The Mitchell Beazley Joy of Knowledge Library

The Modern World

Scientiam non dedit natura semina scientiae nobis dedit

"Nature has given us not knowledge itself, but the seeds thereof."

Seneca

The Joy of Knowledge Encyclopaedia is affectionately dedicated to the memory of John Beazley 1932–1977, Book Designer, Publisher and Co-Founder, of the publishing house of Mitchell Beazley Limited, by all his many friends and colleagues in the company.

The Joy of Knowledge Library

General Editor: James Mitchell
With an overall preface by Lord Butler, Master of Trinity College, University of Cambridge

Science and The Universe	Introduced by Sir Alan Cottrell, Master of Jesus College, University of Cambridge; and Sir Bernard Lovell, Professor of Radio Astronomy, University of Manchester
The Physical Earth	Introduced by Dr William Nierenberg, Director, Scripps Institution of Oceanography, University of California
The Natural World	Introduced by Dr Gwynne Vevers, Assistant Director of Science, the Zoological Society of London
Man and Society	Introduced by Dr Alex Comfort, Professor of Pathology, Irvine Medical School, University of California, Santa Barbara
History and Culture 1 **History and Culture** 2	Introduced by Dr Christopher Hill, Master of Balliol College, University of Oxford
Man and Machines	Introduced by Sir Jack Callard, former Chairman of Imperial Chemical Industries Limited
The Modern World	Introduced by Dr Michael Wise, Professor of Geography, London School of Economics and Political Science
Fact Index A–K	
Fact Index L–Z	

The Mitchell Beazley Joy of Knowledge Library

The Modern World

Introduced by Michael J. Wise, MC, PhD, FRGS
Professor of Geography, London School of Economics and Political Science

MITCHELL BEAZLEY

ISBN 0 85533 112 7

Typesetting by Art Reprographic (London) Limited, England
Photoprint Plates Ltd, Rayleigh, Essex, England

Printed in England by Balding + Mansell

Major contributors and advisers to The Joy of Knowledge Library

Fabian Acker CEng, MIEE, MIMarE; Professor Leslie Alcock; Professor H.C. Allen MC; Leonard Amey OBE; Neil Ardley BSc; Professor H.R.V. Arnstein DSc, PhD, FIBiol; Russell Ash BA(Dunelm), FRAI; Norman Ashford PhD, CEng, MICE, MASCE, MCIT; Professor Robert Ashton; B.W. Atkinson BSc, PhD; Anthony Atmore BA; Professor Philip S. Bagwell BSc(Econ), PhD; Peter Ball MA; Edwin Banks MIOP; Professor Michael Banton; Dulan Barber; Harry Barrett; Professor J.P. Barron MA, DPhil, FSA; Professor W.G. Beasley FBA; Alan Bender PhD, MSc, DIC, ARCS; Lionel Bender BSc; Israel Berkovitch PhD, FRIC, MIChemE; David Berry MA; M.L. Bierbrier PhD; A.T.E. Binsted FBBI (Dipl); David Black; Maurice E.F. Block BA, PhD(Cantab); Richard H. Bomback BSc (London), FRPS; Basil Booth BSc (Hons), PhD, FGS, FRGS; J. Harry Bowen MA(Cantab), PhD(London); Mary Briggs MPS, FLS; John Brodrick BSc(Econ); J.M. Bruce ISO, MA, FRHistS, MRAeS; Professor D.A. Bullough MA, FSA, FRHistS; Tony Buzan BA(Hons) UBC; Dr Alan R. Cane; Dr J.G. de Casparis; Dr Jeremy Catto MA; Denis Chamberlain; E.W. Chanter MA; Professor Colin Cherry D Sc(Eng), MIEE; A.H. Christie MA, FRAI, FRAS; Dr Anthony W. Clare MPhil(London), MB, BCh, MRCPI, MRCPsych; Professor Aidan Clarke MA, PhD, FTCD; Sonia Cole; John R. Collis MA, PhD; Professor Gordon Connell-Smith BA, PhD, FRHistS; Dr A.H. Cook FRS; Professor A.H. Cook FRS; J.A.L. Cooke MA, DPhil; R.W. Cooke BSc, CEng, MICE; B.K. Cooper; Penelope J. Corfield MA; Robin Cormack MA, PhD, FSA; Nona Coxhead; Patricia Crone BA, PhD; Geoffrey P. Crow BSc(Eng), MICE, MIMunE, MInstHE, DIPTE; J.G. Crowther; Professor R.B. Cundall FRIC; Noel Currer-Briggs MA, FSG; Christopher Cviic BA(Zagreb), BSc(Econ, London); Gordon Daniels BSc(Econ, London), DPhil(Oxon); George Darby BA; G.J. Darwin; Dr David Delvin; Robin Denselow BA; Professor Bernard L. Diamond; John Dickson; Paul Dinnage MA; M.L. Dockrill BSc(Econ), MA, PhD; Patricia Dodd BA; James Dowdall; Anne Dowson MA(Cantab); Peter M. Driver BSc, PhD, MIBiol; Rev Professor C.W. Dugmore DD; Herbert L. Edlin BSc, Dip in Forestry; Pamela Egan MA(Oxon); Major S.R. Elliot CD, BComm; Professor H.J. Eysenck PhD, DSc; Dr Peter Fenwick BA, MB, BChir, DPM, MRCPsych; Jim Flegg BSc, PhD, ARCS, MBOU; Andrew M. Fleming MA; Professor Antony Flew MA(Oxon), DLitt (Keele); Wyn K. Ford FRHistS; Paul Freeman DSc(London); G.S.P. Freeman-Grenville DPhil, FSA, FRAS, G.E. Fussell DLitt, FRHistS; Kenneth W. Gatland FRAS, FBIS; Norman Gelb BA; John Gilbert BA(Hons, London); Professor A.C. Gimson; John Glaves-Smith BA; David Glen; Professor S.J. Goldsack BSc, PhD, FINSTP, FBCS; Richard Gombrich MA, DPhil; A.F. Gomm; Professor A. Goodwin MA; William Gould BA(Wales); Professor J.R. Gray; Christopher Green PhD; Bill Gunston; Professor A. Rupert Hall LittD; Richard Halsey BA(Hons, UEA); Lynette K. Hamblin BSc; Norman Hammond; Peter Harbison MA, DPhil; Professor Thomas G. Harding PhD; Professor D.W. Harkness; Richard Harris; Dr Randall P. Harrison; Cyril Hart MA, PhD, FRICS, FIFor; Anthony P. Harvey; Nigel Hawkes BA(Oxon); F.P. Heath; Peter Hebblethwaite MA (Oxon), LicTheol; Frances Mary Heidensohn BA; Dr Alan Hill MC, FRCP; Robert Hillenbrand MA, DPhil; Catherine Hills PhD; Professor F.H. Hinsley; Dr Richard Hitchcock; Dorothy Hollingsworth OBE, BSc, FRIC, FIBiol, FIFST, SRD; H.P. Hope BSc(Hons, Agric); Antony Hopkins CBE, FRCM, LRAM, FRSA; Brian Hook; Peter Howell BPhil, MA(Oxon); Brigadier K. Hunt; Peter Hurst BDS, FDS, LDS, RSCEd, MSc(London); Anthony Hyman MA, PhD; Professor R.S. Illingworth MD, FRCP, DPH, DCH; Oliver Impey MA, DPhil; D.E.G. Irvine PhD; L.M. Irvine BSc; E.W. Ives BA, PhD; Anne Jamieson cand mag(Copenhagen), MSc (London); Michael A. Janson BSc; G.H. Jenkins PhD; Professor P.A. Jewell BSc (Agric), MA, PhD. FIBiol; Hugh Johnson; Commander I.E. Johnston RN; I.P. Jolliffe BSc, MSc, PhD, ComplCE, FGS; Dr D.E.H. Jones ARCS, FCS; R.H. Jones PhD, BSc, CEng, MICE, FGS, MASCE, Hugh Kay; Dr Janet Kear; Sam Keen; D.R.C. Kempe BSc, DPhil, FGS; Alan Kendall MA (Cantab); Michael Kenward; John R. King BSc(Eng), DIC, CEng, MIProdE; D.G. King-Hele FRS; Professor J.F. Kirkaldy DSc; Malcolm Kitch; Michael Kitson MA; B.C. Lamb BSc, PhD; Nick Landon; Major J.C. Larminie QDG, Retd; Diana Leat BSc(Econ), PhD; Roger Lewin BSc, PhD, Harold K. Lipset; Norman Longmate MA(Oxon); John Lowry; Kenneth E. Lowther MA; Diana Lucas BA(Hons); Keith Lye BA, FRGS; Dr Peter Lyon; Dr Martin McCauley; Sean McConville BSc; D.F.M. McGregor BSc, PhD(Edin); Jean Macqueen PhD; William Baird MacQuitty MA(Hons), FRGS, FRPS; Professor Rev F.X. Martin OSA; Jonathan Martin MA; Rev Cannon E.L. Mascall DD; Christopher Maynard MSc, DTh; Professor A.J. Meadows; Dr T.B. Millar; John Miller MA, PhD; J.S.G. Miller MA, DPhil, BM, BCh; Alaric Millington BSc, DipEd, FIMA; Rosalind Mitchison MA, FRHistS; Peter L. Moldon; Patrick Moore OBE; Robin Mowat MA, DPhil; J. Michael Mullin BSc; Alistair Munroe BSc, ARCS; Professor Jacob Needleman; John Newman MA, FSA; Professor Donald M. Nicol MA PhD; Gerald Norris; Professor F.S. Northedge PhD; Caroline E. Oakman BA(Hons. Chinese); S. O'Connell MA(Cantab), MInstP; Dr Robert Orr; Michael Overman; Di Owen BSc; A.R.D. Pagden MA, FRHistS; Professor E.J. Pagel PhD; Liam de Paor MA; Carol Parker BA(Econ), MA (Internat. Aff.); Derek Parker; Julia Parker DFAstrolS; Dr Stanley Parker; Dr Colin Murray Parkes MD, FRC(Psych), DPM; Professor Geoffrey Parrinder MA, PhD, DD(London), DLitt(Lancaster); Moira Paterson; Walter C. Patterson MSc; Sir John H. Peel KCVO, MA, DM, FRCP, FRCS, FRCOG; D.J. Penn; Basil Peters MA. MInstP, FBIS; D.L. Phillips FRCR, MRCOG; B.T. Pickering PhD, DSc; John Picton; Susan Pinkus; Dr C.S. Pitcher MA, DM, FRCPath; Alfred Plaut FRCPsych; A.S. Playfair MRCS, LRCP, DObstRCOG; Dr Antony Polonsky; Joyce Pope BA; B.L. Potter NDA, MRAC, CertEd; Paulette Pratt; Antony Preston Frank J. Pycroft; Margaret Quass; Dr John Reckless; Trevor Reese BA, PhD, FRHistS; M.M. Reese MA (Oxon); Derek A. Reid BSc, PhD; Clyde Reynolds BSc; John Rivers; Peter Roberts; Colin A. Ronan MSc, FRAS; Professor Richard Rose BA(Johns Hopkins), DPhil (Oxon); Harold Rosenthal; T.G. Rosenthal MA(Cantab); Anne Ross MA, MA(Hons, Celtic Studies), PhD, (Archaeol and Celtic Studies, Edin); Georgina Russell MA; Dr Charles Rycroft BA (Cantab), MB(London), FRCPsych; Susan Saunders MSc(Econ); Robert Schell PhD; Anil Seal MA, PhD(Cantab); Michael Sedgwick MA(Oxon); Martin Seymour-Smith BA(Oxon), MA(Oxon); Professor John Shearman; Dr Martin Sherwood; A.C. Simpson BSc; Nigel Sitwell; Dr Alan Sked; Julie and Kenneth Slavin FRGS, FRAI; Professor T.C. Smout; Alec Xavier Snobel BSc(Econ); Terry Snow BA, ATCL; Rodney Steel; Charles S. Steinger MA, PhD; Geoffrey Stern BSc(Econ); Maryanne Stevens BA(Cantab), MA(London); John Stevenson DPhil, MA; J. Sidworthy MA; D. Michael Stoddart BSc, PhD; Bernard Stonehouse DPhil, MA, BSc, MInstBiol; Anthony Storr FRCP, FRCPsych; Richard Storry; Charles Stuart-Jervis; Professor John Taylor; John W.R. Taylor FRHistS, MRAeS. FSLAET; R.B. Taylor BSc(Hons, Microbiol); J. David Thomas MA, PhD; D. Thompson BSc(Econ); Harvey Tilker PhD; Don Tills PhD, MPhil, MIBiol, FIMLS; Jon Tinker; M. Tregear MA; R.W. Trender; David Trump MA, PhD, FSA; M.F. Tuke PhD; Christopher Tunney MA; Laurence Urdang Associates (authentication and fact check); Sally Walters BSc; Christopher Wardle; Dr D. Washbrook; David Watkins; George Watkins MSc; J.W.N. Watkins; Anthony J. Watts; Dr Geoff Watts; Melvyn Westlake; Anthony White MA(Oxon), MAPhil(Columbia); Dr Ruth D. Whitehouse; P.J.S. Whitmore MBE, PhD; Professor G.R. Wilkinson; Rev H.A. Williams CR; Christopher Wilson BA; Professor David M. Wilson; John B. Wilson BSc, PhD, FGS, FLS; Philip Windsor BA, DPhil(Oxon), Roy Wolfe BSc(Econ), MSc; Donald Wood MA PhD, Dr David Woodings MA, MRCP, MRCPath; Bernard Yallop PhD, BSc, ARCS, FRAS Professor John Yudkin MA, MD, PhD(Cantab), FRIC, FIBiol, FRCP.

The General Editor wishes particularly to thank the following for all their support:
Nicolas Bentley
Bill Borchard
Adrianne Bowles
Yves Boisseau
Irv Braun
Theo Bremer
the late Dr Jacob Bronowski
Sir Humphrey Browne
Barry and Helen Cayne
Peter Chubb
William Clark
Sanford and Dorothy Cobb
Alex and Jane Comfort
Jack and Sharlie Davison
Manfred Denneler
Stephen Elliott
Stephen Feldman
Orsola Fenghi
Professor Richard Gregory
Dr Leo van Grunsven
Jan van Gulden
Graham Hearn
the late Raimund von Hofmansthal
Dr Antonio Houaiss
the late Sir Julian Huxley
Alan Isaacs
Julie Lansdowne
Professor Peter Lasko
Andrew Leithead
Richard Levin
Oscar Lewenstein
The Rt Hon Selwyn Lloyd
Warren Lynch
Simon macLachlan
George Manina
Stuart Marks
Bruce Marshall
Francis Mildner
Bill and Christine Mitchell
Janice Mitchell
Patrick Moore
Mari Pijnenborg
the late Donna Dorita de Sa Putch
Tony Ruth
Dr Jonas Salk
Stanley Schindler
Guy Schoeller
Tony Schulte
Dr E. F. Schumacher
Christopher Scott
Anthony Storr
Hannu Tarmio
Ludovico Terzi
Ion Trewin
Egil Tveteras
Russ Voisin
Nat Wartels
Hiroshi Watanabe
Adrian Webster
Jeremy Westwood
Harry Williams
the dedicated staff of MB Encyclopaedias who created this *Library* and of MB Multimedia who made the IVR Artwork Bank.

The Modern World/Contents

Preface

I do not think any other group of publishers could be credited with producing so comprehensive and modern an encyclopaedia as this. It is quite original in form and content. A fine team of writers has been enlisted to provide the contents. No library or place of reference would be complete without this modern encyclopaedia, which should also be a treasure in private hands.

The production of an encyclopaedia is often an example that a particular literary, scientific and philosophic civilization is thriving and groping towards further knowledge. This was certainly so when Diderot published his famous encyclopaedia in the eighteenth century. Since science and technology were then not so far developed, his is a very different production from this. It depended to a certain extent on contributions from Rousseau and Voltaire and its publication created a school of adherents known as the encyclopaedists.

In modern times excellent encyclopaedias have been produced, but I think there is none which has the wealth of illustrations which is such a feature of these volumes. I was particularly struck by the section on astronomy, where the illustrations are vivid and unusual. This is only one example of illustrations in the work being, I would almost say, staggering in their originality.

I think it is probable that many responsible schools will have sets, since the publishers have carefully related much of the contents of the encyclopaedia to school and college courses. Parents on occasion feel that it is necessary to supplement school teaching at home, and this encyclopaedia would be invaluable in replying to the queries of adolescents which parents often find awkward to answer. The "two-page-spread" system, where text and explanatory diagrams are integrated into attractive units which relate to one another, makes this encyclopaedia different from others and all the more easy to study.

The whole encyclopaedia will literally be a revelation in the sphere of human and humane knowledge.

Butler

Master of Trinity College,
Cambridge

The Structure of the Library

Science and The Universe

The growth of science
Mathematics
Atomic theory
Statics and dynamics
Heat, light and sound
Electricity
Chemistry
Techniques of astronomy
The Solar System
Stars and star maps
Galaxies
Man in space

The Physical Earth

Structure of the Earth
The Earth in perspective
Weather
Seas and oceans
Geology
Earth's resources
Agriculture
Cultivated plants
Flesh, fish and fowl

The Natural World

How life began
Plants
Animals
Insects
Fish
Amphibians and reptiles
Birds
Mammals
Prehistoric animals and plants
Animals and their habitats
Conservation

Man and Society

Evolution of man
How your body works
Illness and health
Mental health
Human development
Man and his gods
Communications
Politics
Law
Work and play
Economics

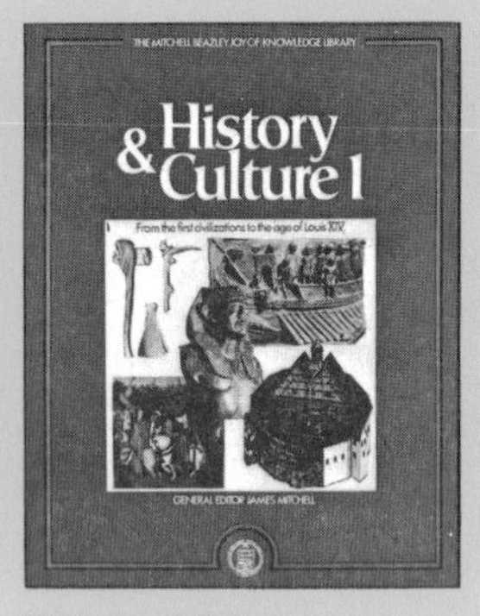

History and Culture

Volume 1 From the first civilizations to the age of Louis XIV

The art of prehistory
Classical Greece
India, China and Japan
Barbarian invasions
The crusades
Age of exploration
The Renaissance
The English revolution

The Modern World is a book of popular general knowledge about the countries of the world. It is a self-contained, self-indexing book and it has its own internal system of cross-references to help you to build up a rounded picture of the subjects it covers.

It is one volume in Mitchell Beazley's ten-volume library of individual books we have entitled *The Joy of Knowledge Library*—a library which forms a comprehensive encyclopaedia.

For a new generation brought up with television, words alone are no longer enough—and so we set out to make the *Library* a new sort of pictorial encyclopaedia for a visually oriented age, a new "family bible" of knowledge which will find acceptance in every home.

Seven other colour volumes in the *Library* are *Man and Society, The Physical Earth, The Natural World, History and Culture* (two volumes), *Man and Machines*, and *Science and The Universe. The Modern World* is arranged alphabetically: the other volumes are organized by topic and provide a comprehensive store of general knowledge rather than isolated facts.

The last two volumes in the *Library* provide a different service. Split up for convenience into A-K and L-Z references, these volumes are a fact index to the whole work. They provide factual information of all kinds on peoples, places and things through approximately 25,000 mostly short entries listed in alphabetical order. The entries in the A-Z volumes also act as a comprehensive index to the other eight volumes, thus turning the whole *Library* into a rounded *Encyclopaedia*, which is not only a comprehensive guide to general knowledge in volumes 1–7 but which now also provides access to specific information as well in *The Modern World* and the fact index volumes.

Access to knowledge

Whether you are a systematic reader or an unrepentant browser, my aim as General Editor has been to assemble all the facts you really ought to know into a coherent and logical plan that makes it possible to build up a comprehensive general knowledge of the subject.

Depending on your needs or motives as a reader in search of knowledge, you can find things out from *The Modern World* in several ways: for example, you can simply browse pleasurably about in its pages haphazardly (and that's my way!) or you can browse in a more organized fashion if you use the map references at the foot of each article that lead you to the atlas section. Or you can locate specific places by using the index to the atlas. Finally, you can set yourself the solid task of finding out literally everything in the book in A to Z order by reading it from cover to cover: in this the Contents List (page 6) is there to guide you.

Our basic purpose in organizing the volumes in *The Joy of Knowledge Library* into two elements—the three volumes of A-Z factual information and the seven volumes of general knowledge—was functional. We devised it this way to make it easier to gather the two different sorts of information—simple facts and wider general knowledge, respectively—in appropriate ways.

The functions of an encyclopaedia

An encyclopaedia (the Greek word means "teaching in a circle" or, as we might say, the provision of a *rounded* picture of knowledge) has to perform these two distinct functions for two sorts of users, each seeking information of different sorts.

First, many readers want simple factual answers to straightforward questions like "What is a rhombus?" They may be intrigued to learn that it is a four-sided plane figure with all of its sides equal and that a square is a rhombus with its interior angles right angles. Such direct and simple facts are best supplied by a short entry and in the *Library* they will be found in the two A-Z *Fact Index* volumes.

But secondly, for the user looking for in-depth knowledge on a subject or on a series of subjects—such as "What has man achieved in space?" short alphabetical entries alone are inevitably bitty and disjointed. What do you look up first—"space"? "astronautics"? "NASA"? "rockets"? "Skylab"? "Mariner"? "Apollo"?—and do you have to read all the entries or only some of them? You normally have to look up *lots* of entries in a purely alphabetical encyclopaedia to get a comprehensive answer to such wide-ranging questions. Yet comprehensive answers are what general knowledge is all about.

A long article or linked series of longer articles, organized

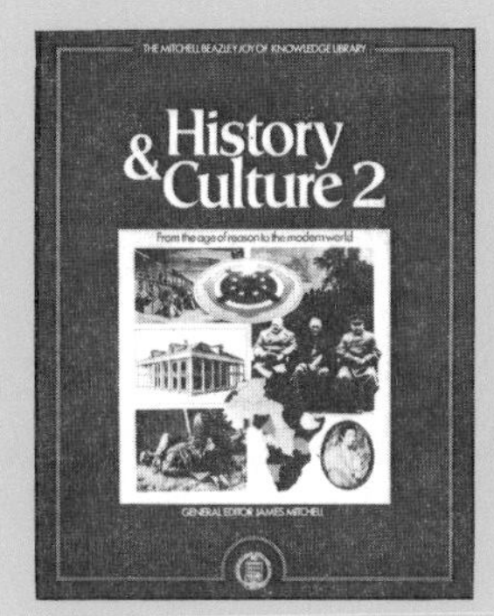

History and Culture

Volume 2 From the Age of Reason to the modern world

Neoclassicism
Colonizing Australasia
World War I
Ireland and independence
Twenties and the depression
World War II
Hollywood

Man and Machines

The growth of technology
Materials and techniques
Power
Machines
Transport
Weapons
Engineering
Communications
Industrial chemistry
Domestic engineering

The Modern World

Flags of the world
Nations of the world
Almanac
Atlas
Gazetteer

Fact Index A-K

The first of two volumes containing 25,000 mostly short factual entries on people, places and things in A-Z order. The Fact Index also acts as an index to the eight colour volumes. In this volume, everything from Aachen to Kyzyl.

Fact Index L-Z

The second of the A-Z volumes that turn the Library into a complete encyclopaedia. Like the first, it acts as an index to the eight colour volumes. In this volume, everything from Ernest Laas to Zyrardow.

by related subjects, is clearly much more helpful to the person wanting such comprehensive answers. That is why we have adopted a logical, so-called *thematic* organization of knowledge, with a clear system of connections relating topics to one another, for teaching general knowledge in the seven general knowledge volumes in the *Library*.

The spread system

The basic unit of all the general knowledge books is the "spread"—a nickname for the two-page units that comprise the working contents of all these books. The spread is the heart of our approach to explaining things.

Every spread in the general knowledge volumes tells a story. It is almost always a self-contained story—a story on how algebra works, for example (pages 34 to 35 of *Science and The Universe*) or on the meaning of myth (pages 206 to 207 of *Man and Society*). The spreads on these subjects all work to the same discipline, which is to tell you all you need to know in two facing pages of text and pictures. The discipline of having to get in all the essential and relevant facts in this comparatively short space actually makes for better results—text that has to get to the point without any waffle, pictures and diagrams that illustrate the essential points in a clear and coherent fashion, captions that really work and explain the point of the pictures.

The spread system is a strict discipline but once you get used to it, I hope you'll ask yourself why you ever thought general knowledge could be communicated, in words and pictures, in any other way.

The structure of the spread system will also, I hope prove reassuring when you venture out from the things you *do* know about into the unknown areas you don't know, but want to find out about. There are many virtues in being systematic. You will start to feel at home in all sorts of unlikely areas of knowledge with the spread system to guide you. The spreads are, in a sense, the building blocks of knowledge. Like living cells which are the building blocks of plants and animals, they are systematically "programmed" to help you to learn more easily and to remember better. Each spread has a main article of 850 words summarising the subject. The article is illustrated by an average of ten pictures and diagrams, the captions of which both complement *and* supplement the information in the article (so please read the captions, incidentally, or you may miss something!). Each spread, too, has a "key" picture or diagram in the top right-hand corner. The purpose of the key picture is twofold: it summarises the story of the spread visually and it is intended to act as a memory stimulator to help you to recall all the integrated facts and pictures on a subject.

Finally, each spread has a box of connections headed "See Also" and, sometimes, "Read First". These are cross-reference suggestions to other connecting spreads. The "Read Firsts" normally appear only on spreads with particularly complicated subjects and indicate that you might like to learn to swim a little in the elementary principles of a subject before being dropped in the deep end of its complexities.

The "See Alsos" are the treasure hunt features of *The Joy of Knowledge* system and I hope you'll find them helpful and, indeed, fun to use. They are also essential if you want to build up a comprehensive general knowledge. If the spreads are individual living cells, the "See Alsos" are the secret code that tells you how to fit the cells together into an organic whole which is the body of general knowledge.

Level of readership

The level for which we have created *The Joy of Knowledge Library* is intended to be a universal one. Some aspects of knowledge are more complicated than others and so readers will find that the level varies in different parts of the *Library* and indeed in different parts of individual volumes. This is quite deliberate: *The Joy of Knowledge Library* is a library for all the family.

Some younger people should be able to enjoy and to absorb most of the pages in *Science and The Universe* on telescopes, for example, from as young as ten or eleven onwards—but the level has been set primarily for adults and older children who will need some basic knowledge to make sense of the pages on thermodynamics or biochemistry, for example.

Whatever their level, the greatest and the bestselling popular encyclopaedias of the past have always had one thing in common—simplicity. The ability to make even

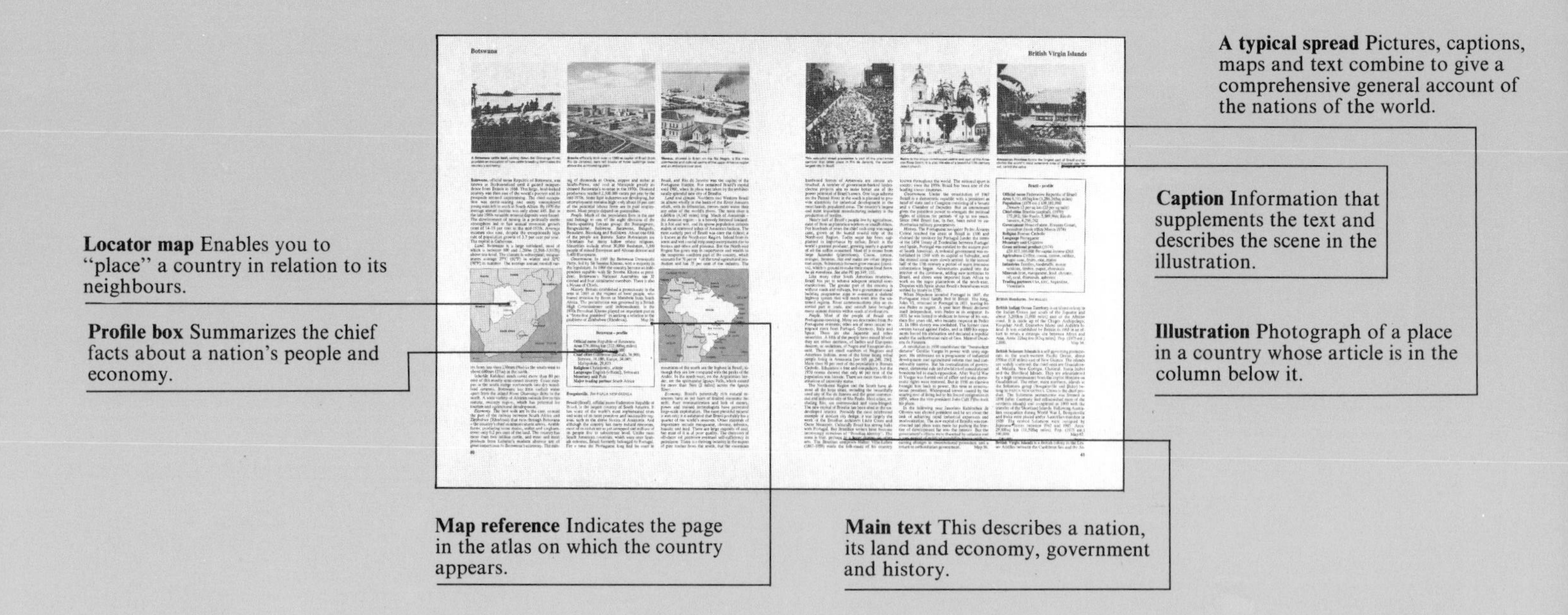

complicated subjects clear, to distil, to extract the simple principles from behind the complicated formulae, the gift of getting to the heart of things: these are the elements that make popular encyclopaedias really useful to the people who read them. I hope we have followed these precepts throughout the *Library*: if so our level will be found to be truly universal.

Philosophy of the Library

The aim of *all* the books—general knowledge and *Fact Index* volumes—in the *Library* is to make knowledge more readily available to everyone, and to make it fun. This is not new in encyclopaedias. The great classics enlightened whole generations of readers with essential information, popularly presented and positively inspired. Equally, some works in the past seem to have been extensions of an educational system that believed that unless knowledge was painfully acquired it couldn't be good for you, would be inevitably superficial, and wouldn't stick. Many of us know in our own lives the boredom and disinterest generated by such an approach at school, and most of us have seen it too in certain types of adult books. Such an approach locks up knowledge instead of liberating it.

The great educators have been the men and women who have enthralled their listeners or readers by the self-evident passion they themselves have felt for their subjects. Their joy is natural and infectious. We remember what they say and cherish it for ever. The philosophy of *The Joy of Knowledge Library* is one that precisely mirrors that enthusiasm. We aim to seduce you with our pictures, absorb you with our text, entertain you with the multitude of facts we have marshalled for your pleasure—yes, *pleasure*. Why not pleasure?

There are three uses of knowledge: education (things you ought to know because they are important); pleasure (things which are intriguing or entertaining in themselves); application (things we can do with our knowledge for the world at large).

As far as education is concerned there are certain elementary facts we need to learn in our schooldays. The *Library*, with its vast store of information, is primarily designed to have an educational function—to inform, to be a constant companion and to guide everyone through school, college and other forms of higher education.

But most facts, except to the student or specialist (and these books are not only for students and specialists, they are for everyone) aren't vital to know at all. You don't *need* to know them. But discovering them can be a source of endless pleasure and delight, nonetheless, like learning the pleasures of food or wine or love or travel. Who wouldn't give a king's ransom to know when man really became man and stopped being an ape? Who wouldn't have loved to have spent a day at the feet of Leonardo or to have met the historical Jesus or to have been there when Stephenson's *Rocket* first moved? The excitement of discovering new things is like meeting new people—it is one of the great pleasures of life.

There is always the chance, too, that some of the things you find out in these pages may inspire you with a lifelong passion to apply your knowledge in an area which really interests you. My friend Patrick Moore, the astronomer, who first suggested we publish this *Library* and wrote much of the astronomy section in the volume on *Science and The Universe*, once told me that he became an astronomer through the thrill he experienced on first reading an encyclopaedia of astronomy called *The Splendour of the Heavens*, published when he was a boy. Revelation is the reward of encyclopaedists. Our job, my job, is to remind you always that the joy of knowledge knows no boundaries and can work untold miracles.

In an age when we are increasingly creators (and less creatures) of our world, the people who *know*, who have a sense of proportion, a sense of balance, above all perhaps a sense of insight (the inner as well as the outer eye) in the application of their knowledge, are the most valuable people on earth. They, and they alone, will have the capacity to save this earth as a happy and a habitable planet for all its creatures. For the true joy of knowledge lies not only in its acquisition and its enjoyment, but in its wise and loving application in the service of the world.

Thus the Latin tag "Scientiam non dedit natura, semina scientiae nobis dedit" on the first page of this book. It translates as "Nature has given us not knowledge itself, but the seeds thereof."

It is, in the end, up to each of us to make the most of what we find in these pages.

General Editor's Introduction

The Structure of this Book

The Modern World is a book about the countries that the peoples of the world inhabit—and to some extent about those peoples themselves, for it is impossible to describe a country without discussing not only the physical nature of the land and its resources but also the language, history, and origins of the people who live in it. But it is not primarily a work of anthropology. It is first and foremost about topography, climate, resources, and only then about the use the peoples of the world have made of their environments. *The Modern World* will tell you almost everything you need to know about what makes a particular country unique—for further information about literature and art, science, philosophy, law, natural history and so on the reader should consult the general knowledge volumes of *The Joy of Knowledge Library* or the two A to Z *Fact Index* volumes.

One general point needs to be made about the statistics in this book. They are taken from the most authoritative and up-to-date sources, but even those sources admit that they cannot be entirely definitive. *The United Nations Demographic Yearbook*, for example, which has been consulted for population figures, is inevitably something like two years out-of-date when it appears. Further, the United Nations, with all its resources, cannot always get accurate figures from the countries themselves. Most Western nations conduct regular and accurate censuses, but some developing nations do not. Similarly, the available figures on gross national product, literacy rate, per capita income and so on can be as accurate only as the sophistication of the bureaucracy producing them will permit. We have made the facts in *The Modern World* as precise as we can, but we have done so in the knowledge that on occasions that precision can be, at best, only an approximation.

Where to start

The articles in *The Modern World* are self-contained and can be read in isolation, but you may find them more significant if when you read them you have in the back of your mind some of the points made in the Introduction (pages 13 to 16) by Michael Wise. Professor Wise begins by pointing out that man's scientific knowledge of the planet he lives on is of only recent origin. Even today it is far from complete—in terms of maps, only six per cent of the land area has been mapped at a scale of 1:31,680 (about 3cm to 1km or 2in to the mile). The need for adequate maps is now urgent: without them scientists cannot plan for a world population that according to one prediction may reach seven thousand million by the year 2000.

Professor Wise also points out that most of the world's land surface is unsuitable for permanent human habitation. "The accounts of the geography and history of countries in this book should be seen against the continuing struggles of societies to adapt to and modify the environmental conditions in which they live," he writes. "Some geographers used to argue that societies are 'determined' in their economic activities—and in other ways— by physical conditions of climate, relief and soil. Today the emphasis is on the opportunities which nations and communities possess through the application of skill and technology, through wise policies and good organization, to make the best of their local natural conditions. But even this optimistic view recognizes that there are severe physical restraints that will be costly to overcome."

The inter-dependence of the countries of the modern world is significant in the story of each country. It is doubtful if today any country is totally self-sufficient.

Plan of the book

There are five main sections in *The Modern World*: the flags of the world, the nations of the world, an almanac, an atlas, and the gazetteer or index to the atlas.

Flags of the World

There are eight pages in *The Modern World* devoted to the flags of 160 independent states. The accuracy of each flag has been checked by one of the world's leading vexillologists, Dr Whitney Smith, of the Flag Research Center, Winchester, Massachusetts, USA. The flags are arranged by regions, beginning with Africa and going on to Europe, South and North America, the Pacific, Asia, the Far East and the Middle East.

Nations of the world

This section forms the heart of *The Modern World*—159 pages describing 231 countries. There are in addition 152 cross-references to major articles: for example, "**Arizona**—*See* UNITED STATES". Accompanying the articles are 170 locator maps, which position a country and show its neighbours, and 467 black-and-white photographs.

The country articles in *The Modern World* are sub-divided, for ease of access to the information that they contain, in much the same way. The longer articles—Australia, for example—begin by discussing the country's land and climate. The article then considers physical resources, industry, agriculture, trade, and transport and communications before covering the people, their cultural life, and their leisure and sporting activities. The structure of the country's government and armed services precede its history. If the country has a federal structure, its states and territories are given and for some Commonwealth countries the article lists the Prime Ministers and Governors-General. Country and city names are given in their familiar English form, followed by the local form in parentheses where appropriate—for example, "Jerusalem (Yerushalayim)" For usage on maps, see below. Most articles end with a cross-reference to the page of the atlas on which a map of the country appears, and articles on major countries also carry a quick-reference "profile" box that contains statistics of area, population, gross national product and so on. In shorter articles, the sub-divisions are fewer, but cover the same topics. Cross-references to other articles in *The Modern World* are set in the body of the article in small capital letters, thus: CHRISTMAS ISLAND.

Almanac

A four-page almanac begins on page 188. Most of it is about time: how it is measured and defined; the time zones the world is divided into; and the secular and religious calendars in current use. There is a map that enables you to find the time differences between any two countries anywhere in the world, a table of important fixed dates throughout the world, and a list of important events that will take place in the next eight years.

The almanac also contains tables listing the world's continents and largest islands, rivers, lakes, deserts, volcanoes, waterfalls and mountains.

Finally, there is a table of international treaties and pacts and their signatories.

Atlas of the world

The 80-page world atlas in *The Modern World* has maps in scales ranging from 1:1,500,000 to 1:60,000,000, with most maps falling between 1:1,500,000 and 1:16,000,000. Eight pages are devoted to the British Isles in the largest scale, 1:1,500,000. Australia, New Zealand and the Pacific receive extensive treatment over ten pages, Canada is covered in a double-page spread, and Africa occupies seven pages. The atlas is arranged with its maps running eastwards from Europe to the Americas.

The gazetteer gives the page on which the entry appears, its latitude and longitude, and keyed information about the entry. The gazetteer, which occupies 48 pages, contains some 20,000 entries.

Most map features are indexed to the largest-scale map on which they appear. Countries, mountain ranges, and other extensive features are generally indexed to the map that shows them in their entirety.

The features indexed are of three types: *point, areal* and *linear*. For *point* features (for example, cities, mountain peaks and dams), latitude and longitude co-ordinates give the location of the point on the map. For *areal* features (countries, mountain ranges, and so on), the co-ordinates generally indicate the approximate centre of the feature. For *linear* features (rivers, canals, aqueducts), the co-ordinates locate a terminating point—for example, the mouth of a river.

Names in the index, as on the maps, are generally in the local form. Many conventional English names are cross-references to the primary map name. For names in languages not written in the Roman alphabet, the locally official transliteration system has been used where one exists. For languages with no one locally accepted transliteration system, notably Arabic, transliteration in general follows closely a system adopted by the United States Board on Geographic Names.

Fuller details about how to use the maps and gazetteer appear on the first page of the atlas section, and on the first page of the gazetteer.

Maps in other volumes

Readers of *The Modern World* should note that there are many specialized maps in other volumes of *The Joy of Knowledge Library*. *The Physical Earth,* for example, contains world relief maps, sea-bed maps, and diagrammatic maps of the world distribution of crops such as rice, maize and wheat. It also contains maps of the earth's crustal plates, of active and extinct volcanoes, world climates, ocean currents, distribution of minerals and resources, deserts, fishing grounds and so on. *Man and Society* contains maps of race and language distribution, and *The Natural World* has maps pertaining to zoogeography. There are innumerable historical maps in both volumes of *History and Culture,* and *Man and Machines* carries maps of such things as communications networks, canals and the spread of iron working.

The Modern World will, I hope, not only stand in its own right as an information source, but also provide essential background to the topics covered in other volumes in the *Library*. It has been argued that both history and morality are largely matters of geography—if so, then this volume of *The Joy of Knowledge Library* ought to explain a great deal of what is covered elsewhere in the *Library*. But whatever the use to which you put this book, I hope you find it entertaining and informative—and above all, helpful.

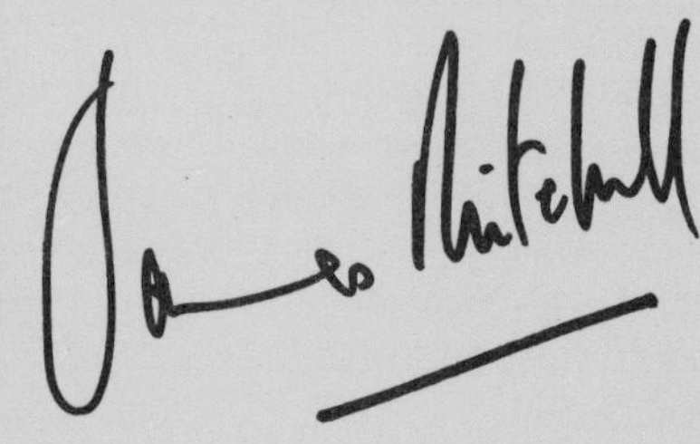

The Modern World

Dr Michael J. Wise

Professor of Geography, London School of Economics and Political Science

According to a Spanish proverb "Geography dictates". As the home of man the Earth provides extremely diverse conditions which greatly influence the location and nature of human activities. But to the variety of the conditions offered by Nature we must link the variety of the thoughts and experiences of men. The combined study of natural with social and economic conditions can then help us to understand the Earth's rich mosaic of life and landscape, and can provide a framework of knowledge that can be employed in devising solutions to the great problems of our time.

The exploration of the Earth and its resources has been a great theme stimulating human endeavour throughout the ages. However, man's scientific knowledge of the planet on which he lives is, in terms of the whole history of mankind, of only recent origin. It is merely two hundred years since Cook's voyages in the Pacific Ocean enabled cartographers to fill major gaps in the map of the world and to dispel myths about the existence of a great southern continent thought necessary to balance the land mass in the Northern Hemisphere. It is only a century or less since exploration of the interiors of Africa and South America supplied knowledge of their main features. Many of the sciences that provide knowledge about the Earth as the home of man – meteorology, geology, botany, economics and political science – themselves became formalized only in the last two centuries. The principles of modern geography owe much to the scientific exploration of such men as Alexander von Humboldt (1769–1859). The year of his death was also the year in which Charles Darwin published *The Origin of Species,* a work of profound importance for both the natural and social sciences whose findings, in the author's own words, would influence "the present welfare and . . . the future success . . . of every inhabitant of this world".

However, in spite of the rapid progress in exploration and science, man's knowledge of the Earth is far from complete – many people believe it has hardly yet begun. The groundwork of the past provides the basis on which modern scientific exploration can continue to reveal more about the qualities of man's environments – the evaluations that can be made about the combinations of natural and man-made conditions in which we live, and about the resources available to maintain and improve standards of life. Knowledge of the oceans and ocean floors is only in its infancy, for example; the search there for valuable resources – as recent discoveries in the North Sea have shown – may well prove rewarding.

Accurate maps of the terrain are essential for environmental planning and, although systematic mapping was begun in some western European countries in the 18th century, it was not until 1891 that geographers adopted a scheme for the uniform mapping of the whole of the Earth's surface on the scale of 1:1,000,000 (1cm to 10km, or about 1in to 16 miles). At the larger scale of 1:253,440 only 72 per cent of the world's land mass has been mapped, and at scales of 1:31,680 (about 3cm to 1km, or 2in to the mile) or better only 6 per cent of the land area is covered. Even the maps that have been made are not always freely available because some countries restrict their publication and distribution. Aerial photography and information from orbiting satellites offer great technical possibilities for filling in gaps in our knowledge and for surveys of land use, crops and resources. But progress will have to be rapid if scientists are to have a satisfactory basis for planning the environment for a world population which by the year 2000 may have risen to some seven thousand million.

The total surface area of the globe is about 510,000,000sq km (196,836,000sq miles) but only a little more than 28 per cent is land. And much of this, at least under present conditions of economics and technology, is unsuitable for permanent human habitation. Areas of permanent ice and snow or permafrost (where the soil or subsoil is permanently frozen) make up about a fifth of the land surface. Another fifth consists of highlands, too rugged or high for settlements that rely on growing crops. Much investment has been made in irrigating arid areas, yet about another fifth can be regarded as desert. In the words of the late Sir Dudley Stamp, Director of the World Land Utilisation Survey and President of the International Geographical Union, "if we regard the habitable lands as those having physical and climatic conditions permitting the growth of crops desired by man, the area available does not cover more than two-fifths of the land surface". But even this proportion is reduced still further by areas of poor soil or, as in equatorial rainforest areas such as Amazonia, by excessive rainfall. Probably not much more than one-third of the land area can be regarded as cultivatable using today's techniques. An optimistic estimate suggests that the total usable land area (now about 40 per cent) could be extended to about 66 per cent. Other estimates present the more pessimistic view that, except for an increase in the utilization of forest land, land use on a world scale may have already reached a maximum.

The accounts of the geography and history of countries in this book should be seen against the continuing struggles of societies to adapt to and modify the environmental conditions in which they live. The great variety of climatic regimes range from the hot and wet equatorial climates of the Amazon and Congo basins to the dry climates of the hot deserts and to the polar climates with their tundra and ice caps. Some climates, particularly the tropical monsoon, impose a discipline and regularity upon choice of crops and the rhythm of farming – societies are heavily dependent on the regularity of the monsoon rains. Others, such as west coast cool temperate climates, offer a wider range of crops and farming methods. Formerly there was much emphasis on the view that societies are "determined" in their economic activities – and in other ways – by physical conditions of climate, relief and soil. Today the emphasis is on the opportunities which nations and communities possess through the application of skill and technology, through wise policies and good organization, to make the best use of their local natural conditions. But even this optimistic view recognizes that there are severe physical restraints that will be costly to overcome.

If all the 3,600 million people in the world were distributed evenly over its land surface, there would be about 24 people for every square kilometre (62 per sq mile). But of course the distribution is far from regular; the great population concentrations in southern, south-eastern and eastern Asia and in the industrialized areas of western Europe and North America stand out clearly on a world map. The increase in the number of people who live in cities has been one of the characteristic features of life in the 20th century. Probably more than a third of the world's population dwells in cities and for the main industrial nations figures of between 60 and 85 per cent are often quoted. In countries such as the United States and Britain, several groups of neighbouring towns and cities have coalesced into conurbations, city regions, or even into a "megalopolis". The megalopolis along the north-eastern seaboard of the United States stretches for about 965km (600 miles) and has at least 37 million people. The proportions of total population living in towns and cities in Asia (13 per cent) and Africa (9 per cent) are low, although there is a significant trend towards the growth of large urban areas, as in Calcutta. The space occupied by cities is probably no more than 1 per cent of the world's land area but the importance of cities as centres of government, skill, communications, trade and innovation is overwhelming.

For these reasons the distribution of population must be related to considerations that are much wider than those merely of physical geography. What matters is how societies make use of locations, and the great French geographer Vidal de la Blache wrote of the concentration of population in western Europe as "an achievement of intelligence and method as much as a natural phenomenon". It is mankind's intelligence and organizing capacity that will be tested if adequate standards of life and environment are to be attained for a world population that recently has been growing at a rate of about 2 per cent per year. Population forecasting is notoriously difficult and it is possible to find very different estimates of the rate at which population is expected to grow. According to one authority, the 3,600 million people of 1970 will become 4,500 million by the year 2000, but another view predicts a rise to at least 7,600 million.

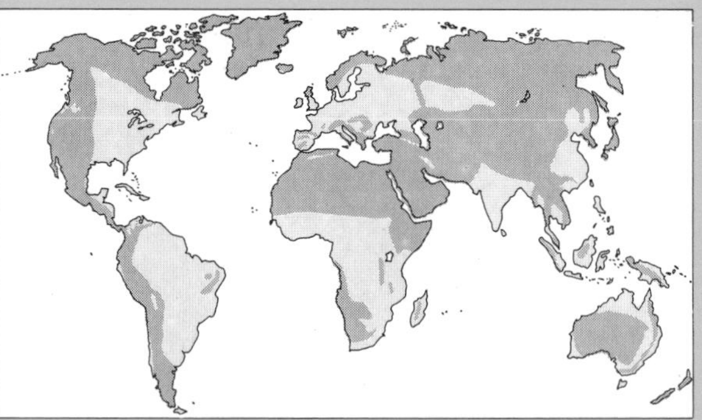

Negative areas of the world – too wet, too dry, too cold or too hilly for permanent human habitation – make up three-fifths of all the land.

How is the greatly increased population to be fed and employed? Another important set of questions relate to what can be done to relieve the undernourishment and poverty of many of the world's population. The countries of the world are often broadly classified as "developed" or "developing". But such differentiation hides as much as it reveals because developing countries are of many different types and at various levels of development. The gap between the developed and developing, as measured by per capita income, continues to widen, much to the worry of politicians and statesmen. But, perhaps, it is more important to discover for each country the nature and causes of poverty, to identify the constraints that prevent or slow down growth, and to discover ways of improving conditions and standards of life. These questions should be borne in mind when reading the articles on individual countries in this volume. Is it not so much the quantity of world resources that is at the heart of the problem as man's ability (or inability) to share them and arrange their distribution? So it is thus important to keep in mind the extent to which countries depend on each other. Very few – if indeed any – nations are totally self-sufficient in terms of their needs for food, raw materials, manufactured goods and specialized services; all have to engage to some extent in trade. The emergence of world trade was greatly advanced by the development of transport systems by sea and rail and, more recently, by road and air. On the broad scale, the great urban and industrial centres could be identified

Feeding the world's population, according to one estimate up to seven thousand million by the year 2000, will be a paramount problem facing mankind.

first in north-western Europe and later also in north-eastern North America. These traded with areas that could provide them with raw materials for their industries and markets for their manufactured products. The pattern changed as countries such as the USSR and United States developed great new industrial regions and as countries formerly dependent only on exporting materials diversified their economies by developing their own industries. In studying individual countries we should also be aware of the attempts that have been made to bring about economic co-operation between nations. The European Economic Community (EEC) is one example and (although it has a different basis) so too is COMECON; there is also, for example, a Central American Common Market (CACM). One of the most influential groupings in recent years has been the Organization of Petroleum Exporting Countries (OPEC), which controls much of the world's oil reserves.

Much recent thinking has focused attention on the global scale of current problems. It is argued by some that if the present trends in world population growth, industrialization, pollution, and rates of food production and resource depletion continue unchanged, the limits to growth on this planet will be reached some time within the next 100 years. The argument continues that the most probable result of a continuation of these trends will be a sudden and uncontrollable decline in population and industrial capacity. If such an outcome is to be avoided, it is thought, men must take action quickly to establish conditions of economic and ecological stability. Such predictions may be found, for example, in the publications sponsored by the Club of Rome and the phrase "The Predicament of Mankind" is evocative of their approach and their conclusions. The ideas rest, broadly, upon assumptions concerning the demands that population growth, urbanization and industrialization will make upon the limited capacity of the biosphere to absorb pollution from increasingly advanced industrial systems; on an awareness of the inter-connections of life systems so that toxic elements (such as mercury) carelessly put to waste in streams or seas may be concentrated in marine organisms with harmful effects to the ecosystem. There are also some at present unanswerable questions. To what extent may the Earth's climate be changed by increased emissions into the atmosphere from the combustion of fossil fuels? How far is the world's industrial system making unsustainable demands upon non-renewable resources? At present rates of consumption known reserves of oil, it is urged, will last for only 30 years, copper for less than 36 years, lead for 26 years or less, mercury for 13, tin for 17 and zinc for 23 or less. Are we, then, using stocks of essential materials at such rates that the expected 7,000 million people in AD 2000 may continue to enjoy a high standard of living?

With such considerations in mind, many have argued the need for an international strategy for the use and development of resources. Recently there has been some recognition of the problem as social and political in nature, as well as technical: the United Nations Conference on the Human Environment held in Stockholm in 1972 studied the problem and the possibilities for action on a comprehensive and worldwide scale. And the United Nations "Habitat" Conference at Vancouver in 1976 studied further the problem of providing adequate human settlements for the increased future population.

However, by no means all the students of the global man-environment problem share the same pessimistic conclusions. Some think that too much emphasis has been placed on certain aspects of the present situation and, especially, that too drastic conclusions have been drawn from extending exponentially current tendencies in the growth of population and use of resources. Insufficient weight may also have been given to man's resources of skill and his ability to solve problems. The adoption of ways of reducing the fertility of human populations, while they would not immediately arrest population growth, would certainly modify longer range forecasts. Others have criticized the estimates of resource availability as based only on the present state of knowledge rather than on the Earth's total potential. Much of the Earth has yet to be explored. Resources, after all, possess only the value placed on them by man, and values change with time, demand and evolving techniques. New "resources" have yet to be discovered, just as new ways of generating energy appear to be possible.

Other criticisms have come from the developing countries which seek in their own economic growth a way of escaping from the problem of poverty and which observe that the problem of environmental quality is a luxury that only advanced industrial nations can afford to recognize. Clearly attempts to take a global view must also incorporate the great world variety of physical, economic and

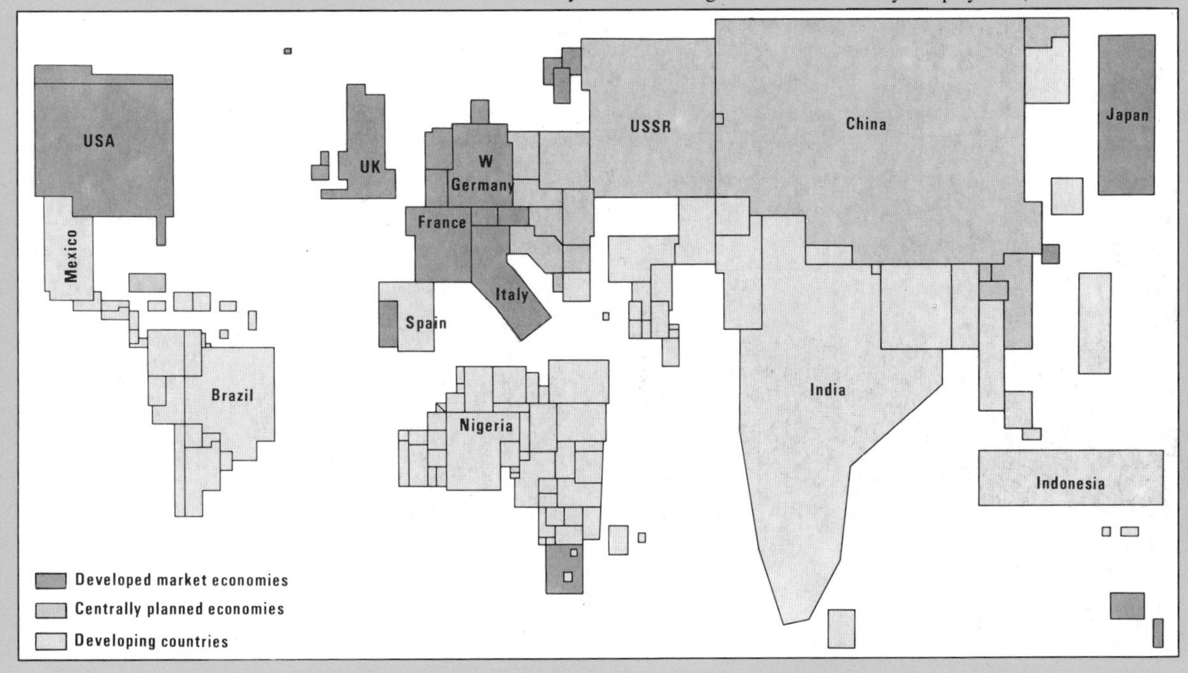

The shape of the world changes when areas of the countries are drawn in proportion to their population densities; the Indian subcontinent dominates the world.

social conditions. And ideas of global "equilibrium", if they are to carry force, must provide for remedies for the worst problems of low life expectancy, poor nutrition and squalid living conditions.

The ideas discussed above have been given new emphasis by the rapid growth of world population, but many of them are by no means new. The idea of conservation, for instance, can be traced back well into the 19th century. One of its early exponents was George Perkins Marsh who, writing in 1864 on *Man and Nature,* used lessons of deforestation, soil erosion and soil exhaustion derived from studies of the Mediterranean lands to warn his fellow Americans of the dangers of over-exploiting the still largely virgin lands of the New World. The view gained strength that the use of land and natural resources should be controlled in the common interest. The likely needs of the future had also to be taken account of and, indeed, the problem of conservation has sometimes been described as "the conflict between the present and the future". This is a view worth discussing: the difficulty, of course, is that we cannot know exactly what value future generations will place on resources which we value today. Resources can fall as well as gain in value. Recent thinking has tended to emphasize the need for carefully considered as well as informed judgements on decisions regarding the use of the natural environment. Thus one leading expert has written that "conservation is all that man thinks and does to soften his impact upon the natural environment and to satisfy all his own true needs while enabling the environment to continue in healthy working order". Many national governments have now enacted laws and established procedures for making what they believe will be wise choices for the use of the environment. How can we proceed to the next stage, of ensuring that wise decisions are made at international and global levels?

A second idea that has had great influence upon recent thinking is that of ecology. This term was first used in 1868 by the German biologist Ernst Haeckel, although it was not until the early years of the present century that scientists began a systematic study of life forms in relation to their habitats. Recognition came of the importance of studying both the interdependence between the members of a plant or animal community and the interdependence between the community as a whole and its environment. The idea of an "ecosystem", a group of living organisms in its relations with its environment, has been used as an integrating concept which helps to bring together the strands of the society-environmental problem. Certainly recent years have brought a greater awareness of social responsibility for the environment.

It should be a central aim of all societies to achieve improved adjustments between man and environments. This will require the persistent and unremitting efforts of all those who study the habitable Earth. Fortunately more and more people are recognizing environmental problems as matters of both national and international concern, and the opportunities available for the application of science and scholarship increase almost daily.

As we saw at the outset, although many aspects remain to be explored, man has begun to learn much more about the Earth as a home. Much progress has also been made in understanding the complex and ever-changing patterns of population and its grouping in towns and cities, of economic activity and trade. Through a study of the histories and activities of peoples and nations we learn more of their attitudes and objectives and of their extremely varied perspectives of the Earth. The Earth as seen from, say, Lagos differs greatly from a view from London or Tokyo. Great tasks lie ahead in accommodating differing aims and actions for the welfare of the complex system of human life on an Earth already much used but still with such great potential.

It is possible to take a pessimistic view of the future and to see ahead for man gross overpopulation and the exhaustion of resources. But I range myself with Dudley Stamp in his view that it is more profitable to see "in the study of earth's riches and the ecology of man a challenge to the fuller understanding which can but lead to the betterment of mankind".

Growing population leads to sprawling urbanization, as in Tokyo (far left), and can result in a "megalopolis" that extends for hundreds of kilometres.

As resources decline, less accessible ones must be tapped. Here an oil drilling platform is being built on land before being floated out to sea.

Flags of the World

In the modern world the flag is a symbol of national or communal identity, but before the rise of modern nation states or even of empires it was more likely to have been a personal symbol – of a king, a noble, or a warlord. The aim of the flag has always been the same: to be an immediately recognizable focal point for people with common military, political or even religious allegiances. It can inspire feelings of cohesiveness and pride in the abstract as well as providing a physical rallying point for soldiers in the uncertainty and turmoil of battle. Flags have become internationally recognized symbols, which can be "read" without the need for language.

The modern flag's ancestral roots lie in the standards of the armies and rulers of the ancient Middle Eastern civilizations. Indeed the oldest flag still in existence is a metal standard from Iran which is probably 5,000 years old. In ancient Egypt the rallying symbol and direct ancestor of the modern flag was a vexilloid, a religious or cultural image set on top of a pole. From the Roman word *vexillum* for the flag or banner carried by soldiers comes the word for the study of flags: vexillology. Later, elements of the modern flag developed as the vexilloid was decorated with or replaced by streamers. Eventually the streamers gave way to the flag: a piece of light fabric that bore an identical emblem on both sides and was attached to a pole.

The earliest flags probably appeared in China and India. The first ruler of the Chou dynasty in China (*c.*1030 BC) was preceded in public by a white flag and this soon came to be closely identified with kingship in its own right. It was an offence even to lay hands on the bearer of the flag; and for it to fall in battle signified defeat. It has been recorded that the early kings of India carried their flags mounted on chariots and elephants.

Asia From China and India cloth flags spread into Burma, Siam and South-East Asia, and to the Middle East. The modern flags of Asia vary greatly in appearance. But even so, their graphic form effectively demonstrates some of the methods by which visual symbols are used to promote a sense of patriotism and unity among the people who willingly give their allegiance to them.

An immediately obvious factor is the use of religious or dynastic emblems. The sun, for example, is a central feature of the Japanese national flag; it represents the Japanese religious belief that the emperor is a direct descendant of the holy sun and hence of prime importance to the nation's welfare. In the Taiwan flag the sun also appears, but there it is a political symbol of the Kuomintang Party which founded the nation of Taiwan.

Both sun and moon form part of the Nepalese flag (the only national flag that is not rectangular). Originally the sun represented the Royal House and the moon stands for the Rana family, which ruled the country until 1951.

As might be expected the basic colour red signifies the political affinities of the Asian Communist countries, although each interprets the colour in its own way. The Chinese People's Republic, for example, shows five stars representing Party and workers. Cultural links can survive political change, however, and this is shown by the appearance of the Chinese yang and yin symbol (good and evil) on the flag of South Korea.

Internal divisions also receive attention in some national flags. The Indian flag includes the country's principal religious groups: the orange of the Hindus, the green of the Muslims and, in the centre, the Buddhist symbol of the wheel. The 14 small stars on the Burmese flag underline the existence in that country of various ethnic groups and the need for living in harmony. It is perhaps significant that the flag of Cyprus ignores its two communities, the Greeks and the Turks, who also fly the flags of Greece and Turkey. The Star of David appears on the Israeli flag, whereas the red star on the flag of North Korea is the international symbol of Communism. Only the central symbols distinguish the flags of Iraq, Syria and Yemen.

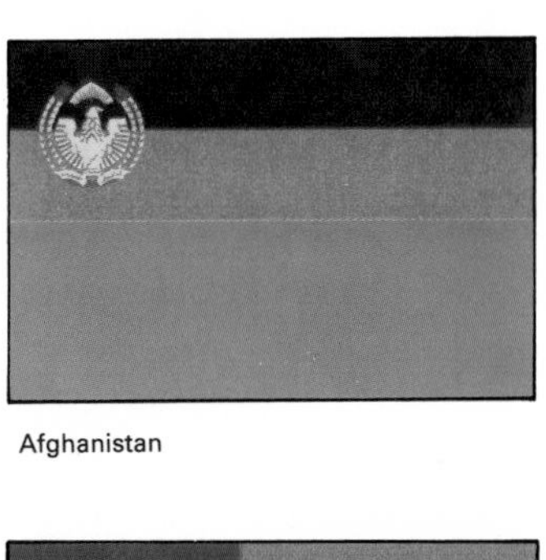

Afghanistan

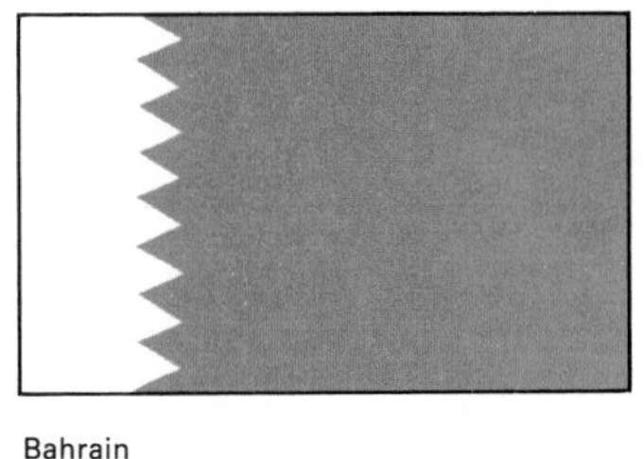

Bahrain

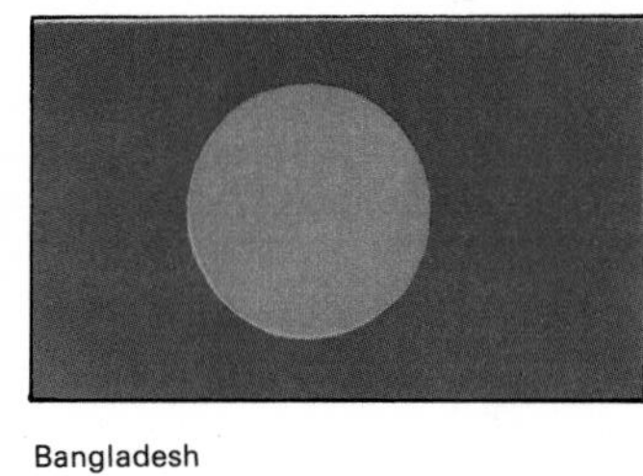

Bangladesh

Bhutan

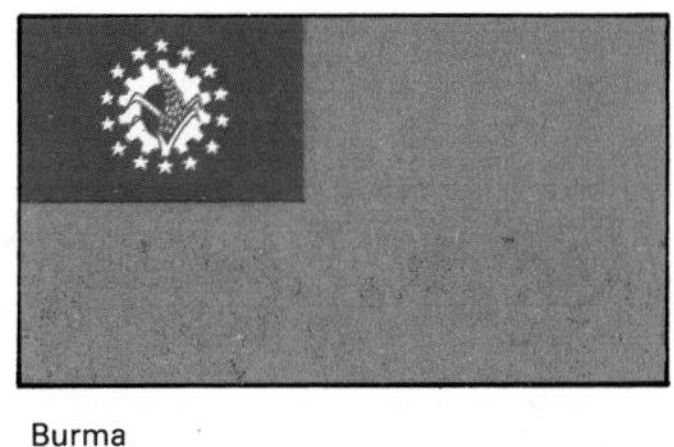

Burma

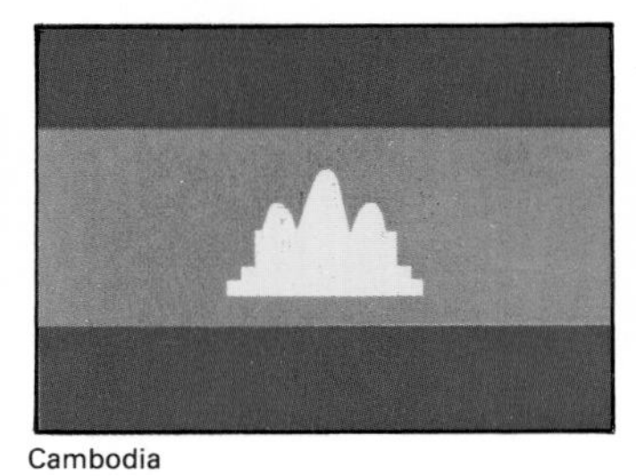

Cambodia

China, People's Republic

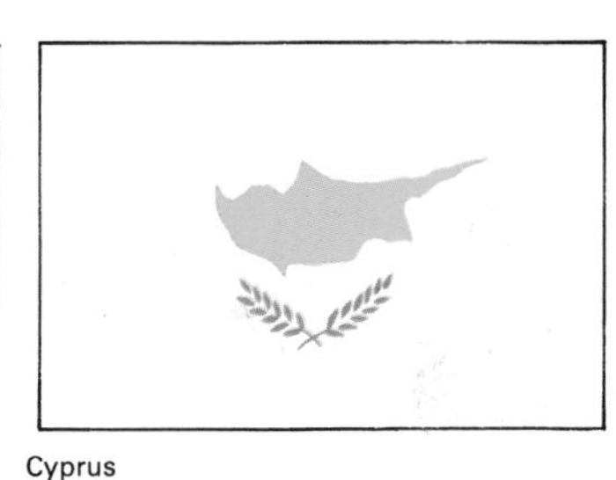

Cyprus

India

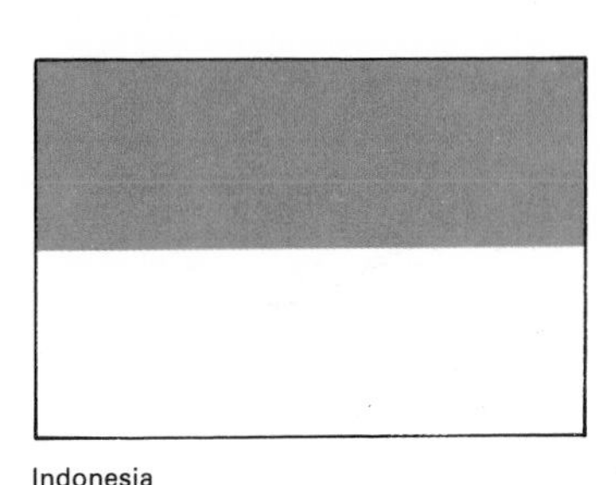

Indonesia

Iran

Iraq

Israel

Japan

Jordan

Korea, North

Flags of the world

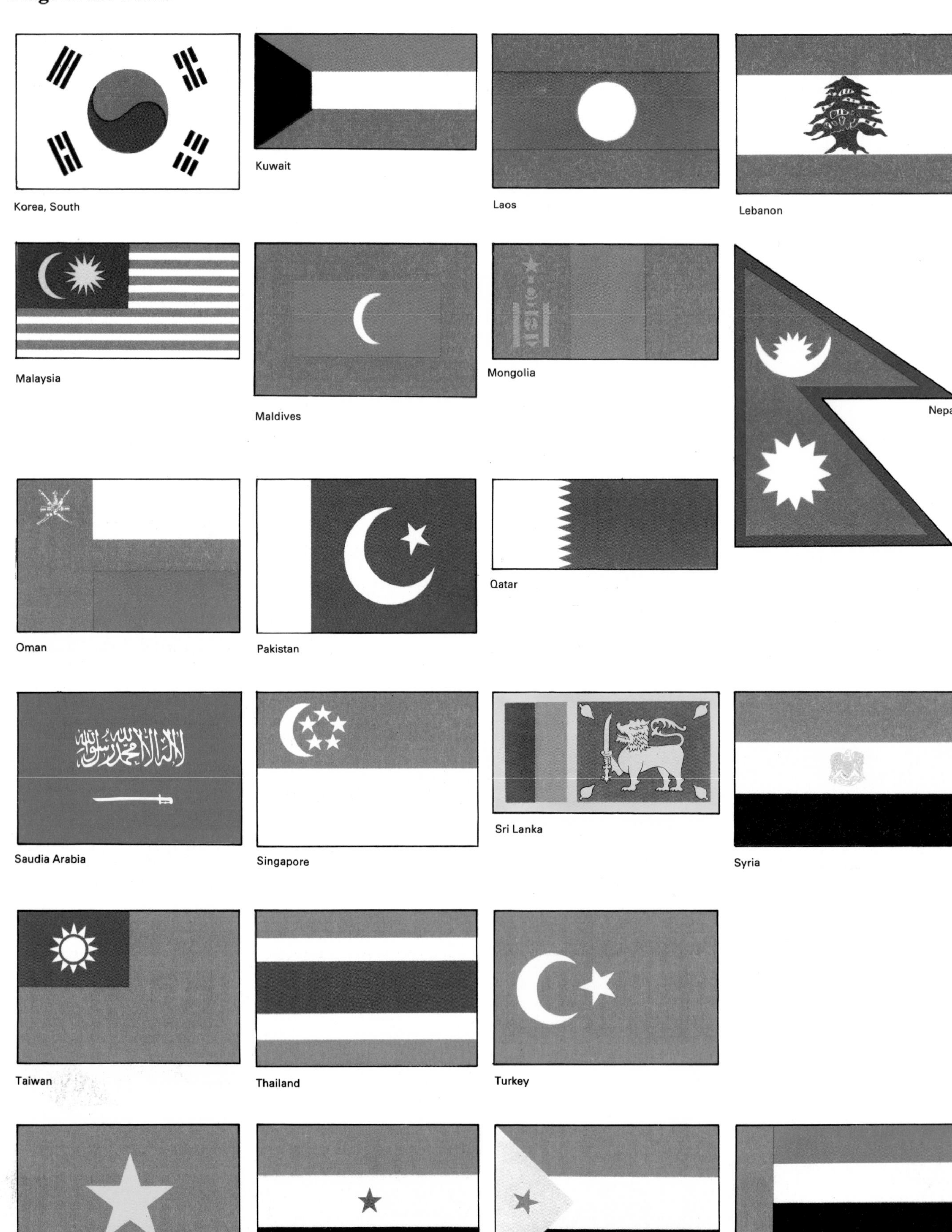

The Americas The country that wields the greatest power in the New World – the United States – also possesses the national flag with the longest and most complicated history. There have been no fewer than 28 different designs to date. The first (1775) was an adaptation of the British Red Ensign; instead of the red background the flag carried 13 red and white stripes to signify the 13 original states and the Union as a whole.

With independence, the British flag was dropped from the corner and replaced by 13 white stars on a blue field. The arrangement of the stars varied greatly over the years. In 1818 Congress decided to add a star for each new state: to date there are 50 such stars. It was not until 1912 that a standard pattern was established for the details of the flag.

Canada's first true national flag was introduced in 1965 to replace the British Red Ensign. The maple leaf, however, has always been Canada's national symbol and is instantly recognizable by people throughout the world.

Farther south, it is possible to discern signs of common colonial backgrounds. Venezuela, Colombia and Ecuador were all Spanish possessions. Their separation from Spain is similarly celebrated in their tricolour flags. In each case yellow represents the independent state's separation by the sea (blue) from Spain (red).

The complicated Brazilian flag is one of the few in the world to carry a message in words (*Ordem e Progresso* – "Order and Progress"). Words are infrequent on flags because they must be "readable" from a distance.

Several flags incorporate national symbols: the Mexican emblem of an eagle fighting a rattlesnake on a cactus refers to an old Aztec legend, and the Barbadian flag symbolizes the island's tropical nature, with Neptune's trident set over golden sand, and blue sky and sea.

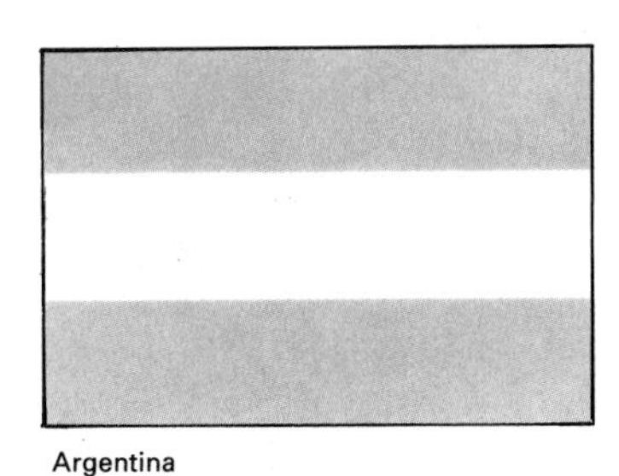

Argentina

Bahamas

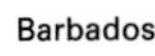

Barbados

Belize (not fully independent)

Bolivia

Brazil

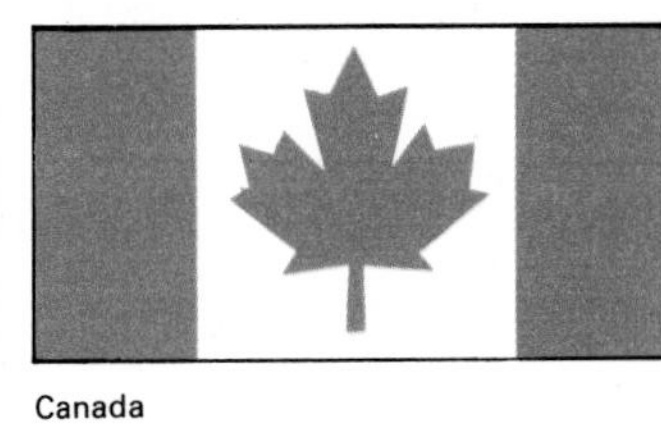

Canada

Chile

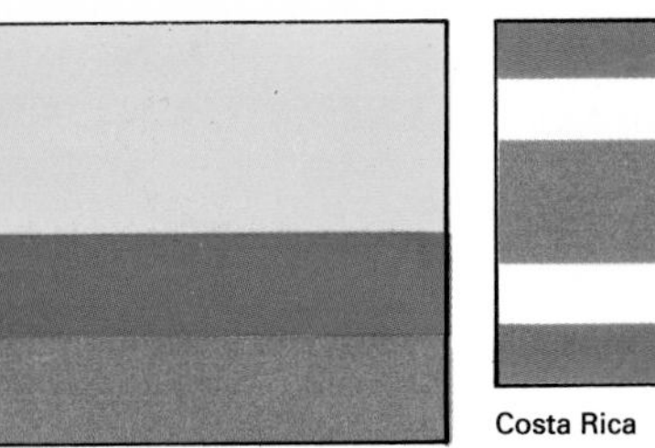

Colombia

Costa Rica

Cuba

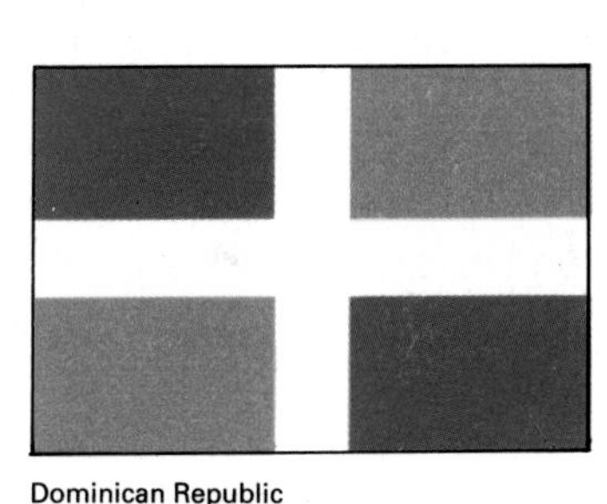

Dominican Republic

Netherlands Antilles (not fully independent)

Ecuador

El Salvador

Grenada

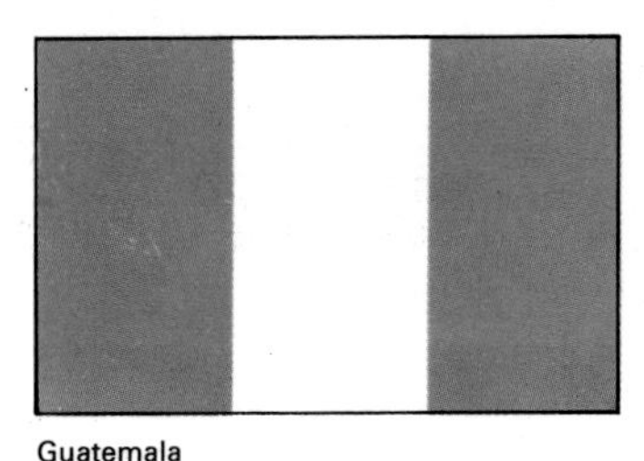

Guatemala

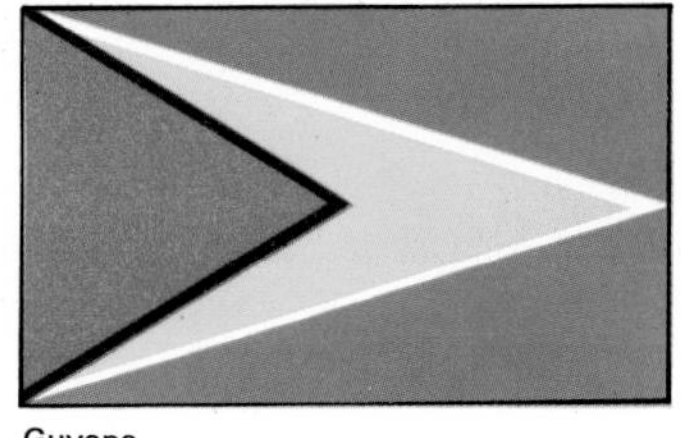

Guyana

Haiti

Honduras

Jamaica

Mexico

Nicaragua

Panama

Paraguay

Peru

Surinam

Trinidad and Tobago

United States

Uruguay

Venezuela

Australasia and Oceania Geographical and historical symbols are curiously mingled in some flags. Both the Australian and New Zealand flags take as their basis the British Blue Ensign which is associated with imperial days. But they also carry a representation of the Southern Cross. This is a salient feature of the night sky in these latitudes and hence implies geographical position. The stars in the New Zealand flag are of different sizes and their positioning is not symmetrical.

The Australian flag was chosen after a public competition in which 30,000 different designs were presented. The seven points of the largest star stand for the six states and the Northern Territory.

The Fijian flag, too, is the same as that used before the country won independence in 1970. Its shield combines English symbols, including the cross of St George, with native products such as bananas and coconuts and the old Fijian image of a dove with an olive branch.

Since gaining full independence from Australia in September 1975, Papua New Guinea continued to use the flag that it had adopted in 1971, with its striking design incorporating five stars and a bird of paradise.

The flag of Western Samoa similarly shows the Southern Cross, which denotes that country's association with New Zealand. The historical and geographical mix occurs even more specifically in the flag of Nauru, the world's smallest republic. It also features a star that has nothing to do with the Southern Cross but represents the island itself, and its positioning represents the island's actual proximity to the equator (the yellow line). The 12 points of the star stand for the 12 ethnic groups that originally inhabited the tiny island.

The influence of the Christian missionaries is recalled by the red cross of the Tongan flag. The red signifies the blood shed by Christ on Calvary.

Australia

Fiji

Nauru

New Zealand

Papua New Guinea

Philippines

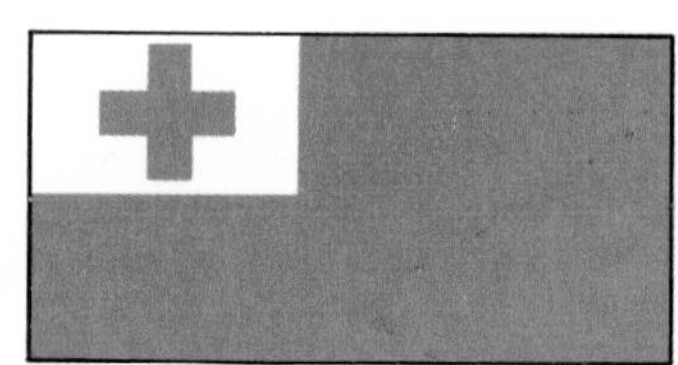
Tonga

Western Samoa

Africa The flamboyant flags of Africa are visible proof that the "wind of change" predicted in the 1960s has indeed blown across the continent. Most of the flags were raised for the first time during the period 1957 to 1975 as the various colonies achieved independence. Many of them are the youngest flags in the world.

Traces of colonialism survive in many African flags. Some are obvious – such as the British Union Flag incorporated in the flag of South Africa. Other colonial influences are more subtle. Most countries that were once French, for example, seem to have adopted the vertical tricolour; many countries with a colonial British background (such as Nigeria and Rwanda) have assumed tricolours with horizontal stripes.

One of the latter countries is Ghana, the first black state to become independent (1957) after World War II. Its flag provided a model for those of many later independent nations. Its colours – the pan-African colours of red, yellow and green first used in 1894 by Abyssinia (present-day Ethiopia), the oldest independent African state – were to become synonymous with African aspirations. Red represents the battle for independence, yellow represents mineral wealth and green signifies the great resources of those countries inherent in forestry, crop raising and other forms of agriculture.

The Ghanaian flag, like many belonging to the emergent African nations, also carries a star that stands for freedom. It may have been inspired by the United States flag, which was the first national flag to make use of symbolic stars.

There is absolutely no doubt, however, of the inspirational source of the Liberian flag. It was a group of former Americans who brought about the founding of Liberia (1822), the first black republic in Africa. The number of stripes on the flag commemorates the number of signatories to the Liberian Declaration of Independence.

During the colonial period Africa was divided up by the European powers in a fairly arbitrary way; frontiers frequently cut ethnic groups in half or bundled together others who had little or nothing in common. In short, the new African leaders have inherited countries that in many cases have little or no historical unity.

It is clear that the emerging nations have looked to their flags for help in changing this. Not only do the flags reflect the bubbling vitality of the newly free; some also seek to call up the African heritage of their peoples. The flag of Kenya is a case in point. The crossed spears and shield were deliberately placed on it to remind today's people of others who, much earlier, sought to defend their homelands. That the attainment of independence was not always peaceful is also reflected in the flag of Mozambique, which includes a modern automatic rifle among its symbols.

The South African flag represents the union of the two British colonies of Natal and Cape Colony with two Boer states, the Orange Free State and the Transvaal, that comprise the modern Republic of South Africa. The flags of the two Boer states appear close alongside the Union Jack on the orange, white and blue tricolour that was originally taken to South Africa by the first Dutch settlers in the 17th century.

Africa also has its share of Muslim nations. One of these, Egypt, spearheaded the struggle for Arab liberation. Its flag, with its pan-Arab colours of red, white and black, also provided a useful model for other Arab countries in the Middle East. Since 1972 Egypt, Libya and Syria, have used the same flags as a sign of religious and political solidarity; many other Arab states also use the red, white and black tricolour as a basis for their flags.

Arab states in Africa, such as Algeria, Mauritania, Morocco and Tunisia, have settled for the traditional simplicities of Islam; a one-colour ground with a star and crescent (symbols of peace and life) or just a star (as on the flag of Somalia). This simplicity is a visible reflection of traditional Islamic distrust of idolatrous images.

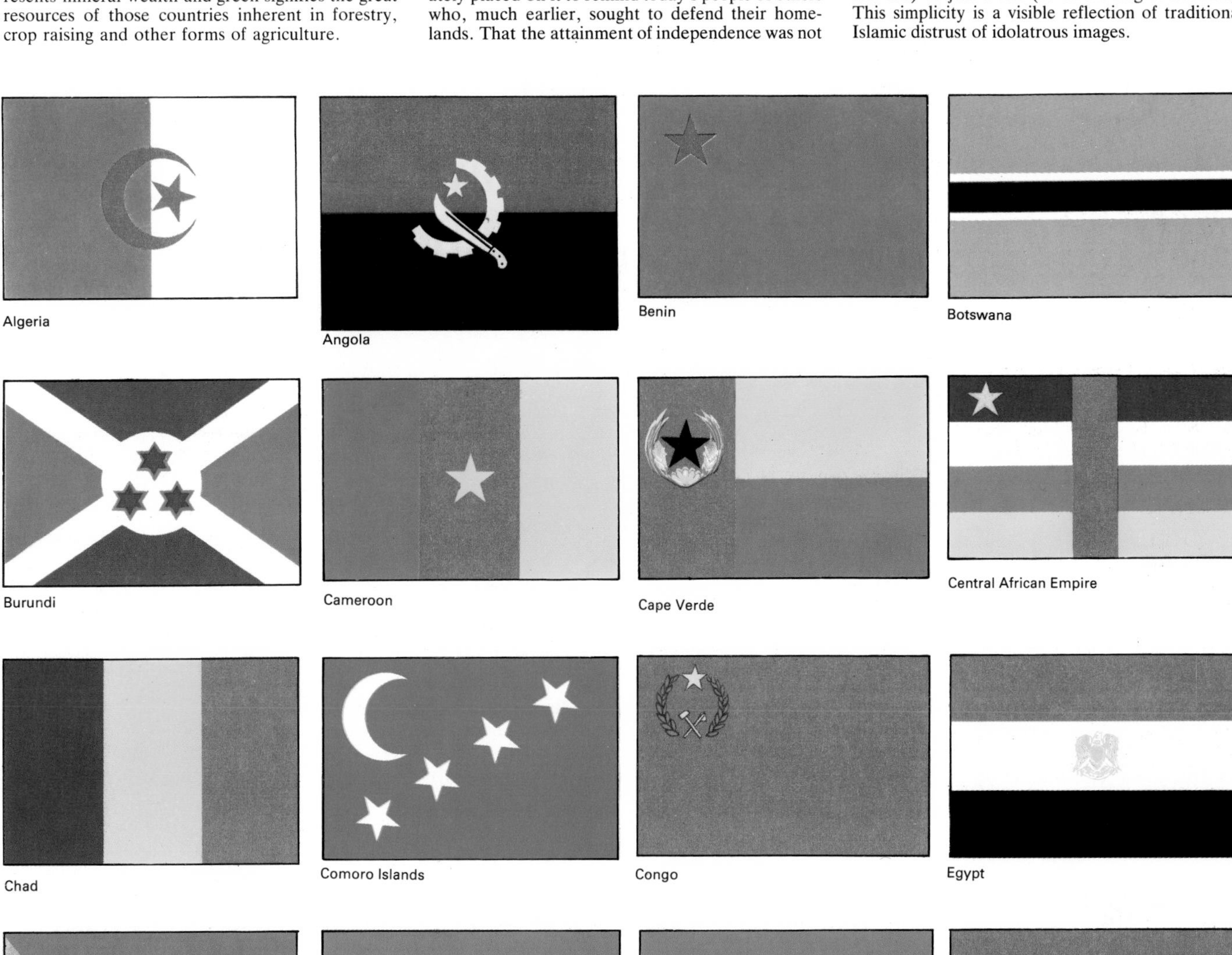

Algeria

Angola

Benin

Botswana

Burundi

Cameroon

Cape Verde

Central African Empire

Chad

Comoro Islands

Congo

Egypt

Equatorial

Ethiopia

Gabon

Gambia

Flags of the world

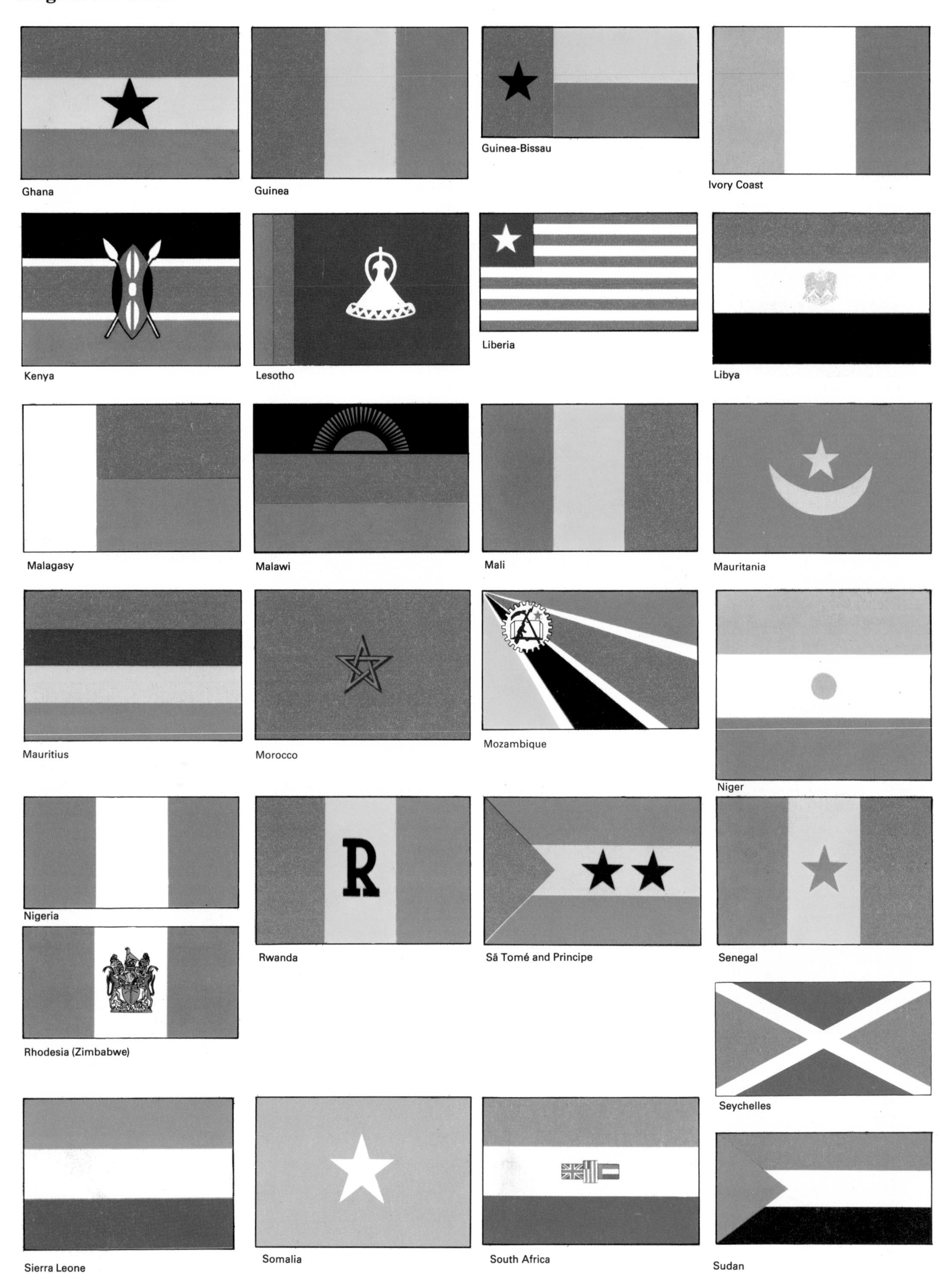

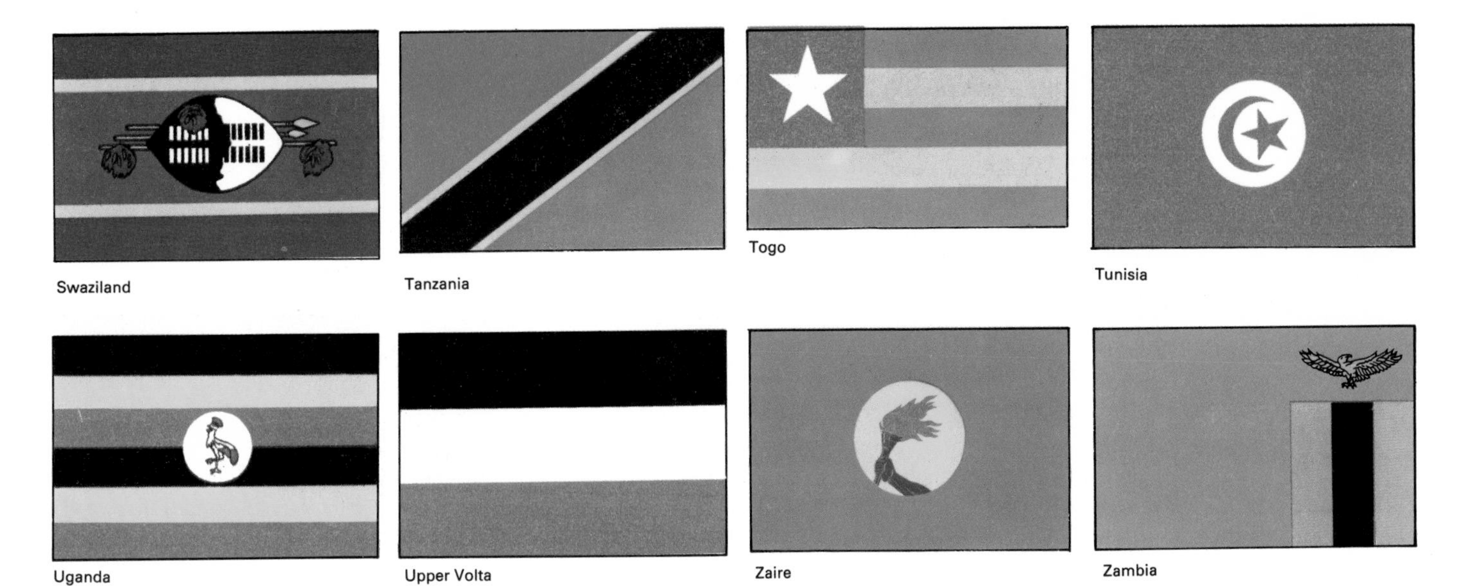

Europe Flags were probably introduced into Europe as a result of the Crusaders' contact with the Saracens. The Christian symbol of the cross, adopted by the Crusaders nearly 900 years ago, has since been profoundly influential in the history of flags and is particularly prominent in those of the Scandinavian countries – Denmark, Finland, Iceland, Norway and Sweden. One of the first flags is said to be that of Denmark which according to legend appeared from the sky in 1219. Its cross has added significance because Denmark once ruled the other countries; the crosses still represent that ancient affiliation. The Swiss flag, with colours reversed to a red cross on a red background, gave rise to the flag of the Red Cross.

The British Union Flag (often called the Union Jack) is a concise history of the United Kingdom; it successively combined the crosses of England, Scotland and Ireland. In 1606 the English cross of St George was combined with the saltire of St Andrew of Scotland to form the first Union Flag. In 1800 the red saltire of St Patrick was added to the flag, and the flag the United Kingdon has remained the same to this day.

Striped flags also have a long history. The tricolour, in red, white and blue, first became a symbol of liberty when it was taken up by The Netherlands in the long fight against Spanish domination during the sixteenth century. It is thought to have first been employed in 1579 (the red was originally orange). The idea received an even bigger boost when the French Revolution adopted the tricolour, this time in a vertical form. The flag of Italy is also a tricolour, but with green instead of blue, and its roots may be dated back to 1796, although the present flag dates from 1946. Only the shape and a subtle difference in the shade of the green stripe distinguishes it from the flag of the Republic of Ireland.

The French and American revolutions introduced great changes in the concept of a flag. With the ensuing growth of nationalism in the nineteenth century, flags became potent political symbols, acting as a focus for nationalist movements.

The history of Germany's flags is exceptionally complex. Since the unification of German states in 1871 the German national flag has been changed four times, until the current black-red-gold design was adopted by both the German states in 1949. East Germany added its state arms in 1959.

Since 1785, red and gold have been the official Spanish colours, although their heraldic source dates back into the thirteenth century. Portugal's flag includes an armillary sphere, a navigational instrument symbolic of the old Portuguese voyages of exploration.

The simple vertical bicolour of the Maltese flag is reputed to date back to 1090 (when the Normans took Malta from the Muslims), although this is in some doubt. In the top left corner of the flag there is the George Cross, which Britain awarded to the island for its bravery in World War II. Monaco, Poland and San Marino have flags with only two horizontal stripes, the red and white of Monaco looking like an inverted form of the white and red of Poland.

The spartan appearance of the flag of the USSR with its famous hammer and sickle has surprisingly not followed the spread of Communist governments. Most European Communist regimes have striped flags of one kind or another. Albania seems to follow the Russian lead, but its two-headed eagle on a red background is a symbol centuries old. Several Communist countries, however, including Albania, Yugoslavia and Romania, include on their flags a red star with a yellow outline, one of the symbols of Communism.

The flag of Andorra is alleged to have been instituted by Napoleon III of France in the 1860s. The tricolour embodies two of the colours from each of the flags of its neighbours – France and Spain – because Andorra is not a sovereign state but a condominium, with some local autonomy, ruled jointly by those two countries.

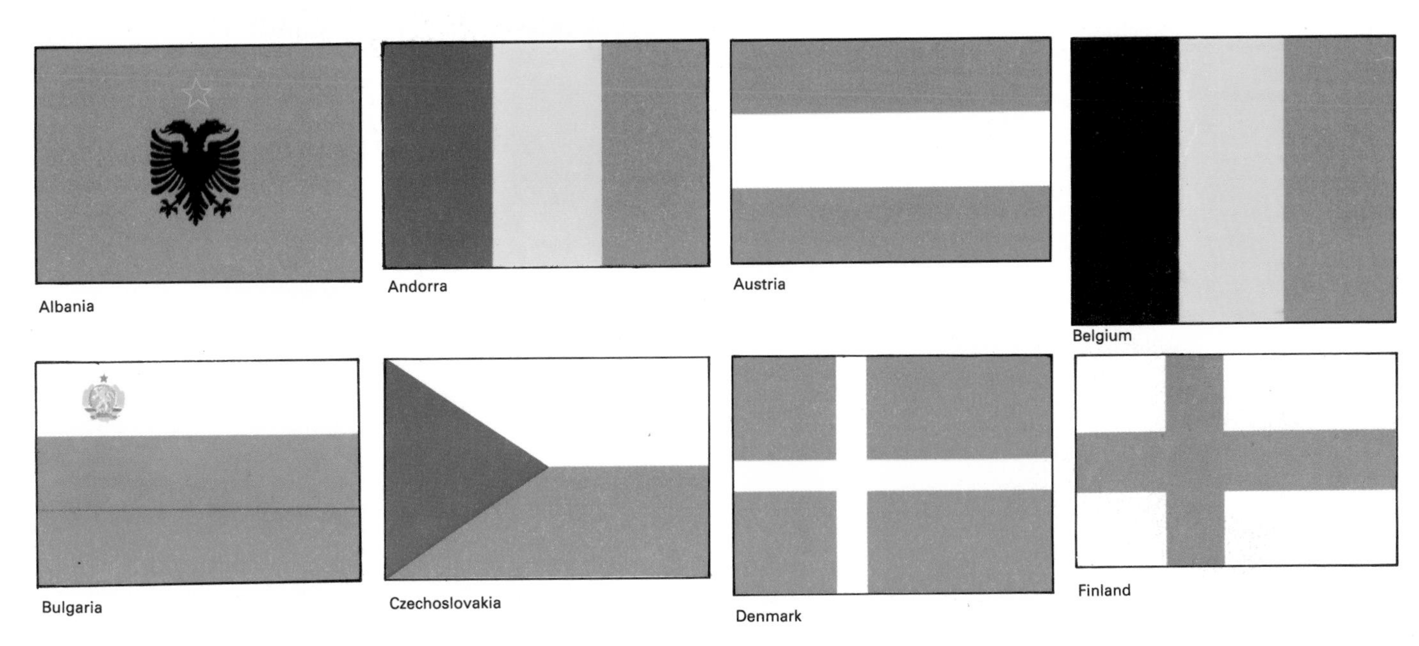

Flags of the world

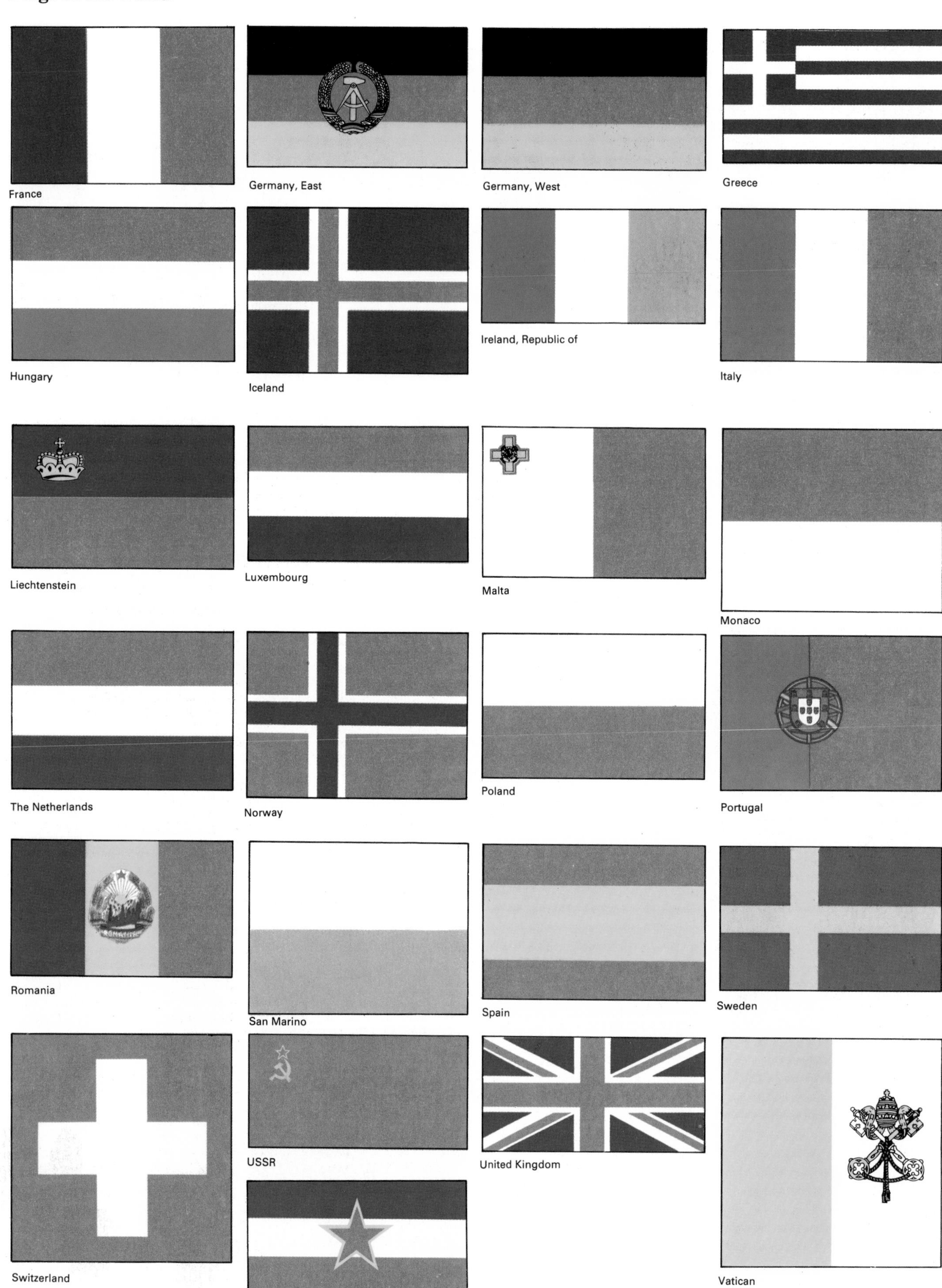

Nations of the world

References to information in other *Joy of Knowledge* volumes use one of the following codes followed by the relevant page number:

MM = Man and Machines
MS = Man and Society
NW = The Natural World
PE = The Physical Earth
SU = Science and the Universe

A page number printed in ordinary numerals indicates that additional information is contained within the text of an article – for example, NW p. 123 is a reference to the text on page 123 of *The Natural World*; a page number printed in italic numerals indicates that additional information is contained in an illustration or its caption – for example, MM p.*123* is a reference to an illustration on page 123 of *Man and Machines.*

The old Muslim section of Kābul, Afghanistan's capital and its largest city, provides a picturesque setting for one of the feast days.

Camels transport crops across Kandahār province, south-eastern Afghanistan; about ninety per cent of Afghanis are farmers or nomadic herdsmen.

Abyssinia. *See* ETHIOPIA.

Afars and the Issas. *See* DJIBOUTI.

Afghanistan (Afghānestān), official name Republic of Afghanistan, is a mountainous, landlocked country in central southern Asia. It is a poor nation with few resources (the average income is only about £30 per person per year); there are no railways and only in the 1970s did industry start to develop. The capital is Kābul.

Land and economy. Three-quarters of Afghanistan is mountainous; the range called the Hindu Kush (meaning Killer of Hindus) extends for 1,000km (620 miles) across the country with peaks up to 6,500m (21,000ft) and the Pamirs in the Wakhan rise to 6,700m (22,000ft). Farming is the chief occupation, followed by 85 per cent of the population, although only 14 per cent of the land is

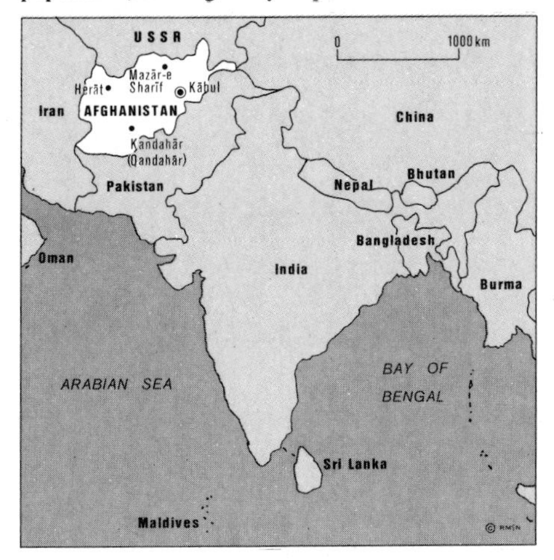

suitable for agriculture (much of the remainder being too dry or even desert). Wheat is the staple diet, and rice, barley and maize are also grown; the main cash crops are cotton and sugar-beet. Afghanistan is famous for its fruits (especially grapes and melons) which, either fresh or dried, are exported to many parts of the world. Farmers raise great numbers of sheep, including the valuable Karakul ("Persian lamb"). Apart from cement works and workshops, such industry as exists is agriculturally based – for example, woollen and cotton textiles and cottonseed oil. Natural gas has been found in the north of the country. Poor quality coal is mined, but the huge deposits of rich iron ore have not yet been exploited.

Foreign aid. Afghanistan pursues a strictly neutralist policy and has accepted substantial development aid from many countries. Principal donors are the United States and the USSR since the 1960s, and Arab oil states and Iran since 1975. Major recent achievements include irrigation schemes, hydro-electric installations and highways.

People. No nationwide census has ever been taken in Afghanistan and so the population can only be estimated. It is unusually mixed because of the many migrations and conquering armies that have swept over this "Cross-roads of Asia" for 2,600 years. The Pathans are the dominant and most numerous group, inhabiting mainly the eastern part of the country, followed by the Tajiks, Uzbeks, Turkmen and Hazararas. There are also about three million nomadic Kuchis who move their flocks of sheep and goats over long distances, travelling regularly between southern Pakistan and the USSR. The Afghans are zealous Muslims, most of whom belong to the Sunni sect. Educated women in the cities no longer wear a veil; purdah (the seclusion of women) is still common, however, in the countryside. Despite a keen demand for education, the level of illiteracy still exceeds 90 per cent. There are no rich people by Western standards and, because the style of dress and diet are much the same for all, glaring contrasts of wealth and poverty are absent. Even the cities show few signs of Westernization. There are two official languages, Pushtu (Pathan) and Dari (Persian).

Government. Afghanistan was "an autocracy tempered by assassination" and by periodic uprisings of the turbulent Pathan tribes until 1964, when a new constitution was adopted whereby King Mohammed Nader Khan became a constitutional monarch. In 1965 the country held the first general election under direct universal adult suffrage for a new National Assembly. This democratic experiment ran into trouble and the executive and the legislature became deadlocked. In 1973 Prince Mohammed Daoud Khan (Prime Minister 1953-63 and cousin of the king), supported by the army, seized power, abolished the monarchy and the National Assembly, and established a republic with himself as president. He rules by means of presidential decrees.

History. Afghan civilization is ancient and its development has been affected by a series of invaders, including Alexander the Great and Genghis Khan. The country is strategically placed for raiding the rich plains of India through the Khyber Pass east of Kābul, and short-lived empires have been centred in Afghanistan by the Bactrian Greeks (3rd-2nd centuries BC), the Buddhist Kushans (1st century AD), the Muslim Ghaznavids (10th-12th centuries AD), Tamerlane (late 14th century) and Babur, who conquered northern India and set up the Mogul Empire in 1526. As a result Afghan culture derives from many sources; there are monumental Buddhist remains at Bamian, and Greek, Persian and Indian influences are evident, although Afghan art is primarily Muslim.

In 1747 Ahmad Shah Durrani, the first national ruler, rebelled against his Persian overlord Nader Shah and created the state thereafter known as Afghanistan. In the 19th century the country became squeezed between the territorial gains of the advancing Russian and British empires. At the end of the Second Anglo-Afghan War in 1880 Afghanistan became a buffer state with Britain controlling its foreign policy until 1919, when it gained full independence. In the early 1970s, President Daoud tended to re-open the dispute with Pakistan over Pathan-inhabited territories lost in the 19th century, but tension was relaxed during 1976. An attempted army coup in November 1976 was discovered before it could be implemented. Maps 40, 42.

Afghanistan – profile

Official name Republic of Afghanistan
Area 647,497sq km (249,999sq miles)
Population (1976 est.) 19,796,000
Density 30 per sq km (77 per sq mile)
Chief cities Kābul (capital) (1973 est.) 318,094; Kandahār, 134,000; Herāt, 62,000
Government Head of state, Mohammed Daoud Khan, president (seized power July 1973)
Religion Islam
Languages Pushtu, Dari
Monetary unit Afgháni
Gross national product (1974) £606,800,000
Per capita income £31
Trading partners USSR, India, USA, Japan

Alabama. *See* UNITED STATES.

Alaska. *See* UNITED STATES.

Albania (Shqiperia or Shqipnija), official name Socialist People's Republic of Albania, is a small independent nation on the Balkan Peninsula in south-eastern Europe. It has been a Communist state since 1945. The country broke its ties with the USSR in 1961 and became the first European country to ally itself closely with China. The capital and largest city is Tiranë.

Land and economy. Albania lies on the Adriatic Sea, bordered by Yugoslavia to the north and east and by Greece to the south. The interior of the country consists of highland plateaus and high mountain ranges such as the rugged North Albanian Alps. The limestone rocks do not support good soil; the most fertile lands are in the valleys of the south. Several westerly flowing rivers descend through narrow, deeply cut valleys to the coastal lowlands. Few people live in these marshy coastal plains, and Durrës is the only sizeable seaport. The geography of Albania largely explains the country's longstanding economic hardship. The Soviet-styled economy is about evenly divided between agricultural production and industrialization. Substantial industrial progress has been achieved since 1945 in a series of five-year plans, particularly in the production of

Many Albanian women still wear traditional clothes, which reflect the influence the Turks had when they ruled the country between the 15th and 20th centuries.

The modern French-built sector of Algiers, capital of Algeria, overlooks the Mediterranean. The city has been a major North African port since the 900s.

Oil and natural gas are Algeria's principal natural resource and earner of foreign revenue; some is processed at oil refineries in the north.

electric power, coal, cement, textiles and petroleum products. Farm output has increased much faster than has the population since 1950. Foreign economic aid is needed each year to offset deficits in the country's trade.

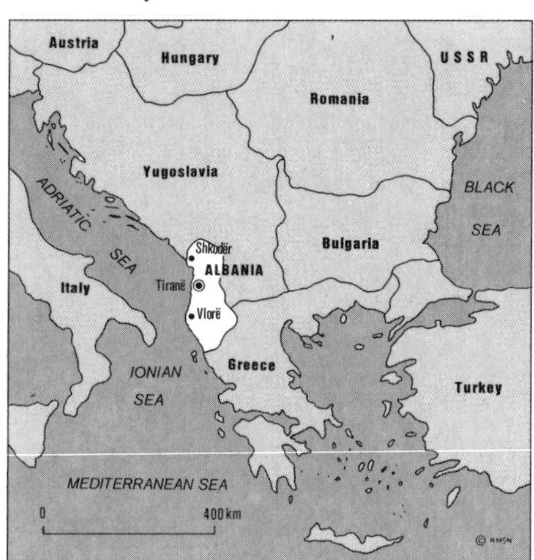

People. The two major tribal groups are the Ghegs, in the north, and the Tosks, in the south. Albanian is spoken by both groups, but the dialect of the Tosks is the official language. Two-thirds of the people are Muslims of the Sunni and Bektashi sects; the remainder are Christians of the Eastern Orthodox and Roman Catholic churches. Religious activities are strongly discouraged.

Albania – profile

Official name Socialist People's Republic of Albania
Area 28,749sq km (11,098sq miles)
Population (1976 est.) 2,549,000
Density 89 per sq km (230 per sq mile)
Chief cities Tiranë (capital) (1973) 182,500; Durrës, 57,000; Shkodër, 55,000
Government Communist, first secretary Enver Hoxha (re-elected 1976)
Religions Muslim, Eastern Orthodox
Language Albanian (official)
Monetary unit Lek
Gross national product (1974) £538,400,000
Per capita income £211
Agriculture Corn, wheat, potatoes, tobacco
Industries Mining, food processing, textiles, petroleum products, cement
Minerals Coal, natural gas, iron, chromium, copper
Trading partners China, Czechoslovakia, Poland, East Germany

Government. The constitution is styled on the Soviet model, and provides for a legislature (the People's Assembly), an executive branch (the Presidium) and an independent judiciary. True power, however, lies with the politburo – the central committee of the Albanian Party of Labour, whose first secretary is commander of all armed forces.

History. Albanians are descended from the Illyrians, who settled in the Balkan Peninsula in the pre-Christian era. For centuries the Albani tribe resisted the influences of the Romans and the Byzantines. Turks gained control over the Albanians in the 15th century, and Islam spread to all but the most mountainous regions. A national revival began in the 19th century and culminated in a declaration of independence in 1912. The Communist Party formed in World War II with Enver Hoxha as its head, seized power in 1944. Albania became a "client state" of the USSR and a member of the United Nations in 1955. The break with the USSR in 1961 severely damaged the country's economy, and China sought to replace the USSR as Albania's chief economic benefactor. Map 26.

Alberta. *See* CANADA

Alderney. *See* CHANNEL ISLANDS.

Algeria (Algérie), official name Democratic and People's Republic of Algeria, is the second largest country in Africa (only Sudan is bigger). Located in the northern part of the continent, the country has a history dominated by invasion and foreign rule from Asia and Europe. Its last colonizers were the French, who left in 1962. Today Algeria is governed by the army-backed National Revolutionary Council from the capital, Algiers, and parts of the land and industry have been nationalized.

Land and economy. Algeria lies in north-western Africa and has a 1,025km (640-mile) coastline on the Mediterranean Sea. It has two clearly defined regions: the northern cultivated region of the Mediterranean littoral and the Atlas Mountains, and the great barren wastes of the Sahara. The country's life is concentrated in the north, which has most of the population and all the large towns. The maritime strip has some of the most fertile farming land in Africa. Fields of wheat, barley and vegetables are interspersed with vineyards and groves of olives and citrus fruits. This region has a moderate Mediterranean climate, with adequate rainfall for agriculture.

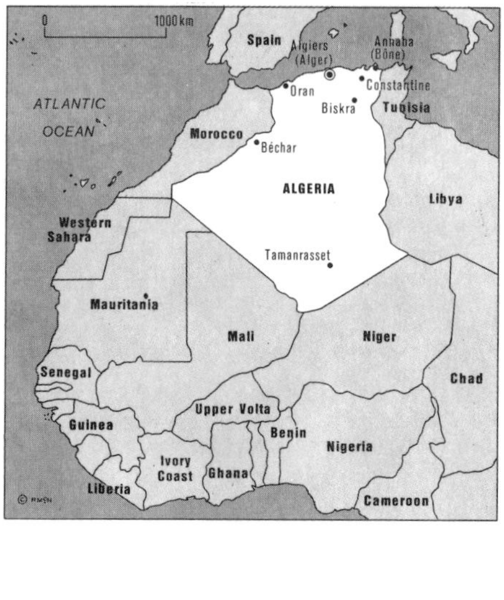

Farther south there are high plateaus covered with grass, and to the south of these are the mountains of the Saharan Atlas. Among them are the Ksour, the Ouled Nail and the spectacular Aurès ranges. The sands, hills and bleak plateaus of the Sahara lie beyond, making up nine-tenths of the country's area. In the south-east the desert rises to the Ahaggar Mountains. The larger oases have small settled communities subsisting on sparse crops (chiefly maize and dates), and there is also a thinly spread nomadic population of herdsmen. The desert is the source of Algeria's most valuable export, petroleum, first located in 1956 and now accounting for 70 per cent of exports. Major deposits of natural gas and iron ore exist in the Sahara, and there are phosphates, coal, zinc and other minerals in the Atlas. The country's industries include the manufacture of textiles, chemicals, steel and food products. Sixty per cent of the population is engaged in agriculture, and in the period 1967-71 a system of worker-managed plants and farms was instituted to develop nationalized properties.

People. Most of the people are Arabic-speaking Muslims, although many also speak French. They are descendants of invaders from Phoenicia, Rome, Arab lands, Turkey and France, intermixed with local Berber tribes. There are about 50,000 foreign technicians and teachers in the country.

Government. A strong, central government controls the country, with policy-making in the hands of the National Revolutionary Council. Rule is by decree.

History. Conquered by Rome after the fall of Carthage in 146 BC and then by the Arabs in the 7th century, the territory was colonized by France in 1830 and annexed as an overseas département with representation in the French Assembly; control remained in French hands. The Algerian push for independence led to a terrorist campaign beginning in 1954, initiated by the National Liberation Front (FLN). In 1962, France signed the Evian accord, which provided for interim economic, cultural and technical relations until a referendum on self-determination could be held. On 1 July 1962 the referendum took place and Algeria was declared independent. In 1965 Premier Ahmed Ben Bella was deposed in a bloodless coup by Col. Houari Boumédienne, and the constitution was suspended. Map 32.

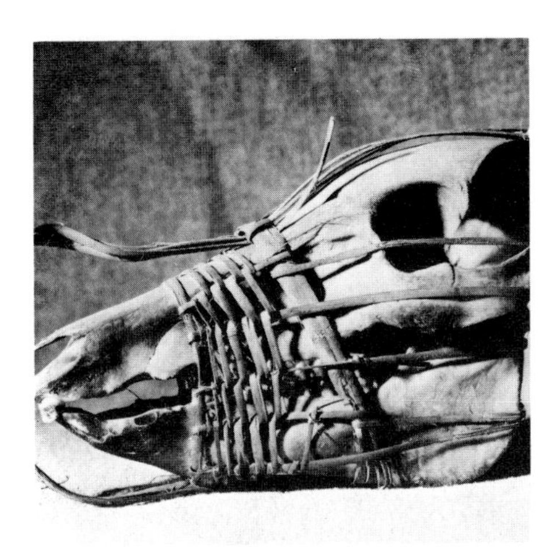

Wild boars are still hunted by bow and arrow on the Andaman Islands. The bow is the Islands' only indiginous weapon and is also used for fishing.

Modern hotels in Andorra la Vella, the largest town and capital of Andorra, have been built to cater for the many tourists who visit the town.

Luanda, capital city of Angola, has a natural harbour with a flourishing port; for 300 years it was the centre of a valuable slave trade.

Algeria – profile

Official name Democratic and People's Republic of Algeria
Area 2,381,741sq km (919,590sq miles)
Population (1975 est.) 16,776,000
Density 7 per sq km (18 per sq mile)
Chief cities Algiers (capital) (1975 est.) 1,179,000; Oman, 327,493; Constantine, 243,600.
Government Military, president Col. Houari Boumédienne
Religion Islam
Languages Arabic, French
Monetary unit Dinar
Gross national product (1974) £4,204,800,000
Per capita income £251
Agriculture Wheat, barley, corn, oats, flax, tobacco, olives, dates, livestock
Industries Wine, iron ore, olive oil, natural gas, petroleum products, leather goods
Minerals Oil and natural gas, iron, zinc, lead, mercury, coal, copper
Trading partners France and other members of the EEC, USA, USSR

American Samoa, the eastern part of the Samoa island group in the southern Pacific Ocean, is a territory of the United States, comprising the islands of Tutuila (site of the capital, Pago Pago), Aunu'u, Rose, Swain's and the Manu'a group. Under the control of local chiefs until about 1860, the islands were granted to the United States in 1899 by a treaty with the former co-administrators, Germany and Great Britain. The 1967 constitution allows for a local legislature to raise money from the islands' income. Exports: canned fish, copra, local craft goods. Area: 197sq km (76sq miles). Pop. (1975 est.) 29,000.

Andaman and Nicobar Islands is an Indian territory in the Bay of Bengal; its capital is Port Blaie, in the Andamans. A home minister in the central Indian government administers the territory, which exports timber and tropical fruits. There are more than 200 islands in the Andaman group and 19 in the Nicobars, which lie 145km (90 miles) to the north. The British made an abortive attempt to colonize the Andamans in the 1790s and made them a penal colony from 1858 to 1945; today's population consists of Negritos and settlers from India. The Nicobars were British territory from 1869 to 1945 (but occupied by the Japanese during World War II) and are populated mainly by people of Mongoloid extraction. Area: 8,293sq km (3,202sq miles). Pop. (1971) 115,133. Map 52.

Andorra is a small independent state (although it still has to pay dues to the president of France and the Spanish bishop of Urgel) situated between France and Spain, high in the eastern Pyrenees. Large flocks of sheep are kept in the high mountain valleys, where the soil is too poor for growing crops. Income from livestock is today supplemented by an increasing tourist trade. Andorra la Vella is the main town, and most of the people are Roman Catholics who speak Catalan. Area: 453sq km (175sq miles). Pop. (1975 est.) 27,000. Map 22.

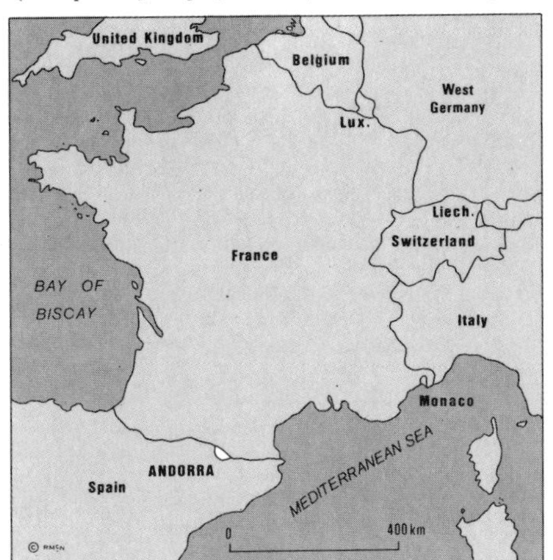

Angola, formerly Portuguese West Africa, is an independent country of western Africa to the south of the Equator. A Bantu nation rich in petroleum, coffee and diamonds, it has been torn by civil war. The capital is Luanda.

Land and economy. Angola is made up of two regions. Angola proper is south of the River Congo, bordered by Zaire to the north and north-east, Zambia to the east and south-west, Namibia (South West Africa) to the south and the Atlantic Ocean to the west. The exclave of Cabinda, a low-lying coastal rain forest, is separated from Angola proper by Zaire. Most of central Angola lies on the Bié plateau, 900-2,100m (3,000-7,000ft) above sea-level. Hydroelectric power is generated by the plateau's falls and rapids. The Moçâmedes desert is in the south. The main rivers are the Cuanza and Cunene, and Luanda and Lobito are the principal ports. Varying altitudes, cold ocean currents, and low rainfall make the north of the country tropical and the south semi-arid. The most important economic development came in 1966 with the discovery of oil off Cabinda and expansion of the Cassinga iron mines. Coffee is the main cash crop, and diamonds are a principal industry.

People. Most Angolans are Bantu black Africans from one of four tribes, the Ovimbundu, Bakongo, Kimbundu and Chokwe. Most speak a Bantu dialect, and Portuguese is common to the country. Roman Catholicism is the state religion, although many tribes retain their own ethnic beliefs.

Government. Warfare has divided Angola into three insurrection groups: the Revolutionary Government of Angola in Exile, the Popular Movement for the Liberation of Angola (MPLA) and the National Union for Total Independence of Angola.

History. The Portuguese explorer Diogo Cão landed on the coast in 1483, befriending the African king. Portugal remained in power, with the exception of Dutch occupation from 1641 to 1648. Angola was a primary source of slaves for Brazilian coffee

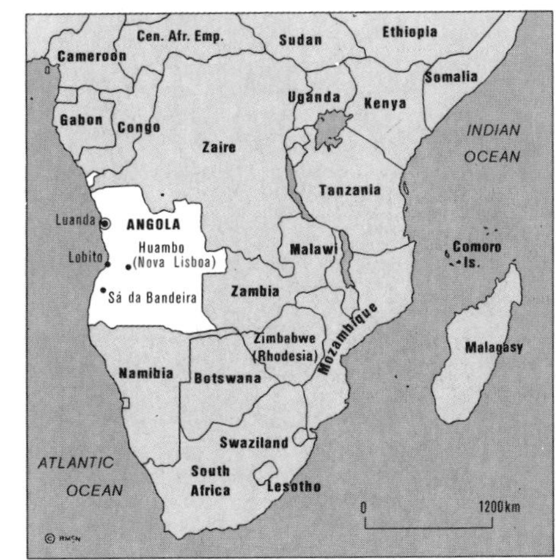

Angola – profile

Official name Angola
Area 1,246,700sq km (481,350sq miles)
Population (1976 est.) 6,761,000
Density 5 per sq km (14 per sq mile)
Chief cities Luanda (capital) (1970) 475,328; Huambo (Nova Lisboa), 89,000; Lobito, 74,000
Government: Transitional (nationalist military parties)
Religions Roman Catholic (state religion) and local cults
Language Portuguese (national)
Monetary unit Escudo
Gross national product (1974) £1,440,000,000
Per capita income £212
Agriculture Coffee, corn, sugar, cotton, wheat, tobacco, sisal, cattle
Industries Food processing, brewing and bottling, cement, glass, paper, cotton, footwear, soap, fishing
Minerals Diamonds, petroleum, iron, copper, manganese, sulphur, phosphates
Trading partners Portugal, EEC, USA

The magnificent Teatro Colón is Buenos Aires' leading opera house. It is the home of the national symphony orchestra and headquarters of the Argentinian ballet.

Part of the musical heritage of northern Argentina is the instrument called the *erke* – a long but light horn with a reedy sound.

Sea birds wade in the shallows at Tierra del Fuego in Argentina's far south; in 1520 Magellan named it "Land of Fire" after local people's cooking fires.

plantations in the New World. After World War II nationalist Angolans sought autonomy, and uprisings continued until a new Portuguese government offered independence in 1975. Soviet-backed forces and Cubans overcame Western-backed factions. The savage civil war prompted most of the white people to leave the country, caused the economy to decline, and made Cabinda seek secession. The country officially became independent on 11 November 1975. By 1976 the Soviet-supported MPLA controlled the government and most of the land area. Map 34.

Anguilla is a small island of the Leewards group in the West Indies. The economy is based on fishing, livestock farming and the production of salt. The island was discovered in 1493 by Christopher Columbus and settled by the British in the 17th century. In 1967 Anguilla joined with Saint Kitts and Nevis to form the self-governing states of Saint Kitts-Nevis-Anguilla but left after a few months, claiming it was being discriminated against. In 1971 it re-adopted British colonial status. Area: 91sq km (35sq miles). Pop. (1971 est.) 6,000.

Antigua is a self-governing island in the Leeward Islands group of the West Indies; its capital is St John's. Occupied by both French and Spanish, it was finally colonized by the British in the late 1600s. Antigua is a coral and volcanic island with natural harbours; it has a US military base. Industries: cotton, sugar cane, tourism. Area: 280sq km (108sq miles). Pop. (1975) 65,525. Map 74.

Argentina, official name Argentine Republic, is a South American country that occupies most of the southern part of the continent. In the Americas, only Canada, the United States and Brazil are larger. Most of the people, Spanish-speaking and Roman Catholic, have European backgrounds. Underdeveloped in some ways, Argentina prospers when labour relations and political stability permit it. The capital is Buenos Aires.

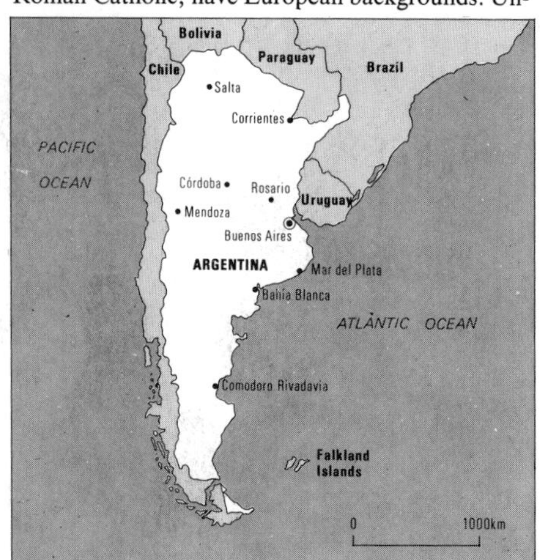

Land and economy. The Andes mountains extend along western Argentina, separating it from Chile. Several peaks rise above 6,100m (20,000ft), and Aconcagua is the highest mountain in South America at 6,960m (22,945ft). Northern Argentina has a hot, humid area, the Chaco, covered with grasses and with only a few trees. Central Argentina is the plains, or pampas, region where livestock and cereal crops thrive in the temperate climate. Southern Argentina, called Patagonia, is cold and mountainous; it is sheep country. Numerous rivers border or cross Argentina, with the Paraná and Uruguay joining to form a wide estuary, the Río de la Plata (River Plate). Minerals, largely unexploited, occur in the mountains. The country is second to Venezuela among petroleum producers of Latin America and has large deposits of natural gas. During the World Wars Argentina industrialized, but still imports many manufactured goods. Exports, including beef, are mainly agricultural.

People. Wandering Indian tribes occupied the land until Spaniards drove them out or mixed with and absorbed them. A large mestizo population resulted; these people of mixed American Indian and European blood became the gauchos, or cowboys, of the pampas. Late in the 19th century, Europeans – particularly Italians, Spaniards, Welsh and Germans – flocked to the country. Their descendants and the mestizos have gradually lost much of their separate identities and most have adopted Roman Catholicism and taken Spanish as a common language. About 68 per cent of the people live in and around the capital, Buenos Aires.

Government. The constitution is patterned on that of the United States. It requires citizens over 18 to vote, and they elect by secret ballot the president and vice-president – who must be Roman Catholics – and members of the Senate and Chamber of Deputies. The governments of the 22 provinces are subordinate to the central government.

History. Under the Vice-royalty of Peru, Argentina suffered by being far from Lima, but in 1776 Spain formed the Vice-royalty of La Plata, with headquarters in Buenos Aires. When Napoleon went to war with Spain in 1808, independence movements spread, and Argentina became free in 1816. Political coups have caused turmoil throughout Argentine history. In 1946, Col. Juan Perón gained popularity with the workers and won the presidency. He became a dictator, but the army overthrew him in 1955. From exile he rebuilt his political influence and regained the presidency in 1973, with his third wife, María Estela (Isabel), as vice-president. He died in 1974, and Mrs Perón became Latin America's first female president. Later, financial scandals, inflation and guerrilla warfare in the provinces brought down her regime and a military junta assumed power in March 1976. Map 78.

Argentina – profile

Official name Argentine Republic
Area 2,776,889sq km (1,072,157sq miles)
Population (1976 est.) 25,719,000
Density 9 per sq km (24 per sq mile)
Chief cities Buenos Aires (capital) (1975 est.) 2,976,000; Rosario, 810,840; Córdoba, 798,663
Government Head of state, president Lt.-Gen. Jorge Rafaél Videla
Religion Roman Catholic
Language Spanish
Monetary unit Peso
Gross national product (1974) £20,041,000,000
Per capita income £779
Agriculture Cattle, wheat, maize, grapes, sorghum, oats, wool, hides, linseed oil
Industries Iron and steel, cars, machinery, textiles, petroleum products
Minerals Petroleum, natural gas, lead, zinc, tin, manganese, iron, copper, beryl
Trading partners EEC, Brazil, USA

Arizona. *See* UNITED STATES.

Arkansas. *See* UNITED STATES.

Armenia. *See* UNION OF SOVIET SOCIALIST REPUBLICS.

Ascension. *See* SAINT HELENA.

Australia, official name Commonwealth of Australia, forms the major part of the continent of Australasia. It is located in the Southern Hemisphere, bounded on the west and south by the Indian Ocean and on the east by the Coral Sea and Tasman Sea, the western limits of the southern Pacific Ocean.

With more than 13 million people and an area of 7,700,000sq km (3,000,000sq miles), Australia is a sparsely populated country in terms of population density – 1.8 people per sq km (4.6 per sq mile), compared with 350 per sq km (910 per sq mile) in England, for example. But it is a highly urbanized country; 85 per cent of the people live in towns and cities, most of which are located along the east coast. Australians have high incomes and consequently enjoy high living standards and their country continues to prosper as it rapidly develops into a major world power.

Since 1901 Australia has been a federation of six states – New South Wales, Queensland, South Australia, Tasmania, Victoria and Western Australia – and two internal territories, Northern Territory and Australian Capital Territory, which lies within New South Wales and is the location of the capital, Canberra (each of these is described in detail later in this article). Before federation each state func-

The fast-growing city of Canberra was originally planned by US architect Walter Burley Griffin in 1911; it replaced Melbourne as federal capital in 1927.

About half of Queensland lies in the tropics; sugar cane is the main crop of this mostly agricultural state of north-eastern Australia.

Newcastle, second city of New South Wales, is the port and industrial centre for the largest coal-mining and processing area in Australia.

tioned autonomously as a separate British colony.

Land and climate. Australia is the lowest of the continents, with an average elevation of less than 300m (1,000ft), compared to the world's average of about 700m (2,275ft). Three-quarters of its area is a vast, ancient peneplain and only about five per cent of the continent is more than 600m (2,000ft) above sea-level [*see* PE p.*107*].

The principal structural division is the Great Western Plateau, which covers virtually the whole of Western Australia, most of the Northern Territory and much of South Australia. Most of it – the "dead heart" of Australia – is practically desert in which nothing grows but sparse grass and scrub [*see* NW pp.204–205].

Average annual temperatures vary considerably from about 27°C (81°F) in the north to 13°C (55°F) in the far south. In January (midsummer) temperatures range from 29°C (84°F) in the north to 18°C (64°F) in the south, whereas in July (midwinter) the corresponding figures are 24°C (75°F) and 10°C (50°F). The highest temperature so far recorded was 53°C (127°F) at Cloncurry, Queensland, in January 1899; the lowest was –22°C (–8°F) at Charlotte's Pass, New South Wales, in June 1945.

Physical resources. Australia has four main economic resources: forests, fish, land and minerals. An estimated 380,000sq km (150,000sq miles) – about 5 per cent of the country's area – are covered with forests that could be exploited commercially. There are also large areas of low-grade woodland of little commercial importance as a source of timber but valuable for the protection they give to the soil. The main commercial forests are in the east and south-east in the wetter parts of the coastal areas [*see* NW p. 210].

In the early 1970s Australia's commercial catches of fish averaged more than 50,000 tonnes per year. The main species caught were tunny (tuna), edible shark, salmon and mullet. Vast areas of land are devoted to farming, mainly the rearing of sheep and dairy cattle and the growing of cereal crops and sugar cane (for details see the section on ***Agriculture*** later in this article). The chief mineral resources are deposits of coal and the ores of iron, copper, nickel and manganese – all products which Australia exports.

Industry. Australia's manufacturing industry grew rapidly after 1900 and the growth accelerated even more after 1960. Before federation (1901) the existing industries in the colonies were limited – generally to the manufacture of such goods as furniture and clothing – and their products were intended only for local markets. Federation removed trade barriers between states, and World War I created a demand for a wider variety of goods, thereby stimulating production. Previously the emphasis had been simply on the repair of imported machinery; after 1914 Australia began to establish its own iron and steel works.

Australia's chief shipbuilding yards are at Adelaide, Brisbane, Maryborough, Melbourne, Sydney and Whyalla. In addition each state has many smaller shipyards in which small steel vessels, such as off-shore oil-rig servicing craft and trawlers, are built.

Large-scale manufacturing of motor vehicles began only after World War II, and even today most manufacturers are subsidiaries of British, American or European companies. Australian models are, however, increasingly being produced, particularly by Holden-GM, and the motor industry is the largest employer in the country.

Australia's heavy engineering industries have expanded enormously since World War II, largely through opportunities created by the growth of other industries, such as the chemical industry and oil refining. Farm machinery, excavating equipment and locomotives are increasingly being made in Australia. Foreign investment in heavy industry, especially by the Japanese since 1945, has made a significant contribution to these developments.

The growth of oil refining is another example of Australia's industrial progress since 1945. There are now ten refineries and at the end of 1972 the distillation capacity of the industry was about 32 million barrels a year. The heavy chemicals industry produces ammonia, chlorine, caustic soda, methanol and sulphuric acid from primary raw materials such as petroleum, coal, salt, sulphur and limestone. From these and other chemicals are derived a wide range of products, chief of which are explosives, plastics and pharmaceuticals.

Mining is of vital importance to the Australian economy. The country is self-sufficient in most minerals (except sulphur and phosphates) and is one of the world's greatest exporters of them. The major mineral is iron ore; by 1980 it is expected that

Rounding up sheep in Australia, the traditionally horse-riding stockman now uses a light trail motorcycle complete with a spare seat for his dog.

Sugar cane, one of Australia's major agricultural products, is grown in south-eastern Queensland and exported from Cairns, the country's chief sugar port.

For every 100 people in Australia there are 40 cars – a surprising statistic that reflects the vast distances people there cover quite casually.

total exports of ore and pellets will be about 100 million tonnes per year. Other minerals exported include coal, bauxite, lead, zinc, rutile and monazite and production of bauxite and nickel is expected to increase.

Agriculture. Output from farming has risen considerably since 1945, although the agricultural work force has declined and droughts (as in 1965-66) have slowed progress. The expansion has been due mainly to wider and more efficient application of scientific advances in soil nutrition and crop management. Australia grows more than enough food for its own needs and derives much foreign currency by exporting the surplus.

Australia is the world's major producer and exporter of wool (producing nearly a third of the world's output). In 1973 there were an estimated 142 million sheep in the country, of which 75 per cent were Merinos, a producer of top-quality wool which has adapted well to the great range of environmental conditions that exist in Australia. Most of the remaining 25 per cent of sheep are crossbreeds, reared principally for the production of fat lambs for meat [*see* PE p. 23]. The export of beef and veal has expanded in recent years, the largest markets being the United States, Britain and Japan.

Grain crops have been planted since the time of the original British settlement around Sydney. Today wheat, barley and oats are grown in all the states. The principal export markets are Egypt, Britain, Japan, India, the Middle East, South America and south-eastern Asia.

Australia's dairy industry is located mainly in the temperate areas of good rainfall: the south-west of Western Australia, the plains around Adelaide and the south-east of South Australia, the eastern districts of New South Wales and the south-eastern coast of Queensland. About 58 per cent of whole-milk production goes for making butter, 10 per cent for cheese and 8 per cent for milk products.

About 95 per cent of the country's sugar production comes from sugar cane grown in Queensland. Output is more than enough for local markets, enabling limited exports to be made to Japan, Britain, the United States and New Zealand. Other crops include fruits (particularly apples), cotton, grapes and tobacco.

Trade. Like so many other aspects of Australia, the pattern of trade has changed greatly since World War I. Britain, the traditional market for Australian products, is no longer the principal trading partner. Japan now holds that position, taking more than 25 per cent of Australian exports. And as Britain concentrates more on her trading links with other European countries, the United States has become the main supplier of imports. Australia trades with nearly 200 countries, and wool, meat and dairy products remain major exports, to which in recent years have been added many minerals and their derivatives.

Transport and communications. In Australia communications have always been difficult and costly because of the vast distances involved and because most of the population is concentrated in the five main state capitals. The country now has one of the most motorized societies, with more than 40 cars for every 100 people. The transport system includes 863,000km (535,000 miles) of roads, 110,000km (68,000 miles) of air routes and 40,000km (25,000 miles) of railways.

Post, telegraph, telephone and railways – the main communications systems – have always been government run, initiated by the colonial governments and maintained ever since by the individual states. The first postal service was established in 1809, the first telegraph in 1854 and the telephone in 1878. Railways did not adopt a standard gauge until after World War II. Qantas Airways is an international airline which flies 10 million passenger kilometres a year to all parts of the world. Internal airlines include Ansett-ANA and the government-owned Trans Australian Airlines.

There are both commercial and state-owned radio and television stations. Newspapers are independent of the government; most are published by large groups that also have major interests in commercial radio and television.

People. The Ice Age stimulated massive movements of human populations in south-eastern Asia between 20,000 and 30,000 years ago. The Aborigines, who have some general physical affinities with ethnic groups in southern India, Sri Lanka and south-eastern Asia, are thought to have travelled to Australia at about this time [*see* MS pp. 31,33].

At the time of the first white settlement, in 1788, there were about 300,000 Aborigines in 500 tribal groups (each with 100 to 1,500 people) distributed over the whole continent. The tribes had a kinship of family units, with inheritance through the male line; most large tribes contained several subdivisions. There were no animals or plants suitable for farming, and the Aborigines lived by hunting and gathering [*see* MS p. 250]. The scarcity of food in the inland areas made the tribes who lived there lead a semi-nomadic way of life, whereas the coastal Aborigines, with better supplies of food and water, made more permanent settlements.

This relatively peaceful life-style was doomed to disintegration by the arrival of white colonists. Tribe after tribe disappeared within a few generations as the new arrivals turned tribal land into stock-raising farms. Many Aborigines were killed in battles with white settlers; others succumbed to diseases introduced by the Europeans, particularly to smallpox, influenza and syphilis.

By the early 20th century the number of Aborigines had decreased to about 50,000 and only in recent years has their birth-rate increased and infant mortality decreased. Today there are about 110,000 Aborigines in Australia and the government is making efforts to provide them with better schools, houses and medical facilities.

The remainder – 99 per cent – of Australia's population derive from other parts of the world, especially Europe. Five other British colonies were established after the founding of New South Wales. Large-scale immigration began with the gold rush of the mid-19th century and many Chinese were admitted to work on the sugar cane plantations of Queensland. Then a "White Australia" policy effectively limited immigration to European people until after World War II. By 1945 the population (which was less than 4 million in 1901) had grown to 7.5 million. In the following 30 years three million more immigrants arrived from more than 50 countries. The birth-rate also rose from 17 per thousand in the 1930s to 20 per thousand in the 1970s. Today's population of nearly 14 million represents a further population increase of 74 per cent since 1945.

As a result, Australia is a multi-national country with nearly one in three of its present population a post-1945 immigrant or the child or spouse of an immigrant. The largest national group is still derived from Britain and Ireland. The second largest group is Italian, followed by Greek, Yugoslav, German, Dutch, Polish and Maltese. The smallest national group living in Australia today is that of the original inhabitants – the Aborigines.

Christianity is the principal religion. The Church of England has nearly 4 million adherents and the Roman Catholic Church 3.5 million; there are also more than a million Presbyterians and a million Methodists. Australia has a free state education system with 7,400 government-run schools as well as many private schools, run mainly by the churches. Attendance at school is compulsory for children between the ages of six and 15 in all states (except in Tasmania, where the school-leaving age is 16). The vast distances between small communities in the outback present particular difficulties in education. Each state capital has a correspondence school to meet the needs of children whose attendance is prevented by distance or illness (more than 20,000 children are educated in this way). In addition Queensland operates the unusual School of the Air, which uses two-way radios to link a central teacher to a number of children on scattered farms. Higher education is financed by the federal government, which abolished all university and college fees in 1974.

Cultural life. The art, literature and music of Australia reflect the mixture of nationalities that makes up its population. Aboriginal myth and ritual find expression in painted sacred symbols (often on bark) and mythological designs carved in stone. Social gatherings called corroborees provide opportunities for dancing and singing, and many myths are preserved in chants.

Cultural development among European settlers took place in three phases. The difficult conditions during the first 100 years of the colonies left little time for leisure. A few writers recorded the great differences the settlers found in Australia – the novel ***Robbery Under Arms*** by T.A. Browne ("Rolf

Sydney Opera House (opened in 1974), which stands on Bennelong Point overlooking Sydney Harbour, was designed largely by the Danish architect, Joern Utzon.

Australians are famous for their love of swimming, surfing and water sports – famous too for well-trained and highly competitive life-savers.

Adelaide is the capital of South Australia. Despite its industrial and commercial importance, its an attractive city with tree-lined streets and squares.

Boldrewood") dates from this period – but the cultural heritage and dominating influence remained that of Britain. Even the people's clothes, totally inappropriate to the Australian climate, remained modelled exactly on those worn by people in Europe.

The second period extended from about 1880 to 1940. The last years of the 19th century were a time of growing nationalism. With this went a sympathy for the struggles of small landowners and the spirit of "mateship", both recurring themes in the literature of the period. During the postwar period Australian artists began to lose their self-consciousness about their environment and turned to more universal themes and the growing complexity of Australian society.

Today the largest patrons of the arts are the federal and state governments. The Australian Council was established in 1973 to assist the arts independently. It is now the central administrative body for a number of boards that deal with various aspects of the arts, such as music, drama and cinema. There is also a board dealing with Aboriginal art.

Many buildings and touring activities are financed by state governments. The New South Wales government, for example, built the A$100 million Sydney Opera House, a complex of theatres and concert halls. The government of Victoria financed the Victorian Arts Centre, and the South Australian government built a similar arts complex in Adelaide (opened in 1973).

Many drama companies receive funds from the Australian Council, which also supports the Australian Opera and the Australian Ballet, both national touring companies. Opera received its greatest stimulus at the turn of the century from the career of Dame Nellie Melba (original name, Helen Mitchell), an obscure Australian girl who became the best-known singer of her time. National pride in her achievement gave Australians, until then largely uninterested in opera, a taste for this art form. The ballet, under the direction of Sir Robert Helpmann, has toured Europe, Asia and North America and in 1973 visited Britain, the USSR and countries of eastern Europe. Music for both opera and ballet is provided by two Elizabethan Trust orchestras. There are six more professional symphony orchestras, one based in each state capital. The Australian Broadcasting Commission (ABC) organizes about 750 concerts every year and there are several distinguished contemporary Australian composers, including Malcolm Williamson (Master of the Queen's Musick), Nigel Butterby, Richard Meale and James Penberthy.

There is also a growing interest in the visual arts, and emphasis is again being placed on Australian artists. There are public galleries in all the state capitals and the Australian National Gallery is being built in Canberra (planned to open in 1978).

The Australian Council has also been helpful to authors and publishers. The book trade – particularly fiction books – was depressed in the early 1970s and the setting up of the Literature Board in 1973 provided a valuable boost with regular government support. In 1974 a public lending rights system was inaugurated under which payments are made to authors and publishers with books in public lending libraries. The best-known Australian novelist is Patrick White, who won the Nobel prize in literature in 1973. Many other writers have followed his lead in exploring Australian themes in universal terms.

Leisure and sport. Australia, despite its relatively small population, does disproportionately well at all kinds of sport. Water sports, particularly swimming and surfing, are popular (much of the population lives on, or within easy reach of, the coast). Until 1902 public bathing was prohibited between the hours of sunrise and sunset. The ban was so ridiculed by William Gooner, the owner and editor of a newspaper in Manly, that it was lifted. Crowds soon flocked to the beaches, many of which are now protected by nets and other measures to lessen the danger from sharks.

Rowing is another popular water sport and the first Australian to win an official world championship in any sport was a sculler, Edward Trickett. In 1875 he beat the British and world champion in a race on the River Thames in London between Putney and Mortlake. Yachting is also keenly followed and winning the Sydney to Hobart race is one of the most coveted prizes.

The two chief field sports are cricket (in the summer) and football (in the winter) and Australia has produced many sportsmen of international renown in both games. In the 1890s the scores of Clem Hill, the second youngest cricketer to play for Australia in a test match, established records that remained unbeaten until Don Bradman's rise to fame in the 1930s. Bradman played for Australia in 37 tests against England, averaging 89.78 runs per innings with 19 centuries – including six scores of more than 200 and two of more than 300.

Four main types of football are played: Australian rules, rugby league, rugby union and soccer (Association football). Soccer is comparatively new to the country but is rapidly gaining in popularity, stimulated by the achievement of the Australian team that reached the World Cup finals in 1974. Rugby union dates from the foundation of the Sydney University club in 1863. Today it is one of the top spectator sports, played mainly in New South Wales and Queensland. Victoria is the state in which Australian rules is most popular.

The long hours of sunshine and settled weather have made tennis a sport that can be played outdoors all the year round in Australia. The country has produced many Wimbledon champions, including Lew Hoad and Rod Laver. Hoad was noted for his power and top-spin backhand shots; he played in Davis Cup matches from 1952 to 1956, won Wimbledon in 1956 and 1957, and then turned professional. Laver won permanent sporting fame with his "grand slam" of the Wimbledon, USA, Australian, French and Italian championships in 1962.

The first athletics meeting in Australia was an intercolonial affair held at the Melbourne cricket ground in 1893 with competitors from New South Wales, Victoria and New Zealand. In the first Olympic Games of 1896 Edwin Flack, a young Australian student in London, privately entered the 800 and 1,500 metres events – and won them both. His victories were officially credited to Britain, although some years later the records were changed to show Flack as a representative of Australia. Since that time athletics has been a regular feature of Australian sporting life. Horse racing began in 1810 with a three-day meeting at Hyde Park, Sydney, organized by officers of the 73rd Regiment. The Australian Jockey Club was founded in 1842 and today the premier event is the Melbourne Cup (the day on which it is held has been made a public holiday).

Australians turn even regular occupations into sporting events. There are chopping contests, in which axemen vie with each other both for speed and accuracy in chopping wood. Sheep-shearing is another example, with yearly contests in all the sheep-farming areas.

Government. The governor-general and the prime minister hold the executive power of the Commonwealth. The prime minister is generally chosen from the political party with a majority in the federal parliament, which consists of two houses: the Senate (with 60 members, serving six-year terms) and the House of Representatives (with about 125 members serving for up to three years). Members of both houses are elected by popular vote by every citizen over the age of 18 — voting is compulsory in Australian elections. The Federal Parliament, which has representatives from all the states and territories, makes the laws that affect the country as a whole on such matters as national finance, foreign policy and defence. In addition each of the six states has its own parliament of two houses (Queensland has only one), which legislates on education, social welfare, law and order and other more local matters. State governments are headed by a governor and a cabinet chosen from the majority party in parliament. Like the governor-general, the state governors take little part in the day-to-day affairs of government.

Political parties. Three parties have representatives in the state and federal parliaments: the Labor, Liberal and National Country parties. There are also other, small parties, such as the Communist and Australian parties. The Labor Party was formed in 1891, having its origins in the trades union movement. It has developed in a similar way to Social Democratic parties in other countries, with an emphasis on nationalism. One of its chief aims is to use federal influence and money to create equality of opportunity for all Australians through improvement in health services and education.

The Liberal Party, formed in 1944, supplanted the United Australia Party. It emphasizes links with Britain, represents private enterprise and seeks to protect the freedom of the individual from govern-

Capt. James Cook's charting of the east coast was the basis of Britain's claim to the vast rich lands of the newly discovered continent.

Alice Springs, near the centre of Australia, has become a popular resort for tourists exploring Northern Territory and the nearby Aboriginal settlement.

Panning was one of the traditional methods of extracting gold from Australia's rivers and streams; gold can still be found in the Yarra River.

ment controls. The National Country Party was founded in 1918, principally to protect the interests of farmers. Never numerically strong, it has nevertheless exerted considerable influence on Australian politics, largely through its coalition with the Liberal Party.

Armed services. During the 19th century Australia was defended by British troops or by Australian troops under British direction. Since federation the country has raised and run its own army, whose structure was radically changed in 1972. The former geographical commands (modelled on the British system) were replaced, over a period of five years, by three major commands: Field, Training and Logistics. Today the army has more than 41,000 men and uses a little more than a third of the defence budget. After the abolition of conscription in 1972 the army became an entirely volunteer force.

The Royal Australian Navy (RAN) has a small but versatile fleet, including an aircraft carrier, destroyers, a boom defence vessel, minesweepers, and training, survey and support ships. The submarine fleet is also being enlarged. More and more of the new ships are being built in Australia, but exchanges of vessels and purchases from abroad continue. The RAN has more than 16,000 men, and uses nearly a quarter of the defence budget.

The Royal Australian Air Force (RAAF) dates from 1921 when it was formed from the small but highly effective Australian Flying Corps of 1915-18 and the postwar Australian Air Corps. Today it has a complement of about 25,000 men, is assigned nearly a third of the defence budget, and is administered by the Air Board. The strike reconnaissance force uses Canberra and Phantom aircraft; the RAAF also has F-111 fighter-bombers bought from the United States in 1973 and uses French Mirage fighters, as well as Hercules, Caribou and Dakota transport aircraft. Maritime work is assigned to Neptunes and Orions, and Sabres and Mirages are used as trainers.

History. Europeans had long assumed that there had to be a large land mass in the south to balance the Northern Hemisphere. But it was many years before they actually discovered Australia, mainly because of the vastness of the Pacific Ocean. In 1601 the Portuguese navigator Manuel de Eredia may have sighted the Australian coast and in 1605-06 the Spaniard Luis de Torres sailed through the Torres Strait from the east and sighted Cape York, the northernmost tip of the mainland. To him the land seemed barren and inhabited by "wild, cruel, black savages" – a reference to the Aborigines. Dutch and other explorers followed, including Dutchman Abel Tasman between 1642 and 1644 and the Englishman William Dampier between 1699 and 1701. But it was not until Capt. James Cook charted the east coast in the 1770s that the size and nature of the continent began to be realized. In the name of the British Crown, he took formal possession of eastern Australia on 23 August 1770.

A British settlement was established in 1788 at Port Jackson (now part of Sydney). This was a penal colony of about 950 people, of whom 736 were convicts transported from Britain in 11 ships (the "First Fleet") under the command of Captain Arthur Phillip. Soon after their arrival the 200 marine guards refused to work any longer, claiming their duties did not extend beyond the outward voyage. Also the expedition had inadequate supplies, and assistance from home was delayed because most of Britain's naval vessels were engaged in the war with France at that time. A special unit, the New South Wales Corps, was enlisted in England to act as police and guards in the new colony. Phillip expected supplies and assistance and was dismayed to discover that when the Second Fleet arrived in 1790 it carried not food and agricultural tools but 750 more convicts (267 had died on the voyage). Despite these setbacks and appalling hardship, however, the colony survived and began exploration and established new settlements, such as that at Parramatta on the Hawkesbury River.

Governor Phillip became ill and had to give up his command in 1792, and control passed to the senior officers of the New South Wales Corps. They rapidly established themselves as a rich and powerful group, and as a source of great vexation to the new governor, John Hunter. The officers wanted convicts assigned to them to work on their land, whereas Hunter wanted to use them as labourers on government land in order to benefit the whole colony. The officers gained monopolies in certain products, especially rum, which became virtually the common currency.

Matters came to a head in 1808 with a rebellion by the army (known as the Rum Rebellion) during which the next governor, William Bligh, was imprisoned in Government House. Yet another governor, Lachlan Macquarie, was sent from England with his own regiment. His established firm rule offered the hope of freedom to the convicts and a chance of becoming farmers in their own right. He established a bank, introduced proper currency, and organized the construction of public buildings and roads.

Little was known of New South Wales beyond a coastal strip about 240km (150 miles) long and 80km (50 miles) wide. Then in 1813 three explorers crossed the Blue Mountains, a barrier around the Sydney settlement that had hitherto proved impassable. It then became possible to occupy the interior of the country, and soon sheep farmers were pushing farther inland to the richer pastures of Victoria and South Australia. The heartland of Australia, however, remained largely unexplored.

Early in the 1820s people began to agitate for better representation in the government. A law of 1823 established a small council which was appointed in New South Wales to advise the governor about new legislation and taxation. Then in the 1830s Australians began to seek a further transfer of power from the British government to the colony. Legislation was enacted in England in 1842 establishing a new council of 36 members, 12 appointed by the British Crown and 24 to be elected in the colony. But true political representation had to wait until after the mass influx of non-convicts that followed the discovery of gold.

In 1851 E.H. Hargraves (who had taken part in the Californian gold rush) went to New South Wales and found gold in workable quantities near Bathurst. Further discoveries were made near Clunes, at Anderson's Creek and then – in great richness – at Ballarat and Bendigo. By the end of 1851 gold worth more than a million pounds had been extracted. The entire political and economic structure of Australia was suddenly and irreversibly changed. Workers flocked to the gold-fields and soon Port Phillip Bay (near present-day Melbourne) was crammed with ships carrying prospectors from all parts of the world. In 1850 the combined population of New South Wales and Victoria was 265,000; ten years later it had grown to 886,400. Transportation to Australia of British convicts was abolished in 1868

By 1857 opportunities for individual prospectors began to decline. Other occupations then rapidly became over-supplied with labour and people turned their attention to the land. Australia's days as a convict colony were virtually over. During the labour surplus of the 1860s unemployment forced the embryo trade unions to adopt a defensive stance. Then, following increasing union membership in the 1880s, strikes began to increase but the powerful government defeated the unions in a series of great strikes between 1888 and 1895.

The financial crises and bank failures that followed the strikes delayed the long-envisaged move towards federation between the separate colonies. This finally came with the Imperial Act of 9 July 1900, passed in each of the colony's parliaments, which resulted in a proclamation which stated that from 1 January 1901 the people of New South Wales, Queensland, South Australia, Tasmania, Victoria and Western Australia (thereafter known as states) would be united in a federal Commonwealth. The central government, which first met in Melbourne, was given certain defined powers, the remainder going to the individual states. In 1908 it was decided to establish a federal capital at Canberra, and parliament first met there in 1927.

The outbreak of World War I in 1914, thousands of miles away in Europe, might have been expected to have had little or no effect in Australia. But the country united in its aim of giving all possible assistance to Britain. More than 416,000 volunteers enlisted in the armed forces, of whom nearly 330,000 served overseas. Most of the Australian contingent served in the Middle East, at the Gallipoli landings in the Dardanelles, and on the Western Front. Nearly 60,000 died as a result of war service. At home, manufacturing industries underwent a spurt in their growth, and it was at this time that the Australian iron and steel industry was developed.

The postwar period found Australians deeply divided over the issue of military service. The prime

The bustling city of Brisbane is Queensland's capital and major trading centre; it is also one of the largest river ports in Australia.

Metals are among Australia's most important natural resources; here aluminium ingots are being melted in a furnace before being made into castings.

Hydroelectric dams in New South Wales are part of Australia's largest hydroelectric project – the Snowy Mountains Scheme – which was completed in 1974.

minister in 1916, William Hughes, proposed the introduction of conscription but the measure was twice rejected in referenda. Many leaders of the Labor Party, who favoured the proposal, were expelled from the party as a result and much popular support swung to the rival Australian Country Party. The issue created a division between urban and rural interests and robbed the labour movement of political influence during the 1920s. Power rested instead with a coalition of the Nationalist and Country parties.

There was a seamen's strike in 1925 and a strike of timber workers in 1929. In that year the government held a referendum on its proposal to abolish state arbitration in industrial disputes in all but the maritime industries. It was roundly defeated and a landslide victory in the resulting election swept James Scullin and the Labor Party into power. The following year the value of exports fell to about half of those of 1928, and all overseas loans ceased. Australia, with its government rendered ineffective by opposition in the Senate, was in the worst possible position to face the depression which was to follow in the 1930s. By the summer of 1932 there were half a million people unemployed. Factories ceased production, shops ran out of stock and then closed, houses remained unlet and unsold, and there were thousands of bankruptcies.

The depression demonstrated to Australia how greatly the country was dependent on external economic conditions. This vulnerability was changed by World War II. A few hours after Britain declared war on Germany in 1939, Australia – with no dissenting voice – did likewise. During 1940 Australian forces were engaged in Greece, Crete and North Africa, where they held the fortress of Tobruk against Rommel's combined German and Italian forces for eight months.

For Australia, the war differed from World War I in two important ways: Japan entered the conflict in 1941 on the side of Germany, and the evolution of aerial warfare now meant that Australia's northern cities were vulnerable to surprise enemy attack. For this reason two Australian divisions were transferred from the Middle East to home bases. Britain's resources were strained to the limit in the European theatre, leaving little military aid available for Australia. The Australian government recognized this and sought closer co-operation with the United States – a significant alliance that was to develop and prosper in the postwar years.

The Japanese advance was halted at sea by American forces in the battles of the Coral Sea and Midway Island. On land they were halted in New Guinea by Australian troops at Gona, Buna and Milne Bay. The campaign to recapture New Guinea and the islands then began. By late 1944 Australia had suffered more than 29,000 casualties. American forces pursued a policy of "island hopping" in their advance towards Japan and the Australians took on the role of isolating and destroying the enemy in Bougainville, New Britain and Borneo.

After the end of the war in 1945 Australia was a transformed country. World War I had strengthened it as a manufacturing country; World War II brought it for the first time into prominence as a nation of international importance. Australia's chief postwar problem was not, as had been predicted, a recurrence of unemployment but inflation. In foreign affairs, the Labor government steered the country to look to the United Nations for security and generally supported those Asian countries that sought independence. The change from Labor to a Liberal-Country party government in 1949 brought a shift of emphasis derived from the new government's opposition to revolutionary Asian nationalism. Later Australian public opinion was sharply divided (as it had been about conscription) over the Vietnam War and the issues it raised.

The Liberal-Country government of Robert Menzies emphasized not only its association with Britain, but also its growing connection with the United States. This was one of the reasons for the 1951 ANZUS treaty (ratified 1952) between Australia, New Zealand and the United States and for Australia's joining the South-East Asia Treaty Organization (SEATO) in 1954, which cemented this alliance against Communism.

Communist influence in the trades unions precipitated a series of strikes in the immediate postwar period that brought 17 years of conservative rule. Divisions within the Labor Party ensured that recent Australian politics have been relatively free from ideological conflict and have been mostly concerned with issues of nationalism and Liberalism, such as the question of conscription or of constitutional relations with Britain. In November 1975 the governor-general, Sir John Kerr, dismissed Gough Whitlam as prime minister. Again the people of Australia have divided views about these topics, although Queen Elizabeth II was well received on her Silver Jubilee tour of Australia in 1977.

States and territories. The Commonwealth of Australia consists of six states, two mainland territories and several external territories. Australia's external territories include CHRISTMAS ISLAND, COCOS ISLANDS, NORFOLK, Ashmore and Cartier Islands, and Heard and McDonald Islands.

The Australian Antarctic Territory (annexed in 1936) covers an area of 6,400,000sq km (2,500,000sq miles) and comprises all islands and territory – except Adélie Land – south of 60° S latitude between 45° and 160°E longitude. The Australian National Antarctic Research Expeditions (ANARE) have carried out exploration and research work there since 1943.

Australian Capital Territory, until 1938 called Federal Capital Territory, is the smaller of the two mainland territories. It consists of an area within the state of New South Wales containing the capital, Canberra. In 1911 most of the area, then known as Yass-Canberra, was surrendered to the Commonwealth by New South Wales. The state then ceded part of the Jervis Bay area in 1915 to complete the present territory and provide it with a port. Nearly all the people live in Canberra. Area: 2,432sq km (939sq miles). Pop. 143,000.

New South Wales is a state in south-eastern Australia, bordering on the Tasman Sea on the east which gives 1,095km (680 miles) of coastline. The surf beaches, Blue Mountains and snow-covered slopes of the Australian Alps are popular tourist at-

Prime Ministers of Australia

Edmund Barton (1901-03)
Alfred Deakin (1903-04; 1905-08; 1909-10)
John Watson (1904)
George Reid (1904-05)
Andrew Fisher (1908-09; 1910-13; 1914-15)
Joseph Cook (1913-14)
William Hughes (1915-23)
Stanley Bruce (1923-29)
James Scullin (1929-31)
Joseph Lyons (1931-39)
Sir Earle Grafton Page (1939)
Robert Menzies (1939-41; 1949-65)
Arthur Fadden (1941)
John Curtin (1941-45)
Francis Forde (1945)
Joseph Chifley (1945-49)
Harold Holt (1965-67)
John McEwen (1967-68)
John Gorton (1968-71)
William McMahon (1971-72)
Gough Whitlam (1972-75)
Malcolm Fraser (1975-)

Governors-General of Australia

John Louis, Earl of Hopetoun (1901-02)
Hallam, Baron Tennyson (1902-04)
Henry Stafford, Baron Northcote (1904-08)
William Humble, Earl of Dudley (1908-11)
Thomas, Baron Denman (1911-14)
Sir Ronald Munro-Ferguson (1914-20)
Henry William, Baron Forster of Lepe (1920-25)
John Lawrence, Baron Stonehaven (1925-31)
Sir Isaac Isaacs (1931-36)
Alexander Gore Arkwright, Baron Gowrie (1936-45)
HRH Prince Henry, Duke of Gloucester (1945-47)
Sir William McKell (1947-53)
Sir William Slim (1953-60)
William Shepherd, Viscount Dunrossil (1960-61)
William Philip, Viscount De Lisle (1961-65)
Richard Gardiner, Baron Casey (1965-69)
Sir Paul Hasluck (1969-74)
Sir John Kerr (1974-)

New multi-storey buildings in Sydney tower over older ones, many of which are being demolished to make way for new, more concentrated development.

The Murray River and its tributaries form Australia's principal water system. Dams create reservoirs, chiefly for irrigation.

Vienna, capital city and province of Austria, was once a cultural and scientific centre. Its historical associations today make it a major tourist attraction.

tractions. The capital is Sydney, a centre of transportation and commerce, with one of the world's finest harbours. The principal industries are the manufacture of iron and steel, textiles, agricultural machinery, cement, paper, petrochemicals and electrical equipment. New South Wales was discovered in 1770 by Captain James Cook and first settled in the region of Botany Bay in 1788. It became part of the Commonwealth of Australia in 1901. Area: 801,430sq km (309,180sq miles). Pop. 4,567,000.

Northern Territory is in central northern Australia between Western Australia and Queensland and bordering on the Timor and Arafura seas. Darwin is the capital and only port. Most of the terrain is desert although there is good pastureland in the north; cattle farming is the chief industry. Northern Territory was first settled in 1820-50. It became part of New South Wales from 1825 to 1863, when it was annexed to South Australia (which became part of the Commonwealth in 1901). Northern Territory became a territory in its own right in 1911. Area: 1,347,525sq km (520,280sq miles). Pop. 71,400.

Queensland, a state in north-eastern Australia, is bordered by the Coral Sea and the Pacific Ocean on the east and by the Gulf of Carpentaria on the north-west. Brisbane is the capital city. It is an agricultural state with sugar cane as the chief crop and cattle as the main livestock. Major industries include dairying, food processing and the mining of copper, coal, lead and zinc. It was first visited by Captain James Cook in 1770 and settled in 1824-43 as a penal colony. It became a colony in 1859 and a part of the Commonwealth in 1901. Area: 1,727,530sq km (667,000sq miles). Pop. 1,799,200.

South Australia is in the central southern part of the continent and includes Kangaroo Island and several smaller islets. The state's terrain is varied and includes deserts, mountains, salt lakes and swampland. Two-thirds of the population lives in the capital, Adelaide. Other important places include Whyalla, location of Australia's largest shipyards, and the Murray River valley, whose vineyards produce the finest Australian wines. The chief industries are mining, metal processing, textiles and food processing. The coast was reputedly visited by Dutch seamen as early as 1627; the first British settlement was established in 1836 and the colony became a state within the Commonwealth in 1901. Area: 984,380sq km (379,760sq miles). Pop. 1,164,700.

Tasmania, formerly Van Diemen's Land, is an island state 240km (149 miles) off the south-east coast of the mainland, separated from it by the Bass Strait. It includes also many islands, chief of which are King Island and the Furneaux Group. The capital is Hobart and the only other large city is Launceston. Mountains rise to 1,617m (5,305 ft) at Mt Ossa. Industries include the manufacture of electrochemicals, metals, paper and textiles. Tasmania was discovered in 1642 by the Dutch explorer Abel Tasman and visited by Captain James Cook in 1777. It was settled by the British as a penal colony in 1803, became a colony in 1825 and was re-named Tasmania in 1853; it was included as a state of the Commonwealth in 1901. Area: 68,332sq km (26,383sq miles). Pop. 392,500.

Victoria is a state occupying the south-eastern part of Australia, south of the Murray River and bounded by the Indian Ocean and Tasman Sea to the south and east. Melbourne, the capital and major port, is the centre of transport, communications and government. The state is noted for the hardwood forests in the highlands; there is also much dairy and livestock farming. Chief industrial products include car components, textiles and chemicals. The region was discovered by Captain James Cook in 1770 and first settled near the present site of Melbourne in 1835. It became a state of the Commonwealth in 1901. Area: 227,620sq km (87,813sq miles). Pop. 3,443,800.

Western Australia is the largest of the states with a coastline 5,550km (3,500 miles) long bordering the Indian Ocean. The capital is Perth, which has a major port at nearby Fremantle. Much of the vast terrain consists of low, dry rock; it is the major gold-producing state in Australia. The chief agricultural products are wheat, wool, meat and dairy produce and iron ore, coal and nickel are mined. The region was visited in 1616 by the Dutchman Dirck Hartog and again in 1688 by the English explorer William Dampier; settlement began in 1829. Western Australia became a state of the Commonwealth in 1901. Area: 2,527,630sq km (975,130sq miles). Pop. 980,000. Map 54.

Australia – profile

Official name Commonwealth of Australia
Area 7,686,848sq km (2,967,892sq miles)
Population (1976 est.) 13,600,800
Density 1.8 per sq km (4.6 per sq mile)
Chief cities Canberra (capital) (1976 est.) 198,700; Sydney, 2,874,380; Melbourne, 2,583,900; Brisbane, 911,000; Adelaide, 868,000; Perth, 739,200
Government Constitutional monarchy
Monetary unit Australian dollar
Gross national product (1974) A$37,416,000,000 (£27,113,100,000)
Per capita income A$2,751 (£1,993)
Agriculture Cereals, sugar cane, livestock (principally sheep), forestry, potatoes, grapes (for wine)
Industries Mining, meat and meat products, iron and steel, dairy products, manufacturing, tourism
Chief exports Iron ore, coal and other minerals, wool, beef, wheat, sugar, dairy products
Trading partners (major) Japan, Britain, USA, China

Austria (Österreich), official name Republic of Austria, is a country of central Europe. It was once the centre of an extensive empire under the royal house of Hapsburg. The Hapsburg emperor ruled not only Austria but also, at various times, Hungary, Bohemia and the other lands that make up present-day Czechoslovakia, and parts of Poland, Yugoslavia, Romania, Italy, Belgium and Luxembourg. Since 1955 Austria has been a sovereign state. Vienna, its capital, is one of the world's most beautiful cities.

Land and economy. Austria is completely landlocked and is one of Europe's most mountainous countries. Ranges and spurs of the Alps in the south-west and centre cover about 70 per cent of the country. The highest peaks are Grossglockner (3,798m; 12,461ft) in the Hohe Tauern and Wildspitze (3,774m; 12,383ft) in the Ötztaler Alps. To the north of the mountains is the broad plain of

the River Danube, at the eastern end of which are the lowlands in which Vienna lies. Farther north the land again rises to the wooded hills of the Bohemian Massif. The River Danube enters the country in the north-west from West Germany and flows eastwards to Czechoslovakia and Hungary. Among its many tributaries are the Inn, Traun, Enns, Ybbs, Letha and Rab. There are many lakes, the largest of which are the Neusiedler on the Hungarian border, the Atter and Traun lakes in the north-west and the Millstätter and Wörther lakes in the south. They are all in regions of great natural beauty, and are popular with tourists.

Austria's climate varies considerably. In the wide Danubian plain summers are hot but winters are often extremely cold. The Alpine areas of the south also have cold winters and much rain.

People in the Alpine valleys live mainly by agriculture. They raise cattle for dairy-farming, as well as sheep and goats. In low-lying areas wheat, maize,

Innsbruck, the capital city of Tyrol province, lies on the Inn River in the Eastern Alps of south-west Austria and is a famous summer and winter resort.

Tyrol, a province in western Austria, has tourism as a principal industry; thousands of people go there to see the hills, woods and, in the south, the alps.

Salzburg, birth-place of the composer Mozart, lies in a picturesque setting on both banks of the Salzach River in Austria, near the West German border.

barley and sugar beet are grown and there are also extensive vineyards producing grapes for wine making. The country's mineral resources include iron, coal, graphite and petroleum. Prosperous manufacturing industries are sited in the north and east; large industrial concentrations have developed around the cities of Vienna, Linz and Graz. Factories there produce textiles, chemicals, iron and steel, machinery, wooden goods and processed foods. The country's largest manufacturing firms (oil and heavy industrial concerns and commercial banks) were nationalized in 1946. Tourism is a major industry: the Tyrol has become one of the world's leading winter sports areas; Salzburg in the Alpine foothills, with its castle and Baroque cathedral, is internationally known for its music festivals; and Vienna's imperial past, its splendid churches and palaces, its famous streets, coffee houses, and wine bowers, and its connection with music – particularly waltz music – give the city a romantic appeal that few others can equal.

People. The Austrians are descended from a variety of stocks, including Germanic, Slavic and Mediterranean peoples. Nearly all are German-speaking, the various dialects being in the main similar to those in Bavaria. About 90 per cent of Austrians are Roman Catholics; 6 per cent belong to Lutheran or Evangelical Protestant churches. Most of the population is concentrated in the cities, Vienna having one-fifth of the entire population.

Government. Modern Austria is a federal republic comprising nine provinces: Vienna, Lower Austria, Upper Austria, Salzburg, Tyrol, Vorarlberg, Carinthia, Styria and Burgenland. The constitution separates the federal government into executive, legislative and judicial branches. Power lies chiefly in the hands of the legislature – the Federal Assembly. This body consists of the Federal Council *(Bundesrat)* and the National Council *(Nationalrat).* The head of government is the chancellor, and he and his ministry are responsible to the National Council. Austria's major political parties are the Socialist Party and the more conservative People's Party.

History. Many settlements existed in the Danube valley in pre-Roman times; the Romans built strongpoints there, including the fortress of Vindobona on the site where Vienna now stands. Part of the territory that is now Austria was a border state of Charlemagne's empire. After the accession of Otto I the Great to the imperial throne in the year 962, *Österreich* (eastern realm) was incorporated in the Holy Roman Empire. Later it was constituted a duchy and in the 1270s passed from its ruler Ottokar II, King of Bohemia, to the personal estates of Rudolph I of Hapsburg, later crowned Holy Roman Emperor. It developed into the centre of the Holy Roman (or German) Empire because from 1438 (when Albert II became emperor) until 1806 the Hapsburg archduke of Austria was almost always elected to the imperial throne.

In 1519, the year Charles V became emperor, Austria was at the summit of its power. It formed the hub of an empire that covered Spain, The Netherlands, parts of Italy and vast territories in the Americas. Ferdinand, successor to Charles, was already King of Bohemia and Hungary, and both of these countries remained under the Austrian ruler until 1918. During the religious wars of the 16th and 17th centuries, Austria remained staunchly Roman Catholic and was at the centre of the Counter-Reformation's opposition to Protestantism in northern Europe. In 1740, the succession of a princess, Maria Theresa, to the Hapsburgs' Austrian possessions resulted in a war – the War of the Austrian Succession – as Austria's power appeared to be on the wane. Napoleon dissolved the Holy Roman Empire by military force in 1806, and the Austrian ruler then took the title Emperor of Austria, a title that continued even when Austria later recovered its possessions. In 1867 the Dual Monarchy of Austria-Hungary was set up.

In 1914 the murder of the Archduke Franz Ferdinand (the heir to the Austrian throne) at the hands of Serbian nationalists escalated into the Europe-wide World War I. At the war's end Austria-Hungary was among the defeated, and the empire was in ruins. Austria was reduced to the small republic it is today. In 1938 Hitler occupied Austria in the *Anschluss* in an attempt to create a single state of Greater Germany. In World War II Austria was again on the losing side and suffered enormous destruction. After the war it was occupied by the victorious Allies for ten years. Gradually it rebuilt its economy, and in recent years the country has maintained an uncommitted position in international politics. Map 19.

Austria – profile
Official name Republic of Austria
Area 83,848 sq km (32,374 sq miles)
Population (1975 est.) 7,523,000
Density 90 per sq km (232 per sq mile)
Chief cities Vienna (capital) (1971) 1,614,841; Graz, 248,500; Linz, 202,874; Salzburg, 128,845
Government Federal republic
Religion Roman Catholic
Language German
Monetary unit Schilling
Gross national product (1974) £13,024,500,000
Per capita income £1,731
Agriculture Wheat, rye, potatoes, dairy products
Industries Chemical products, heavy machinery, vehicles, textiles, electrical equipment
Minerals Lignite, graphite, iron, copper, magnesite, natural gas
Trading partners West Germany, Italy, Switzerland, Britain

Azerbaijan. *See* UNION OF SOVIET SOCIALIST REPUBLICS.

Azores. *See* PORTUGAL.

Bahamas, official name Commonwealth of the Bahamas, is an independent sovereign state consisting of a group of about 700 islands in the Atlantic Ocean south-east of Florida and north-east of Cuba. The chief island (although not the largest) is New Providence, location of the capital, Nassau. The islands, together with approximately 2,000 cays (islets) and coral reefs, extend for about 1,000km (620 miles).

The rocky terrain of the islands provides little opportunity for agricultural development. The subtropical climate has a temperature range of 21–32°C (70–90°F). Tourism is the mainstay of the economy; commercial fishing, salt, rum, cement, oil refining

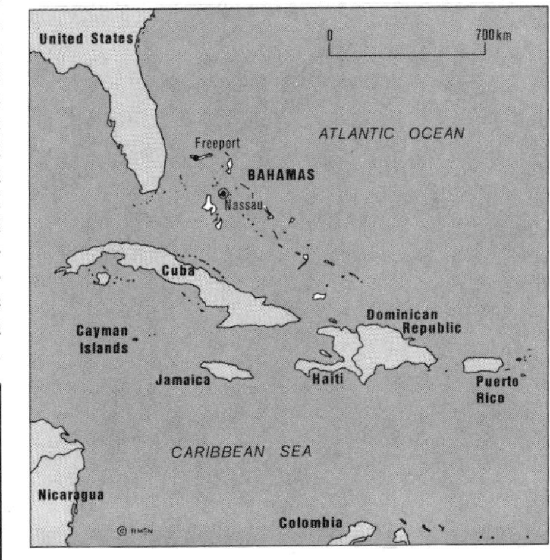

and handicrafts are important industries, which produce most of the country's exports. There is an international airport at Nassau.

About 85 per cent of the native population is African or of mixed Afro-European stock. Most people live on New Providence Island, in or near Nassau, earning a living from the tourist, fishing or handicraft industries. The Baptist Church and the Church of England are the predominant religions (there has been a See at Nassau since 1861). Education is compulsory between the ages of six and 14; Queens College is in Nassau.

The Bahamas were discovered by Christopher Columbus in the 1490s. They were ruled as a British Crown colony from the 18th century. In 1962 the Bahamian political parties demanded a degree of independence and, by 1963, a new constitution had been drawn up providing for a cabinet form of government. In 1973 the Bahamas became an independent nation within the Commonwealth; the

Bahrain derives most of its income from the refining of petroleum, either from its own inland wells or from Saudi Arabia, from where it is shipped or piped.

Low-lying swampy land, such as that of the delta of the River Ganges, makes up much of the land of Bangladesh; it is subject to frequent flooding.

Bangladesh is a poor nation, and the most common forms of transport are man-powered bicycles and heavily laden carts drawn by animals.

monarch of the United Kingdom, represented by a governor-general, remains titular head of state, and a prime minister rules as head of government.

Map 74.

Bahamas – profile

Official name Commonwealth of the Bahamas
Area 13,935sq km (5,379sq miles)
Population (1976 est.) 208,000
Chief Islands Grand Bahama, Great Abaco, Eleuthera, New Providence, Andros, Cat, San Salvador (Watling), Great Exuma, Long, Crooked, Acklins, Mayaguana, Great Inagua
Language English (official)
Government Parliamentary democracy
Monetary unit Bahamian dollar

Bahrain (Al-Bahrayn) is a small island state in the Persian Gulf off the coast of Saudi Arabia, whose economy is almost totally dependent upon petroleum. The nation consists of the islands of Bahrain, Sitrah, Al-Muharraq, Umm Na'sān and about 30 smaller ones. The capital is Manama on Bahrain Island. The terrain is generally subdued and only in the main island does the land rise to more than 100m (328ft). Summers are hot and humid and most rainfall is limited to the cooler, more equable winter months. Because of the lack of surface water much of the supply has to be raised from artesian wells but, once irrigated, the soil is extremely productive. In the irrigated areas the people (who are all Arabic-speaking) grow fruit and vegetables; there is little livestock farming.

Petroleum has been produced from oilfields in the centre of Bahrain Islands for more than 40 years, but the state derives most of its income from refining Saudi oil piped from Dharan to Sitrah. Traditional activities of dhow building and pearling are being replaced by industries serving a more affluent populace. Two major developments of the 1970s have been the smelting of bauxite and the development of Mīnā Salmān as a free port of entry for the southern Persian Gulf. The country became independent in 1971; it is ruled by a royal family assisted by a cabinet, and in 1973 a constitution was adopted.

Map 38.

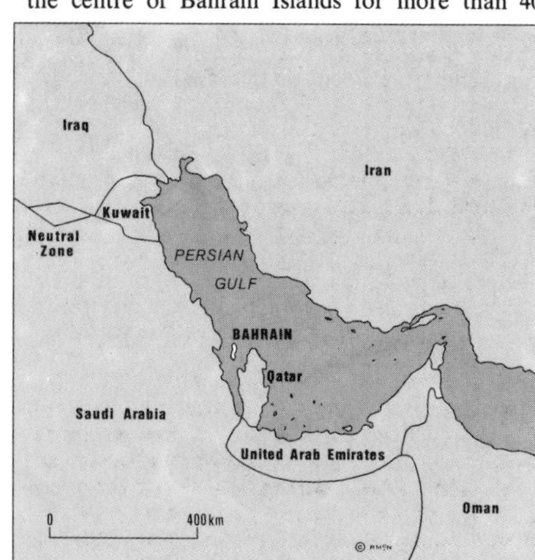

Bahrain – profile

Official name Bahrain
Area 663sq km (256sq miles)
Population (1975 est.) 256,000 *Density* 412 per sq km (1,067 per sq mile)
Chief cities Manama (capital) (1976 est.) 105,400; Al-Muharraq, 41,000; Ar-Rifā, Al, Gharbi, 9,400
Government Emirate; head of state, Sheikh Isā ibn Sulmān Al-Khalīfah
Religion Sunni and Shiite Muslim
Languages Arabic (official), Persian
Gross national product (1974) £247,900,000 *Per capita income* £968
Trading partners Britain, Japan, Saudi Arabia

Bali. *See* INDONESIA.

Bangladesh (Bengal Nation), official name People's Republic of Bangladesh, is an independent nation in Asia on a low-lying alluvial plain formed by the rivers Ganges (Padma), Brahmaputra (Jamuna) and Meghna and their tributaries. Its monsoon climate (which gives it one of the highest rainfalls in the world), location in the cyclone belt and low elevation combine to produce devastating floods. A weak economy has resulted in periodic famines, and it is one of the world's poorest countries. The capital is Dacca.

Land and economy. The world's eighth most populous nation, it is bisected by the Tropic of Cancer. Its coastline borders the Bay of Bengal, with India and Burma adjacent. Most of the land occupies the three great deltas of the rivers Brahmaputra, Ganges and Meghna. As a result it is low-lying and has a hot, humid climate. Only in the south-east on the Burma border does the land rise to any height, and even there the Chittagong hills seldom exceed 90m (300ft) above sea-level. The tropical monsoon climate gives rise to an average annual rainfall of 2,030mm (80in), most of which falls in summer; there is a short dry season in winter. Typhoons, tidal waves and floods often cause much damage and loss of life along the coast; in November 1970 300,000 people were drowned after a tidal wave flooded the delta region. Raw jute and jute manufacture account for 90 per cent of foreign earnings; off-shore oil was found in 1974. More than 80 per cent of the people are farmers and rice is the major crop, but even so the country is not self-sufficient in food; it depends on foreign aid. Poor agricultural yields, disease, famine and a high birth-rate present difficulties not quickly to be overcome.

The chief industrial areas are around Chittagong and Dacca; jute products, paper and leather goods are manufactured and exported. Other exports include cotton textiles (made from imported cotton), tea and fish. The chief import, vital for the country's survival, is food.

People. Ninety-eight per cent of the people are Bengali and speak Bengali, the official language. The remainder include Urdu-speaking Muslim immigrants from India and various tribal units. Islam is the religion of 85 per cent of the people with the balance made up of Hindus, Christians, Buddhists, and followers of ethnic religions.

Government. The 1972 constitution (amended in 1975) was based on nationalism, secularism, socialism and democracy and on a parliament with power in the hands of a prime minister. At present rule is by military decree.

History. A melting-pot of Dravidians, Aryans, Mongolians, Arabs, Persians and Turks, the region of Bangladesh was ruled by Hindu, Buddhist and Muslim dynasties until the British assumed control in the 18th century. In 1947, India and PAKISTAN became separate states. Pakistan was divided into two sections, East and West, and in the mid-1960s Sheikh Mujibur Rahman emerged as the spokesman for East Pakistan autonomy. He became president of the Awami League, was charged with civil disobedience and imprisoned in West Pakistan. Other League leaders fled to India, where they organized a provisional government. Bengali forces defeated Pakistan, and on 26 March 1971 Bangla-

Brussels, Belgium's capital, is also the seat of the European Economic Community; it has several Gothic buildings, art museums and academies.

Ghent, capital city of East Flanders province, is connected with the North Sea by a network of canals and is the chief textile and steel-making centre of Belgium.

Bruges, capital of West Flanders province in north-west Belgium, is a major industrial city and tourist centre sited on an inlet of the North Sea.

desh emerged as a nation. Mujibur Rahman was killed in a coup in 1975 and succeeded by President Abu Sadat Mohammed Sayem. Map 40.

Bangladesh – profile

Official name People's Republic of Bangladesh
Area 142,776sq km (55,126sq miles)
Population (1975 est.) 76,815,000
Density 533 per sq km (1,382 per sq mile)
Chief cities Dacca (capital) (1974) 1,730,253; Chittagong, 889,760
Government Military decree
Religion Islam
Language Bengali
Monetary unit Taka
Gross national product (1974) £3,102,300,000 *Per capita income* £40
Agriculture Jute, rice, sugar cane, tea, oilseeds, fish, timber, cotton
Industries Jute products, cotton textiles, wood products, processed foods
Chief trading partner India

Barbados is an island state in the Windward Island group of the West Indies, east of Saint Vincent. The capital is Bridgetown. It was settled by the British in 1627, became a Crown colony, and was granted independence in 1966. It is a member of the Organization of American States, the Caribbean Free Trade Area and the United Nations. Most of the people are Negroes, who speak English; the main religions are Christianity and ethnic beliefs. Industries include tourism, fishing and the production of sugar cane, molasses and rum. The highest point is Mt Hillaby (336m; 1,104ft). Area: 431sq km (166sq miles). Pop. (1975 est.) 245,000.

Basutoland. *See* LESOTHO.

Bechuanaland. *See* BOTSWANA.

Belgium (België or Belgique), official name Kingdom of Belgium, is a small, densely populated nation of Western Europe. The people are of Flemish, French and German background, and all three languages are spoken (Flemish is similar to Dutch). The country, which is a constitutional monarchy, has one of the world's most highly developed economies and the people enjoy a high standard of living. The capital is Brussels, and Antwerp is the chief port and centre of commerce. Other major cities are Liège and Ghent.

Land and economy. Belgium is a country of coastal dunes and gently undulating plains rising to the hills of the Ardennes in the south. Flanders, the coastal area bordering the North Sea, is a flat, moist plain traversed by many rivers and canals. Belgium's central plain, lying between Flanders and the River Meuse, is a fertile farming region. Most of the

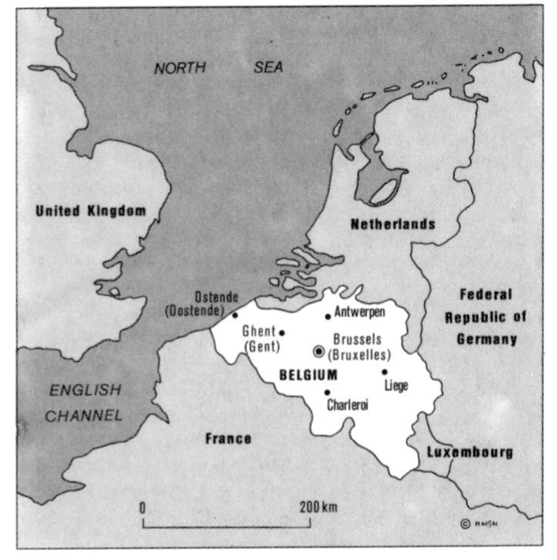

country's industry is also situated there, as are most of the towns. The Ardennes is a rocky, wooded plateau unsuitable for farming. The climate is maritime with damp, foggy winters and mild summers. Industry is focused in the cities along the Scheldt and Meuse rivers. For centuries, Belgium has been famed for its textiles, linens and lace. Other important industries include diamond-cutting, chemicals, coal, metal processing, glass, oil refining, sugar refining and off-shore petroleum. The economy depends upon extensive imports of raw materials and on foreign export markets for its manufactured products. Only 7 per cent of the population is engaged in agriculture.

People. Belgians can be divided roughly into two major language-groups. Most northerners are Flemings, whose official language in Flemish. The Belgians of the south are called Walloons; their official language, Walloon, is a dialect of French. The city of Brussels is officially bilingual, although French is the predominant language. The country is overwhelmingly Roman Catholic.

Government. King Baudouin is the head of state and commander-in-chief of the armed forces. the ministers of government are leaders of the dominant party in parliament, and are appointed by the king. Government policy is carried out by the prime minister and his cabinet. The leading political parties are the Social Christian, Socialist and Liberal. Voting in national elections is compulsory for all Belgians aged 21 years or older.

History. During the Middle Ages many textile manufacturing and trade centres developed in the area that is now Belgium. The French united the small states of the region in the 14th century, and included them in the duchy of Burgundy. After 1477, the Austrian and Spanish branches of the Hapsburgs controlled the area for more than 300 years. Belgium and the Netherlands were united after the Congress of Vienna (1815). The Belgians revolted and founded a constitutional monarchy in 1831 under Prince Leopold of Saxe-Coburg. Belgium colonized Africa's Congo region in the 19th century.

In both world wars the country was overrun and occupied by German forces. Despite Allied air raids and the Battle of the Bulge (the German counter offensive at the end of 1944), Belgian industry was comparatively undamaged at the end of World War II and the country made a speedy economic recovery. It joined the North Atlantic Treaty Organization (NATO) in 1949. King Leopold III, who did not return from Austria until July 1950, was criticized for the way in which he exercised his authority and after a year's reign abdicated in favour of his eldest son, Baudouin. In 1960 Belgium granted independence to its African colony (the Belgian Congo, which became Zaire) and entered a phase of internal strife (between French-speaking and Flemish factions) and unstable government. The creation of Brussels, Flanders and Walloon as equal autonomous regions in 1971 did much to alleviate these problems, with the Church and the monarchy providing unifying influences. Belgium was instrumental in conceiving and forming the European Economic Community (EEC), which has its headquarters there as have NATO and the Supreme Headquarters Allied Powers Europe (SHAPE). Map 18.

Belgium – profile

Official name Kingdom of Belgium
Area 30,513sq km (11,781sq miles)
Population (1975 est.) 9,804,000
Density 32 per sq km (832 per sq mile)
Chief cities Brussels (capital) (1976) 1,074,726; Antwerp, 672,703; Liège, 440,447; Ghent, 224,728
Government Constitutional monarchy
Religion Roman Catholic
Languages Flemish, French, German
Monetary unit Belgian franc
Gross national product (1974) £21,827,200,000
Per capita income £2,226
Agriculture Wheat, barley, sugar-beet, potatoes
Industries Steel, diamond-cutting, chemicals, glass, textiles
Minerals Coal, iron, copper
Trading partners West Germany, France, Netherlands, USA

Belize, formerly British Honduras, is a self-governing British Crown colony on the east coast of central America, on the Caribbean Sea. The capital (since 1970) is Belmopan; much of the former capital, Belize, was destroyed by a hurricane in 1961 but it still contains a third of the population and is the country's chief port. The south is mountainous but

Refining of sugar from locally grown sugar cane is the chief industry of Belize and the sugar produced constitutes the country's major export.

Porto-Novo, Benin, is the capital of the country and a trading and shipping centre founded by the Portuguese in the late 16th century.

The Himalayas, with mountains up to 7,500m above sea-level, dominate the landscape of Bhutan, most of whose people farm the lower slopes.

much of the remainder is lowland with coastal swamps. Dense forests yield timber and fertile grasslands allow the cultivation of sugar cane (the major crop and chief export) and citrus fruits. Most of the people are descended from black Africans or Mayan Indians and they speak Spanish, although English is the official language. Self-government was granted in 1964. In 1977 Guatemala threatened to "take over" Belize. Area: 22,965sq km (8,867sq miles). Pop. (1975 est.) 140,000. Map 72.

Belorussia. *See* UNION OF SOVIET SOCIALIST REPUBLICS.

Benin, formerly a French protectorate called Dahomey and part of the colony of French West Africa, is an independent nation in west Africa, official name People's Republic of Benin. It is a comparatively poor country that has had a succession of military governments since independence, although the development of off-shore petroleum deposits (discovered in 1968) could improve the economy. The capital is Porto-Novo; other large towns include the chief port Cotonou, Abomey, Oüidah and Parakoü.

Land and economy. The coast is hot and humid, and there are two rainy and two dry seasons; average annual rainfall is 810mm (32in). Benin has three plateaus: one inland of the coastal zone is fertile, another to the north-east consists mainly of bare rocks, and a third in the north-west has streams flowing to the rivers Volta and Niger and includes the Atacora mountain range. The eastern part of the country is a plain. The economy is based on subsistence agriculture; half of all exports are palm products and cotton.

People. Benin has various groups of people. Descendants of the Fon, or Dahomey, who established the early kingdom, were trained for civil service by the French. They live in the south and are the best educated (the literacy level is 25 per cent among school-age children). In the north is the Fulani tribe, in the centre and north are the Bariba and Somba, and in the south-east are the Yoruba. Ninety per cent of the population is rural, and 65 per cent practises ethnic religions. French is the common language.

History. Benin's history dates back to three early kingdoms (Allada, Porto-Novo and Dahomey) in the southern area which were under pressure from the northern Kingdom of Abomey in the 16th century. Dahomey became the most aggressive, pushing northwards and selling slaves taken from neighbouring lands. In 1863 the king of Porto-Novo sought French protection against the Dahomey. In 1899 Dahomey was incorporated into French West Africa, which had been formed a few years earlier. The 1956 Overseas Reform Act expanded civil rights, and in 1960 the country became independent (as the Republic of Dahomey; the official name was changed to Benin in 1976). Economic and regional rivalries have caused numerous military coups and changes of government since 1960. In 1970 a charter provided for a three-member presidential council to govern until 1976, but it was overthrown by Lt.-Col. Mathieu Kerekou in 1972. Strained relations with neighbouring Togo were resolved in 1976 by a pact agreeing mutual co-operation in combating subversive elements. Map 32.

Benin – profile

Official name People's Republic of Benin
Area 112,622sq km (43,483sq miles)
Population (1975 est.) 3,112,000
Density 28 per sq km (72 per sq mile)
Chief cities Porto-Novo (capital) (1975 est.) 104,000; Cotonou, 175,000
Government Military
Religions Ethnic, Christianity and Islam
Languages French (official), local dialects
Monetary CFA franc
Gross national product (1971) £158,000,000
Per capita income £55
Agriculture Groundnuts, cotton, cocoa, maize, palm oil, livestock
Minerals Petroleum
Trading partners France and other members of the EEC, USA, Canada

Bermuda, formerly called Somers Islands, is a self-governing British colony in the western Atlantic Ocean, 920km (570 miles) off the United States coast. It consists of about 300 islands, islets and coral reefs (the most northerly in the world). The capital is Hamilton, on Bermuda Island; other inhabited islands include Saint George's, Saint David's, Somerset and Ireland. Bermuda was first colonized when the Englishman Sir George Somers and his expedition were shipwrecked there in 1609. It became subject to the British Crown in 1684 and was granted internal self-government in 1968. There is an American naval and air force base there. Industries: tourism, perfumes, pharmaceuticals, ship repairing, textiles, flowers. Area: 52sq km (20sq miles). Pop. (1975 est.) 56,000. Map 66.

Bhutan is a kingdom in the Himalayas between Tibet (China) to the north and India to the south. The king (Singhi Wangchuk since 1974) rules with the aid of an advisory council and a 130-member national assembly. The official capital is Thimbu although traditionally Punakha, to the north, has this distinction. Eight high mountain ranges cross the country; the highest peak is Kula Kangri, 7,559m (24,784ft). Torrential rainstorms, with an average annual fall of 5,080-6,350mm (200-250in), make it one of the wettest countries in the world.

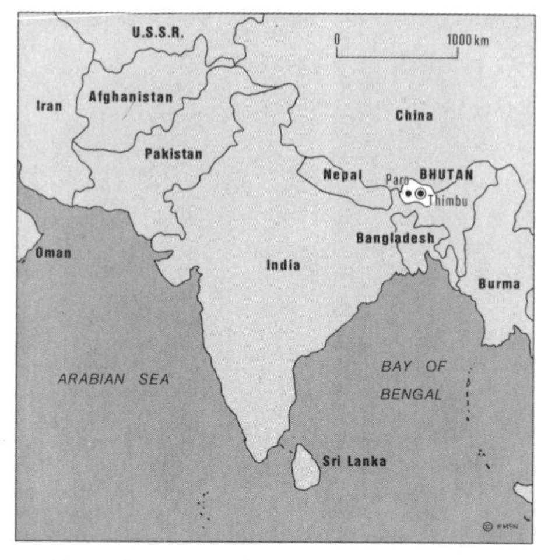

Most of the people of Bhutan are Bhotias, who practise a form of Buddhism closely related to Tibetan Buddhism and speak Dzongka. The chief occupation is farming, with rice as the main crop; yaks, pigs, sheep, cattle and hill ponies are also raised. Craft industries produce goods in metal, wood and leather. The only exports are rice and yak hair (used for making British Guardsmen's bearskins).

By the 16th century, the original inhabitants were conquered by Tibetan armies. China took the country in 1720 and in 1774 opened relations with the British, who occupied the southern region in 1864. Other areas were annexed to India in 1865 and in 1910 Britain signed a treaty making Britain responsible for Bhutan's foreign affairs. When India gained independence in 1949 it took over this role, and Bhutan managed its own internal affairs. The Bhutan – Tibet border was closed in 1960 in the face of threats from China. Area: 47,000sq km (18,147sq miles). Pop. (1976 est.) 1,202,000. Map 40.

Sucre is a major agricultural centre in Bolivia and is the seat of the archbishopric, the supreme court and the national university, San Fransico Xavier.

Mt Illimani is one of the highest peaks of the Cordillera Real in the Bolivian Andes; the top of the mountain is covered with snow throughout the year.

Cochabamba, the second largest city in Bolivia, has many historical buildings, including a convent which has five paintings by the Spanish artist Goya.

Bolivia, official name Republic of Bolivia, is a country in South America. Despite its rich natural resources, it is one of the poorest nations of the continent (the average income is £127 per year). Since 1884, when Chile gained Bolivia's only coastal province in a war, the country has been landlocked. Two-thirds of it is low-lying, but most of the people live on the *altiplano,* a plateau high in the Andes. La Paz, which is effectively (although not officially) the capital, is the world's highest major city. The official capital is Sucre. Bolivia is named after Simón Bolívar, the "Liberator" of north-western South America in the struggle for independence from Spain.

Land and climate. Western Bolivia is occupied by the mountains and bleak tablelands of the Andes, which are there at their widest – 650km (400 miles). Two great Andean ranges extend roughly north-west to south-east enclosing the *altiplano,* which has

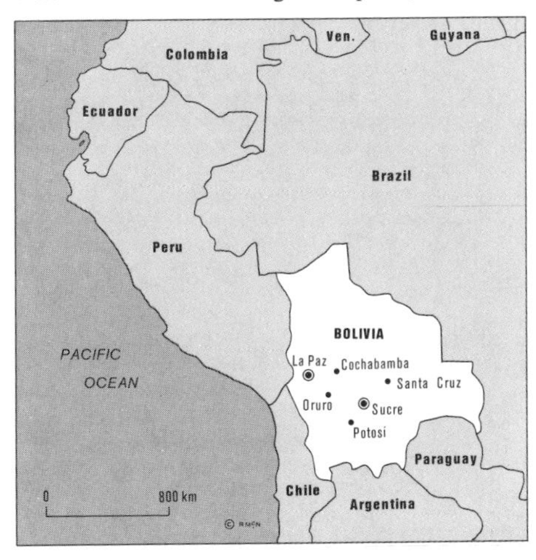

an average altitude of 3,700m (12,000ft). At the north of the *altiplano,* on the boundary with Peru, is Lake Titicaca, the highest navigable stretch of water in the world [*see* PE p. *65*]. Much of the southern part of the plateau is a desert region of salt marshes. Although generally cold and dry, the plateau has the country's best living conditions and its most easily exploited natural resources. Central Bolivia is covered by the forested foothills of the Andes, with steep-sided but often fertile valleys called *yungas.* In this region the climate is warm and humid. Beyond, to the north and east, is an immense lowland region of tropical forests and plains, drained by many great, slow rivers.

Economy. Minerals are the chief source of wealth – they account for nearly 70 per cent of Bolivia's foreign earnings. The most plentiful mineral is tin; its chief centres of production are Potosí and Oruro. Other important minerals include silver, gold, copper, lead, bauxite, tungsten and antimony. Petroleum and natural gas have been discovered near Santa Cruz, and enough petroleum is obtained to meet most of the country's needs. An oil pipeline connects Santa Cruz with the Pacific port of Arica in Chile. The country has little manufacturing industry, partly because of shortage of electric power. There is considerable hydroelectric potential in the fast-flowing rivers of the mountains, but it is as yet insufficiently exploited.

Agriculturally Bolivia is mainly a land of peasant farmers, many of them living at or near subsistence level. After a revolution in 1952 more than 150,000 new peasant holdings were created from former large estates. On the *altiplano,* crops include maize, barley, potatoes and quinoa (a crop that yields edible seeds and leaves); sheep and llamas are also herded. Better farming land is found in the valleys to the east and in the rainy *yungas;* there grapes, cocoa, coffee, rice, bananas and other valuable cash crops are grown.

Poor communications is one of Bolivia's most intractable problems. The building of modern roads and railways is hampered both by the nature of the terrain and by lack of money. But about 19,000km (12,000 miles) of rivers are navigable, and towns along the rivers play an essential part in trade. Exports and imports are routed chiefly through the ports of Antofagasta and Arica in Chile and Mollendo-Matarani in Peru.

People. Bolivia is one of the most thinly and unevenly populated countries of South America. The lowlands of the east are almost uninhabited. American Indians make up some 70 per cent of the population; most of them belong to the Aymará and Quechua tribes. Nearly all live on the land and speak either Aymará or Quechua, the Inca language. About a quarter of the people are mestizos, of mixed Indian and European descent; Spanish-speaking and Roman Catholic, they control government and industry. Education is free and compulsory, but even so more than half the population is illiterate. There are several universities, one of them dating from 1624. As in other parts of the continent, religious holidays and festivals play an important part in the life of the people of Bolivia. The national sport is soccer.

Government. Constitutionally Bolivia is a democratic republic with a president, senate and chamber of deputies elected by adult suffrage. The legislature was, however, suspended in 1971 because of successive coups, and the country reverted to a president and cabinet who rule by decree.

History. In the 13th and 14th centuries the empire of the Incas was extended to include the declining civilization of the Aymará Indians who inhabited a region near Lake Titicaca. The Incas were, in turn, overthrown in 1538 by Spanish conquistadores, led by Hernando Pizarro. The discovery of silver at Potosí in 1545 attracted many adventurers from Spain; other colonists carved out vast estates for themselves. There were numerous Indian revolts against Spanish rule, but it was the settlers who finally won independence for the country. In 1824 Antonio José de Sucre, one of Bolívar's generals, defeated a Spanish army at Ayacucho in Peru, and in the following year an independent republic of Bolivia was proclaimed.

Since independence, the country's history has been one of war and revolution: war has resulted in loss of territory to Chile, Brazil and Paraguay, and there have been more than 180 revolutions. A revolutionary government that came to power in 1952 under the leadership of Dr Víctor Paz Estenssoro nationalized the large mining companies, broke up the large estates, removing economic control from the companies and landowners. A new constitution recognized the rights of the Indians. But continuing economic failure led to a coup by a military junta in 1964. Elections were held in 1966, but constitutional procedure was again suspended after further coups in 1970 and 1971. The present military leaders plan to return the country to democratic government by 1980. Map 76.

Bolivia – profile

Official name Republic of Bolivia
Area 1,098,581sq km (424,162sq miles)
Population (1976 est.) 4,687,618 · *Density* 4.3 per sq km (11 per sq mile)
Chief cities Sucre (official capital) (1975 est.) 106,590; La Paz (seat of government), 660,700
Government Head of state, Gen. Hugo Banzer-Suárez, president (took office Aug. 1971)
Gross national product (1974) £594,000,000 *Per capita income* £127
Agriculture Sugar cane, rice, coffee, cotton, maize, potatoes, wheat
Mining Tin, gold, petroleum, natural gas
Trading partners Argentina, USA, Brazil, West Germany

Borneo (Kalimantan) is a large island in the Malay Archipelago in south-eastern Asia. The third largest island in the world, it is made up of four political units belonging to three countries. The states of Sarawak in the west, and Sabah (north) belong to MALAYSIA; BRUNEI (north-west) is a British protectorate; and Kalimantan state, occupying 70 per cent of the island, is part of INDONESIA. The Dyaks, one of the largest ethnic groups on the island, are among the most primitive peoples in the world today.

In the 7th century Borneo was colonized by Chinese, followed by Malays. Spanish, Portuguese, Dutch and British trade started in the 16th and 17th centuries, and by 1888 Sabah, Brunei and Sarawak were British protectorates, the remainder being claimed by The Netherlands. The present divisions were fixed in 1963. Area: 743,330sq km (287,000sq miles). Pop. (1971 est.) 6,968,000. Map 50.

A Botswana cattle boat, sailing down the Okavango River, provides an indication of how cattle breeding dominates the country's economy.

Brasília officially took over in 1960 as capital of Brazil (from Rio de Janeiro); here tall blocks of hotel buildings tower above the surrounding plain.

Manaus, situated in Brazil on the Rio Negro, is the main commercial and cultural centre of the upper Amazon region and an important river port.

Botswana, official name Republic of Botswana, was known as Bechuanaland until it gained independence from Britain in 1966. This large, land-locked country was then one of the world's poorest and its prospects seemed unpromising. The chief occupation was cattle-rearing and many unemployed young men left to work in South Africa. By 1970 the average annual income was only about £45. But in the late 1960s valuable mineral deposits were found. The development of mining in a politically stable atmosphere led to fast annual economic growth rates of 14–15 per cent in the mid-1970s. Average incomes also rose, despite the exceptionally high rate of population growth of 3.7 per cent per year. The capital is Gaborone.

Land. Botswana is a large tableland, most of which is between 600 and 1,200m (1,968–3,937ft) above sea-level. The climate is subtropical; temperatures average 27°C (81°F) in winter and 32°C (90°F) in summer. The average annual rainfall varies from less than 230mm (9in) in the south-west to about 690mm (27in) in the north.

Infertile Kalahari sands cover more than 80 per cent of this mostly semi-desert country. Grass steppes in the south merge northwards into dry woodland savanna. Botswana has little surface water apart from the inland River Okavango delta in the north. A wide variety of African animals live in this remote, swampy region, which has potential for tourism and agricultural development.

Economy. The best soils are in the east, around the part of the railway between South Africa and Zimbabwe (Rhodesia) that runs through Botswana – the country's chief communications artery. Arable farms, producing some maize, millet and sorghum, cover only 0.2 per cent of the land. The country has more than two million cattle, and meat and meat products from Lobatse's modern abattoir are of great importance to Botswana's economy. The mining of diamonds at Orapa, copper and nickel at Selebi-Pikwe, and coal at Morupule greatly increased Botswana's revenue in the 1970s. Diamond production reached 2,500,000 carats per year by the mid 1970s. Some light industries are developing, but unemployment remains high: only about 18 per cent of the potential labour force are in paid employment. Most people depend on pastoralism.

People. Much of the population lives in the east and belongs to one of the eight divisions of the Bantu-speaking Tswana group: the Bamangwato, Bangwaketse, Bakwena, Batawana, Bakgatla, Bamalete, Barolong and Batlokwa. About one-fifth of the people are literate. Some Botswanans are Christians but many follow ethnic religions. Minorities include about 30,000 Bushmen, 3,500 people of mixed European and African descent and 3,400 Europeans.

Government. In 1965 the Botswana Democratic Party, led by Sir Seretse Khama, won a majority in the legislature. In 1966 the country became an independent republic with Sir Seretse Khama as president. Botswana's National Assembley has 32 elected and four nominated members. There is also a House of Chiefs.

History. Britain established a protectorate in the area in 1885 at the request of local people, who feared invasion by Boers or Matabele from South Africa. The protectorate was governed by a British High Commissioner until independence. In the 1970s President Khama played an important part as a "front-line president" in seeking a solution to the problems of Zimbabwe (Rhodesia). Map 34.

Botswana – profile

Official name Republic of Botswana
Area 576,000sq km (222,000sq miles)
Population (1976 est.) 718,000
Chief cities Gaborone (capital), 36,900; Serowe, 34,186; Kanye, 34,045; Molepolole, 29,623
Religions Christianity, ethnic
Languages English (official), Setswana
Monetary unit Pula
Major trading partner South Africa

Bougainville. *See* PAPUA NEW GUINEA.

Brazil (Brasil), official name Federative Republic of Brazil, is the largest country of South America. It has some of the world's most sophisticated cities and some of its most primitive and inaccessible regions, such as the dense forests of Amazonia. And although the country has many natural resources, most of its riches are as yet untapped and millions of its people live at subsistence level. Unlike most South American countries, which were once Spanish colonies, Brazil formerly belonged to Portugal. For a time the Portuguese king had his court in Brazil, and Rio de Janeiro was the capital of the Portuguese Empire. Rio remained Brazil's capital until 1960, when its place was taken by the architecturally splendid new city of Brasília.

Land and climate. Northern and Western Brazil lie almost wholly in the basin of the River Amazon which, with its tributaries, carries more water than any other of the world's rivers. The main river is 6,669km (4,145 miles) long. Much of Amazonia – the Amazon region – is a heavily forested lowland. It is hot and wet, and its sparse population consists mainly of scattered tribes of American Indians. The most easterly part of Brazil was once the richest; it is known as the North-east Region. Inland from its warm and wet coastal strip steep escarpments rise to broken and often arid plateaus. But the North-east Region has given way in importance and wealth to the temperate southern part of the country, which accounts for 70 per cent of the total agricultural production and has 75 per cent of the industry. The

mountains of the south are the highest in Brazil, although they are low compared with the peaks of the Andes. In the south-west, on the Argentinian border, are the spectacular Iguaçu Falls, which extend for more than 3km (2 miles) across the Iguaçu River.

Economy. Brazil's potentially rich natural resources have as yet been of limited economic benefit. Poor communication and lack of money, power and trained technologists have prevented large-scale exploitation. The most plentiful mineral is iron ore; it is estimated that Brazil probably has a quarter of the world's reserves. Other minerals of importance include manganese, chrome, asbestos, bauxite and lead. There are large deposits of coal, but most of it is of poor quality. The discovery of off-shore oil promises eventual self-sufficiency in petroleum. There is a thriving industry in the export of pine timber from the south, but the enormous

This colourful street procession is part of the pre-Lenten carnival that takes place in Rio de Janeiro, the second largest city in Brazil.

Belim is the major commercial centre and port of the Amazon River basin; it is also the site of a beautiful 17th century Jesuit church.

Amazonas Province forms the largest part of Brazil and includes the world's most extensive area of tropical rain forest, called the selva.

hardwood forests of Amazonia are almost untouched. A number of government-backed hydroelectric projects aim to make better use of the power potential of Brazil's rivers. One large scheme on the Paraná River in the south is planned to provide electricity for industrial development in the most heavily populated areas. The country's largest and most important manufacturing industry is the production of textiles.

Nearly half of Brazil's people live by agriculture, most of them as plantation workers or smallholders. For hundreds of years the chief cash crop was sugar cane, grown on the humid coastal strip of the North-east Region. Today sugar has been supplanted in importance by coffee; Brazil is the world's greatest producer, growing nearly a quarter of all the coffee consumed. Most of it comes from large *fazendas* (plantations). Cocoa, cotton, oranges, bananas, rice and maize are other important crops. Subsistence farmers grow manioc (cassava), which is ground to make their staple food *farinha de mandioca. See also* PE pp.149, 155.

Like many other South American countries, Brazil has yet to achieve adequate internal communications. The greater part of the country is without roads and railways, but a government road-building programme aims to construct a skeletal highway system that will reach even into the untamed regions. River communications play an essential part in trade, and aircraft have brought many remote districts within reach of civilization.

People. Most of the people of Brazil are Portuguese-speaking. Many are descended from the Portuguese colonists; other are of more recent immigrant stock from Portugal, Germany, Italy and Spain. There are also Japanese and other minorities. A fifth of the people have mixed blood: they are either mestizos, of Indian and European descent, or mulattoes, of Negro and European descent. There are small numbers of Negroes and American Indians, most of the latter being tribal people living in Amazonia [*see* MS pp.*249,* 256]. More than 90 per cent of the population is Roman Catholic. Education is free and compulsory, but the 1970 census showed that only 60 per cent of the population was literate. There are more than 60 institutions of university status.

The North-east Region and the South have almost all the large cities, including the beautifully sited city of Rio de Janeiro and the great commercial and industrial city of São Paulo. Most cities, including Rio, are overcrowded and slum-fringed. The new capital of Brasília has been sited in the undeveloped interior. Probably the most celebrated example of modern city design, it was largely the work of the Brazilian architects Lúcio Costa and Oscar Niemeyer. Culturally Brazil has strong links with Portugal. But Brazilian writers have become increasingly conscious of "Brazilian identity". The same is true, perhaps to a lesser degree, in other arts. The Brazilian composer Heitor Villa-Lobos (1887–1959) made the folk-music of his country known throughout the world. The national sport is soccer; since the 1950s Brazil has been one of the leading soccer countries.

Government. Under the constitution of 1967 Brazil is a democratic republic with a president as head of state and a Congress consisting of a Senate and a Chamber of Deputies. But an amendment gives the president power to abrogate the political rights of citizens for periods of up to ten years. Since 1964 Brazil has, in fact, been ruled by authoritarian military govenments.

History. The Portuguese navigator Pedro Álvares Cabral reached the coast of Brazil in 1500 and claimed the territory for Portugal (under the terms of the 1494 Treaty of Tordesillas between Portugal and Spain, Portugal was entitled to the eastern part of South America). A colonial government was established in 1549 with its capital at Salvador, and the coastal areas were slowly settled. In the second half of the 17th century a period of more intensive colonization began. Adventurers pushed into the interior of the continent, adding new territories to Brazil, and slaves were imported from Africa to work on the sugar plantations of the north-east. Disputes with Spain about Brazil's boundaries were settled by treaty in 1750.

When Napoleon invaded Portugal in 1807, the Portuguese royal family fled to Brazil. The king, John VI, returned to Portugal in 1821, leaving his son Pedro as regent. A year later Brazil declared itself independent, with Pedro as its emperor. In 1831 he was forced to abdicate in favour of his son, then five years old, who became emperor as Pedro II. In 1888 slavery was abolished. The former slave owners turned against Pedro, and in 1889 his opponents forced his abdication and declared a republic under the authoritarian rule of Gen. Manoel Deodora da Fonseca.

A revolution in 1930 established the "benevolent dictator" Getúlio Vargas in power with army support. He embarked on a programme of industrial development and agricultural reform that had considerable success. But his centralization of government, dictatorial rule and abolition of constitutional freedoms led to much opposition. After World War II Vargas was forced out of office and some democratic rights were restored. But in 1950 an election brought him back to power, this time as constitutional president. Widespread unrest caused by the soaring cost of living led to his forced resignation in 1954, when the vice-president João Café Filho took his place.

In the following year Juscelino Kubitschek de Oliveira was elected president and he set about the task of achieving rapid industrial expansion and modernization. The new capital of Brasília was constructed and plans were made for pushing the frontier of development far into the interior. But the government's efforts were thwarted by inflation and a new period of political instability began, with recurring changes in constitutional procedure and a return to authoritarian government. Map 76.

Brazil – profile

Official name Federative Republic of Brazil
Area 8,511,695sq km (3,286,365sq miles)
Population (1976 est.) 109,181,000
Density 13 per sq km (33 per sq mile)
Chief cities Brasília (capital), (1970) 272,002; São Paulo, 5,869,966; Rio de Janeiro, 4,296,782
Government Head of state, Ernesto Geisel, president (took office March 1974)
Religion Roman Catholic
Language Portuguese
Monetary unit Cruzeiro
Gross national product (1974) £39,817,100,000 *Per capita income* £365
Agriculture Coffee, cocoa, cotton, rubber, sugar cane, fruits, rice, maize
Industries Textiles, foodstuffs, motor vehicles, timber, paper, chemicals
Minerals Iron, manganese, lead, chrome, oil, coal, diamonds, asbestos
Trading partners USA, EEC, Argentina, Venezuela

British Honduras. *See* BELIZE.

British Indian Ocean Territory is an island colony in the Indian Ocean just south of the Equator and about 3,200km (2,000 miles) east of the African coast. It is made up of the Chagos Archipelago, Farquhar Atoll, Desroches Island and Aldabra Island. It was established by Britain in 1965 in an effort to retain a strategic site between Africa and Asia. Area: 226sq km (87sq miles). Pop. (1975 est.) 2,000. Map 34.

British Solomon Islands is a self-governing protectorate in the south-western Pacific Ocean, about 850km (530 miles) east of New Guinea. The islands are widely scattered; the chief ones are Guadalcanal, Malaita, New Georgia, Choiseul, Santa Isabel and the Shortland Islands. They are administered by a high commissioner from the capital Honiara on Guadalcanal. The other, more northern, islands in the Solomons group (Bougainville and Buka) belong to PAPUA NEW GUINEA. Copra is the chief product. The Solomons protectorate was formed in 1898 (after Germany had relinquished most of the northern islands) and completed in 1900 with the transfer of the Shortland Islands. Following Australian occupation during World War I, Bougainville and Buka were placed under Australian mandate in 1920. The central Solomons were occupied by Japanese forces between 1942 and 1945. Area: 29,800sq km (11,505sq miles). Pop. (1975 est.) 190,000. Map 62.

British Virgin Islands is a British colony in the Lesser Antilles between the Caribbean Sea and the At-

Sofia, apart from being Bulgaria's capital city, is also known for its educational and cultural facilities, together with its many historical buildings.

The valley of the Maritsa River, which flows from Bulgaria and through Turkey to the Aegean Sea, separates the Rhodope and Balkan mountain ranges.

Plovdiv is the second major city in Bulgaria; it has many major industries together with several Orthodox churches and Turkish mosques.

lantic Ocean, ruled by a governor and councils. The colony consists of 36 islands (the remainder of the group, about 70 islands to the south-west, are an American possession known as the VIRGIN ISLANDS OF THE UNITED STATES). The chief islands of the British group are Tortola (site of the capital, Road Town), Anegada and Virgin Gorda. They were acquired from The Netherlands in 1666. The chief sources of income are tourism, fishing and farming. Area: 155sq km (60sq miles). Pop. (1975 est.) 11,000. Map 74.

Brunei is a self-governing British protectorate in north-western Borneo. It is in two main sections, surrounded by the Indonesian state of Sarawak. The capital is Bandar Seri Begawan. Brunei sultans ruled Borneo during the 15th and 16th centuries, but their power diminished after trade with western European countries and by 1888 the island was ruled by Britain. The sultan's power was reinstated

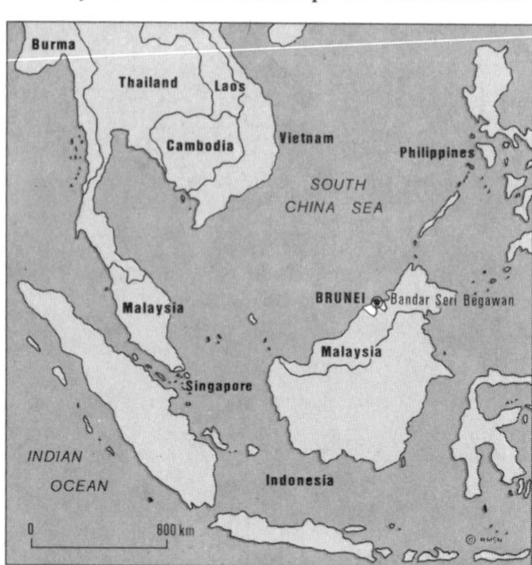

with the formation of an amended protectorate in 1965. Oil was discovered in 1929 and is now the chief export, giving the people of Brunei one of the highest living standards in south-eastern Asia. Other exports include rubber, rice and tropical fruits. Area: 5,765sq km (2,225sq miles). Pop. (1975 est.) 147,000. Map 50.

Bulgaria, official name the People's Republic of Bulgaria, is a country in the Balkan Peninsula in south-eastern Europe. It was under Turkish suzerainty for nearly 500 years before it became an independent kingdom at the beginning of the 20th century. After World War II it was absorbed into the Soviet-dominated bloc of eastern European countries as a Communist "people's republic". Strenuous efforts at industrialization have been made in recent years, but many of the people of this predominantly agricultural country still cling tenaciously to their old ways of life. The capital is Sofia.

Land and climate. The most fertile regions of Bulgaria lie in the basins of two great rivers, the Danube in the north and the Maritsa in the south-east. The broad Danubian plain is exposed to fierce winds from the north and east. In the Maritsa valley the climate tends to be mild. The steep but round-topped Balkan Mountains extend across the country and have only one major pass, the Shipka Pass near their centre. The Rhodope Mountains in the south-west are the historic boundary between Thrace and Macedonia. Some of the mountain scenery is exceptionally beautiful and attracts many tourists, as do the holiday resorts on Bulgaria's Black Sea coast.

Economy. The Communist government that came to power after World War II adopted the idea of a centrally planned economy based on the Russian model. Several five-year plans were promulgated, aimed at a gradual but sustained increase in heavy manufacturing industries and in the production of electric power. Some industries are based on the country's valuable mineral resources, which include oil. The chief oilfields are around Pleven and near Balchik in the north-eastern coastal region. Off-shore oil has also been found.

Most of the arable land has been consolidated into collective, co-operative and state farms. Production has been boosted by mechanization and other modern farming techniques, and new agricultural-industrial towns or complexes have been established. In the Danubian basin the chief crops are cereals: wheat, maize and barley. The milder regions – the mountain valleys of the south, the Maritsa basin and the Black Sea coast – produce also fruits, vegetables, cotton and tobacco. Grapes are grown for wine-making and plums for slivovitz (plum brandy). Damask roses are Bulgaria's most famous crop. They are grown to make attar of roses, an oil used in perfumes.

People. About 90 per cent of the people of Bulgaria are Bulgars, but there is a sizeable Turkish minority and there are other, smaller groups. The country has few large cities – only Sofia and six other cities have more than 100,000 inhabitants – but there are many small towns and about half of the population is now urban. Traditional village life continues, however, and country dwellers keep up the ancient rural crafts of pottery, wood-carving and weaving. They also retain their interest in the country's rich tradition of folk-music and dancing. The Bulgarian language belongs to the Slavonic group (as are Russian and Serbian); it uses Cyrillic script. *See also* MS p.245.

Government. The 1971 constitution provides for a single-chamber National Assembly that elects the executive body, the Council of State. The chairman of the Council of State is in effect the head of state. The elections for the National Assembly return a near-100 per cent majority for the Fatherland Front, a union (mainly) of the Communist Party and the People's Agrarian Union.

History. In ancient times the land that is now Bulgaria was partly in Thrace and partly in Moesia. In the 6th and 7th centuries AD it was settled by Slavs who, in the following century, were conquered by the Bulgars, a people from Asia. The Bulgars became Slavicized, founded an empire and adopted Christianity. By 1018 they were forcibly incorporated in the Byzantine Empire, but a second Bulgarian Empire with its capital at Tŭrnovo was established by Ivan I, the first of the Asen dynasty, in about 1187. But it was short-lived. In the early 14th century it lost its independence to Serbia, and before the end of the century Bulgaria had been added to the empire of the Ottoman Turks.

A revival of Bulgarian nationalism in the 19th century led to a rebellion in 1876, which was put down by the Turks with great severity. Russia took the part of the Bulgarians, and after the Turkish defeat in the Russo-Turkish War (1877–78) tried to create an independent but Russian-influenced Bul-

garian kingdom. The Congress of Berlin (1878), however, reduced the size of the proposed country and made Bulgaria an autonomous principality under Turkish suzerainty. In 1885 Bulgaria seized Eastern Rumelia, part of the territory given back to Turkey by the Congress, and in 1908 Prince Ferdinand of Bulgaria declared the country an independent kingdom with himself as tsar. Bulgaria gained territory in the First Balkan War (1912–13) but lost some possessions in the Second Balkan War (1913). It again lost territory after World War I, during which it fought on the side of Germany and Austria-Hungary. In World War II it again fought alongside Germany (after 1941) but changed sides after a government coup when Soviet troops reached its borders in 1944. In 1946 the monarchy was abolished and the nine-year-old ruler Simeon II left the country. A people's republic was proclaimed with the veteran Communist Georgi Dimit-

The Rhodope Mountains, in the south-west of Bulgaria, include Mount Musala which, at 2,925m (9,598ft), is the highest point in the country.

Buddhism in Burma flourishes both actively and visibly: Rangoon is dominated by the huge gold stupa of the Shwe Dagon Pagoda.

Burma, one of the world's main rice growing areas, is also known for its rich mineral deposits and sources of teak and other hardwoods.

rov as premier. A constitution modelled on that of the USSR was adopted in 1947. Relations with neighbouring Greek and Turkey remained strained and in 1951 160,000 people of Turkish descent were deported. An attempted military coup failed in 1965 and three years later Bulgarian forces helped Soviet troops in the invasion of Czechoslovakia. In 1971 Bulgaria adopted a new constitution. In the mid-1970s the country re-affirmed its alliance to the USSR and to Soviet Communism. Map 26.

Bulgaria – profile

Official name The People's Republic of Bulgaria
Area 110,911sq km (42,823sq miles)
Population (1975 est.) 8,722,000
Density 79 per sq km (204 per sq mile)
Chief cities Sofia (capital), 946,300; Plovdiv, 287,700; Varna, 260,100
Government Chairman of the Council of State, Todor Zhivkov
Gross national product (1974) £6,576,000,000
Per capita income £754
Agriculture Wheat, maize, barley, tobacco, sugar-beet, fruit
Industries Iron and steel, cement, chemicals, textiles
Minerals Iron, lead, zinc, coal, oil
Trading partners USSR, East Germany, Italy

Burma (Myanma), official name The Socialist Republic of the Union of Burma, is a nation in south-eastern Asia. It is a land of small villages and great rivers, mountains and forests, with a small population in a relatively large country. Until 1948 it was part of the British Empire. In recent years, as an independent country, Burma has been strongly nationalistic, determined to shield itself from outside interference. The life of its people is strongly influenced by Buddhism, a philosophy that finds expression in Burma's most famous monument, the graceful, gold-covered Shwe Dagon pagoda in the capital, Rangoon.

Land and climate. Burma is cut off from its neighbours by lofty mountains that extend southwards from China in two branches along the eastern and western (Arakan) sides of the country. On the east they form the large, broken massif of the Shan plateau. Between the east and west ranges is the broad valley of the Irrawaddy River where most of the people live. The Irrawaddy and its largest tributary the Chindwin rise in the northern mountains and flow southwards for more than 1,600km (1,000 miles) to a wide, many-streamed delta on the Bay of Bengal. All the large rivers tend to flood during the rainy season, and for this reason people living on their banks often build their houses on stilts.

Burma has a monsoon climate, with heavy rains

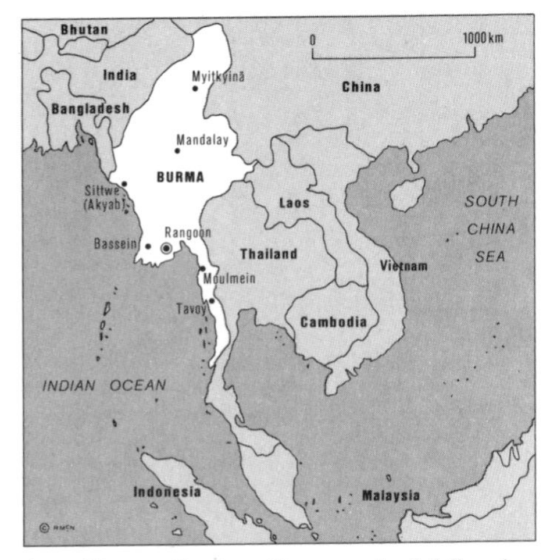

from May to October. But annual rainfall varies greatly from place to place: the coasts of the Arakan and Tenasserim regions may have 5,000mm (200in), whereas the so-called "dry zone" around Mandalay has 1,000mm (40in) or less.

Economy. The country is rich in natural resources, including minerals, although much of it has yet to be surveyed. It has long been known for its gemstones, particularly rubies, sapphires, lapis lazuli and jade. The most important mineral economically is oil; there are also valuable deposits of various metal ores. An under-exploited resource is timber from the forests that cover nearly two-thirds of Burma, although large amounts of teak and ironwood are cut.

The economy is predominantly agricultural. Since independence much farming land has been nationalized and distributed among peasant proprietors. The country is one of the world's leading rice producers, the heaviest crops coming from the Irrawaddy delta. Other major cash crops include rubber, jute, tea and tobacco. River traffic plays an essential part in trade, but the towns of the Irrawaddy valley also have good road and rail connections. Recent economic growth has failed to keep up with the steadily increasing population.

People. Two-thirds of the people are Burmese or Burmans, a Mongoloid people. Their language, Burman, belongs to the Sino-Tibetan group. Several tribal peoples live in the hills. They include the Nagas and the Kachins in the north, the Chins in the west and the Shans in the east. The Karens inhabit a region to the south of the Shans; these two are the largest of the minority groups. Most of the people are Theravada Buddhists, but Buddhism is no longer the state religion. The monasteries and monks are still, however, important elements in the community. Public celebrations, generally of religious inspiration, play a large part in Burmese life. The few towns are mostly small and ancient. The largest city is Rangoon, and the only other cities of any size are the seaport of Moulmein and the former royal capital, Mandalay, which is still the chief religious and cultural centre.

Government. Under a constitution approved by referendum in 1973, Burma is a socialist republic with a president, a People's Assembly and a Council of Ministers elected by the Assembly.

History. The early history of Burma is one of small tribal kingdoms whose peoples were descended from Chinese and Tibetan migrants. In the 11th century unity was imposed by a ruler named Anawrahta, the first of a dynasty that set up its capital at Pagan. It was Anawrahta who introduced Buddhism into Burma. In 1287 Kublai Khan and his Mongols sacked Pagan and disrupted the Burmese kingdom. Lasting unity was not achieved again until 1758, when Alaungpaya created a new kingdom with its capital at Rangoon. The territory of the kingdom extended into north-eastern India and western Thailand. In the 1820s (the First Burmese War) armies of the British East India Company drove the Burmese out of India and took Arakan and part of Tenasserim. Recurring disputes – mostly about trade – between the British and the Burmese (the Second Burmese War) resulted in progressive British annexations of Burmese territory and significant economic development. In 1885 Mandalay (the royal capital since 1860) was taken, and Burma was made an Indian province. It was given a measure of self-government in 1923 as part of India, and in 1937 was made a separate colony with home control of internal affairs.

Burma – profile

Official name The Socialist Republic of the Union of Burma
Area 678,033sq km (261,789sq miles)
Population (1975 est.) 31,240,000
Density 46 per sq km (119 per sq mile)
Chief cities Rangoon (capital) (1973) 2,056,118; Mandalay, 393,000
Government Head of state President U Ne Win
Language Burmese
Religion Buddhism
Monetary unit Kyat
Gross national product (1974) £1,158,000
Per capita income £37
Agriculture Rice, rubber, jute, legumes, tea, tobacco, sesame, millet, groundnuts
Industries Processed food, textiles, tobacco, wood products
Minerals Petroleum, zinc, lead, antimony, tungsten, copper, silver, rubies, sapphires, jade
Trading partners India, Japan, Malaysia, EEC, USA

Ancient rain forest occupies much of Burundi, one of the smallest nations in Africa. It is densely populated and most of the people work on the land.

About one per cent of Burundi's people are pygmies, members of the Twa ethnic group whose stature contrasts markedly with that of the tall Tutsi.

Phnom Penh, the capital of Cambodia, became a key target of the Communist Khmer Rouge in the civil war, which ended when the city was taken in 1975.

When Japan became a belligerent in World War II in 1941, Japanese troops invaded Burma from Thailand, set up a pro-Japanese government and sought unsuccessfully to make Burma their base for an invasion of India. In one of the war's bitterest campaigns the Allies, with Burmese help, defeated the Japanese armies in Burma shortly before Japan's surrender. In 1948 the link with Britain was broken and the Union of Burma became an independent country. Its early years of independence were marred by armed risings of Communist groups and of Karen tribesmen. The Shans and other ethnic groups also reacted against what they saw as "Burmanization" of their territories, but by adopting authoritarian measures the government succeeded in maintaining some stability. Map 52.

Burundi, official name Republic of Burundi, is an independent country in central Africa. It was part of the Belgian-ruled Ruanda-Urundi until 1962, when that territory was divided into RWANDA and Burundi. Small, romote and densely populated, Burundi is one of the world's poorest nations, with an average annual income per person of only £38 in 1974. The capital and only large city is Bujumbura.

Land and climate. Western Burundi lies in the great African Rift Valley and includes part of the River Ruzizi valley and Lake Tanganyika, whose shore is 767m (2,515ft) above sea-level. Mountains between 2,135 and 2,745m (7,000–9,000ft) tower above the Rift Valley. Beyond the mountains, eastern Burundi is a plateau between 1,370 and 1,830m (4,500–6,000ft) above sea-level.

The climate is moderated by the altitude. Annual temperatures average 23°C (73°F) in the Rift Valley, 17°C (63°F) in the mountains and 20°C (68°F) in the eastern plateaus. Rainfall is comparatively low, averaging 760mm (30in) in the Rift Valley, 1,475mm (58in) in the mountains and 1,200mm (47in) in the eastern plateaus, although it is unreliable and long droughts sometimes occur. Grasslands cover the mountains, and moist woodland originally covered the eastern plateaus; but much of the natural vegetation has been destroyed by farmers, particularly in the east.

Economy. Most people are subsistence farmers, especially in the east, where they grow such food crops as beans, cassava, maize, peas, rice and sweet potatoes. The most valuable crop is coffee, which makes up about 75 per cent of Burundi's exports; cotton and tea are also important cash crops. Agriculture productivity is low, partly as a result of severe soil erosion caused by over-intensive farming and cattle-grazing. Some cassiterite (tin ore) is mined, but there is little manufacturing. Because of over-population, many young men used to seek seasonal employment in other countries, although this is discouraged by neighbouring governments.

People. The Kurundi-speaking Hute, most of whom are farmers, form 84 per cent of the population. Most of the remainder are Tutsi, who are strikingly tall Hamitic peoples who live by rearing cattle. For nearly 400 years the Tutsi have ruled the area, and recently Hutu-Tutsi conflict has marred the country's progress. Burundi also has some Twa (pygmies), who form less than one per cent of the population.

Government. In 1966 Burundi was proclaimed a republic and the former prime minister, Col. Michel Micombero, became president. In the 1970s the president ruled with a Council of Ministers and the Political Bureau of Burundi's one remaining political party, UPRONA (the Unity and National Progress Party). Micombero was deposed in 1977 and Lt.-Col. Jean-Baptiste Bagaza, a leading army officer, became president. He dissolved UPRONA and set up a Supreme Revolutionary Council to govern the country.

History. Germany began to rule the area in 1897, but Belgian troops occupied the territory in World War I and in 1919 Belgium was mandated by the League of Nations to rule Ruanda-Urundi. Belgian rule ended in 1962 when Ruanda-Urundi was partioned to form two nations: Rwanda, a republic, and Burundi, a monarchy. In 1966 Tutsi officers deposed the Tutsi *mwami* (king) and established a republic. Power struggles between the Hutu and Tutsi increased. In the early 1970s, President Micombero accused the Hutu of attempting genocide against the Tutsi. In stern reprisals, thousands of Hutu were killed and many others fled from persecution in Burundi to neighbouring countries.

Burundi – profile

Official name Republic of Burundi
Area 27,834sq km (10,747sq miles)
Population (1975 est.) 3,900,000
Chief cities Bujumbura (capital) (1976 est.) 110,000; Kitega, 3,579
Religions Ethnic, Christianity
Languages Kirundi, French (both official)
Monetary unit Burundi franc

California. *See* UNITED STATES.

Cambodia, official name Democratic Kampuchea and formerly the Khmer Republic, is a country in the Indochina peninsula. It was once a kingdom and has an ancient culture that drew much of its inspiration from India. In the north of Cambodia is Angkor Wat, an extraordinary complex of Hindu temples and shrines which is now in ruins but is thought to have been the largest religious structure in the world. In recent years Cambodia has experienced war and bloodshed, and today it is run by a Communist government from the capital, Phnom Penh.

Land and economy. Hills and high plateaus mark the land boundaries of Cambodia, but the greater part of the country is a broad lowland (dominated by the Mekong River) with, near its centre, the Tonlé Sap ("Great Lake"). In the rainy season the Mekong River floods into this lake, which increases vastly in size: its dry-season area of about 2,500sq km (965sq miles) can increase by as much as four times. Rice is grown in the rich silt left behind when the waters recede, and the lake is also an abundant source of fish – a major component of the Cambodian diet. Cambodia has a tropical monsoon climate, with a dry season from November to March or April.

The country's chief natural resources are its forests (covering three-quarters of the land area) and its minerals, which include phosphates and iron ore. Most of the people live by farming and more than half of the cultivated land is used for rice production. The chief export crop is rubber; other cash crops include pepper, maize, soya beans, bananas and cotton. Most industries are on a small scale and are concerned with food processing and the use of timber. Some larger industries, disrupted by the fighting in the early 1970s, are being reconstituted. The chief cities are Phnom Penh, Battambang, Kompong Cham and Kompong Som, the main seaport. Good roads link the main centres of population, and Phnom Penh is connected by railway to Poipet on the Thai frontier and to Kompong Som. The rivers are an important means of communication.

People. Most of the people (85 per cent) are Khmers, but there are sizeable minorities of Chinese, Vietnamese and Chams. There are also hill tribes called Khmer-Loeu. Population statistics are unreliable, and official estimates (*see Profile*) may be as much as a million too large. Most of the people live on the land, many of them in villages along the rivers. The official language is Khmer; about 60 per cent of the population is literate. The country's traditional religion is Theravada Buddhism.

Government. A new constitution came into force in January 1976, and elections were held for a

The Khmer Sacred Dance originated as a result of a Cambodian religion which coexists with pre-Buddhist beliefs in the use of magic to ward off spirits.

Angkor Wat, Cambodia, was created in the 12th century as a monument to the monarch; it is probably the largest religious structure in the world.

Ottawa, the capital of Canada, was founded in 1827 and has several notable buildings including the Anglican and Roman Catholic cathedrals.

People's Representative Assembly. Prince Sihanouk resigned as head of state in April 1976 and the Assembly elected Khieu Samphan in his place (official title, President of the State Presidium). A new government was also appointed.

History. In the 1st century AD a strong state, the kingdom of Funan, developed in the lower Mekong basin. In about the 6th century the Khmer people from the state of Chenla, in the upper Mekong region, extended the boundaries of their country into Funan and present-day Laos, and built up an empire that became the dominant power in south-eastern Asia. The Khmer civilization survived for several hundred years, and scholarship, literature, sculpture and architecture flourished. The greatest monuments to its achievements are the rich and beautiful remains of Angkor the former capital. In the 1400s the Thais took Angkor and the capital was moved to Phnom Penh.

A long period of decline ended in the establishment of a French protectorate in Cambodia in 1863. The country was granted independence within the French Union in 1949, and by 1955 was a fully sovereign kingdom. The leading figure in its political life was then Prince Norodom Sihanouk, who successively occupied the positions of king, head of the government and head of state. Although officially neutral in the Vietnam War, Prince Sihanouk's socialist regime permitted the North Vietnamese to use Cambodian territory in their fight against the South Vietnamese. The country's economic difficulties added to his growing unpopularity and in 1970 he was deposed by the National Assembly, and a Khmer Republic was established. Sihanouk set up a rival government in Peking. A protracted and confused civil war followed: until 1973 American and South Vietnamese troops actively supported the republicans; North Vietnamese troops helped the Sihanoukists and the Communist Khmer Rouge. The war ended in April 1975, when the Khmer Rouge took Phnom Penh. In September 1975 Prince Sihanouk returned to Cambodia as head of state but he resigned the following year and was replaced by Khien Samphan, leader of the Khmer Rouge. In January 1976 the country adopted a new constitution, changing the official name from Cambodia to Kampuchea. Map 52.

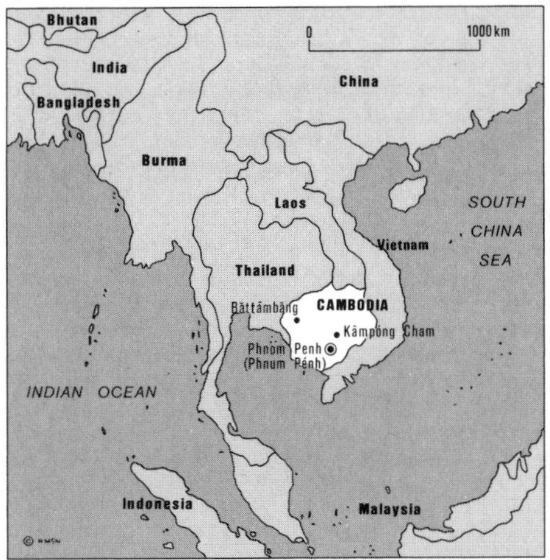

Cambodia – profile

Official name Democratic Kampuchea
Area 181,035sq km (69,898sq miles)
Population (1975 est.) 8,110,000
Chief cities Phnom Penh (capital) (est.) 470,000; Battambang, 43,000; Kompong Cham, 31,000
Government Head of state, Khieu Samphan
Religions Theravada Buddhism, Islam
Languages Khmer (official), French
Monetary unit Riel
Agriculture Rice, rubber, maize, soya beans
Industries Forestry, fishing, mineral processing
Minerals Phosphate rock, limestone, iron ore

Cameroon, official name United Republic of Cameroon, is a nation in western Africa with an extremely diverse population. Historians believe that the Bantu-speaking peoples of Africa originated in this area. The capital is Yaoundé.

Land and economy. Behind the coastal plain, Cameroon is a country of plateaus and volcanic uplands. The volcanic Mt Cameroon, western Africa's highest peak, is 4,070m (13,353ft) above sea-level. The south has an equatorial climate with extreme heat and an extremely high average annual rainfall of more than 9,000mm (354in). The Uplands have a mild climate and the tropical north is dry. Mangrove and rain forests in the south merge into mountain grassland and woodland savanna in the north. Most people are farmers working at subsistence level and the average annual income in 1974 was only £107. The chief exports are coffee, cocoa and bauxite (aluminium ore). Cattle are reared on the northern savanna.

People. The population is divided into more than 200 diverse ethnic groups. Sudanese negroes, Hamitic Fulani (Fulbe) and Arab Choa live in the north. Bantu-speaking peoples, including the Bamiléké, Bassa, Bulu and Fang, live in the south.

Government and history. Cameroon is a republic, consisting of the former French Cameroun and part of the former British Cameroon. It was a German protectorate from 1884 to 1916 but, after World War I, France ruled five-sixths of Cameroon and Britain ruled the rest. They became trust territories of the United Nations in 1946. The French region became independent as the Cameroon Republic in 1960 and the southern part of British Cameroon joined it in 1961 to form the Federal Republic of Cameroon (the north united with NIGERIA). It adopted its present official name in 1972. President Ahmadou Ahidjo rules with a cabinet and an elected National Assembly. Map 32.

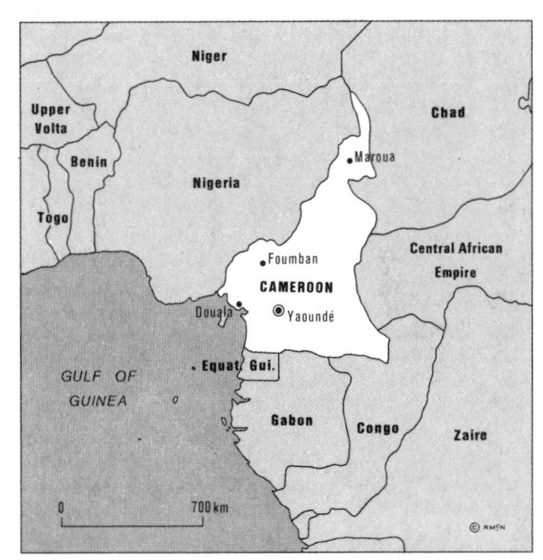

Cameroon – profile

Official name United Republic of Cameroon
Area 475,441sq km (183,568sq miles)
Population (1975 est.) 6,398,000
Chief cities Yaoundé (capital) (1975 est.) 274,400; Douala, 486,000
Religions Ethnic, Christianity, Islam
Languages French, English (both official)
Monetary unit CFA franc

Canada, the second largest country in the world (after the USSR), occupies most of the northern half of North America and has coastlines on three oceans – the Pacific, Arctic and Atlantic. It is so vast that it extends across seven time zones. All its natural features are on a large scale: the lofty, snow-capped ranges of the Rocky Mountains, the rolling fields of wheat on the prairies, the thick forests of the Canadian Shield, and the icy wastes of the Arctic. But it has only one-ninth as many people as its smaller neighbour, the United States, and most of these live in a tiny southern region bordering the Great Lakes and the St Lawrence River. Politically, Canada is a federation and a constitutional monarchy that acknowledges the British Monarch as head of state; it is a member of the Commonwealth of Nations. The capital is Ottawa.

Land and climate. With the exception of the Arctic lands, each of Canada's several distinct natural regions runs broadly north to south. The largest region, comprising almost half the country's land area, is the Canadian Shield [*see also* PE p.103]

Niagara Falls, one of the most famous spectacles in North America, form an international line between Canada and the USA; the falls were discovered in 1678.

Toronto, Canada's second largest city, is the banking and stock exchange centre of the country; it has major rail and port facilities too.

The Royal Canadian Mounted Police, also known as the Mounties, nowadays carry out their duties in mechanized vehicles; horses are reserved for ceremonial occasions.

This rugged upland of granite and other ancient rocks forms – with its United States extension – the core of the North American continent. It is shaped like a horseshoe, with one tip on the northern mainland coast of the Northwest Territories, the other on the northern coast of Quebec, and Hudson Bay in the centre. At the south-western shore of Hudson Bay the rocky slopes descend to a swampy, forested lowland.

To the south-east of the Shield is the most heavily populated region, the lowlands of the St Lawrence River Valley and the peninsula between between Lake Huron and lakes Erie and Ontario. This small but fertile region has most of the major cities. To its north-east is the Canadian section of the Appalachian mountain system, a region of small plains and low hills. Its coastal areas are famous for their long-established fishing communities.

The Interior Plains, to the west of the Shield, belong to the Great Plains of North America, and run in a broad band from the United States border to the shore of the Beaufort Sea. The seemingly limitless prairies that make up their southern part are among the world's most productive agricultural lands, and grow enormous crops of wheat. They rise in three great steps from Manitoba, through Saskatchewan, to Alberta. Beyond them, to the west, are the high peaks of the Canadian Rockies [*see* PE p. *107*] and, farther west still, the precipitous Coast Mountains, one peak of which – Mt Logan (6,050m; 19,850ft) in the Yukon – is the country's highest.

The Arctic north of Canada is remote, icy and barren. For hundreds of years navigators tried to find a route through its maze of island straits so that ships could sail from the Atlantic Ocean, through the Arctic Ocean, and into the Pacific — the so-called North-West Passage. No ship made the passage until the 1900s. Much of this region is north of the tree line, and its landscape is almost entirely treeless. Apart from fur traders and the staffs of weather and scientific stations, the only inhabitants are small communities of Eskimos and American Indians.

Canada's climate tends to the extremes, and most of the country has very cold winters. The mildest area is the Pacific coast; the south-eastern lowlands and coasts are relatively mild, too, but even these may have heavy winter snow. In the Arctic some waters are frozen for up to nine months a year, and permanent ice caps on parts of the eastern islands extend into the sea as sheets of fresh-water ice. Central Canada has short, cool, dry summers. In southern regions the summer is longer and warmer. The highest annual precipitation – as either rain or snow – is in the western and eastern coastal areas. The prairies get most of their rainfall in the summer months.

Physical resources. The Canadian lumberjack is a legendary figure; wood is the country's most obvious resource. Forests cover about 3,000,000sq km (1,160,000sq miles) and more than half of their trees are of commercial value, although in practice many forest areas are almost inaccessible. The chief forest products are pulpwood, timber and plywood. Pulp comes chiefly from the northern coniferous forest belt that stretches right across the country. The pulp is turned into newsprint and it is estimated that half the newspapers in Europe and North America are printed on Canadian paper. High-quality timber is cut in British Columbia, where forestry is the principal industry; the trees there include Douglas fir, hemlock and red cedar. Valuable timber comes also from the deciduous hardwood forests of the south-east. The rivers provide the timber industry with an indispensable means of transport: cut logs are floated downstream, often in the form of giant rafts, to the saw mills or pulp mills. *See also* PE p.219.

Canada's mining industry is among the world's richest and makes a major contribution the country's export trade. Great quantities of minerals are extracted annually and yet untold wealth still lies untouched underground. The largest reserves are thought to be in the Shield, but few regions are without known resources. Canada leads all other countries in the production of nickel, silver, zinc and asbestos, and is among the leading producers of uranium, molybdenum and iron. Other products of prime importance are gold, lead, coal, potash and sulphur.

Canada is also an important petroleum-producing country. The main deposits are in the Interior Plains, in the tar sands along the Athabasca River in Alberta. An extensive network of pipelines carries oil to refineries in southern Canada and the United States and natural gas to major Canadian cities.

Water power is one of Canada's greatest resources, and until the 1950s was almost the only source of electric power. Today hydroelectricity accounts for about 65 per cent of the total, the remainder coming from thermal and nuclear power stations.

Constitution and government. Canada is a federal union consisting of ten provinces and two territories. The country's independent status derives from the British North America Act of 1867 (an Act of the British parliament), which provided that Canada should have a constitution similar in principle to Britain's. As with the British constitution, many accepted rules and conventions are unwritten. The Statute of Westminster in 1931 removed any remaining limitations on Canadian sovereignty. Canada's sovereign is Queen Elizabeth II, who is represented by a governor-general appointed on the recommendation of the prime minister.

The national parliament meets in Ottawa and consists of an upper and a lower house: the Senate and the House of Commons. They have an equal voice in legislation except that the Senate cannot reject finance bills. The Senate consists of 104 senators appointed by the governor-general on the advice of the prime minister, each province and territory having an allotted number of seats. Senators serve until they reach the age of 75. The House of Commons is elected by universal suffrage for maximum terms of five years, and each of its members represents a constituency. The number of constituencies in each province or territory is related to population and is reconsidered every ten years; after the election of 1974 the House of Commons had 264 members. The governor-general invites the leader of the majority in the House to accept the position of prime minister and form a government. The prime minister selects a cabinet from among his supporters in the House, each cabinet minister having charge of a government department.

The highest court of justice in the land is the Supreme Court of Canada, composed of the Chief Justice of Canada and eight judges; it is the final court of appeal in both civil and criminal cases. The next highest court, the Federal Court of Canada, comprises two divisions: a court of appeal and a trial division. The provinces have superior and county courts. In Quebec province justice is based on civil law; elsewhere there is a common law system.

Political parties. The two largest parties are the Liberals and the Progressive Conservatives. The Conservatives are looked upon as the more traditional and "British" party, but in general both parties follow much the same policies. Since the formation of the Dominion of Canada in 1867 the country has had ten Conservative prime ministers and eight Liberal ones. The smaller parties include the New Democratic Party, which is socialist in sentiment, and the Social Credit Party (Ralliement Creditistes). Quebec province has some French-Canadian separatist groups; although these have had little success in national elections, the separatist Parti Québecois won the provincial election in 1976.

Armed services and police. Since 1964 naval, land and air forces have had integrated headquarters and are organized in seven major commands. Mobile Command has ground and air forces (including airborne troops) for the protection of Canadian territory and for meeting Canada's overseas commitments. Maritime Command is primarily responsible for the defence of Canada against sea attack. Air Defence Command shares with United States forces the air defence of the North American continent (through the North American Air Defense Command, NORAD). The Canadian Forces Communications system operates strategic communications for the armed services and also aids the civil authorities in emergencies. And there are Air Transport and Training Commands, as well as a Reserve and Survival Organization with a specific role in case of national emergency. In 1976 the total strength of the armed services was nearly 80,000.

The federal police force is the Royal Canadian Mounted Police (RCMP) but there are also provincial forces in Ontario and Quebec as well as municipal forces in many cities. The working uniform of the "Mounties" is a brown jacket and dark blue breeches, but the famous scarlet jackets and splendid horses are still seen on ceremonial occasions.

Industry. Canada ranks sixth among manufacturing countries and industry is the most important

Saskatchewan consists of vast expanses of unbroken plains which are ideal for large scale mechanized farming of wheat, oats, barley and rye.

Nanaimo, situated on Vancouver Island, has several sawmills and a large pulp processing plant, together with a flourishing fishing industry.

St Lawrence Seaway is an international waterway consisting of a system of canals, dams and locks; it is ice bound for nearly four months of the year.

part of its economy. Its products include motor vehicles and vehicle parts, pulp and paper, processed foods, textiles, industrial and agricultural machinery and equipment, timber, rolling stock, aircraft, plastics and rubber goods, and chemicals. Thousands of people are employed in the smelting and refining of metals, iron and steel milling, petroleum refining and publishing and printing. As with agriculture, industry is concentrated almost entirely in the south, and in particular in Ontario and Quebec.

Toronto and Montreal are the chief manufacturing cities. Toronto has factories producing motor vehicles, electrical equipment, chemicals, textiles and aircraft. Montreal has steel, chemicals, textiles, foodstuffs and petroleum products. Hamilton and Sault Ste Marie are known as iron and steel centres. Windsor makes motor vehicles, London rolling stock, Cornwall chemical products and paper, and Trois-Rivières pulp and paper. The prairie cities also have varied industries, many of them connected with the processing of local agricultural produce. In addition there are engineering industries, including the manufacture of railway equipment, as well as paper and pulp mills, iron and steel mills and oil refineries. In British Columbia much industry is centred on Vancouver and its neighbouring towns.

Agriculture. Only about 7 per cent of the land area is devoted to agriculture; this contrasts with the more than 30 per cent of forest land. But agriculture still plays an extremely important part in the economy. With increased mechanization farms have been declining in number but increasing in size, so that a comparatively small workforce produces an extremely high output. Most farms are concentrated in the prairies and in a relatively small region comprising southern Ontario, southern Quebec and New Brunswick. The prairies have been called "the world's breadbasket". Vast fields of grain stretch away on every side. Farms are large, and the farm buildings stand in isolated clusters. The chief crops include wheat, oats, barley and maize, as well as hay, potatoes, flax and sugar-beet. In the drier areas cattle ranching is important. Other livestock includes dairy cattle and pigs. In recent years, partly as a result of occasional gluts of cereal produce, farmers have tended to diversify.

In southern Ontario and Quebec farms are smaller and mixed farming is common. There farmers enjoy the major advantage of having easily accessible markets. Cereal crops include wheat, oats, maize (chiefly in south-western Ontario) and barley. Vegetables and fruits (including grapes and peaches) are important in some areas, and so is tobacco. There are many dairy farms, especially in Quebec, and other animals are raised for meat.

In the eastern maritime areas cattle are kept for dairy produce, and hay is grown. Potatoes are important in New Brunswick and Prince Edward Island. In the Shield, where the soil is poor and rocky, the northern areas have little farmland and elsewhere many farms operate only at a subsistence level. British Columbia has cattle ranches, dairy farms and sheep and poultry farms. It also has vegetable and fruit farming in specific areas, such as the Okanagan Valley in the south.

Trade. Two feature's of Canada's commerce have caused concern to successive governments. The first is the fact that about half of the country's exports are primary products, and the other is the dominant position of the United States in Canadian industry and trade. Both factors, however, contribute to Canada's prosperity. Apart from the United States – the partner in about 65 to 70 per cent of foreign trade – Canada's chief trading partners are Japan, Britain and West Germany; together they account for about 13 per cent.

Transport and communications. Two transcontinental railway systems serve Canada, the state-owned Canadian National Railways and the Canadian Pacific Railway. There are also numerous small railway companies. As in other parts of the world, the railways play a less important part in public transport than they once did, except for long-distance freight haulage. Computers and standard containers has greatly improved railway efficiency.

An Atlantic to Pacific motor highway, the Trans-Canada Highway, was opened in 1962. It links St John's, Newfoundland, with Victoria, British Columbia, and has a length of 7,700km (4,800 miles). Other great highways include the Alaska Highway from Dawson Creek, British Columbia, to Alaska; all but 480km (300 miles) of its total length of 2,450 km (1,520 miles) is in Canada.

The St Lawrence River provided early explorers with a route into the Canadian interior; today the river is a vital trade route. The Great Lakes-St Lawrence Seaway system [*See* MM p.196] has opened ports on the Great Lakes to all but the largest of ocean-going ships [*See* PE p.60] A series of canals and locks connect a waterway that extends about 3,830 km (2,380 miles) westwards from the Atlantic to Thunder Bay on Lake Superior. At Lake Superior the water level is about 150m (500ft) above that at the mouth of the seaway.

Canada has two international airlines: Air Canada, which is state-owned, and Canadian Pacific Airlines. There are also some domestic airlines and small operators provide air links in the northern regions where an aircraft is often the only feasible means of transport. In the Arctic the snowmobile has proved its usefulness for short journeys, and has been adopted as a patrol vehicle by the RCMP. The dog-sled drawn by huskies, famous in Canadian folk-lore, is still used to some extent by Eskimos and other inhabitants of the cold north.

People. Although there were earlier British attempts at settlement in Canada, the French were the first permanent colonists – in Quebec in the 1600s. The Hudson's Bay Company was founded in 1670, and as early as the 1700s British settlements were established in Newfoundland and Nova Scotia; British immigration to these regions increased after Britain gained Canada from the French in 1763. New Brunswick and southern Ontario were heavily settled in the late 18th century,

Quebec, which has a mainly French-speaking population, is an important port and an industrial, cultural and tourist centre.

Vancouver, an important deepwater port, also has shipyards and a large fish processing industry; it was named after George Vancouver, the British explorer.

The Fraser River, was discovered by Sir Alexander Mackenzie in 1793 and named after Simon Fraser, the Canadian explorer and fur trader.

when about 50,000 United Empire Loyalists moved there from the United States following the success of the revolutionaries in the American War of Independence. In the 19th century hundreds of thousands of immigrants arrived from England, Ireland and Scotland, and at the turn of the century people of many nationalities arrived from mainland Europe, large numbers of them going to the west. After World War II millions more arrived to build a new life in Canada; they travelled not only from Europe, but also from Asia and elsewhere.

Canada's population has quadrupled since the beginning of the century. The contemporary search for a "Canadian identity" is prompted not only by the proximity of the powerful and all-pervading United States, but also by the fact that Canadians remain acutely conscious of one another's ethnic origins. Today French Canadians form about 30 per cent of the population, and have an abiding awareness of their cultural identity and their historic place in the country – an awareness heightened by the fact that most of them live in one province, Quebec. They often resent the dominance of the English language and British-type institutions. Less than half of the English-speaking population is, however, of British stock. The next largest ethnic groups are the Germans and Italians. There are also large Ukrainian, Dutch, Polish, Norwegian, Swedish, Greek, Hungarian, Chinese and Yugoslav communities, and many smaller groupings. Of the indigenous peoples, the American Indians are in the majority: they number about 280,000 and belong mainly to the Algonquin, Huron, Iroquois, Athabascan, Haida and Salish tribes. About 17,000 Eskimos live in scattered communities in the north.

Cultural life and leisure. Most Canadians apart from those of French ancestry tend to identify themselves with the "Anglo-Saxon" culture. As a result some two-thirds of Canadians speak English as their only language, and a mere one-fifth speak only French. About one-eighth speak both languages. But both French and English are used officially – for example, on postage stamps, in the civil service and in the parliament in Ottawa. Television programmes, too, are bilingual. In 1977 the provincial government in Quebec began taking measures to enforce the primacy of the French language throughout the province.

Religion plays a major part in Canadian life, particularly outside the large cities. Most people are Christians, the largest denomination being Roman Catholic – about 45 per cent of the population. The largest Protestant groups are the United Church of Canada, a union of Congregationalists, Methodists and Presbyterians (17 per cent) and the Anglican Church of Canada (12 per cent).

Education is a matter for the provincial governments. It is often linked to the people's religious beliefs, although most schools are interdenominational. Systems of education vary from province to province, and schooling is generally compulsory from the age of six to 14. It is free and coeducational. Except in Quebec, teaching is in English. The Quebec government encourages teaching in French and has embarked on plans to make it compulsory. Special schools for Indians and Eskimos are provided by the federal government, but the majority of Indian and Eskimo children attend ordinary schools. The 60 or so universities include several of international standing, such as the English-language McGill University and University of Toronto, and the French-language University of Montreal and Laval University. Laval, the country's oldest university, dates from 1663.

In the arts, Canadians have always been conscious of three strong outside influences – those of France, Britain and the United States. Writers have been particularly affected, not only because of the near-impossibility of distilling a single Canadian literature from writings in two languages but also because of the pressure of literary output from the United States. But something that is specifically Canadian has come through in writers as varied as L.M. Montgomery *(Anne of Green Gables),* Stephen Leacock *(Literary Lapses),* Mazo de la Roche *(Jalna)* and the poets St-Denys Garneau and E.J. Pratt; John McCrae gained fame for his haunting poem *In Flanders Fields.* Canadian painting has excelled chiefly in depicting landscapes and early Canadian life. The most influential school of painters has been the Group of Seven, who worked in the 1920s and 1930s. The members of the group – Lawren Harris, F.H. Varley, Arthur Lismer, Franz Johnston, A.Y. Jackson, J.E.H. MacDonald and F. Carmichael – used bold design and colour to capture distinctively Canadian landscapes. Earlier, similar paintings had been made by Tom Thompson, a member of the so-called Algonquin school.

Canadians have a reputation for fitness and toughness, probably a legacy of the days when life in the newly settled territories was hard. Today the use of the motor car rather belies this reputation but something of a cult of physical fitness has developed among city dwellers conscious of the dangers of an affluent life style. Canada's favourite spectator sports are fast and often rough. Ice hockey, the most popular team game, demands speed, toughness and skill from players. Lacrosse is another popular game that is hard and fast. It derives from a game played by the Iroquois Indians to train their warriors. Canadian football (similar to American football) is a favourite school and college sport, and professional teams are enthusiastically supported. The diverse Canadian terrain and the vast areas of unspoilt country offer unrivalled opportunities for a host of outdoor pastimes, including hunting, fishing, skiing, snowshoeing and skating.

History. The identity of the first European navigators to reach the coast of Canada is still a matter for conjecture and argument, but it seems that Viking ships sailed along the eastern coast at some time in the 11th century. In 1497 John Cabot, searching for a route to China, sailed from Bristol and touched the North American coast at Newfoundland or Cape Breton Island. The Portuguese explorer Gaspar Corte-Real is believed to have reached Newfounland and Labrador in 1501, and in 1524 Giovanni da Verrazano exlored the coasts of Newfoundland and Nova Scotia. Between 1534 and 1542, the French explorer Jacques Cartier made three journeys to North America: he took possession of Newfoundland for France, named and explored the St Lawrence River and visited the areas where Montreal and Quebec City were later built. Early French attempts at establishing settlements proved unsuccessful, but in 1608 Samuel de Champlain and others founded a well-organized settlement on the north bank of the St Lawrence River; it was named Quebec and was the first permanent European settlement on the Canadian mainland.

The English, too, were active in exploring these new lands and seas. In 1583 Sir Humphrey Gilbert landed in Newfoundland and declared it part of the dominions of Queen Elizabeth I, thus establishing Newfoundland's claim to be the first English colony. Another English navigator, Henry Hudson, entered the great inlet now called Hudson Bay in 1610, and met his death when his mutinous crew cast him adrift in a small boat. In 1629 the English temporarily seized Quebec.

French adventurers and missionaries explored the lands along the St Lawrence and the Great Lakes. Champlain travelled to the lake that is now named after him, found the Ottawa River and saw Lake Ontario. Jean Nicolet made an expedition to Lake Michigan, and the Jesuits Joseph Marie Chaumonot and Jean de Brébeuf reached Lake Erie. Another Jesuit, Isaac Jogues, saw Lake Superior and named Sault Ste Marie. For many early settlers Canada's chief attraction was its abundance of fur-bearing animals, whose pelts brought rich rewards. Most settlements were in the hands of trading companies. But in 1663 King Louis XIV made the French territories in Canada into a royal colony, which was named New France, and thereafter immigration increased and more ordered patterns of life began to develop. The English were also extending their interests. In 1670 a fur-trading company, the Adventurers of England Trading into Hudson's Bay (the Hudson's Bay Company), was granted a charter by King Charles II that gave it territory and exclusive trading rights in Canada. In time the company built forts and trading posts that stretched right across Canada from the Atlantic to the Pacific. The Treaty of Utrecht (1713), ending the War of the Spanish Succession, recognized the British claim to Hudson Bay, Newfoundland and Nova Scotia.

In the early days the French and English settlers looked on the indigenous Indians as the main opponents of their claims in North America: the Indians had fought – sometimes with considerable success – to protect their lands from the greedy invaders. but gradually the French came to see the British as the chief obstacle to French ambitions, and the British were determined to advance their own interests against opposition from any quarter. In 1745 during

Halifax, Canada's principal ice-free Atlantic port, was founded in 1749 and named after the Earl of Halifax, the then President of the Board of Trade.

Nova Scotia, one of the Maritime Provinces of Canada, is highly industrialized; coal is the main industry, followed by fishing and forestry.

Louise Lake was discovered in 1882 and named after Princess Louise; it is noted for its scenic beauty and is surrounded by high peaks and glaciers.

"King George's War" British troops from New England captured Louisbourg, on Cape Breton Island, the strongest French fortress in Canada. It was restored to the French in 1748, but fighting broke out again in 1755 at the beginning of the French and Indian War. The war went badly for the British until 1758, but in that year they again captured Louisbourg and also took other French strongholds. In September 1759 – after the historic battle of the Plains of Abraham in which both British and French commanding generals, Wolfe and Montcalm, were killed – Quebec and Montreal surrendered and all Canada passed into British hands.

The Treaty of Paris (1763) formally recognized British rule in Canada. The Roman Catholic French settlers were not, however, disposed to accept either the disabilities on Catholics then imposed in British territories or the abrogation of French legal and political institutions. To meet their objections the Quebec Act of 1774 (a British statute) allowed them freedom of religion, recognized the dominance of the civil law in the newly created province of Quebec, and defined the province's boundaries. The passing of this law incensed the settlers in the other British colonies in North America, and was one of the "Intolerable Acts" that helped to ignite the revolution. When the American War of Independence began in the following year, the Canadian settlers remained loyal to the British Crown. The revolutionaries were beaten back in their attempt to take Canada, and thousands of "United Empire Loyalists" moved to Canada.

The Indian name *Canada* was not used officially until 1791. In that year the Canada Act divided the province of Quebec in two: the southern part, largely inhabited by American Loyalists, became Upper Canada; the northern part became Lower Canada. In time, Upper Canada formed part of Ontario, and most of Lower Canada was incorporated into the new province of Quebec.

Despite political troubles, the second half of the 18th century and the early 19th century was a period of expansion. In 1778 Captain James Cook explored Nootka Sound; in 1789 Alexander Mackenzie travelled along the Mackenzie River to the Arctic; David Thompson mapped much of the Canada–United States border, and in 1807 crossed the Rockies; and in 1808 Simon Fraser made a 1,300-km (800-mile) journey along the Fraser River, almost reaching the Pacific Ocean.

In the War of 1812 American troops invaded Canada several times; on two occasions they sacked York (now Toronto), but their incursions were easily repulsed. Since the end of that war, the long border between the two countries has been one of the most peaceable and least-policed in the world.

In the early 1800s the northern lands of North America were still trading country. Rivalry between the Hudson's Bay Company and the North West Company of Montreal led to violence that was at length ended by the amalgamation of the two under the name of the Hudson's Bay Company. There was also violence in the east. In 1837 two brief rebellions broke out – one in Lower Canada led by Louis Joseph Papineau and one in Upper Canada led by William Lyon Mackenzie. The British government was alarmed, fearing another war of independence. A new governor-general, Lord Durham, was hurriedly despatched to Canada. In a famous report he recommended the union of Upper and Lower Canada, the improvement of communications to encourage unity, and the granting of responsible government. The British government accepted only part of these proposals: the two Canadas were combined under a single legislature by the Union Act of 1840. But the ensuing disagreement about the way in which the French and British populations were to be represented in the legislature merely worsened the dispute between them. Eventually in the late 1840s the British government accepted that further measures were needed, and various territories were allowed to form their own governments.

In the 1860s the idea of an ordered federation of the British lands in North America gained support. Canadian leaders feared the ambitions of the United States in the west, and it was clear that without a strong central authority in the country it would be almost impossible to build the railways that were essential to prosperity. The foundations of a federal union were laid by a series of conferences – at Charlottetown (1864), Quebec (1864) and London (1866) – and in 1867 the British North America Act was passed by the British parliament and the Dominion of Canada came into being.

At first the dominion consisted of four provinces: Quebec, Ontario, Nova Scotia and New Brunswick. Beyond it to the north and west was the vast extent of Rupert's Land, belonging to the Hudson's Bay Company. On the west coast there were the two colonies of Vancouver Island and British Columbia, and in the east was Newfoundland. Within 15 years, however, all these lands had adhered to the dominion with the exception of Newfoundland, which maintained its separate identity until 1949. The lands of the Hudson's Bay Company were bought.

The government of the new dominion, under the premiership of Sir John A. Macdonald, had many difficulties to overcome. The most intractable was the welding together of disjointed territories and mutually suspicious peoples to form a homogenous country and a united population. A major step forward was the completion in 1885 of the long-awaited transcontinental railway, the Canadian Pacific. It linked Montreal in the east with the Pacific coast of British Columbia, and was built in five years despite the most formidable obstacles which the engineers had to tackle and overcome. As part of his national policy, Macdonald also imposed tariffs to protect Canadian industry and commerce from outside competition, particularly from the United States. In the early 1890s the population was found to be declining in numbers, but by the turn of the century a flood of immigration had begun and the settlement of the west was under way.

The continuing quarrel between French-speaking and English-speaking Canadians flared up again in 1869 with a rebellion of the *Métis* (people of mixed French and Indian blood) in the central lands who feared for their holdings now that the Canadian government had taken over from the Hudson's Bay Company. Under Louis Riel they set up a provisional "government" of their own. The rebellion was quickly suppressed, and the grievances of the *Métis* were met, partly by the creation of the new province of Manitoba. Riel fled to the United States, but in 1884 he returned and led another rebellion in Saskatchewan. He was captured and – despite French protests – hanged.

In 1896 Sir Wilfrid Laurier, a Liberal, became prime minister. He was a French Canadian and did much to bring the people of Canada together. His government increased immigration, some of it from central and eastern Europe, and pushed ahead with the settlement of the prairies. At this period mining developed rapidly in the Shield and many new industries were started. The period has gone into folk-lore because of the gold rush in 1897–98.

Canadian divisions fought in the Allied armies in World War I, and more than 60,000 Canadian soldiers died. Canada's part in the war added to its standing among the nations, and after the war it took an active part in international affairs. In 1926 there was a constitutional crisis when the governor-general refused the request of the prime minister, William Lyon Mackenzie King, to dissolve parliament. The Imperial Conference in London later in the same year, attended by Mackenzie King, declared the autonomy and equality of status of Britain and the dominions in the Commonwealth of Nations, and this declaration was given legislative effect in the Statute of Westminster in 1931.

In September 1939, at the outbreak of World War II, Canada declared war on Germany; Canadian soldiers, sailors and airmen fought on many fronts alongside troops of the other Western Allies. Altogether nearly 1,000,000 Canadians served in the armed forces. After the war, Canada played a major part in helping to rebuild the devastated countries of Europe, and a new wave of immigration began. Canada was by now regarded as a country of affluence and economic stability, the possessor of one of the world's highest living standards. In foreign policy it adopted an independent and moderating position, although participating in NATO and NORAD and stating clearly its place among the democratic nations of the West. The ancient tensions between French-speaking and English-speaking Canadians gained world-wide attention in 1967 when President de Gaulle of France, on a visit to Canada, made his "Vive le Québec libre!" speech, and later when there were terrorist outrages by a minor separatist party in Quebec. Another separatist party, the Parti Québecois, came to power in the provincial elections of 1976 and stated its aim to form an independent Quebec with economic ties to English-speaking Canada.

Montmorency River rises in southern Quebec and flows southwards to the St Lawrence; there is a hydroelectric power station at Montmorency Falls.

Hamilton, Ontario, is at the western end of Lake Ontario, near Niagara; most of the people work in engineering, producing machinery, cars and steel.

Edmonton, the provincial capital of Alberta, is a major market centre for farm products and within its vicinity are oil, gas and coal fields.

Since World War II only two general elections have been won by the Conservatives: in 1957 John Diefenbaker became the first Conservative prime minister for 22 years. But the Liberals were again returned in 1963, with Lester B. Pearson as prime minister. He was succeeded as Liberal leader by Pierre Trudeau, a French Canadian, who took office in 1968 and was re-elected in 1974.

Provinces and territories. The federal union of Canada consists of ten provinces and two territories. Each province has a lieutenant-governor (appointed by the governor-general of Canada in council) and a legislative assembly. An executive council – the provincial government – is headed by a premier, who is the leader of the majority in the assembly. Subject to the over-riding authority of the federal government, the provinces have the power to legislate on such matters as education, property, law enforcement and local finance. The governments of the two territories have more limited powers.

Alberta is a province that was admitted to confederation in 1905. It lies mainly in the Interior Plains and the chief cities are Edmonton (capital), Calgary, Lethbridge and Red Deer. Minerals include petroleum (from near the Athabasca River) and natural gas, and major products are cereals, cattle, processed foods and machinery. Area: 661,188sq km (255,285sq miles). Pop. (1975 est.) 1,770,000.

British Columbia was admitted to confederation in 1871. The province lies in the western mountain region and the chief cities are Vancouver, Victoria (capital) and New Westminster. Minerals include copper and asbestos, and major products come from forestry, fruit and fishing. Area 948,600sq km (366,255sq miles). Pop. (1971) 2,184,621.

Manitoba, admitted to confederation in 1870, lies partly in the Interior Plains and partly in the Shield and the Hudson Bay lowlands. The chief cities of the province are Winnipeg (capital), Brandon and St Boniface. Minerals include nickel (from Thompson), zinc, copper, caesium and tantalite; other products are cereals, livestock, forest products, processed foods and textiles. Area: 650,090sq km (251,000sq miles). Pop (1975 est.) 1,018,000.

New Brunswick, one of the original four provinces confederated in 1867, is in the Appalachian region. The chief cities are Saint John, Moncton, Fredericton (capital) and Bathurst. Mahor minerals are coal, zinc, lead, copper and antimony, and other products include potatoes, dairy products, fish, paper, wood pulp and timber. Area: 73,437sq km (28,354sq miles). Pop. (1971) 635,000.

Newfoundland did not become a province within the confederation until 1949. It consists of the island of Newfoundland and the mainland region of Labrador. The chief cities are St John's (the capital, said to be North America's oldest city) and Corner Brook. Minerals include iron ore (in Labrador), asbestos, zinc, lead and silver; the chief products are fish, paper, wood pulp and ships. Area: 404,520sq km (156,185sq miles). Pop. (1971) 522,104.

Northwest Territories is an icy region that includes most of Canada's Arctic lands – only its southernmost part is outside the permafrost zones. The seat of government is Yellowknife. Minerals include zinc, lead, gold, silver, natural gas and petroleum; the main products are furs, fish and forest products. Area: 3,379,699sq km (1,304,903sq miles). Pop. (1971) 34,807.

Nova Scotia was one of the original four provinces confederated in 1867. It lies in the Appalachian region, and the chief cities are Halifax (capital), Sydney and Glace Bay. The chief minerals are coal, gypsum and salt, and other products include dairy produce, poultry, apples (from the Annapolis valley), processed foods, transport equipment, paper and timber products. Area: 55,490sq km (21,425sq miles). Pop (1975 est.) 822,000.

Ontario is a province, one of the original four confederated in 1867, in the south-eastern lowlands, the Shield and the Hudson Bay lowlands. The chief cities are Toronto (capital), Ottawa (the federal capital), Hamilton, London, Windsor, Kitchener and Sudbury. The chief minerals are nickel, copper, iron, uranium, salt, sulphur and zinc. The province has a wide range of products, including dairy produce, cattle, pigs, poultry, cereals, motor vehicles and parts, iron and steel, meat packing, paper and wood pulp, chemicals, petroleum products, agricultural and industrial machinery, and processed foods. Area: 1,068,587sq km (412,582sq miles). Pop (1976 est.) 8,300,000.

Prince Edward Island, admitted to confederation in 1873, lies in the Appalachian region. The province's chief cities are Charlottetown (capital) and Summerside, and its products include potatoes, pigs, dairy products, lobsters, herring, haddock, boats and building materials. Area: 5,656sq km (2,184sq miles). Pop (1975 est.) 119,000.

Quebec, one of the original four provinces confederated in 1867, is in the south-eastern lowlands and the Shield. Its population is almost entirely concentrated in the southern cities of Montreal, Laval, Quebec (capital), Sherbrooke, Verdun, Hull and Trois-Rivières. Major minerals are copper, iron, zinc and gold, and other products include cattle, dairy produce, oats, maize, potatoes, fish, wood pulp, paper, metals, chemicals, textiles, rolling stock, electrical goods, ships textiles and leather goods. Area: 1,540,687sq km (594,860sq miles). Pop (1974 est.) 6,212,000.

Saskatchewan, admitted to confederation in 1905, is a province mainly in the Interior Plains but partly in the Shield. The chief cities are Regina (capital), Saskatoon, Moose Jaw and Prince Albert. Minerals include petroleum, natural gas and potash, and other products are cereals, cattle, pigs, poultry, wood pulp and furs. Area: 651,900sq km (251,700sq miles). Pop. (1975 est.) 920,000.

Yukon Territory was constituted a separate territory in 1898. It is the cold north-western region, much of it in the permafrost zone; the seat of government is Whitehorse. The many minerals include silver, gold, lead, zinc, cadmium, copper, asbestos and nickel; other products are timber and furs. Area: 536,327sq km (207,076sq miles). Pop. (1976 est.) 21,000.

Map 64.

Prime Ministers of Canada

Sir John A. Macdonald (1867–73; 1878–91)
Alexander Mackenzie (1873–78)
Sir John J. C. Abbott (1891–92)
Sir John S. D. Thompson (1892–94)
Sir Mackenzie Bowell (1894–96)
Sir Charles Tupper (1896)
Sir Wilfrid Laurier (1896–1911)
Sir Robert L. Borden (1911–20)
Aurther Meighen (1920–21; 1926)
W. L. Mackenzie King (1921–26; 1926–30; 1935–48)
Richard B. Bennett (1930–35)
Louis S. St Laurent (1948–57)
John Diefenbaker (1957–63)
Lester B. Pearson (1963–68)
Pierre Elliott Trudeau (1968–)

Canada – profile

Official name Canada
Area 9,976,139sq km (3,851,787sq miles)
Population (1976 est.) 22,831,000
Density 2.3 per sq km (5.9 per sq mile)
Chief cities Ottawa (capital) (1974) 302,345; Montreal, 1,214,355; Toronto, 712,785; Winnipeg, 246,245; Edmonton, 438,150; Vancouver, 426,260
Government Federal union, a constitutional monarchy with a parliamentary system of government
Religions Roman Catholic, United Church of Canada, Anglican Church of Canada
Languages English, French
Monetary unit Canadian dollar
Gross national product (1974) $140,000,000,000 (£58,363,000,000)
Per capita income $6,132 (£2,556)
Agriculture Wheat, barley, oats, rye, potatoes, vegetables, fruit, cattle, dairy products
Industries Motor vehicles and parts, pulp and paper, fishing, processed foods, textiles, industrial and agricultural machinery, rolling stock, aircraft, iron and steel, chemicals, petroleum products
Minerals Petroleum, iron, coal, nickel, silver, zinc, asbestos, uranium, molybdenum, aluminium, gold, lead, potash, sulphur, natural gas
Trading partners (major) USA, Japan, Britain, West Germany

The Panama Canal runs for 64km (40 miles) through the Canal Zone, providing a passage for large ships (up to 306m long) between the Pacific and Atlantic oceans.

Villagers of Azande, near Zemio in the Central African Empire, must be self-sufficient if they are to survive in this sparsely populated country.

Bangui, capital of the Central African Empire, is a port on the Ubangi River that handles nearly all the country's trade – mainly timber, cotton and sisal.

Canal Zone also known as Panama Canal Zone, is a region within PANAMA astride the Panama Canal, administered (since 1903) by the United States. Most of the population are American and the US president appoints the governor (who is also president of the Panama Canal Company). The chief towns, at each end of the canal, are Balboa (western end) and Cristóbal (eastern end). Area: 1,432sq km (553sq miles). Pop. (1975 est.) 44,000. Map 74.

See also EGYPT *(History)* for an account of the Suez Canal Zone.

Canton and Enderbury are coral islands in the Phoenix Islands group in the central Pacific Ocean, 3,220km (2,000 miles) south-east of Hawaii, jointly administered by Britain and the United States. Both nations made conflicting claims for the territory from the end of the 19th century, but signed a 50-year agreement on joint control in 1939. Area: approx. 9sq km (3.5sq miles). Pop (1975 est.) 150. Map 62.

Cape Province *See* SOUTH AFRICA

Cape Verde Islands (Ilhas do Cabo Verde) is an independent republic in the eastern Atlantic Ocean, 620km (385 miles) west of the African coast. The country is made up of about 15 volcanic islands split into two groups: the Windward Islands to the north (including Boa Vista, Sal, São Nicolau, Santa Luzia, São Vicente, Santo Antão, Ilhéu Branco and Ilhéu Raso) and the Leeward Islands (including São Tiago, Maio, Fogo, Brava and the Ilhéus do Rombo). Only Fogo has an active volcano, Cano (2,830m; 9,300ft), the highest point in the group. The capital is Praia, on São Tiago.

Deposits of pozzolana (volcanic dust used in making cement), coal and salt in the mountains are important to the republic's economy. Agricultural products, such as coffee, tobacco, sugar cane, oranges and groundnuts, are susceptible to wide variations in rainfall and droughts are common. Most of the people, of African or Portuguese descent, are Portuguese-speaking Roman Catholics. The islands were held as an overseas province of Portugal from 1495 until they were granted independence in 1975. Area: 4,033sq km (1,557sq miles). Pop (1975 est.) 294,000. Map 2.

Cayman Islands is a group of three islands in the West Indies, about 250km (155 miles) north-west of Jamaica. They are Grand Cayman (location of the capital, Georgetown), Little Cayman and Cayman Brac. The islands were discovered by Christopher Columbus in 1503 and colonized by the British in the late 17th century. The chief industries are tourism, shark and turtle fishing, coconuts and timber. Area: 260sq km (100sq miles). Pop. (1976 est.) 14,000. Map 74.

Celebes *See* INDONESIA.

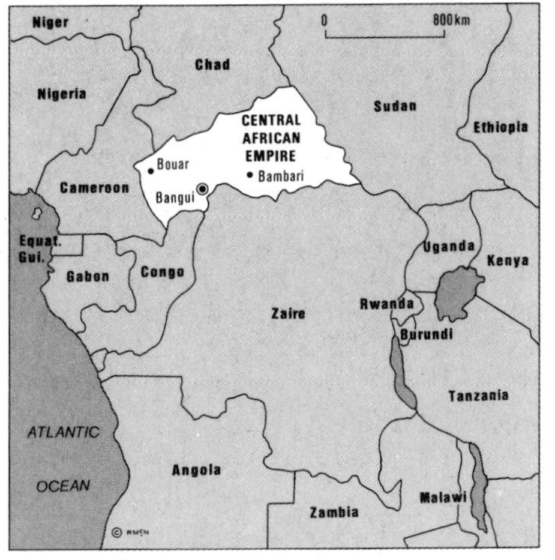

Central African Empire, known as the Central African Republlic until 1976, is a remote landlocked nation in central Africa. The capital, Bangui, has a population of 301,800. It was formerly the French colony of Ubangi-Shari. The capital is Bangui.

The country forms a watershed between rivers flowing north to the Chad basin and those flowing south into the Ubangi system. Landscapes are generally monotonous and most of the land lies at about 610m (2,000ft) above sea-level. The south is hot and humid, with an average annual rainfall of 1,750mm (69in), but the north is drier. Most people are farmers living at subsistence level and the average annual income was only £90 in 1974. The most valuable crops and chief exports are cotton and coffee. Poor communications hamper development.

The largest of the various ethnic groups are the Banda and Baya. Most people follow ethnic religions, but 20 per cent are Christians and there are some Muslims. France ruled the area between 1894 and 1960, when the country became an independent constitutional republic. But a military group, led by Jean-Bédel Bokassa, seized power in 1966. Bokassa was converted to Islam in 1976 and took the name Eddine Ahmed Bokassa. He replaced the government by a military council, declared the country a monarchy and named himself as emperor. Area: 624,977sq km (241,304sq miles). Pop. (1975 est.) 1,800,000. Map 32.

Ceylon. *See* SRI LANKA.

Chad (Tchad), official name Republic of Chad, is a nation in northern central Africa. In 1975 the average annual income was only £33. The capital, Ndjamena, has a population of 224,000.

Landlocked Chad consists mostly of a depression surrounded in the north, east and south by uplands. The highest peak, Emi Koussi, is 3,415m (11,204ft) above sea-level in the northern Tibesti massif. The shallow Lake Chad, in the west, is fed by the Chari and Logone rivers [*see* PE p.*47*]. Temperatures are high and the average annual rainfall is 1,020mm (40in) in the far south. But north of Lake Chad, except for oases, the land is desert. Farming is possible only in the south and the chief export is cotton.

The people in the south are Negroid and to the north there are Sudanese Negroes and Tuareg-Berbers. About half of the people are Muslims; apart from a few Christians, most of the remainder follow ethnic religions. France ruled Chad from 1897 until 1960, when it became an independent republic. Tensions developed as northerners claimed that the government favoured the southerners. In 1975 a military group, led by Brig.-Gen. Félix Malloum, seized control. He survived an assassination attempt in 1976. Area: 1,284,000sq km (495,752sq miles). Pop (1975 est.) 4,030,000. Map 32.

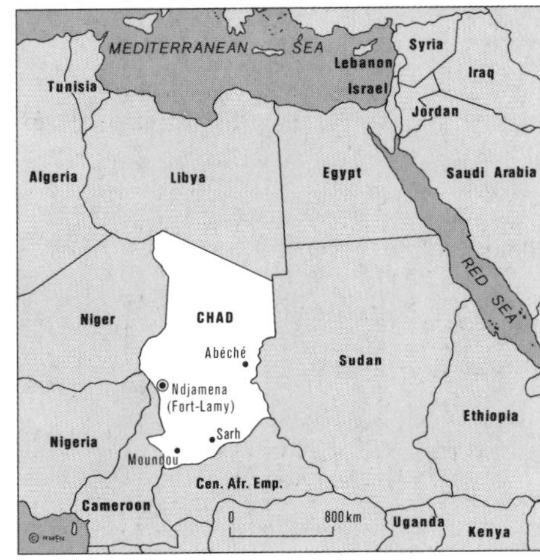

Mont Ougueil Castle looks out towards France from its towering site above the quay at the south-eastern tip of Jersey in the Channel Islands.

One of the world's largest copper mines is situated in Chuquicamata, Chile, on the slopes of the Andes at an elevation of 3,181m (10,435ft).

The Chuquicamata mine is the sole means of livelihood for thousands of people in the town of Chuquicamata, and has been so since it first opened in 1915.

Channel Islands are a group of islands, dependencies of the British Crown and part of the United Kingdom, at the south-western end of the English Channel, about 16km (10 miles) off the coast of France. The main islands are Jersey, Guernsey, Alderney and Sark; smaller islands include Herm, Jethou and Lithou. Jersey, capital St Helier, is the largest island (117sq km; 45sq miles), with a population of 72,700. Guernsey, capital St Peter Port, is about half the size; it lies 40km (25 miles) to the north-west of Jersey and has a population of 51,500. The islands have a mild, sunny climate which makes tourism and agriculture (fruits, vegetables and flowers) the principal industries; Jersey and Guernsey cattle yield high-cream milk.

The Channel Islands are administered as two groups, or bailiwicks, called Jersey and Guernsey (all the islands except Jersey), each with its own lieutenant-governor appointed by the British Crown. The official language of Jersey is French, although most Channel Islanders speak English or a Norman patois (dialect). The islands were part of the Duchy of Normandy in the 10th century, passing to the English Crown with the Norman conquest (1066). Area: 194sq km (75sq miles). Pop. (1975 est.) 126,000. Map 14.

Chile, official name Republic of Chile, is a country that occupies a long strip of land on the Pacific coast of South America; it is more than 25 times as long as it is wide. Although its greatest width is only about 360km (225 miles), it has 12,450km (7,780 miles) of coastline. Its northern region is in the tropics but its southernmost tip Cape Horn – much dreaded by sailors because of the violence of its storms – is only about 640km (400 miles) from Antarctica. In 1973 Chile's revolutionary-minded Marxist president, Salvador Allende Gossens, was overthrown by a military junta which established a regime it described as "authoritarian democracy". The capital is Santiago.

Land and climate. Chile is separated from its neighbours by the tremendous barrier of the Andes mountains. Most of the country has the same dominant physical features, with mountains to the east and ocean to the west, although there are several distinct climatic zones. In the north is a vast desert region which includes the Atacama Desert; the Atacama is 1,150km (720 miles) long, and although extremely dry is seldom very hot because of the moderating influence of the cold Peruvian Current. To the south the desert gradually merges into the fertile central region (which has a Mediterranean-type climate), where most of the people live. Farther south is another mild, though wetter, region of forested hills. The southernmost 1,600km (1,00 miles) of the country is wild and broken, with hundreds of fiords and off-shore islands [*see* PE p.117]. Much of it is in the bleak tableland of Patagonia. The Strait of Magellan divides the mainland from the archipelago of Tierra del Fuego (part of which belongs to Argentina).

Economy. The Chilean economy depends largely on minerals, the largest known deposits of which are in the north. Copper accounts for about 80 per cent of exports. One open-cast copper mine, the Chuquicamata in the Atacama Desert, is the largest in the world. The Atacama is also the source of the nitrates that were once Chile's most valuable product [*see* PE p.*134*]. Other minerals include iron, coal, gold, molybdenum, silver and manganese. Petroleum comes from the Magallanes field on the Strait of Magellan. The forests are also a source of wealth; timber, wood pulp and paper are exported. About one third of the people live by farming. In the early 1970s a state corporation for agrarian reform was set up to establish agricultural settlements. Crop-growing is concentrated in the mild central region and the forest region to its south. The main cereal is wheat; oats, maizebarley and rice are also important. Vegetables include beans, potatoes, peas and lentils. There are many orchards and

groves of fruit trees yielding pears, cherries, apples, peaches and citrus fruits. There are also many vineyards. The chief manufacturing centres are Santiago, Valparaíso and Concepción. A large state owned plant at Huachipato, near Concepción, produces high-quality steel. The cellulose industry is important, and so are the textile and food-processing industries. There are some engineering factories, including vehicle-assembly plants.

Communications are influenced by the unusual shape of the country. Highways link the largest cities: the Chilean section of the Pan-American Highway runs more than half the length of the country. The railway system provides links with Chile's neighbours as well as giving outlets from the interior regions to many ports. Sea transport has been – and still is – important to Chile's life and development, although there are few good harbours. There are five international airports.

People. The first European settlers in Chile were Spaniards, and Spanish culture and traditions have shaped the development of the country. Today the majority of Chileans are mestizos of mixed Spanish and American-Indian blood. In the 18th and 19th centuries there was an influx of Basque, British, German, Swiss and other migrants, some of whom have made large contributions to Chilean life; the British, for example, have taken the leading role in business. Of the indigenous Indian peoples, many Changos still live in the north and there are thousands of Araucanians (about 2 per cent of the total population) in the central areas.

Nine out of ten people are Roman Catholic. Spanish, the official language, is spoken by almost everybody. Education is free and compulsory at primary school level, but some remote areas have no schools. Nearly 100,000 students attend the eight universities.

Chileans take an active interest in the arts, particularly in Santiago. Among internationally known Chileans are the pianist Claudio Arrau, the poetess Gabriela Mistral (real name Lucila Godoy Alcayaga) and the poet Pablo Neruda (Neftali Ricardo Reyes).

History. During the 15th century the Incas of Peru extended their dominions into Northern Chile. Spaniards made overland incursions into Chile in the 1530s, but no settlement was established until 1541 when Santiago was founded by Pedro de Valdivia. The territory was made a dependency of the viceroyalty of Peru. In 1810 the Spanish governor was deposed by revolutionaries, who included Bernardo O'Higgins, the son of an Irish-born viceroy of Peru. The Spaniards re-established their authority, but in 1817 O'Higgins and the revolutionary general José de San Martín led an army across the Andes from Argentina, and the Spanish were decisively defeated at Maipu (1818). O'Higgins became ruler of the Chilean republic, but his authoritarian social and financial reforms provoked the enmity of the landowners and the Church, whose privileges he had curtailed. In 1823 he was deposed.

Chile defeated Peru and Bolivia in a war in 1839, and in the second half of the 19th century fought against Spain (with the aid of Peru and Bolivia) and then again made war on Peru and Bolivia. This last war – the so-called Nitrate War – brought in the mineral-rich northern provinces of Atacama, Arica and Tarapacá, and the port of Antofagasta (leaving Bolivia landlocked).

A period of prosperity during World War I (because of the demand for Chilean nitrates for explosives) was followed by an economic recession and civil disturbance that almost erupted into civil war. Under President Arturo Alessandri industry expanded again, but in 1925 he was forced from office; a military junta led by Carlos Ibáñez del Campo seized power and declared a new constitution. Ibáñez himself was overthrown in 1931 because of his failure to cope with rising unemployment and inflation, and in 1932 Alessandri was back

A third of Chile's population makes a living by tilling the soil; there is some mechanization, but traditional methods of agriculture still prevail.

Canton is a major deepwater port and the market place for China's world trade; the city is also noted for its educational and cultural facilities.

The Yellow River, one of China's great rivers, has a history of devastating flood disasters which are caused by excessive silt depositions.

in the presidency. Chile began an economic recovery that gained momentum during World War II, when the country sided with the Allies. But in 1952 the boom was over, and in the presidential election of that year the old dictator Carlos Ibáñez was recalled.

, Chile suffered a major disaster in 1960 when an earthquake killed more than 5,500 people. In the election of 1964 the moderate Christian Democrat Eduardo Frei Montalva was successful against his Marxist opponent Salvador Allende Gossens; but in 1970, after an inconclusive ballot, the National Congress declared Allende the winner. Allende tried to create a state on the Cuban model, and his measures led to increasing inflation and conflict with congress. In September 1973 he was overthrown by a military coup and he allegedly committed suicide. A military government was formed under the leadership of Gen. Augusto Pinochet Ugarte. It imposed a night-time curfew, outlawed Marxist parties (and later banned all political parties), dissolved the Nastional Congress, and declared its aim to be the stabilization of the economy followed by a gradual return to democracy. Map 78.

Chile – profile

Official name Republic of Chile
Area 756,945sq km (292,256sq miles)
Population (1975 est.) 10,253,000
Density 13 per sq km (35 per sq mile)
Chief cities Santiago (capital) (1975 metropolitan area) 3,263,000; Valparaíso, 238,557
Government Military junta, aiming at "authoritative democracy"; head of state, Gen. Augusto Ugarte
Monetary unit Peso
Gross national product (1974) £3,628,200,000
Per capita income £354
Agriculture Wheat, beans, potatoes, apples, peaches, maize, rice, grapes
Industries Iron and steel, petroleum products, textiles, foodstuffs
Minerals Copper, nitrates, iron, coal, gold, molybdenum, silver, manganese
Trading partners (major) USA, EEC, Argentina

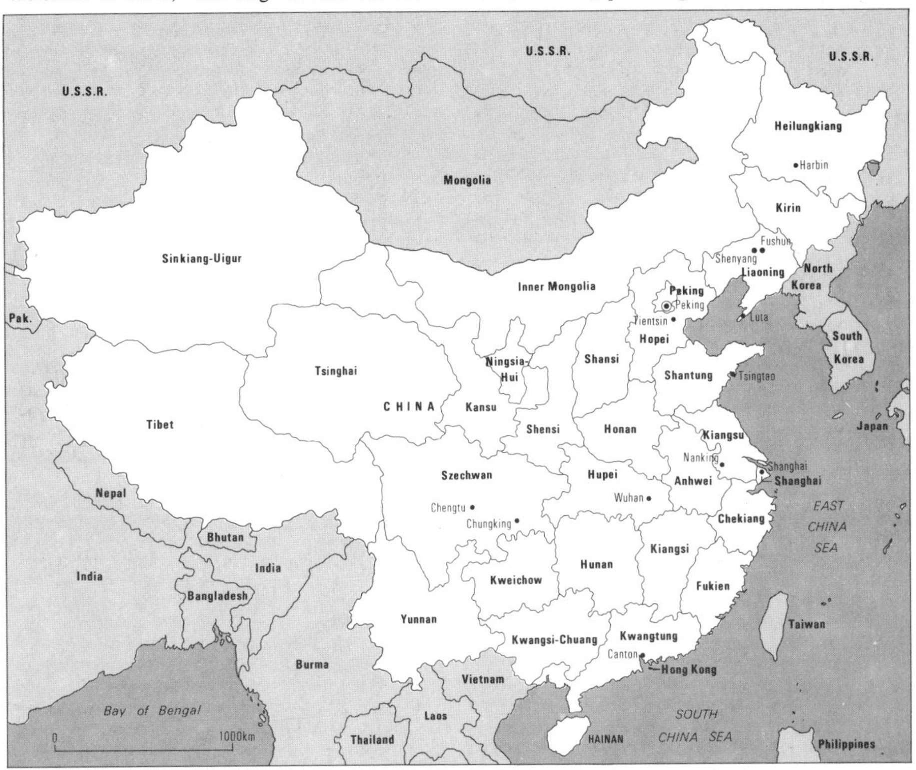

China (Chang-kuo, the Middle Country, or Hua-kuo, the Country of Blossom), official name The People's Republic of China, is the third largest and the most populous country in the world. It also has one of the world's oldest civilizations: its recorded history dates back more than 3,500 years. For much of that time the Chinese tended to shun contact with other peoples, believing strongly in the superiority of their own culture, with its Confucian, Taoist and Buddhist virtues. But in the 20th century China has come to terms with the outside world, the world of technology. It has undergone violent upheavals: the collapse of its imperial monarchy and the establishment of a republic, a protracted war with Japan, and a bitter civil war that ended in a Communist victory and the sweeping away of an age-old way of life. Today China is a super-power in the making, although it is as yet one of the poorest of the world's major countries. The capital is Peking (Bijing).

Land and climate. China occupies more than one-fifth of Asia. It has a 6,800km (4,250 mile) coastline on the Pacific Ocean, and its landward boundaries lie along natural barriers formed by great mountain ranges. In the south-west is the high tableland of Tibet, extending between the snow-covered peaks of the Himalayas in the south and the arid Kunlun Mountains in the north. Beyond the Kunluns are the sand dunes of the Takla Makan Desert. Still farther north are the spectacular mountains of the Tien Shan, separated from the high Altai of Mongolia by the Dzungaria Basin, once the homeland of a Mongolian kingdom. In the north-east of the country are the flat lands of Manchuria (a region now divided into several provinces) the source of many past disputes with Russia and Japan. The Khingan Mountains on Manchuria's west separate it from Inner Mongolia (Chinese Mongolia), part of which lies in the Gobi Desert.

The Great Plain of eastern China is "China proper", the historic region that has most of the cities and for which China's Great Wall was built as protection. The hills of the Chin Ling Shan divide it into two parts, that in the north being watered by the Hwang Ho or Yellow River, and that in the south by the Yangtze-Kiang. Both of these rivers are subject to flooding, caused by melting snows in the mountains in which they have their sources and by the deposition of silt. In the course of centuries great floodplains have been formed in the rivers' lower reaches, providing some of the most fertile land in China. The name of the Hwang Ho refers to the yellow silt it carries; this silt also colours the coastal waters of the Yellow Sea. For part of its course the Yangtze, China's principal river, flows through rocky gorges famous for their stark beauty.

Climatic conditions in China vary greatly because of the country's enormous size and the wide differences of terrain, but most regions are affected by monsoon winds. The interior is subject to the influence of the immense land mass of central Asia, and

The construction of irrigation schemes provides water for soya beans (the major crop), wheat and rice in the Kwantung Province of China.

Shanghai, China's largest city and one of the world's greatest seaports, has many industries including a section devoted to consumer goods.

The Chinese aim to improve irrigation and win back more land for growing the staple crop rice has led to the cultivation of many new paddy fields.

has extremes of heat and cold. The Tibetan Plateau and northern deserts are exceptionally dry. The coastal areas of the south-east have a subtropical climate and are extremely hot in summer. But in the greater part of the eastern lowlands the climate is generally mild; the months from November to February tend to be dry, but May brings heavy rain.

Physical resources. China has large known deposits of minerals even though only a small part of the country has as yet been surveyed for its natural wealth. As industry slowly increases, the demand for coal and oil grows; both are plentiful. China is believed to be the third largest coal producer in the world, the main coalfields being in the north-east. Much of the coal is near the surface and is easily mined. It still provides 80 per cent of the energy used in China, but in recent years considerable progress has been made in the extraction of oil. Some of the main oilfields are in the Dzungaria Basin, and a search for off-shore oil has been started. Increasing industrialization has made the production of electricity a matter of urgency. Many rivers are suitable for harnessing, but to date only a few large-scale hydroelectric installations have been constructed. There are many small local plants.

The most plentiful metal is iron, again found chiefly in the north-east, although immensely rich deposits of iron ore have been located in Hupei, in east central China. The country ranks fourth in iron production, but much of its ore is of low grade. Other metals of importance to the economy are tungsten (of which China is the principal producer), tin, aluminium, antimony, manganese and mercury. Large amounts of salt and graphite are extracted.

Constitution and government. The establishment of the People's Republic of China was proclaimed by Mao Tse-tung in October 1949 after the defeat – so far as mainland China was concerned – of Chiang Kai-shek and the Kuomintang (National People's Party). Then a "Political Consultative Conference of the Chinese People" adopted resolutions that became the basis of the country's constitution, which came into effect in 1954. In 1975 a new constitution was approved. No particular person is designated head of state, although that role is in practice assigned to the chairman of the central committee of the Communist Party – Mao Tse-tung in the first instance. The Communist Party has no special constitutional function; its dominant position is secured by the fact that all key governmental functionaries belong to it. Constitutionally, the chief legislative authority is the National People's Congress, which is made up of elected delegates from the provinces, the autonomous regions, the government-controlled municipalities and the armed forces. It is elected for five years and meets once a year. The executive and chief administrative organ of the Congress is the State Council, which includes a premier and ministers in charge of such matters as planning, foreign affairs, finance and foreign trade.

At provincial level China has 21 provinces (plus Taiwan), five autonomous regions and three government-controlled municipalities. The organs of provincial government, the Provincial Revolutionary Councils, are responsible to the State Council. At the several local levels there are Revolutionary Committees.

The Communist Party has a congress similar in structure to the National People's Congress, and a central committee elected for terms of five years. The central committee appoints a politburo and an inner standing committee. In 1973 Communist Party membership was given as 20,000,000 people. There are eight other parties, but all are members of the United Front.

Armed services. The army is known as the People's Liberation Army and is regarded as descended from the Fourth Red Army that took part in the Long March of 1934. Its members wear green tunics, buttoned up to the neck, with red gorget patches. The soldier's green cap has a five-pointed red star as a badge. All ranks wear the same uniform, without any badges to indicate status. Rank is said to have been abolished: an officer's title indicates his function. All officers are expected to perform regular tasks of manual labour to signify the soldiers' basic equality.

The country is divided into military regions and field armies, each with a commander who has authority over all ground, air, naval and people's militia forces in his region. The army has about 2,800,000 soldiers, and the people's militia is said to number 5,000,000. Although conscription is in force, not all people liable are actually called on to serve. The navy has a strength of about 170,000 men. It is divided into North Sea, East Sea and South Sea fleets, and its vessels – nearly all small – include 60 submarines. The air force, with about 220,000 members, is equipped mostly with planes of Soviet design, said to number about 3,800.

Industry. Apart from cottage industries, which have always been extremely important to the country's life, China in the early 19th century was a land almost devoid of manufacturing industry. Then in the last third of the 19th century an injection of foreign capital gradually established industrial enterprises. After the creation of the republic in 1911 attempts were made to industrialize the economy, and these had some slow success. But many of the factories that were built were lost during the war with Japan, during World War II and during the civil war. The first Communist five-year plan (1953-57) made the development of heavy industry a primary aim, and many factories were built with Soviet financial and technological aid. Later planning had less success. Iron and steel produced in the thousands of small workshops set up during the Great Leap Forward (1958-61) proved to be of too low a quality for use, and in 1960 the Russians withdrew their technical help. The Chinese then diverted manufacturing endeavour into the production of agricultural machinery, lorries and similar equipment, and a policy was adopted of establishing factories in rural areas. In this way, it was hoped, balanced communities would evolve: workers in the communes would be employed in the most productive way, with any one family working in manufacturing and in agriculture. But 70 per cent of industry is concentrated in the north-east and centre – in Liaoning, Hopei, Kiangsu, Hupei and Szechwan.

Despite planning mishaps, China has made genuine advances industrially in recent years. The oldest and most valuable light industry is the making of textiles, most of them containing natural fibres. Iron and steel production is the most important heavy industry; its chief centres are in the east but there are also plants in the north-west. Many other industries are subsidiary to steel production, such as the manufacture of machinery and tools. Much technological effort and skill has gone into industry of military importance, including the production of nuclear weapons. Consumer goods are relegated to second or third place.

Agriculture. Since the change in direction in economic planning in the 1960s, agriculture is seen as being at the centre of material development. In food, the country is self-sufficient – though only barely so. About 35 per cent of the population works on the land (compared with 8 per cent in industry), and altogether about 80 per cent of China's people are peasants. One of the first acts of the Communist government was to confiscate the property of large landowners, and to reallocate it in farming units; it was said to comprise 70 per cent of all agricultural land. Peasant households were arranged in various communal groups to produce food and other crops. Today the rural population is organized into more than 70,000 communes, each consisting of several villages farming communal land and working together as a local government unit. Each commune is run by a committee that decides which crops to grow and allocates the production targets. A working committee carries out day-to-day supervision of the commune, and village committees organize production teams of workers in the fields. The working week consists of six days, each of about nine hours' work. Recent evidence suggests such communes are inefficient.

China is the world's leading producer of rice, the largest crops coming from the plains watered by the Yangtze-Kiang; farmers there have at least two harvests a year, and work laboriously in the paddy fields using the traditional methods. In the northern part of the lowlands and in Manchuria the principal cereal crop is wheat, in the production of which China ranks third in the world. Poor harvests in recent years have resulted in large wheat imports, chiefly from Canada and Australia. Other major crops are vegetables, cotton and tea. In the far south, farmers grow tropical crops including sisal, rubber and citrus fruits. Large herds of sheep and goats pasture on the grasslands of the north and west, but China's commonest farm animal is the pig. Fish is one of the people's chief sources of protein. The large fishing industry depends not only on catches of fish and shellfish from the sea, but also

Fishing is one of the major industries in Chekiang Province, which includes many islands, notably the Chou-Shan Archipelago; the capital is Hangchow.

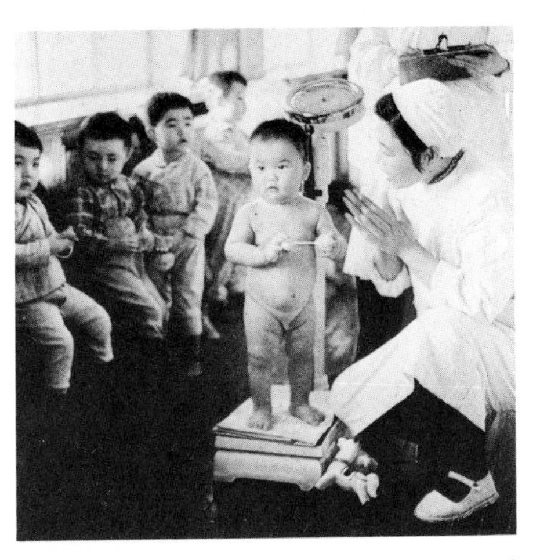

This creche at a Peking factory, illustrates one of the several government supported facilities provided for working mothers in China's capital.

Chinese boys and girls often spend their holidays riding with the mounted guards on the Mongolian pastures, doubling as cattle-herders.

on fish-farming in ponds and rivers [*see* PE p.242].

Trade. Nearly all foreign trade is with non-communist countries, chiefly Japan, Hong Kong, West Germany, Britain, Canada and Australia. Imports include cereals, machinery, aircraft, motor vehicles and rubber. The principal exports include meat and other agricultural products, textiles, tungsten and antimony. No foreign investment is allowed in China.

Communications and transport. The Communist Party and the government foster the notion of a revolutionary people all working towards a common goal, and consequently all means of communication are strictly controlled to ensure that only approved ideas and information are propagated. This control extends to speeches and posters as well as to newspapers, magazines, radio, and television. The most important newspapers are the *People's Daily,* the voice of the Communist Party and frequently of the government, and the *Peking Daily.* Wall newspapers, in the form of posters, carry items of news, exhortations by leaders and local party comments on current affairs. There are also vast numbers of local and institutional news-sheets. Only a minority of people have radios, and very few have television. Postal services are, in the main, quick and efficient. Telephones are used mainly as a means of business communication.

Transport in China tends to be slow. Outside the cities there are few motor vehicles, and loads are carried by boat, on pack animals, in carts, on bicycles or by porters. Recent governments have tried to improve transport, believing that good communications are a prerequisite to industrial expansion. In the early 1970s it was claimed that 100,000km (62,000 miles) of new roads had been built in five years. The highway system is skeletal but extensive, reaching into relatively remote regions; but few roads are metalled. China's railway system has a history going back more than a century when it was built to provide communications inland from the treaty ports, financed by foreign capital and using imported materials. The greater part of it is concentrated in the east where it provides connections between the important towns. Some lines, however, penetrate to the interior, and one links the heavily populated lowlands of the east with Urumchi in the far north-west. There is also a line connecting with the Trans-Siberian Railway. Waterways are an essential part of the transport system. Again, they are confined mainly to the lowlands of the east; there, the larger rivers (in particular the Yangtze) are navigable by passenger steamers and cargo boats, and canals carry a heavy barge traffic. The Grand Canal, started in AD 600, connects Peking with Hangchow, nearly 1,000km (600 miles) away. Internal air services link about 70 cities, and the Civil Aviation Administration operates flights to some overseas cities, including Tokyo, Moscow and Paris. Foreign airlines fly to Peking, Shanghai and Canton.

People. Nearly 90 per cent of China's enormous population is concentrated in about 15 per cent of the land area, chiefly in the east. In the deltas of the Hwang Ho and the Yangtze-Kiang the population density is as great as 800 per sq km (2,000 per sq mile), yet in the west, south-west and north there are vast areas that are almost uninhabited. Out of every five people, four live in rural communes, but in the regions of high population density many live in sprawling cities. China has, in fact, a quarter of the world's biggest cities, and the industrial centre and seaport of Shanghai may have a larger population than any other city in the world.

All Chinese are of the same Mongoloid racial stock, but within this broad grouping they belong to many different peoples. An earlier trend towards cultural uniformity has been reversed under the Communist regime, and now more than 50 "nationalities" are recognized. Overwhelmingly the largest nationality, comprising perhaps 94 per cent of the total population, is that of the Han, who live in the eastern plains and Manchuria. They are descended from the people of the ancient empires, the inhabitants of "China proper". The minority groups are peoples who were incorporated in the empires over the centuries, descendants of the inhabitants of the "Outer Territories". Most of them live in the border lands, or form pockets within the lands of the Han. The largest of these groups, the Chuang, live in the south. Others include the Manchus and Mongols of the north and north-east, the Uigurs and Kazakhs of the north-west, the Tibetans, and the Miaos of the south-east.

Traditionally, life in China centres on the family, although the philosophy of contemporary China strongly emphasizes the over-riding importance of the commune as the cohesive force. In the past the family included remote as well as immediate relatives, sometimes 100 or more people living together as a family unit. As family ties have become less binding, the status of women has changed too. Whereas formerly the woman's place was strictly in the home and her education was directed only to her role as mother and housekeeper, women today have virtual parity with men. For most people, life under the new regime remains as frugal as life under the old. The majority of families live in one or two rooms, sharing washing facilities with others. Food is not plentiful and consists generally of cereals (rice or wheat) and vegetables [*see* PE p.191]. Protein comes mainly from fish, pork and chicken. Clothing is monotonous: for most it consists of the worker's uniform of plain tunic, trousers and soft cap. Consumer goods are few, and such as can be found are usually of poor quality. But although the standard of living is low by comparison with that of Western countries, important advances are being made. Much endeavour is directed towards minimizing the peril and calamitous effects of flood and drought; a precarious self-sufficiency in the essentials of life has been attained.

Cultural life. China has as many languages as nationalities, and hundreds of dialects. People from one part of China are not always able to make themselves understood in other parts. But the majority speak (and most schools teach) a dialect of Mandarin that had its origin in the region of Peking. The word *Mandarin* is not now used, having been replaced by the term *putonghua,* meaning "the generally understood language". Written Chinese is ideographic, employing thousands of characters instead of an alphabet. Each character conveys a complete language entity, a word or a phrase. To aid popular education, the characters have in recent years been simplified and their number greatly reduced. A phonetic alphabet has been introduced, although its application is limited. As a result of these changes and an improved educational system, China's literacy rate is increasing.

Religion seemingly has little influence in contemporary China, except in the outlying regions. It is officially discouraged, and formal religious teaching is not permitted. The traditional Chinese philosophies of Confucianism and Taoism are now held in low esteem; a campaign to eradicate Confucianism was initiated in 1973. Buddhism was formerly pre-eminent in Tibet and Mongolia; during the Cultural Revolution, Buddhist monastries were suppressed, but some were later restored. The fate of Islam, formerly strongest in the west, is hard to assess. Some pockets of Christianity still exist: the Catholic Church in China declared itself independent of Rome in 1958 and formed a Patriotic Catholic Association.

The educational system was greatly disrupted by the Cultural Revolution, when students were in the forefront of the fight to preserve revolutionary ardour. From 1968 schools and colleges began to function again, with increased emphasis on political awareness and the value of manual work. Schooling progresses from primary school (ages seven to 12) through junior middle school (12 to 15) to senior middle school (17 to 19). Then there is a break of two years for practical experience in the army, in a commune or in a factory. Students may then go to university. Before the Cultural Revolution there were 61 universities, and hundreds of other institutes of higher education, such as engineering and medical colleges.

Historically, the arts in China had few points of contact with those in the West. In the classical Chinese tradition, art had the purpose of elevating the mind, and the visual arts in particular had a distinctive delicacy and idealism that often tended towards the exquisite. Brushwork had a range and fineness unrivalled anywhere else in the world. Chinese painting probably reached its peak during the Sung and Yuan dynasties. The ceramic arts, too, were probably at their highest point of excellence in the Sung dynasty, but during the Ming were added qualities of decoration. Contemporary Chinese art retains its elevating purpose – but now that purpose is social and political. The subjects of modern art are found in revolutionary strife and in the commune and factory.

Changchuw is China's major motor car, lorry and tractor production centre; it is also the location of the government-owned film studios.

The cultivation of rice forms part of China's leading occupation; today's methods of harvesting are still the same as they were centuries ago.

The Hall of the Annual Prayers in the Forbidden City of Peking was added in the 19th century to the buildings which comprise the Temple of Heaven.

A feature of contemporary Chinese life is the view held on the constructive use of leisure: leisure time should preferably be spent in improving body and mind. Most communes have cultural houses or "palaces" that offer courses in the arts and in practical subjects, and have facilities for games. The most popular sports are basketball, tennis and volleyball, and China's table-tennis players are internationally famous for their skill. The attainment of physical fitness is encouraged in the whole population: many Chinese engage daily in callisthenics.

History. Artefacts found in the valley of the Hwang Ho indicate the existence of a Stone Age culture in eastern China about 50,000 years ago, and there is considerable evidence of a developing civilization in the years before the 1900s BC when the Hsia dynasty is said to have been founded. According to legend its founder was the hero Yu, but it is doubtful whether such a dynasty really existed. The first certain dynasty was the Shang (dates uncertain, but estimates vary from as early as 1766–1122BC to 1523–1027), during which the people developed farming techniques, made pottery and became skilful in the use of bronze. The Shang dynasty was overthrown by Chou warriors under their leader Fa, known as Wu Wang (King Wu). The Chou came from the western borders (of China proper); their dynasty lasted until about 249 BC and was culturally one of the most important in Chinese history. During it discoveries were made in astronomy, silk weaving was developed, principles of cartography were worked out, and iron came into use. It was also the period in which the philosophers Confucius, Lao Tze and Mencius (Meng Tsz) lived. But the kingdom was weakened by feuding between the territorial lords and it finally collapsed.

By 221 BC the ruler of the Ch'in, a warrior people from the basin of the Wei River, had conquered the quarrelling states, unified China, and established a new dynasty. His name was Ch'eng, but he gave himself the title Shih Huang Ti, meaning "First Emperor". His dynasty lasted for only 14 years but in that time he welded his empire together and laid the foundation of a governmental system that lasted until 1911. He also started the building of the Great Wall to protect his domains, incorporating earlier and smaller structures. He tried to develop a lingua franca for China and, as no friend of past learning, sought (unsuccessfully) to destroy all records of ancient philosophies. The Han dynasty arose from the civil wars that followed the collapse of the Ch'in. Its founder was Liu Pang, and it ruled from 206 BC to AD 220 (with an intermission from AD 9 to 23 during which a usurper, Wang Mêng, occupied the throne, calling his reign that of the Hsin). Under the Han, trade – as far as Rome – flourished, an efficient civil service was developed, the building of a network of canals was begun, and the boundaries of the empire were extended. Art and learning also advanced: the invention of paper and ink made possible a great upsurge in writing and scholarship. An imperial university was founded, and Buddhism was introduced into the country and spread rapidly in the north. But the Han dynasty, in its turn, collapsed, and centuries of disruption and conflicting claims followed. During the San Kuo (three kingdoms) period, the empire was split. It was briefly united under the Tsins (265–420), but again dissolved into a mass of small kingdoms. In 581 Yang Chien, the founder of the Sui dynasty, unified northern China and in 589 made himself master of the south as well. But the attempts of the Sui emperors to add Korea and Manchuria to their dominions ended in disaster. The dynasty fell and was succeeded by the T'ang, which ruled for nearly 300 years (618–907).

Under the T'ang the empire was again enlarged, and the period of T'ang rule is regarded as China's golden age. The governmental system was reorganized and the study of the classics became compulsory for administrators, advances were made in mathematics, much astronomical data was gathered, a technique of printing invented, and an encyclopaedia of history compiled. The arts flourished, too, particularly painting, poetry and essay-writing. Many foreign students attended the schools of Ch'ang-an (Sian) the capital. Despite these accomplishments the rule of the T'ang emperors was uneasy, and more than once they had to desert their capital and even their throne. In the 9th century the Buddhist religion was dislodged from its favoured position in Chinese life and suffered the loss of thousands of temples and shrines. It later made a partial recovery.

At the beginning of the 10th century the empire was again fragmented, but in 960 an official named Chao Kuang-yin ascended the throne as the first of a new dynasty, the Sung, which held power until 1279. During the Sung period there were advances in scholarship, in general education and particularly in the arts. Fine cities were built, agriculture was improved, a more scientific approach to medicine developed and social services were established. Chinese merchants exported their wares to the less civilized parts of the world. Two remarkable innovations of this period were the use of paper money and the use of gunpowder in war. In the dynasty's last years it ruled only in the south; the Hsia and Ch'in kingdoms of the north had been overrun by the Mongols in the 1220s and 1230s. In the 1270s Kublai Khan, grandson of the great Mongol warrior-chief Genghis Khan, established himself on the imperial throne of China and founded a new dynasty called the Yüan. It is the China of Kublai Khan that Marco Polo described so vividly to an

Provinces, Autonomous Regions, and Government-controlled Municipalities of China

All Chinese names, including those of provinces and places, can be rendered in English characters in more than one way. In this list there are many alternative spellings. The first version is an English spelling that reveals the normal pronunciation; the second is a form favoured for international use and appears as the primary spelling on the maps of China in the Atlas at the end of this volume. All population figures are approximate. Map 44.

Anhwei (Anhui) Province, E-central China; capital Hofei (Hefei). Area: 139,970sq km (54,042sq miles). Pop. 35,000,000

Chekiang (Zhejiang) Province, E coast; capital Hangchow (Hangzhou). Area: 101,830sq km (39,317sq miles). Pop. 31,000,000

Fukien (Fujian) Province, SE coast; capital Foochow (Fuzhou). Area: 123,150sq km (47,548sq miles). Pop. 17,000,000

Heilungkiang (Heilongjiang) Province, NE China; capital Harbin (Haerbin). Area: 463,790sq km (179,069sq miles). Pop. 21,000,000

Honan (Henan) Province, E-central China; capital Chengchow (Zhengzhou). Area: 167,090sq km (64,513sq miles). Pop. 50,000,000

Hopei (Hebei) Province, NE China; capital Shihkiachwang (Shijiazhuang). Area: 202,510sq km (78,189sq miles). Pop. 47,000,000

Hunan Province, S-central China; capital Changsha. Area: 210,570sq km (81,301sq miles). Pop. 38,000,000

Hupei (Hubei) Province, E-central China; capital Wuhan. Area: 187,590sq km (72,430sq miles). Pop. 32,000,000

Inner Mongolia (Neimenggu Zizhiqu) Autonomous Region, N China; capital Huhehot (Huhehaote). Area: 1,190,930sq km (459,818sq miles). Pop. 13,000,000

Kansu (Gansu) Province, NW China; capital Lanchow (Lanzhou). Area: 366,625sq km (141,550sq miles). Pop. 13,000,000

Kiangsi (Jiangxi) Province, SE China; capital Nanchang. Area: 164,865sq km (63,654sq miles). Pop. 22,000,000

Kiangsu (Jiangsu) Province, E coast; capital Nanking (Nanjing). Area: 102,240sq km (39,474sq miles). Pop. 47,000,000

Kirin (Jilin) Province, NE China, capital Changchun. Area: 187,070sq km (72,228sq miles). Pop. 1,200,000

Kwangsi-Chuang (Guangxi Zhuang Zizhiqu) Autonomous Region, SE coast;

Grain from the terraced fields of Suchien province is put into sacks and stored in the traditionally straw-built silos of the region.

The Great Wall of China, built originally over twenty-one centuries ago to keep out the marauding Mongol hordes, now welcomes hordes of tourists.

The Yangtze, the longest river in Asia, passes through China's most populated region; the fertile middle basin produces mainly rice.

amazed European audience; he lived at the Mongol court for 17 years. The chief improvement in Chinese life effected by the Yüan dynasty was the reconstruction and extension of the roads and canals. Expeditions were despatched to discover the source of the Hwang Ho, and the imperial observatory was refitted with new and ingenious astronomical instruments. But Kublai Khan had inherited the warlike ambitions of his grandfather. He tried unsuccessfully to conquer Japan, and frequent plundering expeditions were mounted against neighbouring territories. After his death his successors were unable to hold their empire together and finally, in 1368, the Mongols were driven out of China by a peasant-born leader, Chu Yüan Chang, the founder of the Ming dynasty. This dynasty ruled China until 1644; under its rule Peking became the official capital. Today the Ming dynasty is best remembered outside China for its artistic achievements, particularly the elaboration and decoration that characterized its ceramic arts. But the people of the period also have to their credit achievements in education, communications and agriculture. In the early 15th century a Ming emperor, Ch'eng Tsu, sent out naval expeditions that travelled as far as Arabia and Ceylon. Although the purpose was to extract tribute, such interest in the outside world was not common in Chinese rulers, and after Ch'eng Tsu's death the empire relapsed into isolationism. A growing distaste for foreigners was encouraged by encroachments from outside: the Japanese attempted an invasion, and the Portuguese gained a foothold in the south.

In 1644 a new dynasty was established by armies that poured into the old Chinese lands from Manchuria, north of the Great Wall. The purpose of this Manchu "invasion" was ostensibly to recover Peking from the bandit hordes who had seized it. But having taken Peking, the Manchus then pursued and fought the supporters of the Mings until they soon controlled China and a Manchu emperor, Shun-chih, was on the throne as the first of the Ch'ing dynasty — the twenty-second and last imperial dynasty. As a mark of Manchu dominance, every Chinese man was required to adopt the custom of partly shaving his head and wearing a pigtail.

Some of the early Manchu emperors were enlightened rulers, maintaining order, encouraging the arts and endeavouring to advance the country's prosperity. But order gradually succumbed to pressures from within and without. The first troublesome issue was foreign trade. At first the Manchu government resisted attempts by European countries and the United States to obtain trading rights, but eventually it felt obliged to make concessions and some trading rights in Canton were granted to the British. One of the most profitable trades was in opium; when the Manchus barred its import and destroyed British-owned opium stocks in 1839, Britain responded with attacks on coastal cities. A rapid Chinese defeat resulted in the Treaty of Nanking, which ceded Hong Kong to the British and gave British traders access to five Chinese ports. In 1850 the Taiping rebellion against the Manchus broke out, led by a religious visionary who was going to found the *Taiping* ("overpowering peace"). The rebels captured Nanking, but were put down with British and French help. A war about trade led to concessions for France, the United States and Russia, and to Russian intrusion into Manchuria. Other territory was also lost – Annam to the French, Amoy to the Portuguese and Taiwan to the Japanese.

In 1900 the Manchus were again faced with rebellion, the so-called Boxer rebellion, when the secret Society of Harmonious Fists set out on a crusade to rid China of the hated foreigners, who by this time were widespread in the country and were even operating railways. Attacks on foreigners and Chinese Christians culminated in the siege of foreign legations in Peking. By now the Boxers had support at court, including that of the dominant personage there, the dowager empress. The siege was raised by an international army. China was forced to offer indemnity and to permit increased access to foreigners; in the event, much of the indemnity was waived or repaid.

But now a new threat to the Manchu government developed: a rapidly growing republican movement organized from Japan by the revolutionary Sun Yat-sen. In October 1911 soldiers in Wuhan mutinied and captured the arsenal and the mint. The spirit of revolt spread rapidly, and the government was slow and weak in reacting. Province after province turned against the Manchus, a republican government was proclaimed with Sun Yat-sen as president, and in February 1912 the last emperor, the six-year-old Pu Yi, abdicated. The presidency of the republic was transferred to a Manchu officer, Yüan Shin-k'ai, in order to gain the support of the northern provinces. But Yüan set up a dictatorial regime that eventually provoked opposition from Sun Yat-sen's Kuomintang. Yüan embarked on a course of action that would place him on the imperial throne, and the Kuomintang responded by proclaiming a rival government in Canton. But in 1916 Yüan died, and under his successor Li Yuan-hung unity seemed to have been restored.

In 1915 the Japanese had presented China with "Twenty-one Demands" that would in effect have made China tributary to Japan. But this threat receded after World War I (China declared its support of the Allies in 1917) when the major powers guaranteed the territorial integrity of China. But in-

capital Nanning. Area: 220,495sq km (85,133sq miles). Pop. 21,000,000

Kwangtung (Guangdong) Province, S coast; capital Canton (Guangzhou; Kwangchow). Area: 231,480sq km (89,374sq miles). Pop. 40,000,000

Kweichow (Guizhou) Province, S-central China; capital Kweiyang (Guiyang). Area: 174,060sq km (67,204sq miles). Pop. 17,000,000

Liaoning Province, NE coast; capital Shenyang (Mukden). Area: 151,055sq km (58,322sq miles). Pop. 29,000,000

Ningsia-Hui (Ningxia Huizu) Autonomous Region, N China; capital Yinchwan (Yinchuan). Area: 66,435sq km (25,650sq miles). Pop. 2,000,000

Peking (Beijing Shi) Government-controlled Municipality, NE China, the area around the country's capital Peking (Beijing). Area: 8,773sq km (3,386sq miles). Pop. 7,600,000

Shanghai (Shanghai Shi) Government-controlled Municipality, E China. Area: 5,800sq km (2,239sq miles). Pop. 10,800,000

Shansi (Shānxī) Province, N-central China; capital Taiyuan. Area: 157,170sq km (60,683sq miles). Pop. 18,000,000

Shantung (Shandong) Province, NE coast; capital Tsinan (Jinan). Area: 153,360sq km (59,212sq miles). Pop. 57,000,000

Shensi (Shănxī) Province, central China; capital Sian (Xi'an). Area: 195,880sq km (75,630sq miles). Pop. 21,000,000

Sinkiang-Uigur (Xinjiang Weiwuer Zizhiqu) Autonomous Region, NW China; capital Urumchi (Wulumuqi). Area: 1,647,435sq km (636,075sq miles). Pop. 8,000,000

Szechwan (Sichuan) Province, S-central China; capital Chengtu (Chengdu). Area: 569,215sq km (219,774sq miles). Pop. 70,000,000

Taiwan (or Formosa) Regarded by the People's Republic of China as part of its territory, but claimed independence in 1949; capital T'aipei. Area: 35,962sq km (13,882sq miles). Pop. 16,324,000. *See* TAIWAN.

Tibet (Xizang Zizhiqu) Autonomous Region, SW China; capital Lhasa (Lasa). Area: 1,222,070sq km (471,841sq miles). Pop. 1,400,000

Tientsin (Tianjin) Government-controlled Municipality, NE China. Area: 6,205sq km (2,395sq miles). Pop. 4,000,000

Tsinghai (Qinghai) Province, W-central China; capital Sining (Xining). Area: 721,280sq km (278,486sq miles). Pop. 2,000,000

Yunnan Province, S-central China; capital Kunming. Area: 436,380sq km (168,486sq miles). Pop. 21,000,000

China's industrialization dates from the development of heavy industries, particularly those based on iron and steel; here iron castings are being made.

A grand parade in Kwangsi-Chuang province displays in military ranks – but without military equipment – the might and power of China's agricultural force.

Bogotá, Colombia, has several universities, many museums and churches, and a cathedral; it was named after a Chibcha Indian chief called Bacatá.

ternally China was in chaos. The official government in Peking was unable to control the welter of rival war-lords who were campaigning in the north, and the reactivated "rebel" Kuomintang government in Canton had similar problems in the south. Moreover a new contender had entered the lists – the Chinese Communist Party, founded in 1921. The Soviet government engineered an alliance between the Kuomintang (the Nationalists) and the Communists, and after Sun's death in 1925 his successor Chiang Kai-shek launched an attack on the northern war-lords and the official government. Before long he was in control of the Yangtze lowlands; agitation against foreigners, imperialists and property owners was skilfully used to win popular support. But a split developed between the Nationalists and the Communists, and Chiang began a purge of his former Communist allies; many were killed.

By 1928 the greater part of China, including the former capital of Peking, was in Chiang's hands.

China – profile

Official name The People's Republic of China
Area 9,560,948sq km (3,691,482sq miles)
Population (1975 est.) 838,803,000
Density 88 per sq km (227 per sq mile)
Chief cities Peking (Bijing, the capital) 7,600,000; Shanghai, 10,800,000; Tientsin (Tianjin), 4,000,000; Shenyang (Mukden), 2,800,000; Wuhan, 2,560,000; Chunking (Chongqing), 2,300,000; Canton (Guangzhon; Kwangchow), 2,200,000.
Government Communist people's republic
Religion Organized religion is discouraged; the major historic religions and philosophies are Buddhism, Confucianism, Taoism, Islam, Christianity
Language (official) Mandarin Chinese, called *putonghua* ("the generally understood language")
Monetary unit Yuan
Gross national product (1974) £105,059,000,000
Per capita income £125
Agriculture Rice, wheat, cotton, tea, vegetables, tobacco, soya beans, sisal, rubber, citrus fruits, fish products, meat
Industries Iron, steel, textiles, oil products, machinery, pharmaceuticals, fertilizers, building materials
Minerals Coal, iron, manganese, tungsten, antimony, molybdenum, oil, asbestos, graphite, salt, mercury
Trading partners Canada, Australia, Japan, Hong Kong, Singapore, West Germany, Britain

The Communists were, however, still in existence. Thousands of peasants and deserters from various forces had been recruited to form an army, and a Communist government had been set up in Kiangsi in the east. But under nationalist pressure the Communists, led by Mao Tse-tung and Chu Teh, decided to move to Yenan (Fushih) in the central north, where they established a stronghold. This was the famous "Long March" (1934): of 100,000 people that set out, only 20,000 arrived.

While the Chinese were fighting each other, the Japanese had in 1931 seized Manchuria and made it into a separate puppet state, Manchukuo, with the former emperor Pu Yi as its ruler. Then in 1937 a local clash between Chinese and Japanese troops heralded a new Japanese advance into Chinese territory. Nationalists and Communists united to oppose them but were forced to fall back, and the Japanese established another puppet regime in eastern China. When in 1941 the Japanese attacked American and British territories, the Sino-Japanese conflict became merged in the general conflict of World War II, and nationalist China was accepted as one of the major Allies. At the end of the war, China regained her lost lands. But a new situation developed in Manchuria. When the Red (Soviet) Army, which had taken Manchuria from the Japanese, withdrew its troops in 1946, the Chinese Communists took over from them. Civil war in China between Communists and Nationalists ensued. Despite American help, Chiang Kai-shek suffered defeat after defeat; his army was exhausted, money had become worthless, and support for the Communists grew every day. Eventually the Nationalists retired to the island of Taiwan (Formosa). In October 1949, the Communist People's Republic of China came into being, with its capital in Peking and Mao Tse-tung at its head. Chou En-lai was appointed premier. Several countries gave quick recognition to the new regime but others, including the United States, continued to recognize the Nationalists in Taiwan as the rightful government of China. The People's Republic was not given the Chinese seat in the United Nations until 1971.

In 1950 Tibet was forcibly incorporated into China's territory. In the same year Chinese troops were sent to help the Communist North Koreans in the Korean War. After 1959 the Chinese army was also involved in border skirmishes with Indian troops. And before the end of the 1950s, disagreements with the USSR became apparent. In 1963 there was a definite split between the two great Communist countries. The Chinese accused the Russians of "revisionism" and disputed the Soviet claim to world leadership of the Communist nations: China's foreign policy has since been directed towards diminishing Soviet influence in the world. Revisionism was again in the air in 1966–68, when China was engulfed in the Great Proletarian Cultural Revolution – a struggle between the disciples of Mao, who believed that the major consideration was the purity of the country's revolutionary ideals, and others who argued that China's primary goal should be modernization. For years the country was in a ferment of student demonstrations, mass meetings, arguments and even beatings. Millions of students joined Red Guard brigades to protect the revolution. The "little red book" (*Quotations from Chairman Mao Tse-tung*) was brandished on every side. In January 1976 the greatly respected premier Chou En-lai died, and in September Mao also died. Hua Kuo-feng, who had succeeded Chou En-lai as premier, was elected chairman of the central committee of the Communist Party in Mao's place, also retaining the premiership. His appointment signified a belated victory for the "moderates" in the government. Four "radical" members of the politburo were arrested, accused of plotting a coup. The four, referred to in the newspapers as "the gang of four", included Mao's widow, Chiang Ching. In July 1976 earthquakes in Hopei province caused much damage and estimates of the number of killed range between 100,000 and 700,000 people.

Christmas Island is the name of two islands. The one in the Pacific Ocean is part of the GILBERT AND ELLICE ISLANDS (the other Christmas Island is in the Indian Ocean, 320km (200 miles) south of Java). The Pacific island, annexed by Britain in 1888, is now worked as a British copra plantation employing most of the population. Both Britain and the USA carried out nuclear tests on the atoll in the late 1950s and early 1960s, which led to an unresolved dispute over sovereignty. Area: 575sq km (222sq miles). Pop, (1968) 367. Map 50.

Cocos Islands, also called Keeling Islands, are a group of 28 coral islets in the Indian Ocean, 1,200km (750 miles) south-west of Java. Since 1955 they have been administered by Australia. The islands were discovered by William Keeling in 1609 and settled in the 1820s. They were annexed by Britain in 1857 and incorporated into the Straits Settlements in 1903. Only West Island, Home Island and Direction Island are inhabited. The chief products are copra and coconuts. Area: 13sq km (5sq miles). Pop. (1975 est.) 1,000.

Colombia, official name Republic of Colombia, is a country in the north-western corner of South America. It is rich in natural resources, with fertile soil and important deposits of precious metals, emeralds and petroleum. Coffee is the mainstay of the economy. Colombia's modern development has, however, been seriously hampered by the *Violencia,* long periods of internal violence which began in 1948. The capital is Bogotá.

Land and climate. The western part of Colombia is dominated by three great ranges of the Andes, the Western, Central and Eastern Cordilleras, which are separated by the deep troughs of Cauca-Patía and Magdalena. The larger, eastern, part stretches south of the Equator and is a vast plain, watered by the tributaries of the rivers Orinoco and

A little Colombian girl plays with her toys, a mixture of ancient and modern; the traditional beads and bangles — and a torch battery.

Cartagena is capital of Bolivar department, Colombia. It is a port from which petroleum, sugar and other locally produced goods are exported.

Antioquia is a province in north-western Colombia, at the northern end of the Cordillera Occidental; the chief city of the region is Medellín.

the Amazon. The country has a wide range of climate, caused mainly by variations in altitude, but there is little seasonal variation in any particular locality. Bogotá, 2,640m (8,661ft) above sea-level, has a cool, moist climate, with temperatures averaging about 14°C (57°F). Barranquilla, on the Caribbean coast, averages 28°C (82°F) and is fairly dry. The swampy Pacific coast is one of the wettest areas in the Americas.

Economy. Agriculture is the chief industry, but only a small percentage of the land is under cultivation. Coffee often accounts for 60 to 70 per cent of export trade; Colombia ranks second only to Brazil in world production, and leads in the output of mild coffee. Other important crops include sugar cane, rice, potatoes, maize, bananas and cotton. Beef cattle (23 million) are reared on the grassy plains. Colombia's gold output is the highest in South America; the country has the largest deposit of platinum in the world, and it also mines most of the world's

emeralds. Petroleum, once second to coffee in export value, is now becoming increasingly important to Colombia's internal economy as the country tries to industrialize, diversify and improve its poor roads and railways.

People. With more than 23 million inhabitants, Colombia is rapidly overtaking Argentina as South America's second most populous country (to Brazil). Most of the people live in the valleys of western Colombia. About 68 per cent are mestizo (of mixed European and American Indian descent), 20 per cent white (mainly Spanish descent), 7 per cent American Indian (the native people), and 5 per cent Negro. The official language is Spanish and the state religion is Roman Catholicism. About 70 per cent of the people depend on agriculture for a living, but most of them own little land. Lack of good communications has resulted in isolated groups of people with a strong local identity and has made for feuding between villages and even families. Bandits descend on remote communities to pillage crops and obtain food.

Pre-Conquest Indians showed a high degree of artistic ability in their gold and silver jewellery, and modern Colombian art shows both Indian and European influences. Bull-fighting is a favourite sport and baseball and soccer are also popular.

Government. Colombia is a presidential democracy with a congress of two houses: a Senate (112 members) and a House of Representatives (199), each elected for four years. The president, who is also in office for four years but cannot serve consecutive terms, is elected by popular vote.

History. Before the coming of the Spanish conquistadors, the region was occupied by several groups of Indians, notably the Chibcha in the mountainous interior and the Carib along the northern coast. In about 1500 the Spanish explorer Alonso de Lugo, who accompanied Columbus on his second voyage, visited what is now Colombia. In 1538 Gonzalo Jiménez de Quesada founded an outpost later to be called Santa Fé de Bogotá, and the whole area was named New Granada after his home province. As well as Colombia, New Granada eventually came to include what is now Panama, Ecuador and Venezuela, and became a viceroyalty in 1718 (with a break between 1723 and 1739). Revolution and liberation (by Simón Bolívar) led to independence in 1819, and then the breakup of the union (1830). Panama did not break away until 1903, after which Colombia enjoyed a period of relative tranquillity under both Conservative and Liberal governments up to 1948.

The assassination of Jorge Eliécer Gaitán, the left-wing mayor of Bogotá, led to an explosion of rioting in the capital that took three days to quell and cost the lives of 2,000 to 3,000 citizens. This developed into the countrywide *Violencia*, which raged fiercely for ten years and continued sporadically for another ten, with more than 200,000 deaths. In 1953 Gustavo Rojas Pinilla assumed power and became president but was ousted in 1957 by a military junta. Parity between the traditional Liberal and Conservative parties was maintained from 1958 by a National Front system that ran for 16 years, under which the president was chosen from each party in turn. Parity in congress and all government departments was planned to continue until 1978. In 1974 the first presidential election not subject to the National Front system was won by the Liberal candidate, Alfonso López Michelsen. Terrorism was again rife in the mid-1970s, with large outlying areas virtually run by groups of bandits, ostensibly non-political but alleged to be Communist-assisted. To combat guerrilla action (kidnapping and assassination of businessmen and diplomats), strikes, rioting, student unrest and the occupation of land by peasant organizations, the president declared a state of siege in June 1975, which lasted a year. In 1976 he promised to accelerate the programme of land reform begun 15 years earlier.

Colombia – profile

Official name Republic of Colombia
Area 1,138,914sq km (439,735sq miles)
Population (1975 est.) 23,542,000
Density 21 per sq km (54 per sq mile)
Chief cities Bogotá (capital) (1976 est.) 3,153,000; Medellín 1,070,924; Cali, 898,253
Government Head of state, Alfonso López Michelsen, president (elected 1974)
Religion Roman Catholicism (official)
Language Spanish
Monetary unit Peso
Gross national product (1974) £4,957,300,000
Per capita income £211
Agriculture Coffee, cocoa, sugar cane, rice, potatoes, cotton, maize, bananas, tobacco, livestock
Minerals Platinum, gold, silver, emeralds, copper, lead, mercury, petroleum, iron, coal, salt
Industries Metal products, industrial chemicals, petrochemicals, cotton yarn, textiles, processed foods, beverages
Trading partners (major) USA, West Germany, Japan

Colorado. *See* UNITED STATES.

Comoro Islands, official name Comores State, is a group of volcanic islands off the eastern coast of Africa between Mozambique and Madagascar. The main islands are Grande Comoro, Anjouan and Mohéli. Most of the people are Muslims and farming is the chief industry. France annexed the islands in the 1800s and early 1900s. In 1975 they became

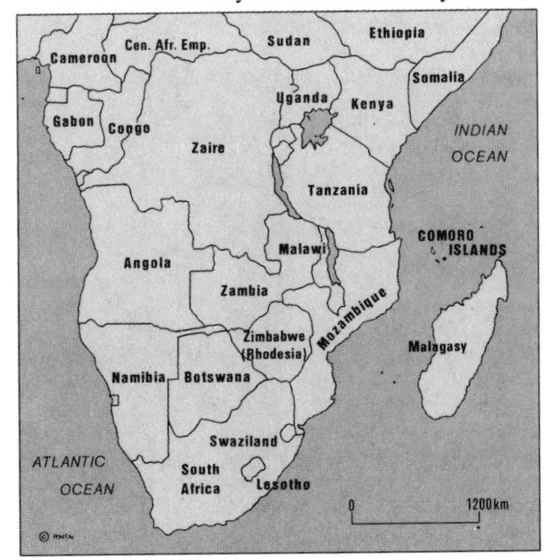

The Congo River, now known as the Zaire, takes its modern name from the word *nzari,* which in various African dialects means simply "river".

Sir Henry Stanley's second expedition (1879-84) traced the Congo River from its source to the sea; three years later he helped to establish the Congo.

Sugar refining is a key industry in Costa Rica and sugar, along with coffee, cocoa and bananas, is one of the country's major exports.

an independent republic but one island, Mayotte (predominantly Christian) remained a French overseas département. The capital of the republic is Moroni. Area: 1,797sq km (694sq miles). Pop. (1975 est.) 292,000. Map 34.

Congo, official name People's Republic of the Congo and formerly French Congo, is an independent nation in western central Africa. (The neighbouring nation, formerly the Belgian Congo, is now ZAIRE). Congo was proclaimed a Communist state in 1970. It established links with China, Cuba and the USSR and has pursued radical policies, including land redistribution and reduction in the power of local chiefs. But Congo has also retained contacts with the West, especially France. The capital is Brazzaville.

Land and climate. A narrow coastal plain is crossed by the River Kouilou whose tributary, the Niari, flows through Congo's chief farming region. Behind the coastal plain the land rises, reaching more than 800m (2,626ft) in the Batéké plateau near the Congo-Gabon border. The north consists of low ridges separated by broad, swampy river valleys occupied by tributaries of the Ubangi and Zaire rivers, which form most of Congo's eastern border.

The climate is hot and wet. In much of the country temperatures average 24°C (75°F) all the year round, but the coastal plain is kept cooler by the cold off-shore Benguela Current. Northern Congo has an average annual rainfall of 1,780mm (70in). The coastlands are drier, with 1,220mm (48in). Rain forest and swamps cover more than half of Congo. Woodland savanna flourishes in the drier south east, but the coastal plain is almost treeless.

Economy. Most Congolese are farmers living at subsistence level. They grow such crops as cassava, plantains (cooking bananas), rice, sweet potatoes and yams. Timber and timber products are the most valuable exports. The chief commercial crops, which are also leading exports, are sugar cane, cocoa, coffee and tobacco. Plantation agriculture is especially important in the Niari valley, which lies on the railway between Brazzaville and Pointe-Noire, the country's chief port. Industrial diamonds and potash are exported and petroleum is extracted on the coast. Manufacturing is relatively well developed and Congo has considerable potential for hydroelectricity. Further economic progress will, however, depend heavily on foreign aid.

People. Most people live in the south and about 40 per cent of the population is in urban areas. Congo has various Bantu-speaking groups, the largest of which is the Kongo, who live in and around Brazzaville. About 12,000 Europeans live mostly in the towns and there are about 12,000 pygmies. More than 50 per cent of the people follow ethnic religions, but there is also a large Christian community. Literacy is estimated at 20 per cent.

Government. Congo's one political party, the Congolese Labour Party, introduced a new constitution in 1972 and it was approved in a referendum in 1973. The country is governed by a State Council, a Council of Ministers and an elected, 115-member National Assembly.

History. From 1880 to 1960, when it became independent, Congo was ruled by France. In 1960 a former priest, Fulbert Youlou, became president but was deposed in 1963 by a left-wing group under Alphonse Massamba-Débat. Massamba-Débat was removed by a military group in 1968 and Marien Ngouabi became president. Some opposition to the government's radical policies, communal differences and tension between civilian and military leaders all contributed to further unrest in the 1970s. In March 1977 Ngouabi was assassinated and Massamba-Débat executed for his alleged role in the assassination. The new president Col. Joachim Yhombi-Opango announced that a Military Committee would assume full powers. Map 34.

Congo – profile

Official name People's Republic of the Congo
Area 342,000sq km (132,046sq miles)
Population (1975 est.) 1,345,000
Density 4 per sq km (10 per sq mile)
Chief cities Brazzaville (capital) 289,700; Pointe-Noire, 141,700
Government Self-proclaimed Communist
Religions Ethnic, Christianity
Language French (official)
Monetary unit CFA franc
Gross national product (1974) £204,000,000
Per capita income £152
Agriculture Cassava, cocoa, coffee, groundnuts, palm oil, rice, sugar cane, timber, tobacco, yams
Industries Brewing, cement, chemicals, flour, sugar, textiles, wood products
Minerals Industrial diamonds, petroleum, potash
Trading partners France, West Germany, Netherlands, USA, Italy

Congo, Republic of the. *See* ZAIRE.

Connecticut. *See* UNITED STATES.

Cook Islands, formerly called Hervey Islands, are a group of about 15 self-governing islands and atolls (under New Zealand sovereignty) in the southern Pacific Ocean, about 3,000km (1,870 miles) north-east of New Zealand. The chief of the Northern Cook (or Manihiki) Islands is Manihiki, and the main island in the Southern (or Lower) Cook Islands is Rarotonga, location of the capital Avarua. Some of the islands were discovered by Capt. James Cook in 1773 and Rarotonga was discovered in 1823 by the British missionary John Williams; the others remained unexplored until the 1920s. They became a British protectorate in 1888 and were annexed by New Zealand in 1901. The chief products are copra and citrus fruits. Area: 240sq km (93sq miles). Pop. (1975 est.) 25,000. Map 62.

Corsica. *See* FRANCE.

Costa Rica, official name Republic of Costa Rica, is a nation in Central America between Nicaragua (to the north) and Panama; the capital is San José. It has a long history of stable, democratic government. The climate varies from hot and humid on the coastal plains to the more temperate conditions on the central plateau, where most of the population lives. A mountain range along the axis of the country includes several active volcanoes, such as Irazú (erupted 1964) and Arenal (erupted 1968). The

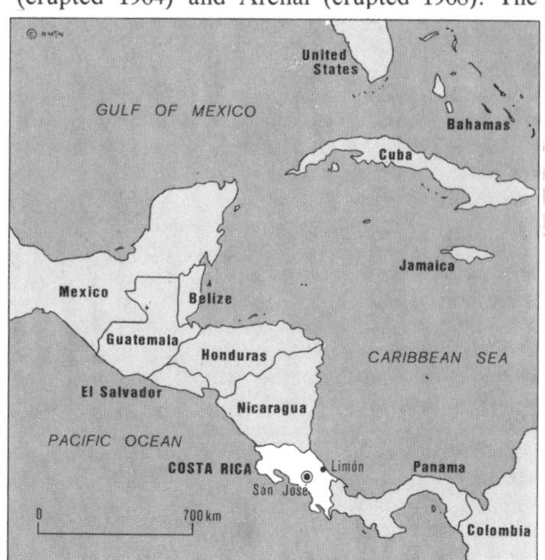

Havana, capital of Cuba, is the largest city and chief port in the Caribbean; it is also a popular tourist resort and factories make the famous cigars.

The produce of this vegetable farm near Havana proves the semitropical climate of Cuba does not benefit only the country's vast sugar plantations.

At a modern agricultural school in Havana, great stress is laid on the practical applications of studies in the technology of agriculture.

chief products are coffee (first cultivated in Costa Rica), cocoa, bananas and sugar cane, which are exported in exchange for fuel and manufactured goods. Most of the people are Spanish-speaking Roman Catholics.

Costa Rica was conquered by the Spanish in the 1560s and ruled from Guatemala. The country declared independence in 1821 and a year later became part of the Mexican Empire. From 1823 to 1838 it was a member of the Central American Federation, and then it became a republic. Revolutions in 1917 and 1948 temporarily halted democratic rule. Area: 50,898sq km (19,652sq miles). Pop. (1975 est.) 1,968,000. Map 74.

Crete. *See* GREECE.

Cuba, official name Republic of Cuba, is an island country in the West Indies. Originally under Spanish rule, it came under American influence in the early 1900s and then became a Communist state after the revolution of 1959, led by Fidel Castro. Its economy depends largely on the cultivation of sugar cane and on Soviet aid. The island is renowned for its great natural beauty. The capital is Havana (La Habana).

Land and climate. The largest island in the West Indies, Cuba has a much-indented coastline of about 3,200km (2,000 miles). The coasts are fringed with coral islands and reefs, forming bays and lagoons. The many excellent harbours include Havana, Santiago de Cuba, Guantánamo, Cienfuegos and Matanzas. About three-fifths of the country is gently rolling land with wide, fertile valleys and plains. The rest is hilly or mountainous, rising to 2,400m (7,874ft) in the south-east along the Sierra Maestra range. The climate is tropical in the lowlands but cool in the hills. There is a rainy season from May to October, with an annual average of about 1,270mm (50 inches) in Havana. Hurricanes are common, especially in the west.

Economy. Two-thirds of Cuba's cultivated land is devoted to sugar cane, and sugar accounts for about 80 per cent of exports. By-products of sugar, such as molasses and rum, are also important exports, as are tobacco and tobacco products (the famous Havana cigars) and nickel, which is one of many minerals mined in the eastern part of the island. Cuba imports machinery, chemicals, transport equipment and an estimated 40 per cent of its food requirements, mainly from Communist-bloc countries. Despite a high level of support and economic investment from the USSR, Cuba's economy has made little progress since the Communist take-over in 1959. The government has carried out programmes of land reform and has nationalized nearly all commercial enterprises. State farms or cooperatives occupy about 70 per cent of the cultivated land, and the private smallholders who farm the remainder are also government-controlled to some extent.

People. Nearly three-quarters of the people are white (of Spanish descent) and most of the rest are Negro or mulatto. Spanish is the official language and English is widely understood. Education is free and compulsory, and the government claims to have completely eliminated illiteracy. There is no state religion but most of the people are Roman Catholics. About half the working population is employed in agriculture. Cubans are by nature a cheerful people and on the whole have accepted, with remarkably good grace, the deprivations of the revolution. Nevertheless more than 650,000 Cubans have gone into exile since the Castro take-over, most of them to the United States.

Government. Since the suspension of the constitution in 1959 government has been by decree of the politburo, headed by the premier and nominally appointed by the president. The Communist Party is the only authorized political party. A new socialist constitution came into force in 1976, providing for a National Assembly of Popular Power.

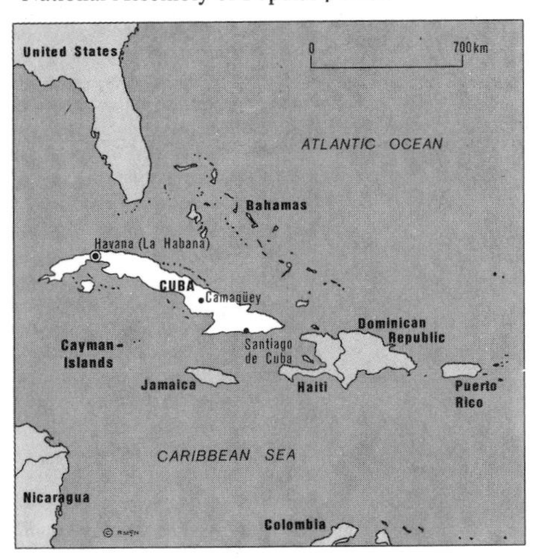

History. Christopher Columbus landed on Cuba in 1492 and claimed it for Spain. It became Spain's richest colony in the West Indies and, apart from the British occupation of Havana in 1762-63, remained Spanish for more than 400 years. The mysterious sinking of the USS *Maine* in Havana harbour in 1898 resulted in the short Spanish-American War, in which the Spaniards were defeated and gave up all claims to Cuba. The United States formally withdrew in 1902, but continued to exert an important influence on Cuba, especially on its economy. In 1933 Fulgencio Batista led an army revolt that made him the leading power in the country, and he eventually seized the presidency in 1952.

Opposition to the corrupt Batista dictatorship intensified in 1956 under the leadership of a young lawyer, Fidel Castro. Batista finally fled in 1959 in the face of guerrilla action, and Castro became premier in February of that year. The new government, dominated by left-wing extremists, immediately initiated a programme of sweeping economic and social reforms. It also executed hundreds of dissidents and ousted moderates from office. In 1960 the government took over all private enterprise, including more than $1,000 million worth of American-owned property. The United States broke off relations with Cuba in 1961, and Castro disclosed his alliance with the Soviet bloc. Thousands of Cubans fled to the United States and, financed and encouraged by the American CIA, led an abortive invasion force that landed in the Bay of Pigs, in southern Cuba. In 1962 the United States ordered a total embargo on Cuban exports. The USSR began building missile bases in Cuba and precipitated a world crisis. President Kennedy demanded the removal of this threat on 22 October 1962, confronting Soviet ships with the US Navy, and the Russians withdrew. Continued Soviet influence developed a stranglehold on the Cuban economy, but American-Cuban relations showed signs of softening in the early 1970s and the United States began to lift its embargo on trade. The country adopted a new constitution in 1976 which gave the people a greater voice in the government. Map 74.

Cuba – profile

Official name Republic of Cuba
Area 114,524sq km (44,218sq miles)
Population (1974 est.) 9,194,000
Density 80 per sq km (208 per sq mile)
Chief cities Havana (capital) (1974 est.) 1,838,000; Camaguey 178,600; Santiago de Cuba 277,600
Government Dr. Fidel Castro Ruz, president of the Council of State
Religion Roman Catholicism is widely practised, but the state recognizes no religion
Languages Spanish (official), English
Gross national product (1974) £2,470,100,000
Per capita income £269
Agriculture Sugar cane, tobacco, rice, maize, coffee, cotton, citrus fruits
Industries By-products of sugar, tobacco and tobacco products, fishing, dairy products
Minerals Nickel, copper, chromite, manganese, iron
Trading partners (major) USSR, other Communist countries

Cyprus (Greek, Kypros; Turkish, Kibris) is an island republic in the Mediterranean Sea. The rocky but beautiful land of Cyprus, with its wooded mountains, ancient ports and groves of fruit trees, has been torn by strife in the second half of the 20th

Cyprus gained independence in 1960, marked by this monument in Nicosia — the scene of bitter fighting shortly before, and again in the Turkish invasion.

Prague, the capital of Czechoslovakia, dates from the 9th century and has old buildings on hills astride the River Vltava and modern industrial development.

Czechoslovakia, a highly industrialized country, also has extensive forests and a highly developed and efficient agricultural industry.

century because it is the home of two seemingly irreconcilable communities. The capital is Nicosia.

Land and economy. The dominant features of the landscape are the Troodos Mountains in the west-central region and the Kyrenia Mountains along the northern coast. Inland from the latter range is a fertile lowland plain, the Mesaoria. The island has a typical Mediterranean climate with mild, wet winters and hot, dry summers. Minerals – found chiefly in the Troodos massif – are important to the economy; they include iron, asbestos, chromite and copper (from which the island is said to derive its name). But Cyprus is predominantly agricultural; wheat and barley are grown, and there is an export trade in oranges, grapefruit, lemons and potatoes. Grapes are produced for wine. The island has no heavy industry but there are many light manufacturing industries, most of which produce consumer goods. Until the violence of 1974 tourism was a profitable source of income. Cyprus has no railways

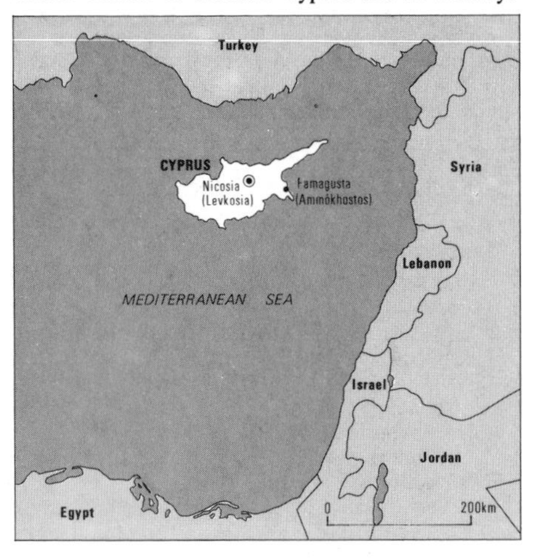

and depends on road communications and coastal boat traffic. Nicosia is in the Mesaoria but other important towns – Famagusta, Larnaca, Limassol, Paphos and Kyrenia – are all situated on the coast.

People. The two main communities differ in origin, traditions, language and religion. There is no intermarriage and little social contact between them. The Greeks (about 78 per cent of the population) belong mainly to the Greek Orthodox Church and in recent times have been the more prosperous community. The Turks (18 per cent) are Muslims and are descended from settlers who arrived after the Turkish conquest of 1571. Education is free and compulsory at primary level. The official languages are Greek and Turkish; English is widely spoken.

Government. The constitution of 1960 provided for a Greek Cypriot president, a Turkish Cypriot vice-president and a mixed House of Representatives. In practice this arrangement led to continuous dispute between the two communities. After the Turkish invasion of 1974 the leaders of the Turkish Cypriot community rejected the 1960 constitution and unilaterally declared the formation of a Turkish Cypriot state within a federal republic of Cyprus.

History. Early settlers on the island included Phoenicians and Greeks, and at various times Cyprus was ruled by Assyria, Egypt, Persia and Greece. In 58 BC it became a Roman province. Richard I of England took the island from the Byzantines in 1191 and gave it to Guy de Lusignan, who founded a dynasty that ruled until 1489, when it was captured by Venice. The Turks ousted the Venetians in 1571. In 1878 the Ottoman Turks permitted the setting up of a British administration in Cyprus. At the beginning of World War I Britain made the island a Crown colony.

In 1950 the archbishop of Cyprus, Archbishop Makarios, organized a plebiscite in which 96 per cent of Cypriots were said to have voted for *Enosis* (union with Greece). Later, after the Greek Cypriot movement EOKA began a campaign of violence to further this aim, the archbishop was deported and held in custody (1956-57). But eventually all parties to the dispute, including the Turkish Cypriots who opposed *Enosis,* agreed at a London conference that Cyprus should be an independent republic: Britain, Greece and Turkey guaranteed its independence. Britain retained sovereignty over certain military bases. In 1960 Archbishop Makarios became the first president of the new state, which elected to remain in the Commonwealth of Nations. But ethnic and cultural differences remained, with continuous friction between the two main communities and the need for peacekeeping forces – at first from the guarantor powers and later from the United Nations. In 1974 a coup inspired from Greece temporarily ousted President Makarios. Shortly afterwards Turkish troops invaded Cyprus, ostensibly to restore order. They occupied much of the northern part of the island, and 200,000 Greek Cypriots fled to the south or emigrated. The Turkish troops consolidated their hold on the northern areas and supported the establishment of a Turkish Cypriot Federated State. In 1975 the United Nations called (unsuccessfully) for the withdrawal of Turkish troops. In 1976 tentative talks were held in Vienna between the two factions. Makarios died in 1977. Area: 9,251sq km (3,571sq miles). Pop. (1976 est.) 639,000. Map 38.

Czechoslovakia (Československo), official name Czechoslovak Socialist Republic, is a country in central Europe. Its two major groups of people, the Czechs and the Slovaks, have a long history of domination by powerful neighbours. They became a prosperous, industrialized nation after World War I, but first the Germans and then the Russians subjugated the country, now a Soviet satellite.

Land and climate. Czechoslovakia is a landlocked country with three natural regions: Bohemia in the west, Moravia in the centre and Slovakia in the east. Bohemia is a basin-shaped plateau ringed with mountains. The capital of the country, Prague (Praha), stands on the River Vltava on low-lying land in the centre of the region. To the south-west are the wooded hills of the Bohemian Forest. Morava slopes from the highlands of the north and west to a low, hilly, fertile region drained by the River Moravia, which flows into the Duna (Danube) on the southern border. The southern part of Slovakia lies in the Danube basin and is extremely fertile, although most of Slovakia is mountainous, dominated by ranges of the Carpathians. They rise to more than 2,500m (8,200ft) in the scenic High Tatra, the highest peak being Gerlachovka, 2,663m (8,737ft). Czechoslovakia has cold, dry winters, with average January temperatures of -7°C (20°F). The summers are warm, with average July temperatures of 20°C (68°F) and some extremely hot days. Rainfall is heavy in most of the mountainous regions.

Economy. Czechoslovakia is one of the most highly industrialized countries of central Europe. Industry, which is state-owned, accounts for about two-thirds of the national income and employs more than a third of the working population. Most of the important industries are in Bohemia and Moravia, and include the manufacture of iron and steel, chemicals, cars, textiles and shoes. There are abundant deposits of hard and soft coal and some iron ore, but much raw material has to be imported. Agriculture provides about 15 per cent of the country's income and employs about 20 per cent of the working population. Nearly all the cultivated land is run by the state or by co-operative farms. The chief crops include sugar-beet, wheat, potatoes, barley and hops for brewing beer, and other cereals. Dairy farming is also important and Czechoslovakia has nearly 5 million cattle and more than 6 million pigs. Forests cover 35 per cent of the land, mainly spruce (50 per cent), beech and pine. The timber provides pulp for paper-making, which is a major industry. Nearly 70 per cent of Czechoslovakia's trade is with other Communist countries. Its major exports are arms and machinery, industrial consumer goods, and iron and steel. It imports machinery, raw materials and fuel. Trade with non-Communist countries is increasing, the chief of these new trading partners being West Germany, Austria and Britain.

People. Czechs, most of whom live in Bohemia and Moravia, make up about 64 per cent of the population and Slovaks form another 30 per cent. About 4 per cent of the people are Magyars (Hungarians), who live mainly in Slovakia, and there are smaller minorities of Germans, Poles, Ukrainians and Russians. Czech and Slovak, the two official languages, are only slightly different from each other and are mutually understood. About 80 per cent of the people are Roman Catholics and there is also a large Protestant (Hussite) community. There is no state religion, but the churches are under state control. Education is free between the ages of six and 15; where possible children remain at the same

Ceske Budejovice is noted for its inner town, with an arcaded square, and its famous breweries; the city was founded in the 12th century.

Brno is an industrial centre which is particularly known for its woollen industry and armaments factories where the famous Bren gun was designed.

Slovakia, with its rich, fertile soil, has numerous farms, vineyards and orchards which make a major contribution to Czechoslovakia's economy.

school for the whole nine years. They may then choose to go to secondary school for another three years to continue their general education or for vocational studies. There are six universities, including the Charles University in Prague (1348).

In the arts, Czechoslovakia's greatest contributions have been in music and literature. The tradition of music is strong and goes back to the plainsong chants and folk melodies of the Middle Ages. Despite persecution by the Hapsburgs, Czech composers and instrumentalists became widely known in the 1600s. The growing feeling of nationalism in the 1800s inspired such composers as Bedřich Smetana (1824-84), the pioneer of Czech opera (he wrote *The Bartered Bride*), Antonín Dvořák (1841-1904), Leoš Janáček (1854-1928) and Gustav Mahler (1860-1911). The Reformation provided the inspiration for early Czech literature, seen in the writings of the Bohemian religious reformer Jan Hus (John Huss) (*c.* 1369-1415), who spread the teachings of John Wycliffe, and the Moravian theologian and educator John Amos Comenius (1592-1670). Later writers included Jaroslav Hašek (1883-1927), whose multi-volume novel *The Good Soldier Schweik* satirized military bureaucracy; Karel Čapek (1890-1938), whose satirical play *Rossum's Universal Robots* introduced the word "robot" into the English language; and Franz Kafka (1883-1924), whose posthumously published novels have had an immense influence on Western literature. The people of Czechoslovakia are famous for their traditional handicrafts, which include elaborate embroidery and needlework, weaving and pottery. They are also a sports-loving people and have been successful in international soccer, athletics and ice hockey.

Government. Since 1 January 1969 Czechoslovakia has been a federal socialist republic. The supreme organ of the state is the Federal Assembly, which consists of two equal chambers: the Chamber of Nations and the Chamber of the People. The chief executive is the president, elected by the Assembly for a five-year term. The premier and his cabinet are appointed by the president, but are responsible to the Assembly. The republic consists of two equal nations, the Czech Socialist Republic and the Slovak Socialist Republic, each governed by a National Council which elects 75 members to the Chamber of Nations. The Chamber of the People has 200 deputies, elected by national suffrage. The primary source of power is the Communist Party of Czechoslovakia. Communists head the other political parties, trade unions and youth organizations, which are all incorporated in the National Front. Since 1971 only one candidate for the Federal Assembly is allowed to stand in each constituency.

History. The Czechoslovak state came into existence in 1918 but the history of its peoples dates from the nomadic Slavs (including the Czechs) who occupied the area in the 6th century AD and had set up a group of states by the 7th century. These were absorbed into Charlemagne's empire in about 800, and when it broke up the Slavs formed the kingdom of Greater Moravia. This was destroyed in 905 by the Magyars, who ruled the eastern region of Slovakia for the next thousand years. The Czechs founded the kingdom of Bohemia, which flourished from the 10th to the 16th century under several efficient monarchs, including Charles IV (who became Holy Roman Emperor in 1346 and made Prague an imperial capital).

Czech nationalism was revived in the 15th century by the religious teachings of Jan Hus. Bohemia was ruled by Polish kings from 1471 to 1526. But then a Hapsburg, Ferdinand of Austria, ascended to the throne. The Protestant Czechs rebelled in 1618 but were defeated in 1620 and were then ruled for almost 300 years as part of the Austrian Empire. In 1848 the peoples of Bohemia and Slovakia, together with the Magyars of Hungary, revolted against the Austrian emperor but were ruthlessly put down, although the Magyars achieved equality in 1867 when the dual monarchy of Austria-Hungary was created.

With the collapse of the Hapsburgs in 1918 Czechoslovakia declared itself an independent republic. The leader of the Czech nationalists, Thomas Masaryk (who had set up a National Committee in exile at the start of World War I), was elected first president. The country soon established itself as a liberal democracy with a prosperous economy. But there were serious problems, especially with the minority of 3,500,000 Germans who lived in the Sudetenland, on the western border. Following Nazi pressure, Britain and France signed the 1938 Munich agreement and persuaded the Czechs to give up the Sudeten territory to Germany. As a result, Czechoslovakia lost its western defences and was divided into separate Czech and Slovak states. In 1939 the Prague government dismissed the Slovak government, who appealed to Hitler for help and so the Germans invaded Prague (breaking the Munich agreement) and occupied the rest of the country.

The Allies organized a Czech government-in-exile under Eduard Beneš, who came to power in 1945 when Soviet troops liberated Czechoslovakia. The Communists won 38 per cent of the vote in the 1946 elections, becoming the strongest single political party; Beneš became president. Backed by threats of Soviet intervention they took complete control of the government in 1948, Beneš resigned, and the Communists set up one of the harshest regimes in the Communist block – with purges, "show trials" and executions. A movement towards liberalization in the 1960s gained impetus when Alexander Dubček became party leader in 1968. He resisted Soviet demands to halt his programme of reforms, but Communist-bloc troops invaded Czechoslovakia in August 1968 [*see* MS p.*264*]. Dubček was replaced in 1969 by the Soviet-backed Gustáv Husák, and Dubček's supporters were purged from the party. The Czech reform movement continued, however, and the voices of Dubček and his followers were still being heard well into the 1970s. Map 18.

Czechoslovakia – profile

Official name Czechoslovak Socialist Republic
Area 127,869sq km (49,370sq miles)
Population (1976 est.) 14,862,000
Density 116 per sq km (301 per sq mile)
Chief cities Prague (capital) (1974) 1,095,615; Brno, 343,860; Bratislava, 328,765
Government Head of state, Gustáv Husák, president (re-elected 1976)
Religions Roman Catholic, Protestant (Hussite)
Languages Czech, Slovak (both official)
Monetary unit Koruna
Gross national product (1974) £20,200,900,000
Per capita income £1,359
Agriculture Sugar-beet, cereals, potatoes, cattle, pigs
Industries Iron and steel, paper-making, chemicals, cars, textiles, shoes
Minerals Coal, iron ore
Trading partners USSR, East Germany, Poland, West Germany

Dahomey. *See* BENIN.

Delaware. *See* UNITED STATES.

Denmark (Danmark), official name Kingdom of Denmark, is a monarchy in western Europe, consisting of a peninsula and more than 100 islands. It is

Copenhagen, Denmark's capital, is a major fishing and naval port and the country's chief commercial, industrial and cultural centre.

Århus, one of Denmark's oldest cities, is a cultural centre and is also noted for its museum of early Danish houses and a 12th century Cathedral.

Helsingør, an industrial centre and fishing port, is also the site of Kronberg castle, now a maritime museum and a venue for Shakespeare's *Hamlet.*

a small country, noted for its beautiful rolling farmlands and famous for its dairy produce. The capital is Copenhagen (København). The Danes are a peaceful people – but they have not always been so. A thousand years ago Danish Vikings terrorized Europe, and for centuries afterwards the Danes were regularly at war with the Swedes, British, Germans and many other nations. Today Denmark is one of the most efficient agricultural countries in the world. Lacking the natural resources of their Scandinavian neighbours, the Danes have nevertheless become highly industrialized, with a fine reputation for engineering and shipbuilding.

Land and climate. Denmark consists of the peninsula of Jutland (Jylland), which accounts for approximately 70 per cent of its land area, and about 500 islands, of which only about 100 are inhabited. Copenhagen lies on the largest island, Zealand (Sjael-land). Most of the country is made up of low-lying, gently undulating plains and hills. The

highest point is Yding Skovhøj, in central Jutland, a hill rising only 173m (568ft) above sea-level. Long sandy beaches are a feature of the western coast of Jutland, fiords a feature of the eastern coast. Denmark has a mild, damp climate, uniform over most of the country because of the absence of natural barriers. In Copenhagen the average January temperature is -1°C (30°F) and for July it is 18°C (64°F). Rainfall is low, but fog is common on the west coast, especially in winter. GREENLAND, the largest island in the world, is a province of Denmark and the FAEROES, a group of islands in the northern Atlantic Ocean, are also part of the kingdom.

Economy. Although Denmark is known primarily as an agricultural country, it now exports more manufactured goods than farm products. The largest industries are food-processing, engineering, iron and steel goods, and chemicals. Other important industries include shipbuilding, textiles, clothing, beverages, and high-quality furniture and silverware. The country is poor in natural resources, so large amounts of raw materials and fuel have to be imported. Nearly 10 per cent of the working population is engaged in agriculture, and about 70 per cent of the land is farmed. Agriculture is organized on a co-operative basis, and the co-operatives are united in national federations. The chief activity is raising pigs and cattle. With more than 7 million pigs and 3 million cattle, Denmark is the world's third-largest exporter of meat. It leads the world in the export of pork and bacon; butter and cheese are also important export products [*see* PE p.228-234]. Denmark is one of the leading producers of barley, and other crops include potatoes, wheat, oats and sugar-beet. The fishing industry has become a basic part of the Danish economy, and tourism also provides a considerable contribution.

People. The population of Denmark is almost entirely Scandinavian, the only minority group being about 30,000 people of German descent just inside the border with West Germany. The Danish language is similar to Swedish and Norwegian, and publications can be read without translation in many other parts of Scandinavia. The National Lutheran Church is the established Church; about 94 per cent of the population belongs to it. Education is free and compulsory from the ages of seven to 16, and most younger children go to kindergarten. Denmark has four universities, including the University of Copenhagen, founded in 1479. An interesting aspect of Danish education is the "folk high school", of which there are about 80. These are private schools which receive financial support from the state and provide courses in Danish culture and government. The first was founded in 1844, its aim being to attract students from rural areas and give them some feeling for literal and cultural values. That the Danes today have a real interest in culture and art is in no small way due to the influence of these schools. Folk art is extremely popular, and much new Danish design combines modernity with tradition. Danish furniture design is world famous, and quality cabinet-making (especially in Copenhagen) dates back 400 years.

Apart from Hans Christian Andersen (1805-75), author of the famous fairy tales, the Danes have produced several other notable writers, including three Nobel Prize winners. And the books of the philosopher and theologian Søren Kierkegaard (1813-55) have had a profound influence on modern thinking, and led to the development of existentialist philosophy. Few Danish artists or composers have won much fame outside their own country, although the sculptor Bertel Thorvaldsen (1770-1844) won international acclaim with his statue of *Jason* in Rome. The buildings and designs of architect Arne Jacobsen (1901-71) may be seen in many parts of the world. Danish life in general is conducted at an easy pace, and Danes take their sport (which is nearly all amateur) leisurely and without too much competitiveness. Soccer is the national sport, and bicycling and sailing are popular pastimes. The famous national dish, or snack, is ***smørrebrød,*** an open sandwich prepared almost as a work of art.

Government. Denmark is a monarchy, with a constitution founded on the *Grundlov* (charter) of 1953. Legislative power is invested jointly with the monarch and the one-house parliament, the *Folketing,* which is elected by and from citizens of age 20 or over. Executive power is exercised by the monarch through ministers. The Folketing is made up of 135 members elected by proportional representation in 17 districts, 40 additional seats divided among parties that have not obtained sufficient returns at the district elections, and two members each from Faeroes and Greenland. Members serve four-year terms.

History. The Danes were a Germanic people who settled in the area in about AD 250, living in small communities governed by local chieftains. They formed a loose confederation of states which were united in about 950 by King Harald "Bluetooth", who spread Christianity throughout Denmark. Danish Vikings had been plundering European coastal towns since the 8th century, and after the unification of Denmark these raids became missions of conquest. Erik "the Red" colonized Greenland in 982 and Harald's son, Sweyn "Forkbeard", conquered much of Britain in 1013. His son, Canute ("the Great", reigned 1014-35), added Norway in 1028. Valdemar ("the Great", reigned 1157-82) began to build up an empire based on the Baltic Sea, which at its height included much of northern Germany. Later, civil wars and struggles with the increasingly powerful cities of the Hanseatic League weakened the country. But the country gradually regained strength, and in 1397 Queen Margaret united Denmark, Norway and Sweden in the Union of Kalmar, with power centred in Denmark. This lasted until 1523, when Sweden broke away.

During the 17th and 18th centuries, Sweden defeated Denmark in a number of wars and won much territory. Finally, as part of the peace settlement of 1814 at the end of the Napoleonic Wars (in which Denmark had sided with France), Denmark ceded Norway to Sweden. In 1849 Frederik VII introduced a liberal constitution and created a two-house parliament. In a brief war with Prussia in 1864 Denmark lost the German duchies of Schleswig and Holstein. At about this time, Denmark began to develop into an industrialized nation and introduce many social, political and agricultural reforms.

After World War I, in which Denmark remained neutral, it granted independence under its own sovereignty to Iceland (which was still a colony) and recovered northern Schleswig after a plebiscite. In April 1940 Nazi Germany invaded Denmark. The Danes, hopelessly outmatched, surrendered after a few hours to avoid unnecessary bloodshed and found themselves under total foreign occupation for the first time in their history. At first the Danes were allowed to manage their own affairs, but in the face of increasing anti-German feeling and sabo-

Tivoli Gardens is one of the most favoured spots in Copenhagen and consists of a large amusement park and areas of recreation; it was opened in 1843.

Bananas are the main crop of farmers in Dominica; the chief variety grown there has a large, pendulous flower which gives rise to small green fruits.

Spanish influence lasting for nearly 400 years is evident in the architecture of many public buildings in the Dominican Republic; most people speak Spanish.

tage, the Germans took over the government in August 1943. The Danes organized a secret Freedom Council to aid the Resistance, who stepped up their activities and also carried out the remarkable operation of evacuating most of the country's 7,000 Jews to Sweden before the Nazis could seize them. The Allies liberated Denmark on 5 May 1945.

After the war Denmark continued its political and economic reforms, abolishing the upper house of parliament in 1953 and further developing manufacturing industries. Denmark was a founder member of EFTA in 1960, but left with Britain to join the EEC in 1973, after a referendum. The mid-1970s, however, saw Denmark struggling with one of the highest rates of inflation in Europe, rising unemployment, and a serious balance of payments deficit and associated economic problems. Map 17.

Denmark – profile

Official name Kingdom of Denmark
Area 43,068sq km (16,625sq miles)
Population (1975 est.) 5,059,000
Density 117 per sq km (304 per sq mile)
Chief cities Copenhagen (capital) (1975 est.) 5,059,000; Århus, 245,212
Government Head of state, Queen Margrethe II (succeeded 1972)
Religion National Lutheran Church
Language Danish
Monetary unit Krone (plural kroner)
Gross national product (1974) £12,558,800,000
Per capita income £2,482
Agriculture Pigs, cattle, barley, potatoes, wheat, oats, sugar-beet
Industries Food-processing, engineering, chemicals, fishing
Trading partners West Germany, Sweden, Britain, Norway, Netherlands, France

District of Columbia. *See* UNITED STATES.

Djibouti, formerly French Territory of the Afars and the Issas, is an independent nation on the east coast of Africa near the southern end of the Red Sea. It is a small country which achieved complete independence from France only in 1977. About 90 per cent of the land is stony desert, allowing little agriculture, and more than half the people follow a nomadic way of life, raising cattle, sheep, goats and donkeys. There is some manufacturing industry in and around the capital Djibouti.

There is an extremely small rainfall (50-500mm; 2-20in a year), allowing a few farmers to grow dates or cultivate market gardens. Temperatures are high, averaging 30°C (85°F) on the coast. Industries include meat packing and the production of salt, hides and skins.

Most of the people are members of the Afars and the less numerous Issas tribes, who frequently disagree politically and economically. Less than 20 per cent of the population is European or Arab, with a few Ethopians and Indians. French is the offical language, although most people speak Afar or Somali, the language of neighbouring SOMALIA.

In 1862 France obtained Obock on the Somali coast, and gradually extended southwards to Djibouti after making agreements with Afar and Issa chiefs. In 1896 France signed with Britain, Italy and Ethiopia treaties that defined the boundaries of French Somaliland. From 1957 to 1967 the colony gradually gained autonomy; in 1967 it voted to remain a French possession and adopted the name the Afars and the Issas. In a referendum in May 1977 the people voted for complete independence and changed the country's name to Djibouti, immediately recognized by Somalia. The new prime minister was Hassan Gouled. Area: 22,000sq km (8,494sq miles). Pop. (1976 est.) 226,000. Map 38.

Dominica is a self-governing island, an associate state of Britain, in the Windward Islands group of the West Indies. The capital and main port is Roseau. Most of the people, descended from African slaves, speak a French dialect, although the official language is English; there are still some Carib Indians. Agriculture is the chief occupation, and farmers produce bananas, citrus fruits, copra and tobacco.

Dominica was discovered by Christopher Columbus in 1493. Following rival French and British claims during the 1700s, the island was made a British possession by the Treaty of Paris (1783). In 1871 Dominica and other islands to the north were formed into the Federation of the Leeward Islands colony. Then in 1940 Dominica was transferred to the Windward Islands group. It achieved self-government in 1967. Area: 750sq km (290sq miles). Pop. (1975 est.) 75,000. Map 74.

Dominican Republic is an independent nation occupying the eastern two-thirds of the island of Hispaniola in the British West Indies (the remainder of the island is HAITI). The capital is Santo Domingo. The mountainous centre of the country includes Pico Duarte (3,175m; 10,467ft), the highest peak in the West Indies. Minerals, particularly bauxite and nickel, are becoming increasingly important exports, although agriculture still dominates the economy. The chief crops are sugar cane, coffee, cocoa, fruits and tobacco, all of which were affected by a severe drought in 1975. The people are predominantly mulatto and most of them are Spanish-speaking Roman Catholics.

Hispaniola was discovered in 1492 by Christopher Columbus who a year later founded the first settlement in the new world at Isabela (now ruined); in 1496 his brother Bartholomew established Santo Domingo as the Spanish capital in the West Indies. By 1697 Spain had ceded the western part of the

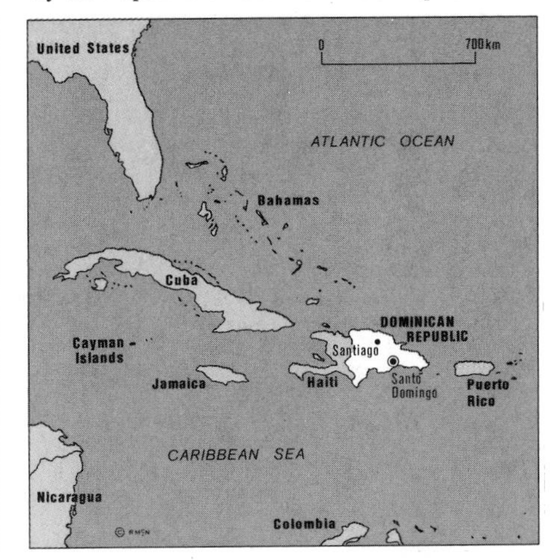

island (Haiti, then called St Domingue) to France, and ceded the rest under the terms of the Treaty of Basel (1795) to be ruled jointly by a French-Haitian administration. Haiti declared independence in 1804 and claimed Santo Domingo but Spanish rule was soon re-established, followed by Haitian. The Dominicans revolted against the Haitians in the 1840s and founded a republic which survived.

The United States took over most of Dominica's financial control from 1905–41 because the country was virtually bankrupt. A military dictatorship was formed in 1930 and war with Haiti nearly broke out in 1937. The dictator (Raphael Trujillo Molina) was assassinated in 1961 and free elections the following year returned Juan Bosch as president. He was overthrown by a military coup d'etat in 1963. Joaquín Balaguer became head of state in 1966 and was re-elected in 1970. Area: 48,442sq km (18,703sq miles). Pop. (1975 est.) 4,697,000. Map 74.

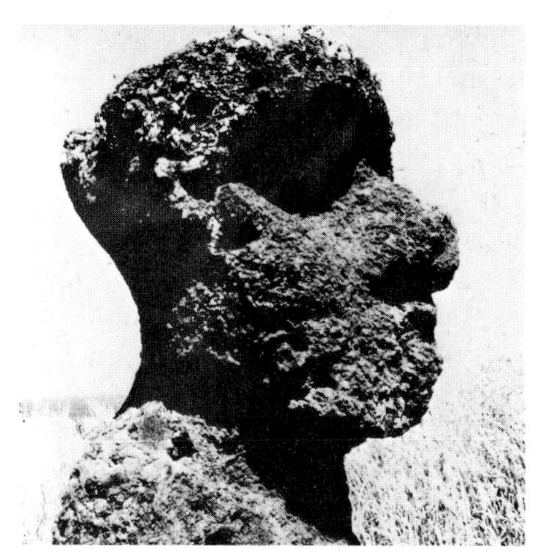

Easter Island is famous for its statues which were carved from tufa, a soft volcanic stone, and date from Polynesian ancestors of about 600 years ago.

The ancient city of Quito, capital of Ecuador, is dominated by its cathedral which was built in the 17th century by Spanish colonists.

Cotopaxi, in the mountains of Ecuador south of Quito, is the world's highest active volcano; its frequent eruptions have caused much damage.

Easter Island (Isla de Pascua) is an island belonging to Chile in the south-eastern Pacific Ocean, about 3,500km (2,174 miles) off the west coast of South America. It was so called because it was discovered on Easter Day in 1772, by the Dutch navigator Jakob Roggeven; it was annexed by Chile in 1888. Most of the island's inhabitants are Polynesian farmers. It is famous for its large stone statues standing up to 12m (39ft) tall, whose origins are still a subject of speculation by anthropologists. Chile has made Easter Island an historic monument. Area: 119sq km (46sq miles). Pop. (1970 est.) 1,600. Map 2.

Ecuador, official name Republic of Ecuador, is a nation in north-western South America. It is a colourful country, bisected from north to south by ranges of the Andes, in whose valleys more than half the people live. It is among the poorest countries of South America, with an economy depending largely on the export of bananas, although recently

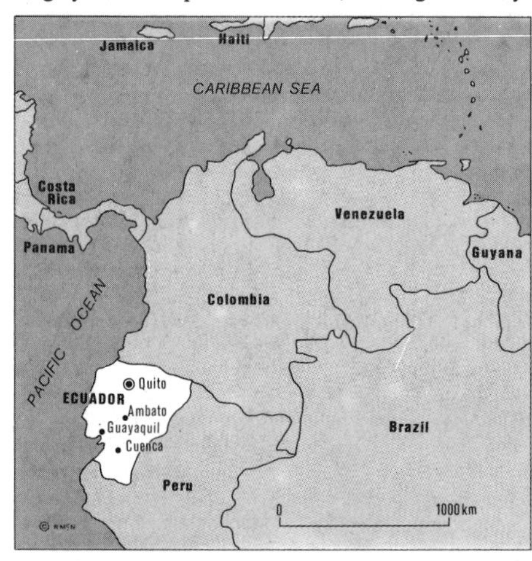

discovered oilfields are being developed. It has a long history of political instability, and in one 23-year period had 22 different presidents or ruling juntas. The capital is Quito, which lies on the Equator (*ecuador* is Spanish for "equator").

Land and climate. Ecuador has a simple geographical division into three distinct regions: a coastal strip (Costa), the Andes mountains (Sierra) and the eastern lowlands (Oriente). The Costa is low-lying and green, and from 80 to 160km (50-100 miles) wide. It is hot and humid, with average temperatures of 28°C (83°F). The majestic peaks of the Andes include about 30 volcanoes; the highest, Chimborazo, rises 6,272m (20,577ft) above sea-level. Cotopaxi, at 5,896m (19,344ft) the highest active volcano in the world, stands 65km (40 miles) south of Quito. The climate in this region alters with altitude. Quito has a mild climate because it lies among the mountains at an altitude of 2,750m (9,022ft). The largely uninhabited Oriente (also claimed by Peru) is a tropical lowland, a jungle and forest region watered by several large tributaries of the River Amazon. The GALAPAGOS ISLANDS, famous for their unique wildlife, belong to Ecuador; they lie in the Pacific Ocean, about 1,000km (620 miles) off the coast.

Economy. Bananas account for 46 per cent of Ecuador's exports, and another 36 per cent is shared by coffee and cacao. Rice and sugar are also exported, and crops grown on the Costa and Sierra for local use include potatoes, maize, barley and wheat. Cattle and sheep provide dairy products, meat and wool. The discovery and development of oilfields in the Oriente region in the late 1960s offers considerable possibilities not only to satisfy the country's own needs but also to provide a much-needed boost in exports. The vast forests are rich in valuable timber but are undeveloped commercially, although Ecuador is the world's leading producer of balsa wood.

People. About 40 per cent of the people are American Indians, 40 per cent mestizo (of mixed Indian and European descent), 10 per cent white (of Spanish descent) and 10 per cent Negro or mulatto. Spanish is the official language, but a large number of Indians neither speak nor understand it. The Indians who live in the highlands are descended directly from the Incas, and still follow the customs of their ancestors and speak the Quechuan language. Because of the geography, they are separated from those that live in the Oriente – said to be some of the fiercest left in the world today. It is said that no member of the Aucas has ever become civilized, and the Jívaros are renowned as the head-shrinkers of the Amazon.

There is no state religion and there have been long periods of anticlerical rule, yet Ecuador is one of the most predominantly Roman Catholic countries in the world. Less than one per cent of the population owns more than half the land; most of the people are poor and in the provincial areas scrape a bare living from the soil. About 60 per cent of the people live in the highlands and most of the remainder in the coastal lowlands. Communications between the regions are poor. The Pan American Highway runs through the country from north to south, but there is little contact between peoples of the three regions.

Government. Nominally a presidential democracy, Ecuador has such a history of political instability that no 20th-century president had served his whole four-year term until Galo Plaza Lasso (1948-52). It has had several constitutions, mostly based on a congress with a Senate and a Chamber of Deputies. But government is frequently by dictatorship, either presidential or military.

History. Indian civilizations inhabited parts of what is now Ecuador 2,000 years before the Incas conquered the kingdom of Quito in about 1470, and made it their northern capital. The Spaniards under Francisco Pizarro overthrew the Inca Empire in 1533, and Ecuador became a Spanish colony in 1534. It was ruled as part of New Granada (now Colombia) under the viceroy of Peru. Revolt against Spanish rule grew strong in the early 1800s, and the country won its independence in 1822 when the Spaniards were routed at the Battle of Pichincha by Marshal Antonio José de Sucre. Ecuador was united with Colombia and Venezuela in the Republic of Greater Colombia by Simón Bolívar, but broke away in 1830 to become a separate republic. Weak government and border disputes have plagued Ecuador and slowed its development. Neighbouring countries have seized land – Brazil in 1904, Colombia in 1916 and Peru in 1942. The dispute with Peru is still bitterly resented in Ecuador. The constitution of 1946 was suspended following a military coup in 1963. But the four-man junta set up was itself overthrown in 1966, giving way to a temporary president, and a new constitution was announced in 1967. In 1968 José María Velasco Ibarra, who had been elected for four previous terms (serving a total of eight years) and had withdrawn into exile in 1961, was elected president for the fifth time. Following continual trouble between students and security forces, he assumed dictatorial powers in 1970 but was deposed again in 1972. A National Military Government was formed under army commander-in-chief Brig.-Gen. Guillermo Rodriguez Lara, but he in turn was replaced by a three-man military junta in 1976. Map 76.

Ecuador – profile

Official name Republic of Ecuador
Area 270,670sq km (104,506sq miles)
Population (1976 est.) 7,305,000
Density 26 per sq km (67 per sq mile)
Chief cities Quito (capital) (1974) 597,000; Guayaquil, 814,100
Government Military junta (assumed power January 1976) led by Vice-Admiral Poveda Burbano
Religion Roman Catholicism, ethnic
Language Spanish
Monetary unit Sucre
Gross national product (1974) £1,367,500,000
Per capita income £187
Agriculture Bananas, cacao, coffee, cereals, sugar, vegetables, cattle
Industries Forestry (balsa wood, kapok, rubber), chemicals, hats, nails, soap
Minerals Petroleum, gold
Trading partners (major) USA, Japan, West Germany

Egypt, official name Arab Republic of Egypt, is a country located most in north-eastern Africa, but with part (the Sinai Peninsula) in south-western Asia. Egypt is often called the "land of the desert

Alexandria, Egypt, dates from 332 BC and is named after its founder, Alexander the Great. Its harbours are among the finest in the Mediterranean.

The Sphinx is a mythical beast of ancient Egypt; thousands of sphinxes were built and the most famous one is the Great Sphinx at Al Jizah (Giza).

The Colossus of Memnon is situated near the temple of Luxor, the greatest monument of antiquity in the city; the name was derived from Greek mythology.

and the river" – the Sahara desert, which covers most of the country, and the River Nile, close to which most of the people live. It was on the banks of the Nile that a great civilization grew up 5,000 years ago. The ancient Egyptians established a strong empire noted for its scientific and cultural achievements, to which the pyramids still stand as monuments. Most of the people, however, have always been poor, scraping a living from the fertile soil of the Nile, whose flow is now regulated by the Aswān High Dam. Modern Egypt, capital Cairo, is a socialist state struggling to modernize itself. Since 1948 it has been in continual conflict with its neighbour ISRAEL.

Land and climate. The Nile divides the deserts of Egypt into two: the Libyan Desert to the west and the Arabian Desert to the east. The Sinai Peninsula, across the Gulf of Suez, is also desert and has the highest point in Egypt, Jabal Katrīnah (2,637m; 8,652ft) [*see* PE p. *52*]. Almost all of Egypt's farm-

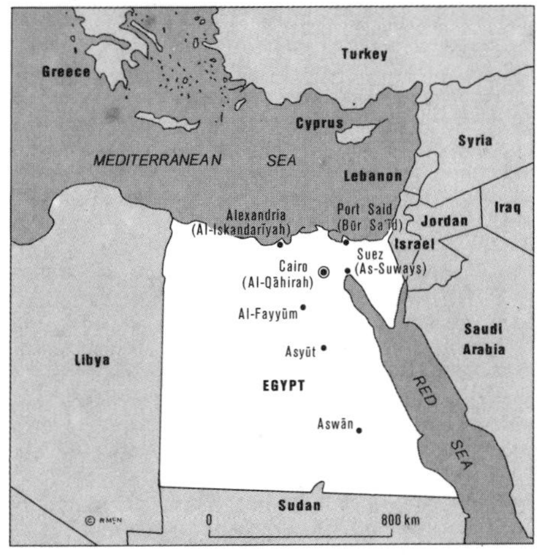

land lies in the vicinity of the Nile, which for the whole of its course through Egypt (about 1,545km; 960 miles) has no tributaries. The southern part of the Nile valley has been turned into the huge Lake Nasser by the Aswān High Dam [*see* MM p. *198*]. Just downstream of Cairo the Nile enters its delta, dividing into several branches [*see* PE p.*46*]. Egypt has hot summers, with clear skies, bright sunshine and daytime temperatures of 30-40°C (86-104°F), which fall dramatically at night. The winters are mild and in the spring there is an occasional hot, dry, southerly wind known as the *khamsin*. There is an almost total absence of rain.

Economy. Most of Egypt's working population is engaged in or dependent on agriculture. Cotton is the chief crop and raw cotton accounts for nearly half of all exports; cotton yarn and goods make up a further 14 per cent. Other crops include sugar cane, maize, rice (6-7 per cent of exports) and wheat. The USSR is Egypt's chief customer. Egypt's oilfields, on each side of the Gulf of Suez, are small by Middle East standards. But production was rising in the mid-1970s and a major exploration effort was mounted in 1976. Another major source of foreign earnings are tolls charged to vessels passing through the Suez Canal. Many tourists go to Egypt to see the ancient remains, and these provide another source of income [*see* SU p.*21*]. Imports, which include cereals, machinery, equipment and chemicals, are chiefly from the United States, France and the USSR. The economy is conducted largely on socialist lines. Nearly all cultivation is carried out by peasant farmers, and no family may own more than 100 *feddâns* (42 hectares, or 103.8 acres). There has been a major effort to step up industrialization, and most industries have been nationalized.

People. Nearly all the people live in the Nile valley or delta, or on oases – that is, on about 3½ per cent of Egypt's total land area. There are three distinct groups of people: the fellahin, peasant farmers mostly descended from the original northern inhabitants; the bedouin, nomadic Arabs of the deserts; and the Nubian people of the southern Nile valley, of mixed Arab and Negro descent. The fellahin are the largest group. They are poor people living in crowded conditions, who spend most of their lives working. Only about a seventh of the bedouin are real nomads; the rest are tent-dwellers who live on the outskirts of the cultivated Nile valley. City life in Egypt is much like rural life, with cramped living conditions. Many fellahin have moved into the towns to find work. The cornerstone of Egyptian social life is the family which, particularly in the villages, is extended to include grandparents, uncles, aunts and cousins. Women and children are expected to treat men with great respect.

About 98 per cent of the people speak Arabic, the official language. The Nubians speak their own African languages, which have no written forms. The religion of Egypt is Islam; more than 90 per cent of the people are Sunni Muslims. There are also more than a million Christians, mostly Copts. For centuries Islam has been closely linked with society and the law in Arab countries, but since the early 19th century Egypt has led the movement to loosen these ties, a movement that has accelerated since the 1952 revolution. Nevertheless Islam is still the major force in the lives of most Egyptians, particularly in rural areas.

Education is free at all levels, and compulsory between the ages of six and 12. Illiteracy, which was 93 per cent at the beginning of the century, had been reduced to 65 per cent by 1970. Egypt's three leading universities – Cairo, Ain Shams (the state university, in Cairo) and Alexandria – are among the largest in the world, with more than 50,000 students each. The government has played an active part since 1952 in encouraging the development of the theatre, the visual arts and sport. The most popular sports include soccer, basketball and boxing. Backgammon is also a favourite pastime.

Government. The constitution of 1964 defines Egypt as a democratic socialist state. A People's Assembly of 360 members, half of whom must be workers or peasants, is elected by universal suffrage for five-year terms. The president of the republic is nominated by the Assembly and confirmed by plebiscite for a six-year term. He may appoint up to ten additional members to the Assembly. He holds executive power and appoints the prime minister. There is only one political party.

History. Egypt's history dates back to about 3200 BC, when the already highly civilized kingdoms of Upper and Lower Egypt were united by King Menes. A great culture was developed by the rulers of what is known as the Old Kingdom, the 3rd to 6th dynasties of pharaohs. Their capital was Memphis. Local rule replaced central government from about 2258 to 2050 BC, when Egypt moved into a new period of stability (the middle Kingdom, with a capital at Thebes). The 18th Dynasty marked the beginning (1570 BC) of the New Kingdom, the golden age of Egypt's imperial power.

At its height, under such empire-builders as Rameses II (reigned 1304-1237 BC), Egyptian rule reached the River Euphrates. But from about 1200 BC the empire began to decline. It was conquered by the Persians in 525 BC and then in 332 BC by Alexander the Great, who founded Alexandria as the capital. One of his generals, Ptolemy, founded the dynasty that took his name. The last of the Ptolemies was Cleopatra, who failed to prevent Rome annexing Egypt in 30 BC. It came under the rule of the Eastern Empire of Byzantium when the Roman Empire was divided (AD 395), and in 642 was conquered by the Arabs, who introduced Islam.

Islamic rule lasted for more than 1,150 years, during which time Egypt was governed by several different powers: the Fatimid Dynasty (969-1171), who built a new capital, Al-Qāhirah (Cairo); the Aiyubid Dynasty (1171-1250), whose founder, Saladin, fought the Crusaders; the Mamelukes (1250-1517), who were originally slaves and saved Egypt from the Mongols as well as the Crusaders; and the Ottoman Turks (1517-1805), who added Egypt to their empire while allowing the Mamelukes local control. In 1798 the French under Napoleon Bonaparte invaded Egypt and captured Alexandria, but were forced out in 1801 by an Anglo-Turkish alliance. An Albanian mercenary with the Turkish army, Muhammad Ali, seized power in 1805, and this event marked the birth of modern Egypt.

The building of the Suez Canal, opened 1869 [*see also* MM p.*196*] plunged Egypt into debt and led to British intervention, occupation (1882), and the appointment of a consul-general (1883-1907). Britain made Egypt a protectorate during World War I. Nationalism grew and the British granted Egypt independence as a kingdom in 1922, eventually withdrawing its troops in 1936 (except from the Canal Zone). In World War II, British troops had to surround the palace to ensure King Farouk's co-

Egypt is predominantly an agricultural country, especially in the Nile valley and delta where the principal crop (cotton) is produced.

The Houses of Parliament, the legislative seat of the United Kingdom, has many historical features including the Victoria Tower – 102m (336ft) high.

Aldeburgh, once a flourishing fishing and ship building centre, is now known for its world famous music festival held there every June.

operation and eventually won their decisive North African victory at El Alemein (1942).

A military coup staged by a movement called the Free Officers and led by Col. Gamal Abdel Nasser deposed Farouk in 1952 and declared Gen. Muhammad Neguib president of the new republic (1953). Nasser ousted Naguib in 1954 and became president. British troops withdrew from the Canal Zone in 1956, and Nasser nationalized the canal. He prevented Israel from using the canal in October 1956; Israel invaded the Gaza Strip and Sinai, and Britain and France bombed Egyptian bases, before the United Nations brought the fighting to an end. Nasser became the leader of a movement to unite Arab nations and set up the short-lived United Arab Republic (UAR) with Syria (1958-61). A major achievement during his presidency was the building of the Aswān High Dam, mainly using Soviet financial and technical aid [*see* MM p.*198*]. Nasser's continual confrontation with Israel led to the Six-Day War (1967), in which Israel inflicted a crushing defeat on the Arab countries and captured the GAZA STRIP, the eastern bank of the Suez Canal and Sinai, and the canal was blocked by sunken ships.

Nasser died in 1970 and was succeeded by Anwar as-Sadat. Sadat resented Soviet attempts to dictate Arab strategy and expelled their military mission from Egypt in 1972. Another fierce war with Israel (1973) was inconclusive, and Sadat decided to put faith in the "step-by-step" diplomacy of Henry Kissinger, the American secretary of state. The Suez Canal was reopened in 1975. Aid from other Arab nations and from the United States helped Egypt through a crisis in 1976 caused by inflation. Map 32.

Egypt – profile

Official name Arab Republic of Egypt
Area 1,001,449sq km (386,659sq miles)
Population (1976 est.) 38,067,000
Density 38 per sq km (98 per sq mile)
Chief cities Cairo (capital) (1976 est.) 6,133,000; Alexandria, 2,259,000; Giza, 853,700; Suez, 368,000
Government Head of state, Anwar Sadat, president (elected 1970)
Religion Islam
Language Arabic (official)
Monetary unit Egyptian pound
Gross national product (1974) £4,312,000,000
Per capita income £113
Agriculture Cotton, sugar cane, maize, rice, wheat, livestock
Industries Petroleum, chemicals, textiles, iron and steel
Minerals Petroleum, natural gas, iron ore, phosphates, coal
Trading partners (major) USSR, USA, France

Éire. *See* IRELAND.

El Salvador, official name Republic of El Salvador, is a nation on the south-western coast of Central America bordered by Guatemala to the west and Honduras to the north and east. The capital is San Salvador and most of the people, of mixed European and American-Indian descent, are Spanish-speaking Roman Catholics.

Land and economy. Two mountain ranges cross the country from east to west and create fertile upland plains. The tropical climate is modified by the elevation. Agriculture is the chief occupation, with coffee and sugar cane as the main crops. Industrialization is progressing rapidly and cotton cloth is produced for export.

Government. El Salvador is a republic, with a president who is elected every five years but is ineligible for immediate re-election; a National Assembly of Deputies elected for two-year terms; and a Supreme Court and lesser courts. Everyone over the age of 18 may vote.

History. Spain conquered the country in 1524 and established the first permanent settlement. El Salvador won its independence from Spain in 1821 and came under Mexican control until 1823, when it became a member of the Central American Federation. It regained its autonomy in 1839 but since World War II has been politically unstable, with frequent coups and revolutions and a rapid succession of presidents. Area: 21,393sq km (8,260sq miles). Pop. (1976 est.) 4,240,000. Map 74.

England, official name Kingdom of England, is the largest nation within the UNITED KINGDOM of Great Britain and Northern Ireland. It has a long history of freedom and independence – the last successful invasion was by the Normans in 1066. England became a great sea power in the 16th century, and sent out explorers and traders, missionaries and colonists to all parts of the world and helped to build the largest and most far-flung empire in history. As a result, the English language is one of the most widely spoken tongues in the world today (after Chinese), being the chief language of countries as far apart as the United States, Canada, South Africa, Australia and New Zealand. In 1976 the population of England was 44.6 million (about 78 per cent of the total United Kingdom population); the capital, London, had a population of 7,028,200.

England has produced many of the world's great scientists, from Isaac Newton, who ushered in the English Age of Reason, to the inventors who made possible the Industrial Revolution. England has also produced an extraordinary number of great writers – poets, playwrights, novelists and historians – including William Shakespeare, whose plays are renowned, studied and translated throughout the world. England's form of government and legal system have been adopted by many other countries, and its people's ideals of democracy have had worldwide influence. England was the first country to become highly industrialized, and as the nucleus of Britain became one of the world's leading nations, both economically and politically.

Land. England covers an area of 130,362sq km (50,333sq miles). It has a complex landscape, which in general slopes from the north and west down to the south and east. The variety and richness of the landscape is remarkable in a land of such a relatively small area. The mountains of the north and west are geologically old. There are three distinct highland areas: Cumbria, or the Lake District [*see* PE p.115] in the north-west; the Pennine Chain, which extends from the Scottish border to nearly half-way down the country and is often called the "backbone of England"; and the south-western peninsula. The lowland areas, in contrast, are made up of alternate scarps and vales; limestones and clays make up the rest of the country. In general, the highland regions are wetter and colder than the lowlands.

The Lake District is the wettest region in England. An ancient mountain area with deep valleys rises to England's highest point, Scafell, which stands 978.5m (3,210ft) above sea-level. The individual mountains tower above lakes such as Windermere and Derwentwater, providing England's most scenic tourist attraction. Across the Eden valley from the Lake District stand the Pennines, which extend farther south, flanked by plains. The Cheviots, to the north, are separated from the Pennines proper by the Tyne valley and extend beyond the Scottish border. The south-western peninsula of Cornwall and Devon rises to more than 600m (2,000ft) on Dartmoor, a high moorland region.

Lowland England comprises the Midlands, East Anglia and the scarplands of the south-east. The Midlands, the geographical heart of England, lie south of the Pennines and consist of a large plateau broken by rolling hills and fertile valleys. They lie in the centre of a large drainage system (the River

The 12th century castle at Alnwick stood in ruins for 200 years before it was restored by the Dukes of Northumberland in the 18th and 19th centuries.

Beachy Head attracts many tourists during the summer season; the lighthouse at its foot sends its beam 10km across the English Channel.

A part of Camden on London's Regent's Canal, which consisted of derelict warehouses, has now been restored and has art galleries and restaurants.

Severn on the west, Trent on the north, Ouse on the east and Thames on the south). South of the Midlands region Salisbury Plain is the centre of the downs, a system of chalk hills, which extend westwards to reach the sea at Dorset. Eastwards, the North and South Downs reach the sea at Dover and Beachy Head. The chalk hills north of the downs include the Chilterns and the East Anglian Heights, east of which is East Anglia. This is the most extensive arable region in England, covered as it is largely with fertile glacial drift. Between the Chilterns and the North Downs lies the Thames, or London, Basin, which is now dominated by the urban spread of Greater London.

The Thames is England's longest river, flowing 338km (210 miles) from the Cotswolds to its estuary on the North Sea [*see also* MM pp.202-3]. The Severn, which is the longest river in Britain, flows for 354km (220 miles) in a great curve from its source in Wales, across the English border to its outflow in the Bristol Channel.

England has several offshore islands including the Isle of Wight, just off the Hampshire coast, and the Scilly Islands, a group of 40 small islands in the south-west about 40km (25 miles) from Land's End.

Climate. The most striking feature of England's climate is its extreme variability. In general it is mild with few extremes, although long periods of settled weather are uncommon. Westerly winds may bring mild weather from the Atlantic Ocean at any time. England lies in a low-pressure belt between the Arctic and the Tropical areas of high pressure. When air masses from these meet over the Bay of Biscay, the resulting depressions tend to move in a north-easterly direction, crossing central England. A mild, windy and wet period is followed by a warm, dull spell or by cool, squally showers. Sometimes high-pressure areas from the Azores expand to cover the whole of England, producing clear and sunny weather. England's coasts often experience gales, and fog is common inland.

Average temperatures in winter vary little over the country. In January, generally the coldest month, temperatures average about 4.5°C (40°F) in both London and the north. The south is warmer than the north in summer, with a July average of 17°C (63°F) as compared to 15.5°C (60°F).

Rainfall is evenly distributed throughout the year, with more falling in the west than in the east. It also increases with altitude. Much of the rain comes in long spells of steady drizzle associated with the warm fronts of depressions. The annual average ranges from 500mm (20in) in parts of the south-east to more than 2,500mm (100in) on the mountains of the Lake District.

Natural resources. About three-quarters of England's land area is farmland. With the exception of coal, England is poor in natural resources. The largest deposits of coal are found along the Pennines, particularly in Durham, the Yorkshires and Nottinghamshire. Natural gas fields are being exploited in the Yorkshires and on the North Sea coast. There are also substantial deposits of low-grade iron ore in the Midlands. Various materials are mined in the south-west, including china clay, used in the paper and pottery industries, limestone for cement, and tin and copper ores. The shallow coastal waters provide excellent fishing grounds, particularly in the North Sea.

Agriculture. About 40 per cent of England's land is arable, another 25 per cent permanent pasture, and 10 per cent rough grazing land. The chief crops are barley and wheat, most of which are grown in the drier, eastern part of the country. Other important crops include potatoes, fodder crops (turnips, swedes) sugar beet and oats. Most of the farms average about 45 hectares (111 acres), and many owners practise mixed farming. Britain uses more than 500,000 tractors and 60,000 combine harvesters.

Market gardening is an important factor in England's agricultural economy, with growing areas near most of the large conurbations. This specialized cultivation of vegetables and soft fruits is generally practised on smallholdings, although in some areas (such as the London Basin) holdings range from 20 to 120 hectares (50-300 acres) and are completely mechanized.

Livestock products make the main contribution to England's agricultural output. Dairying is the cornerstone of the agricultural industry and is practised throughout the country, particularly along the lowlands in the west Midlands; beef cattle are also raised. Hill sheep are bred in the southern uplands, the Lake District, the Pennines and the south-western moors. A few pigs and poultry are kept on most farms and are raised in large numbers on some specialized holdings [*see also* PE pp.224-237].

Fishing. Britain's sea-fishing industry is among the most important in the world, and British fishing vessels range over a wide area. The principal grounds include the North and Irish seas, the Faeroes, off Iceland and as far afield as Labrador and Newfoundland. About 950,000 tonnes of fish are landed each year, mainly herring, cod, haddock, plaice and hake; 61,000 tonnes of shellfish are caught, principally oysters, crabs and lobsters. England's chief herring ports are Lowestoft and Great Yarmouth, whereas Hull, Grimsby and Fleetwood are the main trawler ports for white fish.

Industry. Most of England's manufacturing industries grew up in the coalfield areas. With the increasing use of oil, gas and nuclear energy, new industries developed in the south-east, particularly in the London area. Today engineering industries are the most important single sector, and these include shipbuilding and marine engineering, aircraft, motor vehicles, textile machinery, electrical engineering and electronics products. The change from iron to steel as the chief shipbuilding material in the 1890s led to the development of shipyards close to steel-producing centres such as those near the rivers Tyne, Wear and Tees. The industry was later also established at Birkenhead and Barrow.

The motor-vehicle industry, like shipbuilding, is an assembly industry and is controlled by a few large companies, supplied by specialized component manufacturers. Britain's main centres of the motor industry are in England, particularly in the West Midlands (Birmingham, Coventry, Wolverhampton), Dagenham, Luton and Oxford. The government has also encouraged the development of a car-manufacturing centre on Merseyside. The materials for the industry, such as sheet steel and other metals, are produced in Sheffield, Birmingham, Tees-side, Cheshire, Wales and Scotland. Tractors are made in several places, chiefly Coventry and Basildon.

The chief centres of the aircraft industry are in the Midlands, Greater Manchester and the south-west (Bristol), where engines and parts are manufactured, and in the east and south, where there is space available for assembly and testing. English textiles have always enjoyed a high reputation. The woollen industry is dominated by West Yorkshire and cotton by Lancashire. Man-made fibres such as rayon, nylon and Terylene have a much wider distribution. Allied to the textile industries are the manufacture of clothing, knitwear and carpets, which use both natural and man-made fibres. England also has important chemical industries, including heavy chemicals (alkalis, acids and salts), petro-chemicals and plastics. Glass, pottery, rubber and paper are also manufactured.

Mining. Coal accounts for about 90 per cent of Britain's mineral production. In the mid-1970s

The iron bridge at Coalbrookdale was the first of its kind to be built; the semicircular arch measures 43m (140ft) and it was built in 1777–79.

Dartmoor is a wild picturesque area in the county of Devon which attracts many tourists during the summer; a civilian prison is sited there too.

The largest public school in England is Eton College, founded in 1440, where the famous wall game (a form of football) takes place.

about 125 million tonnes of coal were being produced annually, nearly all by the 250 National Coal Board mines (this represented, however, less than half the production of 60 years before). Britain's domestic gas supplies were converted to the use of natural gas in the 1970s, most of it coming from the North Sea gasfields, which began production in 1967. The North Sea oilfields, with their vast reserves, began production in the mid-1970s. Most of England's high-grade iron ore has been worked out.

Trade and commerce. England has the most highly developed economy in Britain and produces most of its industrial and farm products. England's trade, however, may not be considered individually, but should be regarded as part of Britain's economy as a whole. Britain is one of the world's leading trading and manufacturing nations. It is poor in natural resources and does not produce enough food to meet all of its requirements. As a result, foreign trade is of paramount importance, and Britain's economy relies heavily on exports.

Britain's chief exports are machinery and electrical appliances (29%), transport equipment (12%) and chemicals (11%). Its chief imports, broadly classified, include machinery and transport equipment (19%), mineral fuels and lubricants (18%), food (16%), raw materials such as wood, wood pulp, ores and fibres (8½%), and other manufactured goods. Britain's imports have nearly always been greater in value than its exports – by as much as 40 per cent in 1974, a figure that was soon reduced to 20 per cent. The country is able to reduce this balance of payments deficit somewhat with earnings from "invisible" exports, chiefly services such as banking, insurance and tourism.

Britain trades mainly within the EEC, which it joined in 1973, and with the United States. Exports to EEC countries make up 32 per cent of Britain's total, the leading customers being West Germany (6½%), France (6%), The Netherlands (5½%), Ireland (4½%) and Belgium and Luxembourg (4½%). The United States is the biggest single customer (9%); EFTA countries still account for a large share (13%) and so does the Commonwealth (16%). The pattern of imports is similar, with the EEC accounting for 37 per cent (West Germany 8¼%, The Netherlands 7¾%, France 6¾%). Again the biggest single supplier is the United States (9½%); EFTA countries supply 12½% and Commonwealth countries 13½%. In addition large quantities of oil are still imported from Arab countries.

Transport. In the mid-1970s Britain's merchant fleet of more than 3,000 registered ships represented nearly 10 per cent of the world's total tonnage, third in size after Liberia and Japan. There are about 300 ports, of which London, Liverpool and Southampton are the busiest. Britain has 150 civil and 50 military airports, and ranks second only to the United States in the number of passenger-kilometres flown. The chief airline is the state-owned British Airways, and there are about 30 private airlines [*see also* MM p.186]. Hovercraft services operate on the coasts and across the English Channel. Britain has more than 18,500km (11,500 miles) of railways, mostly state-owned, and more than 330,000km (205,000 miles) of roads, including 1,879km (1,168 miles) of motorways. About 90 per cent of passenger travel and 75 per cent of freight movement is by road. [*See also* MM pp.184, 192-3].

Communications. Britain has the world's third-highest newspaper circulation per head of population, and some of the highest circulation figures for individual newspapers. The nine leading daily papers have a combined circulation of more than 14 million, including 4 million for the *Daily Mirror* and 3.4 million for the *Sun*. The *News of the World*, a Sunday paper, has a circulation of 5.6 million [*see also* MM pp.210-11].

Radio and television services are controlled by two bodies, the British Broadcasting Corporation (BBC) and the Independent Broadcasting Authority (IBA), both public corporations independent of the government. The BBC is financed mainly by revenue from television licences (which are compulsory) and from selling programmes overseas. In 1976 there were 17.8 million television licences in force, including 8.6 million for colour receivers. The BBC, which inaugurated a regular television service in 1936, broadcasts two national television channels, an information channel (Ceefax), four radio channels and has several local radio stations. The IBA is responsible for one commercial television channel (shared by several regional companies) and a number of local commercial radio stations. Commercial broadcasting is financed by advertising revenue and the sale of programmes overseas; there is no direct sponsorship. Britain has 19 million telephones, more than any other country except the United States and Japan. The internal telephone and telegraph services are run by the Post Office, which also carries some 11,000 million items of mail a year. The world's first public postal systems originated in England in the 15th and 16th centuries, and Britain issued the first postage stamps in 1840.

Government. Britain is a constitutional monarchy with a parliamentary government, as described in the article on UNITED KINGDOM. England has the status of a kingdom within the United Kingdom. Of the 635 members elected to the House of Commons, 516 come from English constituencies. Members of Parliament serve terms of up to five years. In 1974, the total United Kingdom electorate was 39,798,899, of which 32,769,792 lived in England. The minimum voting age is 18; women received equal voting rights in 1928.

Legislation may be initiated in either the House of Commons or the House of Lords, but usually originates in the Commons. Each bill has three readings in the Commons before it is referred to the Lords, which may return it with amendments or suggestions. In practice the Lords can delay a bill, but cannot prevent its becoming law after it has been passed three times by the Commons. Executive power is vested in the cabinet, which is headed

English Counties

County	Area sq km	[sq miles]	Population (1976)
Avon	1,346	[520]	920,200
Bedford	1,234	[476]	491,700
Berkshire	1,255	[485]	659,000
Buckinghamshire	1,882	[727]	512,000
Cambridgeshire	3,409	[1,316]	563,000
Cheshire	2,329	[899]	916,400
Cleveland	583	[225]	567,900
Cornwall	3,546	[1,369]	407,100
Cumbria	6,808	[2,629]	473,600
Derbyshire	2,631	[1,016]	887,600
Devonshire	6,711	[2,591]	942,100
Dorset	2,654	[1,025]	575,800
Durham	2,436	[941]	610,400
Essex	3,674	[1,419]	1,426,200
Gloucester	2,642	[1,020]	491,500
Hampshire	3,782	[1,460]	1,456,100
Hereford and Worcester	3,926	[1,516]	594,200
Hertford	1,634	[631]	937,300
Humberside	3,512	[1,356]	846,600
Kent	3,732	[1,441]	1,448,100
Lancashire	3,040	[1,174]	1,375,500
Leicestershire	2,553	[986]	837,900
Lincoln	5,886	[2,273]	524,500
London, Greater	1,580	[610]	7,028,200
Manchester, Greater	1,284	[496]	2,684,100
Merseyside	646	[249]	1,578,000
Norfolk	5,356	[2,068]	662,500
Northampton	2,367	[914]	505,100
Northumberland	5,033	[1,943]	287,300
Nottinghamshire	2,164	[836]	977,500
Oxfordshire	2,612	[1,008]	541,800
Salop	3,490	[1,347]	359,000
Somerset	3,450	[1,332]	404,400
Staffordshire	2,716	[1,049]	997,600
Suffolk	3,807	[1,470]	577,600
Surrey	1,679	[648]	1,002,900
Sussex, East	1,795	[693]	655,600
Sussex, West	2,016	[778]	627,400
Tyne and Wear	540	[208]	1,182,900
Warwick	1,981	[765]	471,000
West Midlands	899	[347]	2,743,300
Wight, Isle of	381	[147]	111,300
Wiltshire	3,481	[1,344]	512,800
Yorkshire, North	8,309	[3,208]	768,500
Yorkshire, South	1,561	[603]	1,318,300
Yorkshire, West	2,039	[787]	2,072,500

Evesham, with its quaint shops and narrow passage-ways, is the centre of the Vale of Evesham, which is known for its fine market gardens.

The seaside resorts of Kent attract thousands of tourists every summer; one of the favourite pastimes is the old English game of bowls.

Lavenham is one of the most picturesque towns in Suffolk, with fine medieval timber houses; centuries ago it was headquarters of the local wool trade.

by the prime minister. (For a list of British prime ministers since 1714, *see* UNITED KINGDOM).

English local government is run by county councils and district councils. There are 39 non-metropolitan counties, six metropolitan counties, and Greater London, which has a different structure (it is divided into 32 boroughs). Within the counties there are 296 non-metropolitan and 36 metropolitan districts. County and district councillors are elected by their local electors every four years, and annually elect one of their number as chairman (in a district with city or borough status, the title is mayor, or in some places lord mayor). The relationship between the various types of council is not hierarchical, but one of specialization. Local government is financed by a system of local rates (levied as a property tax) and grants from the central government.

For more efficient administration, the counties of England were reorganized in 1974, resulting in the creation of the six metropolitan counties and of Avon, Cleveland, and Humberside. Some of the traditional counties were absorbed, amalgamated or changed in name to form the following: Cumbria (Cumberland, Westmorland and Furness area of Lancashire), Hereford and Worcester, Cambridgeshire (Huntingdonshire and the Soke of Peterborough; Cambridgeshire and the Isle of Ely), Leicestershire (absorbed Rutland), Salop (Shropshire). Smaller boundary adjustments were made between certain other counties, some counties were subdivided (Sussex and Yorkshire) and some dropped the suffix "shire" from their names (e.g. Hertford, Lincoln). The county of Middlesex was abolished in 1965, when most of it was absorbed in the new area of Greater London, although it still exists, for example, as a cricket county and a postal area. A full list of today's counties is given in the accompanying table.

Judiciary. English common law has formed the basis of the legal systems of most of the English-speaking world. Two factors help to ensure a fair trial: the independence of judges (who are appointed for life or until retirement and are outside the control of the executive) and the participation of a jury of 12 private citizens in all important criminal, and some civil, cases. A majority verdict (at least ten to two) has been in operation since 1967. The death penalty was abolished in 1965. The legal systems of Scotland and Northern Ireland differ in some respects from that of England and Wales.

There are three sources of law as administered in the law courts: statute law, common law and equity. The various courts of law in England and Wales include the magistrates' courts, or petty sessions, which are criminal courts of the lowest jurisdiction presided over by unpaid laymen (justices of the peace) who are empowered to try non-indictable offences and to commit criminal offenders for trial. County courts try most civil actions and are presided over by a single, paid judge. Crown courts, which sit at various centres and try criminal cases, are presided over by high court judges, circuit judges or recorders, according to the status of the court. Decisions of lower courts may be appealed against in higher courts, the ultimate court of appeal being the House of Lords. The strength of the police in England and Wales in the mid-1970s was about 100,000 (including nearly 5,000 women).

Armed forces. In the mid-1970s the British armed forces had a regular complement of about 350,000 made up of 175,000 Army, 98,000 Royal Air Force, 70,000 Royal Navy and 7,000 Royal Marines. The Army personnel include about 5,000 women and there are in addition about 74,000 people in the Territorial and Army Volunteer Reserve. All servicemen and servicewomen are volunteers, serving for periods of up to 22 years. Britain is a member of NATO [*see* p.190] and spends approximately £5,000 million a year on defence. Its armoury includes nuclear weapons, which can be delivered by aircraft or by land-based or submarine-launched missiles.

The Army in Britain is organized in district commands; there are overseas commands for the Near East, Hong Kong and the British Army of the Rhine (BAOR), with additional garrisons in Belize and Gibraltar. British troops in Northern Ireland are under direct control of the Ministry of Defence.

There are three home commands of the Royal Air Force – Strike, Training and Support – and overseas commands include RAF Germany, with smaller units in Gibraltar and Hong Kong. In 1975 the RAF had about 2,000 aircraft, including 200 helicopters. Among the types flown by Strike command are Vulcan bombers; Buccaneer strike aircraft; Harrier, Jaguar, Lightning and Phantom fighters; Belfast and Hercules transports; and Puma and Wessex helicopters. Bloodhound surface-to-air missiles are available for defence.

The number of Royal Navy vessels has decreased steadily in recent years – for instance, 11 cruisers were sold or scrapped between 1959 and 1967. In 1975 there were about 100 warships, including one aircraft carrier, two command carriers, two cruisers (helicopter carriers), ten destroyers, 30 submarines and 58 frigates, and about 250 support and maintenance vessels.

People. Most English people are descended from peoples who invaded the British Isles from the mainland of Europe between the 8th century BC and the 11th century AD, including Celts, Romans, Angles, Saxons, Jutes, Danes and Normans. England is the second most crowded major country in Europe (after The Netherlands). With almost 45 million people it has a population density of nearly 350 people per sq km (about 900 per sq mile). About 80 per cent of the people live in towns. More than 40 per cent live in the seven metropolitan counties, the largest of which is Greater London, with a population of more than 7 million and an area of 1,580 sq km (610 sq miles).

Language. English is the official language, but it is spoken with a great variety of accents in different parts of the country, departing most prominently from standard English progressively northwards from the Midlands. There are numerous dialects, each with its own characteristic words, phrases and pronunciation, which can identify the speaker as coming from a certain county.

Religion. The Church of England (Anglican), one of the Reformed Churches, is the established Church in England, of which the British monarch is titular head. The spiritual head of the Church is the Archbishop of Canterbury, the "primate of all England". There is one other archbishop (York) and 41 bishops. About two-thirds of the children born in England are baptized by the Church of England. There is complete freedom of worship, and minority religions (in England and Wales) include about 4 million Roman Catholics, 560,000 Methodists, 500,000 members of other Free Churches, and more than 400,000 Jews.

Education. Schooling in England and Wales is free and compulsory between the ages of 5 and 16. The education service is national but is administered by local education authorities. A certain amount of free nursery education is available for children under five years old, and a plan was published in 1972 to expand nursery schooling so that in ten years there would be schooling available for 90 per cent of four-year-olds and 50 per cent of three-year-olds, generally on a half-day basis. About half the primary schools teach children from the ages of five to eleven, the others teaching either infants (ages five to seven) or juniors (eight to eleven).

Secondary education has been a controversial subject in England since the gradual introduction of comprehensive schools, which provide courses for pupils of all abilities and aptitudes. By 1976 about 70 per cent of children were receiving their secondary education in comprehensives. Some areas still maintained the traditional grammar school, with its selection on ability and its emphasis on an academic education (ages 11 to 16 or 18), with secondary modern schools providing a general education with a practical bias (11 to 16 and over). Direct-grant grammar schools (operating on government grants and independent of local authorities) were phased out after 1976, either joining the maintained sector as comprehensives or becoming totally independent and charging fees. England's public schools, which are large independent private schools, include Eton College (1440) and Harrow School (1571).

There are about 600 institutions of further education in England and Wales, including about 30 polytechnics (technical colleges). They provide a wide spectrum of courses, ranging from instruction in shorthand to degree and postgraduate work. There are also more than 6,500 evening institutes offering mainly courses in leisure activities. England has more than 30 universities offering degrees. The largest is London University; it has colleges distributed over a wide area and places for more than 40,000 internal students. The oldest and most famous universities are Oxford and Cambridge, both established in the 13th century. The Open Universi-

Buckingham Palace has been the residence of British sovereigns since 1837 and has nearly 600 rooms; it was built in 1703.

London's Hyde Park has many attractive features including the Serpentine, an artificial lake constructed for Queen Caroline in 1730.

A popular London attraction is Tower Bridge over the River Thames; it was originally operated by hydraulic power derived from steam.

ty (founded in 1969), financed by the central government and by students' fees, awards its own degrees and gives tuition by means of television and radio broadcasts, correspondence textbooks, summer schools and more than 250 local study centres. Any British resident of age 21 or over may apply to join, and no formal qualifications are required for entry. It has up to 50,000 students at a time.

Social services. Britain has been known as a "welfare state" since the end of World War II, because of the great expansion in state aid that took place at that time. The National Insurance Act (1946) came into operation in 1948, along with the National Health Service. National Insurance is collected as a tax related to a person's earnings. It provides a range of benefits, including financial aid for unemployment, sickness, maternity or bereavement and a retirement pension. There are also benefits for industrial injuries, disablement, and death of a family member. The Health Services (administered separately for each county or borough) include a wide range of hospital, specialist, general (medical, dental, ophthalmic and pharmaceutical), community and school services, mostly free (adults have to pay towards the costs of drugs, spectacles and dental treatment). Other benefits of the welfare state include allowances towards the cost of housing for those in need.

Culture. The Arts Council of Great Britain was established in 1946 to develop and improve the understanding and performance of the arts and increase their accessibility to the public. It receives an annual grant-in-aid from the government, and helps to support and encourage both organizations and individuals in the fields of drama, music, dance, writing, painting, sculpture and photography.

There are about 200 professional theatres in England, about a quarter of them in London, which is a great centre of the arts. The National Theatre, which consists of three separate theatres (the Olivier, the Lyttelton and the Cottesloe), was opened in 1976 on the South Bank of the Thames.

It is in literature, particularly in drama, that the English have excelled (from the time of the old English epic poem *Beowulf,* by an anonymous poet in about the 8th century AD). The late Middle Ages produced several important writers, the foremost being Geoffrey Chaucer (*c.* 1340–1400), whose *Canterbury Tales* is regarded as a masterpiece. The Elizabethan Age was the golden age of English literature – the time of William Shakespeare (1564–1616), now universally acknowledged as one of the greatest dramatists the world has ever known. His 36 plays include historical dramas, tragedies, comedies and fantasy romances, and he also perfected the sonnet.

Although Shakespeare's works overshadow all other literary works of his age, there were other fine writers at about that time, including Francis Bacon (1561–1626), who introduced the essay form into English literature; the playwrights Christopher Marlowe (1564–93) and Ben Jonson (*c.* 1572–1637); and the poets Edmund Spenser (*c.* 1552–99), whose *The Faerie Queene* was the first epic in Modern English verse and John Donne (*c.* 1572–1631). John Milton (1608–74) wrote his masterpiece *Paradise Lost* after becoming totally blind, and John Bunyan (1628-88) pioneered the development of the novel with *Pilgrim's Progress,* a religious allegory.

The Restoration period saw the emergence of more fine writers in all fields, such as the poet, critic and dramatist John Dryden (1631–1700), the philosophers Thomas Hobbes (1588-1679) and John Locke (1632–1704), the diarists Samuel Pepys (1633-1703) and John Evelyn (1620-1706), and the dramatists William Wycherley (*c.* 1640–1716) and William Congreve (1670–1729). There followed the Age of Reason, which produced two masters in the Irish satirical writer Jonathan Swift (1667–1745) and the poet Alexander Pope (1688–1744).

The modern style of novel began to emerge in the 18th century, with Daniel Defoe (1660–1731), who wrote *Robinson Crusoe,* and Samuel Richardson (1689-1761), whose *Pamela* is regarded as the first true English novel. Other leading novelists of the period included Henry Fielding (1707–54) and the Irishman Laurence Sterne (1713–68). A major influence on English literature at this time was the critic Samuel Johnson (1709–84), who compiled a famous dictionary. Johnson's literary circle included the Irish playwright, novelist and poet Oliver Goldsmith (*c.* 1730–74), Richard Brinsley Sheridan (1751–1816), an Irish-born dramatist noted for his satirical comedies, and the historian Edward Gibbon (1737–94), who spent 20 years writing *The History of the Decline and Fall of the Roman Empire.* The poet and artist William Blake (1757–1827) was a forerunner of the romantic movement. He rebelled against all conventions, and drew on his powerful imagination to produce such works as *Songs of Innocence* and *Songs of Experience.*

The first wave of romantic poets included William Wordsworth (1770–1850), Samuel Taylor Coleridge (1772–1834) and Robert Southey (1774–1843), and these were followed by a second wave, Lord Byron (1788–1824), Percy Bysshe Shelley (1792–1822) and John Keats (1795–1821). The romantic age also produced two great novelists, the Scottish Walter Scott (1771–1832), famous for his historical romances, and Jane Austen (1775–1817), who described certain aspects of English life in her day in such novels as *Pride and Prejudice.*

The Victorian Age in literature followed, with the poetry of Lord Tennyson (1809–92) and Robert Browning (1812–89), the historical works of Thomas Babington Macaulay (1800–59) and Thomas Carlyle (1795–1881), and the novels of Charles Dickens (1812–70), William Makepeace Thackeray (1811–63), George Meredith (1828–1909), Charlotte Brontë (1816–55), Emily Brontë (1818–48), George Eliot (1819–80), Thomas Hardy (1840–1928), Robert Louis Stevenson (1850–94) and Rudyard Kipling (1865–1936). The outstanding playwright of this period was the Irish-born Oscar Wilde (1854–1900), a controversial and unconventional figure famed for his biting wit. Another Irishman, George Bernard Shaw (1856–1950), was to become one of the greatest of modern playwrights.

Other leading 20th-century writers include the poets T.S. Elliot (1888–1965, born in the United States), W.H. Auden (1907–73), Walter de la Mare (1873–1956), Robert Graves (1895–), and the "war poets" Rupert Brooke (1887–1915), Wilfred Owen (1893–1918) and Siegfried Sassoon (1886–1967); the novelists Joseph Conrad (1857–1954, born in Poland), H.G. Wells (1866–1946), Arnold Bennett (1867–1931), John Galsworthy (1867–1933), W. Somerset Maugham (1874–1965), E.M. Forster (1879–1970), Virginia Woolf (1882–1941), D.H. Lawrence (1885–1930), T.E. Lawrence (1888–1935), G.K. Chesterton (1874–1936), Aldous Huxley (1894–1963), J.B. Priestley (1894–), Evelyn Waugh (1903–66), George Orwell (1903–50) and Graham Greene (1904–); and the playwrights Christopher Fry (1907–), Noel Coward (1899–1973), John Osborne (1929–) and Harold Pinter (1930–). These, and many others of different nationalities, have added to the wealth of English literature.

Although England's art cannot match its vast literary output through the ages, its museums and art galleries are among the finest in the world. In London alone there are the British Museum (with its incomparable collection of art from ancient Egypt, Greece and Rome); the National Gallery (an outstanding comprehensive collection of European painting); the Tate Gallery (modern art) and the Victoria and Albert Museum (applied arts from all places and periods). In addition, many private art collections are on show throughout the country.

It was not until the 18th century that English painters began to develop their own individual styles. William Hogarth (1697–1764), with his engravings and paintings, was the first great English master of social caricature. Thomas Gainsborough (1727–88) and Joshua Reynolds (1723–92) also painted in a highly personal style, and George Stubbs (1724–1806) was unsurpassed as a painter of animals. Two great landscape painters emerged in the 19th century, Joseph Turner (1775–1851) and John Constable (1776–1837). The leading 20th-century artists include painters L.S. Lowry (1887–1976), Ben Nicholson (1894–) and Graham Sutherland (1903–) and sculptors Jacob Epstein (1880–1959), Henry Moore (1898–) and Barbara Hepworth (1903–75).

England is rich in fine architecture, seen particularly in its churches and country houses. The 12th-century Durham Cathedral, a great masterpiece of early Norman architecture, is the first example of rib-vaulting in Europe. Gothic, introduced from France in the late 12th century, soon acquired an English flavour, as seen in Lincoln and Salisbury cathedrals. The English style of Perpendicular Gothic appeared in the mid-14th century (e.g., the

The Tower of London was originally a fortress, became a royal residence in the Middle Ages, was used as a prison and now contains the crown jewels.

Lowestoft is the easternmost town in the country; it is a popular seaside resort and has fishing, shipbuilding and food processing industries.

Stonehenge, huge blocks of standing stones on Salisbury Plain, is evidence of England's early history visited by thousands of tourists annually.

choir in Gloucester Cathedral). The great houses of the Elizabethan and early Stuart period reflected the wealth and pretentions of the gentry. A neoclassical style emerged in the 17th century, developed by two great English architects, Inigo Jones (1573–1652), who brought English Renaissance architecture to a dramatic maturity (the Banqueting House, Whitehall), and Sir Christopher Wren (1632–1723), who rebuilt St Paul's Cathedral and 51 City churches after the Great Fire of London. Two colleagues of Wren, Nicholas Hawksmoor (1661–1736) and Sir John Vanbrugh (1664–1726), were highly original Baroque architects.

The Georgian style (1725–1800) made extensive but modest use of brick and stone. The predominant English style of the early 19th century was Regency and its greatest exponent was John Nash (1752–1835), with fine examples in London's Regent's Park. Of England's modern architects, the most versatile and perhaps the most controversial has been Sir Basil Spence (1907–76), who designed the new Coventry Cathedral.

Until this century, England produced few classical composers of world renown. Early English music was closely bound up with the Church, and several notable composers emerged in the 16th century, including Thomas Tallis (*c.* 1505–85) and William Byrd (1543–1623). With the Restoration, masques and operas became popular and the period produced one of England's greatest composers, Henry Purcell (1659–95). But after Purcell it was a German, George Frideric Handel (1685–1759), who dominated the English music scene – he became a British subject in 1726. His English contemporary was William Boyce (1710–79), and various other European composers visited or worked in England, such as J.C. Bach, Joseph Haydn and Felix Mendelssohn. It was not until Edward Elgar (1857–1934) gained popularity in the late 1800s that England had another composer of international repute, although the light operas of Arthur Sullivan (1842–1900), with librettos by W.S. Gilbert (1836–1911), won lasting fame at home.

Then in the 20th century several English composers achieved worldwide acclaim, including Frederick Delius (1862–1934), Gustav Holst (1874–1934), Ralph Vaughan Williams (1872–1958), Arthur Bliss (1891–1975), William Walton (1902–) and Benjamin Britten (1913–76). English orchestras and conductors have also achieved universal regard during this century.

In the 18th century, the design and manufacture of furniture and pottery won England international acclaim. The three great furniture designers were Thomas Chippendale (1718–79), George Hepplewhite (died 1786) and Thomas Sheraton (1751–1806). Josiah Wedgwood (1730–95) and Josiah Spode (1754–1827) both produced beautiful chinaware, industries still important today.

Science and invention. Britain is a leading centre for scientific research, with some of the world's finest laboratories and medical schools. Its oldest scientific institution is the Royal Society, founded in 1660. Since then British scientists have made important advances and discoveries in many disciplines. In medicine, William Harvey (1578–1657) discovered the circulation of blood, Edward Jenner (1749–1823) developed vaccination, Joseph Lister (1827–1912) founded antiseptic surgery, and Alexander Fleming (1881-1955) discovered penicillin. Robert Boyle (1627–91, born in Ireland) founded modern chemistry, Joseph Priestly (1733–1804) first isolated and identified many gases (including oxygen) and John Dalton (1766–1844) proposed the atomic theory. Regarded by many as the greatest scientist of all is Isaac Newton (1642–1727), who formulated the laws of motion and gravitation, invented calculus and discovered properties of light and colour. The Industrial Revolution maintained its impetus through a great many British inventions, notably those of James Watt (1736–1819), who revolutionized the steam engine, and George Stephenson (1781–1848), who pioneered railways.

Following in Newton's footsteps, Michael Faraday (1791–1867) founded the science of electromagnetism and invented the first generator, and James Clerk Maxwell (1831–79) predicted the existence of electromagnetic waves. Charles Darwin (1809–82) caused immediate controversy with his theory of evolution, but paved the way for much modern scientific thought. And in the early 20th century, Bertrand Russell (1872–1970) revolutionized mathematical thought; he was also the leading philosopher of his day and won the Nobel prize for literature in 1950. Many British scientists of the 1900s have won Nobel prizes and made important discoveries and inventions, including television, radar, the jet engine and the hovercraft.

Food and drink. England has a reputation for plain and simple cooking. Fried fish and chips is still a popular "take-away" meal, despite growing competition from Chinese, Indian and American rivals. Most of the best restaurants, however, specialize in European or Oriental cuisine. Certain provincial areas are known for specialities, such as black pudding or pigs' trotters in the north and Cornish pasties in the south-west. Home-made meat or fruit pies are perhaps among the best of the traditional English dishes, and there are a variety of puddings.

Tea has been a popular drink for hundreds of years, and coffee is becoming almost as widely drunk. Many people drink beer, generally in public houses (pubs), which are found throughout the country. There is an enormous variety of beers, the most popular kinds being "on draught" – that is, pumped straight from the barrel [*see* PE p.204]. Cider, an alcoholic drink made from apples is a speciality of south-western counties.

Leisure and sport. The British have originated several sports and introduced them to other countries. They have produced world champions in such diverse sports as athletics, bowls, boxing, cycling, fencing, horse riding, ice skating, motorcycling, motor racing, snooker, speedway, swimming, table tennis, tennis and yacht racing. The national sports are soccer and cricket, and rugby (League and Union) is popular. Many people play golf and tennis, and other leisure activities include angling, hiking and sailing. The traditional sports of fox hunting and shooting (of game) are still practised, although many people are opposed to blood sports.

History. Even after the sea separated Britain from the mainland of Europe, Stone Age man could easily cross the English Channel. Most early migrations from the mainland were peaceful, but the warlike Celts of the Iron Age began invading the islands from the 8th to the 4th century BC, and tribes such as the Gaels and the Britons settled in various parts. The Romans first invaded England when Julius Caesar sailed from Gaul (France) in 55 and 54 BC. But trouble in Gaul forced him to withdraw, and the Romans did not return until AD 43, when the emperor Claudius sent armies to defeat the Celtic tribes and conquer Britannia, as it was called. Queen Boudicca (Boadicea) led the Iceni in a brave but unsuccessful revolt in AD 61.

The Romans ruled England as a province for about 350 years, and built an impressive network of roads and forts. In the north they built two walls across the country, Hadrian's Wall (AD 120s, from the Tyne to the Solway Firth) and the Antonine Wall (AD 140s, from the Firth of Clyde to the Firth of Forth).

England prospered, however, and towns sprang up round the Roman camps; Londinium (London) began to develop as a port. Roman soldiers and traders brought Christianity to England. When the Roman Empire declined, towards the end of the 4th century, the legions were withdrawn, and England was left at the mercy of the Picts from Scotland, the Scots from Ireland, and the Angles, Saxons and Jutes, who began invading the coast from northern Germany and Denmark. In the mid-5th century, these seafaring tribes began to establish settlements in the south and east, and the region became known as Angle-land. Despite resistance (led possibly by the legendary King Arthur), the Britons were pushed out to the north and west.

Christianity reappeared when St Augustine arrived from France and converted Ethelbert, King of the Jutes, in 597. By the end of the 7th century, seven Anglo-Saxon kingdoms were established – Sussex, Essex, Wessex, Kent, East Anglia, Mercia and Northumbria. In the 9th century, the Danes conquered all the kingdoms except Wessex, whose king, Alfred (the Great), defeated them in 886. But after his death in 899, the Danes gradually extended their territory, which was called the Danelaw, until finally the Dane Canute conquered Wessex and became ruler of all England.

Canute's kingdom collapsed soon after his death in 1035, and the Saxons ruled until 1066, when Edward (the Confessor) died with no direct heir. Harold, Earl of Wessex, became King, but was defeated by William, Duke of Normandy, at the Battle of Hastings. William the Conqueror, as he be-

Three typical "settlements" in England are the village, the industrial town and the cathedral city; this is the unspoiled village of Finchingfield, Essex.

One of the most highly industrialized cities in England is Newcastle-upon-Tyne; it has five bridges, including one built in 1871 by Redheugh.

The cathedral in Salisbury is one of the most beautiful in the country and is a fine example of Early English Gothic; it was founded in 1220.

Rulers of England

This table lists the rulers of England from Egbert through to Elizabeth I. The monarchs after the end of her reign (1603), who also ruled Scotland, are given in a table which appears in the article on UNITED KINGDOM.

Saxons	
Egbert	828–39
Ethelwulf	839–58
Ethelbald	858–60
Ethelbert	860–66
Ethelred I	866–71
Alfred (the Great)	871–99
Edward (the Elder)	899–924
Athelstan	924–39
Edmund I	939–46
Edred	946–55
Edwy	955–59
Edgar	959–75
Edward (the Martyr)	975–78
Ethelred II (the Unready)	978–1016
Edmund II (Ironside)	1016
Danes	
Canute (or Cnut)	1016-35
Harold I (Harefoot)	1035–40
Hardicanute	1040–42
Saxons	
Edward (the Confessor)	1042–66
Harold II	1066
House of Normandy	
William I (the Conqueror)	1066–87
William II (Rufus)	1087–1100
Henry I (Beauclerc)	1100–35
House of Blois	
Stephen	1135-54
House of Plantagenet	
Henry II	1154–89
Richard I (Coeur de Lion)	1189–99
John (Lackland)	1199–1216
Henry III	1216–72
Edward I (Longshanks)	1272–1307
Edward II	1307–27
Edward III	1327–77
Richard II	1377–99
House of Lancaster	
Henry IV	1399–1413
Henry V	1413–22
Henry VI	1422–61; 1470–71
House of York	
Edward IV	1461–70; 1471–83
Edward V	1483
Richard III	1483–85
House of Tudor	
Henry VII	1485–1509
HenryVIII	1509–47
Edward VI	1547–53
Mary I	1553–58
Elizabeth I	1558–1603

came known, was the last successful invader of Britain. He established strong central rule, and divided most of the land among Norman nobles, who continued the feudal system that had already existed for centuries. The Anglo-Saxon peasantry, however, retained their language and many of their customs, except in the presence of their Norman overlords when they had to learn and use new words. Although the Normans initially spoke French and asserted their foreign ways, the two peoples eventually became united. In 1085 William ordered a survey of land and other possessions throughout the kingdom, and the results were recorded in 1086 in what became known as the *Domesday Book* [*see* MS p.*261*]. This enabled the king to apply taxation directly, instead of through the feudal lords.

The Norman monarchs and their successors, the Plantagenets, sought to extend their authority over the Church and the mighty barons. In the reign of Henry II (1154–89) the quarrel with the Church reached its peak and led in 1170 to the murder of Thomas à Becket, Archbishop of Canterbury. In 1215 the barons, supported by the Archbishop (Stephen Langton), compelled King John to grant the charter of English liberty, the Magna Carta.

The reign of Edward I (1272–1307) saw the beginnings of parliament (although Simon de Montfort's "parliament" of 1265 is usually considered to be the first). Like earlier kings, Edward held meetings with leading nobles and churchmen, but he enlarged these to include representatives of the whole country. The "Model Parliament" of 1295 set a pattern for later parliaments. Edward had begun a war with the Welsh in 1277, and in 1282 brought Wales under English control. He also began his campaign against the Scots (1296) and declared himself King of Scotland. But the Scots resisted the English advance and beat Edward II at Bannockburn (1314).

Rivalry between the English and the French had been growing since the reign of Henry II, who ruled half of France through marriage or inheritance. But most of this land was lost by later rulers. The Hundred Years' War (1337–1453) between the two countries began when Edward III landed an army in Normandy, and the conflict quickly resolved itself into a struggle for land. The English won famous victories at Crécy (1346, under Edward III), Poitiers (1356, under his eldest son, Edward, the Black Prince) and Agincourt (1415, under Henry V). Henry and his successors made further conquests, but the dramatic appearance of Joan of Arc in 1429 turned the tide, and by 1453 Calais was England's only remaining possession in France.

Meanwhile, great changes were taking place at home. The Black Death (bubonic plague), which reached England in 1348, took a heavy toll on lives. Feudalism was declining, because many lords began to prefer rents to feudal service. Wat Tyler led an unsuccessful peasants' revolt in 1381, and the English were awakening to a greater sense of identity.

Towards the end of the war in France, a struggle for the throne began to develop in England which led to the Wars of the Roses (1455–85), between the House of York (emblem, white rose) and the House of Lancaster (red rose), which ended when Henry Tudor defeated Richard III and the Yorkists at the Battle of Bosworth Field (1485). As Henry VII, he united the two houses by marrying Elizabeth of York.

The Tudor dynasty ruled for more than a hundred years. Henry VIII's quarrel with the Church of Rome led to the English Reformation and the formation of the Church of England. Mary I's reign re-established Roman Catholicism briefly, but one of the first acts of Elizabeth I was to restore the Church of England. England became one of the most powerful nations in Europe during the reign of Elizabeth I (1558–1603). She successfully blocked French and Spanish designs on the English throne, and in Church affairs she established a compromise between the extremes of Rome and the Protestantism of Luther and Calvin. She challenged the might of Spain, and in 1588 an English fleet defeated the Spanish Armada. English seamen, such as Sir Francis Drake and Sir Walter Raleigh, explored the coasts of the Americas and, by raiding and trading, took home some of the spoils and established England's presence in the New World. English literature blossomed with the works of Shakespeare and others, and great advances were made in education and scholarship. In 1600 English merchants formed the East India Company. And despite inflation and economic upheaval at home, there was an upsurge of national enthusiasm, centred largely on the person of the Queen. Elizabeth never married. She was succeeded by her cousin, James VI of Scotland, who ruled the two countries as separate kingdoms. They were eventually united in 1707 by the Act of Union.

For the history of England since 1603, *see* UNITED KINGDOM. Map 8.

Equatorial Guinea, official name Republic of Equatorial Guinea, is an independent nation in western Africa. It consists of mainland Río Muni and the island of Macías Nguema Biyoga (formerly Fernando Póo). The capital, Malabo (formerly Santa Isabel), has a population of 19,869.

Most of Río Muni is forested and underdeveloped; coffee and timber are its chief products. Macías Nguema Biyoga is a fertile volcanic island, with plantations producing cocoa and coffee. About 75 per cent of the people live in Río Muni, the largest group being the Fang; the official language is Spanish. Macías Nguema Biyoga has a mixed population, including Bantu-speaking people, Creoles and migrant plantation workers.

Spain took Fernando Póo in 1778 and Río Muni in 1885, governing them as overseas provinces of Spain until 1968. They were then joined to form an independent republic. The president, Francisco Macías Nguema, was made president for life in 1972. Area: 28,051sq km (10,830sq miles). Pop. (1976 est.) 316,000. Map 34.

Ethiopia's modern and tourist-conscious capital, Addis Ababa, lies on a high but well-watered plateau at the very centre of the country.

The headquarters of the OAU are located in Addis Ababa; since 1963 the OAU Centre has entertained many African delegations from member countries.

The technique of weaving home-produced wool was probably taken to the Faeroe Islands by the early Scandinavian settlers.

Estonia. *See* UNION OF SOVIET SOCIALIST REPUBLICS.

Ethiopia, formerly called Abyssinia, is an ancient empire in north-eastern Africa. Most of its emperors, including Haile Selassie I (who was deposed in 1974), claimed descent from the biblical King Solomon and the Queen of Sheba. Ethiopia has been a bastion of Christianity surrounded by often hostile, non-Christian forces since the AD 300s and has attracted the attention of Europeans since the 1400s. The capital is Addis Ababa, since 1963 the headquarters of the Organization of African Unity (OAU).

Land and climate. The heart of Ethiopia consists of two highland regions separated by an arm of the East African Rift Valley. This valley, which contains several large lakes and the River Awash, broadens northwards in the eastern Danakil plains. The northern highlands, which are crossed by the

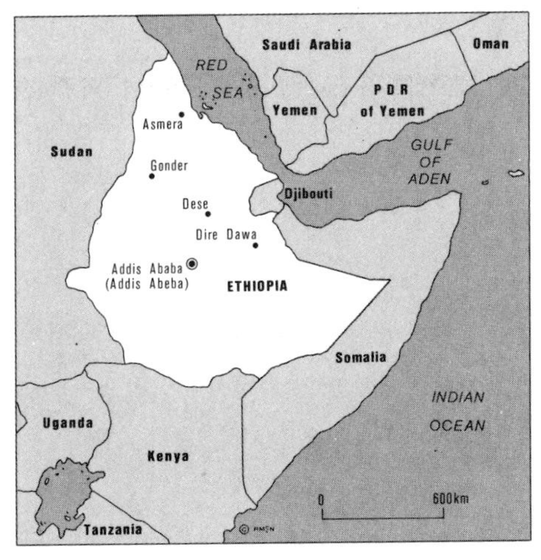

Blue Nile (or Abbay) [*see* PE p.*114*], are the highest, reaching 4,620m (15,158ft) at Ras Dashen. The southern uplands gradually descend to the arid, south-eastern plateaus, which border Kenya and the Somali Republic.

The climate is largely determined by altitude. The highest zone, the *dega,* is more than 2,750m (9,023ft) above sea-level, and has a cool climate. It is grassy but has few trees. Between about 1,830 and 2,750m (6,004-9,022ft) is the *woina dega* (wine highlands), which has an annual average temperature of 22°C (72°F) and an annual rainfall of between 500 and 1,500mm (20-59in). This is a grassy zone, with many trees, and it contains Ethiopia's most fertile land. Below 1,830m (6,004ft) the land becomes increasingly hot and arid, with vegetation ranging from dry savanna to desert. This region, called the *quolla,* has an average annual temperature of 26°C (79°F) and less than 500mm (20in) of rain per year. The rainfall is unreliable and long droughts occur, causing great hardship to the local people.

Economy. Coffee grown in the *woina dega* is the country's most valuable crop, accounting for more than half of Ethiopia's exports. Much coffee grows wild and it was in Ethiopia that the plant probably originated. Another fifth of Ethiopia's exports comes from oilseeds, oil nuts and oil kernels, fruits and vegetables, and meat, hides and skins. A little gold and some salt are mined. Addis Ababa and Asmera are the chief manufacturing centres.

People. Most Ethiopians are Hamitic in origin, having long faces and thin noses and lips. Intermixing has, however, led to varying degrees of Negroid features. There are three main language groups. Semitic languages, including Amharic, are spoken by about half of the people; these languages were introduced from Arabia. Cushitic languages, such as Galla and Somali, are spoken in the south-east and Nilotic tongues predominate in the south-west. About half of the people are Coptic Christians. The isolated Ethiopian Church is unique, displaying Hebraic, Semitic and African influences. Islam is followed mostly in the south and east, with ethnic religions being practised in the south-west. In Eritrea the people are divided almost equally into Christians and Muslims. Literacy is estimated at about 10 per cent.

Government. Since Emperor Haile Selassie was deposed in 1974 a Provisional Military Administration Council, called the Dergue, has ruled the country. This Council has announced its aim of making Ethiopia a people's Democratic Republic, with an elected People's Revolutionary Assembly, but no timetable has been fixed for these changes.

History. In the 4th century Ezana, King of Aksum in northern Ethiopia, was converted to Christianity. After Aksum declined, Islam became a powerful force on the coast in the 7th century and the Christian kingdom survived only in the inaccessible interior highlands. The Christian kingdom endured long periods of isolation until the 1520s, when Portuguese explorers arrived. Shortly afterwards the Portuguese helped Ethiopia to resist a major Muslim onslaught.

In the late 1800s Italy declared Ethiopia a protectorate but the Ethiopians defeated an Italian army at the Battle of Adowa (Adwa) in 1896. But Italy continued to hold Eritrea on the Red Sea coast. After 14 years as regent Haile Selassie became emperor in 1930. In 1935–36 Italy seized Ethiopia, but the Italians surrendered to Ethiopian and Allied forces in 1941. After 1941 Haile Selassie steadily introduced reforms in this essentially feudal country. He established an elected parliament but remained an absolute ruler.

Eritrea was federated with Ethiopia in 1952 and was fully incorporated in 1962, despite opposition from the Eritrean Liberation Front. Haile Selassie's slow pace of reform led to widespread discontent. The removal of Haile Selassie led to disunity and threats of secession from several parts of the country, especially in Eritrea where an armed rebellion was mounted. Because of the difficult military situation, the left-wing military government was unable to implement many of its reformist aims. Disorder has led to the imprisonment and execution of many opponents of the regime, which in 1977 began to accept military aid from the USSR. Map 38.

Ethiopia – profile

Official name Ethiopia
Area (including Eritrea) 1,221,900sq km (471,776sq miles)
Population (1975 est.) 27,946,000
Density 23 per sq km (59 per sq mile)
Chief cities Addis Ababa (capital) (1975 est.) 1,161,267; Asmera, 317,950
Government Military council, led by Brig.-Gen. Teferi Benti
Religions Coptic Christianity, Islam
Language Amharic (official)
Monetary unit Ethiopian dollar
Gross national product (1974) £1,089,700,000
Per capita income £39
Agriculture Barley, cattle, coffee, hides and skins, maize, millet, sugar cane, tobacco, wheat
Industries Brewing, bricks, cement, food processing, fuel oils, shoes, sugar, textiles
Minerals Gold, salt
Trading partners USA, Italy, West Germany, Japan, Britain

Faeroe (or **Faröe**) **Islands** (Faerøerne) are a group of 22 volcanic islands in the northern Atlantic Ocean between Iceland and the Shetland Islands; the capi-

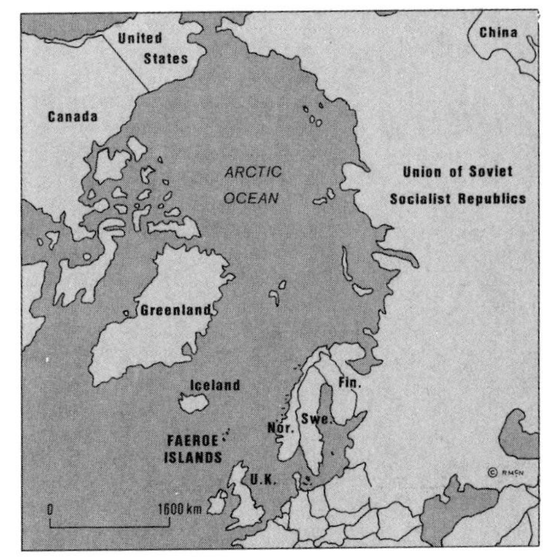

Fijian dancers dress in decorative costumes to give expression to their culture. Despite outside influences most of their customs have survived.

Helsinki, capital city of Finland, was founded in 1550 and was mostly destroyed in 1808 by fire. The first Finnish railway linked Helsinki and Hämeenlinna.

Turku was Finland's capital city until 1812. It remains important for its culture, commerce and industry, and is also the largest port in the country.

tal is Tórshavn. Seventeen of the islands, which are high and rugged with little vegetation, are inhabited. They have belonged to Denmark since 1380 but became a British protectorate temporarily during World War II after the German occupation of Denmark. The parliament of the Faeroes (the Lagting) declared independence in 1946, but a new parliament reversed the decision. They were granted home rule in 1948 and since 1953 have sent two members to the Danish parliament. Area: 1,399sq km (540sq miles). Pop. (1975 est.) 41,000. Map 6.

Falkland Islands (Islas Malvinas) are a group consisting of two large and about 200 small islands in the southern Atlantic Ocean, 600km (373 miles) east of Argentina. They comprise a British Crown colony (although Argentina and Chile also claim them) which has its own dependencies such as South Georgia, the South Orkney Islands and the South Shetland Islands. The main islands are West Falkland and East Falkland, location of the capital, Stanley. The Falklands are rocky and windswept but have good pastures for the large flocks of sheep. Most of the people are farmers, fishermen or hunters of the seals that abound in the waters round the islands. Attempts are being made to exploit extensive beds of seaweed for making milled dried kelp (a fertilizer and source of potash). There is a British Antarctic Survey station at King Edward Point in South Georgia.

The Falkland Islands were probably first sighted in 1592 by the English navigator John Davis and they were charted by the Dutch sailor Sebald de Weert in 1600. The first landing on the islands was made by the Englishman Capt. John Strong, who named them after Viscount Falkland, Treasurer of the Royal Navy. They were occupied at various times by Argentines, French and Spanish until permanent British occupation began in 1832. Area: 11,961sq km (4,618sq miles). Pop. (1975 est.) 2,000. Map 78.

Fiji is an independent republic made up of more than 800 islands and islets in the southern Pacific Ocean. The main islands are of volcanic origin and include Viti Levu (location of the capital, Suva), Vanua Levu, Taveuni, Kandavu and Ovalau. Most indigenous Fijians are of Melanesian stock, with Tongan Polynesians forming a minority group, but since 1945 both groups have been outnumbered by descendants of Asian Indians taken to the islands as labourers by the British to work on the sugar plantations. This is reflected in the local religions, Methodism and Hinduism; the chief language is English. The main products are copra, sugar, rice, bananas, gold and manganese.

Fiji was discovered by Abel Tasman in 1643 and visited by Captain James Cook in 1774. Tribal wars and exploitation by Europeans caused much damage before the islands were annexed by Britain in 1874. In 1970 Fiji became an independent nation with dominion status within the Commonwealth. Area: 18,272sq km (7,055sq miles). Pop. (1975 est.) 573,000. Map 62.

Finland (Suomi), a republic in northern Europe, is a country of lakes and forests. The vast coniferous forests of pine and spruce that cover more than two-thirds of Finland are the mainstay of its economy. The 60,000 lakes make up a tenth of its total area. The Finns are a proud people who find fortitude in adversity. After defeat in World War II, and faced with a huge war indemnity, they made a remarkable recovery to build up a thriving economy. The Finns are known for their love of books, the theatre and sport. Finnish architecture and design are admired throughout the world for their colour and ingenuity. The capital is Helsinki.

Land and climate. The great ice sheets that advanced and retreated over northern Europe during the Ice Ages scooped out hollows that have since

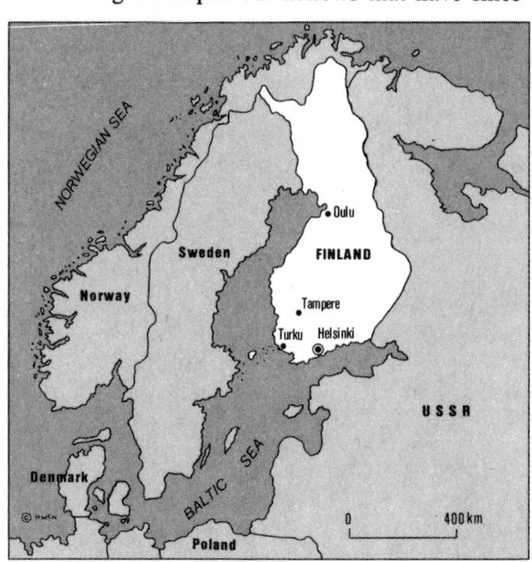

filled with water and formed an intricate system of lakes and rivers. Most of the country is low-lying, but outside the forests the landscape is rugged and broken. There are mountains in the north, where Finland extends into Lapland, and the country's highest point (Haltiatunturi) stands 1,324m (4,344ft) above sea-level on the border with Norway. There are thousands of off-shore islands. Because of the proximity of the Gulf Stream, Finland's climate is not as severe as that of other regions in the same latitude. The winters are long and cold, the summers warm and sunny. July temperatures in Helsinki average 18°C (64°F) and in Lapland 12°C (54°F). In February, the coldest month, Helsinki averages -6°C (21°F), Lapland -12°C (10°F). Rainfall is heaviest in the south-west, with an annual average of about 650mm (26 inches).

Economy. Industry accounts for 95 per cent of all Finland's exports. Forest-based industries, which make up more than half the exports, include the manufacture of paper (especially newsprint) and paper-board, wood pulp, shaped or worked wood, veneers, plywood and sawn timber. Since World War II Finland has built up its other manufacturing industries, which include metal processing (steel, copper, nickel), engineering, chemicals and shipbuilding. Important design industries include furniture, fashion, textiles, jewellery and glass. About 9 per cent of Finland's land area is under cultivation. The principal crops are hay, oats, barley, wheat, potatoes and rye. The farms, on average, are small. Most farmers keep cattle, and dairy products include butter and cheese. Finland's chief imports are fuel and raw materials, machinery and equipment. Finland became an associate member of EFTA in 1961, and signed a free trade agreement with the EEC in 1973. A pipeline was completed in 1973 to supply Soviet natural gas to Finnish industry.

People. The Finnish language (*suomi*), spoken by about 93 per cent of the people, is unlike any other Scandinavian language; it belongs to the same family as Hungarian. About 7 per cent of the people speak Swedish, the other official language; they live chiefly near the southern and western coasts. There are about 2,500–3,000 Lapps, who live in the far north. About 93 per cent of the people belong to the evangelical Lutheran Church and 1.4 per cent to the Orthodox Church of Finland, both state churches. The Finns have one of the highest literacy rates in the world. A comprehensive school (*peruskoulu*) system was initiated in 1968 for completion in the early 1980s. After *peruskoulu* (seven-16 years) students may go on to the *lukio*, a sixth-form college or high school. There are eight universities and several other institutions of higher education. Because of the difficulties of language, few Finnish writers are known outside their own country. In 1835 Elias Lönnrot published his collection of folk poetry and ballads (the *Kalevala*). Now recognized as one of the world's great works of epic poetry, it helped to inspire the national awakening (it also inspired Henry Longfellow's poem *The Song of Hiawatha*). Other Finnish masterpieces include *The Seven Brothers,* by Aleksis Kivi (1834–72), and *The Unknown Soldier,* by Väinö Linna (1920–). Finland has a thriving professional theatre, which is heavily state-subsidized, and most Finns participate in amateur theatricals. The country's outstanding composer, Jean Sibelius (1865–1957), was inspired by the late 19th-century romantic movement. Finnish architects enjoy world-wide esteem. Eliel Saarinen (1873–1950), who moved to the United States in 1922, had a considerable influence on the development of American architecture, particularly the skyscraper, and Alvar Aalto (1898–1976) influenced European architecture particularly building techniques. Finland is a great sporting country. The main winter sport is cross-country skiing; ice hockey, swimming and car rallying are also popular. The Finns have produced several of the world's outstanding long-distance runners, notably Paavo

Finland is traditionally an agricultural country. It is composed largely of forests, and has a labyrinth of lakes, rivers and canals serving as waterways.

The Cathedral of Sainte-Croix is one of Orleans' finest buildings. Destroyed by the Huguenots in 1568, it was rebuilt between the 17th and 19th centuries.

Cannes, on the south coast of France, has been a fashionable resort since the mid-1800s. It is also the home of a major international film festival.

Nurmi, who dominated the Olympics in the 1920s, and Lasse Viren, who did the same in the 1970s.

Government. Finland has a one-chamber parliament of 200 members, elected for four years by proportional representation. The president is elected for a six-year term by a college of electors. He has far-reaching powers, particularly in foreign affairs. There are usually seven or eight political parties represented in parliament, and no party has ever had an overall majority. There has been an average of about one cabinet a year.

History. The early Finns, probably migrants from the southern Baltic in the 1st century AD, lived as independent tribes for hundreds of years. In about 1157 King Erik IX of Sweden led a crusade into Finland, but as well as introducing Christianity he gained a foothold in Finland, which eventually became a province of Sweden. Swedish rule lasted

Finland – profile

Official name Republic of Finland
Area 337,009sq km (130,092sq miles)
Population (1976 est.) 4,734,000
Density 14 per sq km (36 per sq mile)
Chief cities Helsinki (capital) (1975) 495,287; Tampere, 165,621; Turku, 163,752
Government Head of state, Dr Urho Kekkonen, president (elected 1956)
Religion Christianity
Languages Finnish, Swedish (both official)
Monetary unit Markka
Gross national product (1974) £8,269,200,000
Per capita income £1,747
Agriculture Hay, oats, barley, wheat, potatoes, rye, cattle
Industries Timber, wood products, paper, metal processing, chemicals, ceramics, engineering, shipbuilding
Trading partners (major) USSR, Sweden, West Germany, Britain

until 1809 (and many places in Finland still have alternative Swedish names). Then Russia, who had already occupied the south-eastern part, seized the rest of the country and made Finland a grand duchy. With the outbreak of the Russian Revolution in 1917 the Finns declared their independence. A short but savage civil war in 1918 between the Russian-backed "Reds", attempting to establish a socialist state, and the nationalist "Whites" resulted in a victory for the Whites, under Carl Gustav Mannerheim (with German military assistance). Finland became a republic in 1919.

The Finns refused Soviet demands for certain strategic territory in 1939 and the USSR declared war. After heroic resistance under Mannerheim, the Finns finally yielded in 1940. They lost more land to the USSR in a further war (1941–44), and in the final peace treaty of 1947 ceded the important Karelia isthmus. They lost a third of their hydroelectric capacity to the USSR, a fifth of their railways, their outlet to the Arctic, and several mines and factories; in addition they were faced with a reparations bill for goods to the value of £60 million. The northern third of the country had been devastated by the Germans, and 13 per cent of the population had to be resettled. Yet within ten years Finland's economy had recovered, and the Finns went on to become a highly prosperous nation. They signed pacts of mutual security with the USSR, while endeavouring to avoid involvement in the European East-West political divisions. Since the war the Communists have regularly won about 25 per cent of the seats in parliament. The government was weakened in 1976 by a deepening divergence between the Communists and the other coalition partners over how to tackle economic problems caused by inflation. Map 6.

Florida. *See* UNITED STATES.

Formosa. *See* TAIWAN.

France, a republic on the western seaboard of the European mainland, is the largest country in Europe with the exception of the USSR. For centuries France has been at the centre of Western culture, a leader in the arts and sciences. Its capital, Paris, is one of the most beautiful cities in the world. It has produced fine artists, writers and musicians, and is a place where both aspiring youth and established masters of France and other countries have lived and worked. Paris is also renowned as the centre of modern Western fashion. France has an international reputation for excellent cuisine and produces some of the world's finest wines. For hundreds of years France was also a major world power. But much of it was devastated by the two world wars and later drained by costly fighting in its colonies, most of which eventually won or were granted independence. France became a major power in the EEC and, in a remarkable transition from a largely agricultural to a prosperous industrialized nation, achieved what many regard as an economic miracle.

Land. France is low-lying in the north and west, with highlands in the south and east. In the centre of northern France lies the Paris basin, which occupies a third of the country's area. The Loire is the longest river, rising on the edge of the Massif Central (Central Plateau) and flowing 1,020km (634 miles) to the Bay of Biscay. The Massif Central, which contains old volcanic peaks rising to more than 1,500m (5,000ft) above sea-level, occupies about a seventh of France. The country's highest mountains are the spectacular Alps in the south-east, topped by Mont Blanc (4,810m; 15,781ft), the highest peak in Europe outside the USSR [*see* PE pp. *44, 45, 107, 117*]. Another great range of mountains in the south-west, the Pyrenees, separates France from Spain. Marseille, France's largest port, lies at the mouth of the River Rhône, on the Mediterranean [*see* PE p. *45*]. The eastern part of this coast is the Côte d'Azur, the French Riviera, with its alternating bays and headlands, its numerous small harbours, steep cliffs and sandy beaches. About 160km (100 miles) south-east of this coast lies Corsica, a large hilly island with a rocky coastline; it is a metropolitan département of France.

Climate. France is the only country with the three major European climates – maritime, continental and Mediterranean. The west and north-west have a maritime climate similar to that of the British Isles, with mild winters, cool summers and most rain in autumn. The Massif Central and eastern France share a moderate continental climate, with cold winters and wet, warm summers. The Vosges and Alps experience heavy winter snows. The south coast enjoys a Mediterranean climate, characterized by mild winters and dry, hot summers, modified oc-

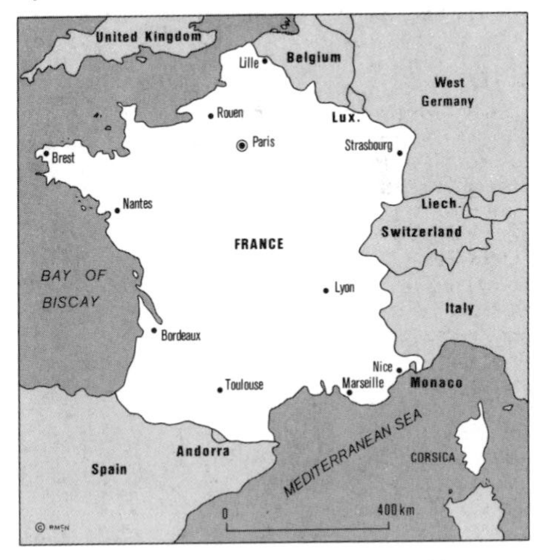

casionally by a cold, brisk mistral which blows across southern France from the Alps. France also experiences pronounced local variations of climate, particularly in the Alps.

Physical resources. France's richest natural resource is its soil. About 90 per cent of the land is productive – a third of it being under cultivation, a quarter permanent pasture and a quarter forest. Lorraine has the greatest reserves of iron ore in Europe, and there are large deposits of bauxite in the south-east. There are several coalfields, although they do not yield enough for the country's needs and coal has to be imported. Potash is mined in Alsace, and deposits of natural gas and sulphur at Lacq (in the Pyrenees) are being developed.

Agriculture. More than a tenth of the country's labour force works on the land, and France produces more than enough food to support its population. Most of the country's arable land is divided

The city of Lyons, south-eastern France, has been modernised. But there are still many historic buildings, notably the Gothic Cathedral of St Jean.

The Romanesque Church of St Michel d'Aiguilhe overlooks the ruins of the 11th-century chapel at Le Puy, in southern France.

The Hôtel de Ville, Compiègne, is a fine example of late Gothic architecture. Built in the 1400s, it has a decorative façade topped by an elegant belfry.

into small farms, a relic of the Napoleonic inheritance law under which estates were divided equally among children. The number of farms has been reduced, however, from 4 million in 1930 to 1½ million in the 1970s. More than half the agricultural land is used to produce cereal crops, and France ranks in the first ten of the world's leading producers of wheat, barley, maize and oats, as well as of potatoes (grown mainly in the north-east and Brittany) and sugar-beet (northern Paris basin). Rice is grown in the Camargue, which produces enough to meet all France's requirements.

Grapes are one of the country's most important crops, and nearly 2½ per cent of all agricultural land is used for vineyards. France is second only to Italy in wine production, regularly producing some 60 million hectolitres (1,300 million gallons) a year [*see* PE pp. 200, 202, 207]. Many fruits are grown, including apples (for making cider as well as for eating), peaches, pears, grapes (for eating also), plums, apricots, cherries, olives, almonds, walnuts and chestnuts. Half the value of France's agricultural produce comes from its herds of beef and dairy cattle. France ranks second (to the USSR) in the production of butter, third in cow's milk, third in cheese (including the famous Camembert, Brie and Roquefort) and third in meat (chiefly beef and veal). The country has about 24 million cattle, 12 million pigs and 10 million sheep.

Fishing. France has more than 30,000 fishermen and 13,000 fishing vessels, which operate off the French coasts or as far afield as Newfoundland and Iceland. They catch fish such as cod, herring, mackerel and tuna as well as oysters, lobsters, shrimps and other crustaceans and shellfish [*see* PE p. *242*].

Industry. Some 27 per cent of the labour force now works in manufacturing. The chief industries include steel (27 million tonnes a year), pig iron (more than 20 million tonnes), cars (more than 3 million), other mechanical engineering products, electrical appliances, heavy chemicals and plastics, and textiles (cotton, synthetic fibres, wool, silk). The chief industrial areas are located near Paris and the ports and mining areas, and new centres have grown up where hydroelectric power has been developed in the mountain areas and the Massif Central. The aircraft industry is based chiefly at Toulouse and Marseille. As well as producing a whole range of military aircraft, France co-operated with Britain to build Concorde, the world's first supersonic airliner. Also, France is traditionally famous for such luxury goods as perfumes, porcelain (Limoges, Sèvres) and lace. France's scenery and cultural heritage make tourism important.

Economy and trade. At the end of World War II (1945) France and its economy were in near ruins. The industrial areas had been devastated, communications in the war zones seriously damaged, and what machinery remained was old (because prewar policies had allowed France's industries to deteriorate). A postwar reconstruction and modernization policy, however, soon turned France into one of the world's leading industrial nations. The policy, called the National Plan, is renewed every few years. The government has considerable control over the economy because of its highly centralized administration, and is able to enforce its guidelines.

By the mid-1970s France had become the world's fourth largest exporter, and ranked third in the export of arms. The values of imports and exports are roughly equal, and the balance of payments is favourable because of considerable revenue from tourism and capital investment abroad. France's chief exports include machines, chemicals, cars, and iron and steel. EEC countries receive 54 per cent of exports (West Germany 18%, Italy 12%, Belgium and Luxembourg 11%, Britain 7%, The Netherlands 5%). Other leading customers include Switzerland (6%) and the United States (5%). France's chief imports are machinery, foodstuffs, oil and chemicals. EEC countries are the leading suppliers with 48 per cent (West Germany 19%, Belgium and Luxembourg 10%, Italy 7%, The Netherlands 6%); the United States provides 8 per cent.

Transport and communications. France has one of the finest transport systems in the world. Its railway system is the third largest (after the United States and the USSR), with some 37,000km (23,000 miles) of track and modern trains that have broken world speed records. The major highways (*routes nationales*) total more than 73,000km (45,000 miles), including more than 2,000km (1,250 miles) of motorways (*autoroutes*). France has a merchant fleet of some 500 ships, 70 per cent of which are cargo vessels. There are three national airlines, including Air France, one of the world's largest; France ranks third after the United States and Britain in the number of passenger-kilometres flown per year. There are nearly 100 daily newspapers with a combined circulation of more than 11 million. Organization and development of broadcasting became the responsibility of the Public Broadcasting Establishment in 1974. There are three television channels and three radio stations, and some 14 million television sets.

Government. The 1958 constitution (amended 1962), which established the Fifth Republic, greatly extended the president's powers. Elected by universal suffrage for a seven-year term, the president nominates and may dismiss the prime minister and other members of the government; he can dissolve the National Assembly; and he appoints people to all military and civil offices. Parliament consists of two houses: the National Assembly, with 490 members elected for five-year terms, and the less powerful Senate, with 283 members elected for nine-year terms (a third elected every three years). Ministers may not be members of parliament, and the prime minister is assumed to have the Assembly's confidence unless an Opposition censure motion (signed by a tenth of the deputies) gets an absolute majority. A Constitutional Council is responsible for supervising all elections and referenda. There are several political parties, and the French political scene has a long tradition of confusion. It has undergone a remarkable simplification, however, under the Fifth Republic, and elections are now usually fought among a few groups, the chief being the Gaullist *Rassemblement pour la République* (formerly UDR), a left-wing group of Socialists and Radicals, the Communist Party, Independent Republicans (supporters of the Gaullist government), Reformers, and the Centre Democratic Party.

For administrative purposes, metropolitan France (including Corsica) is divided into 95 *départements.* France's former colonies – MARTINIQUE, GUADELOUPE, REUNION and FRENCH GUIANA – have the status of overseas départements. The départements were regrouped in 1964 into 22 programme regions for purposes of national planning. Each département is administered by a prefect, appointed by the central government, and is divided into *arrondissements* (more than 300 in all), *cantons* (more than 3,200) and finally *communes* (nearly 38,000), the basic unit of local government. Most communes (more than 33,000) have fewer than 1,500 inhabitants, although 334 have populations exceeding 20,000. Each commune is governed by an elected municipal council headed by a mayor, except in Paris, which is administered uniquely as a city-département by a 90-member council.

Judiciary. The principles of French law laid down under the Revolutionary and Napoleonic regimes have largely survived, but a thorough reorganization of the courts took place in 1958-59. Minor cases are dealt with in more than 450 *tribunaux d'instance* (civil) and *tribunaux de police* (criminal), each tribunal under a single judge. The next level is composed of some 170 *tribunaux de grande instance* (civil) and *tribunaux correctionnels* (criminal), usually with three judges for each tribunal. There are some 30 courts of appeal. More serious crimes are tried by the courts of assizes, which sit in every département and are composed of a president, two other magistrates and a jury of nine people.

Armed forces and police. The president of the Republic excercises command over the armed forces and has direct responsibility for the nuclear striking force. Military policy is determined by the Supreme Defence Council, which is composed of the prime minister and certain other ministers. The effective strength of the armed forces is about 500,000 (army 330,000; navy 70,000; air force 100,000). National service of 12-15 months is compulsory. France withdrew from the military side of NATO in 1966. The *gendarmerie* (effective strength about 70,000) is an integral part of the army, but also co-operates with the civil administration in maintaining public order.

People. There are several regional differences in language and tradition, especially between the people of the Paris area (nearly a sixth of the population) and the rest of the country. All the people speak French (which derives almost entirely from Latin), but there are several minorities that also cling to their own language: the people of Brittany,

The château at Chaumont, north-east France, was built during the 10th century; for about 100 years, until 1329, it was the home of the counts of Champagne.

The 13th-14th century Cathedral of Saint-Nazaire looks over the old town of Beziers which lies on the River Orb in the Languedoc region.

The magnificent gardens of Versailles, just outside Paris, were designed and laid out by the Frenchman André Le Nôtre during the reign of Louis XIV.

of Celtic origin, who speak Breton; the Basques, of the western Pyrenees; and the Catalans, of the eastern Pyrenees, who speak a Romance language resembling Provençal (a language that has retained a hold in Provence). In addition a German dialect is common in Alsace, Flemish around Dunkirk, and an Italian dialect in Corsica. French is the official language of more than 20 countries (particularly in the West Indies), and its structure and vocabulary are carefully supervised by the French Academy (founded 1635). There is no state religion, but most of the people are Roman Catholics; 90 per cent are baptized as such, but active participation is about 20 per cent and varies from region to region. There are about 800,000 Protestants and 500,000 Jews.

The French have a comprehensive system of social security, and state insurance provides financial benefits for sickness, unemployment, maternity and industrial injury. There is a health scheme in which patients pay for treatment and then claim back all or a part of the cost. The French are renowned the world over for their cuisine, and every region has its own specialities. To the French gastronomy is an art, and complementing the food is the inevitable wine. Each Frenchman consumes an average of about 135 litres (30 gallons) of wine a year. The most popular sport is soccer, and rugby flourishes in the south-west. Boules, a game of bowls that can be played on any patch of rough ground, is popular throughout the country. The annual Prix de l'Arc de Triomphe, held at Longchamp, is the richest horse-race in the world, and Frenchmen enjoy a weekly gamble called the *tiercé*, in which they have to forecast the first three horses in a special Sunday race. The sporting event for which France is perhaps best known is the Tour de France cycle race. This massively sponsored event is eagerly followed every summer for more than three weeks as teams of professional cyclists from many countries race round the roads of France, watched by hundreds of thousands of spectators and millions more on television.

Education. The French educational system is highly developed and centralized. The Ministry of National Education determines the course and teaching methods of primary and secondary schools throughout the country. Education is free and compulsory from the ages of six to 16. Free nursery schools (two to six) are also provided. Instruction in primary schools and certain classes of *lycées* consists of three courses: preparatory (one year), elementary (two) and intermediary (two). The entrance examination to secondary schools (*lycées* and *collèges d'enseignement*) has been abolished. All children enter the first cycle (11-15), but a special commission decides whether they have the ability to take a further course which may lead eventually to the *baccalauréat* examination at the *lycées.*

University education is undergoing a reform begun in the late 1960s, by which the 23 state universities, with their separated disciplines, are being gradually replaced by 61 smaller universities with many departments. Tuition is free and students have a voice in administration. A considerable amount of higher education is available outside the universities in the famous *grandes écoles,* such as the École Polytechnique for engineers and the École Normale Supérieure for the training of teachers and future administrators.

Culture. Before World War II Paris was known as the art capital of the world and attracted not only budding artists but also, over the years, great painters from other countries, such as Van Gogh and Picasso. Of Paris's many distinguished museums and art galleries, the Louvre is perhaps the most celebrated in the world, for its architecture as well as its contents. The French have been to the fore in most of the art movements since the 18th century. Jean-Antoine Watteau (1684–1721) and Jean-Honoré Fragonard (1732–1806) led the Rococo movement, Jacques-Louis David (1748–1825) and his pupil Jean Ingres (1780–1867) Neoclassicism, Eugène Delacroix (1798–1863) Romanticism, which grew out of Neoclassicism, and Gustave Courbet (1819–77) Naturalism. French artists also led the Impressionist movement in the 1870s – Édouard Manet (1832-83). Claude Monet (1840–1926), and Pierre Renoir (1841–1919). A whole series of movements followed: Paul Cézanne (1839–1906) Post-Impressionism, Paul Gaugin (1848–1903) Synthetism, and Georges Seurat (1859–91) pioneered Divisionism. Edgar Degas (1834–1917) painted ballet dancers and Henri de Toulouse-Lautrec (1864–1901) posters of café and theatre life. Henri Matisse (1869–1954) was the leading artist of the Fauve group and another Fauvist, Georges Braque (1882–1963), led the development of Cubism with Picasso. Auguste Rodin (1840–1917) was the leading sculptor of the late 19th century.

French architecture has flourished since the Carolingian dynasty. Romanesque art reached its peak after AD 1000, and many monasteries still survive as fine examples. Gothic art originated in France. Great Gothic cathedrals include Notre-Dame, Paris (1163), Chartres (1194), Rheims (1211) and Amiens (1220). France is also famous for its *châteaux* (castles) the finest of which were built around the Loire valley in the 16th century, and for the palaces of Versailles (17th century) and Fontainebleau (16th), which are superb examples of the high standard of decorative art achieved in France in that period.

The French have produced much of the world's finest literature, inspiring the literature of other nations, and they bestow great honours on their own authors. The first important French literature emerged in the 11th and 12th centuries in the form of *chansons de geste,* long epic poems such as *La Chanson de Rolande.* The Renaissance produced the coarse satire of François Rabelais (*c.* 1494–1553) and the philosophical essays of Michel de Montaigne (1533–92). With the patronage of Louis XIV, the 17th century was a golden age for French literature. The outstanding writers included René Descartes (1596–1650) and Blaise Pascal (1623–62), both philosophers and mathematicians, and the three masters of classical drama, Pierre Corneille (1606–84), Molière (1622–73) and Jean Racine (1639–99). Outstanding writers of the 18th century included Voltaire (1694–1778), who attacked the establishment and injustices in brilliant satirical prose, and Jean Jacques Rousseau (1712–78), whose works also championed liberty. The *Encyclopédie,* edited by Denis Diderot (1713-84), included contributions from almost every major writer. Romantic writers of the 19th century included novelist, dramatist and poet Victor Hugo (1802–85), the popular novelist Alexandre Dumas (1802–70) and Honoré de Balzac (1799–1850), who had a great influence on later novelists. The realism and naturalism of the later 19th century is typified by the novels of Gustave Flaubert (1821–80) and Émile Zola (1840–1902) and the short stories of Guy de Maupassant (1850–93), in contrast to the symbolism of poets such as Stéphane Mallarmé (1842–98), Paul Verlaine (1844–96) and Arthur Rimbaud (1854–91). The great contributions of the 20th century have been by novelist Marcel Proust (1871–1922), whose masterpiece *À la Recherche du Temps Perdu* is regarded by many as the greatest of all French novels, and André Gide (1869–1951), who wrote novels, plays and essays. Other outstanding writers of the 20th century include Jean-Paul Sartre (1905–), leader of the Existentialist movement, and Albert Camus (1913–60), both novelists and dramatists, the playwrights Jean Anouilh (1910–), Jean Giraudoux (1882–1944) and Jean Cocteau (1889–1963), who also wrote novels, essays, poems and films; and the "theatre of the absurd" dramatists such as Irish-born Samuel Beckett (1906–), Jean Genet (1910–) and Romanian-born Eugène Ionesco (1912–).

France has a rich musical tradition going back to the 17th century, but few composers have used the larger musical forms. An outstanding exception is Hector Berlioz (1803–69), a master of orchestration. Many foreign composers settled in Paris, including Polish-born Frédéric Chopin (1810–49) and Belgian-born César Franck (1822–90), who had been a brilliant student at the famous Paris Conservatoire, and later Russian-born Igor Stravinsky (1882–1971), a great influence on young French composers. French music of the 19th century was mainly operatic, produced by such composers as Charles Gounod (1818–93), German-born Jacques Offenbach (1819–80), Léo Delibes (1836–91), Georges Bizet (1838–75) and Jules Massenet (1842–1912). Later composers include Gabriel Fauré (1845–1924), Claude Debussy (1862–1918) and Maurice Ravel (1875–1937), loosely referred to as the Impressionist group, and another group known as "the Six", which included Darius Milhaud (1892–1974) and Francis Poulenc (1899–1963).

The French tradition of opera and ballet is maintained in Paris by the famous Baroque-style Théatre National de l'Opéra and the Opéra-Comique. The

A traditional sword dance is performed in the town of Biarritz for the enjoyment of the many tourists who flock into this fashionable French resort.

Some remarkable megalithic monuments are found near Carnac, north-west France. Made of granite, huge menhirs (erect stones) stand in long rows outside the village.

The medieval walled town of Aigues-Mortes, south-east France, was built by Louis IX as an embarkation port for two crusades; today it is still remarkably intact.

former was the ballet centre of the world in the 18th century, and the 20th century revival began in Paris in 1909, when Diaghilev founded the Ballet Russe. France's theatrical life is concentrated in Paris, which has about 50 professional theatres, including the Comédie-Française (founded in 1680 by Louis XIV), with a largely classical repertoire, and the Théatre National Populaire, which specializes in modern drama.

History. France was inhabited by Celts and other peoples in ancient times. Roman armies began to invade the region, which they called Gallia (Gaul), in about 200 BC, and Julius Caesar conquered the native kingdoms between 58 and 51 BC. The Gauls adopted Roman customs and language (Latin) and prospered under Roman rule for hundreds of years. The Romans built towns, roads, aqueducts and theatres [*see* MM p.*200*]. In the 5th century AD the Roman Empire began to decline and its border defences began to crumble. The Franks – from whom France got its name – were the most successful of the Germanic tribes that invaded from the east. Clovis, who defeated the Romans in 486 at Soissons, founded the Merovingian dynasty and adopted Christianity. Charles Martel (the Hammer) became virtual ruler in the 730s and his son, Pepin (the Short), became king in 751 and founded the Carolingian dynasty. His son, Charlemagne, expanded the kingdom far beyond what is now France, and was crowned Emperor of the Romans in 800 by Pope Leo III. But the Carolingian dynasty declined after Charlemagne's death, and Hugh Capet started the Capetian dynasty (987–1328).

The monarchs increased their authority gradually at the expense of the nobility, and also regained lands ruled by English kings. The Hundred Years' War (1337–1453) with England was a catalogue of disasters for France until they rallied under the inspiration of Joan of Arc, eventually emerging from the Middle Ages as the leading nation-state of Europe. In the early 17th century French power and influence expanded under Cardinal Richelieu, minister to Louis XIII, and the reign of Louis XIV (1643–1715) was the golden era of the monarchy.

The French Revolution (1789–99) destroyed the power of the monarchy, and the First Republic was established in 1792. The infamous Reign of Terror began in 1793, when the leaders of the Jacobins (such as Robespierre and Danton) sent hundreds of people to the guillotine. Meanwhile Napoleon Bonaparte, who became a general in 1794, was winning victories abroad, and in 1799 he overthrew the revolutionary government (which had been taken over by moderates in 1795) and seized control of France. He was crowned Emperor in 1804, and France dominated European politics in the Napoleonic era. He ruled until 1814 and then again for a few months until his defeat at Waterloo in 1815.

A restoration of the monarchy was followed by the Second Republic (1848–52) and then the Second Empire (1852–70), when Napoleon's nephew Louis Napoleon declared himself Emperor Napoleon III. His adventurous foreign policy was at first successful but he declared war against Prussia in 1870 and the ensuing humiliating defeat brought the Second Empire to an end. The Third Republic was proclaimed, and successive governments established a colonial empire. In 1907 France allied itself with Russia and England in the Triple Entente.

During World War I the Germans occupied about a tenth of France and some of the worst battles of the war were fought on French soil. Although a victor, France emerged from the war seriously weakened economically and with some 10 per cent (1.3 million) of its able-bodied male population lost. The French, however, were slow to realize that they could no longer stand alone as a great power, and a succession of weak governments left France unprepared for the German onslaught in 1940 and they signed an armistice after little more than a month's fighting. The Germans occupied northern and western France, while the unoccupied zone in the south became an authoritarian French state, with its capital at Vichy. But Gen. Charles de Gaulle set up a "Free French" government in London, and a Resistance movement started in France.

When France was liberated in 1944, De Gaulle formed a provisional government but resigned in 1946, opposing the new constitution as not providing strong enough executive powers. The Fourth Republic was beset by political troubles at home and colonial revolts abroad. Fighting since 1947 to retain their empire in Indochina, the French finally withdrew in 1954 after defeat at Dien Bien Phu. Later that year the Algerian Revolt broke out. Nevertheless, with American aid and a series of national plans, the country struggled to its feet again and France was a prime mover in the formation of the EEC in 1957.

De Gaulle emerged from retirement in 1958 and established the Fifth Republic, with greater power for the president and less for parliament. He was elected president for a seven-year term. France withdrew from Algeria in 1962 and De Gaulle began to assert France's position internationally – making it a nuclear power as well as a dominant force in the EEC. With France no longer at war, industrial production accelerated and the economy flourished, particularly under Georges Pompidou, who became president when De Gaulle resigned in 1969. But tensions at home, which erupted in 1968 with revolts from both students and workers, were still evident, and inflation, unemployment and an ailing economy faced Valéry Giscard d'Estaing when he became president in 1974 [*see* MS p.*285*]. A member of the Independent Republican Party (supporters of the Gaullist government), he introduced measures to stimulate the economy. Map 20.

Planning Regions of France

Region	Population
Alsace	1,543,000
Aquitaine	2,614,000
Auvergne	1,371,000
Basse-Normandie	1,339,000
Bourgogne	1,605,000
Brittany	2,681,000
Centre	2,200,000
Champagne	1,365,000
Corsica	220,000
Franche-Comté	1,093,000
Haute-Normandie	1,621,000
Languedoc-Roussillon	1,830,000
Limousin	761,000
Lorraine	2,396,000
Midi-Pyrenees	2,332,000
Nord	3,938,000
Paris	9,896,000
Pays-de-la-Loire	2,825,000
Picardie	1,719,000
Poitou-Charentes	1,568,000
Provence-Côte d'Azur	3,731,000
Rhône-Alpes	4,872,000

France – profile

Official name French Republic
Area 547,026sq km (211,207sq miles)
Population (1976 est.) 52,904,000
Density 97 per sq km (250 per sq mile)
Chief cities Paris (capital) (1975) 2,289,800; Marseille, 889,028; Lyon, 527,890; Toulouse, 370,796
Government Head of state, Valéry Giscard d'Estaing, president (elected 1974)
Religions Roman Catholicism, Protestantism, Judaism
Language French
Monetary unit Franc
Gross national product (1974) £116,414,500,000
Per capita income £2,200
Agriculture Wheat, barley, maize, oats, potatoes, sugar-beet, rice, grapes, soft fruits, nuts, cattle, pigs, sheep
Industries Wine-making, beef and dairy products, fishing, steel, pig iron, motor vehicles, aircraft, electrical and mechanical engineering, chemicals, plastics, textiles, luxury goods, tourism
Minerals Iron ore, coal, potash, natural gas, sulphur
Trading partners (major) West Germany, Belgium and Luxembourg, Italy, USA, Netherlands, Britain

Moorea, 12 miles off the coast of Tahiti in French Polynesia, has great scenic beauty, with eroded volcanic peaks and sandy coral beaches.

Bamboo, in the tropical regions of Africa, can be used for a multitude of purposes: building, furniture, as writing material, and even as food.

Banjul, capital of The Gambia, is situated at the mouth of the Gambia River in West Africa; it is the nation's chief port and was formerly known as Bathurst.

French Guiana (La Guyane française), official name Department of French Guiana, is a French overseas département on the north-eastern coast of South America. It was once the site of the infamous Devil's Island penal colony. Much of the country is covered by uninhabitable tropical forest and most of the small population lives along the coast, with 75 per cent of the people in or near Cayenne, the capital. The country's economy depends on France, although exploitation of the large bauxite deposits could change this situation.

The region was explored by the Spanish in about 1500 and settled by the French in 1604. After being controlled at various times by Dutch, English and Portuguese, the colony became a permanent French possession in 1817. The people, most of whom are French-speaking, have had full French citizenship since 1848 and from 1870 have sent two deputies to Paris. French Guiana became a French département in 1946. Area: 91,000sq km (35,135sq miles). Pop. (1975 est) 60,000. Map 76.

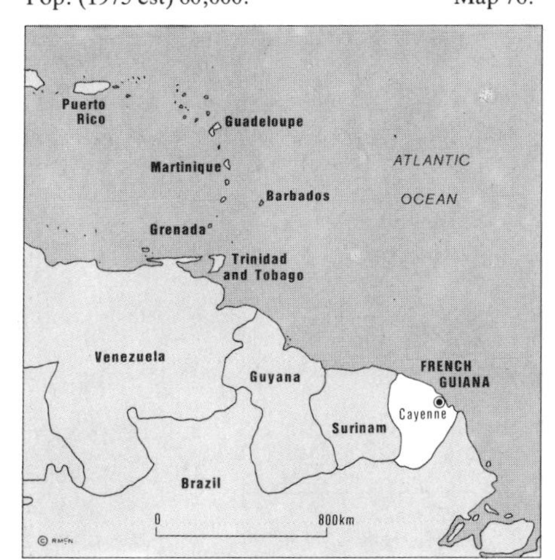

French Polynesia is a group of about 105 islands in the southern Pacific Ocean. An overseas territory of France, it includes the Society Islands (location of the capital, Papeete, on Tahiti), Marquesas Islands, Austral Islands, Tuamotu Islands [*see* PE p.*56*] and Gambier Islands. The principal sources of income are tourism and the export of tropical fruits and copra. Most of the islands were discovered in 1767 by the English navigator Samuel Wallis but a French claim was made a year later by Louis de Bougainville. They were visited by Capt. James Cook in 1769. Most of the islands came under French protection before 1850 and the whole group became a French colony in the 1880s; in 1946 all the indigenous population was granted full French citizenship, with a representative in the National Assembly. Area: approx. 4,000sq km (1,500sq miles). Pop. (1975 est.) 135,000. Map 62.

Frisian Islands are a group of about 30 low-lying islands in the North Sea 5 to 32km (3-20 miles) off the coast of Denmark, The Netherlands and West Germany, to which countries they belong. The inhabited West Frisians (Dutch) include Ameland, Schiermonnikoog, Terschelling, Texel and Vlieland; and the East Frisians (West German) include Baltrum, Borkum, Juist, Langeoog, Norderney, Spiekeroog and Wangerooge. Among the North Frisians are the German-held Amrum, Föhr, Pellworm and Sylt and the Danish islands of Fanø, Manø and Rømø. The few permanent inhabitants raise cattle and sheep; many of the islands are holiday resorts. Map 18.
See also PE p.*44*.

Gabon, official name Gabonese Republic, is an independent country in western central Africa. By African standards it is an affluent nation, with an average annual income per person of £534 in 1973. It owes its wealth to its minerals, especially petroleum from the coast. The capital is Libreville.

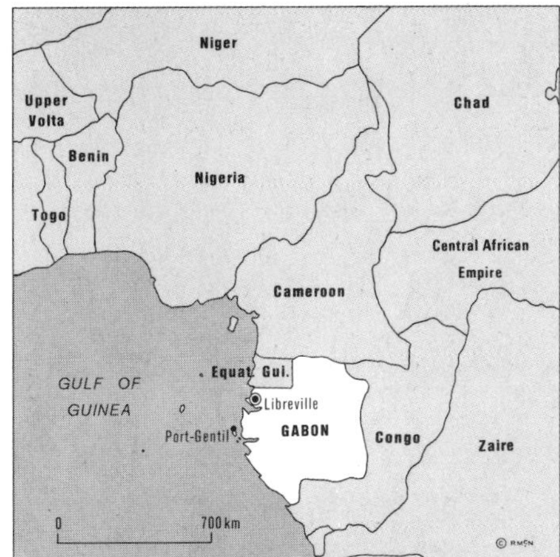

Most of Gabon lies in the basin of the River Ogooué, which cuts through high coastal mountain ranges. Inland is a dissected tableland, between 600 and 900m (1,968-2,952ft) above sea-level. The climate is hot and wet; annual temperatures average 25°C (77°F) and the rainfall everywhere exceeds 1,520mm (60in) per year, reaching 3,800mm (150in) in the north-west. Rain forest covers most of Gabon.

The chief subsistence crop is cassava and some cocoa, coffee and palms (for palm oil) are grown. Exports are, however, dominated by minerals and timber. Petroleum accounts for half of the value of the exports and manganese, thorium and uranium are important. Timbers include ebony, mahogany and okoumé. Most people belong to one of 40 Bantu-speaking groups, largest of which is the Fang. Minorities include about 12,000 Europeans and a few thousand pygmies.

French influence in the area began in the late 18th century, and Gabon became part of the French Congo in 1889. In 1910 Gabon was incorporated into French Equatorial Africa. From 1913 Lambaréné in Gabon became known as the site of the missionary hospital of the Nobel peace prize winner Albert Schweitzer who died in 1965. From 1946 it was administered as a French overseas territory. Since independence in 1960, Gabon has been fairly stable, although French troops put down an attempted coup in 1964. Upon assuming the presidency in 1967, Omar Bongo declared Gabon a one-party state. It is now ruled by the president and the elected National Assembly. Map 34.

Gabon – profile

Official name Gabonese Republic
Area 267,667sq km (103,346sq miles)
Population (1975 est.) 1,155,800
Chief city Libreville (capital) (1975 est.) 169,200
Religions Christianity, ethnic
Language French (offical), Bantu languages
Monetary unit CFA franc

Galápagos Islands are a group of about 15 volcanic islands and many small islets belonging to Ecuador on the Equator in the eastern Pacific Ocean, 1,045km (650 miles) west of Ecuador. The larger islands have vegetation only on the upper mountain slopes, the remainder of the terrain being barren lava. The Galápagos were discovered in 1535 by the Spanish navigator Tomás de Bertanga and claimed by Ecuador in 1832. Following the visit of the British naturalist Charles Darwin in 1835, the islands became known for their unusual range of wildlife, including giant tortoises whose name (in Spanish) became the islands' [*see* NW pp.32-33, *229*]. Area: 7,845sq km (3,029sq miles). Pop. (1973 est.) 4,205.

Gambia, official name Republic of The Gambia, is Africa's smallest nation, forming a narrow enclave within SENEGAL. Its economy is based almost entirely on groundnuts and groundnut products, which together account for 95 per cent of the country's exports. The average annual income per person was only £53 in 1973.

Land and climate. Gambia is a narrow country about 320km (200 miles) long, varying in width from 24 to 48km (15-30 miles) on either side of the Gambia River, which rises in the Fonta Djalon plateau in Guinea and flows through Gambia towards the Atlantic Ocean for the last 475km (295 miles) of its course. The capital Banjul (formerly Bathurst) stands on a coastal island. East of the sandy coast are low-lying grasslands which are flooded in the wet season, with mangrove swamps on the lower reaches of the Gambia River. Sand-

Former emperors of Germany were crowned at the 15th-century cathedral of St Bartholomew in Frankfurt am Main; the tower is 95m (312ft) high.

Ulm developed as a port and crossing-point of the River Danube (Donau). The tower of the 14th century minster was added in the late 19th century.

One of the oldest German cities is Regensburg, which is situated at the confluence of the Danube and Regen rivers; a Roman gate built in AD 179 still remains.

stone hills and low sandstone plateaus, between 30 and 50m (98-164ft) above sea-level, lie farther inland. The climate is tropical and rainfall averages between 760 and 1,140mm (30-45in) per year.

Economy. Groundnuts are the main cash crop, grown mostly on small farms in the inland sandstone regions. Around the estuary palm kernels and nuts are grown and fish provide an extra source of protein. Rice cultivation is developing and some cattle are reared. Mining is insignificant and manufacturing accounts for only two per cent of the gross national product.

People and government. Gambians are divided into five main groups, the largest of which is the Mandingo. About 80 per cent of the people are Muslims; of the remainder most are Christians and others follow ethnic religions. Gambia became a republic in 1970. The president, Sir Dawda Jawara, rules with a cabinet. Parliament consists of 32 elected members, four chiefs elected by the Chiefs in Assembly and some nominated members.

History. Britain's influence in Gambia dates from 1588 when English merchants first obtained trading rights in the area. Between 1765 and 1783 the country formed part of the Crown colony of Senegambia. The Gambia River was a major route for slave traders. The traffic in slaves was formally abolished in 1807 and Gambia was a British colony from 1888 until 1965, when it became independent. Since then it has co-ordinated its defence and foreign policies with Senegal. Moves to amalgamate the two countries have failed, mainly because of the widely differing British and French traditions. Map 32.

Gambia – profile

Official name Republic of The Gambia
Area 11,295sq km (4,361sq miles)
Population (1975 est.) 524,000
Chief city Banjul (capital) (1975 est.) 42,689
Religion Islam
Language English (official)
Monetary unit Dalasi

Gaza (or **Ghazzah) Strip** is a small area on the south-eastern coast of the Mediterranean Sea bordering on Israel and until 1967 administered by Egypt. It was part of Britain's Palestine Mandate after World War I and by 1947 Britain intended to give up the area. The peace settlement after the 1948-49 Arab-Israeli War made it an Egyptian possession; it was occupied by Israel after the 1967 war and is populated mainly by Arab refugees. Area 370sq km (140sq miles). Pop. (1971) 365,000. Map 38.

Georgia. *See* UNION OF SOVIET SOCIALIST REPUBLICS.

Georgia. *See* UNITED STATES.

Germany was a former nation in central Europe that since 1949 has been divided into two independent countries, East Germany (the German Democratic Republic) and West Germany (the Federal Republic of Germany). Some territory was also transferred to other countries, principally Poland and the USSR. By a treaty of 1972, East and West Germany recognized each other's sovereignty.

A unified German nation existed from the end of the Franco-Prussian War in 1871 to the end of World War II in 1945. Twice in the 20th century, 1914–18 and 1939–45, Germany was defeated in world wars. This article deals with the history of Germany up to 1945. For its subsequent history, *see* GERMANY, EAST and GERMANY, WEST.

In prehistoric times Germanic tribes inhabited the forests of northern Europe. Later, as the Celtic cultures in central Europe declined, the Germans spread northwards and southwards, eventually encroaching upon the borders of Imperial Rome. Caesar and Tacitus both found that even the skilful and disciplined Roman legionaries were tried by the courage and audacity of these marauders, who to them seemed also barbarous and treacherous.

By the 5th century the Franks had firmly established themselves in what is now France and in the Rhineland. The Frankish tribes were united by Clovis (*c*.466–511), founder of the Merovingian dynasty. By accepting Christianity in about 491 he established a link with Rome that three centuries later led the pope to crown the Frankish Charlemagne as Roman Emperor. Following Charlemagne's death, the German lands east of the Rhine passed as a separate kingdom to his grandson, Louis II (the German); it is he who is now regarded as the first ruler of Germany. His kingdom was later split between his three sons, but reunited again by the emperor Arnulf. Arnulf's son, Louis III (the Child; the last of the Carolingians to rule in Germany) died in 911 at the age of 18.

For the following three centuries the German empire – successor to the Roman Empire (although no longer including most of Gaul) – wanted to incorporate the heartland of its ancient heritage. The continuing domination of Italy became the chief preoccupation of the imperial policy makers. Rome was the centre of the Christian faith, which the emperors believed was their God-given duty to protect. But the emperors' assertions of their claims often led to conflict with the popes, especially over the ill-defined division between spiritual and temporal authority over the clergy.

During the 13th century the German cities became rich and they resented interference in their affairs by local territorial magnates, so they formed leagues to protect their interests. One great association of trading cities was the Hanseatic League, which was centred on Lübeck and had more than 100 members from the Low Countries to Poland. At the height of its power, the 13th to 15th centuries, it developed monopolies, regulated customs and put down piracy in the North Sea and the Baltic.

The coronation of Rudolf of Hapsburg in 1273 after a prolonged period of political disorder made him the first king of the remarkable Hapsburg family. By the time Albert of Hapsburg, Duke of Austria and King of Hungary, came to the throne in 1438 it was already a tradition for the imperial crown to be offered to the head of the house of Hapsburg. Moreover the "election" of the emperor still took place in Frankfurt-on-Main, but the trend of power had already moved away from (what is now) Germany and towards Austria.

The empire's lack of unity and the conflict of interest between the emperor and the princes became increasingly apparent during the Protestant Reformation. Although Charles V allied himself with the pope in trying to suppress Martin Luther and his adherents, several princes gave the Protestants their support, and some increased their strength by enriching themselves with the lands and other properties of the Church.

After the end of the Peasant's War in 1525, Charles V – who was far more accustomed to military successes against the French or the Turks – resolved to maintain the unity of the Church. He duly defeated an alliance of Protestant princes and free cities known as the League of Schmalkalden. But in 1555 the Protestant and Roman Catholic rulers decided to settle their differences peaceably. The Peace of Augsburg declared that Protestantism would be permitted in any state and that Protestant states should have equal status with Catholic states, thus affording formal recognition of a new kind of disunity within the empire.

The cities and regions of Germany suffered worst in the Thirty Years War, which began as a civil war in Bohemia and then developed into a protracted and complex series of skirmishes between Roman Catholic Europe and Protestant Europe. Some of the countries participating did so more from motives of territorial gain than of religious conviction, however, and at the Peace of Westphalia in 1648 France emerged as the strongest power in Europe and took possession of Alsace and Metz, leaving the Holy Roman Empire to exist in name only.

The end of the Holy Roman Empire was finally brought about when, after the emperor in Vienna abandoned his title of German Emperor to become Emperor of Austria in 1806, Napoleon Bonaparte seized territories that belonged to Austria and Prussia. He organized Bavaria, Württemburg, Saxony, Westphalia and Baden into a Confederation of the Rhine under his protection. (Eventually this same Confederation together with Prussia was to play a major part in Napoleon's final defeat.)

After the upheavals of the Napoleonic Wars, many Germans strongly desired unity, but few of the larger states were prepared to lose their sovereignty. The Congress of Vienna (1814–15) eliminated hundreds of petty states by incorporating them with other territories, thus arriving at a final total of 39 German states. But although a German Confederation was established with a *Bundes-*

The half-timbered Renaissance buildings of the old quarter of Hanover were destroyed during World War II; they have since then been carefully restored.

Trier is an ancient city, now in West Germany, which still retains much of its Roman architecture, including this bridge over the River Mosel.

Dresden has many buildings with historic associations, including the Zwinger Museum, now restored after being damaged in World War II.

tag (a federal assembly with no legislative powers) at Frankfurt-on-Main, it had no effect in terms of the unification of Germany. This was because the Confederation was dominated by the forceful and conservative Austrian chancellor, Prince Metternich, who wished to preserve the status quo.

Prussian leadership of a united Germany separate from Austria became the goal of Otto von Bismarck, appointed chief minister under William II of Prussia in 1862. He unconstitutionally dissolved the parliament that had refused the king funds to reform and enlarge the army, thereafter levying taxes for that purpose. He then embarked on three wars in the quest for a German Empire. First, together with Austria, Prussia defeated Denmark and gained the territory of Schleswig. Then in 1866 Bismarck accused Austria of breaking the Gastein Convention that had ended the Danish War; in the Seven Weeks War that followed, Prussia crushingly defeated Austria at Sadowa and annexed Hanover, Hessen, Nassau, Schleswig-Holstein and the free city of Frankfurt. In 1867 a new North German Confederation, without Austria, was set up with the King of Prussia as president. Finally in 1870 war was joined between Prussia and France – a war desired by France in order to curb Prussian pretensions, and desired by Prussia to demonstrate its strength and to rouse German nationalism in the southern German states. Bismarck's forces won and, in the peace treaty with France, Germany gained Alsace and Lorraine.

The new German Empire grew in prosperity and confidence. Its legislature now consisted of the *Reichstag* (a lower house with no authority in foreign or military affairs) and the *Bundesrat* (an upper house dominated by the Prussian chancellor Bismarck). Consolidation of the Empire seemed to Bismarck to be his next task, which he set about fulfilling by means of alliances with other countries and by discouraging within Germany those interests he believed hostile (among which he included Roman Catholicism).

But these alliances did not prevent the outbreak of World War I. When Austria-Hungary declared war on Serbia in 1914, Russia mobilized and moved troops to its western borders. Germany immediately declared war on Russia and its ally France. German armies fought their way through Belgium (causing Britain to enter the conflict) and northern France. But with British help the French held them at the River Marne, and although Russia capitulated in 1918, the length of the war together with the entry of the United States into the arena in 1917 made it impossible for Germany and its exhausted allies (Austria-Hungary, Bulgaria and Turkey) to prevail. In late 1918 there was open revolt inside Germany; on 9 November William II abdicated and Germany was proclaimed a republic. Two days later Germany formally surrendered to the Allies, signing the peace treaty in 1919 at the Palace of Versailles in the same Hall of Mirrors in which the German Empire had been born 49 years earlier.

Under the terms of the Treaty of Versailles, Germany lost all its colonies; its army was restricted to 100,000 men; the south bank of the Rhine was temporarily occupied by the victors; and France recovered Alsace-Lorraine. In 1921 reparations were provisionally assessed at £12,000 million.

Friedrich Ebert was the first president of the "Weimar Republic". His main problem – apart from occasional attempts by the extreme left or right to seize power (including one attempt by Adolf Hitler, leader of the fledgling National Socialist German Workers' Party, in 1923) – was ungovernable inflation and the resultant chaotic economic situation. At one time 20,000 million marks were equivalent to only one pound sterling.

Germany's fortunes seemed to improve, however, on the election of the wartime hero Field-Marshall Paul von Hindenburg to the presidency in 1925. The economic situation was being brought under control, and the Locarno Pact of 1925 between Europe's major powers brought warmth rather than bitterness to international affairs. But the world depression again devastated the German economy, and extremists took advantage of the general misery and discontent. The National Socialists (Nazis) and Communists increased their strength in the Reichstag to the extent that a government could not be formed without Nazi support, and Hitler was therefore appointed chancellor by Hindenburg. An election was called for March 1933, but in February the Reichstag was destroyed by fire; the Nazis blamed the Communists and a state of emergency was proclaimed. The elections were held in an atmosphere of Nazi intimidation.

In the new house the Nazis held 288 seats out of 647; with Nationalist support (52 seats) this gave them a narrow majority. Hitler then manoeuvred the passing of an Act giving the government power to rule by decree, independent of the Reichstag and the president – and when Hindenburg died in 1934, the presidency was abolished altogether. Hitler became the *Führer* (leader) of the "Third Reich" (the first had been the Holy Roman Empire, the second Bismarck's empire). [*See also* MS *p.277.*]

In 1938 Germany annexed Austria – the so-called *Anschluss.* Hitler also made clear his intention of seizing the Sudetenland, the German-speaking part of Czechoslovakia. In the Munich Agreement, Britain and France agreed to this seizure provided it marked the end of Germany's ambitions. But six months later Hitler took the rest of Czechoslovakia as well. Then on 1 September 1939 his armies invaded Poland and quickly occupied it, except for the eastern part which was taken by the USSR. On 3 September France and Britain (followed immediately by the countries of the Commonwealth) declared war on Germany, and World War II began.

Initially the German forces had swift and crushing successes. Then in December 1941 Japan's attack at Pearl Harbor brought the United States into the war against Germany. From then on it became a losing war of attrition for the German forces: defeated in Africa and Italy, and slowly driven back into Germany by the Red Army in the east and (after D-Day, 6 June 1944) by the British, Americans and others in the west. The German homeland was devastated from the air. In April 1945 Hitler committed suicide, and in May his successor, Grand Admiral Karl Doenitz, agreed to surrender.

Germany was divided into four zones of occupation by the major victorious powers – British, French, Soviet and American. Berlin, in the Soviet zone, was similarly divided into four sectors. Nazis accused of war crimes or other crimes against humanity were put on trial, and the Germans set about the task of rebuilding their lives and their ruined country. Map 18.

Germany, East, official name German Democratic Republic (*Deutsche Demokratische Republik*), is a Communist country of northern Europe, formed after World War II from the north-easterly part of GERMANY that comprised the occupation zone of the Red Army. The rest of pre-war Germany – the part occupied by the Western Allies – is now a separate country called West Germany (*see* GERMANY, WEST). Today East Germany is the richest Communist nation in Europe apart from the USSR, but in 1945 it was the most devastated part of a war-ruined country. The city of Berlin lies within the territory of East Germany, but only East Berlin ("Democratic Berlin") belongs to it. East Berlin is the country's capital.

Land and climate. The eastern boundary of the country is the line of the rivers Oder and Neisse. The parts of Prussia and Silesia lying east of this line were transferred to Poland by the Potsdam Conference of 1945, a segment of East Prussia (including the capital Königsberg, now Kaliningrad) going to the USSR. The greater part of East Germany is within the northern European plain, and slopes gently downwards towards the north and north-west. Inland from the sandy Baltic coastlands is the low Mecklenburg plateau with lakes and wooded hills.

The southern, mountainous region of the country has many of the most important industrial cities, including Erfurt, Jena, Leipzig, Zwickau, Karl-Marx-Stadt and Dresden. One of the peaks of the Harz Mountains, in the west of this region, is the Brocken (1,142m; 3,747ft), famous in German legend for the witches' sabbath held there on *Walpurgisnacht* (the eve of 1 May), named after the English missionary St Walpurgis (Walburga). In the south-west are the thickly wooded hills of the Thuringian Forest. The Erzgebirge (Ore Mountains), in the south-east, mark the border with Czechoslovakia. One of Europe's great rivers, the Elbe, flows south-east to north-west across the country. As in most of Germany, the climate is generally mild, but winters in the east can be extremely cold.

Economy. East Germany produces about one-third of the world's lignite (brown coal); some of it is exported, but much is used at home in industry, for domestic heating and for generating electricity.

The horizon at Warnemünde, East Germany, is spiked with cranes and harbour machinery; this Baltic port has a shipbuilding yard and is also a resort.

Much of East Berlin has been rebuilt since 1960; the imposing blocks overlooking Alexander Square are apartments with modern shops at ground level.

Eisenach, East Germany, has had some world famous residents; Martin Luther lived in this house (1498–1501) and J. S. Bach was born in Eisenach in 1685.

Apart from this, the country is poorly endowed with minerals. Copper ore is found in the Harz Mountains, and small amounts of uranium, cobalt and bismuth in the Erzgebirge. Some black coal and iron ore are mined, and there are considerable deposits of potash.

Before the division of Germany, the whole country depended greatly for food on the farmlands of the east. Agriculture is still of major importance although it employs only 13 per cent of the working population. Most of the land is farmed by collectives; the estates of the *Junkers,* the former landowning class, have been broken up and the land redistributed. The most fertile land is in the south and the north; the central region is sandy, but produces fair yields of rye. The other principal crops are sugar-beet, potatoes, wheat, barley and oats. Although most of the country's farming land is classified as arable, many farmers raise livestock. The most common farm animal is the pig; there are also large numbers of cattle and poultry. [*See also* PE pp. 230, 235.]

East Germany, like West Germany, has to import most of its industrial raw materials. Industry is almost entirely state controlled; and it accounts for two-thirds of the national income. Steel manufacture and its associated industries (concentrated along the River Elbe) are of primary importance; the major products are heavy machinery, machine tools and motor vehicles. The chemical industry has a wide range of products, including fertilizers, dyes, synthetic rubber and industrial chemicals; electrical engineering, textile and food-processing industries are also important. Well-known smaller industries include the making of cameras and other optical equipment at Jena and of porcelain at Meissen ("Dresden china"). East Germany's chief trading partners are the USSR, other Communist countries of eastern Europe and West Germany.

The airline Interflug operates international services, chiefly to other countries of the Communist bloc. Internal communications need modernization; many highways are badly surfaced, and much railway rolling stock is old.

Government. Under the constitution of 1968, the *Volkskammer* (people's chamber) of 500 deputies is the supreme organ of state power. It elects the *Staatsrat* (council of state), the *Ministerrat* (council of ministers), the National Defence Council, the Supreme Court and the procurator-general. In practice, power lies in the hands of the politburo of the Socialist Unity Party of Germany (SED). There are some other small political parties, but they all belong to the National Front, which follows the SED line. The chairman of the *Staatsrat* acts as head of state. For regional government the former *Länder* (states) of Mecklenburg, Saxony-Anhalt, Brandenburg, Saxony and Thuringia have been divided into 14 *Bezirke* (districts).

People. East Germany lies in the strongly Protestant part of the old Germany. The constitution permits freedom of religion, but in the face of intense official disapproval the exact extent of religious affiliation is hard to determine. In the 1964 census 59 per cent of the population declared itself Protestant and 8 per cent Roman Catholic.

Education is free and compulsory between the ages of six and 15. The basic school is the polytechnical high school; a further two years' schooling at an extended polytechnical high school is optional. Entrance to one of the 54 universities is determined by a board which, among other criteria, takes account of the applicant's proficiency in Marxist-Leninist studies. Karl Marx university, formerly Leipzig university (founded in 1409), is the oldest.

Between 1949 and 1961 about 3,000,000 East Germans left their country to live in West Germany – even more had left the region between the end of World War II and 1949. The strengthening of the frontier (including the building of the Berlin Wall) put an end to this traffic; even today, the border guards open fire on anybody trying to cross the

frontier without permission. The standard of life has gradually improved since the beginning of East Germany's own, smaller "economic miracle" in the 1960s. In the rebuilding of the historic cities an effort has been made to recover something of their former character, but more common are buildings in the "Russian grandiose" style or in rather bleak functional designs.

Despite the basic harshness of the regime, many of the distinctive traditions and habits of German life still continue. The theatre is as popular as elsewhere in German lands, and in political cabarets performers are permitted some freedom of satire. Restaurants have introduced many dishes from the USSR and other countries of the Eastern bloc, but the old-fashioned pork and *wurst* (sausage) dishes remain extremely popular. Several good beers are made in the nationalized breweries, and some wine is made in the south-east. As in West Germany, physical fitness is highly regarded. Thousands of sports clubs provide facilities for adults as well as children, and East Germany has a high reputation in international sports events.

History since 1945. (For the history of the country before 1945, *see* GERMANY.) The most important political figure in the early years of East Germany was the veteran Communist leader Walter Ulbricht, who had left Germany for Moscow in 1933. In 1949 he became deputy premier and in 1960 chairman of the *Staatsrat.* Ulbricht saw his task as the rebuilding of the country and the fashioning of its institutions in the correct Marxist-Leninist mould. In 1950 the eastern frontier was fixed permanently on the Oder-Neisse line by a treaty with Poland.

In 1953 there were workers' uprisings in East Berlin and other cities. They were put down by the Red Army, but in 1955 the USSR recognized East German sovereignty, although still kept its forces in the country. The economy began to improve in the 1960s, the decade that started with the sealing of the western border and the building of the Berlin Wall. At this time 30,000 people a month were crossing to the West, a manpower loss that the economy could not stand. A softening of the Soviet line towards West Germany was reflected in East German receptivity to West Germany's *Ostpolitik* of Willy Brandt, particularly after Ulbricht was succeeded by Erich Honecker as the dominant figure in the party in 1971. East and West Germany signed a treaty in 1972 agreeing the basis of their mutual relationship, and tensions between the two countries relaxed considerably in the following years. Map 18.

East Germany – profile

Official name German Democratic Republic
Area 108,178sq km (41,610sq miles)
Population (1976 est.) 16,850,000
Density 156 per sq km (403 per sq mile)
Chief cities East Berlin (Democratic Berlin, capital) (1976 est.) 1,098,200; Leipzig, 574,400; Dresden, 506,100
Government Communist republic
Religions (1969) Protestant (80%), Roman Catholic (10%)
Language German
Monetary unit GDR mark
Gross national product (1974) £25,162,400,000
Per capita income £1,493
Agriculture Sugar-beet, potatoes, cereals, pigs, cattle, poultry
Industries Iron and steel, heavy machinery, motor vehicles, electrical equipment, chemicals, textiles, processed foods, optical products, ceramics
Minerals Lignite, potash, copper, uranium, bismuth, iron
Trading partners (major) USSR, Poland, Czechoslovakia, West Germany

The River Rhine, the principal river in Europe, rises in the Swiss Alps before emptying into the North Sea, and is one of the world's busiest rivers.

The major winter sports centre in West Germany is Garmisch-Partenkirchen, a picturesque town that was the scene of the 1936 Olympic winter games.

Erich Mendelsohn (1887–1953), a German architect, was noted for his prominent and imaginative use of glass in strongly horizontal compositions.

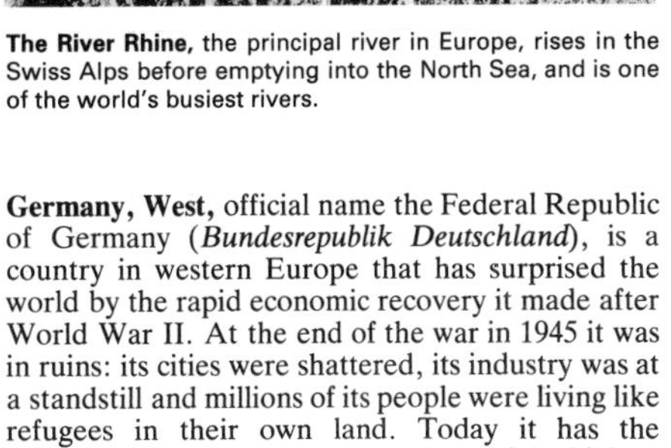

Germany, West, official name the Federal Republic of Germany (*Bundesrepublik Deutschland*), is a country in western Europe that has surprised the world by the rapid economic recovery it made after World War II. At the end of the war in 1945 it was in ruins: its cities were shattered, its industry was at a standstill and millions of its people were living like refugees in their own land. Today it has the strongest economy in Europe and one of the highest living standards in the world. The capital is Bonn.

The country is only part of prewar Germany – the part that was occupied by the Western Allies at the end of the war. The zone that came under Soviet control is now a separate country called East Germany (*see* GERMANY, EAST). In building up their country, the leaders of West Germany have sought, successfully, to erase the stigma of militarism that was attached to Germany in the 75 years between the foundation of the Weimar Republic and the fall of Hitler. In the EEC, West Germany has forged close links with other European nations.

Land and climate. The north of the country lies in the vast northern European plain that extends from the Low Countries across the Germanies and Poland into the USSR. Some of the plain is high and bleak, but there is good farming land in its southern part. The Baltic coastal areas are generally sandy, with many inlets. The lands bordering the North Sea are distinctively different; they are broken and marshy and are reinforced with dikes in many places. The offshore waters are shallow, with low, sandy islands (*see* FRISIAN ISLANDS).

South of the plain are the central uplands, again part of a physical region that stretches across the continent. The uplands have some of Germany's most beautiful and varied scenery. In the west are the wooded hills and characteristic round crater lakes of the Eifel (which continues into Belgium as the Ardennes). To the east of the Eifel is the valley of Germany's great river and one of Europe's most important waterways, the River Rhine. In the northern section of the Rhine basin is the industrial landscape of the *Ruhrgebiet* – a place of crowded, busy cities. The central part of the river valley is the Rhine of legend and romance – of the Nibelungen and the Lorelei Rock – with its high, castle-crowned sides covered with vineyards and thick woods. To the south of the Eifel is another valley famous for its vines and its picturesque villages, that of the River Mosel (Moselle).

The most southerly part of the country is a highland region. Its mountain ranges include the Schwarzwald (Black Forest), named after the dark fir woods that clothe its slopes. To the extreme south is the German section of the Alps; although relatively low it has some of the most spectacular of alpine scenery. One of its peaks, the Zugspitze (2,963m; 9,722ft), is the highest in Germany.

Almost all of West Germany has a mild climate that becomes colder towards the east and south. In alpine areas the winters can be severe. The highest rainfall is in the mountains.

Physical resources. The development of heavy industry in Germany was based on the abundant coal resources, both hard coal and lignite (brown coal). The industrial towns of the Ruhrgebiet are situated in one of the world's largest coalfields. Hard coal is also mined in the Saar, and lignite in Lower Saxony, Hessen and Bavaria. Germany also has iron ore, chiefly in Lower Saxony, but most of it is of poor quality. There are small deposits of ores of other metals, including copper, zinc, lead and tin. About one-tenth of the country's oil requirements are home produced; the chief oilfields are, again, in Lower Saxony. Other important minerals include carnallite and rock salt.

Germany's forests are a national asset both for their beauty and their commercial value. They are carefully tended and tree cutting and replacement are strictly regulated. Today coniferous trees – pines and firs – predominate, but beeches, limes and birches are also common. The production of timber is a major industry, and wood is also important for many of the crafts – carving and the making of toys, clocks and other wooden or partly wooden goods – for which Germany has long been known.

Constitution and government. The West German republic grew out of the amalgamation of the three zones occupied by the military forces of Britain, France and the United States after World War II. It attained full sovereignty on 5 May 1955 when the occupation formally ceased, although the republic was proclaimed earlier in 1949 after the Berlin airlift. A constitutional democracy, the country is a federation of 11 *Länder* (states), one of which, West Berlin, has not yet been incorporated.

The republic has a president, who is elected by a special procedure for a term of five years. The legislature consists of a lower house, the *Bundestag,* and an upper house, the *Bundesrat.* The Bundestag has 518 members (including 22 non-voting members from Berlin) elected for a term of four years by direct universal suffrage. The president nominates as chancellor (premier) the leader of the party or group that has a majority in the Bundestag, and the Bundestag ratifies (or rejects) his choice. The chancellor selects a cabinet of ministers, usually from among the members of the Bundestag.

The Bundesrat consists of representatives of the governments of the *Länder,* each *Land* having three seats or more, according to population. Laws adopted by the Bundestag are submitted to the Bundesrat, which has limited powers of veto. The major political parties are the Social Democratic Party (SPD), the Christian Democratic Union (CDU), the Christian-Social Union (CSU) and the Free Democratic Party (FDP).

Justice is administered by federal courts and courts of the *Länder.* The highest courts are the Federal Constitutional Court (the supreme court) and the Federal High Court, both at Karlsruhe.

Armed services. Under the treaties establishing the country's sovereignty, West Germany undertook to contribute to Western defence within the framework of NATO. The *Bundeswehr* (federal armed force) has a total strength of about 500,000 men. The navy has a number of submarines as well as small surface craft. The *Luftwaffe* (air force) is equipped with modern aircraft organized almost entirely in defensive units.

Agriculture. West Germany depends to a considerable extent on imported foodstuffs. A major reorganization of agriculture has taken place since World War II because much of the most productive farming land of prewar Germany is now in East Germany. In general the farms are small and the prevalent practice of strip farming does not permit efficient cultivation. Government schemes to improve efficiency have, however, had considerable success. The chief cereals grown are wheat (principally in the south), barley, oats and rye (in the north). Sugar-beet, the major crop after cereals, is grown in Lower Saxony and in part of the Rhineland. Large crops of potatoes are produced in many

parts of the country, particularly in Bavaria and Lower Saxony. The valleys of the Rhine and the Elbe are known for their orchards. In Bavaria, famous for its beer, there are many hop farms [*see* PE p.204]. Another drink for which Germany is renowned is wine. Most German wines are white, and some of them are among the finest white wines in the world. The principal vineyards are in the valleys of the rivers Rhine, Mosel (Moselle), Saar and Ruwer (Ruhr) and in Franconia [*see* PE p.200, 202].

Livestock is raised in most parts of the country, but is of major importance in the north and in the southern highlands. The most common farm animal is the pig; pork figures prominently on any German menu. Much land is also devoted to cattle and poultry farming, but there are few sheep and goats. The dairy-farming industry is one of the most advanced in Europe, and Germany makes many good cheeses

Dusseldorf is a major industrial and commercial centre; the Goethe Museum (over 20,000 manuscripts) and an art academy are sited there too.

One of the many attractions to tourists visiting West Berlin is the Kurfurstendamm, an elegant street comprising shops and theatres.

Apart from being West Germany's busiest port, Hamburg has large shipyards and other manufacturing industries; the city has a university (founded 1919).

and other milk products. *See also* PE pp.228-235.

Industry. German industry concentrates on productivity and high quality goods. This is particularly true in heavy engineering, chemical production and precision engineering. The greatest concentration of heavy industry is in the *Ruhrgebiet* in North Rhine-Westphalia (which includes the rich industrial cities of Essen, Bochum, Duisburg, Gelsenkirchen, Oberhausen and Dortmund), where the chief industry is iron and steel; other steel centres are the Saar and the neighbourhoods of Bremen and Salzgitter in Lower Saxony. Many other major industries are associated with steel production. The most important of them is the manufacture of motor vehicles – not only motor cars, for which Germany has long had a reputation of functional design and reliability, but also heavy goods vehicles and buses. Other associated industries are the making of industrial machinery, tools and fine cutlery. Germany is Europe's leading producer of electrical goods.

Next in importance to iron and steel production is the chemical industry. The largest chemical plants are in the *Ruhrgebiet* and at various points along the Rhine, including Leverkusen (near Düsseldorf) and Ludwigshafen. The industry makes an immense variety of products: industrial chemicals, drugs, dyes, fertilizers and plastics. Oil refining and the manufacture of petroleum products have increased in importance in recent years, and the textile industry – the country's oldest industry – is still one of the most important in Europe. It produces both natural and synthetic fabrics.

West Germany has a leading place in several industries requiring particular skills and precision, such as the making of musical instruments, watches and porcelain, and in printing and cartography. Food processing gives employment to millions of workers; major sectors of this industry include sugar refining (from sugar-beet), meat and dairy products, and flour milling.

West Germany is one of the few countries in Europe that has suffered from a shortage of workers to meet the needs of its industrial expansion. Millions of "guest workers" have been recruited to man its factories, many from Turkey and Italy.

Trade and economics. West Germany is the fourth largest industrial power in the world (after the United States, the USSR and Japan) and, economically, the strongest member of the EEC. It imports chiefly food and raw materials and exports manufactured goods. Its principal trading partners are France, The Netherlands, the United States, Belgium and Luxembourg, Italy and the USSR.

Transport and communications. The busiest express rail routes run in a general north-south direction, along the Rhine from the Ruhr and Cologne to Basle and from Hamburg to Munich. The trains – particularly the Trans-Europe-Express (TEE) trains – are clean and comfortable; several major international routes cross the country. Railway services are complemented by extensive bus services. In the cities trams are common; like some of the buses, they often consist of two coaches linked together. The famous network of *Autobahnen* (motorways) – one of the world's wonders in the 1930s – is still among the best highway systems anywhere, and is being constantly extended.

The inland waterway system is one of Europe's great trade arteries. Its main component is the Rhine, which has its outlet to the sea at the port of Rotterdam in The Netherlands and which is linked by canal to several other rivers. The immense Rhine barges carry heavy cargoes of the raw materials upon which West Germany's factories depend. Passenger ships carry tourists and also operate scheduled services between the Rhineland towns. Other busy waterways include the Dortmund-Ems Canal, connecting the Ruhr to the North Sea port of Emden, and the Kiel Canal across Schleswig-Holstein, which provides a direct passage from the North Sea to the Baltic. The national airline is Deutsche Lufthansa. State-run radio and television stations broadcast national as well as regional programmes; they co-operate with each other through a joint consultative body.

People. Until the 19th century Germany was an association of a large number of kingdoms and principalities, many of them with long histories of mutual rivalry and often enmity. Although the pressures and habits of this century have encouraged the emergence of a "national mix" that is primarily and essentially German, the legacy of the age-old fragmentation still persists. In a political sense it is enshrined in the ten *Länder* (excluding Berlin), each of which has its own identity and some independence. Throughout the country there are marked differences in customs, speech, and even in physical appearance; regional loyalties are strong.

The population is unevenly distributed. The chief concentrations are in the Rhine valley from the *Ruhrgebiet* to Ludwigshafen and Stuttgart, in the Saar, around the great cities of the north, and around Munich. Although prices are relatively high in West Germany, the German worker is more prosperous than any other in Europe. One worker in three is a woman.

Education. The traditional German regard for scholarship and education is still a marked feature of life in West Germany, and the use of academic and professional distinctions is more common than in most countries. Full-time education is compulsory between the ages of 6 and 16. After *Volksschule* (first level) to the age of ten, there is a choice of further schools according to ability. The higher types are the *Gymnasium,* which offers science, languages and the classics and which provides a nine-year course, and the *Mittelschule,* which offers a six-year course in more practical studies. Children who leave school at the age of 16 are obliged to attend lessons part-time until the age of 18. There are a number of institutions for vocational training and a wide variety of other kinds of specialized instruction. More than 500,000 students attend the 57 universities, of which the oldest include the universities of Heidelberg (1386) and Cologne (1388).

Cultural life and leisure. The German language is probably the most widely spoken in Europe. Within West Germany there are many dialects and styles of speech – often a subject for jokes. Broadly, there are two principal speech forms: *Hochdeutsch* (High German), spoken in the uplands and mountains of the centre and south and *Plattdeutsch* (Low German), spoken in the northern lowlands. But for several hundred years the written language almost everywhere has been Hochdeutsch; it is the language not merely of literature but of books and newspapers generally.

In religion, slightly less than half the people are Protestants and about 45 per cent are Roman Catholics. The largest Protestant body is the Evangelical Church, which has 21 constituent churches, mainly Lutheran. There is a small Jewish community, all that has survived the Nazi pogroms.

Tradition and local pride are strong in Germany, and nearly every town and village has its festival during which the local way of life is celebrated. Some festivals, such as the Oktoberfest in Munich, the pre-Lent celebrations in Cologne and Mainz and the wine festivals of the Rhine and Mosel valleys, attract thousands of visitors. Many villages have their own bands, which play at weekends and at local celebrations. Competitions to find the best band in an area are eagerly contested.

In the 19th century, the Germans acquired a reputation for painstaking and rather unimaginative scholarship. But although thoroughness is undoubtedly a facet of the German character, the Germans are essentially a creative people. The arts, particularly music, literature and drama, play an important part in their lives. Few towns of any size are without a theatre, and there are more than 70 symphony orchestras. Germany's greatest contribution to the arts has probably been in music. Its composers include Johann Sebastian Bach, Ludwig van Beethoven, Johannes Brahms, Robert Schumann and Richard Wagner. Today, various towns have international festivals associated with particular composers: Ansbach with Bach, Bonn with Beethoven and Bayreuth with Wagner.

In other arts Germany's contribution has also been immense. The heroic epic the *Nibelungenlied* written about 1200, provided inspiration for Wagner. Wagner also found inspiration in the epic *Parzival,* written by a medieval minnesinger (from *minne,* meaning "love"); the most famous minnesinger was Walther von der Vogelweide. Among the great names in German literature are Johann Gottfried Herder, Johann Wolfgang von Goethe, Heinrich Heine, Friedrich von Schiller and Christoph Wieland. Writers of the 20th century include Thomas Mann and Bertold Brecht. Pre-eminent in German philosophical writings are Arthur Schopenhauer, Georg Hegel and Immanuel Kant.

In the visual arts, some of the most enduring achievements have been in architecture. Germany is a land of magnificent churches, castles, guildhalls

Munich, in West Germany, is a highly industrialized and commercial centre with many buildings and churches dating from the 14th century.

Lindau, built on an island in the Lake of Constance (Bodensee), is a popular summer resort with buildings dating back to the 16th century.

Bananas form part of the staple diet of Ghanaians. The fruit is sold in open markets (typical in the tropics) by bunches rather than by weight.

and other public buildings. Probably the most famous among German artists is Albrecht Dürer. Matthias Grünewald is known for the sometimes almost unbearable realism of his Crucifixion scenes, Lucas Cranach shows the landscapes of Renaissance Europe and Hans Holbein the Younger is among the most renowned of portrait painters. Tilman Riemenschneider made sculptures of delicate realism. An important German contribution to the visual arts in the 20th century was the school of art called the Bauhaus founded by Walter Gropius in Weimar in 1919. It has had a profound influence on modern design throughout the Western world.

German names also figure prominently in the history of science, discovery and invention. Examples are Johannes Gutenberg, the inventor of modern printing; Johannes Kepler, the astronomer; Gottfried von Leibniz, the mathematician; Conrad Röntgen, discoverer of X-rays; Karl Benz, maker of the first practical motor car; and Max Planck, who laid the foundations of the quantum theory.

Some ten million Germans belong to the German Sports Union, and a belief in the importance of physical exercise is deeply ingrained. Walkers, usually carrying walking-sticks, are a common sight in country areas, and it is generally possible to buy a local map showing the most pleasant routes for a walker to follow. Shooting and climbing are also popular leisure occupations, and, on a less energetic level, bowling. The most common spectator sport is soccer. Cycling, athletics, ice-skating and swimming all also have their followers.

History since 1945. (For the history of the country before 1945, *see* GERMANY.) Following World War II, the area now covered by West Germany was divided into three zones occupied by troops from Britain, France and the United States. In 1949 the Federal Republic was formed by the amalgamation of the occupied zones, although sovereignty had to wait until 1955, when the USSR formally ended the war with Germany. As time passed the division of Germany hardened into an historic fact. The difference in ideology between East and West, the development of two differently based and successful economies and the armed frontier (of which the Berlin Wall is part) have all reinforced their status as two separate nations.

The first chancellor of West Germany was Konrad Adenauer, a Christian Democrat, who headed the government until 1963. His great achievement was the expansion and stabilization of the economy. Not the least of his problems was the absorption of some ten million refugees from the east. The country received much aid from the United States, and its economic position was further strengthened by membership of the European Coal and Steel Community and, later, the EEC.

In 1963, Adenauer was succeeded by Ludwig Erhard, heading a CDU-FDP coalition, and in 1966 by Kurt-Georg Kiesinger in a coalition of the CDU, CSU and SPD. In 1969 the Christian Democrats were defeated, and a FDP-SPD coalition government was formed under Willy Brandt, a Socialist and former mayor of West Berlin. Brandt sought for practical agreement with the Communist bloc – the *Ostpolitik* begun when he was foreign minister. He met East German leaders, signed non-agression treaties with the USSR and Poland and was party to a 1971 agreement on access to Berlin. He resigned in 1974 when one of his aides was discovered to be a Soviet spy; he was succeeded by Helmut Schmidt. Schmidt was returned to power in the election of 1976.

The West German Länder. West Germany is divided into 11 *Länder* (states), each with its own constitution, legislature and government. Two are the autonomous cities of Bremen (*Freie Hansestadt Bremen*) and Hamburg (*Frei und Hansestadt Hamburg*). The others are as follows:

Baden-Württemberg, capital Stuttgart. Area: 35,751sq km (13,804sq miles). Pop. (1974) 9,226,200.

Bavaria (Bayern), capital Munich. Area: 70,547sq km (27,238sq miles). Pop. (1975) 10,849,100.

Hessen, capital Wiesbaden. Area: 21,112sq km (8,151sq miles). Pop. (1975) 5,576,100.

Lower Saxony (Niedersachsen), capital Hannover. Area: 47,426sq km (18,311sq miles). Pop. (1974 est.) 7,264,840.

North Rhine-Westphalia (Nordrhein-Westfalen), capital Düsseldorf. Area: 34,057sq km (13,149sq miles). Pop. (1974) 17,217,800.

Rhineland-Palatinate (Rheinland-Pfalz), capital Mainz. Area: 19,835sq km (7,658sq miles). Pop. (1974) 3,688,100.

Saarland, capital Saarbrücken. Area: 2,569sq km (992sq miles). Pop. (1974 est.) 1,103,300.

Schleswig-Holstein, capital Kiel. Area: 15,678sq km (6,053sq miles). Pop. (1974 est.) 2,584,300.

West Berlin Comprises the British, French and United States sectors of the city; not yet formally incorporated as a *Land;* laws passed by the federal legislature require formal adoption by the Berlin House of Representatives. Area: 408sq km (158sq miles). Pop. 2,024,000. Map 18.

West Germany – profile

Official name Federal Republic of Germany
Area 248,577sq km (95,976sq miles)
Population (1976 est.) 61,000,000
Density 248 per sq km (642 per sq mile)
Chief cities Bonn (capital) (1976) 283,900; West Berlin, 2,024,000; Hamburg, 1,738,800; Munich, 1,317,000; Cologne, 1,017,200
Government Federal union, with democratically elected upper and lower houses
Religions Protestant (49%), Roman Catholic (45%)
Language German
Monetary unit Deutsche Mark
Gross national product (1974) £156,102,600,000
Per capita income £2,532
Agriculture Cereals, sugar-beet, potatoes, fruit, hops, grapes (for wine), vegetables, pigs, cattle, poultry
Industries Iron and steel, motor vehicles, industrial machinery, chemicals, electrical goods, textiles, processed foods, optical equipment, cutlery, ceramics
Minerals (major) Coal, iron, potash, petroleum
Trading partners (major) EEC, USSR, USA

Ghana, a nation in western Africa known as the Gold Coast before it became independent in 1957, is the world's leading cocoa producer. The name *Ghana* was the name of a large medieval West African empire, situated to the north and north-west of present-day Ghana, whose capital is Accra.

Land and climate. Apart from the south-west, Ghana occupies the drainage basin of the Black Volta, White Volta and Oti rivers. These rivers now flow into the man-made Lake Volta, which covers 8,420sq km (3,251sq miles). The lake lies behind the dam at Akosombo, where hydroelectricity is produced. The Volta basin is separated from the south-west by the Kwahu escarpment. The south-west is an area of low plateaus, ridges and fertile basins, whose rivers flow south into the Gulf of Guinea.

The climate in the south is equatorial and the average annual temperature at Accra is 27°C (80°F); parts of the interior are even hotter. In the south, the rainfall decreases east of Accra and increases west of the capital. The forested south-west has an average annual rainfall of more than 2,030mm (80in). But the south-east, where grassland and

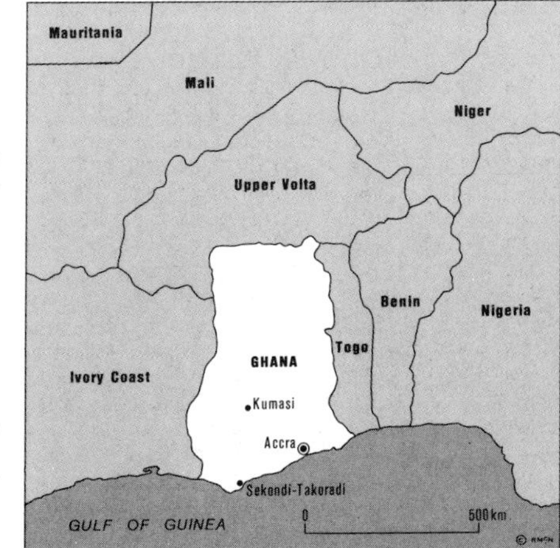

A modern block in Accra – Ghana's capital and largest city – forms part of the new horizon in this young and rapidly developing country.

A herd of goats are driven in search of water over the desert-like terrain of Boeotia, a region of ancient Greece which lies north of the Gulf of Corinth.

Athens, named after Athena, its patron goddess, is capital and largest city of Greece. Ancient Athenian culture has greatly influenced the rest of the world.

scrub are the dominant types of vegetation, has less than 890mm (35in). The rainfall generally decreases inland, averaging about 1,020mm (40in) in the north-east. The rainfall in the interior is also less reliable than on the coast and droughts sometimes occur. Savanna is the characteristic landscape of northern and central Ghana.

Economy. More than 70 per cent of the people live in rural areas and depend on farming. The chief food crops are cassava, groundnuts, guinea corn, maize, millet, cooking bananas, rice and yams. Cocoa is grown mostly in the south-west, Ghana's chief economic region and most thickly populated area. It is the chief crop, accounting for just over 50 per cent of all exports. The second most valuable export is timber and timber products. Livestock are reared in the south-east and in the central and northern savanna. Fishing is also important. Ghana's chief minerals are gold, diamonds, manganese and bauxite. Aluminium is produced at the industrial port of Tema, using power from the Volta River project. By African standards, Ghana is fairly prosperous, with a national income per person of about £180 a year in 1974. But an over-ambitious development programme, aimed at rapid industrialization, increased Ghana's external debts to unmanageable proportions in the early 1960s. Since then, Ghana has been struggling to solve its economic problems. In the 1970s, the government launched an "Operation Feed Yourself" campaign in an attempt to increase food production.

People. The people are divided into about 50 groups. The largest, the Akan, includes the Ashanti, Fante and Twi in the south and south-west. In the north, the people have mixed to some extent with Hamitic peoples. Ethnic religions are followed by 38 per cent of the people, Christianity by 43 per cent and Islam by 12 per cent.

History. The Portuguese began trading along the coast in the 1470s, but later the area became a centre of competition between several European powers, who built coastal forts for trading. British influence became dominant in the late 1800s. The coast became a British colony in 1874 and, after strong resistance, Ashanti was annexed in 1901 and the northern territories became a British protectorate.

After World War I Britain was mandated to rule part of neighbouring German Togoland, the rest going to France. At independence in 1957, British Togoland was integrated with Ghana. In 1960 Ghana became a republic and the prime minister Dr Kwame Nkrumah became president. Increasingly, Nkrumah and his regime were accused of dictatorial rule, corruption and economic mismanagement. In 1966 Nkrumah was deposed, to spend the rest of his life in exile. The military rulers restored the civilian government in 1969 but, following further economic crises, a second military coup occurred in 1972. The National Redemption Council, led by its chairman Col. Ignatius Acheampong, was reorganized in 1975 as the Supreme Military Council. Map 32.

Ghana – profile

Official name Ghana
Area 238,537sq km (92,099sq miles)
Population (1975 est.) 9,866,000
Density 41 per sq km (107 per sq mile)
Chief cities Accra (capital), 848,000; Kumasi 249,000; Sekondi-Takoradi, 161,000
Government Military council
Religions Ethnic, Christianity, Islam
Language English (official)
Monetary unit Cedi
Gross national product (1974) £1,765,000,000
Per capita income £179
Agriculture Cassava, cocoa, coffee, livestock, maize, millet, nuts, palm-oil, cooking bananas, rice, rubber, tobacco, wood, yams
Industries Aluminium smelting, brewing, cement, cocoa, food processing, wood products
Minerals Bauxite, diamonds (mainly industrial), gold, manganese
Trading partners Britain, USA, West Germany, Japan, Netherlands, France

Gibraltar is a British Crown colony at the north-western end of the Rock of Gibraltar, on the tip of the Iberian Peninsula at the entrance to the Mediterranean Sea. The free port and town of Gibraltar makes up the whole of the colony, which has large military installations employing some of the small local population and, traditionally, many workers from the neighbouring Spanish town of La Línea. Most of the people are of Portuguese, Spanish or Italian descent but English is the official language. Spain continues to make claims for the territory, and in the late 1960s closed its border with the colony. The places of many of the Spanish workers have been taken by immigrants from Morocco and Portugal.

Gibraltar's history dates from its capture by the Moors in 711. The Spanish attempted to re-take it but did not establish permanent possession until 1462. The British, using a force of seamen and marines, captured Gibraltar in 1704 (an action chosen as the only battle honour on the colours of the Royal Marines) and have resisted all attempts to recapture it since. Modern techniques of warfare have lessened Gibraltar's strategic importance, but a 1967 referendum reaffirmed the population's desire to maintain links with Britain. Area: 6.5sq km (2.5sq miles). Pop. (1975 est.) 27,000. Map 22.

Gilbert and Ellice Islands comprise about 40 islands near the Equator in the central Pacific Ocean, 4,500km (2,800 miles) north-east of Australia. There are four main groups – the Gilbert Islands, Ellice Islands, Phoenix Islands and Line Islands – and Ocean Island 386km (240 miles) to the west of the Gilberts. Ocean Island was annexed by Britain after it was found to be composed of phosphate rock. The islands are a British Crown colony, with the capital at Tarawa in the northern Gilbert Islands. Most are low-lying atolls; copra, mother-of-pearl and phosphate are produced for export. The islands were declared a British protectorate in 1892 and became a colony in 1915. The Ellice Islands (Tuvalu) were constitutionally separated in October 1975. Area: 970sq km (375sq miles). Pop. (1975 est.) 66,000. Map 62.

Gold Coast. *See* GHANA.

Great Britain. *See* UNITED KINGDOM.

Greece (Hellas), official name Hellenic Republic, is a nation on the southern part of Balkan Peninsula in the eastern Mediterranean. It is a country of mountains, islands and peninsulas. The capital is Athens (Athínai). Greece is a leading maritime nation, with about 3,000 liners, freighters, tankers and other ships, but its agriculture and industry are still largely underdeveloped. Magnificent civilizations flourished in Greece in the 1st millennium BC and the impressive ruins throughout the country, together with its sunny climate, attract a considerable tourist trade. Greece was under foreign domination for hundreds of years, and has suffered much political turmoil in the 20th century.

Land and climate. The Pindus Mountains make up the central backbone of mainland Greece, separating Epirus from Thessaly and Macedonia. Composed mainly of limestone and with peaks rising to 2,500m (8,200ft), they form an effective barrier to east-west communication. Thessaly, the only extensive plain in Greece, is surrounded by mountains. These include Mt Olympus, the mythological home of the gods, which stands 2,917m (9,571ft) above sea-level near the coast and is the highest mountain in Greece. The hundreds of islands (166 are inhabited) make up almost a fifth of the country's land area. The largest are Crete (Kríti), Rhodes (Rhódhos), Chios (Khíos), Lesbos (Lésvos) and Corfu (Kérkira). Most of them are rocky, and many have vine-covered southern slopes. The climate is predominantly Mediterranean, with hot, dry summers and mild, wet winters. Most rivers dry up in summer because of lack of rain. There are sharp local variations in climate because of differences in altitude.

Economy. About 40 per cent of the working population is engaged in agriculture, although less than a third of the land is arable. The chief crops are cereals (wheat, barley, maize), tobacco, currants, grapes, olives, cotton, tomatoes, sugar-beet and citrus fruits. Various mineral deposits are being increasingly exploited, including lignite, bauxite, iron ore, chromite, magnesite and nickel. Oil was struck in 1963 in west-central Greece.

The Eastern Orthodox monastery on Mount Athos encloses a large community which in former times was accorded administrative independence.

In a busy street on Kérkira, or Corfu (second largest of the Ionoan islands in north-western Greece), pedestrians shop without the hazards of motor cars.

The terrace of lions towers above the ruins of a city which flourished about 500 BC on Delos – a small island of the Greek archipelago in the Aegean Sea.

From 1960 the scale of manufacturing was increased and a quarter of the working population is now engaged in industry. The main industries are textiles, chemicals and food processing. The chief exports are tobacco, textiles, metals, manufactured goods and dried and fresh fruit. Greece became an associate member of the EEC in 1962, and its chief source of imports are other EEC countries (50%), particularly West Germany (21%) and Italy (9%). The main imports include machinery, transport equipment, raw materials and meat, and the leading suppliers are EEC countries (43%), including West Germany (16%), Italy (8½%) and France (7%), and the United States (9%). Imports exceed exports by 2½ times in value, a trade gap effectively closed by invisible exports, especially tourism and merchant shipping. The 3,000 vessels of the merchant fleet rank fifth in the world in gross tonnage, with another 1,300 registered under foreign flags. The chief ports are Piraeus (Piraiévs) and Salonika (Thessaloníki).

People. Greece has only few and small racial minorities (such as Turks and Albanians). There are two forms of modern Greek: *katharevousa,* a conscious revival of the classical tongue, and *demotiki,* the commonly spoken language which has been taught in the schools since 1974. More than 97 per cent of the people belong to the Greek Orthodox Church, the state religion, and some 1½ per cent are Muslims. All education is free and is compulsory from the ages of six to 12 (to be extended to 15 in the 1980s). The chief universities are at Athens and Salonika, each with more than 20,000 students. Illiteracy has been reduced from 72 per cent to 10 per cent since World War II, as the population moves steadily from the farming villages to the cities.

The family is the dominant social unit, and a Greek's first loyalty is to his kinsmen. Greek emigrants regularly send large sums of money home to their families. There is still much poverty in some rural areas, and the staple diet rarely contains meat. Some of the more sophisticated dishes of urban areas have, however, spread internationally, such as moussaka, a pie with layers of potato, minced meat, cheese and aubergine. Greece is also known for retsina, a type of wine to which resin is added. Most families in rural areas own a small vineyard and use the village grape-press for making their own retsina. Soccer is the most popular sport. The first modern Olympic Games were held in Athens in 1896, and Greece has a special place in Olympic ceremonies.

Government. The 1975 constitution provided for a parliamentary democracy with a president as head of state. Elected by parliament for a five-year term, the president has considerable political power. He represents the state in relations with other nations; he appoints the prime minister, other ministers and judges; and he can veto bills passed by parliament. The one-house parliament has a maximum of 300 deputies (200 minimum) elected to four-year terms by popular vote. The chief political parties are the New Democracy Party (72% of seats in 1975), Centre New Union Forces (20%), the Pan-Hellenic Socialist Movement (5%) and the United Left (3%), a Communist coalition. Local government is carried out by 52 prefectures (*nomes*). Greece withdrew its military obligation to NATO in 1974. The strength of the armed forces is 161,000, including 112,000 conscripts (military service of up to two years is compulsory), and there is a gendarmerie of 30,000 and a National Guard of 70,000.

History. The recorded history of Greece goes back to 776 BC, the date of the first ancient Olympic Games, but legends and traditions of earlier civilization have been remarkably well corroborated by archaeological discoveries on Crete and at Mycenae. Greek civilization – with its achievements in art, architecture, science, philosophy, literature and democracy – reached the peak of its glory in the 5th century BC, particularly in the city-state of Athens. Philip of Macedonia conquered Greece in 338 BC, and his son Alexander (the Great) spread Hellenistic civilization far and wide. By the middle of the 2nd century BC, however, Greece had declined to the status of a Roman province. Christianity began to take hold in Greece, particularly among the poor. In the 4th century AD Greece became part of the Byzantine Empire, which in the 13th and 14th centuries began to break up into small states.

After the fall of Constantinople (the Byzantine capital) to the Turks in 1453, Greece became a province of the Ottoman Empire. The Turks allowed the Greeks religious freedom and considerable local self-government. In the late 18th century, as a result of increasing prosperity and education, the Greeks began to experience a national reawakening, which led to a revolt in 1814 and to the Greek War of Independence (1821-30). The Turks, even with the help of the Egyptians, could not subdue the Greeks. Eventually Britain, France and Russia intervened and crushed the Turkish-Egyptian fleet at the Battle of Navarino, off the Peloponnese Peninsula (1827), and the same European powers recognized Greek independence in the Protocol of London (1830).

Otto I, a Bavarian prince, became the first King of Greece in 1833 and 11 years later Greece became a constitutional monarchy, with an area less than half of today's. For several decades Greece sought to acquire foreign territories inhabited by Greeks. By the end of the Balkan Wars (1912-13) and World War I (1914-18), Greece had added Thrace, Crete and many other islands, a part of Macedonia, and southern Epirus. An expedition into Turkish Asia Minor was repelled in 1922, however, and claims to territory were finally settled by an exchange of populations (Treaty of Lausanne, 1923). A military revolt in 1923 forced George II from the throne, and Greece declared itself a republic in 1924.

Greece – profile

Official name Hellenic Republic
Area 131,944sq km (50,944sq miles)
Population (1975 est.) 9,046,000
Density 69 per sq km (178 per sq mile)
Chief cities Athens (capital) (1971) 867,023; Salonika, 345,799; Piraeus, 187,362
Government Head of state, Constantine Tsatsos, president (elected 1975)
Religions Christianity, Islam
Language Greek
Monetary unit Drachma
Gross national product (1974) £7,555,600,000
Per capita income £835
Agriculture Cereals, tobacco, currants, grapes, olives, cotton, tomatoes, sugar-beet, citrus fruits
Industries Shipping, textiles, chemicals, food processing
Minerals Lignite, bauxite, iron ore, chromite, magnesite, nickel, oil
Trading partners West Germany, Italy, USA, France, Britain

The next decade was marked by political confusion, economic weakness and coups before the monarchy was restored in 1935. But the country was ruled by a military dictatorship (1936-41), established by Gen. John Metaxas. In World War II the Italians invaded Greece (late 1940). The Greeks beat back the superior Italian forces, but succumbed to the German airborne invasion of Crete in 1941. In the face of starvation and mass executions, the Greeks organized a strong resistance. When they were eventually liberated in 1944 by British and Greek troops, a civil war developed between the rival Communist and right-wing factions, and Greece found itself a battleground for East-West ideologies. With American aid, the Communist guerrillas were defeated by 1950.

At Delphi, in the Greece of pre-Hellenistic times, the temple of the famous oracle was dedicated to Python; later deities were Apollo and Dionysus.

A polar current flows southwards down Greenland's east coast, carrying dangerous ice; the west coast enjoys the warmer North Atlantic Drift.

Guatemalan women sell their wares at a market; about forty-five per cent of the Guatemalan population are pure Indians descended from the Mayan tribes.

After a period of reconstruction and economic development, chiefly under the conservative governments of Constantine Karamanlis, a handful of army officers known as "the colonels" seized power in 1967 and suspended parliamentary government and most civil liberties. After an unsuccessful attempt to overthrow the military junta, King Constantine II fled into exile. In 1968 George Papadopoulos emerged as the leader of the junta and took the title first of premier and then (in 1973) of president, when Greece was once again proclaimed a republic. The dictatorship's repressive actions met with almost worldwide disapproval. Papadopoulos was ousted in November 1973 in a bloodless coup, in the wake of student riots. The military regime finally resigned in July 1974, following an unsuccessful attempt to take political control of CYPRUS.

Karamanlis returned from exile and was sworn in as premier of Greece's first civilian government since 1967. Parliamentary elections were held in November, when Karamanlis's New Democracy Party won 216 of the 300 seats. A referendum in December voted by two to one against a restoration of the monarchy. In 1976 the EEC agreed to consider Greece's application for full membership of the community. Map 26.

Greenland, in the north-western Atlantic Ocean, is the largest island in the world; the capital is Godthaab (Godthåb). It belongs to Denmark and most of it, covered with ice (thickness up to 3,000m; 10,000ft) and uninhabitable, lies north of the Arctic Circle [*see* PE p.188]. Most of the people are Eskimos, who live mainly along the south-west coast hunting seals and catching fish [*see* MS p.*252*]. Fiords indent the coastline and there are many offshore islands. Glaciers moving down to the coast produce icebergs that drift southwards into the Atlantic Ocean [*see* PE p.116]. The only trees, stunted and shallow-rooted, grow near the south coast; grasses, mosses and lichens make up the remainder of the vegetation. There is some mining for coal and other minerals such as quartz, mica and cryolite, and there is an American air base at Thule.

The island was discovered in about 960 by Erik (the Red) who deliberately chose the optimistic name Greenland (Danish *Grønland*) to attract settlers. By the 12th century there was a colony of about 10,000 people which in 1261 came under the control of Norway, but by 1400 most of the settlements had been deserted. Colonization began again in 1721 under the Norwegian Hans Egede, "the Apostle of Greenland", who founded Godthaab. The Congress of Vienna (1815) re-established Danish sovereignty and colonial status ended in 1953. Since then Greenland has been administered by a ministry of the Danish parliament (to which Greenland sends two members) through a local governor and administrative council. Area: 2,175,600sq km (840,000sq miles). Pop. (1975 est.) 49,500. Map 4.

Grenada, official name State of Grenada, is an independent republic of the Commonwealth in the south-eastern Caribbean Sea, 137km (85 miles) north of Trinidad. It is made up of Grenada, the southernmost of the Windward Islands and location of the capital (Saint George's), and the smaller islands of the southern Grenadines. The economy depends almost entirely on agriculture, with cocoa, bananas, coconuts, sugar and nutmegs as the chief crops. More than half the people are of African descent; and most of the rest are of mixed African-East Indian. The official language is English.

Grenada was discovered by Christopher Columbus in 1498 and colonized by the French in 1650. It was ceded to Britain in 1783 and remained a British colony until 1974, when it was granted independence. Area: 344sq km (133sq miles). Pop. (1975 est.) 105,000. Map 74.

Guadeloupe is an overseas département of France in the Leeward Islands, eastern West Indies. It comprises the islands of Basse-Terre (Guadeloupe proper, location of the capital, Basse-Terre) to the west, Grande-Terre to the east and various smaller islands. Most of the people, who speak a French dialect, are of African descent. The chief occupations are the production of sugar cane, rum and bananas; tourism is also a major industry. Guadeloupe was discovered in 1493 by Christopher Columbus and settled by the French in 1635. It was a French colony until 1946, when it attained département status. Area: 1,780sq km (687sq miles). Pop. (1975 est.) 354,000. Map 74.

Guam is an unincorporated United States territory, the largest of the Mariana Islands in the western Pacific Ocean, 2,600km (1,617 miles) east of Manila. Much of the non-forested land is occupied by US military bases, leaving room only for small farms. Its civil airport is a staging post for aircraft flying between south-eastern Asia and North America. More than half the population are Chamorros (people of mixed Spanish, Micronesian and Filipino descent) and many work for the US military authorities. Guam was discovered in 1521 by Ferdinand Magellan and belonged to Spain until the end of the Spanish-American War in 1898, when it passed to the United States. Area: 541sq km (209sq miles). Pop. (1975 est.) 104,000. Map 62.

Guatemala, official name Republic of Guatemala, is a nation of Central America between the Atlantic and Pacific Oceans. It is one of the world's leading producers of coffee. The capital is the city of Guatemala.

Land and economy. The Sierra Madre Mountains, many of volcanic origin, run parallel to the Pacific Ocean in the south, branching off into four principal ranges in the north. The extinct volcano Tajumulco (4,211m; 13,816ft) is the highest peak in

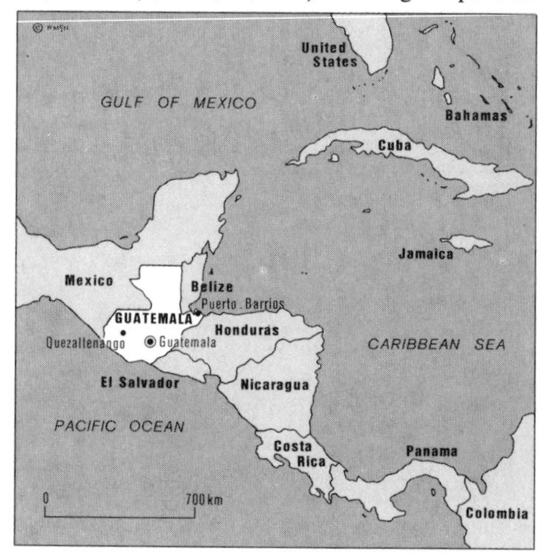

Central America. A plain about 48km (30 miles) wide extends the length of the Pacific side for about 320km (200 miles) from Mexico to El Salvador. The climate is tropical but moderated by the elevation. Although the economy is basically agricultural (coffee and bananas are the chief exports), mining has become increasingly significant. The principal ports, Puerto Barrios on the Atlantic and San José on the Pacific, are connected with the capital by a transcontinental railway.

People. More than half the people are of American-Indian origin, and most of the remainder are Mestizos (of mixed Spanish and Indian descent). Most are Roman Catholic and speak Spanish.

Government. The republican government is made up of three departments; executive, with a president elected for four years and ineligible for re-election for another 12 years; legislative, with a National Congress elected every four years; and

The great pyramid-temple at Tikal, Guatemala. displays all the characteristics of Classical Mayan architecture; Tikal was one of the largest Mayan ceremonial centres.

Guinea has some of the world's largest bauxite deposits, from which aluminium can be processed to be exported to America and Western Europe.

Both Guinea-Bissau and Guyana rely quite considerably for export revenue on the coconut palm – every part of which may be put to use

judicial, with a Supreme Court and lesser courts. Every literate citizen more than 18 years old must vote in elections; voting is optional for people who cannot read and write.

History. Guatemala was the home of the Mayan Empire for 1,000 years until conquered by the Spanish in 1524. From 1821 to 1823 it came under Mexican control, after which it became a member of the Central American Federation. It re-established its independence in 1839. Throughout its history, Guatemala has been characterized by political upheaval and revolution. Justo Rufino Barrios attempted to re-establish a Central American Union in the late 19th century, but failed. Communists and anti-Communists battled for control after World War II, and the United Fruit Company, a US firm, played a major role in domestic affairs. It owned 95 per cent of the banana plantations and was the largest employer in the country. In 1953 the Guatemalan government nationalized the plantations; the United States intervened, and the government was overthrown. In the 1970s Guatemala pressed its claims to the neighbouring British colony of Belize, and threatened war twice. In February 1976 the nation was rocked by a severe earthquake that left more than 16,000 people dead and nearly a fifth of the population homeless. Map 74.

Guatemala – profile

Official name Republic of Guatemala
Area 108,889sq km (42,042sq miles)
Population (1973) 5,540,000
Density 51 per sq km (132 per sq mile)
Chief city Guatemala (capital), 717,300
Government Republic; head of state President Kjell García
Religion Roman Catholic
Language Spanish (official)
Monetary unit Quetzal
Gross national product (1974) £1,286,300,000
Per capita income £232
Agriculture Coffee, cotton, bananas, sugar cane
Industries Food, beverages, tobacco
Minerals Zinc, lead, nickel
Trading partners USA, El Salvador, West Germany, Japan

Guernsey *See* CHANNEL ISLANDS.

Guinea, official name Republic of Guinea, is an underdeveloped nation in western Africa. In the early 1970s the average income was only about £45 per year. The capital is Conakry.

Land. Behind the coastal lowlands the land rises to the Fouta Djallon highlands. The south-west is also a highland zone, but in the north-east the land descends to the interior Niger plains. The Niger, Gambia and Senegal rivers all rise in Guinea. The

Guinea – profile

Official name Republic of Guinea
Area 245,857sq km (94,925sq miles)
Population (1975 est.) 4,416,000
Chief city Conakry (capital) (1974) 412,000
Religions Islam, ethnic, Christianity
Language French (official)
Monetary unit Guinea franc

climate is hot and wet and the coast has about 4,300mm (169in) of rain per year. The highlands are rather less rainy and cooler. About 1,525mm (60in) of rain fall on the Niger plains.

Economy. More than 80 per cent of the people work in farming. Coffee, palm products and bananas are all exported, although 65 per cent of export income comes from bauxite and aluminium.

Guinea also has reserves of diamonds and iron ore. Manufacturing contributes only three per cent of the gross national product.

People and government. Most people are Negroid, the largest group being the Fulani (or Peul). More than 60 per cent of the people are Muslims. Most of the others follow ethnic religions, although two per cent are Christians. Guinea is governed by its president, elected for seven-year terms, and a cabinet. The 75 members of the National Assembly belong to the Parti Démocratique de Guinée, the only political party.

History. In the 18th century, the northern part of the region was part of the empire of Ghana. Guinea became a French colony in 1891. In 1958 the French withdrew after the people had voted to become independent. President Sékou Touré asked Communist nations for help. Today, Guinea maintains good relations with both East and West. Map 32.

Guinea-Bissau, official name Republic of Guinea-Bissau, is a country in western Africa that was known as Portuguese Guinea until independence in 1974. The capital, Bissau, has 25,000 people.

This low-lying country is highest on the border with Guinea, where it reaches about 300m (984ft). The climate is hot and wet, with between 2,030 and 3,300mm (80-130in) of rain per year on the forested coast. Inland, the woodland savanna is drier. Farming is the main occupation; more than 80 per cent of the exports come from groundnuts and groundnut products, palm products and copra.

From 1879 the country was a Portuguese colony. After a bitter armed struggle starting in 1962, the country became independent in 1974. It is now ruled by the president (Luis Cabral) and the State Council. The National Popular Assembly has 120 elected deputies. Portuguese remains the official language. Area: 36,125sq km (13,948sq miles). Pop. (1976 est.) 534,000. Map 32.

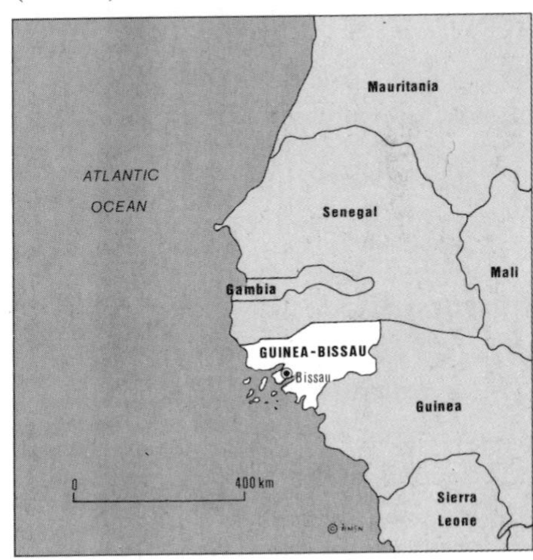

Guyana, official name Republic of Guyana and formerly British Guiana, is an independent nation in north-eastern South America between Venezuela and Surinam. The capital is Georgetown. The people of Guyana have a wide variety of origins: about half are of East Indian descent, a third are of African origin, and most of the remainder are American Indian or of mixed descent. More than 90 per cent of the population lives in the flat coastal region. The rest of the land is covered by tropical forest and savanna, making it difficult to gain access to the country's rich bauxite deposits. Guyana's economy depends mainly on agriculture, with sugar cane, rice and coconuts as the chief crops. The country is governed by a president (Arthur Chung) and a 53-member National Assembly, whose members are elected by proportional representation for terms of four years.

The region was first explored by the Spanish in

Haitian women make their way to market; most Haitians are of African descent and make their living either by subsistence farming or by working on plantations.

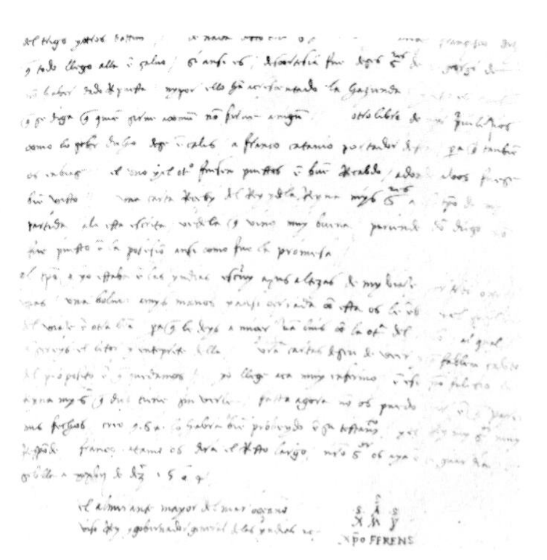

Eight years of writing letter after letter to the rulers of Spain finally paid off for Christopher Columbus, and he sailed – to discover Hispaniola.

Sampans provide homes for many people in Hong Kong; despite the housing programmes of the last twenty years Hong Kong still remains chronically overcrowded.

1499 and settled in *c.* 1620 by the Dutch, who ceded it to Britain in 1815. The British introduced large numbers of African slaves and East Indians to work on the sugar plantations. Guyana achieved independence in 1966 and became a republic within the Commonwealth in 1970. In 1976 the government nationalized much of industry and took over the running of schools. Area: 214,970sq km (83,000sq miles). Pop. (1975 est.) 794,000. Map 76.

Haiti, official name Republic of Haiti, is an independent nation comprising the western third of the Caribbean island of Hispaniola (the remainder of the island is occupied by the DOMINICAN REPUBLIC). The capital is Port-au-Prince and the country includes the off-shore islands of Tortuga and Gonâve. It is the oldest independent Negro republic and one of the most densely populated countries in the world. Most of the people are descended from African slaves and speak a dialect of French. Roman Catholicism is the state religion, although voodoo is still widely practised. Two-thirds of the land area is rough, mountainous terrain unsuitable for cultivation, and the majority of the people try to eke out an existence by subsistence farming in the remaining area, growing rice, coffee, sugar cane, bananas and tobacco. Bauxite and copper ores are mined and exported.

Hispaniola was discovered in 1492 by Christopher Columbus and divided in 1697 when Spain ceded part of it to France. The slave population, taken from Africa to work in the plantations, gained independence in 1804 and gave the country the American Indian name Haiti – "the Land of Mountains". Between 1822 and 1844 Haiti also held the Spanish-speaking part of the island (now the Dominican Republic), which had broken away from Spain in 1821. From 1843 to 1915 there were more than 20 dictatorships in Haiti. François Duvalier (known as Papa Doc) was elected president in 1957, and his term of office was extended for life by the 1964 constitution. He died in 1971 and was succeeded by his son, Jean-Claude Duvalier, also for a life term. Area: 27,750sq km (10,714sq miles). Pop. (1975 est.) 4,584,000. Map 74.

Hawaii. *See* UNITED STATES.

Hispaniola is the second largest island in the West Indies, between Cuba (to the west) and Puerto Rico (east). The DOMINICAN REPUBLIC occupies the eastern two-thirds of the island and HAITI makes up the remainder. The island was discovered in 1492 by Christopher Columbus who called it Española, the first Spanish colony in the New World and the starting point for European settlement in the West Indies and Central America. In 1697 part became the French colony of Santo Domingo, incorporated by Haiti when it gained independence in 1804; the Dominican Republic was established in 1844. Area: 76,480sq km (29,559sq miles). Pop. (1975 est.) 9,281,000. Map 74.

Holland. *See* NETHERLANDS.

Honduras, official name Republic of Honduras, is a nation in Central America between Guatemala (to the west) and Nicaragua (south-east). The capital is Tegucigalpa. It has a damp, tropical climate with most of the rain falling to the north of the central mountain ranges, which rise to 2,740m (9,000ft) above sea-level. The land descends to forests and coastal swamps in the east. About 90 per cent of the population are Spanish-speaking mestizos, of mixed European and American-Indian descent, and most of the people work in agriculture, growing such crops as bananas (accounting for 50 per cent of all exports), coffee, sugar cane and rice. Beef cattle are

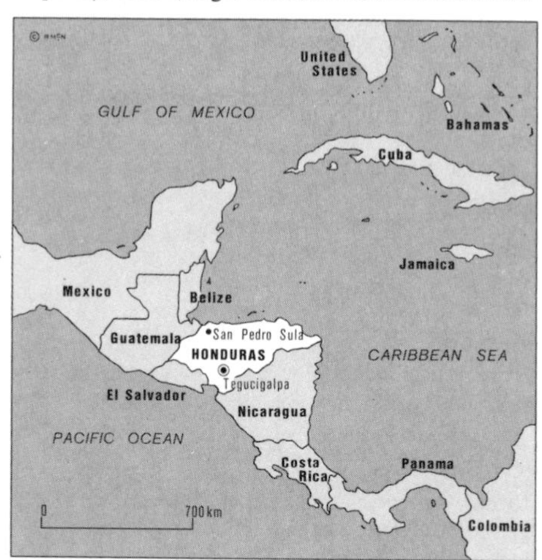

also raised. A poor transport network is a severe handicap and prevents the country from exploiting its mineral resources, which include silver, lead, zinc and off-shore petroleum.

The country once had a rich Mayan culture, which reached its peak in the 4th century AD. It was conquered and settled by the Spaniards in the 1500s (after being sighted by Christopher Columbus in 1502). Honduras broke away from Spain in 1821 and then joined the Central American Federation until 1838. Frequent revolutions have marked its subsequent history – in 1972 an army coup overthrew the government. In 1976 the head of state was Col. Juan Castro. Area: 112,088sq km (43,277sq miles). Pop. (1975 est.) 3,037,000. Map 74.

Hong Kong is a British Crown colony in southern China, 145km (90 miles) south-east of Canton. The colony is made up of Hong Kong Island, Kowloon

Peasants draw water for horses from a primitive well or shaduf in Hortobagy, eastern Hungary. In this region much of the traditional steppe way of life has survived.

About half the people of Hungary live in towns or cities; Budapest's commuter traffic crosses to and fro in a tunnel under the Danube each day.

Although Hungary's best natural resources are rich soil and continental climate, nearly half the total work force is in industry and construction.

Peninsula, the New Territories on the mainland and about 230 islets in the South China Sea. The capital is Victoria on Hong Kong Island. Since 1949, with the establishment of the Communist regime in China, more than 3,000,000 Chinese have emigrated to Hong Kong and live within a 104sq km (40sq mile) area, continuing a tradition that began in the early days of the colony. It is a major centre of world commerce in the Far East because of its good harbour, administration and absence of customs dues (it has been a free port since its foundation). The many industries include agriculture, tourism, fishing, mining, textiles, plastics, electrical and electronic goods, shipping, printing and publishing.

Hong Kong Island was ceded in perpetuity to Britain by the Treaty of Nanking in 1842; the peninsula was acquired in 1860 and the New Territories in 1898 (leased for 99 years). Area: 1,036sq km (400sq miles). Pop. (1975 est.) 4,367,000. Map 48.

Hungary (Magyar Népköztársaság), official name Hungarian People's Republic, is a landlocked nation in central Europe. It is a member of the Communist bloc, but has historical and cultural ties with Western Europe. The capital is Budapest.

Hungary was once joint ruler of the Austro-Hungarian Empire, and one of the most authoritarian states in Europe, with a landowning aristocracy that dominated the peasants and repressed non-Magyar peoples. Today its Communist rulers also keep a firm grip on the country's life. But Hungarians have retained a fierce independence, as shown during the uprising in 1956.

Land and climate. The country consists mainly of two rolling plains, the Little Alfold in the north-west and the Great Alfold, which stretches across central and eastern Hungary and is divided into three parts by the rivers Duna (Danube) and Tisza. Ranges of mountains in the north include Hungary's highest peak, Mt Kékes, which rises to 1,015m (3,330ft) above sea-level in the Matra Mountains. In the west, south of the Little Alfold, lies the Bakony Forest and Lake Balaton which, with an area of 635sq km (245sq miles), is the largest lake in central Europe. Hungary has a continental climate with long, dry summers – averaging 21°C (70°F) in July – and severe winters – averaging 1°C (30°F) in January.

Economy. Formerly a predominantly agricultural country, Hungary began to industrialize in the late 19th century, a process that has accelerated since World War II. More than a fifth of the working population is still employed on the land, however, and Hungary's fertile soil is its greatest natural resource. More than half the land is arable and most of the remainder is forest, pasture or meadow. After the war the government divided the large farms into small plots and gave them to farm workers, but later organized them into collectives. Irrigation schemes of the 1950s increased the area of cultivatable land. The chief crops are maize, wheat, sugar-beet and potatoes. Fruit and garden produce are also important, and vines are cultivated on more than two per cent of the land for producing such high-quality wines as Tokay. The chief farm animal is the pig, of which there are about 8 million. Hungary has relatively few mineral resources. There are substantial deposits of bauxite, and coal and lignite are also mined. Some oil and natural gas has been found and is being developed.

About 45 per cent of the total work force is now engaged in industry and construction, which contributes about 57 per cent to the economy (as opposed to agriculture's 18 per cent). The principal exports include machinery, vehicles (especially buses), chemicals (mainly pharmaceuticals), metals, food and food products. The leading customers are the USSR (32%), East Germany (10%) and Czechoslovakia (9%). Hungary's imports include crude oil, raw cotton and motor vehicles, and its chief suppliers are the USSR (28%), West Germany (9½%) and East Germany (9½%).

People. About half the population is urban, almost a fifth living in Budapest. Nearly all the people are Hungarian (Magyars) and speak the Hungarian language. There are small minorities of Germans, Slovaks, Gypsies, Romanians, Serbs and Croats. There is no state religion and all religions have equal standing, although the state keeps a close watch on all religious activity. The Roman Catholic Church was extremely powerful until 1949 and today about two-thirds of the people are Roman Catholics and most of the remainder are Protestants, chiefly Calvinists. There is a Jewish population of about 100,000 but nearly half a million Hungarian Jews were murdered by the Nazis during World War II.

Education is free and compulsory between the ages of 6 and 16; kindergartens are provided for three- to six-year-olds and there are public nurseries for younger children. Basic school ends at age 14, after which secondary schooling is available at general schools, technical schools and apprentice-training schools attached to factories or co-operatives. There are four general universities and 14 specialized universities. The government is rapidly expanding the educational programme.

Hungarians have a great love of literature and music. The best known Hungarian dramatists include Imre Madách (1823-64) for his epic masterpiece ***The Tragedy of Man*** and Ferenc Molnár (1878-1952) for his witty comedies. The most widely read novelist has been the prolific and inventive Mór Jókai (1825-1904). Hungary is renowned for its tradition of folk-music, which inspired many of the works of such great composers as Franz Liszt (1811-86), Béla Bartók (1881-1945) and Zoltán Kodály (1882-1967). Gypsy violin playing and folk-dances such as the csárdás are well-known features of the country, as is the national dish goulash, a stew of beef, vegetables (especially onions) and paprika. Hungarian chefs have an international reputation.

Soccer is the most popular sport and the Hungarian national side of the early 1950s is regarded by many as the greatest football team of all time. Hungary has a fine record in the Olympic Games, excelling in such sports as fencing and water polo. Other major activities include swimming, athletics and winter sports, and table-tennis.

Government. The People's Republic was established in 1949 with a Soviet-type constitution, vesting power in a parliament which elects a Presidential Council. The Council, usually represented by its president, acts collectively as head of state. But it is responsible to parliament, which takes the form of a National Assembly consisting of 352 deputies elected for five-year terms. The only political party, the Hungarian Socialist Workers' Party, plays a major role in government through its Central and Political committees – the politburo. Hungary was a founder member of the Warsaw Pact in 1955. The armed forces number 105,000 regulars; there are also 20,000 border guards and a Workers' Militia.

History. Hungarians are descended from the Magyars, nomadic horsemen who migrated across the Urals in the late 9th century AD and defeated the tribes of the Hungarian plains (such as the Goths and Huns). They established a dynasty that lasted for 400 years. St Stephen (reigned *c.* 1000-1038 as Stephen I) introduced Christianity and centralized feudal rule, making Hungary a powerful state. Following a disastrous defeat at Mohács in 1526 by the Ottoman Turks, who laid waste much of the country (and controlled most of it until the late 1600s), the Hungarian nobles elected Ferdinand I of Austria as king. Austrian power in Hungary was shaken by their defeat in 1866 at the hands of Prussia, and in 1867 the dual kingdom of Austria-Hungary was established, giving Hungary equal rights. Hungarian rule proved to be intolerant, however, and it led to suppression of the non-Magyar peoples (Czechs, Slovaks and Serbs) who lived mainly in the border provinces.

Sopron, in the west of Hungary, still retains at its centre the original medieval walled city, although the modern town has now spread far around.

Most of Iceland's coast is rugged and indented by numerous gulfs and fiords. Only in the south of the country is the coast flat and unbroken.

Reykjavík is Iceland's capital and centre of its cod-fishing industry. It was founded in 1874 and today about two-fifths of the population lives there.

The Hapsburg Empire collapsed at the end of World War I, and a Hungarian republic was proclaimed. It was soon supplanted by a Communist regime under Béla Kun, but after only five months his government was brought down with the aid of Romanian intervention (1919), and a new government was formed under the regency of Admiral Miklós Horthy. Peace was made with the Allies, and by the Treaty of Trianon (1920) Hungary gave up more than two-thirds of its territory, including Transylvania (to Romania), Croatia (to Yugoslavia) and Slovakia (to Czechoslovakia). As a result, Hungary lost valuable mineral deposits, 80 per cent of its forests and a third of its population. The country was left economically weak, and its foreign policy between the wars was focused on the recovery of its lost provinces and led to the signing of military pacts with Nazi Germany and Italy.

At first these treaties proved beneficial when Germany seized Czechoslovakia and part of Romania (1938-40) and returned some of Hungary's former territory. Hungary then joined the Axis powers and entered World War II, declaring war on the USSR in 1941. But they suffered heavy losses, and subsequent attempts by Hungary to sign a separate armistice with the Allies provoked German military occupation of the country in 1944. Soviet troops displaced the Germans and occupied Hungary from 1944 to 1945. The Hungarians were allowed to set up a provisional government, and by the terms of an armistice with the Allies gave up the territory they had acquired since 1938. Elections were held in 1945, in which only a few Communist candidates were returned, and a new republic was proclaimed in 1946. Then, with Soviet backing, the Communists seized power in 1947-48, and a Soviet-type constitution was adopted in 1949.

In 1956 a large-scale uprising broke out in Budapest. Soviet troops entered Hungary and crushed the revolt, despite vain but courageous resistance, especially in the streets of Budapest. Thousands of Hungarians were killed, and nearly 200,000 fled to the West. The Soviet-backed János Kádár was installed as premier, and as First Secretary of the Communist party maintained continuous power. The Kádár regime has proved to be progressive, while taking care not to alienate the USSR. It remains faithful to the Soviet party line, although Hungarians are now allowed to travel abroad and there has been a relaxation of censorship. In 1968, the government initiated the "New Economic Model", a policy designed to decentralize industry and increase productivity. Some private enterprise has been fostered, trade with the West has been doubled, and there has been a significant amount of Western investment. With a consumer-oriented economy and increased tolerance for cultural freedom, Hungary emerged in the late 1970s as one of the freest of the "Eastern bloc" nations. Map 18.

Hungary – profile

Official name Hungarian People's Republic
Area 93,030sq km (35,919sq miles)
Population (1975 est.) 10,619,000
Density 114 per sq km (298 per sq mile)
Chief cities Budapest (capital) (1976 est.) 2,071,000; Miskolc 193,000; Debrecen, 177,000
Government Head of state, Pál Losonczi, president of Presidential Council (elected 1967)
Religions Roman Catholicism, Protestantism, Judaism
Language Hungarian
Monetary unit Forint
Gross national product (1974) £9,576,900,000
Per capita income £902
Agriculture Maize, wheat, sugar-beet, potatoes, grapes, pigs
Industries Machinery, motor vehicles, chemicals, food processing
Minerals Bauxite, coal, lignite, natural gas, oil
Trading partners USSR, East Germany, Czechoslovakia, West Germany, Poland

Iceland (Island) is an independent island nation in the northern Atlantic Ocean, just south of the Arctic Circle and 290km (180 miles) east of Greenland. It includes several other smaller islands. The capital and only large city is Reykjavík. Iceland's economy is dominated by the fishing industry, which the country has felt in recent years to be threatened by the activities of foreign trawlers. But it is a prosperous country and incomes are generally higher than the European average.

Land and economy. Much of Iceland is of volcanic origin, with a level lava desert 610m (2,000ft) above sea-level, ice fields, glaciers and lakes. The 200 or so active volcanoes include Hekla (1,490m; 4,890ft), and geysers and lakes of boiling mud are evidence of continuing volcanic activity [*see also* PE p.*31*]. The highest point is the glacial peak Öraefajökull (2,120m; 6,950ft). Only about 0.5 per cent of the land is cultivated. Damp, cool summers characterize a climate tempered by the Gulf Stream. Winter temperatures average –1°C (30°F).

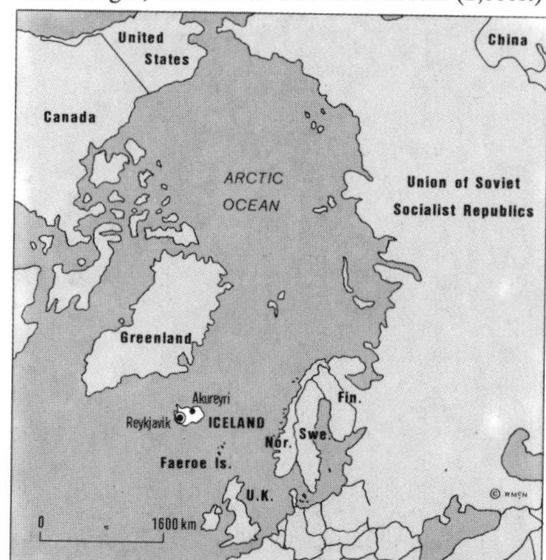

Fishing employs 14 per cent of the population, another 14 per cent works in agriculture, and 30 per cent is engaged in manufacturing and construction. Fish products form nearly 80 per cent of all exports.

People. Icelanders are descended from Norwegian settlers; the Icelandic language has remained almost unchanged since the 12th century and is consequently similar to Old Norse. The literacy rate is 99.9 per cent, the highest in the world. About 97 per cent of the population belongs to the Evangelical Lutheran Church.

Government. Iceland is a constitutional republic with an elected president and a parliament (called the Althing).

History. Iceland was settled by Norwegians in the 9th and 10th centuries, and the ruling chiefs established a republic and the Althing, said to be the oldest parliament in the world. In 1262 Norway took control of the island and, along with Norway, it passed to Denmark in the 14th century. In the early 19th century the rise of nationalism brought demands for independence, and home rule and sovereignty were granted under the Danish crown in 1918. During World War II Iceland depended on the United States for defence, and a plebiscite in 1944 established it as an independent republic. A

Iceland – profile

Official name Republic of Iceland
Area 103,000sq km (39,768sq miles)
Population (1975 est.) 218,000
Density 2.1 per sq km (5.5 per sq mile)
Chief city Reykjavík (capital) (1975 est.) 84,900
Government Constitutional republic
Religion Evangelical Lutheran
Language Icelandic
Monetary unit Krona
Gross national product (1974) £521,400,000
Per capita income £2,392
Agriculture Fish, potatoes, turnips, hay, cattle, sheep
Industries Fish canning and freezing, aluminium smelting, cement, ammonium nitrate, diatomite, clothing, shoes, chemicals, fertilizers, hydroelectric power
Minerals Natural hot water, skeletal algae, perlite
Trading partners USA, Britain, West Germany, USSR

In northern India much of the heavy work, such as shifting rock and soil for building a new road, is still done almost entirely by manual labour.

Fatehpur-Sikri, founded in 1569 to honour the muslim saint, Shaikh Salim Chishti, is the only nearly complete Mogul city in India.

The capital of India since 1931, New Delhi is built in stark contrast to the old city which still retains its original commercial functions.

conservative coalition government allows the United States and NATO to have military bases on the island. In the mid-1970s the presence of foreign fishing fleets near Iceland led to clashes, particularly with British vessels. In September 1972 Iceland extended its fishing limits from 22 to 93km (12 to 50 nautical miles) and in October 1975 it extended them even farther to 370km (200 nautical miles). This action, taken to protect and conserve fish stocks, severely strained relations between Iceland and the other fishing nations involved. Map 2.

Idaho. *See* UNITED STATES.

Illinois. *See* UNITED STATES.

India (Bharat), official name Union of India, is a republic of southern Asia. It occupies the greater part of the Indian subcontinent, the vast triangular peninsula that extends southwards into the Indian Ocean [*see* PE p.*53*]. India is the home of about one-seventh of the world's population, and has one of the oldest civilizations in the world. Two of mankind's major religions, Hinduism and Buddhism, developed within its borders. Over the centuries many great Indian empires have risen, run their course and fallen. The most recent was the British Indian Empire, which came to an end in 1947. India is a member of the Commonwealth of Nations. The capital is New Delhi.

Millions of India's people live in busy industrial and commercial cities – some of the most crowded cities in the world. Many more live in the half million small villages and wrest their living from the soil. For them life's greatest problem is obtaining sufficient to eat: for in India – with its periodic devastating droughts and floods – famine is an ever-present threat. Before making plans to better the conditions of the poorer classes, the government has first to balance the relationship between production (both agricultural and industrial) and the enormous and rapidly increasing population.

But despite its problems, India is a land of great physical and cultural wealth: of vast rivers and high mountains, of forests and gardens, and a land of art, music and subtle philosophy. It is a land of contrasts, particularly between wealth and poverty.

Land and climate. India is divided naturally into three main regions: the triangular Deccan plateau, sometimes called "the peninsula"; the northern plains; and the Himalayas. The Deccan plateau is highest in the south and west and tilts downwards towards the east. In the north, east and west it has mountainous rims. The northern rim is the Vindhya Range, following the line of the Narbada River, which has often formed a cultural and political frontier in India. On its eastern and western edges, the Deccan rises to ranges of hills called the *Ghats,* which converge towards the southern tip of the peninsula. The Eastern Ghats are low and fringed by a wide coastal plain. The Western Ghats extend for about 1,300km (800 miles); at their southern end are the Nilgiri Hills – the Blue Hills – famous for their hill stations, such as Ootacamund. The Western Ghats are bordered by a narrow coastal plain, on the north of which Bombay stands.

The northern plains, which stretch right across the subcontinent, are the most extensive alluvial lowlands in the world. They have India's best agricultural land and its highest density of population. Three large rivers cross the plains: the Indus, the Ganges and the Brahmaputra, all three of which rise in the Himalayas. The Indus is the river from which India gets its name: Aryan invaders called it *Sindhus* ("river"). It was on its banks that the first known civilization developed – in the cities of Mohenjo-daro and Harappa. The Indus rises in Tibet and flows across north-western India and Pakistan to reach its mouths on the Arabian Sea near Karachi. The Ganges is the holy river of the Hindus; one of the cities on its banks, Vārānasi (Benares), is a place of pilgrimage. The Ganges rises on the southern slopes of the Himalayas and flows eastwards through India and into Bangladesh. It has a wide, many-mouthed swampy delta called the *Sundarbans* on the Bay of Bengal. The third great river, the Brahmaputra, rises in south-western Tibet and flows through north-eastern India and Bangladesh before meeting the Ganges in the Sundarbans.

The Himalayas, the highest mountains in the world, rise above India's northern frontier, extending from the valleys of the Indus in the west to those of the Brahmaputra in the east – about 2,400km (1,500 miles). Only the foothills project into India; the highest is Nanda Devi (7,816m; 25,645ft).

The sea is the dominating climatic influence, not the land mass of central Asia, from which India is sheltered by the Himalayas. The climate varies considerably from place to place because of the diversity of surface features and the great distances involved. The heaviest rainfall ever recorded anywhere in the world in one year was 26,466mm (1,042in – nearly 87ft) in 1860-61 at Cherrapunji, in Assam in the north-east, but the Thar Desert in the north-west is one of the driest places in Asia.

There is, however, a general pattern of seasons: hot, rainy and cool. The hot season begins in March and lasts until the rains come in about the middle of June. The rainy season continues for about three months; then the south-west monsoon winds blow in from the Indian Ocean and bring torrential rain to a large part of the country. As the monsoon comes to an end, temperatures rise and the air becomes humid. The season that follows is cool in most places, but can be bitterly cold in the mountainous districts of the north.

Physical resources. India has immense deposits of iron ore, much of it of high quality, estimated as about a quarter of the world's total reserves. The deposits are chiefly in the east and south, and are as yet under-exploited. There are also large reserves of ores of manganese and aluminium – India ranks among the three leading producers of each. The most plentiful and most valuable mineral, however, is coal, which is mined principally in Bihar, West Bengal and Madhya Pradesh. Other important minerals are copper ore, gold and dolomite (calcium and magnesium carbonate); India produces four-fifths of the world's mica. Petroleum, natural gas and sources of nuclear energy (uranium and thorium) are of increasing importance.

Constitution and government. The constitution came into effect on 26 January 1950. In it India is described as a union of states which comprises 22 states; Sikkim was incorporated in the union by the 38th amendment to the constitution in 1975. The states have autonomy in a number of matters, but certain responsibilities are fulfilled by the central government; these include foreign affairs, defence, communications, coinage and customs. There are also nine union territories, each of which has an administrator who acts on behalf of the president. The president is the head of state and holds office for

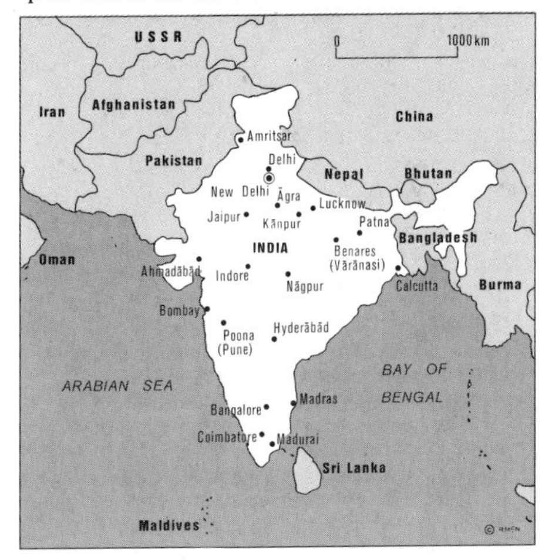

five years; he can be re-elected. He is appointed by an electoral college consisting of the members of the central legislature and the state legislatures. Certain fundamental rights are enshrined in the constitution. "Untouchability", one of the social evils of the caste system, has been abolished and any attempt to put it into practice is an offence.

The central legislature has two houses. The upper house, the *Rajya Sabha* (Council of States), has up to 250 members. Most are representatives of the states and are elected by the members of the state legislatures; a third retire every second year. The lower house, the *Lok Sabha* (Council of the People), has up to 500 members directly elected by universal adult suffrage from constituencies in the states. In addition, it has up to 25 members that represent the union territories. It is elected for a term of six years (if not dissolved sooner). The legislature meets in New Delhi.

The largest political party is the Congress Party,

The high price obtainable for tiger skins has promoted the illegal hunting of these animals. Only the killing of proven maneaters is allowed in law.

The Palace of the High Court of Justice in Madras is a fine example of the legacy of public buildings left by the British in India.

Meditating Buddhists provide a tranquil contrast to Calcutta's turbulent streets. The city is severely overcrowded and homelessness is endemic.

which won all the central elections until 1977. In that year it was defeated when an alliance was formed between the Janata Front and the Congress for Democracy.

The highest court is the Supreme Court of India, which has the final voice in constitutional matters and is also the final court of appeal. Immediately below it are several high courts.

Armed services. The army is organized in four commands – southern, western, central and eastern. Its authorized strength is 826,000 men. Officer cadets are trained at the Indian Military Academy, Dehra Dūn. The navy comprises about 30,000 officers and ratings; vessels include an aircraft carrier and two cruisers, as well as destroyers, frigates, submarines and escort vessels. It has an air arm. The air force has three operational commands – western, central and eastern – and has about 100,000 members. Many of its combat squadrons are equipped with MiG-21s, either Soviet-built or manufactured in India under licence.

Agriculture. Three-quarters of India's people live on the land, and agriculture accounts for about 45 per cent of the national income. Since independence the government has tried to effect a fairer distribution of land and to relieve peasants of their historic burden of debt, but legislation introduced to achieve these ends seems to have been only marginally effective. Efforts have also been made to increase the yields of crops, which have always been low. Improved strains of seeds have been made available, irrigation work has been put in hand and the use of fertilizers has been explained and encouraged. As a result, output has slowly increased, despite the disastrous drought of 1965–66.

The most important food crop is rice, of which about 4,000 varieties are grown; about a quarter of all the cultivated land is used for growing rice. The next most important cereal crop is wheat, which is grown chiefly in the north and centre. There are also large crops of Indian millet (*bajra*), maize and many kinds of vegetables. The chief cash crop is tea, of which India is the world's largest producer. More than a million people are employed in tea cultivation, and tea accounts for a quarter of all India's exports. The country is also the largest producer of pepper, groundnuts and sugar cane. Other major cash crops are cotton, jute, rubber, coffee and tobacco. Of the vast numbers of livestock, the largest categories are cattle (more than 160 million) and buffaloes (55 million). Cows are regarded as sacred by Hindus and are not raised for beef, although about a third of them are used as work animals for such tasks as pulling ploughs and carts and for turning millstones. Both cows and buffaloes provide milk. There are also large numbers of sheep, goats, horses and poultry.

Industry. Recent government planning restricts expansion in certain industries to state corporations; such industries include the manufacture of iron and steel, shipbuilding and the mining of coal, iron, manganese and gypsum. The largest industry is the making of textiles, chiefly cotton, which employs more than a quarter of all factory workers. Its main centres are in the west-coast cotton-growing states of Mahārāshtra and Gujarat. The traditional centre of the industry is Bombay, but Ahmadābād is also of major importance. Jute processing is a long-established industry, centred on Calcutta.

The government has treated the expansion of the steel industry as one of its priorities. Since 1911 Jamshedpur (Bihar) has had some of the world's largest iron and steel mills, but in the 1960s and 1970s several other plants were opened, many of them also in the north-east. Associated industries have expanded with the growth in steel production, and India now manufactures a wide range of engineering products, including machine tools, rolling stock, diesel engines, aircraft and electronic equipment. Oil refining is of growing importance, as is the manufacture of chemicals.

Despite the wide scope of India's large-scale industry, only about five million people are employed in it, whereas the older village industries employ about 20 million. The government has also invested in this local sector of the economy. The chief single product is cotton cloth but some finer fabrics are woven and many kinds of craft goods are made, often to folk designs. "Backyard factories" make light engineering products, such as bicycles and sewing machines.

Trade. India has difficulty in balancing its imports and exports, but sometimes the trade gap is narrowed by foreign aid (such as imports of wheat from the United States). The chief exports are jute, tea, iron ore, cotton and leather; the chief imports are petroleum, cereals, machinery, chemicals, transport equipment and copper. India's principal trading partners are the United States, the USSR, Britain, Japan, Iran and West Germany.

Transport and communications. The railway system, one of the largest in the world, is state owned and is divided into nine zones. It has broad-, metre- and narrow-gauge track. There are about 1,100,000km (680,000 miles) of roads suitable for motor vehicles, but less than half is metalled and thousands of villages are several miles from the nearest permanent road. Carts drawn by bullocks and other draught animals are an indispensable means of transport outside the cities. The national highways, the best roads, total about 24,000km (15,000 miles) and are linked with the international routes of the Economic Co-operation Administration Far East (ECAFE). Air transport is state owned: Air India operates long-distance international flights, and Indian Airlines flies over domestic and short-haul international routes. There are more than 80 airports, of which four are international (Bombay, Calcutta, Delhi and Madras).

People. Although 90 per cent of the people of India are Caucasian in origin, all of the major racial groups of mankind are represented in the country if the Negritos – the Negroid pygmies of the Andaman Islands – are included. The Caucasians include both the Dravidians (today represented by the Tamils and others of the south) and the Aryans who began their invasions of India in the 1500s BC and are thought to have driven the Dravidians out of the northern lands. Many later groups of invaders also settled on the plains of the north. Most northern people have lighter skin and a heavier stature than people of the south. India also has many tribespeople, about seven per cent of the population. The origin of some of the tribes is obscure, but those who live in Assam and other mountainous regions of the north-east are of Mongoloid stock. Veddoid tribes in the south are thought by some anthropologists to be related to the Australoid peoples of Australia.

In population India ranks second only to China, and its population is still rising rapidly – nearly 25 per cent in the years 1961 to 1971. The chief reason for this increase is the declining death-rate, the result of improvements in preventive medicine – for example, in the fight against malaria – and in the availability of medical treatment. But life expectancy is still only about 47 years for men and 46 for women. Economically, the rise in population compounds India's difficulties, since it swallows up increases in agricultural and industrial production. The government provides inducements for people to have smaller families, has made family planning facilities widely available and has encouraged sterilization in some circumstances.

The greatest densities of population are in Delhi (2,738 per sq km; 7,091 per sq mile) and Chandīgarh (2,257 per sq km; 5,846 per sq mile) in the north, and the lowest in Assam (6 per sq km; 16 per sq mile) in the north-west.

Education. About 30 per cent of the population is literate; the literacy rate for men is twice as high as that for women. The Indian constitution makes the provision of free and compulsory primary schooling a national aim, and many of the states have passed legislation with the same end in view. But for financial reasons the attainment of this goal must be far ahead: the money for teacher training and school building is not available, and in millions of families even young children have to contribute work or earnings to help to provide food. At present about 80 per cent of children in the 6-11 age group go to school, and about 35 per cent of those in the 11-14 group. Urban areas are much better served than rural ones. In higher education the situation is quite different. India has nearly 100 universities and hundreds of research and specialized-training institutions. In this, as in other aspects of life, there is an immense gap between the standards of the richer people and those of the poorer sections of society.

Cultural life. The official language of India is Hindi, the most widely spoken language. It is written in the Devanagari script. English still has a special place as a language of convenience. Of the country's scores of languages (with 1,500 dialects), 15 are listed in the constitution as having particular

The followers of Buddhism have built many lavish temples in India as well as dome-shaped memorial shrines called stupas, which are often topped by ornate towers.

Central to the Sikh religion, the Golden Temple is situated in the Punjabi city of Amritsar, which was the 19th century centre of the Sikh Empire.

The Taj Mahal, Uttar Pradesh state, was completed in 1648 as a mausoleum for Mumtaz Mahal, the favourite wife of the Mogul Emperor, Shah Jahan.

historical and cultural importance: Assamese, Bengali, Gujarati, Hindi, Kannada, Kashmiri, Malayalam, Marathi, Oriya, Punjabi, Sanskrit, Sindhi, Tamil, Telugu and Urdu. They belong to two main language families, Indo-European and Dravidian. The Indo-European languages predominate in the north. They include Assamese, Bengali, Gujarati, Punjabi, Hindi and Urdu. Urdu is written in Arabic script and incorporates many Persian and Arabic words. A simpler, spoken form of the same language is called Hindustani. The Dravidian languages, which predominate in the south, include Kannada, Malayalam, Tamil and Telugu.

According to its constitution, India is a secular state, but its population is overwhelmingly Hindu (about 510 million people; 85 per cent). The largest minority religion is Islam (67.5 million; 11 per cent). At the time of independence many Muslims moved to Pakistan; most of those that remain live in the north. There are also large minorities of Christians (16 million), Sikhs (12 million), Buddhists (4 million) and Jains (3 million). The influential Parsee community numbers about 120,000. Certain cities have particular religious associations. To the Hindus, Vārānasi (Benares), Allāhābād, Nāsik and Puri are sacred, and the Sikhs consider Amritsar to be a holy city.

The distinctive features of Indian life are closely bound up with the teachings and customs of Hinduism. Hinduism is unusual among the great religions of the world in that it had no founder or founding teacher but has been formed from many religious and cultural strands over thousands of years. It has a very large number of sects, but all Hindus hold sacred the *Vedas* – collections of hymns (*Samhitas*), religious instructions and metaphysics (*Brahmanas* and *Upanishads*) – which date from the time of the Aryans. For hundreds of years the *Vedas* were passed on by word of-mouth, but eventually they were written down in Vedic, the parent language of Sanskrit. Two other writings of great importance in Hinduism are the epic poems the *Ramayana* and the *Mahabharata*.

The caste system, now forbidden, was deeply rooted in Hindu society. In its origin it was based partly on a division of labour, but it gradually developed into a system of rigid social compartments, each with its own rules, customs and sometimes even beliefs. Caste membership depended on ancestry, perhaps connected with place of birth and occupation. The 3,000 or so different *jatis* (castes) were divided into four major groups: *Brahmins*, the priests and scholars; *Kshatriyas*, the warriors or rulers; *Vaisyas*, the traders; and *Sudras*, the cultivators. The Indian constitution of 1950 forbids any form of discrimination based on caste and outlaws the idea of "untouchability" – the condition of the millions of outcastes who were formerly condemned for life to the lowest and most unpleasant tasks. *See also* MS p.*265*.

Most Indians live in small villages: of the 575,000 villages, 300,000 have fewer than 500 inhabitants. Village houses are generally made of mud and straw, and have at the most two rooms. The chief items of household furniture are a bed – a wooden framework laced in with interwoven string – and a few pots for cooking and water. Food is scarce, and the main item of diet is a kind of bread made from rice, wheat or seeds; vegetables are also eaten. Few Hindus eat meat and none eats beef. Water is drawn from the village well, and cattle dung is dried and burned for heating. A few of the richer villagers may live in small brick houses, and a village usually has a shop or two and a schoolroom.

The peoples of India have a long and varied – even exuberant – artistic history, one rich in indigenous invention and one also that has drawn inspiration from many outside sources. The age-old love of design and colour can be seen in even the poorest temples in remote areas. Southern India is particularly rich in the gigantic, dark, pyramidal Hindu temples of the Dravidians, each rising to its narrow peak in elaborately carved stages. Indian sculpture dates back as far as the subcontinent's known history: statuettes showing considerable artistic skill have been found in the ruins of Mohenjo-daro and Harappa. Most Hindu sculpture is concerned with living things, particularly the human body, but despite a Greek presence in northern India the Indian artist's view of human beauty has not been influenced by the Greek ideal that permeates Western art. Some of the finest ancient sculptures are found in the many extraordinary cave temples, such as the Hindu caves on Elephanta Island near Bombay and the Buddhist caves at Ajanta near Aurangābād.

In the AD 1500s the Moguls introduced a distinctive art and architecture – one that, because it was Muslim, permitted no representation of the human form. Its most famous example is the Taj Mahal at Agra (the work of a Persian architect), a building that some people consider to be the most beautiful in the world [*see* MS p.*184*].

Indian music has its own tradition, said to be linked to philosophy and religion, with the emphasis more on melodic effects than on harmonic variation. The best known stringed instruments include the vina and the similar but simpler sitar. The tamboura, also stringed, is played to provide a drone accompaniment. Percussion instruments include the tabla, a pair of small kettle-drums that are played with the fingers. Indian dancing is a highly developed art form. Each of the classical dances tells a story by means of graceful and stylized movements, involving also the hands and fingers.

History. The earliest recorded Indian civilization flourished in the Indus valley in about 2000 BC. Archaeologists have uncovered the remains of two great cities of this civilization, Mohenjo-daro and Harappa, which reveal a complex and ordered way of life. The people of the Indus valley evidently engaged in commerce, and it is assumed that they had contact with the civilizations of Mesopotamia.

What happened to the civilization of the Indus valley is not known; it may have fallen to the onslaught of the Aryans, who invaded India in about 1500 BC. From their settlements, Hindu culture developed, although some of the features of Hinduism seem to have been taken over by the Aryans from the Indus valley civilization. As the Aryans extended their region of settlement, the indigenous peoples, the Dravidians, withdrew southwards.

The first important kingdom established by the Aryans was that of Magadha. During the reign of the Magadha king Bimbisara (540-490 BC) the Buddha – Siddhartha Gautama – turned away from the vanities of his princely life and began the teachings that developed in Buddhism. Another religious philosophy founded then was Jainism.

Alexander the Great invaded India in 326 BC. He captured much territory, but after his departure his successors were driven out by Chandragupta Maurya, who made himself king of Magadha and built a vast dominion called the Mauryan Empire, which had its capital at Patna. This was the first of the great empires that attempted to bring all India under one rule. Maurya's grandson was the Emperor Asoka, the greatest of ancient India's rulers.

The Mauryan Empire lasted for only 50 years after Asoka's death (*c.* 232 BC). Two centuries of disorder followed, with many invasions and the founding of many kingdoms. At this time Tamils emigrated from the south of India to settle in Indonesia and the Malay Archipelago, taking their Hindu culture with them. In AD 319 a strong kingdom was formed in Magadha and Oudh by Chandragupta I, the first of the Gupta dynasty. His successors conquered in turn Assam, the Deccan, Gujarat and the land of the Mahrattas.

After 606 a new empire was created by Harsha, King of Thanesar, who established his capital at Kanauj and gradually extended his rule over most of northern India. The Kanauj Empire lasted until the Muslim conquest of the late 12th century. The Muslims (from Afghanistan) established a sultanate with its capital at Delhi. By the 14th century the Delhi sultanate included almost the entire subcontinent. Its power ended with the sack of Delhi by Tamerlane and his Tartar hordes in 1398. The sultans lingered on as local rulers until the Battle of Panipat (1526) when Babur, another Muslim invader from Afghanistan, defeated the then sultan and founded the Mogul Empire.

Europeans had travelled to India at about the same time as the Moguls. In 1498 Vasco da Gama landed at Calicut, and shortly afterwards the Portuguese founded a trading colony at Goa. As well as the Portuguese, the British, French and Dutch were all competing for the riches of India. Queen Elizabeth I granted a charter to the British East India Company in 1600 for a monopoly of trade in the Eastern Hemisphere. The company dealt mainly in textiles and tea. To protect its trade it built forts in India, recruited an army and handed out "subsidies" to local rulers – and the Mogul emperor – for favours granted.

The Ajanta caves in Maharashtra state, India consist of a series of Buddhist chapels cut out of the hillside. The interiors are decorated with frescoes.

The city of Hyderabad which was in the 16th century the capital of the Kingdom of Golconda is now the principal city of Andgra Pradesh state.

Varanasi, formerly Benares, situated on the banks of the Ganges is the Hindus's holiest city. It is estimated to contain 15,000 temples.

The East India Company lost much ground, however, when the governor of the French settlement of Pondicherry, the Marquis Dupleix, set out in the 1740s to establish French supremacy in India by a mixture of warfare and diplomacy. But Dupleix came up against the military genius of the Robert Clive, and eventually was recalled to France in disgrace. Clive's defeat of the Nawab of Bengal at Plassey (1757) was the greatest of many victories that established the British Empire in India.

In 1774 Warren Hastings was made the first governor-general of India, but the administration of the now vast Indian possessions was still in the hands of the British East India Company. The British conquests continued into the mid-19th century: Sind was taken in 1843 and the Punjab in 1849. But in February 1857 two mutinies occurred in the East India Company's army when sepoys (soldiers) refused to bite cartridges thought to be greased with sacred cow and pig fat. In May another mutiny erupted at Meerut, becoming a full-scale rebellion.

Soon the whole of north-central India was at war. The rebels took Delhi and declared the Mogul emperor Bahadur Shah II ruler of all India. The Mahratta leader Nana Sahib massacred the British garrison and colony at Kānpur (Cawnpore), although many princes maintained their support for the British. The rebellion – called the Indian Mutiny or the Sepoy Rebellion – was put down within a year, but it changed the whole course of events in India. Bahadur Shah II was deposed and exiled, and the British government took over all the territories of the East India Company. They were arranged in provinces, each with a governor. Other parts of the subcontinent, the so-called princely states, were left in the hands of their rulers but under overriding British control. A few small French and Portuguese colonies also remained. Then in 1876 Queen Victoria was made Empress of India, represented there by a viceroy. The capital of British India was Calcutta from 1833 to 1912, when it was transferred to the splendid new city of New Delhi (officially inaugurated in 1931).

Nationalist sentiment grew in strength towards the end of the 19th century. The Indian National Congress, founded in 1885 as an organization working for political and economic advance, soon hardened into a body demanding radical change. By 1910 its moderate members were asking for dominion status for India, and its militant members were demanding *swaraj* (independence). The Congress was paralleled by the Muslim League, founded in 1906 to advance Muslim interests. Meanwhile the British administration was taking cautious measures to involve Indians in government. The outbreak of World War I in 1914 temporarily put an end to the growing ferment in India.

In 1920 Mohandas Karamchand Gandhi became the leader of the Indian National Congress. Gandhi was a lawyer and had gained fame in South Africa by organizing a campaign of non-violent civil disobedience that had forced the government to change laws discriminating against Indians. He now led several similar *Satyagraha* campaigns in India; millions of people boycotted schools and courts and obstructed streets and railways. To many people the Mahatma ("great-souled") was a saint.

In 1934 Gandhi withdrew from his position in the Congress, but his disciple Jawaharlal Nehru was appointed leader in his place. Also in 1934 Mohammed Ali Jinnah became leader of the Muslim League. The League grew in strength, and voiced its doubts about the position of Muslims in the new India planned by the Congress. In 1940 it demanded a seperate Muslim state, for which the name *Pakistan* ("land of the pure") came to be used.

When World War II broke out in 1939, the Muslim League supported Britain but the Indian National Congress refused its support without immediate self-government. The Japanese entered the war in 1941, and Japanese armies soon stood on the frontier of India. The British government made another attempt to gain the support of the Indian leaders, and proposed dominion status for India after the war. This suggestion was rejected. When Gandhi called for another campaign of civil disobedience, he and other Congress leaders were interned. One extreme section of the Congress gave help to the Japanese. Meanwhile the Indian Army played an important part in the victorious Allied campaign in North Africa and in the later campaigns that recaptured Burma.

In 1945 and 1946 several conferences took place between British and Indian leaders to discuss the form independence should take. They were held against a background of mounting violence between Hindu and Muslim factions. The Congress leaders would not agree to the partition of India, as demanded by the Muslim League. In 1947 the viceroy Lord Mountbatten and Indian political leaders agreed that to stop the violence India must be partitioned. On 14 August the dominion of Pakistan came into being, and on 15 August the dominion of India.

Partition was accompanied by great transfers of population and by violence in which hundreds of thousands of people died. The rulers of the princely states in the new India agreed to relinquish their rights, and accept incorporation into the new country. Fighting between India and Pakistan about Kashmir (where the majority of people were Muslim although the ruler was Hindu) was ended by a ceasefire arranged by the United Nations. A year after independence, Gandhi was assassinated by a Hindu fanatic who believed the Mahatma was destroying Hinduism.

The first governor-general of independent India was Lord Mountbatten, and in 1948 he was succeeded by Chakravarti Rajagopalachari. A Constituent Assembly governed the country pending the establishment of agreed organs of government; its leading figure was Jawaharlal Nehru. The Assembly ratified a new constitution, which came into force on 26 January 1950. The general election of 1952 returned the Congress Party to power with a large majority; Jawaharlal Nehru remained as prime minister. The only Indian territories still outside the republic were the small French and Portuguese colonies, but by 1962 India had gained control of these also.

Nehru (who died in 1964) was succeeded by Lal Bahadur Shastri. Then when Shastri died in 1966 his place was taken by Mrs Indira Gandhi, Nehru's daughter. She embarked on a Socialist programme, but two of her measures (to nationalize the banks and to cease payments to the former princes) were declared unconstitutional. In an election in 1971, however, she obtained the two-thirds majority needed to amend the constitution, and the government gained much popularity by its forceful intervention in East Pakistan, which led to the creation of BANGLADESH. But the government's popularity then waned and there was growing civil disorder.

In 1975 an opposition leader, Joyaprakash Narayan, secured a court conviction of Mrs Gandhi for electoral corruption. She responded in June 1975 with an Emergency Declaration giving the government proscriptive powers: thousands of opposition politicians were imprisoned and censorship was imposed. In February 1976 the term of the legislature was extended for a further year. When an election was called in March 1977 Mrs Gandhi released

India – profile

Official name Union of India
Area 3,275,833sq km (1,266,479sq miles)
Population (1975 est.) 598,097,000
Density 183 per sq km (472 per sq mile)
Chief cities New Delhi (capital) (1971) 301,800; Bombay, 5,970,575; Delhi, 3,287,883; Calcutta, 3,148,746; Madras, 2,469,449
Government Federal democracy, constitutionally described as a union of states
Religions Hindu (85%), Muslim (11%)
Languages Hindi (official), English
Monetary unit Rupee
Gross national product (1974) £33,756,400,000
Per capita income £56
Agriculture Rice, wheat, tea, sugar, vegetables, rubber, coffee, tobacco, groundnuts
Industries Textiles (cotton), jute products, iron and steel, machinery, electronic equipment, rolling stock, processed foods, chemicals, aircraft, petroleum, bicycles, sewing machines
Minerals (major) Iron, manganese, aluminium, coal, mica, gypsum, salt
Trading partners (major) USA, Japan, USSR, Iran, Britain, West Germany

Bombay is the most important sea-port on the west coast of India, being the principal outlet for the agricultural produce of the rich Deccan Plateau.

Sumatra, the second largest of the Indonesian islands, is also the most westerly of the main islands; the interior is only sparsely populated.

Celebes is the most mountainous of the Indonesian islands, and is heavily forested in the interior. The population of the island is mainly Malay.

the imprisoned opposition leaders. The election resulted in a resounding defeat for the Congress Party and Mrs Gandhi. A coalition government was formed, with Morarji Desai as premier. Map 40

States and Territories The Union of India consists of 22 states, each with its own constitution, legislature and government and each having limited autonomy. There are nine centrally administered territories.

States
Andhra Pradesh Capital, Hyderābād; area 276,814sq km (106,878sq miles); pop. (1971) 43,502,708
Assam Capital, Dispur; area 78,523sq km (30,318sq miles); pop. (1971) 14,625,152
Bīhar Capital, Patna; area 173,876sq km (67,134sq miles); pop. (1971) 56,353,369
Gujarat Capital, Ahmadābād; area 195,984sq km (75,669sq miles); pop. (1971) 26,697,475
Haryana Capital, Chandīgarh (with Punjab); area 44,222sq km (17,074sq miles); pop. (1971) 10,036,808
Himachal Pradesh Capital, Simla; area 55,673sq km (21,495sq miles); pop. (1971) 3,460,434
Jammu and Kashmir Capital, Srīnagar; area 222,236sq km (85,805sq miles); pop. (1971) 4,615,176
Karnataka Capital, Bangalore; area 191,773sq km (74,044sq miles); pop. (1971) 29,263,334
Kerala Capital, Trivandrum; area 38,864sq km (15,005sq miles); pop. (1971) 21,347,375
Madhya Pradesh Capital, Bhopāl; area 442,841sq km (170,981sq miles); pop. (1971) 41,449,729
Mahārāshtra Capital, Bombay; area 307,762sq km (118,827sq miles); pop. (1971) 50,412,235
Manipur Capital, Imphāl; area 22,356sq km (8,632sq miles); pop. (1971) 1,072,753
Meghalaya Capital, Shillong; area 22,489sq km (8,683sq miles); pop. 1,011,699
Nāgāland Capital, Kohīma; area 16,527sq km (6,381sq miles); pop. (1971) 516,449
Orissa Capital, Bhubaneswar; area 155,782sq km (60,147sq miles); pop. (1971) 21,944,615
Punjab Capital, Chandīgarh (with Haryana); area 50,376sq km (19,450sq miles); pop. (1971) 13,551,060
Rājasthān Capital, Jaipur; area 342,214sq km (132,129sq miles); pop. (1971) 25,724,142
Sikkim Capital, Gangtok; area 7,298sq km (2,818sq miles); pop. (1971) 208,609
Tamil Nadu Capital, Madras; area 130,069sq km (50,220sq miles); pop. (1971) 41,103,125
Tripura Capital, Agartala; area 10,477sq km (4,045sq miles); pop. (1971) 1,556,822
Uttar Pradesh Capital, Lucknow; area 294,413sq km (113,673sq miles); pop. (1971) 88,341,144
West Bengal Capital, Calcutta; area 87,853sq km (33,920sq miles); pop. (1971) 44,312,011

Union territories
Andaman and Nicobar Islands Seat of administration, Port Blair; area 8,293sq km (3,202sq miles); pop. (1971) 115,133
Arunachal Pradesh Seat of administration, Shillong; area 83,578sq km (32,269sq miles); pop. (1971) 444,744
Chandīgarh Eventually to be the capital of the Punjab alone when a new capital is completed for Haryana; area 114sq km (44sq miles); pop. 257,251
Dādra and Nagar Aveli Seat of administration, Silvassa; area 491sq km (190 sq miles); pop. 74,170
Delhi Area 1,485sq km (573sq miles); pop. (1971) 4,065,698
Goa, Daman and Diu Seat of administration, Panjim; area 3,813sq km (1,472sq miles); pop. (1971) 857,711
Lakshadweep Seat of administration, Kavaratti Island; area 32sq km (12sq miles); pop. 31,810
Mizoram Area 21,087sq km (8,142sq miles); pop. 400,000
Pondicherry Capital, Pondicherry; area 480sq km (185sq miles); pop. (1971) 471,707

Indiana. *See* UNITED STATES.

Indonesia, official name Republic of Indonesia, is the largest nation in south-eastern Asia. It consists of more than 3,000 islands, extending some 5,150km (3,200 miles) between the Indian and Pacific oceans and crossing three time zones. The Islands are among those once known as the *Indies*, the goal that Christopher Columbus was seeking when he reached America in 1492. They include the famous Spice Islands – today called the Moluccas (Maluku) – the source of the Oriental spices that were so highly valued in Renaissance Europe. For 150 years until 1949 the main islands of Indonesia formed part of the Netherlands East Indies. Today it includes parts of the second and third largest islands in the world: the Indonesian part of New Guinea is called West Irian (Irian Jaya) and Indonesian Borneo is called Kalimantan. The largest entirely Indonesian islands are Sumatra (Sumatera), Java (Djawa), Celebes (Sulawesi) and Timor. The capital is Djakarta (formerly Batavia), on Java.

Land and climate. West Irian has high, forested mountain ranges in the centre and swampy coastal lowlands. Kalimantan is thinly populated and covered with thick jungles and swamps. Sumatra has a high, volcanic mountain range extending along its south-western coast. The rest of the island is an alluvial lowland, with immense areas of swamps. Java is somewhat similar in formation: a belt of volcanic mountains rises along its southern coast and to the north is a broad plateau and a fertile alluvial plain. Celebes has a mountainous and thickly forested interior, but its valleys are fertile.

All the islands are subject to monsoons, and are generally hot and wet. But no part of Indonesia is far from the sea and so temperatures are seldom extreme, even on the Equator.

Economy. The country has for long been among the world's leading producers of tin; today, however, petroleum is a more valuable export (it was the most important strategic objective of the Japanese campaigns in 1941–42). Other minerals exported include bauxite, coal, and ores of nickel, iron, copper, manganese and gold; salt is also mined. The forests that cover large areas of the islands are another natural source of wealth. They contain many valuable timber trees, chief of which are teak, ebony and the scented sandalwood.

Some two-thirds of the working population is engaged in agriculture, either producing subsistence food crops or working on plantations and commercial smallholdings. The chief crop is rice, some of it produced by the wasteful method of clearing and planting land and then leaving it fallow after a few seasons. Other food crops are maize, cassava, soya beans and sweet potatoes. Cash crops include tea, rubber, copra, spices, coffee and sugar. These are

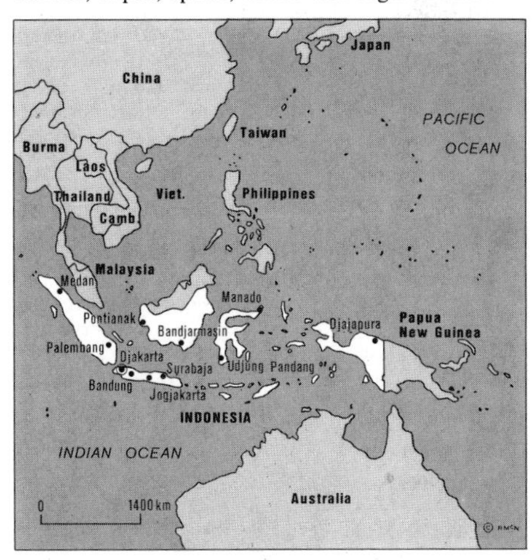

traditionally plantation crops, but since independence governments have encouraged farmers to grow them on smallholdings.

Manufacturing industry is poorly developed. Many factors restrict industrial expansion: the extremely irregular distribution of population, poor communications and shortage of power. There is some shipbuilding, oil refining and assembly of motor vehicles and bicycles. Chemicals, paper and textiles are made, and food and tobacco processing give employment to many. The oldest industry is probably textile-making, including batik (fabric printed by a wax process) for which Java is famous.

A national shipping line operates ferry services between the islands. There is also a state railway company, which operates mainly on Java. Garuda Indonesian Airways (GIA) flies to several Asian countries and to The Netherlands.

Government. Supreme power is in the hands of

Djakarta, capital and largest city of Indonesia, stands on the Liwung River in the northern plain of Java. The city was formerly called Batavia.

Most Indonesian children are of Malay ancestry but in school they are taught the official Indonesian language, which was developed from Sumatran Malay.

A caravanserai (inn for the accommodation of caravans) stands isolated in the harsh, desert landscape of Iran – the second largest country of the Middle East.

the People's Consultative Assembly (MPR), which meets at least once every five years. It includes the House of People's Representatives, which has 460 members (360 of whom are elected).

The president and prime minister is Gen. Suharto, who governs with a cabinet; he gained power by means of the army. In 1968 he was appointed president by a provisional People's Consultative Assembly, and in 1973 was reappointed for a further seven years by the first regular meeting of the Assembly.

People. Two-thirds of Indonesia's population lives on Java. Most are of Malay or Polynesian stock, although there are many distinct ethnic groupings, such as Javanese and Sundanese in Java, Balinese in Bali, Dyaks in Borneo, Papuans in West Irian and Ambonese in the Moluccas. There are also large Chinese (3 million) and Arab (800,000) minorities. There are more than 200 languages and dialects, but the official language *bahasa Indonesia* – a variant of Malay and originally a trading lingua franca – is widely understood. The literacy rate is 70 per cent, and there are several universities.

Most Indonesians are Muslims. The largest religious minorities are Hindus (6 million), Christians (6 million) and Buddhists (1 million). Throughout Indonesia there is an undercurrent of ethnic beliefs and cults; there is also a great diversity of cultures. In some places – West Irian, for instance – life is still at a primitive level; in others – for example, Java and Bali – there is an old and sophisticated civilization. The Hindu culture of Bali has given rise to the island's distinctive dance and drama, renowned for its colour and grace [*see* MS p.205]. Puppet plays are a popular form of entertainment, and Java and Bali have their own type of orchestra, the *gamelan*, composed of various percussion instruments, flutes and *rebabs* (two-stringed instruments like cellos).

Most people live in villages, but their ways of life vary from island to island. Javanese villages are generally prosperous and well-built; in Kalimantan, the Dyaks still live in the communal long-houses of their head-hunting ancestors. Many village people wear the traditional sarong – a length of cloth, often brightly coloured, wound round the body. Most of the cities (seven of the ten largest) are on Java. Many people on Java have only one name.

History. Some of the Indonesian islands have been inhabited since prehistory. Monks and traders from India introduced Hinduism and Buddhism to the larger islands in the first centuries AD. A strong Buddhist kingdom was centred on Sumatra between the 7th and 9th centuries; the famous shrine of Borobudur, still extant, dates from this period.

Islam was first introduced by Arab traders and became the dominant religion by the end of the 16th century, although no strong Muslim state emerged. Portuguese traders were the first Europeans to arrive, followed by the Dutch and British. By the early 17th century the Dutch East India Company had triumphed over its rivals and ruled the main islands. In 1799 the Netherlands East Indies was formed when the Dutch government took over the company's territories. Occasional rebellions against Dutch rule had little success.

In the 1800s two major volcanic eruptions devastated Indonesian islands – Tambora in 1815 and Krakatoa in 1883 [*see* PE p.31, *31*].

An Indonesian nationalist party was formed under the leadership of Achmad Sukarno in the 1920s. In 1945, after the Japanese occupation of the islands in World War II, Sukarno and Mohammed Hatta declared Indonesia an independent republic; the Dutch reluctantly agreed to independence. West Irian (then Dutch New Guinea) did not become part of the new state until 1963. Sukarno's government expelled people of Dutch ancestry, and set about removing foreign influences. He had to deal with rebellion inside Indonesia from those who opposed Communist tendencies in his government, and he staged a desultory conflict with the newly created Malaysia in 1964 over Malaysia's possession of Sarawak and Sabah (northern Borneo).

In 1965 the army thwarted an attempted Communist coup and in the following year the army minister, Gen. Suharto, took over power. Suharto became acting president in 1967 and president in 1968; he began establishing good relations with Indonesia's neighbours and restoring the country's shattered economy. Some foreign capital has returned, and there have been achievements in building up industry. In 1975 Indonesia took over Portuguese Timor. Then in 1976 the government discovered that the state-owned oil company Pertamina was virtually bankrupt and had to borrow money abroad to keep the company going, again undermining the economy. Map 60.

Indonesia – profile

Official name Republic of Indonesia
Area 1,919,400sq km (741,080sq miles)
Population (1977 est.) 138,133,500
Density 72 per sq km (186 per sq mile)
Chief cities Djakarta (capital) (1977 est.) 6,178,500; Bandung, 1,114,000
Government Republic; limited democracy
Religion Islam (85%)
Language Bahasa Indonesia (official)
Monetary unit Rupiah
Gross national product (1974) £7,948,700,000
Per capita income £58
Agriculture Rice, maize, cassava, tea, copra, sugar, coffee, rubber, spices
Industries Shipping, petroleum products, textiles, processed foods, tyres
Minerals Petroleum, tin, bauxite, coal, nickel, iron, copper, manganese, gold, silver
Trading partners Japan, Australia, Singapore, USA, Britain

Iowa. *See* UNITED STATES.

Iran (Irân), official name Imperial Kingdom of Iran, is a country in south-western Asia whose history goes back to the Persian Empire of 2,500 years ago. It is still sometimes called Persia. Over the centuries, the Persian civilization suffered under many invaders, including the Greeks led by Alexander the Great and the Mongols of Genghis Khan. Today, Iran is classed as a developing country, but it draws enormous revenues from petroleum. The shah (king) has embarked on a programme of modernization and industrial expansion, using authoritarian methods that have been much criticized. The capital is Teheran (Tehrān).

Land and climate. Most of the country is a high plateau, rimmed by the Elburz Mountains in the north and the Zagros Mountains in the south-west. The central part of the plateau consists of two vast deserts, the Dasht-e Kavir (Great Salt Desert) and the Dasht-e Lūt (Great Sand Desert). Mountain streams form salt ponds and marshes in the Dasht-i-Kavir during the wet season. In the hot season these dry out, leaving sparkling deposits of salt on the desert's surface. A region of particular importance to the country is the oil-rich Khuzistan plain at the head of the Persian Gulf in the west. On the interior plateau, the weather is extremely hot in summer and extremely cold in winter. In the south the climate is more temperate, but there is still a distinct hot season.

Economy. Iran is the largest exporter of petroleum in the Middle East. Until the 1950s the oil industry was largely British-controlled; in 1951 it was nationalized. The resulting international dispute brought oil exports to an end, but in 1954 it was settled by an agreement between the National Iranian Oil Company and an American-sponsored international consortium of oil companies. Iran's other mineral resources include iron, copper, lead and chromite.

More than half the people live by agriculture, although only one-seventh of the land area is cultivatable. Agricultural output is being improved by irrigation and mechanization. The chief crop is wheat; other important crops include barley, rice, sugar-beet, vegetables, fruits, cotton and gums. Wool from sheep and goats is also a major product. The government is fostering industrial expansion; manufactures include textiles, processed foods, motor vehicles, petrochemicals and iron and steel. Village industries are still of major importance: they are renowned for their fine Persian rugs and their pottery and jewellery. Road and rail networks connect the larger cities. Iranian National Airlines operates on domestic and international routes.

Government. The country is a constitutional monarchy. The Constitution, many times amended, dates from 1906. The legislature has two houses: a Senate of 60 members (some elected, some nominated by the shah) and the *Majlis* (Chamber of Deputies), whose members are elected for four-year

Tehran, or Teheran, is the capital of Iran and Tehrān province. The city lies in the north of Iran near Mount Damavand and is famous for its trade in fine carpets.

An ornate, jewel-like mosque is one of many fine buildings embellishing Isfahan (or Esfahān), the ancient and picturesque city on the Zayandeh River in Iran.

Baghdad is one of the most ancient centres of civilization and culture; the modern capital of Iraq now has a population of nearly three million.

terms. The shah can dissolve either house, and can return finance bills to the Majlis for reconsideration. In 1975 he abolished the two-party system, and announced the formation of a single new party, the *Rastakhiz* (National Resurgence Movement).

People. The people are of mixed descent, the various strains in their ancestry including Iranians (from central Asia), Kassites, Elamites, Medes, Persians, Arabs, Mongols and Turks. The sizeable minority of tribal pastoral nomads includes Kurds, Lurs, Turkomans and Baluchi [*see* MS p. *29*]. The most widely spoken language is Farsi (Persian), which uses many Arabic words and is written in Arabic script. But perhaps as much as half of the population uses other languages, including Kurdish and Turkish. The literacy rate is about 40 per cent. A Literacy Corps, formed in 1963, brings basic schooling to rural areas. The country has eight universities, the largest being Teheran University. Most people follow the state religion, the Shiite sect

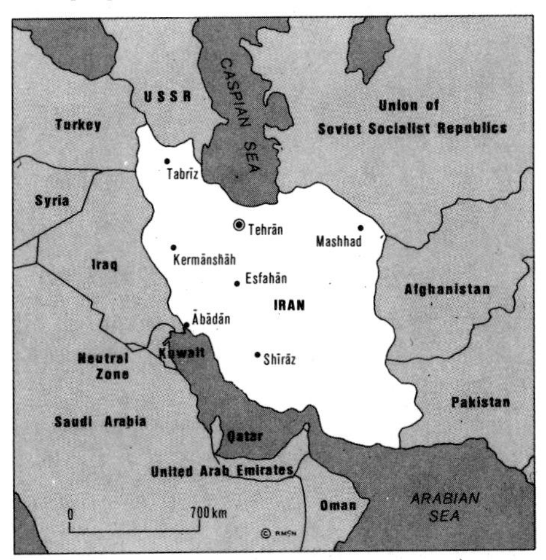

of Islam called the *Ithna-'Ashariyya*. The largest minority group belongs to the Sunnite sect of Islam, and there are small numbers of Zoroastrians, Christians, Jews and Bahais.

Teheran has been the country's capital since 1788. Many other cities have a long history and their monuments testify to former magnificence. The most interesting is probably Isfahan (Esfahān), which was the capital of the Seljuks and again of the Safavids in the 17th century. It contains some of the world's greatest architectural treasures, including the wonderful imperial mosque of Shah Abbas I with its turquoise majolica dome and exquisite mosaics. Shiraz (Shīrāz), also a former capital, has splendid buildings too; one of them is the tomb of the great 14-century poet Hafiz.

History. The earliest Persian kingdom was that of the Elamites between the 1100s and 600s BC. Then the Medes established a kingdom, but were overthrown by the Persian Cyrus the Great (about 550 BC), who extended his rule over most of Asia Minor. His descendant Darius ruled an empire that stretched from the Danube to northern India. He was defeated by the Greeks at Marathon (490 BC); in 480 BC the Greeks vanquished his son, Xerxes, at the naval battle of Salamis.

Under Alexander the Great, the Greeks overran Persia (about 330 BC); they were ousted by the Parthians (about 250 BC). The Persian Sassanid dynasty overthrew the Parthians in the AD 200s and ruled Persia for 400 years. They defeated invading Roman armies, but were conquered by the Arabs, who introduced Islam into the country. Under the Arabs and the Seljuks who succeeded them (about 1037) Persia became a centre of art and learning, but in 1221 it fell to Genghis Khan and his Mongols. In 1738 Nader Shah made himself king of Persia; he conquered Afghanistan and sacked Delhi. In the late 1700s the Qajar dynasty came to power and ruled despotically until 1906, when Shah Muzaffer-ud-Din granted the country a Constitution.

In 1925 the last Qajar shah was deposed by the Majlis, and the prime minister, Reza Khan, was elected shah in his place, with the title Reza Shah Pahlavi. Reza Shah abdicated in 1941, and his son Mohammad Reza Shah Pahlavi took his place [*see* MS p. *264*]. In 1963 the shah launched his "White Revolution" for the modernization of the country, and the government began to split up large estates and redistribute the land to farmers. Map 38.

Iran – profile

Official name Imperial Kingdom of Iran
Area 1,648,000sq km (636,292sq miles)
Population (1975 est.) 33,957,000
Density: 21 per sq km (53 per sq mile)
Chief cities Teheran (capital) (1974 est.) 3,931,000; Isfahan, 605,000; Mashhad, 592,000; Tabrīz, 510,000
Government Constitutional monarchy; limited democracy
Religion Shiite sect of Islam (90%)
Language Farsi
Monetary unit Rial
Gross national product (1974) £15,008,500,000
Per capita income £442
Agriculture Wheat, barley, rice, cotton, gums, tobacco, sheep, goats
Industries Petroleum products, textiles, processed foods, motor vehicles, carpets
Minerals (major) Petroleum, iron, copper, lead
Trading partners (major) West Germany, USA, Britain, USSR

Iraq, official name Republic of Iraq, is an Arab country of south-western Asia, at the head of the Persian Gulf. Two great rivers, the Tigris and the Euphrates, flow across it enclosing the ancient region of Mesopotamia – the region where the complex civilization of Sumeria developed more than 5,000 years ago. Iraq draws large revenues from petroleum, of which it is one of the largest producers in the Middle East. But since 1958, when its last king and its prime minister were assassinated, it has been plagued by violence. The capital is Baghdad.

Land and climate. Iraq has three well-defined natural regions: the desert of the west, the central lowlands and the north-eastern mountains. The desert – largely barren steppe land – is part of the Syrian Desert. To its east are the lowlands crossed by the Tigris and Euphrates, which join to form the Shatt al Arab some 190km (120 miles) from Iraq's short coastline on the Persian Gulf. Mesopotamia, "the land between the rivers", consists of a fertile alluvial plain in the south, with a low plateau, al Jazira ("the island"), in the north. The rugged high-

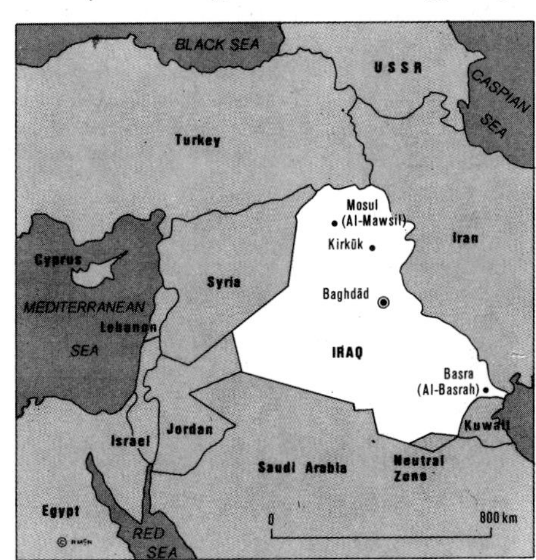

lands of north-eastern Iran form part of Kurdistan, which is shared between Iraq, Iran and Turkey. Nearly all of Iraq is hot and dry except in the winter, which can be very cold in mountainous areas.

Economy. Petroleum is the chief source of wealth. The largest output comes from oilfields at Kirkuk in the north-east; there are also fields near Mosul in the north and Basra in the south-east. Crude oil from Kirkuk can be piped across the Syrian Desert to ports in Syria and Lebanon, but the operation of the pipeline has been disrupted by disputes with the Syrian government. The oil industry is state owned, and a petrochemical industry is being developed.

Despite its place as a leading oil producer, the country is predominantly agricultural, and the area of cultivatable land is being extended by irrigation. On the southern plains the chief crops are wheat, barley, millet, maize and vegetables. Other cereals,

The ornately vaulted ceiling of the palace in Baghdad dates back to the time of the last of the Abbasid caliphs who ruled in the 13th century.

The River Liffey is the chief river in the east of the Republic of Ireland and flows through Dublin, the capital city, before entering the sea.

All is peaceful now on the Grand Parade, Cork, Ireland – but in the nationalist disturbances of 1920, the mayor was murdered by his constabulary.

tobacco and fruit are grown in the north-east. Dates and cotton are other important crops. Most industries are small, although the government has produced national development plans to encourage industrial expansion. Manufactures include processed foods, textiles and building materials.

Road and rail communications are fairly well developed. One rail route from Basra near the Persian Gulf through Baghdad to Mosul – the Baghdad Railway – connects with the Syrian railway system and thus provides through routes to Egypt and Europe. There are international airports at Baghdad and Basra.

People. Most Iraqis are Arabs, but there is a large Kurdish minority (about 18 per cent) in the north-east. The official language is Arabic; the Kurds speak their own language, Kurdish. About 20 per cent of the population is literate. Education is free at primary and secondary levels but is not compulsory. Iraq has five universities, of which the largest is the University of Baghdad. Nearly three-quarters of the people are Muslims, belonging chiefly to the Sunni sect; there is a large Shiite community in the Basra region.

The people have a reasonably high standard of living, and in most areas food is plentiful. Meat (chiefly beef and mutton) is often cooked in the form of kebab (squares of meat skewered and roasted) and is accompanied by bread or rice. Village houses are generally small and built of mud bricks. More than half the population is urban and most of them live in the cities and towns of the Tigris and Euphrates valleys. Baghdad, the largest city, rose to power and eminence as a centre of art and scholarship in the 9th century under the caliph Hurun ar-Rashid – remembered in many stories in *The Thousand and One Nights.* Mosul, the commercial centre of the north, was once the capital of an independent kingdom (it gives its name to the cloth muslin). There are many famous archaeological sites in Iraq, including those of Ur "of the Chaldees" and Nineveh, and it also has (at Al-Qurna) the legendary site of the Garden of Eden.

Government. A provisional constitution of 1964 describes Iraq as an Arab, Islamic, independent sovereign republic based on democracy and socialism. Its aim is the unity of all Arab peoples. The Ba'ath (Arab socialist) Party, ousted in late 1963, seized control again in a coup in 1968. The president, backed by the military, rules through a Revolutionary Command Council.

History. The early civilizations of Mesopotamia included those of the Sumerians, Akkadians, Amorites, Assyrians, Babylonians and Chaldeans. In 539 BC the region that is now present-day Iraq became part of the Persian Empire, and later it was invaded by the Romans and Parthians. In the 7th century AD it was taken by the Arabs. The Turks seized it in the 16th century and it remained part of the Ottoman Empire until taken by the British in World War I.

In 1920 Britain was given a League of Nations mandate over Iraq but the country became independent in 1932. It was then a monarchy: Faisal ibn Husein, a son of the sherif of Mecca, had been elected king as Faisal I in 1921. In the 1950s the government initiated a programme of social and economic reform, financed mainly by royalties from foreign oil interests in Iraq.

Then in 1958 the monarchy was overthrown by a group of army officers. King Faisal II (the grandson of Faisal I), the Crown Prince and prime minister Nuri as-Said were assassinated and a republic proclaimed. The leader of the revolt, Gen. Abdul Karim Kassem, ruled until February 1963, when he was himself overthrown and executed in another (Ba'athist) military coup. The Ba'athists were ousted in yet another coup in November 1963.

Iraq took part in the Arab war against Israel in 1967, and in the following year the Ba'athists returned to power after a coup led by Gen. Ahmed Hassan al-Bakr. His government completed the nationalization of the oil industry and established close relations with the USSR. It also took drastic military action to end the long-standing revolt of the Kurdish nationalists. In the 1970s Iraq took a hard line with regard to the Arab-Israeli conflict, refusing to accept any political solution. Map 38.

Iraq – profile

Official name Republic of Iraq
Area 434,924sq km (167, 924sq miles)
Population (1975 est.) 11,124,000
Density 26 per sq km (66 per sq mile)
Chief cities Baghdad (capital) (1974 est.) 2,800,000; Basra, 370,000; Mosul, 293,100
Government Authoritarian one-party government with army backing
Religion Islam: Sunni (74%), Shiite (22%)
Language Arabic (official)
Monetary unit Iraqi dinar
Gross national product (1974) £4,444,400,000
Per capita income £400
Agriculture Wheat, barley, millet, maize, vegetables, tobacco, cotton, dates
Industries Processed foods, textiles, petrochemicals, building materials
Minerals (major) Petroleum, rock sulphur
Trading partners West Germany, Japan, Britain, France, USA

Ireland, Republic of, (Éire) is a country that occupies five-sixths of Ireland, the westernmost island of the British Isles (the remainder is NORTHERN IRELAND, a part of the United Kingdom). The country's tranquil appearance belies its turbulent history. Overwhelmingly Roman Catholic, Ireland's people were persecuted by Protestant English rulers from the 16th century; in the 19th century about a million people died in a potato famine, and many more emigrated to the USA.

The Catholic southern part of Ireland gained its independence from Britain as the Irish Free State in 1922, and became a republic in 1949. Partition from Protestant-dominated Northern Ireland is still a bone of contention for many, but a programme of industrialization begun in the 1950s and entry into the EEC in 1973 have made for closer ties with Britain and renewed hope for a solution to the political problems.

Land and climate. The central part of the country is a lowland plain that extends from Dublin in the east to Galway in the west. It consists of gently rolling farmlands, pastures and woods, and includes the extensive Bog of Allen. Surrounding the plain is a broken fringe of low mountains, most of which rise near the coasts and rarely exceed 900m (2,950ft). The principal ranges include the Wicklow Mountains in the south-east, topped by the 926m (3,039ft) Lugnaquilla; the Mountains of Kerry in the south-west, where Macgillycuddy's Reeks rise to 1,040m (3,414ft) at Carrantuohill, the highest peak in Ireland; the Mountains of Connemara and Mountains of Mayo in the west; and the Mountains of Donegal in the north-west.

Peat bogs cover about a sixth of the country, mostly to the west of the River Shannon. The Shannon, the longest river in the British Isles, is an important inland waterway. It rises in north-central Ireland and flows south and south-west for about 385km (240 miles) before emptying into the Atlantic Ocean through a wide estuary nearly 95km (60 miles) long. It passes through several *loughs* (lakes), including Ree and Derg. Another important river is the Liffey, which rises in the Wicklow Mountains and flows into the Irish Sea at Dublin. Most of the loughs are in the west, including the Lakes of Killarney, renowned for their scenic beauty, and the wild, dark loughs of Connemara.

The west coast is indented with numerous bays and long inlets, and there are hundreds of offshore islands. High cliffs of solid, exposed rock line parts of the coast, such as the spectacular Cliffs of Moher which extend for 8km (5 miles) along the coast of County Clare, or the even higher cliffs farther north, in Donegal Bay, which drop 600m (nearly 2,000ft) almost vertically to the sea. The other coasts are less rugged, and provide many excellent harbours, such as Dublin Bay and Cork Harbour.

Ireland has a temperate climate, with mild winters, warm summers and plentiful but gentle rain brought by ocean winds. Temperatures average about 15°C (59°F) in summer and 5°C (41°F) in winter. The rainfall varies from less than 750mm (30 inches) at Dublin, on the sheltered east coast, to more than 1,300mm (50in) on the west coast.

Natural resources. Ireland's most valuable resource is its rich farmland. There is little coal, so Ireland has been forced to develop its extensive resources of peat, which is dug from bogs in the lowlands [*see* PE p.*137*]. A state-sponsored organization, *Bord na Mona* (Turf Board), produces both machined and milled peat, mostly for the Electricity

Crofters' cottages on Malin Head, the most northerly point of Ireland, are built facing squarely into the prevailing wind and weather.

The quaint old houses of Limerick give the city a familiar charm. The most populous city on Ireland's west coast, it is bisected by the River Shannon.

Peat, which is dug from bogs, is an important fuel for heating and cooking in the Republic of Ireland because few oil and coal resources exist.

Supply Board. For domestic fuel, peat is used in the form of briquettes. Substantial deposits of lead, zinc, silver and copper have been discovered and are also being developed.

Economy. When Ireland became independent in the 1920s, more than 90 per cent of its exports were agricultural and Britain was the only market. Since the late 1940s, when the Industrial Development Authority was set up, Ireland has been changing from an agrarian to an industrial society. The problem of power supply was solved by the invention of machines to utilize peat as a fuel and by the harnessing of the country's water power by building hydroelectric plants. Since 1959 government programmes for economic expansion have been implemented, and foreign industrialists and capital have been welcomed. Offshore oil and natural gas reserves are being investigated.

Agriculture is still a major factor in the economy, and it provides a sixth of the national income and

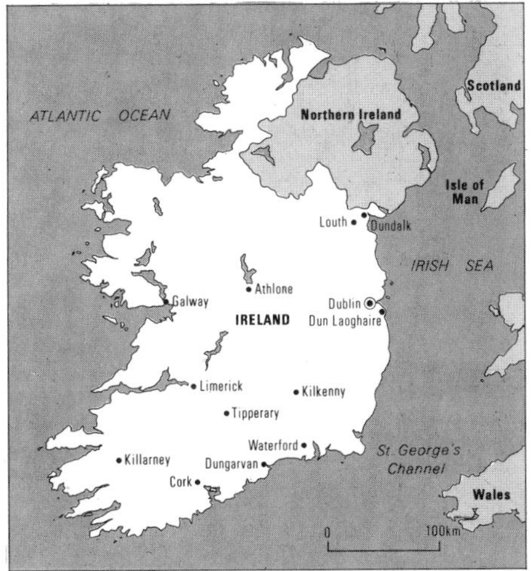

more than 45 per cent of exports. But today more workers are employed in industry than on the land. Ireland produces most of its own consumer goods, and exports a wide range of products.

Economic prospects were brightened when Ireland joined the EEC in 1973. Its chief exports are beef, chemicals, machinery, dairy products and eggs, textiles and fabrics, and livestock. Britain is still by far the biggest customer with 56 per cent, followed by the United States (9%) and West Germany (6%). Ireland's imports include machinery and transport equipment, petroleum and petroleum products, chemicals, grain and other foodstuffs, and textiles. The leading suppliers are Britain (47%), West Germany (8%), the United States (6½%) and France (5%). Imports usually exceed exports by as much as 30-40 per cent, a gap narrowed to an extent by the considerable earnings from tourism.

Agriculture. Nearly two-thirds of the land consists of pastures and meadows, and more is used for rough grazing; crops are grown on only about 7 per cent. About a quarter of the working population is engaged in agriculture, and most farms are family concerns. An average holding is about 16ha (40 acres), and mixed farming is practised, with the emphasis on livestock. The chief sources of income are cattle, milk and pigs. More than 7 million cattle are raised, mainly for beef on the central lowlands and in the east, with dairy cattle predominant in the south. Some 4 million tonnes of milk is produced annually, and butter and cheese are also important. There are nearly a million pigs, and bacon is a major product; there are also about 4 million sheep, mainly in the west. Other products include poultry and eggs. Barley is grown on more than half the cultivated land, and is used for animal feed and in brewing and distilling [*see* PE p.205, 206]. Other important crops are wheat, oats, potatoes, sugar-beet and turnips.

Fisheries. Sea fishing has expanded steadily with the setting up of fishermen's co-operatives for marketing and the establishment in 1977 of a separate Ministry of Fisheries to develop the fishing industry. The chief catches are of cod, whiting, mackerel, haddock and plaice. Salmon and trout fishing are important commercially and as a tourist attraction.

Industry. About 30 per cent of the working population is engaged in manufacturing, mining or construction industries. There is little heavy industry. Most of the manufacturing is concentrated in and around Dublin, and the chief industries are food processing, building materials, clothing, textiles, chemicals, metals, electrical products, tobacco and alcoholic beverages. Waterford glass and Irish whiskey have worldwide reputations, and the Guinness brewery (in Dublin) is one of the largest breweries in the world.

Transport and communications. Public transport is run by the state-sponsored Córas Iompair Éireann. There is 2,190km (1,361 miles) of railway track, mostly single-line; the principal network links Dublin with the other major cities. There is one large airline, Aer Lingus-Irish International Airlines, which operates internal and international services. The chief airports are Shannon, Dublin and Cork. The Irish merchant fleet has about 50 vessels of more than 100 gross tons. The main international ports are Dublin, Cork, Limerick and Waterford. Broadcasting is operated by Radio Telefis Éireann, a statutory public body whose income is derived from both advertising and licence fees. There are more than half a million television sets. British broadcasts can also be received in most parts of Ireland. There are seven daily newspapers (five published in Dublin and two in Cork) and several weeklies, which are in competition with the Irish editions of English newspapers.

Education. School is compulsory from the ages of six to 15, although most children attend from the age of four. Primary education to the age of 12 is mostly at the so-called national schools, which are in effect denominational parish schools.

Post-primary education takes several forms. Secondary schools are privately controlled (usually by religious bodies or boards of governors) but are subsidized and overseen by the Department of Education. Pupils take an Intermediate Certificate at 15 or 16 and may go on to a Leaving Certificate at 17 or 18 in five recognized subjects. Vocational schools provide general and practical training for employment (especially technical), and the cost of running them is shared by the Department of Education (two-thirds) and local authorities. Comprehensive schools and community schools, both state-financed, are a more recent development, designed to provide post-primary schooling in areas in which it was inadequate. Community schools are an amalgamation of existing secondary and vocational schools. There are two universities: Trinity College, Dublin (founded in 1592), with more than 4,000 students; and the National University of Ireland (founded in 1908 in Dublin), with colleges at Dublin, Cork, Galway and Maynooth (St Patrick's, a Catholic seminary with the status of a "recognized" college), and a total of some 17,000 students.

People. Most of the people are descended from the Celts, who began settling in Ireland about 2,400 years ago. From them have been inherited a lasting ethnic identity and a distinctive culture, influenced through the years by succeeding settlers – Vikings, Normans and the English. The population is half rural and half urban, although only two cities (Dublin and Cork) have more than 100,000 inhabitants. Emigration has long been a factor affecting Ireland's population. The Irish abroad (emigrants and their descendants), chiefly in Britain and the United States, outnumber the Irish in Ireland by about ten to one. Only in the 1960s, with increased industrialization, was the tendency for emigration to outstrip the natural population increase finally reversed.

Freedom of religion is guaranteed in Ireland, but the Roman Catholic Church (to which 94 per cent of the people belong) plays an important role in the life of the people and the running of the country. This is especially so in education, social welfare and morality (the constitution prohibits divorce, the sale of contraceptives is illegal, and there are strong censorship laws for books and films). About 4 per cent of the people are Protestants, some three-quarters of them belonging to the Church of Ireland and only about an eighth to the Presbyterian Church (which is the major denomination in Northern Ireland).

The first official language of the country is Irish (or Gaelic), with English as the second. The revival of Irish has been a national policy since the first days of independence; it is taught in schools and a proficiency in Irish is an essential requirement for success in examinations. Yet English is the everyday language of the Irish people, spoken with their characteristic accent and even used in the Gaeltacht, the coastal region in the west where Irish is spoken now by fewer than 80,000 inhabitants.

The Irish people in general lead more simple lives

Dublin, capital of the Irish Republic, means "dark pool" in Gaelic; the city stands on the banks of the River Liffey and has many fine-looking buildings.

The art of thatching is a diminishing trade, but here in Sligo, Republic of Ireland, a craftsman performs his trade on one of the cottages in the town.

Co. Killkenny is noted for its cattle raising and agricultural production; one of its main rivers, the Barrow, flows into the St George's Channel.

than most other Europeans. There is little variety in their food, which is plainly cooked, although nourishing. Potatoes are still prominent in the Irish diet, especially with meat, onions and other vegetables in Irish stew. The traditional alcoholic beverage is Guinness stout, and there are about 14,000 public houses in the country. A sixth of the people, however, have pledged never to drink alcohol, so that total abstainers outnumber drinkers in Ireland.

The Irish are fond of the outdoor life, and hunting, shooting, angling and golf are popular pastimes. There is a keen interest in all aspects of horse-racing, from breeding to betting. Irish-bred racehorses have become important to the export industry, and Irish-trained horses are among the best in Europe. One horse, Arkle, became a national hero and is widely regarded even outside Ireland as the greatest steeplechaser in the history of the sport. There are about 30 race-courses, including the Curragh, in County Kildare, the headquarters of Irish racing. Among the races held at the Curragh is the Irish Sweeps Derby which, thanks to the contribution of prize-money from the Irish Hospitals Trust since 1962, has become one of the richest and most important races in Europe. The Hospital Sweepstakes was set up in 1930 to finance the hospitals. With the full approval of the Irish government it runs lotteries on leading Irish and English races (contributing prize-money to some of them) and sells tickets in the United States, Australia and throughout Europe.

Two exclusively Irish sports are Gaelic football, a cross between soccer and rugby, and hurling (with its women's counterpart, camogie), a fast and furious game akin to hockey. The All-Ireland Championship finals of these sports attract up to 80,000 spectators to Croke Park, Dublin. They are controlled and administered by the Gaelic Athletic Association (GAA), founded in 1884 to preserve the traditional Irish games. Other peculiarly Irish sports include a game akin to handball, played in a four-walled court, and road bowling, in which iron bowls are hurled along country roads in Cork. And despite a former (but largely ineffective) GAA ban on "foreign" sports which was rescinded in 1971, soccer and rugby are also popular, and in rugby (as well as in hockey, cricket and sometimes golf) Ireland fields a united team composed of players from both the Republic and Northern Ireland. An Irish peer, Lord Killanin, became president of the International Olympic Committee in 1972.

Culture. Examples of Ireland's cultural heritage may be seen scattered throughout the country in the form of Celtic crosses. The earliest of these elaborately carved sandstone monuments are attributed to the master masons that St Patrick took with him to Ireland. The 7th and 8th centuries are regarded as Ireland's golden age. While the rest of Europe was suffering a cultural decline, scholars from throughout the known world travelled to study in Ireland's monasteries. Excellent works of art were produced, including illuminated manuscripts such as the famous *Book of Kells* and ecclesiastical metalwork [*see* MS p.*223*]. Architecture flourished, and Ireland has a huge collection of early Christian monuments, including round towers and crosses.

The Irish are also known for their lively folk-music and dances such as jigs and reels. But it is for its literature that Ireland is more famous. In literature the golden age continued until the 10th century and produced lyrical poems, sagas and romances, including such heroic tales as *The Cattle Raid of Cooley* and the stories of the deeds of Finn MacCool. From the 16th century onwards some Irish authors began to write in English, and they number among them such distinguished writers as satirist Jonathan Swift (1667-1745), who wrote *Gulliver's Travels;* poet, novelist and playwright Oliver Goldsmith (1731-74); and playwright Richard Brinsley Sheridan (1751-1816).

At the end of the 19th century there was an outburst of creative activity, born out of the attempt of a group of young Irish writers to awaken the people to the wealth of their native culture. Among its leaders were the poets William Butler Yeats (1865-1939) and A.E. (George Russell, 1867-1935). One development of the Irish Literary Revival, as it is called, was the founding in 1893 of the Gaelic League by Douglas Hyde (who later became the first president of the Republic) to restore Gaelic as the official language of Ireland. It also led Yeats, the dramatist Lady Isabella Gregory (1852-1932) and others to establish in 1904 Dublin's famous Abbey Theatre, which for the following 30 years was to be a storm centre of dramatic controversy. Its first major playwright was John Millington Synge (1871-1909), whose masterpiece *Playboy of the Western World* caused riots in the audience when it was first performed in 1907. The works of Sean O'Casey (1884-1964), including *Juno and the Paycock*, had a similar effect in the 1920s. The movement and the mood of rebellion in Ireland at the time proved a source of inspiration to the novelists and short story writers Seán O'Faoláin (1900-), Liam O'Flaherty (1897-) and Frank O'Connor (1903-66).

The great restoration of Irish culture made an immeasurable contribution also to English literature. The major Irish novelist, James Joyce (1882-1941), dissociated himself from the revival, although his masterpiece *Ulysses* and most of his other work is steeped in the experience of his youth in Ireland. A writer who gained wide acclaim only after his death was Flann O'Brien (1912-66), for his comic novel *At Swim-Two-Birds*. Later writers to draw on their youthful Irish experiences include playwright and wit Brendan Behan (1923-64), who because of his revolutionary activities spent much of his youth in prison, and novelist Edna O'Brien (1936-), whose uninhibited novels are banned in Ireland.

Irish-born writers who made their reputations elsewhere include three great playwrights: Oscar Wilde (1854-1900), whose epigrams are among the most oft-quoted in the English language; George Bernard Shaw (1856-1950), most of whose plays (more than 50) are comedies of ideas and have become classics; and Samuel Beckett (1906-), whose plays, mostly written in French and translated by himself into English, are considered major works in the theatre of the absurd.

Politics and government. The chief political parties are the *Fianna Fáil* (Republican Party), which was founded in 1926 and in government for all but six years between 1932 and 1973; the *Fine Gael* (United Ireland Party), founded in 1933 by an amalgamation of the Cosgrave and Centre parties and the National Guard; and the Labour Party.

The Republic of Ireland is a sovereign, independent, democratic state. The *Oireachtas* (parliament) consists of the president and two houses, the *Dáil Éireann* (House of Representatives) and the *Seanad Éireann* (Senate). The president is elected by direct suffrage (the minimum voting age is 18) for a seven-year term, and is the head of state. Most of his powers are formal or ceremonial: calling parliament into session, signing laws and appointing ministers nominated by the government.

The chief law-making body is the Dáil, composed of 144 members elected for five-year terms by direct suffrage (using proportional representation on the basis of the single transferable vote). The head of government is the *Taoiseach* (prime minister), appointed by the president on the nomination of the Dáil; he is the leader of the majority party or coalition in the Dáil. He selects a cabinet of seven to 15 members, not more than two of whom may be members of the Seanad.

The Seanad is composed of 60 members: 11 nominated by the prime minister, six by the universities, and 43 elected from five specially constituted panels established on a vocational basis (culture and education, agriculture, labour, industry and commerce, and public administration) by some 900 electors drawn from national and local government.

For local administration, the country is divided into 27 county councils (see the table) and four county borough councils. Within the counties are seven boroughs, 49 urban districts and 28 town commissions. All members of these councils are elected on a system of proportional representation. County and county borough councils vary in size from 15 to 46 members and are administered by a manager rather than by committee. Nominated by a Local Appointments Commission, the manager is a paid officer and is ultimately responsible to his council but in practice exercises considerable authority in his day-to-day decisions.

The legal system is based on that of England, drawing on common law and the statutes of parliament. Justice is administered by Courts of First Instance and the Supreme Court, which is the court of final appeal and consists of the chief justice and four other judges. The president may, on the advice of the Council of State, refer certain bills passed in parliament to the Supreme Court for a decision regarding its compatibility with the constitution.

The Aran Islands, situated on the west coast of the Republic of Ireland, are noted for their weaving industry and in particular for the famous Irish tweed.

Round towers are a feature of early Irish architecture, although not all are as well preserved as on this tiny church in County Wicklow.

Tourists on the Aran Islands are shown round in these characteristic horse-drawn carts; on the three islands are many prehistoric remains.

The Courts of First Instance consist of the High Court and circuit and district courts. The High Court, made up of a president and seven judges, has full original jurisdiction concerning all matters and questions of law or fact, civil or criminal. The circuit and district courts have limited powers. All criminal cases except those dealt with summarily by a justice in the district court are tried by a judge and a jury of 12 people. Juries also serve in many civil cases in the High Court. The jury must be unanimous in criminal cases, but the agreement of nine members is sufficient in civil cases. There is also a Court of Criminal Appeal, which consists of a Supreme Court judge and two High Court judges.

Constitutionally, the president is the supreme commander of the armed forces, but the minister of defence exercises practical control. The standing strength of the armed services is about 13,000 and includes a small navy and air force. Military service is voluntary, and there is a Reserve Defence Force.

History. The earliest settlers in Ireland probably arrived from mainland Europe in about 6000 BC. There is no evidence of a large-scale invasion by the Celts, but they began to arrive from Gaul in about 400 BC and gradually built up a Gaelic civilization, subduing and assimilating the native Picts and Erainn. Pre-Christian Ireland was divided into five kingdoms, identifiable with the four present-day provincial units plus Meath.

St Patrick introduced Christianity to Ireland in 432, and the people readily accepted the new religion. Many great monasteries were founded, and Ireland developed into a centre of Gaelic and Latin learning. The comparative peace of this period was broken in 795 with the first Viking raids, and attacks continued all round the coast and farther inland until the Vikings were decisively defeated at Clontarf in 1014. The Normans arrived in Ireland in 1170, and by the 14th century controlled most of the country. But their allegiance to England diminished as they intermixed with the Irish, and by the late 15th century effective control of the Crown was limited to the Dublin area.

In 1534 Henry VIII set out to re-establish England's influence. At first he tried persuasion and when this policy did not work (his break with Rome ensured its failure), force. The Tudors were determined to establish Protestantism in Ireland and resorted to the persecution of Irish Catholics. Mary tried to strengthen English rule by seizing land and replacing the landholders with "planted" English settlers. The Stuarts and Cromwell continued this so-called plantation.

There were frequent rebellions, because the Irish bitterly resented British rule. In 1688 they revolted in support of a dethroned English king, the Roman Catholic James II, but were defeated by William of Orange at the Battle of the Boyne (1690). The British granted an independent (all-Protestant) parliament to Ireland in 1782, but a further rebellion by Wolfe Tone and his United Irishmen in 1798 led to its abolition and to Union with Britain in 1801.

Provinces and counties of Ireland

Connacht Galway, Leitrim, Mayo, Roscommon, Sligo
Leinster Carlow, Dublin, Kildare, Kilkenny, Laoighis, Longford, Louth, Meath, Offaly, Westmeath, Wexford, Wicklow
Munster Clare, Cork, Kerry, Limerick, Tipperary (North and South), Waterford
Ulster (part) Cavan, Donegal, Monaghan

The population of Ireland was growing rapidly, but there was great poverty and nearly half the $8\frac{1}{2}$ million inhabitants depended almost entirely on potatoes for their food. A potato blight in the mid-1840s caused widespread famine. About a million people died of starvation and another $1\frac{1}{2}$ million emigrated to the United States. Discontent grew and the demand for Home Rule became stronger. Reforms were slow in coming and, although Home Rule was eventually granted in 1914, Protestant Ulster refused to accept it and it was suspended at the outbreak of World War I.

The Irish supported Britain in the war. But the Irish Republican Brotherhood (a secret organization set up to achieve a completely independent republic), led by Patrick Pearse and James Connolly and with help of an armed force called the Volunteers, staged an insurrection in Dublin in 1916 – the Easter Rising. Although it received little support at first, severe British reprisals (the execution of 15 leaders) created sympathy for the republican movement, which gained control of the Sinn Féin Party and won 73 of Ireland's 105 seats in the British parliament in 1918. The new members met in Dublin and called themselves the Dáil Éireann.

There were further widespread uprisings, and the British authorities reinforced their armed police in Ireland with two less disciplined forces known as the Black and Tans and the Auxiliaries. The British government signed a truce with the Sinn Féin in 1921, and in 1922 an unsatisfactory treaty was agreed in which independence was granted to the Irish Free State, which was given Dominion status. But the six northern counties of Ulster, which were about two-thirds Protestant, remained in the United Kingdom as Northern Ireland. The new country drifted into a bitter civil war as uncompromising republican elements – chiefly the Irish Republican Army (IRA), which had emerged when the Volunteers were reorganized – took up arms, but fighting died out in 1923. The two factions continued in political opposition, with Eamon de Valera leading the Sinn Féin and William Cosgrave leading the group that supported the treaty. De Valera boycotted the Dáil until 1927. In 1932, as leader of Fianna Fail he became president of the Executive Committee, and began to break all links with Britain.

The Irish Free State was renamed Éire in 1937 and in 1949, largely through the efforts of prime minister John Costello (1891-1976), it became the Republic of Ireland and withdrew from the Commonwealth. Militant republicans were still not satisfied with partition; the IRA (which had been outlawed in 1936) carried on a campaign of violence in Ulster between 1956 and 1962, and the Irish government interned several IRA suspects. Seán Lemass, who became prime minister in 1959, stressed economic expansion and attempted a realistic compromise with Northern Ireland. His successor, Jack Lynch, also urged moderation, but further "troubles" involving Northern Ireland in the late 1960s made it difficult for him to maintain this policy. In 1976, after the IRA assassination of the British ambassador in Dublin, the Irish government brought in new anti-terrorist measures, with wider powers of detention and increased penalties. Following criticism from the government, President Caroll O'Daly resigned in October 1976 and the subsequent election of Patrick Hillery was unopposed. Map 8.

Prime ministers of Ireland

William T. Cosgrave	1922-32
Eamon de Valera	1932-48
John A. Costello	1948-51
Eamon de Valera	1951-54
John A. Costello	1954-57
Eamon de Valera	1957-59
Seán Lemass	1959-66
Jack Lynch	1966-73
Liam Cosgrave	1973-

Ireland – profile

Official name Republic of Ireland
Area 70,282sq km (27,136sq miles)
Population (1976 est.) 3,163,000
Density 45 per sq km (117 per sq mile)
Chief cities Dublin (capital) (1971) 567,900; Cork, 128,235; Limerick 57,371
Government Head of state, Dr Patrick Hillery, president (elected December 1976)
Gross national product (1974) £3,132,500,000
Per capita income £990
Agriculture Cereals, potatoes, sugar-beet, livestock
Industries Dairy products, meat and meat products, tobacco, cereal products, metal processing, motor vehicles, electrical machinery
Exports Livestock and meat, manufactured goods, machinery and vehicles, chemicals
Trading partners (major) Britain, USA, West Germany, France, Netherlands

Jerusalem, the holy city of Jews, Christians and Muslims, is also the capital of Israel; it has a combination of biblical and modern architecture.

The market at Akko (Acre) in north-west Israel is a place of bustling activity. The inhabitants, once mainly Arab, are now predominantly Jewish.

Nazareth, in northern Israel, was the home of Jesus Christ during his early youth; today it is the largest town in the country with an all-Arab population.

Ireland, Northern. *See* NORTHERN IRELAND.

Irian Barat. *See* INDONESIA.

Isle of Man, a possession of the British Crown, is a hilly island in the Irish Sea off the north-west coast of England; the capital is Douglas. The island was occupied by Vikings in about 800 AD and was a dependency of Norway until 1266. It belonged to the earls of Salisbury and Derby from the 14th century until 1735, and has been a dependency of the British Crown since 1765. Man has its own parliament (an assembly called the Tynwald). The governor is appointed by the English monarch. A local Celtic language (Manx) is dying out and most of the people speak only English. The island uses English currency but has minted coins to its own design and sells its own postage stamps. The Tourist Trophy (TT) motorcycle races are held on the island each year. A race of tailless cats, called Manx cats, probably originated on the Isle of Man. Tourism and agriculture are important industries. Sheep and cattle are raised on the hills and other products include cereal and root crops, fruit, flowers and vegetables. The other main industries are dairying, fishing and quarrying. Area: 572sq km (221sq miles). Pop. (1975 est.) 59,000. Map 12.

Israel, official name the State of Israel, is an eastern Mediterranean republic that comprises most of the region known historically as Palestine. Most of its people are Jews. Israel enshrines the centuries-old ambition of Jews throughout the world to re-establish a Jewish state in their ancient homeland. Since it came into being in 1948 it has had to face the hostility of its Arab neighbours and of refugees from Palestine who oppose the creation of a Jewish country in what they regard as Arab territory. This hostility has led to several wars, in the course of which Israel has occupied parts of EGYPT, JORDAN and SYRIA. The Israeli people – mostly immigrants or the children of immigrants – have worked hard to build up their country, constructing modern towns and villages, turning wasteland into productive farms and developing prosperous industries. The capital is Jerusalem.

Land and climate. The southern part of the country, more than half the total area, is the barren Negev. It is a wedge-shaped region which narrows to a point at the Gulf of Aqaba and changes gradually from rough steppe land to hot desert [*see* PE p.*52*]. The long coastal plain along the Mediterranean is widest in the south; its fertile northern part, from Tel Aviv-Jaffa to the heights of Mt Carmel (546m; 1791ft) is called the Plain of Sharon. Another fertile plain – Esdraelon – lies east of Mt Carmel in the valley of the Kishon River. It stretches to the Sea of Galilee (also called Lake Tiberias) and the depression through which the River Jordan flows on its way to the Dead Sea, which is 394m (1,292ft) below sea-level. Only the south-western shore of the Dead Sea is in Israel.

Most of the eastern regions of Israel are hilly, from Galilee in the north, through the hills of Samaria and Judaea, to the mountains of the Negev. Israel's highest point, Mt Meron (1,208m; 3,963ft) is north-west of Galilee; the Golan Heights lie to the north-east. The Samarian and Judaean hills form part of the plateau on which Jerusalem is built.

Israel's climate is generally hot in summer and mild in winter ranging from 24–32°C (75–90°F) in August to 7–16°C (45–60°F) in January, depending on altitude and distance from the Mediterranean Sea. Most rain falls in December, January and February, but it does not exceed about 1,000mm (40in).

Economy. The most valuable natural resources are the mineral deposits of the Dead Sea, chiefly salts of potassium, sodium and magnesium. Copper, rock phosphates, manganese, glass sand, kaolin, iron ore, petroleum and natural gas are found

in the Negev. Agricultural production relies on irrigation. Most farms are co-operatives, such as the *kibbutzim* (collective farms in which all property is held in common and labour is shared) and *moshavim* (co-operative settlements of smallholders), which are practical examples of the determination of the Israelis to establish a Jewish state. The principal agricultural region is the Plain of Esdraelon; the coastal plain is also fertile, as are the valleys. Farmers grow winter wheat and barley, as well as millet, maize and sorghum. The wide variety of fruits include oranges and grapefruits, which are exported in large quantities. In Galilee olives, bananas and tobacco are produced; some central hilly areas and parts of the coastal plain have vineyards. The chief industrial crops are cotton and sugar-beet. Livestock include large numbers of poultry, cattle, sheep and goats, and in several areas carp and trout are raised in artificial ponds.

Sympathizers in other countries have made considerable investments in Israel's industry. Major industries include diamond cutting and polishing, food processing, metalworking, and the manufacture of textiles, chemicals, plastics, glass, precision instruments and electrical goods. Tourism is an important source of foreign earnings. In trade the value of imports far exceeds that of exports.

Internal communications are mainly by road, although there are about 1,000km (620 miles) of railway. The Israeli airline El Al operates services to cities in Europe, North America, Asia and Africa.

People. Israel has two major groups of people: the Jewish majority and the minority Arab community of about 400,000 people. Under the Law of Return passed by the *Knesset* (parliament) in 1950, any Jew who wishes to settle in Israel must be granted an immigrant visa. Since 1948 about 1½ million Jews from a hundred different countries have entered Israel.

One of the major difficulties in building up the state has been the unification of a community formed of peoples from widely different cultural backgrounds. The official language is Hebrew, and Arabic is recognized for the benefit of the Arab community. In religion Judaism has a special status, but the law ensures freedom of worship for people of all faiths. Despite the central place of religion in the life of Israelis, only about a third of the people consider themselves practising members of any faith. *See also* MS pp.218, 222.

Education is free and compulsory between the ages of five and 14. There are seven institutions of university status, including the Hebrew University of Jerusalem (founded in 1925), the Technical Institute of Haifa and the religious university, Bar-Ilan, at Ramat Gan.

All men between the ages of 18 and 29 have to serve for a time in the armed forces; the same obligation applies also to unmarried women under 26. And after military service, everyone is on reserve for a long period.

Social life in Israel is Western in habits and outlook. It is also predominantly urban: four out of five Israelis live in towns. The chief centres of population are Jerusalem (declared the capital in 1950), Tel Aviv-Jaffa and Haifa. All three cities have many new buildings, often with architecture that is original in concept and high in quality. There is a strongly developed cultural and artistic life, particularly in music. All sport in Israel is amateur; spectator sports include soccer and basketball; tennis, swimming and athletics are also popular.

Government. Israel is a republic, whose head of state is the president. The Knesset is a one-chamber legislature whose 120 members are elected by universal suffrage for four-year terms. The leader of the majority party in the Knesset becomes prime minister and forms a government. Israel has many political parties; two major party groupings are the left-wing Mapam – the United Workers' Party – and the nationalist Likud.

History. The first practical step towards the establisment of modern Israel was the formation in the

Traditional handicrafts still flourish in parts of Israel. This Jewish silversmith uses skills that his family has probably practised for generations.

An aerial view of Florence, the birthplace of the Italian Renaissance, highlights one of the city's ornaments, the Brunelleschi dome of the cathedral.

Mt Etna, off the east coast of Sicily, is the highest active volcano in Europe. Eruptions, the most recent in 1971, have left 260 craters on its slopes.

Israel – profile

Official name State of Israel
Area (excluding captured territory) 20,700sq km (7,992sq miles)
Population (1976 est.) 3,483,000
Density 168 per sq km (436 per sq mile)
Chief cities Jerusalem (capital) (1974) 344,200; Tel Aviv-Jaffa, 357,600; Haifa, 225,000
Government Republic, representative democracy
Religions Judaism, Islam
Languages Hebrew, Arabic
Monetary unit Israeli pound
Gross national product (1974) £4,765,000,000
Per capita income £1,368
Agriculture Citrus fruits, cereals, vegetables, cotton, sugar-beet, olives, bananas, grapes, figs
Industries Diamond polishing, food processing, textiles, chemical products, plastics, electrical goods, precision instruments
Minerals Copper, phosphates, magnesium, manganese, glass sand, kaolin, petroleum, natural gas
Trading partners Britain and other EEC countries, USA, Canada

late 19th century of the Zionist movement, which aimed to create a Jewish homeland in Palestine. In the Balfour Declaration of 1917 the British promised help, as long as the rights of non-Jewish Palestinians were protected. Britain received a League of Nations mandate over Palestine in 1920. It strictly controlled Jewish immigration and continued this policy after World War II in spite of intensified demands for a Jewish state.

After much violence and bitterness, the United Nations proposed that Palestine be partitioned between Jews and Arabs. The British mandate ended on 14 May 1948 and the independent state of Israel was proclaimed. It was immediately attacked by several Arab countries; after intervention by the United Nations, uneasy peace was restored and armistice agreements signed.

Increasing pressure on Israel reached its climax in 1956 (when Egypt nationalized the Suez Canal). Israeli, British and French troops invaded Egyptian territory, but later withdrew. War between Israel and its neighbours broke out again in 1967 – the Six-Day War – and by its end Israeli troops had occupied the west bank of the River Jordan, the Sinai Peninsula and the Golan Heights.

In October 1973, Egyptian troops launched a surprise offensive against Israeli positions on the Suez Canal – the so-called Yom Kippur War. They achieved initial successes that demonstrated a change in the balance of military power between Israel and the Arab nations. Through mediation by the United States, a measure of Israeli-Egyptian disengagement was negotiated in December 1973 and again in late 1975. But there was no diminution in the hostility towards Israel by the Palestine Liberation Organization (PLO), who continued terrorist activities both within and outside Israel. In July 1976 Israeli troops rescued Jewish hostages from an airliner hi-jacked by Palestinians at Entebbe in Uganda. In elections in May 1977, there was a swing of opinion to the right, and the left-wing Mapam was defeated for the first time. Map 38.

Italy (Italia), official name Italian Republic, is a Mediterranean country of great variety and with an unsurpassed history of culture. It was united under the Romans, but for nearly 1,400 years after the fall of the Roman Empire was divided at various times into many kingdoms, city-states, duchies, principalities and papal states. This political fragmentation is reflected today in the vast regional differences, especially between north and south. Northerners live in an industrialized society and are heirs to the great traditions of the Renaissance. Southerners are making slow progress with limited resources on poorer land, with extremes of climate and much poverty.

The country as a whole has made a remarkable transition from an agrarian to an industrial economy, after the trauma of Fascism under Mussolini and the devastation of World War II. Italy's revival has, however, been hampered by uncertain government and a series of uneasy coalitions against the strongest Communist party in Western Europe (which received more than a third of the votes cast in the 1976 election).

In spite of its long history of disunity, Italy has produced people who have made great contributions to the advancement of civilization. In addition to the vast artistic heritage (left especially by the geniuses of the Renaissance), there are the journeys of Marco Polo, Columbus and Amerigo Vespucci, the scientific discoveries of Galileo, Alessandro Volta, Guglielmo Marconi and Enrico Fermi.

Land and climate. Northern Italy consists of the great alluvial plain of the Po valley bounded in the north by the uninterrupted ranges of the Alps. The highest mountains in Italy, the Alps rise to more than 4,500m (14,750ft). Mont Blanc (Monte Bianco) straddles the Franco-Italian border and stands 4,810m (15,781ft) high [*see* PE p.*117*]; the loftiest peak entirely within Italy is the Gran Paradiso (4,061m; 13,324ft), in the Graian Alps. The Apennines, the "backbone" of Italy, range down the country for 1,125km (700 miles), averaging about 1,200m (3,900ft) in height. To the west of the Apennines, in central Italy, is a coastal plain. It includes the hilly region of Tuscany and the reclaimed Pontine Marshes near Rome, and extends to just south of Naples.

The most important river is the Po, which is also the longest (673km, 418 miles) and rises in the Alps. There are several large lakes noted for their scenic beauty, including Maggiore, Como and Garda. Italy's territory includes the two largest islands in the Mediterranean: Sicily and Sardinia. Sicily is dominated by Mt Etna, at 3,296m (10,814ft) the highest volcano in Europe and one of the world's most active. Within Italy's borders are two of the smallest independent countries in the world: the VATICAN CITY, which lies wholly in Rome, and the republic of SAN MARINO in the north-east on the slopes of the Apennines.

The north-western, Riviera coast has mild winters, but winters in the Po valley are cold. Towards the south the climate becomes more Mediterranean, with mild or warm winters and moderate rainfall. Temperatures in Rome average 24°C (75°F) in July and 9°C (48°F) in January.

Economy and resources. Since World War II Italy has changed rapidly from a mainly agricultural country to an industrialized one. Agriculture accounted for 32 per cent of the gross national product in 1950, but contributed less than 10 per cent 25 years later; in contrast industrial production has increased by more than 300 per cent since the war. Yet Italy is not rich in natural resources. It is a leading producer of mercury, zinc and sulphur, but its deposits of coal and iron are inadequate for its own needs. Natural gas and oil, discovered since the war, are being exploited, especially in Sicily.

Industry is concentrated in the north, and about 45 per cent of the working population are engaged in manufacturing, construction, mining and quarrying industries. The major industries include textiles (especially silk), chemicals, machinery, motor vehicles, oil refining and food processing. Italy is famous for its marble, quarried mainly in the coastal mountains of Tuscany.

About 17 per per cent of the working population is employed on the land, of which 31 per cent is

The hilly Abruzzi region of central Italy is a poor, agricultural area, with small-scale farming of olives, grapes, sugar-beet and tobacco.

The Sassarian people of Sardinia exhibit their ethnic consciousness by dressing up in traditional folk costume for the photographer.

The leaning tower of Pisa, built in the late 12th century, has a height of 55m (180ft); owing to subsidence it is out of perpendicular by 5m (16ft).

arable, 17 per cent pasture and 21 per cent forested. About 80 per cent of the farms and smallholdings are privately owned. The chief agricultural products are sugar-beet, wheat, rice, vegetables, grapes, olives and citrus fruit. In most years Italy is the world's leading producer of wine, much of it of a high quality [*see* PE pp.200, 202]. Livestock is also raised, including more than 8 million each of cattle, sheep and pigs. Italy is a leading producer of cheese, including the famous Gorgonzola and Parmesan [*see* PE pp.229–235].

Italy's chief exports include machinery and motor vehicles, textiles, footwear and clothing, chemicals and petroleum products, and food and wine, and its principal customers are West Germany (18½%), France (13%) and the United States (6½%). The main imports are crude oil (more than 20%), minerals, machinery, iron and steel, copper and meat. Italy's chief suppliers are West Germany (17%) France (13%), the United States (9%) and Saudi Arabia (6%). Italy's adverse balance of trade is considerably reduced by income from tourism. With a merchant fleet of about 4,000 ships, totalling more than 10 million gross tonnes, Italy ranks among the first ten of the world's maritime nations. The chief ports are Naples and Genoa.

Transport and communications. A feature of Italy is its fine roads, in particular the motorways (*autostrade*). A network of more than 5,250km (3,266 miles) of motorways covers the country and is planned to total 6,146km (3,823 miles) when completed. The motorways – especially the Autostrada del Sole (Highway of the Sun), which runs from Milan through Bologna, Florence, Rome and Naples and is being extended to the "toe" of Italy – are not only feats of civil engineering, but are landscaped to blend with the spectacular hilly countryside. There are also more than 20,000km (12,500 miles) of railways, including some 16,000km (10,000 miles) of state-operated track, about half of which is electrified. The chief airline is the largely state-owned Alitalia, and there are more than 20 international airports. The state-controlled Radiotelevisione Italiana broadcasts on three radio channels (plus regional programmes) and two television channels. There are about 80 daily newspapers.

People. There is a marked difference in Italy between the people of the industrialized north and centre, where culture and ways of life are akin to other countries of Western Europe, and the largely agricultural south, where the people are relatively poor and are closer in many ways to people of northern Africa.

The Catholic Church has an enormous influence on the lives of the people and on the state, enjoying a privileged position with regard to education and legislation concerning marriage and the family. (The Church lost a long-term struggle in 1974, however, when the people voted by a three to two majority to retain a three-year-old law permitting divorce.) About 98 per cent of the people are Roman Catholics. Other denominations are permitted.

The official language is Italian, a Romance language that developed from the Tuscan dialect of Latin. There are numerous regional dialects, and languages spoken by minorities include French in the Val d'Aosta, German in the South Tyrol, Slovene and other Slav tongues in Friuli-Venezia Giulia, Greek and Albanian in Apulia and Sicily, and Sard in Sardinia.

Education is free and compulsory from the ages of six to 14; the curricula are standardized by the Ministry of Education. A shortage of classrooms makes schools in some areas operate double shifts. After three years in lower secondary schools children take an examination for entry into higher secondary schools (classical, scientific, or various technical institutes). Italy has more than 40 major universities and higher institutes, such as Bologna (founded in about 1200), Padua (1222), Naples (1224) and Genoa (1243). Some are extremely large – Rome has more than 100,000 students and Naples and Milan each have more than 80,000. University courses last four to six years and most examinations are oral. The faculty system of teaching predominates and there are no separate colleges.

Italian cooking has a worldwide reputation and each region has its own special dishes. The staple food is *pasta*, a wheat dough that includes eggs; it takes various forms, such as spaghetti and macaroni. Herbs and spices are freely used, and olive oil, tomatoes, mushrooms and peppers are common ingredients. Italy is also famous for its ice creams and cheeses; the chief beverage is wine.

The most popular sports are soccer, which has an enthusiastic following in most major cities, particularly Milan and Turin, and cycle racing, in which the Gran Giro d'Italia holds the whole country's attention for three weeks every summer. Italians have a fine reputation in motor racing and motorcycling both as drivers and as manufacturers, and other sports at which they excel internationally include fencing, show-jumping, alpine skiing and bobsleigh. Boccie, a game like bowls, is a popular pastime of older men.

Culture. Italy is a treasure-house of art, going back 2,700 years to the Etruscan age. Few Western countries have such a long history of art and architecture. The bold art of the Etruscans can still be seen in the hill towns of central Italy; in the south, outstanding Doric temples stand as monuments to the colonizing Greeks; the grandeur of the Roman Empire can be seen in every major city, especially in Rome itself; and there are many fine examples of later periods, such as early Christian, Byzantine and Romanesque art.

The forerunners of Renaissance art were the Florentine painters Giovanni Cimabuse (*c*.1240-*c*.1302) and his great pupil Giotto (1267–1337). And it was also in Florence that the Renaissance proper got under way, with the sculptors Lorenzo Ghiberti (1378–1455) and Donatello (*c*.1386–1466) and the architect Filippo Brunelleschi (1377–1466), soon followed by the painter Fra Angelico (1387–1455) and the sculptor Luca Della Robia (*c*.1400–82). The Renaissance reached its peak in the late 15th and early 16th centuries, when Sandro Botticelli (1444–1510) painted the *Birth of Venus*, Leonardo da Vinci (1452–1519) *The Last Supper* and *Mona Lisa* (among many other masterpieces), and Michelangelo (1475–1564) completed the magnificent murals in the Sistine Chapel.

Outstanding artists in the transition from Renaissance to Baroque were Correggio (*c*.1494–1534) and Caraveggio (1569–1609). Baroque sculpture owes its impetus to Giovanni Bernini (1598–1680) and architecture to Bernini and Francesco Borromini (1599–1667). Andrea Palladio (1518–80) was a Renaissance architect whose classical villas and houses influenced styles for hundreds of years.

Italian literature dates from about 1200, when the first lyric and religious poetry and chronicles appeared in the language, which gradually began to replace Latin in formal literature. About 100 years later one of the world's greatest works of literature was produced in Italian, the *Divina Commedia* by the Tuscan poet Dante Alighieri (1265–1321). Drama saw the development of a new, improvised comedy form, the *commedia dell' arte*, which gave the world such characters as Harlequin, Punchinello and Columbine. Later Italian writers of note include poet, dramatist and novelist Gabriele d'Annunzio (1863–1938); Nobel prize-winners such as playwright and novelist Luigi Pirandello (1867–1936) and poets Eugenio Montale (1896–) and Quasimodo (1901–68).

Italy was the birthplace of opera, and the Italians are great opera lovers. Milan's La Scala is the most famous opera house in the world, and many places have yearly open-air performances, as at the Baths of Caracalla in Rome and the ancient amphitheatre in Verona. Claudio Monteverdi (1567–1643) is regarded as the father of this musical form and his first opera, *Orfeo* (1607), was an important milestone in the history of music. Among Italy's other great operatic composers were Gioacchino Rossini (1792–1868; *The Barber of Seville*), Gaetano Donizetti (1797–1848; *La Favorita, Don Pasquale*), Vicenzo Bellini (1801–35; *La Sonnambula, Norma*); Giuseppe Verdi (1813–1901; *Rigoletto, Il Trovatore, La Traviata*); and Giacomo Puccini (1858–1924; *La Bohème, Tosca, Madama Butterfly*).

Orchestral and chamber music were also greatly enhanced by Italian composers, including Alessandro Scarlatti (1660–1725) and his son Domenico (1685–1757), Antonio Vivaldi (*c*.1675–1741) and the virtuoso violinist Niccolò Paganini (1782–1840). Other great names in Italian music include the master violin-maker Antonio Stradivari (*c*.1644–1737), the operatic tenors Enrico Caruso (1873–1921) and Beniamino Gigli (1890–1957) and the conductor Arturo Toscanini (1867–1957).

In modern times Italy's film-makers have won a worldwide reputation, and since World War II Rome has become, after Hollywood, the cinematic

The excavation of Pompeii, buried by Vesuvius in AD 79, began in the eighteenth century; the temple of Apollo is one of the ruins now visible.

The Appian Way was the first military highway built by the Romans. It was started in 312 BC and originally ran for 212km (132 miles) from Rome to Capua.

From dawn to nightfall the narrow and hilly streets of Naples are alive with all aspects of the life of the urban Italian poor.

capital of the world. The leading directors have included Vittorio de Sica (1902–74), Roberto Rossellini (1906–), Michelangelo Antonioni (1912–) and Federico Fellini (1920–).

Government. Under the 1948 constitution, Italy is a democratic republic with legislative power in the hands of a two-chamber parliament – the Chamber of Deputies (630 members) and the Senate (315), both elected for five-year terms. The head of state is the president, elected for seven years by a joint session of Chamber and Senate reinforced by regional delegates. The president appoints the prime minister (president of the Council of Ministers). The chief political parties are the Christian Democrat Party, a middle-of-the-road party founded in 1943 and openly anti-Communist; the Communist Party, the largest in Western Europe; the Italian Socialist Party; and a neo-Fascist party, the Italian Social Movement. The reorganization of the Fascists is forbidden by the constitution. In the 1976 elections, ten parties won seats in the Chamber of Deputies (election is by proportional representation), most votes going to the Christian Democrats (38.7 per cent) and the Communists (34.4 per cent).

Italy is divided into 20 regions, each with a considerable degree of autonomy. They include five regions with special status (Sicily, Sardinia, Val d'Aosta, Trentino-Alto Adige, and Friuli-Venezia Giulia), which have their own governments. For direct administration under the central government, the regions are subdivided into more than 90 provinces, each under a prefect, and for local government the provinces are further divided into more than 8,000 communes.

Italian law is based on Roman law and is regulated by a system of Codes. Judges are appointed and promoted by means of competitive examinations. There are several grades of courts. Most communities have their own police force. The national force is the Public Security Police, under the overall control of the Ministry of the Interior, with a *quaestor* in each city; linked with these are the *Celere*, the riot police. The *Carabinieri* are an élite and popular force, some 80,000 strong, that belongs to the army. They are a mounted force which patrols every street in the country and are well organized, with a proud tradition. The army has a total strength of more than 300,000, the navy 42,000 and the air force 70,000. National service is for between 12 and 24 months, according to the force.

History. Little is known of the Stone and Bronze Age peoples who inhabited the Italian peninsula thousands of years ago, but by the 500s BC there were four distinct cultures: the Celts in the north; Etruscans in the centre; hill tribes (including the Romans) and cities in the interior; and Greek settlements from Naples to Sicily. The establishment of the Roman republic is traditionally dated at 509 BC, the year in which the last of the legendary Roman kings, Tarquinius Superbus, was exiled. The Romans won domination of the Mediterranean by defeating Carthage in the Punic Wars (264–146 BC) and then built a powerful empire, which at its greatest extent in the early 2nd century AD stretched from Britain and Spain in the west to the Persian Gulf in the east.

Italy became the centre of the known world and enjoyed a period of order and peace until the fall of the Western Empire in AD 476, when the country reverted to the fragmented divisions of pre-imperial times. It was to remain disunited for nearly 1,400 years. Barbarian tribes – first the Ostrogoths and then the Lombards – dominated the north and centre, while the Byzantines held Rome and the south until the late 6th century.

The popes, who had increased their political influence enormously and had resisted Lombard attempts to take Rome, finally enlisted the help of the Franks to subdue the Lombards. In 800 Pope Leo III crowned the Frankish king Charlemagne as Roman Emperor. This alliance gave the popes Rome and central Italy – the papal states – and revived the idea of empire in the West. But it remained no more than an idea, for there ensued an era of power politics between the popes and the emperors. Two parties arose, the Guelphs (or Welfs), who were anti-empire (although not always propapalist), and the Ghibellines (Weiblingers), who were pro-empire. The powerful, almost independent city states that had grown up since the 11th century (such as Florence, Milan, Venice, Genoa, Pisa and Naples) were plagued by the Guelph-Ghibelline confrontation for hundreds of years, a situation complicated by countless other struggles between and inside states and between noblemen and townspeople.

Italy became the battleground for other European powers, which began to take advantage of the country's disunity as they became stronger. First Spain defeated France in a series of wars (1521–59) and brought almost all of Italy under its control. Spain then gradually lost it to Austria, which held sway for most of the 18th century. Then in 1796 Napoleon seized Italy back again for France.

French rule lasted less than 20 years yet had a profound effect because it unified much of the country under the same form of government and the same laws, army and monetary system. For a short period (1805–15) there was even a Kingdom of Italy (in the north). As as result, when the old order was restored by the Congress of Vienna in 1815 (with even firmer control by Austria), the possibility of a united Italy was still in the minds of the people.

In 1861 Italy was declared a kingdom under Victor Emmanuel II of Sardinia. With the addition of Venice in 1866 (with the help of Prussia, which defeated Austria) and Rome in 1870 (French troops who had been guarding Rome were withdrawn to fight Prussia), the jigsaw was complete and Italy was finally united as one kingdom. The pieces were not, however, well-fitting, with not only the major north-south differences but also a considerable residue of regional rivalry, and Italy found itself struggling to establish a unified and universally acceptable government.

Italy entered into the Triple Alliance with Austria-Hungary and Germany in 1882, hoping to establish a colonial empire in northern Africa. But Italy also coveted Trieste and Trentino (Austrian territory), which is why it eventually entered World War I on the side of the Allies (1915). After prolonged struggles and some serious reverses, Italy finally gained its territorial objectives in Austria, but was dissatisfied with the other postwar settlements. It emerged from the war in poor shape politically and economically, a condition ripe for the spread of Fascism led by Benito Mussolini [*see* MS p.*264*]. The king gave him power in 1922, and from 1927 he ruled the country as a dictator. He built up Italy's military strength and conquered Ethiopia (Abyssinia) in 1935–36. His grandiose plans of Mediterranean dominance led him into an alliance (the Axis) with Nazi Germany, although he entered World War II in 1940 only when he felt confident of an Axis victory. The Italians suffered defeats on all fronts, first from the Allies and then, having negotiated an armistice and declared war on Germany, they suffered Nazi occupation. Italy emerged from the war devastated and dispirited.

In 1946 the monarchy was abolished after a referendum, and a new democratic republic set up. Italy's first postwar leader was Alcide De Gasperi whose party, the Christian Democrats, was the largest in the state. De Gasperi led eight coalition

Italy – profile

Official name Italian Republic
Area 301,245sq km (116,311 sq miles)
Population (1975 est.) 56,483,000
Density 188 per sq km (486 per sq mile)
Chief cities Rome (capital) (1976 est.) 2,856,300; Milan, 1,732,500; Naples, 1,224,300; Turin, 1,202,900; Genoa, 807,100
Government Head of state, Giovanni Leone, president (elected 1971)
Language Italian
Monetary unit Lira (plural lire)
Gross domestic product (1974) £66,794,900,000
Per capita income £1,183
Agriculture Wheat, barley, oats, rye, maize, sugar-beet, potatoes, grapes (for wine), olive oil, citrus fruits, livestock
Industry Textiles, chemicals, minerals, iron and steel, sugar refining, food processing, fishing
Chief exports Fruit and vegetables, textiles and clothing, iron and steel, machinery and electrical goods
Trading partners West Germany, France, USA, Netherlands, Britain, Belgium

The volcanically-formed Pontine (or Ponza) Islands, off the west coast of Italy, served as a prison for political prisoners during the Fascist years.

The relic of a once elegant mansion from the French colonial era stands decaying at Grand Bassam, Ivory Coast, despite extensive building development.

Abidjan, capital of the Ivory Coast, has developed into one of Africa's most modern cities; it is a port on a lagoon off the Gulf of Guinea.

governments up to 1953, and managed to establish programmes of industrial expansion and agrarian reform. Italy was a founder member of NATO in 1949 and of the EEC in 1958.

Economic progress was remarkable, but the development of education and social and health services did not keep pace, and the continuing instability of the government led to unrest. From 1969 to 1972 economic growth was seriously slowed by wave after wave of strikes, with riots or demonstrations by students, workers, pensioners and civil servants. The world oil crisis of 1974 led to inflation and unemployment, and further unrest. In April 1976 the thirty-third postwar government collapsed amidst charges of political corruption and in a climate of economic chaos. Despite Communist gains, the Christian Democrats retained a plurality in the June elections, and formed another minority government. A severe earthquake, in which 1,000 died, occurred in May 1976 in the northern province of Friuli. Map 24.

Italian Regions

Region	Area sq km	[sq miles]	Capital
Abruzzi	10,794	[4,168]	L'Aquila
Apulia (Puglia)	19,347	[7,470]	Bari
Basilicata	9,992	[3,858]	Potenza
Calabria	15,080	[5,822]	Reggio Calabria
Campania	13,595	[5,249]	Naples
Emilia	22,123	[8,542]	Bologna
Friuli-Venezia Giulia	7,846	[3,029]	Udine
Lazio	17,203	[6,642]	Rome
Liguria	5,413	[2,090]	Genoa
Lombardy	23,834	[9,202]	Milan
Marches	9,692	[3,742]	Ancona
Molise	4,438	[1,714]	Campobasso
Piedmont (Piemonte)	25,399	[9,807]	Turin
Sardinia	24,090	[9,301]	Cagliari
Sicily	25,708	[9,926]	Palermo
Trentino-Alto Adige	13,613	[5,256]	Trento
Tuscany	22,992	[8,877]	Florence
Umbria	8,456	[3,265]	Perugia
Val d'Aosta	3,262	[1,259]	Aosta
Veneto	18,368	[7,092]	Venice

Ivory Coast (Côte d'Ivoire), official name the Republic of Ivory Coast, is a nation in western Africa. Since it became independent in 1960, it has enjoyed political stability and has achieved the high economic growth rate of nearly eight per cent per year. It now ranks third among the world's coffee producers and is second only to Ghana in cocoa production. Timber is another major export and manufacturing has been expanding rapidly. The average national income per person in 1974 was £197.

Land and climate. The south-western coast is rocky, but that to the south-east is lined with sand bars, enclosing lagoons. The capital, Abidjan, was built on a lagoon but a canal links it to the sea. Inland of the coast the southern third of the country is a flat, forested plain. The northern two-thirds is a savanna plateau, reaching about 1,220m (4,003ft) above sea-level. In the south the average annual temperature is 27°C (81°F); and there is little seasonal variation. The rainfall totals 2,350mm (93in) per year in the south-west and 1,960mm (77in) in the south-east, although the central coastlands are drier. Most of the country has an equatorial climate; the remainder, in the north-west, is tropical and there are greater temperature variations, with rainfall averaging about 1,400mm (55in) per year.

Economy. The country's chief exports are coffee (35 per cent of the total), timber (24 per cent) and cocoa (22 per cent). Cotton, pineapples, bananas and rubber are also exported. Some diamonds and manganese are mined, but fuels are imported to power the fast-increasing manufacturing industries. Industrial production, which was insignificant in 1960, was increasing by more than 20 per cent per year in the early 1970s. It now accounts for more than 13 per cent of the gross national product. The chief industrial centre and port, Abidjan, has a petroleum refinery and a great variety of mainly foreign-owned factories.

People. The mostly Negroid people are divided into more than 60 ethnic and language groups. The Akans, including the Anji and Baule, are the chief group in the densely populated south-east. Some 64 per cent of the people follow ethnic religions, 23 per cent Islam and 13 per cent Christianity.

Government. The country is ruled by the president and a cabinet. The National Assembly has 100 members elected for five-year terms. There is only one political party.

History. French influence began in the 18th century. Coastal forts were built in the 1840s, but the area was not proclaimed a colony until 1893. After independence in 1960, President Félix Houphouët-Boigny maintained stability and attracted the Western foreign investment required for the country's economic advancement. Map 32.

Ivory Coast – profile

Official name Republic of Ivory Coast
Area 322,463sq km (124,503sq miles)
Population (1975) 6,673,000
Chief city Abidjan (capital), 650,000
Language French (official)
Monetary unit CFA franc

Jamaica is an independent nation of the Commonwealth that occupies an island in the West Indies, 145km (90 miles) south of Cuba. The capital is Kingston. It is a comparatively poor country with low wages and high unemployment.

Land and economy. Mountains cover 80 per cent of the land, which has a tropical maritime climate in the path of the trade winds. The economy depends on tourism, mining, light engineering, construction and agriculture; bauxite and alumina are the chief exports [*see* MM p.*34*] and Jamaica is one of the world's major producers of these aluminium ores. Principal crops are sugar cane for rum and molasses [*see* PE p.207], bananas and citrus fruits. Tourism attracts 400,000 visitors a year and, although declining, is the second largest source of foreign income. Even so, poor home production, strikes and unemployment caused a crisis in the mid-1970s.

People. More than 90 per cent of the inhabitants of Jamaica are of African descent, and the remainder are mulatto, Asian, Indian or Chinese. English

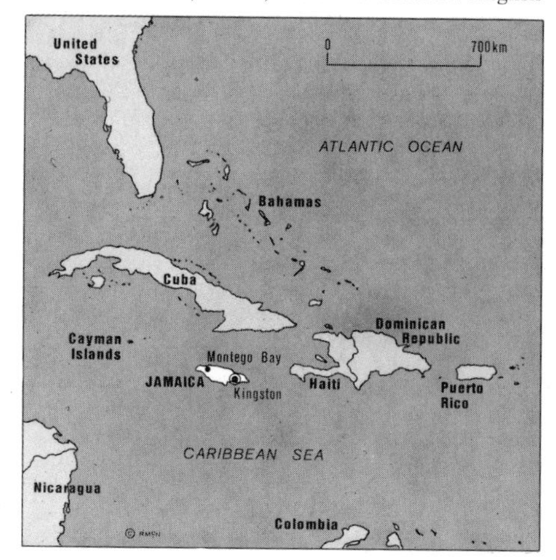

At Oracabessa in northern Jamaica, the hot climate with its heavy rainfall contributes to highly productive banana plantations; much of the fruit is exported.

Fujiyama, a volcano which last erupted in 1707, is Japan's highest mountain (3,776m). The shrine at its peak is visited by thousands of pilgrims annually.

Oyster farming in the Amakusa archipelago, whose islands are too mountainous to support agriculture, provides Japan with a valuable trade in pearls.

is the official language but many speak a dialect form, creole English. Most people work in agriculture which, however, accounts for less than half the national income. Primary education is free, and literacy is estimated at 85 per cent. The Anglican Church is the predominant religion.

Government. The Jamaican constitution, signed in 1962, set up a British-style parliamentary system of government. The British Crown appoints a governor-general, but executive power resides in the cabinet, led by an elected prime minister (in 1976 Michael Manley), an elected House of Representatives and an appointed Senate. The unit of local government is the parish, headed by an elected councillor.

History. Jamaica was discovered in 1494 by Christopher Columbus and occupied by the Spanish until 1655, when British forces captured the island. Sugar production made it an important possession. After a long period of colonial rule, Jamaica began to seek independence in the 1930s. In 1945 the Jamaica Labour Party formed the first government elected by popular vote. Jamaica joined the West Indies Federation in 1958 but withdrew when it gained independence in 1962. In 1968 it joined the Caribbean Free Trade Association, whose members made Jamaica a loan of nearly £50 million in 1976 to help the country in its economic difficulties. Map 74.

Jamaica – profile

Official name Jamaica
Area 10,962sq km (4,232sq miles)
Population (1975 est.) 2,029,000
Density 185 per sq km (479 per sq mile)
Chief city Kingston (capital) 550,100
Government Parliamentary system within the Commonwealth; governor-general, Florizel Glasspole
Religion (major) Anglican
Language (major) English
Monetary unit Jamaican dollar
Gross national product (1974) £970,100,000
Per capita income £478
Agriculture Sugar cane, bananas, citrus fruits
Industries Tyres, chemicals, clothing, food products
Minerals Bauxite, alumina
Trading partners Britain, USA, Canada

Japan (Nihon or Nippon), is an independent nation consisting of four large islands and about 3,000 small ones ranging in a long arc off the coast of eastern Asia. The capital is Tokyo. For hundreds of years the Japanese government pursued a policy of isolation and the country was a mystery to the outside world. Having lost territory through their defeat in World War II, the Japanese have achieved a remarkable economic recovery to become in less than 20 years the world's third greatest industrial nation (after the United States and the USSR). The Japanese tendency to study and emulate Western ideas was once ridiculed. Today their products – from cameras to cars, television sets and supertankers – enjoy a worldwide reputation for quality.

Land and climate. Japan's four main islands (Hokkaidō, Honshū, Shikoku and Kyūshū) are strung out off the coast of eastern Asia for about 2,100km (1,300 miles), They are extremely mountainous, being the upper part of a massive range that rises off the floor of the Pacific Ocean. There are about 200 volcanoes, nearly 60 of them are active, including the world's largest active crater Aso-san [*see* PE p. *53*]. The highest peaks are in the Japanese Alps, on Honshū, rising to the famous extinct volcano Mt Fuji (3,776m; 12,389ft). The islands are perched on the edge of a great Pacific trench, an unstable geological location that produces frequent earthquakes, particularly around the Bay of Tokyo.

There are many fast-flowing mountain rivers, which provide excellent irrigation and considerable hydroelectric power. The longest river is Ishikari (443km; 275 miles), on Hokkaidō. There are also many small lakes, some lying in the craters of extinct volcanoes, and numerous hot springs. The coastline is deeply indented, with hundreds of bays and inlets that provide excellent harbours.

The climate varies from cool and temperate in the north, with cool summers and cold winters, to subtropical in the south, with long hot summers and mild winters. In Tokyo (the capital) temperatures average 26°C (78°F) in August and 3°C (37°F) in January. There is plentiful rain throughout the country, mostly in excess of 1,000mm (40 inches) per year and considerably more in the south (3,000mm; 120 inches). Japan is subject to frequent destructive typhoons, which bring heavy rain and often cause severe flooding.

Economy and resources. With only meagre natural resources, Japan's postwar economic recovery is a triumph largely attributable to the determination and hard work of its people. Between 1952 and 1960 the country's gross national product rose at an average rate of 8 per cent a year, and in the next decade increased by more than 10 per cent a year. Water is Japan's only abundant physical resource; hydroelectric power contributes about half the country's electrical energy, in the production of which Japan ranks third in the world. Japan is the leading shipbuilding nation, annually producing half the world's new ships, and also leads the world in the manufacture of motorcycles, electronic equipment, radios, cameras, watches and sewing machines, and ranks second to the United States in the production of motor vehicles and television sets.

Most of Japan's raw materials have to be imported, especially fuel. Although Japan has substantial deposits of coal, only about a quarter can be used for industrial purposes. A wide range of minerals is mined but only a few, such as lead, zinc and sulphur, in sufficient quantities to meet Japan's basic needs. Copper and aluminium ores are also important, and there are valuable deposits of gold and silver. About 25 per cent of the total work force is engaged in manufacturing industries, which contribute 37 per cent of the gross national product, the heavy and chemical industries predominating. They include the manufacture of cement (Japan ranks third in the world), iron and ferrous alloys (second) crude steel (third), synthetic rubber (second) and newsprint (third). Textiles and food and tobacco products are also important.

Only about 15 per cent of Japan's land is arable, but agriculture is highly efficient and the yield per hectare is the highest in the world [*see* PE p.172]. About 13 per cent of the labour force works on the land, and the average land holding is less than 1ha (2 acres). Rice is grown on nearly half the arable land, and Japan is self-sufficient in this, its staple food. Large acreages of potatoes are also grown.

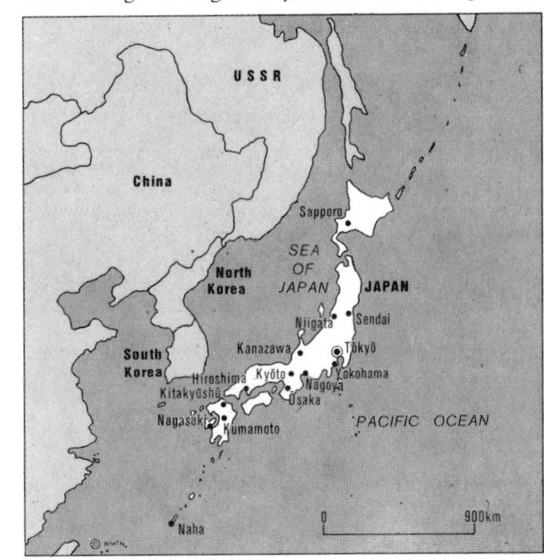

Other crops include fruit, tea, tobacco, hemp and mulberry (for silkworms). Some wheat and barley are grown, but Japan has to import more than 70 per cent of its cereals (excluding rice) and fodder crops. About two-thirds of the country is forested, and products include timber, charcoal, wood pulp and paper.

Nearly 4 million cattle and 8 million pigs are raised. In all, Japan produces about 85 per cent of its own food. Fish is an important source of protein in Japan, and the country is a leading fishing nation. The main catches are Alaska pollack, chub, mackerel, and many species of crustaceans and shellfish. Japan is the world's foremost whaling nation.

Japan's chief exports, with approximate percentages in the mid-1970s, include iron and steel (19%), motor vehicles (13%), shipping (10%), electrical and electronic products (10½%), other machinery (11%), chemicals (7%) and textiles (5½%).

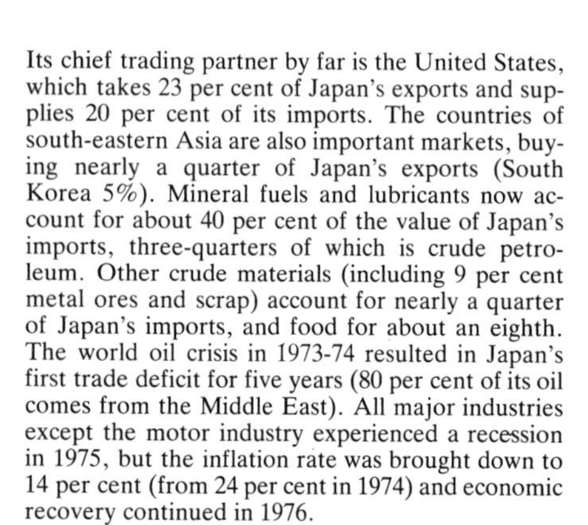

Fukushima prefecture, a mountainous and agricultural region of Japan, produces large crops of rice and soya beans and is also famous for its horse-breeding.

The Imperial Palace in the heart of Tokyo is the residence of the Japanese emperor. In front of it stands the Niji-bashu, or "double bridge".

Japanese painting and engraving reached its peak of perfection in the early 19th century; this example is by Katsushika Hokusai (1760-1849).

Its chief trading partner by far is the United States, which takes 23 per cent of Japan's exports and supplies 20 per cent of its imports. The countries of south-eastern Asia are also important markets, buying nearly a quarter of Japan's exports (South Korea 5%). Mineral fuels and lubricants now account for about 40 per cent of the value of Japan's imports, three-quarters of which is crude petroleum. Other crude materials (including 9 per cent metal ores and scrap) account for nearly a quarter of Japan's imports, and food for about an eighth. The world oil crisis in 1973-74 resulted in Japan's first trade deficit for five years (80 per cent of its oil comes from the Middle East). All major industries except the motor industry experienced a recession in 1975, but the inflation rate was brought down to 14 per cent (from 24 per cent in 1974) and economic recovery continued in 1976.

Transport and communications. Japan has a highly developed railway system, with about 27,000km (16,800 miles) of track (mostly electric), more than a fifth of which is privately owned. The world's longest underwater railway tunnel, the Shin Kanmon Tunnel (18.6km; 11.6 miles), connects Honshū with Kyūshū. A 54km (33.5-mile) tunnel, the Seikan, is under construction between Honshū and Hokkaidō. The Japanese National Railways operates the world's fastest scheduled service, the "New Tokaido", from Ōsaka to Okayama – a distance of 160km (100 miles) covered in 58m minutes at an average speed of 166km/h (103mph). The first monorail system was opened in 1964 to connect Tokyo airport with the centre of the city. Four cities have underground railways, and in Tokyo personnel are employed specifically to pack the passengers in during the rush periods. There are about 30,000km (18,660 miles) of "national" roads, and the number of vehicles in use in the mid-1970s (26 million, including nearly 16 million passenger cars) had quadrupled in ten years.

Japan's merchant fleet of some 40 million gross tonnes is second only to Liberia's in size and has more ships (9,932 in 1975) than any other nation's. The Japanese oil tanker *Nissei Maru* of 238,517 gross tonnes was the world's heaviest when launched in 1975. A car-transporting ship, *Polar Ace* (built in Japan in 1977), resembles a floating multistorey car park and can carry 400 cars for exporting countries. The chief Japanese ports are Yokohama, Kōbe, Tokyo, Nagoya and Ōsaka. The principal airlines are Japan Airlines and All Nippon Airways, and there are large international airports at Tokyo and Ōsaka. The Japan Broadcasting Corporation has three radio and two television channels, and numerous companies operate commercial radio and television. About 98 per cent of all households own television sets, a total of more than 25 million. There are about 42 million telephones in Japan (only the United States has more), and daily newspapers sell nearly 60 million copies.

Government. In accordance with the constitution of 1947 the emperor is the symbol of the state, but has no governing powers. Legislative power rests with the two-chamber Diet, which consists of the House of Representatives, with 511 members elected to four-year terms by universal suffrage (minimum voting age 20), and the House of Councillors, with 252 members (half elected every three years to six-year terms). Executive power is vested in the Cabinet, which comprises a prime minister and 11 to 16 ministers of state. The major political parties are the Liberal-Democrats, a conservative, free-enterprise party that has ruled for almost the whole postwar period, and the Socialist Party.

For local administration Japan is divided into 47 prefectures, 561 cities and more than 2,000 towns and 800 villages. Each city, town and village elects its own mayor and one-house assembly. There is a central police force some 25,000 strong, controlled by the national government through a five-man commission. All judiciary power is invested in the Supreme Court, which consists of the chief justice and 14 other judges, appointed in the first place by the Cabinet but thereafter having to seek reelection every ten years. There are eight regional high courts, and various local courts.

Although the constitution renounces war and the use of armed force, it does not exclude self-defence. For this reason Japan's army is organized as a "Ground Self-Defence Force", with an authorized strength of 180,000; the navy (Maritime Self-Defence Force) is 42,000 strong and the air force (Air Self-Defence Force) 45,000. Military service is voluntary. Almost all the military equipment is supplied by the United States, and there are about 58,000 American military personnel based in Japan, mostly on Okinawa.

People and culture. The Japanese are a Mongoloid people originating from mainland Asia. There has been little mixture for the past thousand years, so there is virtually complete identity of race and nation. The only minority group indigenous to Japan is a Caucasoid people called the Ainu, whose ancestors were among the first occupants of the land. They now number about 15,000 and most of them live on Hokkaido. More than 600,000 Koreans live in Japan. Japan's population distribution is 72 per cent urban; 15 cities have more than half a million inhabitants, eight more than a million.

The two major religions are Shintō and Buddhism, each with more than 80 million adherents; millions of people practise both. There are about 800,000 Christians. Shintō ("the Way of the Gods"), a belief based on myths and legends, was the state religion and was used by the government to support the idea of the emperor's divinity. After World War II, state support for Shintō was banned, and religious teaching is now forbidden in state schools. Many new religions have grown up since the war, based on beliefs of other major religions.

Japanese is the official and universal language of the country. A number of dialects exist and there are also different styles of speaking, according to the social situation, such as intimate, polite, honorific and impersonal. Spoken Japanese has no affinity with spoken Chinese, but the Japanese, who had no written language of their own, adopted written characters from Chinese in about the 6th century AD. Although these have been codified into syllabaries called *kana,* Japanese is still a complex language in written form. Nevertheless literacy is high, nearly 99 per cent.

Education is free and compulsory between the ages of six and 15 (planned to be increased to 18), and all institutions are co-educational. There are kindergartens, mostly privately controlled, for three- to five-year-olds, elementary schools (6-12), and lower (12-15) and upper (15-18) secondary schools. There are seven main state universities and several private universities. The University of Tokyo (founded in 1877) has more than 18,000 students. In all more than 1,800,000 students are enrolled in higher education. After World War II the Japanese accepted an entirely new approach to education initiated by the occupying American authorities, which involved dropping official dogma.

The Japanese have developed one of the most distinctive cultures of any modern civilization. This is evident in everyday life in their clothing, their customs, and particularly in their homes. Houses are open and airy, with sliding paper panels instead of interior walls, thick straw mats, charcoal braziers for heating and deep oval porcelain or wooden bathtubs for relaxing in after a day's work. The ritual of the evening bath is almost a national cult. After it, in most homes, the people don the traditional kimono and sit on cushions round a low table for dinner. A typical Japanese meal consists of rice, slices of various kinds of fish (often raw), soup, pickles and fruit. The ritual tea ceremony is uniquely Japanese and regarded by them as a form of art. It derives from Zen Buddhism, which has had a profound influence on Japanese art and culture. Other influences have been Shintō, Chinese and, in the 19th century, Western art.

Japanese painting began as an expression of Buddhism, adopting Chinese techniques. Picture scrolls began to develop in the 12th century; hanging scrolls, with ink washes, in the 14th; and colour printing from wood blocks in the 18th – the most admired of the Japanese arts outside Japan. The most famous examples of Shintō influence in art are the Ise shrines. The best modern Japanese architecture reflects the simplicity of Shintō and the power of Zen, and is well illustrated by the work of Tange Kenzō (1913-), who designed Tokyo's Roman Catholic cathedral and the National Gymnasium. Japanese music is also characterized by its simplicity, the chief instruments being the zither, lute, flute, horn, gongs and drums. Western music is also appreciated, both classical and popular. In the theatre, the Japanese created two unique forms of drama, *nō* and *kabuki.* Nō plays (which evolved in the 14th century) are formal, stylized, slow-moving dramas, with masks and music, based on historical themes. Kabuki, dance dramas that developed in

Sendai, a bustling commercial and industrial city manufacturing chemicals, metal goods and silk yarn, is also the seat of Tohuku University.

The O-dori Promenade in Sapporo, the centre for Japan's winter sports, is decorated during the annual Snow Festival with giant ice sculptures.

Kamakura, situated on Sagami Bay, is remarkable for having more than 80 shrines and temples, despite a population of fewer than 150,000 people.

the 17th century from nō, are livelier and less formal; there are only male actors, and performances last several hours. Another traditional form of Japanese drama is the *bunraku*, a puppet theatre now confined to Ōsaka. The Japanese also enjoy Western drama and cinema.

Japanese literary forms include *haiku*, a type of verse with a maximum of 17 syllables, influenced by Zen. The most widely-read Japanese book is probably *The Tale of Genji*, a story about the life and loves of an emperor's son written by an 11th-century court lady, Murasaki Shikibu. The Japanese are also known for several minor arts, including *bonsai* (the cultivation of miniature trees), *bonkei* (miniature gardens on trays) and *ikebana* (flower arrangement). They are also famous for wood-carving, ceramics, enamels and embroidery.

The national sport of Japan is *sumo*, a ritualistic style of wrestling practised by huge men [*see* MS p. *304*]. Japanese martial arts such as judo and karate are now popular throughout the world, and kendo has also spread. After sumo, the most popular sport is baseball, which was introduced to Japan in 1873 by an American schoolteacher. Soccer and rugby union are gaining in popularity, and visiting European teams attract large crowds of spectators. The Japanese excel at table tennis and gymnastics, and golf, mountaineering and skiing are popular outdoor pursuits. Japan was the first Asian country to be host for the Olympic Games (Tokyo, 1964) and also the Winter Olympics (Sapporo, 1972).

History. According to legend, Jimmu, a direct descendant of the sun-goddess, became the first emperor of Japan in 660 BC. Little is known, however, about Japan's early history. The Japanese are thought to be descendants of peoples who arrived from mainland Asia in the 1st century AD and subjugated the native Ainu. The present emperor is believed to be a direct descendant of the House of Yamato, the dynasty that dominated Japan from about the 3rd century AD. The Japanese established a colony at the southern end of Korea in the 4th century, which lasted more than 200 years. Korean and Chinese craftsmen and artisans began to settle in Japan, bringing new skills and ideas, such as rice cultivation and the Chinese form of writing.

The emperors gradually lost authority, and power passed from the throne to nobles and then to families of warrior knights called *samurai*. One such family, the Minamoto, gained power in 1185. Their leader, Yoritomo, set up a new form of government called the *Bakufu* (camp office), based at Kamakura, away from the imperial capital. In 1192 the emperor granted him the title *shōgun* (commander-in-chief). Shōguns, or their advisers, controlled Japan for nearly 700 years, always in the name of the emperor. This warrior class was the dominant force in the country. In the late 13th century they successfully repelled two Mongol invasions, but the 14th to 16th centuries were marked by civil wars.

The first Europeans to land in Japan were Portuguese sailors in 1542, soon to be followed by Catholic missionaries (St Francis Xavier spent two years in Japan) and by Portuguese traders who established a small community on Kyūshū. In the 1590s Japan was united again, under Toyotomi Hideyoshi, who also launched two abortive invasions of Korea. After his death another powerful figure, Tokugawa Ieyasu, gained control, taking the office of shōgun and making Yedo (modern Tokyo) his headquarters. It was the Tokugawa shōgunate that began the policy of isolation, fearful that foreign influence might stir up internal trouble. Christianity was suppressed, traders expelled, and the Japanese forbidden to leave the country. The only contact with the outside world was a limited concession allowing the Dutch and Chinese to use Nagasaki for trading. This isolationist policy lasted for 200 years and developed into a general mistrust of all things foreign. Then the United States, eager for power in the Pacific, sent a naval squadron under Commodore Matthew Perry to open trade relations with Japan in 1853. He returned in 1854 to complete a treaty with the reluctant Japanese, and the English, Russians and Dutch followed.

A new constitution was established in 1889 and a two-chamber Diet instituted, but the armed forces were independent of the Cabinet. Japan's imperialist expansion began with the war with China (1894-95), in which China was forced to recognize Korea's independence, and Japan gained Formosa (now Taiwan), the Liaotung peninsula and Port Arthur (in Manchuria), although it had to hand Port Arthur back to Russia in 1898. Japan was now a world power, and signed an alliance with Britain in 1902. Even so its defeat of Russia (1904-05), which was forced to give up Port Arthur and other territories, astonished the world. The Japanese gained control of Korea and annexed it formally in 1910.

Japan made further gains as a result of World War I, which it entered as Britain's ally, acquiring German possessions in Asia and the Pacific. The war also brought an economic boom to Japan, but the postwar years saw the collapse of some of its newly won markets, some unsuccessful foreign adventures, the disastrous Tokyo-Yokohama earthquake of 1923 (in which more than 90,000 people were killed) and economic failure hastened by the world recession.

Angered by the West's refusal to accept the Japanese as equals and with tension mounting with China, the country's military leaders took advantage of a growing feeling of nationalism to renew their imperialistic ambitions. Their troops occupied Manchuria in 1931 and they initiated a campaign of terror at home that gave them full control of the government. In the late 1930s Japan began to overrun China. It joined in alliances with Germany and Italy, and in 1941 entered World War II by bombing American military bases at Pearl Harbor and attacking Hong Kong, Malaya and Singapore. Initially the Japanese won many battles, and extended their control over vast areas of the Pacific and south-eastern Asia by mid-1942. Then they were defeated by the Americans in the Pacific (in the Battle of Midway) and this proved to be the turning point of the war. They gradually lost ground on all fronts, and finally surrendered after the devastation of Hiroshima and Nagasaki by atomic bombs in August 1945 [*see* MM p. *180]*.

The Allied occupation of Japan (1945-52) was supervised almost entirely by the Americans under Gen. Douglas MacArthur, with the willing co-operation of most of the Japanese. A new constitution came into force, establishing a democratic government. Ratification of the San Francisco peace treaty (signed by Japan in 1951 with 48 nations) ended the occupation, and a security treaty with the United States ensured American protection.

Japan's economic revival gained impetus from the industrial boom resulting from the Korean War (1950-53), and from the mid-1950s advanced at an astonishing rate. In politics, however, things did not run so smoothly; student unrest and discontent in the Socialist Party erupted in 1960 over the terms of a new security treaty with the United States. The conservative government survived the crisis, as well as frequent allegations of corruption.

Eisaku Sato became prime minister in 1964, and had the satisfaction of negotiating the return of Ōkinawa (from US control) before he retired in 1972. His successor, Kakuei Tanaka, restored diplomatic relations with Peking and severed ties with Taiwan. He also set about solving such problems as

Japan – profile

Official name Japan
Area 377,535sq km (145,766sq miles)
Population (1976 est.) 112,550,000
Density 298 per sq km (772 per sq mile)
Chief cities Tokyo (capital) (1975) 8,642,000; Ōsaka, 2,980,487; Yokohama, 2,778,975; Nagoya, 2,079,694; Kyōto, 1,461,050
Government Symbol of state, Hirohito, emperor (succeeded 1926)
Religions Shintō, Buddhism
Language Japanese
Monetary unit Yen
Gross national product (1974) £182,000,000,000
Per capita income £1,617
Agriculture Rice, potatoes, sweet potatoes, wheat, barley, fruits, tobacco, tea, livestock, forestry
Industries Coal, iron and steel, petrochemicals, fishing, shipbuilding, textiles, cement, chemicals, optical and electronic instruments, motor vehicles
Exports Machinery, iron and steel, motor vehicles, ships, chemicals
Trading partners (major) USA, Australia, Indonesia, Arab oil countries

Amman, capital of Jordan, was once the major Greek and then Roman city of Philadelphia; surviving Roman remains include this large amphitheatre.

Traditional methods of reaping grain crops by hand do not prevent Jordan from being, in years of good harvests, self-sufficient in agriculture.

Olive orchards, which flourish in the Jordanian uplands, yield the country's sixth largest crop, after grains, legumes, citrus fruit and tomatoes.

inflation, pollution, overpopulation of the cities and underpopulation of rural areas, while avoiding militarism. But before he could get to grips with these, he found himself involved personally in a political scandal. He resigned in 1974 and was succeeded by Takeo Miki.

Meanwhile the Liberal-Democratic Party was being split by rival factions. They failed to win an overall majority in the House of Representatives in 1976 but were saved by the support of a number of independent members. Prime Minister Takeo Miki was forced to resign and was replaced by Takeo Fukuda, who set about reuniting his party. He also called for a world economic summit meeting of industrially advanced countries, which he attended when it took place in London in May 1977. Map 46.

Java. *See* INDONESIA.

Jersey. *See* CHANNEL ISLANDS.

Jordan (Al-Urdunn), official name the Hashemite Kingdom of Jordan, is an Arab monarchy in the Middle East in an historic but largely desert region. Its territory extends into several lands with biblical associations, including Palestine, Ammon, Edom and Moab. Since 1948 it has been involved in the conflict between ISRAEL and Arab countries; its territory west of the River Jordan has been occupied by Israel, and in 1970 it had a brief civil war. The capital is Amman.

Land and climate. Jordan's western border is only about 20km (12 miles) from the Mediterranean Sea (from which it is separated by Israel), but the country's only port is Aqaba, on the Gulf of Aqaba (an arm of the Red Sea). The River Jordan flows north-south through the Ghor Depression in the west to the Dead Sea. Really a salt lake, the Dead Sea is the lowest point on the Earth's land mass – 396m (1,299ft) below sea-level. Highlands to the west of the Depression include the Judaean Hills in the south and Samaria in the north. To the east are the barren stony uplands that form three-quarters of Jordan. The country's climate is generally hot and dry, ranging from 37°C (99°F) in summer to 5°C (41°F) in winter. The heaviest rainfall is in the western highlands – 510 to 760mm (20-30in) – with less than 50mm (2in) in the barren uplands.

Economy. Rock phosphate is the most important mineral resource and accounts for about a third of all exports. Other minerals include potash (from the Dead Sea), marble and manganese ore. An oil refinery at Zerqa (Az-Zarqā'), fed by a branch pipeline of the Trans-Arabian Pipeline (TAPline), provides most of Jordan's needs.

Much of the country is unsuitable for agriculture; in most desert and semidesert areas the only inhabitants are nomadic bedouin who herd sheep, goats and camels. The most fertile land is west of the Jordan and on the edge of the eastern uplands. Although the area irrigated by the East Ghor Canal is being extended, many places are still dependent on the sparse, uncertain rainfall. The chief cereal crops are wheat and barley. Legumes are also grown, and there is a valuable export trade in vegetables and fruits – particularly citrus fruits and grapes; olives are another valuable crop.

Several industries have been established in the cities. They include food processing and the making of textiles, electrical goods and building materials. Tourism is still an important source of income (the main attractions being archaeological sites such as Petra), in spite of Israel's occupation of the historic region west of the River Jordan. The main cities are linked by highways, and there are good road connections with Damascus in Syria and Baghdad in Iraq. A railway from Damascus passes through the cities of Amman and Ma'ān, and has a branch to Aqaba. Aircraft of Royal Jordanian Airlines (ALIA) operate several international services.

People. Most Jordanians are of Arab stock, although there are deep social and cultural distinctions between various communities: the bedouin of the desert, the Westernized city-dwellers, and the refugees from Israel and the Israeli-occupied west bank of the Jordan. Most of the people are Muslims, 80 per cent belonging to the Sunni sect. The official language is Arabic. Primary education is free where schools are available and there is a University of Jordan (founded in 1962). About 40 per cent of the people are literate.

The population is most concentrated in the Israeli-occupied highlands west of the River Jordan. Amman is on this side of the river; its modern part is well planned and has many fine buildings. Until the Arab-Israeli war of 1967, Jordan's second largest city was its section of Jerusalem. Since the end of World War II the country's most intractable social problem has been how to provide shelter, food and work for the half million or so Palestinian refugees, many of whom still live in camps.

Government. Jordan is governed as a constitutional monarchy, with a two-chamber legislature. A constitutional amendment of November 1974 gave the king power to dissolve the legislature, and this authority was immediately exercised. In February 1976 the legislature was reconvened for long enough to approve another amendment enabling the king to postpone elections indefinitely.

History. Amman was once the chief city of the Ammonites and later became the most southerly city of the Decapolis – the Roman federation of ten cities. For the 400 years before World War I Jordan was part of the Ottoman Empire. After Turkey's defeat in World War I, the emirate of Transjordan was established in the eastern uplands of present-day Jordan as part of the British mandated territory in Palestine and neighbouring lands. Its first emir was Abdulla ibn Hussein, a Hashemite. Britain established a Transjordanian army, the famous Arab Legion.

In 1946 Transjordan became an independent kingdom with the Emir Abdullah as its king. On the day Britain ended its mandate west of the River Jordan and the state of Israel was proclaimed, the Arab Legion crossed the river and occupied most of the part of Palestine that had been designated Arab territory by the United Nations. In 1950 this region was incorporated in the Hashemite Kingdom of Jordan, as the country then became known. Jordan had to absorb great numbers of Arab refugees from Israeli territory. Abdullah was assassinated in 1951; his son Talal succeeded him but was shortly afterwards deposed on the grounds of mental illness. Talal's son, Hussein, then became king.

In the Six-Day War of 1967, Jordan's west bank territory was seized by Israel. On the east bank, *fedayeen* (commando) organizations that had de-

Jordan – profile

Official name Hashemite Kingdom of Jordan
Area (including Israeli-occupied territory) 97,740sq km (37,737sq miles)
Population (1975 est.) 2,702,000
Density 28 per sq km (72 per sq mile)
Chief cities Amman (capital) (1974) 598,000; Zerqa, 226,000
Government Constitutional monarchy
Religion Islam (Sunni, 80%)
Language Arabic
Monetary unit Jordanian dinar
Gross national product (1974) £444,400,000
Per capita income £164
Agriculture Wheat, barley, vegetables, fruit
Industries Foodstuffs, textiles, electrical goods, building materials
Minerals Phosphates, potash, manganese
Trading partners USA, Britain, West Germany, Syria, Lebanon.

Jordan's desert wastes are occasionally broken by an oasis, but the only really fertile areas are uplands on the west bank of the River Jordan.

Nairobi is one of Africa's fastest-growing cities. Its modern buildings, most of which are made from local stone, give it a distinct Western appearance.

Mombasa, Kenya's chief port, still retains the strongly Oriental flavour given to it by the 12th-century Arab founders of the city.

veloped in the Palestinian refugee camps for war against Israel became a challenge to Hussein's authority. The Jordanian army was sent into action against them in 1970, and in the following year they were forced across the border into Syria. Hussein found himself at odds with other Arab states over his refusal to accept the Palestinian Liberation Organization (PLO) as the official voice of the Palestinian Arabs; but in 1975 he reluctantly gave recognition to the PLO's claims. In 1976 Jordan was the only Arab country to support Syria in its action of sending troops to fight on the non-Christian side in the civil war in LEBANON. Map 38.

Kalimantan. *See* INDONESIA.

Kampuchea. *See* CAMBODIA.

Kansas. *See* UNITED STATES.

Kazakhstan. *See* UNION OF SOVIET SOCIALIST REPUBLICS.

Keeling Islands. *See* COCOS ISLANDS.

Kentucky. *See* UNITED STATES.

Kenya, official name Republic of Kenya, is a nation in eastern Africa, famous for its abundant wildlife and scenic national parks. Tourism is second only to coffee, the chief export, as a source of revenue. But the reduction of natural habitats and unauthorized hunting threaten the future of wild animals in Kenya, as they do in other African countries. The capital is Nairobi.

Land and climate. Inland of the narrow coastal plain the terrain rises steadily to the East African plateau, which reaches about 2,100m (6,890ft) above sea-level. Some highlands and extinct volcanoes, such as the ice-capped Mt Kenya, rise above the plateau. Mt Kenya (5,199m; 17,057ft) is Africa's second highest peak [*see* NW p.*197*]. The plateau is divided by the impressive African Rift Valley, which is between 60 and 80km (37–50 miles) wide and more than 900m (2,953ft) deep in places. The Rift Valley encloses lakes Nakuru and Naivasha in the south and Lake Turkana (formerly Rudolf) in the north. Lake Victoria in the west occupies a shallow depression which is not part of the Rift Valley.

The climate is affected by the altitude. Temperatures on the hot, humid coast average 27°C (80°F) all the year round. But the annual temperature range on the south-western plateau is 10-20°C (50-68°F); the mild conditions there proved highly attractive to European settlers. The coast has about 1,000mm (39in) of rain each year and the south-western plateau and Lake Victoria regions have more than 1,250mm (49in). But the Rift Valley floor is arid, with only 760mm (30in), and rainfall also decreases from south to north; about 85 per cent of Kenya is subject to long and severe droughts. The vegetation ranges from patches of forest on the coast to dry scrub in the south-east and desert in the north-east.

Economy. Most people depend on agriculture, although only about 12 per cent of the land can be cultivated. The chief commercial crops are coffee (30 per cent of all exports), tea (about 14 per cent), sisal (4 per cent) and pyrethrum (3 per cent). Livestock is important, and meat and meat producers account for another 3 per cent of exports. Mining is comparatively unimportant, but manufacturing has been steadily expanding and diversifying, especially in Nairobi and Mombasa. The oil refinery at Mombasa processes imported oil, and petroleum products now account for about eight per cent of all exports. Commercial activities and tourism are developing, increasing Kenya's invisible earnings.

Since independence Kenya has attracted much foreign aid and investment, partly as a result of its political stability and partly because it encourages capitalist enterprise. Between 1964 and 1973 the gross domestic product increased by 56 per cent. In the mid-1970s the rate of growth declined to about 3.5 per cent per year, which is about the same as the high annual population increase. This comparative decline was caused partly by severe droughts and partly by rises in petroleum prices.

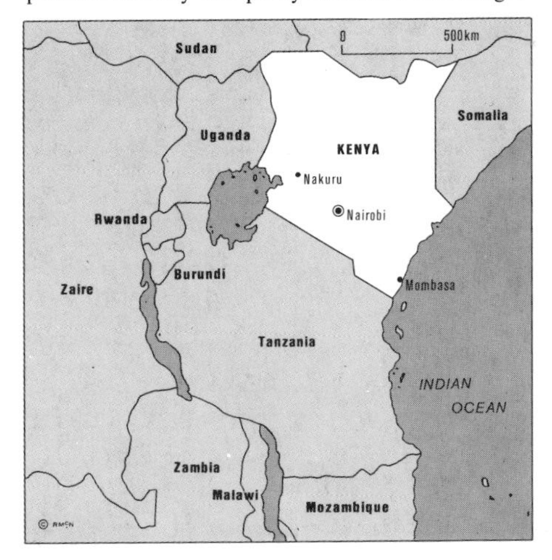

People. Kenya has more than 40 ethnic groups. The largest are the Kikuyu (21 per cent), the Luo (14 per cent) and the Luhya (13 per cent). There are 139,000 Asians, mostly in commerce, 40,000 Europeans, most of whom own farms or work in the professions or in businesses, and 28,000 Arabs, who are also in commerce. About one-third of Kenya's people is Christian and one-fifth is Muslim. Literacy is estimated at 27 per cent. Swahili, the first language of a small coastal group, is the lingua franca; with English, it is an official language.

Government. Kenya is a constitutional democracy with an elected National Assembly which elects the president. The president, who is commander of the armed forces, rules with a cabinet.

History. The Portuguese arrived in eastern Africa in 1498 but Arabs expelled them in the 1700s. Britain made the coast of Kenya a protectorate in 1895 and built a railway inland from Mombasa to Nairobi and Kisumu. By 1906 Britain controlled the interior and, in 1920, Kenya became a British colony.

Large parts of the fertile south-western highlands were reserved for European occupation. This led to discontent, especially among landless Kikuyu. From 1952 British troops fought against the Mau Mau, a Kikuyu secret society with nationalist aims. The state of emergency ended in 1960 and the Mau Mau's alleged leader, Jomo Kenyatta, was released from detention in 1961. He was elected prime minister in May 1963 and Kenya became independent in December 1963. A year later it became a republic and Kenyatta, popularly known as *Mzee* (old man), became president. Kenyatta's policies included the redistribution of land by buying European estates and dividing them into African smallholdings. He encouraged rapid economic development within a mixed economy. The plan to set up an East African federation has not been achieved and tensions have arisen between Kenya, Tanzania and Uganda.

Kenya – profile

Official name Republic of Kenya
Area 582,646sq km (224,960sq miles)
Population (1975 est.) 13,349,000
Density 23 per sq km (59 per sq mile)
Chief cities Nairobi (capital) (1975 est.) 700,000; Mombasa, 255,400
Government Republic; head of state President Jomo Kenyatta
Religions Ethnic, Christianity, Islam
Languages English, Swahili (both official)
Monetary unit Kenya shilling
Gross national product (1974) £1,145,300,000
Per capita income £86
Agriculture Coffee, cotton, cereals, hides and skins, maize, meat, pyrethrum, sisal, sugar, tea
Industries Brewing, cement, dairy products, food processing, petroleum products
Minerals Salt, soda ash
Trading partners Britain and other EEC countries

Khmer Republic. *See* CAMBODIA.

Kirgizia. *See* UNION OF SOVIET SOCIALIST REPUBLICS.

Korea (Choson) is a peninsula that juts out from the Chinese mainland towards Japan. Since 1948 it has

Draught animals outnumber tractors in North Korea where, despite attempts to increase mechanization in agriculture, traditional methods are still used.

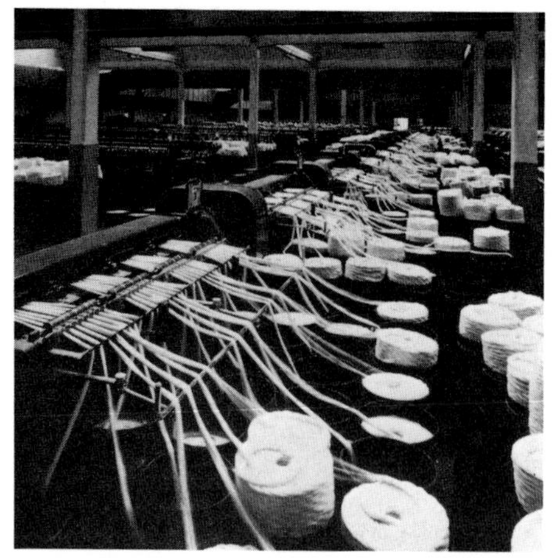

Industry has developed rapidly in South Korea in recent years; textiles, particularly cotton yarn and cloth, are exported to many overseas countries.

Kuwait is one of the world's smallest countries and one of its richest. Its vast wealth is based on oil, some of which is processed at local refineries.

been partitioned into two countries: the Communist nation of North Korea and the independent republic of South Korea. The first Korean nation was Koguryo, founded in the north in about 100 AD. Two southern states were the kingdoms of Paekche (from about 250 AD) and Silla (350 AD). By the 7th century Silla had conquered the others and unified the peninsula. Mongols took over by 1260, to be displaced in turn in 1392 by the Yi dynasty, which ruled until Korea was annexed by Japan in 1910. Partition came in 1945 with the Japanese defeat in World War II; Soviet troops occupied the northern part of the peninsula and Americans the south. Two separate nations were formally established in 1948.

Korea, North (Chosŏn Minjujuŭi In'min Konghwaguk), official name Democratic People's Republic of Korea, is an independent nation of north-eastern Asia. It occupies the northern part of a large peninsula between the Sea of Japan to the east and the Yellow Sea. North Korea was established in 1948 and is governed by the Communist Labour Party (KLP) through a premier and the People's Assembly from the capital, P'yŏngyang.

Land and economy. Moderately high mountain ranges and hills separate valleys and small plains, leaving only about 15 per cent of the land suitable for cultivation. The country has extreme summer and winter temperatures. Its mineral resources are highly developed and it is among the world's leading producers of tungsten, graphite and magnesite. Agriculture is based on collective farming, with rice and maize as the main crops.

People. North Koreans' racial origins are Tungusic (mixed Mongol and Chinese). Between 1925 and 1940 many South Koreans worked in the industrial regions of the north. They returned south after 1945 when the peninsula was divided under American and Soviet control. Korean is the official language; there are two writing styles, one phonetic and one based on Chinese characters. Confucianism was the dominant religion until 1945.

History. Korea was a semi-independent state with links with China until conquered by Japan in 1910. After Japan's defeat in World War II, Korea was partitioned (1945) into American (to the south) and Soviet zones. Numerous efforts to reunite the countries failed and in 1948 the Soviets established the Democratic People's Republic of Korea in the north. In 1950 North Korean forces invaded South Korea. The United States and forces from 15 other members of the United Nations aided the south, and Communist countries assisted the north. The Korean War ended in stalemate in July 1953. Since that time North Korea's premier Kim Il Sung has maintained strong links with China and the USSR and remained hostile to American influences in South Korea. For instance a military alert was ordered in August 1976 after two US officers (working for the United Nations) were killed by North Korean border guards. Area: 120,538sq km (46,540sq miles). Pop. (1975 est.) 15,852,000. Map 44.

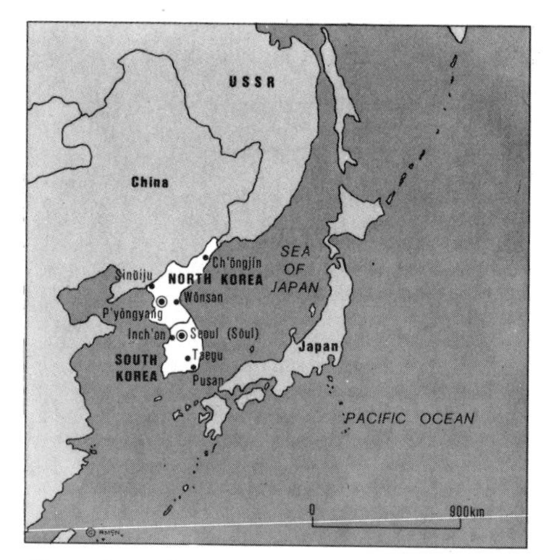

Korea, South (Taehan-Min'guk), official name Republic of Korea, is an independent nation of north-eastern Asia, occupying the southern half of the Korean Peninsula. The country was established in 1948 with its capital at Seoul. It is a racially mixed society which aligns itself with non-Communist countries rather than with its immediate neighbour, North Korea.

Land and economy. South Korea is a mountainous country, with most of its harbours on the west and south coasts. The climate is hot and humid in the summer, dry in the winter. Poor in natural resources, lacking skilled workers and densely populated, South Korea still suffers from the economic after-effects of the Korean War (1950-53). About 25 per cent of the gross national product comes from agriculture, fishing and forestry; manufacturing contributes another 25 per cent. Only about 20 per cent of the land is cultivated, with rice the chief crop. A few animals are raised for meat but fish is the principal source of protein in the diet. Tungsten and coal are the chief minerals, some of which are exported. About 25 per cent of the country's budget is spent on defence.

People. Many Koreans are of Tungusic (mixed Mongol and Chinese) descent and most of the population lives in the southern valleys or in or near Seoul, in the north. The chief language is Korean, although many people also speak English. The racial mixture is reflected in the three main religions: Buddhism, Shamanism and Christianity, although organized religion has only a weak hold on the people.

History. The whole Korean Peninsula was a semi-independent nation, under Chinese influence, until the Japanese conquest in 1910. After World War II and Japan's defeat (1945), Korea was partitioned into a Soviet-occupied zone (to the north) and an American zone. Attempts at unification failed and in 1948 the separate nations of North and South Korea were established. With Communist backing North Korea invaded the south in 1950; the United Nations went to the aid of South Korea, but the war ended in stalemate three years later. South Korea's first president, Syngman Rhee, was deposed in 1960 after riots about alleged irregularities in an election. A military coup led by Maj.-Gen. Park Chung Hee seized power in 1961 and, despite assassination attempts, President Park consolidated his authority and remained the country's leader. Area: 98,484sq km (38,025sq miles). Pop. (1975 est.) 34,688,000. Map 44.

Kuwait (Al-Kuwayt), official name State of Kuwait, is a small independent Arab nation in the north-eastern Arabian Peninsula at the northern end of the Persian Gulf. A leading oil producer, it is one of the world's wealthiest countries. It is named after its capital, Kuwait.

Land and economy. Except for the Al-Jahrah Oasis and a few fertile regions in the south-east and coastal areas, the country is almost entirely desert. Nearly all the population is concentrated in the cities. Kuwait has about 15 per cent of the world's petroleum reserves and oil dominates the economy, which received a tremendous boost with the sharp increases in oil prices in 1974. The government has used its huge financial resources to create a welfare state in which there are no taxes, and medical care, education and social security are all free. It also makes loans to other Arab nations.

People. Most of the people are Arabs; Arabic is the official language and Islam the official religion. More than half of the people are literate and, with expanded free educational facilities, the literacy rate is rapidly increasing. Almost half of the population are not Kuwaitis and have no representation in the government.

Government and history. Kuwait is a constitutional monarchy, governed by an emir and a 50-member

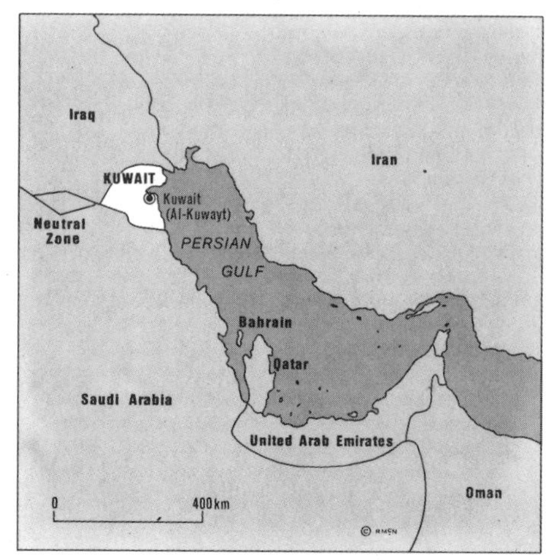

For much of its length, the Mekong River flows in or along the borders of Laos; here a woman washes clothes in the shallows on the river's bank.

Laotian women sit with their wares in a market street in Luang Prabang, the economic centre of northern Laos. Most of the city's shops are Chinese-owned.

The Lebanese town of Sayda, now of little economic importance, occupies the site of ancient Sidon, one of the great trading ports of the Phoenicians.

National Assembly. Founded in 1756 by members of the al-Sabah dynasty, the state is still ruled by the family today. By a treaty of 1899 Britain administered foreign affairs and protected territorial rights. Kuwait became independent in 1961, and shortly afterwards Iraq laid claim to the country. Through a protection agreement with Britain, Kuwait requested and received British troops to forestall the Iraqis; since then relations between Kuwait and Iraq have improved. Kuwait adopted a constitution and joined the United Nations in 1963, but the constitution was suspended in 1976 after the government resigned. Area: approx. 17,000sq km (6,560sq miles). Pop. (1975) 994,837. Map 38.

Laos (Lao), official name People's Democratic Republic of Laos, is an independent nation in south-eastern Asia. Its development has been hampered by its difficult geography and climate and by a poor economy. For six centuries a monarchy, it was conquered by Pathet Lao Communist forces in 1975. The capital is Viangchan (formerly Vientiane).

Land and economy. Landlocked and located in the centre of the south-east Asian peninsula, Laos borders on five nations: China (to the north); Vietnam (east); Cambodia (south); Thailand (south and west), including 800km (500 miles) along the Mekong River; and Burma (north-west). It is a mountainous country covered with jungles and has no access to the sea. Its three-season climate is monsoonal. Natural resources are largely unexplored, and 85 per cent of the population works in subsistence agriculture. The chief exports are tin, timber and coffee; almost all manufactured products are imported.

People. Laos' sparse population is concentrated along the Mekong River valley. The Lao majority are descended from a south-western Chinese people, the Thai, who migrated there in the 13th century. Mountain tribes without a common language or tradition inhabit the central and southern regions. Theravada Buddhism is the principal religion; the mountain tribes follow various ethnic religions. Lao is the dominant language although French is used in schools, and many tribes speak dialects not yet studied. The literacy rate is about 25 per cent. Laos has been ruled as a Communist people's republic since 1975. The king and other old regime leaders were retained as advisers.

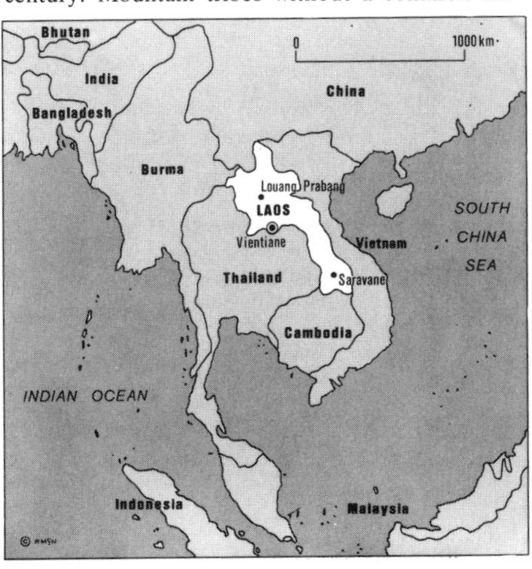

History. United in the 14th century under King Fa Ngum, Laos was the object of centuries-long invasions by neighbouring countries. Siam ruled the country in the 19th century, France from 1893 (when Laos was part of Indochina), Japan in World War II, and France again in 1946, when Laos was granted independence within the French Union. In 1953 Laos became a sovereign nation and there began a three-way struggle for power among Communists (Pathet Lao), right-wing factions and "Neutralists". An end to hostilities in 1961 was followed by the formation of a coalition government under the neutralist prime minister Prince Souvanna Phouma. In 1964 the Communist forces, with aid from Communist troops from North Vietnam, seized Laotian territory and in 1971 the United States bombed their supply line through Laos (the so-called Ho Chi Minh Trail). In 1973 a coalition government was formed with the Pathet Lao. In 1975 the coalition was dismantled and the Kingdom of a Million Elephants was succeeded by the Communist Pathet Lao regime, with Prince Souphanouvong as president and Kaysone Phomvihan as premier. Area: 236,800sq km (91,428sq miles). Pop. (1976 est.) 3,381,000. Map 52.

Latvia. *See* UNION OF SOVIET SOCIALIST REPUBLICS.

Lebanon (Al-Lubnan), official name Republic of Lebanon, is a small Middle Eastern nation bordering Israel and Syria. In recent years the country has been threatened by serious internal divisions. Although the official language is Arabic, Lebanon differs from other Arab countries in that about 40 per cent of the people cherish strong Christian traditions. The capital is Beirut.

Land and climate. Lebanon has three main regions: the narrow coastal area in the west, backed by a mountain range which parallels the coast; the interior plateau, including the fertile Bekaa (Biqa) valley; and the Anti-Lebanon mountains in the east. The western mountains include Lebanon's highest point – 3,088m (10,131ft) above sea-level east of Tripoli. The Bekaa region is watered by the River Litani, which flows south-westwards to the Mediterranean Sea, and the River Orontes, which flows northwards to Syria. The Mediterranean climate, generally with cool, moist winters and hot, dry summers, varies with altitude. Beirut on the coast has average temperatures between 16 and 38°C (61–100°F) and an average annual rainfall of 920mm (36in). The mountain regions are cooler and mostly wetter; in winter, snow covers many of the mountain slopes.

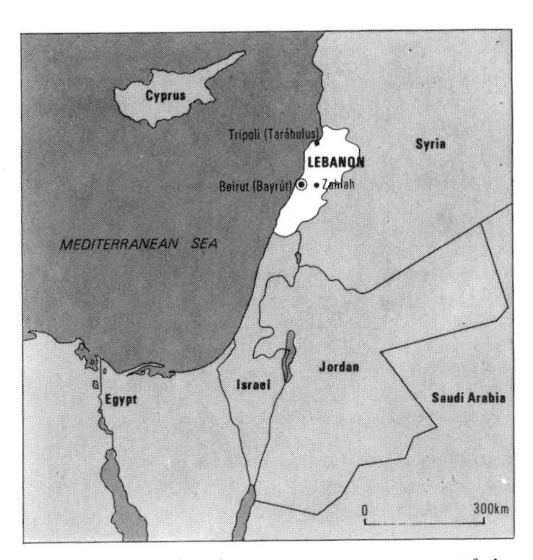

Economy. Lebanon is well known for the industry and ingenuity of its businessmen. Under settled conditions, trade, international finance, banking and tourism account for two-thirds of the gross national product. About half of the people are farmers and the most fertile regions are along the coast and in the Bekaa region. Fruits, vegetables and cereals are important crops and many goats are reared. The country lacks major mineral resources and most manufacturing utilizes imported raw materials. Oil pipelines from Iraq and Saudi Arabia terminate on Lebanon's Mediterannean coast.

People. About 90 per cent of the people are Arabs, although there are Armenian, Assyrian, Kurdish and other minorities. In 1976 Lebanon also had about 400,000 Palestinians. The largest Christian group is the Maronite. The Arabs belong to three Muslim sects – Sunni, Shi'ite and Druse. The literacy rate (86 per cent) is the highest in the Arab world.

Government. Lebanon is a republic with a president, cabinet and an elected Chamber of Deputies in which Christians and Muslims have had equal representation since 1976. Offices in the government are divided between the main religious groups.

History. The region was the homeland of Phoenician traders who flourished between the 12th and 9th centuries BC. Christian Monothelites (Maronites) colonized Lebanon in the AD 500s and, from the early Middle Ages. European (especially French) influences were strong in the region.

Modern Lebanon was created in 1920 when the League of Nations mandated France to govern it. In 1941 the country was proclaimed independent, and French troops were withdrawn in 1946. In the 1948 Arab-Israeli War Lebanon fought with other Arabs against the Israelis. After the war, many Palestinian

Corinthian columns of a 4th-century temple in Sur stand witness to the time when the city, then Tyre, was a prosperous port in the Roman empire.

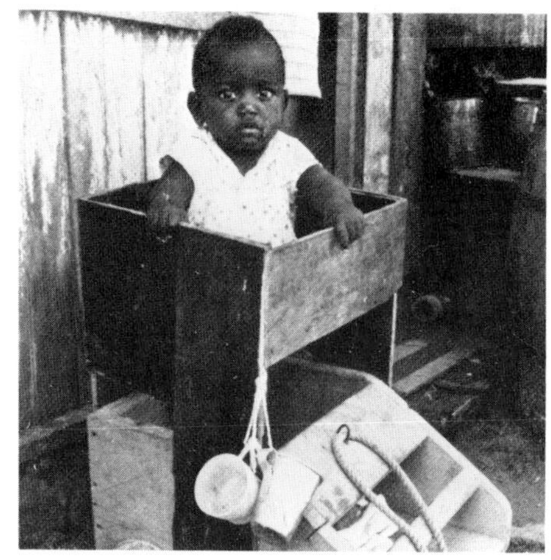

Lesotho is a poor African nation and women lucky enough to find work take their children with them – perhaps, as here, in a home-made play pen and pram.

Subsistence agriculture and ethnic customs continue to play their part in Liberian life, although modern Monrovia is dependent on a growing iron industry.

refugees settled in Lebanon. Their political views were alien to many Lebanese; fighting occurred between the Lebanese army and the Palestinians in 1969 and in 1972-73. During 1975 and 1976 about 45,000 people were killed in a civil war between private Christian and Muslim armies. The war forced the resignation of the Christian president Suleiman Franjieh, who was replaced by Elias Sarkis. Nearly half a million people fled the country. Maps 38, 39.

Lebanon – profile

Official name Republic of Lebanon
Area 10,400sq km (4,015sq miles)
Population (1975 est.) 2,869,000
Density 276 per sq km (715 per sq mile)
Chief cities Beirut (capital), 702,000; Tripoli, 175,000
Government Parliamentary republic
Religions Islam, Christianity
Language Arabic (official)
Monetary unit Lebanese pound
Gross national product (1974) £1,410,300,000
Per capita income £492
Agriculture Apples, cereals, citrus fruits, grapes, olives, tobacco, vegetables
Industries Cement, fertilizers, food products, leather goods, publishing, textiles, tobacco
Minerals Iron ore, lignite, limestone
Trading partners France, USA, Britain, West Germany

Leeward Islands are a group of West Indian islands in the Lesser Antilles, between the Atlantic Ocean and the Caribbean Sea. The major islands or island groups are the Virgin Islands of the United States, Guadeloupe, Anguilla, Antigua, Saint Kitts-Nevis, Montserrat and the British Virgin Islands, each of which has a separate article in this book. The islands of Saba and Saint Eustatius, in the NETHERLANDS ANTILLES, are also sometimes included in the Leeward group. *See also* WEST INDIES. Map 74.

Lesotho, official name Kingdom of Lesotho, was called Basutoland until 1966. This small African nation is surrounded by South Africa and Transkei, a self-governing Bantustan whose independence in 1976 was not recognized by Lesotho. But Lesotho's economic survival is largely dependent on South Africa. The capital is Maseru.

Land and climate. Most of Lesotho is 1,550m (5,085ft) above sea-level. It includes the lofty Drakensberg range, which is the rim of the interior plateau of southern Africa; the highest point is 3,482m (11,424ft). Flatter land occurs in the west and in the southern Orange River valley. The climate is warm and moist in summer and cold and dry in winter; rainfall varies with altitude.

Economy. Lesotho is one of Africa's poorest countries – the average annual income in 1974 was only about £60. Most of Lesotho's land is infertile and only 13 per cent of the country is farmed. Beans, maize, peas, sorghum and wheat are leading food crops, but food has also to be imported. The chief exports are cattle, mohair and wool; some diamonds are also exported, but there is a large adverse balance of trade. Manufacturing is on a small scale, but money sent home by the 40 per cent of Lesotho's male workers who are employed in South Africa is a major source of revenue.

People and government. Most people belong to the Basotho group. Small minorities include some South African whites, Coloureds and Asians. About 70 per cent of the people are Christians and Lesotho has a high literacy rate of about 85 per cent. In 1966 Lesotho became an independent constitutional monarchy, but the constitution was suspended in 1970. In 1974 a National Assembly was formed with nominated members. The head of state is King Moshoeshoe II and the prime minister is Chief Leabua Jonathan.

History. The Basotho are descended from refugees displaced by Zulu and Matabele wars who were united by Chief Moshoeshoe I in the 1820s. The territory came under British protection in 1868 and, as Basutoland, became a British protectorate in 1884. Since it became independent in 1966 its development has been hampered by economic problems, political conflict and strained relations with Botswana and Swaziland. Map 36.

Lesotho – profile

Official name Kingdom of Lesotho
Area 30,355sq km (11,720sq miles)
Population (1976 est.) 1,214,000
Chief city Maseru (capital) 15,000
Religion Christianity
Languages English, Sesotho (both official)
Monetary unit South African rand

Liberia, official name Republic of Liberia, is a nation in western Africa that was created by the American Colonization Society in the 1800s as a home for freed slaves. Although independent since 1847 it still retains close links with the United States – for example, the American dollar, at par with the Liberian dollar, has been in circulation since 1947. The capital is Monrovia.

Land and climate. The coastal plain is between 20 and 100km (12-62 miles) wide. Inland there are forested plateaus and grassy highland regions. The climate is hot and humid – temperatures average between 21 and 26°C (70-79°F) all the year round. The rainfall is heaviest on the coast, averaging between

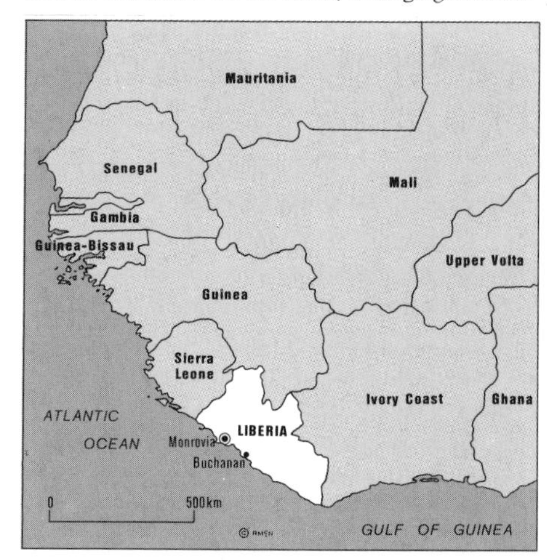

2,500 and 4,060mm (98-160in) per year. Inland the rainfall averages 1,780mm (70in) per year.

Economy. Until the early 1960s rubber dominated Liberia's economy. But by 1973 iron ore had become the chief product, accounting for 72 per cent of all exports, whereas rubber contributed only 14 per cent. Other exports include timber, diamonds, coffee, palm kernels and cocoa. The chief food crops are cassava and rice. In 1974 the average annual income was £167. One important source of revenue comes from Liberia's merchant navy. The ships are foreign-owned, but they are registered in Liberia because of its low taxation and lenient inspection policies.

People and government. Liberia has 16 main ethnic and language groups, the chief languages being Bassa and Kpelle. The official language, English, is spoken by the powerful Americo-Liberian minority. This group, numbering less than 50,000, are de-

Cyrenaica, one of three provinces which came together as independent Libya in 1951, is dotted with ruins from its days of Roman splendour.

The discovery of oil fields in the 1950s, like this one at Zelten in the Sahara, has made the petroleum industry the heart of Libya's economy.

The medieval castle of the princes of Liechtenstein, now an art gallery famed for its Dutch masterpieces, overlooks the nation's capital city, Vaduz.

scendants of freed slaves. Until recently they dominated the political and social life of the country. Today, however, attempts are being made to involve the indigenous people in government, although about 90 per cent of the people are illiterate. Liberia's constitution resembles that of the United States; executive power is divided between the president (William Tolbert, Jr), the Senate and the House of Representatives, and justice is administered by the Supreme Court.

History. The first settlement for freed slaves was at Cape Masurado (now Monrovia) in 1822. After Liberia became independent in 1847 its progress was hampered by internal disorder, hostility from European colonial powers and severe economic problems. Modern development began in 1926 when the American Firestone Rubber Company began operations there. By 1945 rubber accounted for more than 95 per cent of the exports. Recently the mining and export of iron ore to Japan and West Germany has enabled Liberia to make steady economic progress. Map 32.

Liberia – profile

Official name Republic of Liberia
Area 111,370sq km (43,000sq miles)
Population (1976 est.) 1,750,000
Chief city Monrovia (capital) (1974) 180,000
Religions Ethnic, Christianity, Islam
Language English (official)
Monetary unit Liberian dollar (= US dollar)

Libya (Lībiya), official name Libyan Arab Republic, is an independent nation on the northern coast of Africa. The capital is Tripoli. For centuries a possession of various Mediterranean countries, it gained independence in 1951 through action by the United Nations. Oil is its chief commodity, and its government is a military dictatorship.

Land and economy. Libya is a country of 95 per cent desert or semidesert, with a coastline 1,760km (1,100 miles) long on the Mediterranean Sea (to the north). The highest point is 2,286m (7,500ft) in the southern mountain area. Farming is possible only along the narrow coast, on the slopes of two northern hill areas and in a few oases. Libya has no permanent rivers; only 2 per cent of the land is arable, 4 per cent is used for grazing. The *ghibli*, a hot, dust-filled wind from the south, blows in the spring. Oil is Libya's main product; it dominates the economy, accounting for 99 per cent of all exports. Limited by meagre rainfall, Libya is not self-sufficient in foodstuffs, although subterranean sources of water are being tapped. Lack of skilled workers has limited industrial expansion.

People. Because of Libya's topographical features, 90 per cent of the people live on less than 10 per cent of the land, primarily in the coastal regions; 20 per cent live in the largest cities, Tripoli and Benghazi. The population is a mixture of Arab and Berber; nomadic or semi-nomadic tribes, the Tebou and Tuareg, live in the south. There are nearly 200,000 foreigners in Libya, most of whom are Egyptian. Islam is the dominant religion (Sunni Muslim) and Arabic is the official language. The literacy rate is 30-35 per cent. Under a provisional constitution, the government is operated by a 12-man Revolutionary Command Council (RCC). The chairman of the RCC is the chief of state.

History. Occupied in succession since ancient times by the Phoenicians, Greeks, Romans, Vandals and Ottoman Turks, Libya spent centuries under foreign rule. The region was an Italian possession from 1911 and in 1934 was formed as a colony by the joining of Tripolitania and Cyrenaica. It was one of the major battlegrounds between Allied and Axis forces in World War II. In 1947 King Idris I combined with the Allies and liberated the country. On 21 November 1949 a United Nations resolution called for Libyan independence as soon as possible, making it the first country to achieve such status through United Nations action. On 24 December 1951 it declared itself a constitutional monarchy under King Idris. A military coup overthrew his regime in 1969 and the new leader, Col. Muammar el-Gaddafi, abolished the monarchy and announced the formation of the Libyan Arab Republic. Its goal of confederation with other Arab states has not yet been realized, and relations with Egypt remain strained. Map 32.

Libya – profile

Official name Libyan Arab Republic
Area 1,759,540sq km (679,358sq miles)
Population (1975 est.) 2,444,000
Chief cities Tripoli (capital), 245,000; Benghazi, 140,000
Religion Islam
Language Arabic
Monetary unit Libyan dinar

Liechtenstein, official name Principality of Liechtenstein, is a small independent state on the eastern bank of the River Rhine between north-eastern Switzerland and western Austria. The capital is Vaduz. Liechtenstein is an alpine country with terraced slopes, suited for fruit trees, vines and dairy cattle; farmland on the fertile Rhine plain yields cereals and vegetables. Before World War II the economy depended primarily on agriculture; it now prospers with modern industries, including machinery, textiles, processed foods, furniture, pottery, pharmaceuticals, wine and touris.

The native inhabitants of Liechtenstein are descendants of the Alemanni, a German tribe. By the early 1970s, due to increased economic development and the need for labour, a third of the population consisted of foreigners. The state religion is Roman Catholicism, and German is the national language.

Liechtenstein is a constitutional monarchy with a democratic and parliamentary base. The constitution (1921, amended in 1972), calls for a parliament of 15 members. Switzerland maintains the country's postal, telephone and telegraph services, and handles diplomatic relations with other states; it uses Swiss currency. Social legislation is modelled on that of Switzerland, and education is compulsory up to the secondary school level. Liechtenstein has no army and a police force of only 40 men; Swiss frontier guards are stationed at the border in accordance with a customs treaty of 1924.

History. Liechtenstein was part of the Roman province of Rhaetia, which passed in the mid-5th century to a Germanic tribe. In 1396 the county of

Dependence on imports for its raw materials has kept Schaan, Liechtenstein's chief manufacturing community, free from industrial pollution.

Despite its diminutive size, the Grand Duchy of Luxembourg is a member of the UN, UNESCO, NATO and the EEC, and is a founder member of Benelux.

Malagasy, like many countries in east and south-east Africa, has a minority community of Asians, mostly immigrants from India and Indonesia.

Vaduz was placed under the suzerainty of the Holy Roman emperor; in 1699 the lordship was sold to the Austrian Prince Johann Adam von Liechtenstein, and in 1719 the territories were given the title Imperial Principality of Liechtenstein. Liechtenstein was a member of Napoleon's Confederation of the Rhine from 1806 to 1815, and then part of the German Confederation until 1866, when it achieved independence. Map 18.

Liechtenstein – profile

Official name Principality of Liechtenstein
Area 157sq km (61sq miles)
Population (1976 est.) 24,000
Chief city Vaduz (capital) (1975 est.) 3,900
Religion Roman Catholicism
Language German
Monetary unit Swiss franc

Lithuania. *See* UNION OF SOVIET SOCIALIST REPUBLICS.

Louisiana. *See* UNITED STATES.

Luxembourg (Letzeburg), official name the Grand Duchy of Luxembourg, is an independent state in western Europe. It is ruled by a hereditary monarch as the head of state and a premier as the head of government, who is reponsible to a Chamber of Deputies. The capital is Luxembourg.

The country is divided into two topographical regions: the heavily forested and elevated Ardennes plateau to the north (sometimes known as the *Oesling*) and the fertile Bon Pays in the south. The south-west is part of the rich Luxembourg-Lorraine iron-mining area, and Luxembourg is a major producer of iron and steel. Other industries include chemicals, cement, tanning, textiles, agriculture, wine, slate, tourism, postage stamps and banking. The people, strongly Roman Catholic, earn high wages and there is little inflation.

The government operates as a democratic parliament; the Council of Government headed by the prime minister is responsible to the Chamber of Deputies. All laws and decrees are brought before a 21-member Council of State, advisers appointed for life by the grand duke. Jean became grand duke in 1964, and is sovereign and chief of state; Gaston Thorn took office in 1974 as prime minister.

Founded in 963 as a fief of the Holy Roman Empire, the country was made a duchy by John of Luxembourg, King of Bohemia, in 1354. After occupation by France (1684-97), Spain (1697-1714) and Austria (1714), it was formally ceded to France by the Treaty of Campo Formio in 1797. It was made a grand duchy in 1815 and at the same time joined the German Confederation of the Rhine (with its fortress garrisoned by Prussians). Luxembourg's neutrality was confirmed by the London Conference (1867). During both world wars its neutrality was violated by German occupation, forcing Grand Duchess Charlotte to establish a government-in-exile in London. After liberation by Allied troops in 1944, Luxembourg's policy of neutrality was abolished and military service initiated (abolished in 1967). It became a member of the United Nations in 1946 and of NATO in 1949. Luxembourg signed a treaty in 1948 with Belgium and The Netherlands creating Benelux, which became fully effective in 1960. Map 20.

Luxembourg – profile

Official name Grand Duchy of Luxembourg
Area 2,586sq km (998sq miles)
Population (1975 est.) 357,000
Density 138 per sq km (358 per sq mile)
Chief cities Luxembourg (capital) 78,000, Esch-sur-Alzette, 28,000
Religion Roman Catholicism
Languages Letzeburgesch, French, German
Monetary unit Luxembourg franc
Gross national product (1974) £850,400,000
Per capita income £2,382

Macao (Macão, Macau) is a small Portuguese overseas province in south-eastern China, on the south China Sea 64km (40 miles) west of Hong Kong. The colony consists of the Macao peninsula and the islands of Taipa and Colôane. It was visited in 1497 by Vasco da Gama and the city of Macao settled by the Portuguese in 1557. It prospered in the 18th and 19th centuries as one of the two Chinese ports (the other was Canton) open to foreign trade. Macao was declared a free port and made independent by Portugal in 1849, but this status was not recognized by China until 1887. Following the establishment of China as a Communist country in 1949 thousands of refugees entered Macao, but entry has been denied them since 1967. The principal industries are fishing, textiles and tourism. Area: 15.5sq km (6sq miles). Pop. (1975 est.) 271,000. Map 48.

Madagascar. *See* MALAGASY.

Madeira. *See* PORTUGAL.

Maine. *See* UNITED STATES.

Majorca. *See* SPAIN.

Malagasy, official name Malagasy Democratic Republic and formerly known as Madagascar, is an island nation about 400km (250 miles) east of the African mainland. The island is still known geographically as Madagascar, but the nation is Malagasy. Its capital is Antananarivo (Tananarive).

Land and climate. The narrow, forested eastern coastal plain is hot and wet. Inland the central plateau covers about two-thirds of the country. This plateau, of savanna and steppe, has mild temperatures and an average annual rainfall of 1,000 to 2,030mm (39-80in). The western zone consists of low plateaus and plains. The north-west is wet, but the south-west is semidesert.

Economy. About 80 per cent of the people are farmers, although only 3 per cent of the land is cultivated. The chief exports are coffee, sugar, rice, vanilla, cloves and clove oil. Two-thirds of the country is given over to pasture, although the cattle are of poor quality. Mining is limited but manufacturing is expanding. The average annual income in 1974 was only £77 per person.

People. The people are descended from Africans (from the mainland) and Indonesians, who settled in the area more than 1,000 years ago. One of the two official languages, Malagasy, is of Indonesian

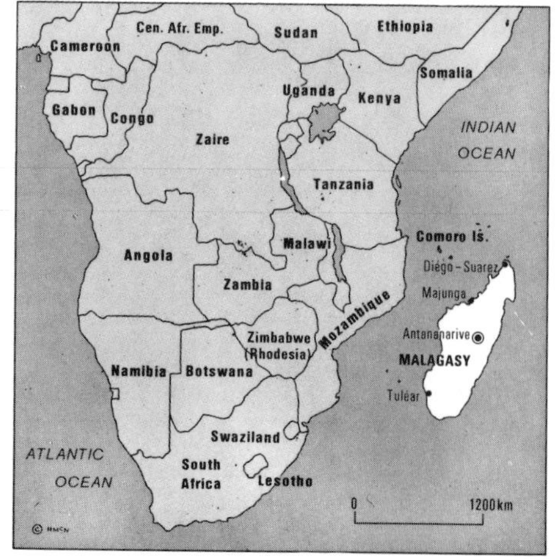

Although it occupies the fourth largest island in the world, Malagasy has no well organized fishing industry and only small boats are used.

Nearly all of Malawi's people are Bantu-speaking black Africans, most of whom work in agriculture raising cereal crops such as maize and millet.

A row of shops comprising a typical Malaysian street includes a coffin shop, a hairdresser's and a dentist-photographer's jostling for position.

origin. There are 18 ethnic groups, of which the largest is the Merina. More than 50 per cent of the people follow ethnic religions, about 35 per cent are Christians and 5 per cent are Muslims. The literacy rate is about 50 per cent.

Government. Under the constitution introduced in December 1975 the country is ruled by the president, Capt. Ratsiraka, the Supreme Revolutionary Council and a People's National Assembly.

History. The first Europeans to reach the island of Madagascar were Portuguese navigators in the 16th century. In the early 1800s, about two-thirds of the country was united under a Merina monarch. France declared the territory a protectorate in 1885. The country became fully independent in 1960; a military group took control in 1972. Capt. Didier Ratsiraka became president in 1975 and declared his support for revolutionary socialist policies. In 1976, however, he disbanded Power to the Underlings, a left-wing revolutionary group. Map 34.

Malagasy – profile

Official name Malagasy Democratic Republic
Area 587,045sq km (226,658sq miles)
Population (1975 est.) 8,044,000
Chief city Antananarivo (capital) (1975 est.) 438,000
Government Socialist republic
Religions Ethnic, Christianity, Islam
Languages French, Malagasy (both official)
Monetary unit Malagasy franc

Malawi, official name Republic of Malawi (including much of the territory formerly called Nyasaland), is a landlocked nation in east-central Africa. It was renamed Malawi (after Maravi, an African empire of the 1500s) when it became independent in 1964. The capital is Lilongwe.

Land and climate. More than 20 per cent of Malawi is covered by water. The African Rift Valley extends through the country and includes Lake Malawi (called Lake Nyasa in neighbouring Tanzania) and the Shire River valley in the south. Lake Chilwa in the south-east is an inland drainage basin. The western highlands are between 1,525 and 1,830m (5,000-6,000ft) above sea-level. Malawi's highest peak, Mt Mlanje (3,000m; 9,843ft), is in the highlands east of the Shire valley. The great variation in altitude gives rise to a wide variety of physical and climatic conditions; for example, the lowlands are hot and humid, whereas the highlands are cooler and well watered.

Economy. Malawi is a poor country — the average annual income in 1974 was only £56. Most people are farmers, and crops such as tobacco and tea account for nearly 80 per cent of all exports. Mining is unimportant but manufacturing is expanding. Many workers seek employment in other countries, especially in South Africa, Zambia and Zimbabwe (Rhodesia). The number of migrant workers probably exceeds 300,000, and they supply valuable revenue for Malawi.

People and government. Most of the people are Bantu-speaking black Africans, including the Tumbuka in the north, the Nyanja-Chewa in the centre and south, and the Lomwe and Yao in the south-east. Minorities include about 11,000 Asians and 7,000 Europeans. Malawi is a one-party state. The president, Dr Hastings Kamuzu Banda, is also head of the government and there is an elected National Assembly.

History. There was a Malawi kingdom from the 15th to the late 18th centuries which, at its height, conquered much of modern Mozambique and Zimbabwe (Rhodesia). In 1891 Britain established the British Central African Protectorate, renaming it Nyasaland in 1907. Between 1953 and 1963 Nyasaland was federated with Northern Rhodesia (Zambia) and Southern Rhodesia (Zimbabwe). The federation broke up because the Africans feared domination by the Europeans in Southern Rhodesia. Nyasaland became independent as Malawi in 1964 and was made a republic in 1966. In 1971 Dr Banda was made president for life. Map 34.

Malawi – profile

Official name Republic of Malawi
Area 118,484sq km (45,747sq miles)
Population (1975 est.) 5,044,000
Chief cities Lilongwe (capital), 87,000; Blantyre-Limbe, 181,000; Zomba, 20,000
Religions Ethnic, Christianity, Islam
Languages English (official), ciNyanja, ciTumbuku
Monetary unit Kwacha

Malaysia, a country in south-eastern Asia, is composed of the member states of the former Federation of Malaya (now sometimes called West or Peninsular Malaysia) together with the former Borneo territories of Sarawak and Sabah (now East Malaysia). All these territories were once ruled or controlled by Britain. Malaysia is still a member of the Commonwealth and is one of the most prosperous countries in Asia, much of its wealth coming from rubber and tin. It has not yet succeeded in removing the mutual suspicion and mistrust of its two largest groups of citizens, the Malays and the Chinese, and memories still linger of the protracted and costly civil war of the late 1940s and the 1950s, when government forces fought Chinese Communist guerrillas in the Malayan jungles. The capital is Kuala Lumpur.

Land and climate. Peninsular Malaysia occupies most of the Malay Peninsula. At its centre is a heavily forested mountain mass made up of short, parallel ranges; the mountains are flanked by plains extending to the coast. Many rivers flow from the mountains, the largest being the Perak, the Pahang and the Kelantan. In East Malaysia narrow coastal plains give way in the interior to sparsely populated regions of mountains and thick rain forest. Mt Kinabalu (4,101m; 13,455ft) is Malaysia's highest mountain; it rises in Sabah, the most northerly part of East Malaysia. Most towns and other settlements are on the coastal plains, but there are also villages along the river valleys, principally in those of the Rajang in Sarawak and the Kinabatangan in Sabah. The tropical maritime climate of Malaysia is strongly affected by the monsoons, but nearly all parts of the country are hot and humid. The highest temperatures occur on the coasts, where they vary from a minimum of 22-25°C (72-76°F) to a maximum of 31-33°C (88-92°F). Rainfall averages between 152 and 406mm (60-160in) and is highest in Perak in Peninsular Malaysia.

Economy. Peninsular Malaysia is predominantly agricultural; the chief crop is rubber, which is grown on more than 60 per cent of the cultivatable land. Malaysia is one of the world's leading producers of natural rubber. Rubber plantations were once large commercial undertakings, but recently an increasing number of peasant farmers have also planted rubber trees [*See* MM p.54]. Another valuable cash crop is palm oil, of which Malaysia is the largest producer and leading exporter. Although much farming land (about one-fifth of the total) is devoted to rice growing, the yields are relatively poor. Other crops include tea, sago, coconuts, pepper, coffee and tobacco.

Malaysia is the world's largest producer of tin ore (cassiterite), found chiefly in Perak and Selangor in the western part of the peninsula. Mining of iron ore and bauxite are also important, petroleum has been found in Sarawak, and some gold is extracted. Another major natural resource is timber. It is especially important to the economy of East Malaysia, which has little land suitable for farming.

A Sunday market draws potential buyers from all the various ethnic communities to be found in Sabah, a Malaysian state on the island of Borneo.

Children play in a stream as a farmer leads water buffaloes towards the paddy fields, in the dense tropical vegetation of Malaysia.

Rubber, tapped from the tree's bark, is one of Malaysia's principal exports; rubber trees (from South America) were introduced in the late 1800s.

Manufacturing industries have expanded since independence, but most of them are in Peninsular Malaysia and are still concerned with the processing of Malaysia's mineral and agricultural raw materials – tin, rubber, copra, and so on. But there is also an increasing number of other industries, chiefly the manufacture of machinery and chemicals. Communications are well developed on the western side of the peninsula, near the chief commercial cities. Port Swettenham (near Kelang) and Penang are important and busy seaports. Malaysian Airlines System (MAS) operates domestic routes, and flies to other Asian countries, Britain and Australia.

People. Population is unevenly distributed throughout the federation: East Malaysia, with 60 per cent of the country's land, has only 20 per cent of its people. In Sarawak the Dyaks are the largest ethnic group but there are also many Chinese. Sabah has Chinese and Malay minorities, but most of its people belong to such tribes as the Kadazans, Bajuas and Bruneis. In Peninsular Malaysia about half the people are Malays, who are the dominant community in the federation, 34 per cent are Chinese and 9 per cent Indian. The two latter communities are descended from immigrant workers who settled in Malaya in the late 19th and early 20th centuries. Many Chinese work in business and the professions.

The national language is Bahasa Malaysia (Malay); in Sarawak English is also recognized as an official language, but its use is to be reconsidered. Many Chinese and Indian people use their own languages as well as Bahasa Malaysia. Iban, the language of the Sea Dyaks, closely resembles Bahasa Malaysia. Only a minority of the population is literate; the literacy rate is highest in the cities of the peninsula, where there are the best educational facilities. The several institutions of university status include the University of Malaya at Kuala Lumpur and the University of Science at Penang. Islam is the official religion and is the faith of more than half the population, chiefly the Malays. The Constitution safeguards religious freedom and there are about 2,500,000 Buddhists in the Chinese community and nearly 800,000 Hindus.

Most of the important cities, including Kuala Lumpur, are in the western lowlands of the peninsula. Although the Chinese community is predominantly urban, about 60 per cent of the population lives in *kompongs* or other villages. Most village houses are made of wood and many are built on stilts. In Sarawak some villages have traditional Dyak longhouses in which several families live.

Government. The country is a federal parliamentary democracy composed of 13 states: Sarawak, Sabah and the 11 peninsular states of Johore, Kedah, Kelantan, Malacca, Negri Sembilan, Pahang, Penang, Perak, Perlis, Selangor and Trengganu. Kuala Lumpur is a federal territory. Malaysia's head of state is the *Yang di-Pertuan Agong* (Supreme Head of the Federation), who is elected for a term of five years from among the rulers of the princely Malay states. The legislature has two houses – the *Dewan Negara* (Senate) and the *Dewan Rakyat* (House of Representatives).

History. For about 500 years until the 13th century the Malay Peninsula was ruled from Sumatra as part of the Hindu and Buddhist kingdom of Sri Vijaya. Later it was ruled by Javanese and Thais and then by the Malay rulers of Malacca. In the 15th century Islam was introduced into Malacca and it spread throughout the peninsula. The Portuguese took Malacca in 1511, and a period of trade rivalry followed between them, the Dutch and the British East India Company.

Starting with settlements in Penang (1786) and Singapore (1824), the British gradually established their dominance commercially and politically, although not without a struggle with Thailand, which claimed rights in several Malay states. The British colony of Straits Settlements was established in 1826 and in 1914 Johore, the last Malay state to hold out, accepted British rule.

During World War II Malaya was occupied by the Japanese: the strongest opposition to the new invaders came from Malayan Communist guerrillas. In 1948, three years after the end of the war, the guerrillas staged a revolt against the government of the Federation of Malaya, which the British had formed as a stage on the road to independence. "The Emergency", lasted until 1960.

The federation became independent in 1957 under the leadership of Tunku Abdul Rahman. In 1963 the larger Federation of Malaysia was formed by the addition of the states of Sarawak, North Borneo (Sabah) and Singapore; SINGAPORE, however, seceded from the federation in 1965. The creation of Malaysia led to a confrontation with the Philippines and INDONESIA, whose president, Achmad Sukarno, declared his intention to destroy it and unsuccessfully attempted invasion.

In 1969 the mutual mistrust of the Malay and Chinese communities erupted into violent communal riots; order was eventually restored by a National Operations Council under Tun Abdul Razak. In the following year Razak succeeded Abdul Rahman. Razak's government set out to improve relations between Malaysia and its neighbours, and also moved the country to a more neutral position in relation to the great Communist powers, China and the USSR. Razak died while he was on a visit to London in 1976 and was succeeded by Datuk Hussein bin Onn. Map 50.

Malaysia – profile

Official name Malaysia
Area 329,740sq km (127,313sq miles)
Population (1975 est.) 11,900,000
Density 36 per sq km (93 per sq mile)
Chief cities Kuala Lumpur (capital) (1975 est.) 557,000; Penang (George Town) 270,000; Ipoh 247,700
Government Federal parliamentary democracy
Religions Islam, Buddhism
Language Malay (official)
Monetary unit Malaysian dollar
Gross national product (1974) £3,252,100,000
Per capita income £273
Agriculture Rubber, palm oil, rice, tea, sago, coconuts, pepper, coffee, tobacco
Industries Foodstuffs, steel, machinery, fertilizers
Minerals Ores of tin, iron, aluminium, petroleum
Trading partners Japan, Singapore, EEC, USA, USSR, China, Saudi Arabia, Thailand

Maldives (Diveh Raajje), formerly Maldive Islands, is an independent island republic in the Indian Ocean, about 645km (400 miles) west of Sri Lanka. The country consists of 12 groups of islands together totalling nearly 2,000 coral islands and islets, of which about 220 are inhabited. The main island, and capital, is Male. Maldives has a hot, humid climate. Its economy is based almost entirely on fishing, with bonita as the chief export; tourism is being promoted as an additional source of income. Most of the people are Muslims of Arab or Aryan descent living at subsistence level. The main language is Diveli (similar to Sinhalese), although some people speak Arabic. The Portuguese, the first Europeans to arrive, ruled the area from 1558 to 1573. The Dutch gained control in the 17th century and were ousted by the British who took possession from 1887 until independence in 1965. The republic was proclaimed in 1968. Area: 298sq km (115sq miles). Pop. (1975 est.) 119,000. Map 2.

Much of Mali is arid desert which will support no agriculture or livestock; even in more fertile areas only subsistence farming is largely practised.

The interior of Malta is rocky, with little productive land and no trees; even so the growing of cereal crops and vegetables is the main industry.

Iron ore is the chief resource and export of Mauritania. Most of the mines are owned by foreign interests (principally France and Britain).

Mali, official name Republic of Mali, is a poor, landlocked nation in western Africa. It was called French Sudan until it became independent in 1960, when it was renamed after a great medieval African empire. The capital is Bamako.

Most of the country is flat, hot and arid. North of Timbuktu the land merges into the Sahara. Only the south, which is drained by the River Niger, has sufficient rain for farming and even this region suffers from extensive droughts (as in 1974). About half of the exports come from livestock, although fish, cotton, oil seeds and groundnuts are also important. The average annual income — £34 in 1974 — is one of the world's lowest. Negroid peoples live in the south and Tuareg nomads roam the arid north. The chief religion is Islam and the official language is French.

The medieval Mali Empire was succeeded by the equally powerful Songhai Empire of Gao. Moroccan invaders broke it up in 1590 and the region ex-

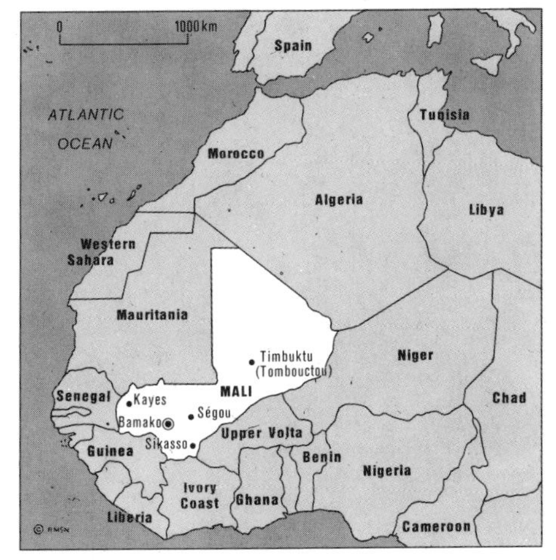

isted as separate states until unified under Islam in the 19th century. But by the end of the century the country had been taken by the French and incorporated into French West Africa. After gaining independence from France in 1960, Mali suffered severe economic problems and political unrest. A military group, led by Lt. Moussa Traoré, took control in 1968. President Traoré rules with a National Liberation Committee. A constitution intended to restore civilian rule in 1979 was approved in 1974. Area: 1,240,000sq km (478,764sq miles). Pop. (1976 est.) 5,842,000. Map 32.

Malta, official name Sovereign State of Malta, is an independent island republic in the Mediterranean Sea, about 96km (60 miles) south of Sicily. It includes the islands of Malta, Gozo and Comino; the capital, Valletta, is on Malta. Until the 1960s Malta, with its airport and good harbours, had great strategic value for Britain, under whose rule it functioned until independence in 1964. The islands are overpopulated and in recent years there has been severe unemployment.

Malta has a fairly flat terrain, no rivers and little rainfall. The hillslopes are terraced and 25 per cent of the population works in agriculture growing such crops as potatoes, wheat, onions, beans, oranges, cotton, grapes and cumin seeds; livestock raising is also important. Industries include tourism, ship repairing and the manufacture of lace, rubber products, buttons, gloves, hosiery and textiles. The make-up of the population reflects the many peoples who have inhabited the islands; the main languages are English, Italian and Maltese.

Formerly a Phoenician and Carthaginian colony, Malta was taken by the Romans in 218 BC. In 1530 it was given to the Knights of St John (Knights Hospitallers) by the Hapsburg emperor Charles V. It was held by Napoleon from 1798 until 1800, when it was taken by the British. In the 19th century Malta prospered as a principal British naval base and, as a British Crown colony, suffered great hardship when besieged by Axis forces during World War II. In 1942 King George VI awarded Malta's whole population the George Cross.

In 1962 a new constitution allowed for a 50-member legislature voted by proportional representation. Defence and foreign affairs were controlled by the British governor and a high commissioner. Malta became a fully independent member of the Commonwealth in 1964 as a constitutional monarchy under the British Crown; it became a republic in 1974. It underwent economic decline in the 1970s with the running-down of British naval interests; the country (under Prime Minister Dom Mintoff) planned to be independent of rental payments for the naval bases by 1979. Area: 316sq km (122sq miles). Pop. (1976 est.) 301,000. Map 24.

Man, Isle of. *See* ISLE OF MAN.

Manitoba. *See* CANADA.

Martinique is an overseas French département in the Windward Islands, West Indies, made up of a volcanic island which is the largest in the Lesser Antilles. Fort-de-France is the capital and chief trade centre. The highest peak of the rugged northern mountains is Mont Pelée, an active volcano 1,397m (4,583ft) high which erupted with great violence in 1902. Rain forests in the north slope down to plains and coastal valleys where farmers grow crops such as sugar cane, pineapples and tobacco. Agriculture and tourism are the chief industries. Most of the people, who speak French, are Negroes or mulattos (of mixed African and European descent). Martinique was discovered in 1502 by Christopher Columbus and inhabited by Carib Indians until they were displaced by French settlers after 1635. It became a French département in 1946. The main exports are sugar, rum, fruits, cocoa, tobacco and vegetables. Area: 1,100sq km (425sq miles). Pop. (1975 est.) 363,000. Map 74.

Maryland. *See* UNITED STATES.

Massachusetts. *See* UNITED STATES.

Mauritania (Mauritanie), official name Islamic Republic of Mauritania, is a country in western Africa. It has borders with Algeria, Mali, Senegal and Western Sahara. The capital, founded after independence, is Nouakchott (pop. 103,500).

Sandy Saharan plateaus cover much of the land and the chief farm region is the Senegal River plain in the south-west. The rainfall is low, decreasing from south to north. Most people work on the land. Fishing is also important, although the chief resource and export is iron ore.

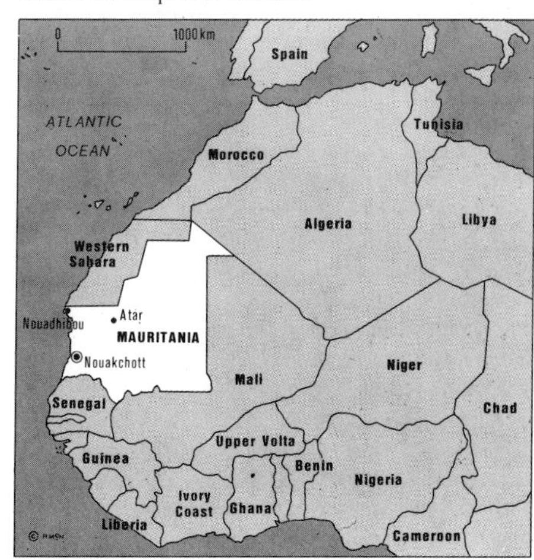

Ixtacihuatl, a dormant volcano in central Mexico, last erupted in 1868. It has three summits, the highest of which reaches 5,286m (17,352ft).

Glazed tilings, characteristic of the Spanish colonial style, are to be seen even in parts of the architecturally ultra-modern Mexico City.

Taxco, with its cobbled streets, gabled roofs and white adobe walls, is a fine example of a Spanish colonial town; modern buildings are banned.

The people are Muslims, 80 per cent of whom are of Arab-Berber descent, the others being Black Africans. Arabic is the official language and the literacy rate is 12 per cent. Mokhtar Ould Daddah is president and the head of the government.

France ruled Mauritania from 1903 to 1960, when it became an independent republic. In 1976 Mauritania took control of the southern third of WESTERN SAHARA (formerly Spanish Sahara) to the north, but guerrillas – members of the Popular Front for the Liberation of Saharan Territories (Polisario) – opposed partition of the country. Area (excluding part of Western Sahara): 1,030,700sq km (397,953sq miles). Pop. (1975 est.) 1,318,000.

Mauritius is an independent island nation in the Indian Ocean about 805km (500 miles) east of Malagasy. Rodriguez island and the Agalega and Cargados Carajos groups are dependencies. The capital is Port Louis. Mauritius is a member of the

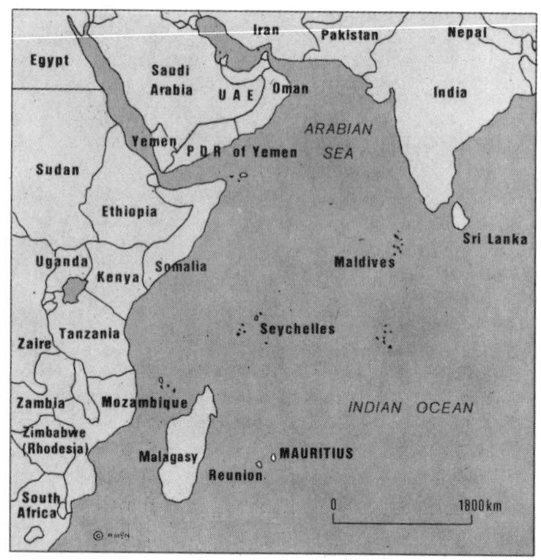

Commonwealth, with a governor-general and a 70-member legislative assembly. Sugar, the principal product, accounts for about 90 per cent of the export trade, with tea as a secondary crop. Most of the people speak English and belong to one of four ethnic groups: Indian, European, Malagasy and Chinese. The official languages are French and English. Mauritius was first visited in 1510 by the Portuguese, and the Dutch twice tried to establish settlements there. Permanent settlement was made by the French in 1722, and in 1810 the island was captured by the British, who controlled it until it gained independence in 1968. Area: 2,046sq km (790sq miles). Pop. (1975 est.) 899,000. Map 34.

Mexico (México), official name United Mexican States, is a republic in North America. Once the home of the great Maya and Aztec civilizations, and then a colony of Spain for 300 years, it won its independence in 1821. Today it is one of the most rapidly developing Latin American nations. But with a growing population (which exceeded 62 million in 1976), it faces severe economic and political problems, and many Mexicans still exist at subsistence level. The capital is Mexico City (Ciudad de México) which, at 2,380m (7,808ft) above sea-level, is one of the highest capitals in the world.

Land and climate. Mexico's land mass is a southwards extension of the North American cordilleras mountain system. This gives rise to an extremely varied terrain, including high mountains, broad plateaus, deep valleys, jungle lowlands and coastal plains. In general terms the country consists of a large central plateau – the *altiplano* – which rises from 900m (2,950ft) above sea-level in the north to 2,400m (7,875ft) in the south, and is bordered on all sides except the north by ranges of the Sierra Madre mountains. There are several active volcanoes, including Popocatépetl (5,452m; 17,887ft) and Ixtacihuatl (5,286m; 17,352ft), whose snow-capped peaks tower over Mexico City, and Citlaltépetl (Mt Orizaba) (5,700m; 18,701ft), Mexico's highest peak [*see also* PE p.31]. In the west is the long peninsula of Baja (Lower) California, a region of mountains and deserts [*see* PE p.*61*]. The Yucután Peninsula, in the south-east, is a low limestone plateau; it has no rivers, and there are tropical rain forests in the south. The chief river in Mexico is the Rio Grande (Río Bravo del Norte), which forms much of the border with the United States. Climate varies with altitude, being subtropical up to about 900m (2,950ft) then temperate up to 1,800m (5,900ft), above which it becomes alpine. Rainfall varies from 50mm (2in) in the north-west to 3,000mm (120in) in the jungles of Tabasco (south-east). Throughout the country, most rain falls in the summer.

Resources and economy. Only about 15 per cent of the land is cultivatable, and two-fifths of that requires artificial irrigation. Nevertheless 40 per cent of the working population is employed on the land and produces about a quarter of the national income. The introduction of modern farming methods is increasing production. The chief subsistence crops include maize and wheat; cotton, sugar, fruit, vegetables and coffee are leading exports. The Yucatán Peninsula produces more than half the world's sisal hemp and nearly all the chicle (the base for chewing-gum). Livestock, raised mainly on non-arable land, include cattle (28 million) and sheep (5 million). Mexico's vast forest resources are strictly controlled because of wasteful tree-felling in the past. They include some of the finest commercial woods, and resins and turpentine are also produced. Coastal fishing is on the increase, and shrimps are an important export.

Mexico is rich in minerals. It is one of the world's leading producers of silver, and there are considerable resources of natural gas, coal, copper, sulphur, gold, lead, uranium and zinc. The country became self-sufficient in oil in the mid-1970s. An industrial expansion programme trebled manufacturing exports in the first half of the 1970s, and 80 per cent of Mexico's consumer goods are now home-produced. The country's chief trading partner is the United States, which accounts for about half of both exports and imports. Revenue from tourism helps to reduce the large trade gap. Despite the difficult terrain, Mexico has good road and rail networks; it also has an excellent air service.

Government. Mexico is a federal republic divided into 31 states (each with its own constitution) and one federal district. The president of the republic has full executive powers and is elected for a maximum of one six-year term. Congress consists of a Senate (with 64 members elected for six-year terms) and a Chamber of Deputies (with about 200 members elected for three-year terms). The "official" political party, the Partido Revolucionario Institucional (PRI), has always won a majority of seats.

People. About 60 per cent of the people are mestizos (of mixed American Indian and Spanish descent), 30 per cent are Indians, and most of the remainder are of European descent. Spanish is the official language, although about a million Indians speak only their own languages. Mexico is predominantly Roman Catholic (about 97 per cent), and there are strict state laws regarding religion and the clergy. Education is free, secular, and compulsory (up to the age of 15) and is run entirely by the state, which spends on it about a sixth of the national budget. Illiteracy, which stood at 52 per cent in 1946, was reduced to 24 per cent in the following 25 years. There are more than 40 universities; the National Autonomous University of Mexico has nearly 100,000 students.

Mexicans have a fine tradition of art and architecture dating back to the time of ancient Indian civilizations. The Aztec and Maya built stone temples on flat-topped pyramids and decorated them with murals and sculptures. Many still stand near Mexico City and at Chichén Itzá, in Yucatán. The Spaniards built highly decorated churches, especially in the 18th century. Today's architects often combine ancient Indian designs with modern constructional methods. Some of the best-known Mexican paintings are the murals depicting the story of the revolution painted on public buildings in the 1920s.

The staple food of the Mexicans is maize, usually ground into meal and cooked as a *tortilla* (a thin pancake), either plain or in various forms with fillings. Kidney beans, squashes and rice are also major parts of the diet, and chillies give Mexican food its characteristic hot flavour. National sports are bullfighting and soccer, for which Mexico held the 1970 World Cup finals. Other popular sports include jai alai, baseball, swimming, diving, athletics and volleyball.

History. Before the arrival of the Spanish conquistadors in the early 16th century, Mexico was the home of a succession of American Indian cultures, including the Olmec (1200-100 BC), Zapotec (AD 1-300), Maya (flourished 300-900), Toltec (900-1200) and Aztec (from 1200) [*see* MS pp.*212, 214*]. A

Toluca is set in the fertile central Mexican plain. It is famous for its traditional craftsmanship in pottery, embroidery and basket-weaving.

The cathedral church of Santo Domingo in Oaxaca, Mexico, is a national monument that is lasting evidence of the Spanish influence in the country.

Monaco is located on the Mediterranean Riviera, between Nice and San Remo; the guaranteed good weather makes tourism the major industry.

Spanish expedition under Hernán Cortés landed in 1519 and founded Veracruz. The Spaniards killed thousands of Aztecs, destroyed their capital (Tenochtitlán) and took the empire.

The colony was named New Spain in 1535 and lasted 300 years. The Spanish-born people (*peninsulares*) retained power in the government and Church; the Creoles (people of Spanish ancestry born in the New World) held only minor posts; mestizos were free workers; and the exploited Indians fared little better than the Negro slaves imported from Africa.

The Mexican War of Independence, which began with a creole rising in 1810, was largely a guerrilla struggle which eventually resulted in independence in 1821. Mexico became a republic two years later. Torn by internal disputes for the next 50 years, the country lost half its territory (Texas, California, New Mexico and Arizona) to the United States (1845-48). Internal unrest, civil war and occupation by France (1863-67) plagued Mexico until a strong ruler emerged in Porfirio Díaz, a mestizo general. He seized power in 1876 and ruled as dictator, with one gap of four years, until deposed in 1911. He gave Mexico law and a certain amount of stability – but at the cost of political and social oppression.

After the overthrow of Díaz, a result of the Revolution of 1910, the revolutionaries continued to fight bitterly among themselves. Eventually Venustiano Carranza gained power, and a new constitution was proclaimed in 1917 giving the government control over education, the Church and all mineral resources; land was to be restored to the Indians and mestizos. The revolutionary programme was continued by succeeding presidents. The National Revolutionary Party, founded in 1929, was reorganized in 1938 and again in 1946 as the PRI. President Lázaro Cárdenas (1934-40) carried out vigorous land reforms (redistributing about 18 million hectares among the Indians), strongly supported the labour unions and, through a policy of compulsory nationalization, took over the foreign-controlled oil industry.

Mexico's modern industrial growth began in 1940 and has been the speediest in Latin America. But a rapidly increasing population has meant that half Mexico's people still live in poverty. President Luis Echevarría Álvarez (held office 1970-76) attempted to introduce a greater measure of democracy (under mounting pressure from the poor peasants and the student movement), but was opposed by conservative politicians and businessmen. When José López Portillo became president in December 1976 he also faced several other problems, including a massive trade deficit, mounting foreign debts, an ailing currency and the guerrilla activities of the Communist League (formed in 1973). Map 72.

Mexico – profile

Official name United Mexican States
Area 1,972,547sq km (761,600sq miles)
Population (1976 est.) 62,329,000
Density 32 per sq km (82 per sq mile)
Chief cities Mexico City (capital) (1975 est.) 8,591,750; Guadalajara, 1,560,805
Government Head of state President José López Portillo (elected 1976)
Religion Roman Catholicism
Languages Spanish, Indian languages
Monetary unit Peso
Gross national product (1974) £24,841,900,000
Per capita income £399
Agriculture Maize, wheat, sorghum, sugar cane, beans, fruits, cotton, livestock
Industries Petroleum, cement, iron and steel, other metals, coal, chemicals, fishing, tourism
Trading partners USA, West Germany, Japan

Michigan. *See* UNITED STATES.

Midway is a United States territory in the central Pacific Ocean, about 2,000km (1,243 miles) northwest of Hawaii. There are two small coral islands, Eastern and Sand, which were discovered by American navigators in 1859 and annexed to the United States in 1867. A civilian air base was built there in 1935 and an American naval base established in 1941. It was the scene of the Battle of Midway, an important Allied victory in which American carrier-borne aircraft defeated Japanese naval and air forces in June 1942. It is still an important military base. Area: 5sq km (2sq miles). Pop. (1975 est.) 2,000. Map 62.

Minnesota. *See* UNITED STATES.

Mississippi. *See* UNITED STATES.

Missouri. *See* UNITED STATES.

Moldavia. *See* UNION OF SOVIET SOCIALIST REPUBLICS.

Moluccas. *See* INDONESIA.

Monaco, official name Principality of Monaco, is a European sovereign state on the northern shore of the Mediterranean Sea near Italy and bordering on France. Monaco-Ville, the capital, occupies one geographical region; the others are the port district of La Condamine and the resort area of Monte Carlo. The chief source of income (55 per cent) is tourism, with light industry contributing a further 25-30 per cent. Native Monegasques are outnumbered eight to one by other nationalities, chiefly French who make up about half the population. Most of the people are Roman Catholics.

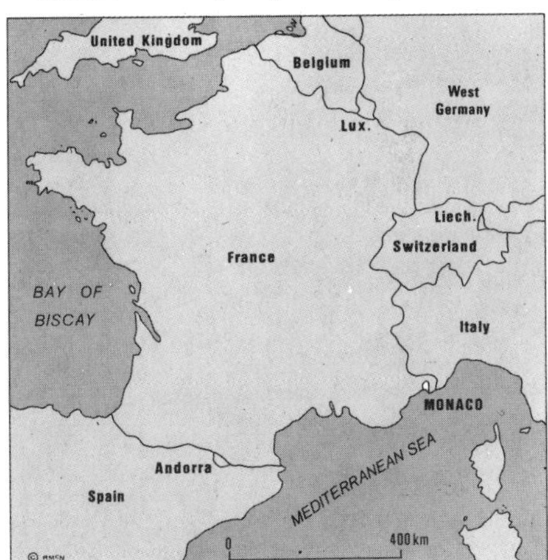

Monaco was originally settled by Phoenicians, was then taken by Rome and became christianized in about AD 100. It had several rulers before coming under the Grimaldi family of Genoa in the 13th century; it is now ruled by their French descendants. France assumed protection of Monaco in 1860 and should the ruler have no male heir, the principality would become an autonomous state of France. The present ruler, Prince Rainier III, married American actress Grace Kelly in 1956; in 1958 they had a son (Prince Albert). Area: 1.5sq km (0.6 sq miles). Pop. (1975 est.) 25,000. Map 20.

Mongolia (Mongol Ard Uls), also called Outer Mongolia and official name Mongolian People's Republic, is a landlocked nation of central Asia lying between China and the USSR. The capital is Ulan Bator. A vast plateau with extensive grasslands embraces the heart of the country; part of the Gobi Desert occupies the south and in the north the

Montserrat has some strikingly rugged scenery and the islanders make use of any suitable soil to raise crops, the chief of which is cotton.

The City Gate of Fez in Morocco is part of a former regional capital made up of an ancient city and a medieval city connected by walls.

There are few customers for the basket-sellers in northern Morocco during the winter months; the main tourist season is from February to April.

Altai Mountains rise to more than 4,000m (13,000ft). More than 80 per cent of the land is pasture and the agricultural economy relies mainly on herd animals; most people work as herdsmen on collective farms. Major minerals include coal, tungsten and copper. The country is governed by a nine-member Presidium (chairman Yumzhagiyen Tsedenbal) chosen from the elected People's Great Hural of Deputies.

In the 13th century, under Genghis Khan's leadership, Mongolia conquered most of Asia and much of Europe. The empire collapsed in the 14th century and came under Chinese rule. With Russian backing, Mongolia declared its independence in 1911 after the Chinese Revolution and became a republic in 1924 (not recognized by China until 1946). In 1966 a 20-year friendship treaty between Mongolia and the USSR reinforced their stand against China. Area: 1,565,000sq km (604,247sq miles). Pop. (1976 est.) 1,489,000. Map 44.

Montana. *See* UNITED STATES.

Montserrat is an island in the Leeward Islands, West Indies, between the Atlantic Ocean (to the east) and the Caribbean Sea. The capital and chief port is Plymouth. The land, of volcanic origin, is mountainous and intensively cultivated.The chief agricultural product is cotton, most of which is exported; some cereal crops are also grown. Montserrat was discovered in 1493 by Christopher Columbus and settled by the English in 1632. After being held briefly by France it was returned permanently to Britain in 1783. It became a member of the Leeward Islands colony and later of the Federation of the West Indies, but in 1966 rejected self-government and Montserrat is now part of the Caribbean Free Trade Association. A severe earthquake in October 1975 damaged much property. Area: 98sq km (38sq miles). Pop. (1975 est.) 13,000. Map 74.

Morocco (Al-Maghrebia), official name Kingdom of Morocco, is a country that occupies the north-western shoulder of Africa. It is one of the most urbanized of African nations and about 38 per cent of the people live in towns. The capital is Rabat.

Land and climate. The land is dominated by the rugged Atlas ranges – Africa's only true fold mountains apart from the Cape ranges of South Africa. The Atlas mountains are divided into three main regions. The northern Rif Atlas, bordering the Mediterranean Sea, reaches a height of 2,456m (8,058ft) above sea-level. To the south is the Middle Atlas and, farther south still, the High Atlas, which contains Morocco's highest peak, Jebel Toubkal (4,165m; 13,665ft). The Anti-Atlas lies to the south and east of the High Atlas; it is an uplifted rim of the Saharan plateau. Beyond the Anti-Atlas the land slopes down to the Sahara [*see also* PE p.*121*]. In central Morocco, plateaus separate the Atlas ranges from the narrow coastal plains. Other lowlands include the fertile Rharb-Sebou region, south-west of the Rif Atlas, and the Moulouya valley in the north-east.

Northern Morocco has a Mediterranean climate. Tangier has average temperatures of 11°C (52°F) in the coldest month and 29°C (84°F) in the warmest. The rainfall, most of which falls in winter, averages 810mm (32in) per year. To the south, the Atlantic coast is cooled in summer by the cold Canaries current. The central Moroccan coast has an average annual temperature range of 14 to 20°C (57-68°F). The rainfall averages only 330mm (13in) per year. The mountains are cooler and north- and west-facing slopes are wetter.

Economy. Agriculture is the chief activity. Leading export crops include citrus fruits, tomatoes, legumes and other vegetables. The chief food crops are cereals, particularly barley and wheat. Grapes are grown and wine is made. Pastoral farming is practised, especially in upland areas, and camels, cattle, horses, goats and sheep are reared. Forest products include cork and wood pulp. Sea fishing (mainly for sardines) is important, the chief centres being Agadir, Casablanca, Essaouira and Safi.

Morocco is the world's largest producer of phosphates, which account for more than a half of all exports. In 1976 it acquired the northern two-thirds of WESTERN SAHARA (formerly Spanish Sahara), to the south. This barren desert territory contains, at Bu Craa, the world's largest phosphate reserves. Other minerals are the ores of lead, iron, manganese and zinc, with some coal and petroleum. Manufacturing includes traditional handicrafts and the processing of agricultural products and minerals, although manufactured goods and fuels are major imports. Nearly 1½ million tourists visited Morocco in 1973.

People. Most Moroccans are Arabs, but Berbers (most of whom live in mountain regions) make up about 30 per cent of the population. The state religion is Islam and the official language is Arabic, although French and Spanish are commonly spoken in the cities and town. Europeans, mainly Frenchmen and Spaniards, once formed a substantial minority, but their numbers have declined. In the 1971 census only 146,000 foreigners were recorded in Morocco. The literacy rate is 15 per cent.

Government. The 1972 constitution vests final civic and religious authority in the king. The parliament has 240 deputies, 180 being elected by universal suffrage and 60 by local government electoral colleges. The king appoints ministers, approves legislation and may dissolve parliament.

History. Morocco came under successive Carthaginian, Roman, Vandal and Berber rule, until the Arabs converted the region to Islam in the early AD 700s. It became part of the great Moorish Empire which included the Iberian Peninsula. The present Alouite dynasty, which has ruled Morocco since 1649, claims descent from the Prophet Mohammed.

In 1912 France took most of Morocco, except for

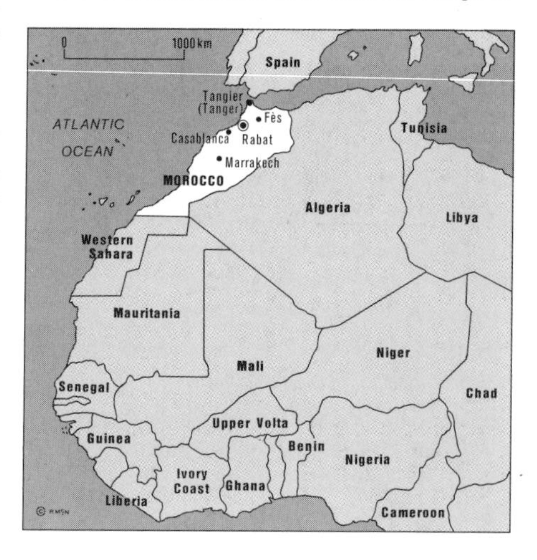

three areas ruled by Spain – Ifni; a desert strip in the far south; and the northern coast, not including Tangier, which became an international zone. Nationalists opposed foreign rule with growing force after World War II. Independence was finally achieved in 1956 for all of Morocco except for some Spanish areas (which joined Morocco later) and Ceuta and Melilla, which remain Spanish garrisons. Also in 1956 the international status of Tangier was ended. King Mohammed V died in 1961 and was succeeded by his son Hassan II. In 1963 fighting broke out along Morocco's disputed south-eastern border with Algeria.

The 1976 partition of Western Sahara was agreed by Spain, Mauritania and Morocco, in consultation with chiefs and community leaders in Western Sahara. Morocco has since been strongly opposed by the nationalist Popular Front for the Liberation of Saharan Territories (Polisario). Polisario claims

Morocco has some of the richest phosphate deposits in the world; mostly calcium phosphate, they are used as fertilizers and for making phosphorus.

The Zambezi River descends several spectacular waterfalls before flowing to the sea in Mozambique; the last 465km (290 miles) are navigable.

Mozambique achieved independence in mid-1975, since then the population has seen some violent changes; their love for tribal dancing will never change.

that Western Sahara should be an independent nation and has repeatedly launched guerrilla attacks against Moroccan bases there. In 1977 relations with Algeria deteriorated when Algeria supported Polisario guerrillas in Western Sahara. Map 32.

Morocco – profile

Official name Kingdom of Morocco
Area 458,730sq km (177,116sq miles), not including the part of Western Sahara taken in 1976
Population (1975) 17,305,000
Density 38 per sq km (98 per sq mile)
Chief cities Rabat (capital), 525,000; Casablanca, 1,395,000; Marrakech, 305,000
Government Monarchy
Religion Islam
Languages Arabic, Berber
Monetary unit Dirham
Gross national product (1974) £2,965,800,000
Per capita income £171
Agriculture Barley, wheat, fruits, olives, legumes, livestock
Industries Petrochemicals, cement, ceramics, fertilizers, processed food, leather goods, textiles, wine, tourism
Minerals Phosphates, coal, manganese, lead, iron ore, petroleum, zinc
Trading partners France, West Germany, Italy, Spain, USA

Mozambique (Moçambique), official name the People's Republic of Mozambique, is a nation in south-eastern Africa. It became independent in 1975 after a long guerrilla war between African nationalists and Portuguese troops. The capital and chief port, Maputo, was known as Lourenço Marques until 1976.

Land and climate. Mozambique is a long, Y-shaped country with the most extensive plains of any southern African nation. About 40 per cent of the land is less than 180m (590ft) above sea-level. The coastal plain is narrowest in the north and widest in the centre and south, where it is crossed by the Zambezi, Save and Limpopo rivers. Inland are plateaus with highlands along the border with Zimbabwe (Rhodesia). The northern interior upland is an extension of the Tanzanian plateau. Its climate is tropical, with average annual temperatures of 23°C (72°F) in the south and 27°C (81°F) in the north. Rainfall is greatest in the uplands. The coastline is well watered, but the rainfall diminishes rapidly inland and most of the broad coastal plain is fairly dry.

Economy. About 90 per cent of the people live by subsistence farming, and grow such food crops as cassava, maize and rice. Cattle-rearing is kept at a low level because of the presence of the ubiquitous tsetse fly, which causes the disease nagana in cattle (and sleeping sickness in man). The chief exports are cashew nuts, copra, cotton, sugar and tea. Mining is unimportant, except for some coal, but the country possesses great potential for hydro-electricity, especially from the Cabora Bassa dam being constructed on the Zambezi. At present there is only small-scale manufacturing in Beira and Maputo, although large-scale industrialization is planned. Invisible earnings are important to the economy. They include revenue from transit trade from Malawi, South Africa, Swaziland, Zaire, Zambia and, until the borders were closed in 1976, Zimbabwe. Substantial revenue also comes from workers who are employed abroad, especially in South Africa.

People. Most people are Bantu-speaking black Africans. In the north the main group is the Makua-Lomwe, and the Tonga predominate in the south. Nearly everyone follows ethnic religions, although

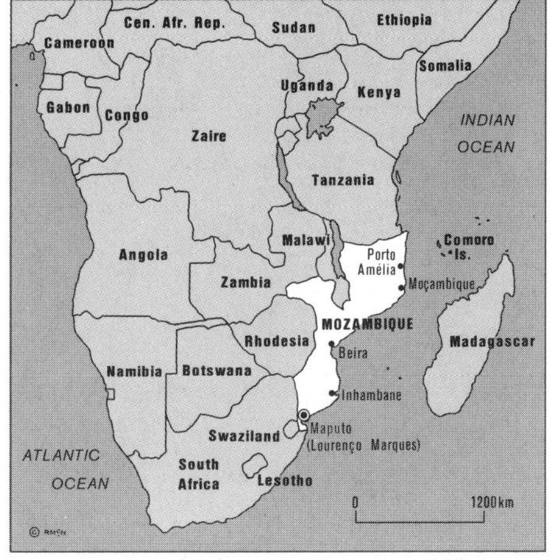

there are some Muslims in the north and Christians in the south. Almost all the 200,000 Portuguese left Mozambique in 1975. Most of the population is illiterate but, since independence, great efforts have been made to raise the literacy rate.

Government. Under the 1975 constitution, the president of the nationalist movement FRELIMO (Frente de Libertação de Moçambique) is president of the country. The constitution also made allowance for a 210-member People's Assembly. In early 1977, FRELIMO announced that elections for the Assembly would be held in February 1978.

History. Portuguese influence in the area began with the visit of Vasco da Gama in 1498 and was gradually extended over the next four centuries. In 1891 Mozambique's boundaries were settled by a treaty with Britain and in 1910 it was formally made a Portuguese colony. In 1951 Portugal designated Mozambique an overseas province. The Portuguese did not pursue segregationist policies, but to achieve full citizenship Africans had to obtain a Portuguese education and adopt a European lifestyle. Education was not free and opportunities were extremely limited. As a result, few Africans achieved citizenship and, in 1964, an armed rebellion began.

Following a coup in Portugal in 1974, the Portuguese quickly ended the war by granting Mozambique independence on 25 June 1975. FRELIMO took control and most Portuguese left the country. Led by President Samora Machel, the government launched a radical socialist programme of nationalization of hospitals, land and schools; the take-over of abandoned and rented properties; and the establishment of collectives, co-operatives and communal, self-help villages. Political education was undertaken by a network of groups intended to spread radical ideas. Mozambique also assisted Zimbabwean guerrillas (who were fighting Rhodesian troops) and, in March 1976, the country closed its borders with Zimbabwe. Economic contacts with South Africa continued, despite Mozambique's denunciations of its racial policies. Map 34.

Mozambique – profile

Official name People's Republic of Mozambique
Area 783,030sq km (302,328sq miles)
Population (1976 est.) 9,454,000
Density 12 per sq km (31 per sq mile)
Chief city Maputo (capital), 799,400
Government Socialist republic; head of state President Samora Machel
Religions Ethnic, Christianity, Islam
Languages Portuguese, Bantu languages
Monetary unit Conto
Gross national product (1974) £1,329,000,000
Per capita income £144
Agriculture Cashew nuts, cassava, copra, cotton, maize, rice, sisal, sugar, tea, tobacco
Industries Alcohol, cement, food, textiles
Minerals Coal, gold
Trading partners (1974) Portugal, South Africa, West Germany, USA, Britain

Namibia, also known as South West Africa, is a territory governed by South Africa. The name Namibia was adopted by the United Nations in 1968, although South Africa did not accept it. Since 1946 Namibia has been the centre of an international dispute concerning the legality of South Africa's control. In 1976 a constitutional committee in Namibia, which was approved by South Africa, announced that Namibia would be independent by the end of 1978. The capital is Windhoek.

Land and climate. The coastal region, averaging 130km (81 miles) in width, consists of the barren and uninhabited Namib Desert [*see* PE p.*47*]. The

Large areas of Namibia, including the region called the Kalahari sandveld, are arid desert that supports only stunted bushes, scrub and sparse grass.

The main square of Katmandu is overlooked by the 16th-century temple which gives the city its name (*kath* meaning 'wood', *mandir*, 'temple').

The canals of Amsterdam are perhaps the city's most famous feature, together with the astonishing number of bicycles during the rush hour.

land rises inland to the central highlands – the largest region. The highest point is 2,483m (8,146ft) above sea-level, near the capital Windhoek. North and east of the central highlands is the Kalahari bushveld – a semi desert. Namibia is one of the world's most arid countries. The Namib is almost rainless and the Kalahari sandveld has between 100 and 250mm (4-10in) of rain per year. The best-watered region is the central highland. The north has about 560mm (22in) of rain per year and the south 350mm (14in).

Economy. Cattle are reared in the northern part of the central highlands, but sheep are more important in the drier south – karakul pelts and meat being important exports. There is also some crop farming in the northern parts of the highlands. But Namibia's chief resources are minerals, including gem-quality diamonds (60 per cent of the mineral exports), lead, tin, uranium, vanadium and zinc. Namibia also has a sizeable fishing industry.

People and government. Non-whites form 88 per cent of the people, the Ovambo being the largest ethnic group. A few Bushmen, southern Africa's original inhabitants, still roam the Kalahari sandveld. But the whites are the chief minority, forming 12 per cent of the population. They control the economy and elect members to the local Assembly, although South Africa has final authority.

History. Namibia was first explored in the 15th century by Portuguese and Dutch expeditions, and the British and Germans built missions there in the 18th century. Between 1884 and World War I the territory was a German protectorate. Then in 1914-15 South African troops conquered the territory. In 1919 the League of Nations granted South Africa a mandate to administer Namibia. But after World War II the United Nations replaced the mandate system with the trusteeship agreements. South Africa, alone among the United Nations, did not accept this change. From the 1950s many of South Africa's racial laws were applied in Namibia, despite protests from the United Nations. In 1971 the International Court of Justice ruled that South Africa's administration of Namibia was illegal.

In 1975 a multi-racial constitutional committee, including representatives of all ethnic groups, was set up in Namibia to discuss the future. The nationalist party SWAPO (South West African People's Organization), which the United Nations recognized as representative of the majority of the people, refused to join the committee. SWAPO argued that it had been set up to ensure continuing South African rule. In 1976 the committee announced that Namibia should become independent by the end of 1978. But guerrilla activity, which had begun in 1966, continued. Map 36.

Namibia – profile

Official name Namibia
Area 823,327sq km (317,887sq miles)
Population (1975 est.) 883,000
Chief city Windhoek (capital) 77,000
Religions Ethnic, Christianity
Languages Afrikaans, English, German (all official)
Monetary unit South African rand

Natal. *See* SOUTH AFRICA.

Nauru (Naoero), formerly Pleasant Island, is an island republic of the Commonwealth in the western Pacific Ocean, south of the Equator and west of the Gilbert Islands. The economy is based on extensive deposits of phosphates (discovered in 1900), which are exported mainly to Australia and New Zealand. Most of the native inhabitants are Polynesians, of Micronesian and Melanesian descent. Nauru was discovered in 1798 by the British navigator John Hunter. Annexed by Germany in 1888, it came under League of Nations mandate to Australia after World War I. It was occupied by the Japanese in World War II, after which it was made a trusteeship of the United Nations until 1968, when it became the world's smallest independent republic. Area: 21sq km (8sq miles). Pop. (1975 est.) 8,000. Map 62.

Nebraska. *See* UNITED STATES.

Nepal is an independent kingdom in central Asia between China (to the north) and India. One frontier cuts through Mt Everest (8,848m; 29,030ft), the highest mountain in the world. The capital is Katmandu. The country comprises three major regions: a central lowland region with arable land, called the Terai; a central mountainous area, including the populated Katmandu valley; and a high mountainous region in the north which extends to the Himalayas [*see* PE p.*52*]. Two-thirds of the nation's income is from agriculture; less than 1 per cent of the population works in industry. The people come from Mongoloid and Indian backgrounds.

In early times Nepal was a collection of independent principalities, such as Gurkha, established west of Katmandu in about the 10th century by Rajput warriors. By the 18th century they were united under one rule. The Chinese checked the expansion of the Gurkhas in 1792 and in 1814-16 Nepal clashed with Britain over the positioning of the Nepal-India border. In 1846 the Rana family assumed power and its descendants continued to rule until the revolution of 1951. A constitution was drawn up in 1959, followed by a return to Rana rule. In 1962 King Mahendra (reigned 1952-72) dissolved parliament and assumed absolute power, ruling through various councils responsible only to the king. Mahendra was succeeded by his son, Prince Birendra, in 1972. Area: 140,797sq km (54,362sq miles). Pop. (1976 est.) 12,904,000. Map 40.

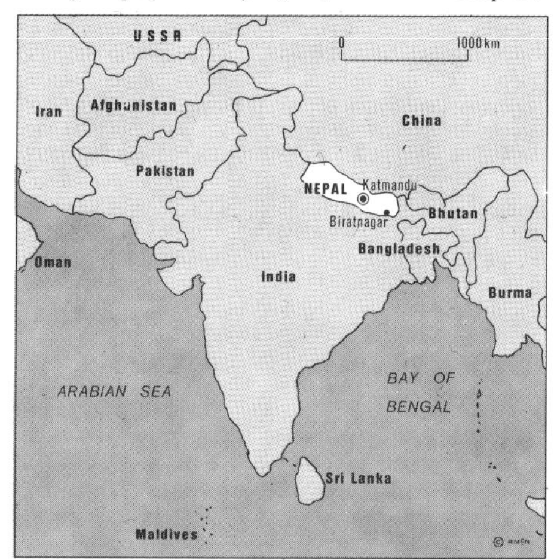

Netherlands, official name Kingdom of The Netherlands, is a small country in north-western Europe. Its name means "the low country", and it is one of the lowest-lying and flattest countries in Europe. About two-fifths of the land is below sea-level, protected from the encroachment of the sea by more than 2,000km (1,245 miles) of dykes and dams. The Netherlands is one of Western Europe's most prosperous and advanced countries – among the continent's leaders both in agriculture and in industry. It is joined with Belgium and Luxembourg in the Benelux Customs Union and is a member of the EEC. The country is often called *Holland*, a name that properly belongs only to two of its western provinces. The capital is Amsterdam, although the seat of government is The Hague ('s-Gravenhage).

Land and climate. The long low coastline is fringed with sand dunes, many of which have been reinforced over the centuries to afford protection from

Dykes and artificial drainage have turned the Zuider Zee, the great arm of water that used to cut into The Netherlands, into reclaimed land.

Rotterdam's Europoort is a complex of quays, container warehouses and harbour basins, that handle more than 250 million tonnes of cargo a year.

Much of Amsterdam is built on reclaimed land; building space is at a premium and so many people have their homes, or even businesses, afloat.

the sea. Off the northern part of the coast a chain of narrow islands, the West Frisian (or Wadden) Islands, enclose the shallow Wadden Sea, part of the former Zuider Zee. In the south-west are the islands and peninsulas of Zeeland. This "delta region" encompasses the mouths of three great rivers – the Rhine, Maas and Schelde. Behind the central coastal strip is Holland (now divided into the provinces of North and South Holland). Much of Holland, and much of Friesland and Groningen in the north, is *polder* – land formerly under the sea, swamp or fresh-water lakes, that has been reclaimed by the building of dykes [*see* PE p.*155*]. The largest area of polder is around the IJsselmeer, a large artificially created fresh-water lake. The IJsselmeer was formerly part of the Zuider Zee, a great bay that was cut off from the sea (1927-32) by the building of a 32km (20-mile) dam, the Afsluitdijk, across its mouth between North Holland and Friesland. Further dykes were built within the Zuider Zee to reclaim areas of polder. The deeper central part has been left as the IJsselmeer; it is fed by the River IJssel. *See also* PE p.*44*.

The eastern part of the country is somewhat higher than the rest, rising in places to about 300m (1,000ft). It is a region of forests, sandy ridges and peat bogs. In many places the generally infertile soil is being improved to make it productive farmland.

The climate throughout the country is moderate, with warm summers and mild winters. Summer temperatures average between 16 and 18°C (60-65°F) and in winter temperatures average –1°C (30°F). The average precipitation (rain and snow) is about 850mm (33in). The weather tends to be changeable, and the western regions are known for their frequent high winds.

Economy. Dutch agriculture is extremely efficient. In a small country with a large population, farmland is at a premium and some 65 per cent of all the country's land has been put to agricultural use. Livestock farming – chiefly for dairy produce – is of major importance: about half of the farming land is under grass or other fodder crops. Three-quarters of the country's cattle are the black-and-white Friesians. The chief dairy products are butter and cheese, the most popular Dutch cheeses being Edam and Gouda. Poultry farming is also important; The Netherlands is the world's largest exporter of eggs. Some 7 million pigs are kept, largely for the production of bacon (*see also* PE p.224-231).

Potatoes are the principal crop by weight, and after them sugar-beet. Many farmers concentrate on high-yield crops, especially fruit and vegetables. Some of them, particularly in South Holland, specialize in greenhouse horticulture; their exports include grapes and tomatoes. The famous bulb fields are confined chiefly to a small area in Holland – the "Bollenstreek" between Leiden and Haarlem.

The Netherlands has huge reserves of natural gas, discovered in the 1960s at Slochteren in the province of Groningen, in the north. It also has considerable deposits of petroleum. Other raw materials of importance are salt, chalk and peat. In the first half of the 20th century the most valuable mineral resource was coal, but the coal mines were gradually worked out and finally closed in 1975.

Since the end of World War II there has been a rapid expansion of manufacturing industry and many new industries have been introduced. The production of iron and steel and non-ferrous metals such as aluminium is of increasing importance. Engineering industries include the manufacture of motor vehicles, aircraft, bicycles, industrial machinery and electrical goods. In electronics, Dutch products include radios, television transmitters and radar equipment. Other major industries are the manufacture of building materials, chemicals (including petrochemicals), textiles – one of the country's traditional industries – clothing, processed foods and tobacco products; printing is also important. The ship- and boat-building industry is one of the oldest in Europe – Peter the Great went to The Netherlands from Russia in 1698 to observe Dutch ship-building techniques. Diamond cutting and polishing, a craft centred on Amsterdam and one for which The Netherlands was once famous, is today of relatively small importance.

Communications are well developed. As might be expected, the country has a good system of inland waterways; there are some 4,800km (3,000 miles) of navigable rivers and canals, of which nearly half can be used by vessels of up to 1,000 tonnes. Rotterdam-Europoort, near the mouth of the *Nieuwe Waterweg* (New Waterway) connecting Rotterdam with the sea, is by far the largest and busiest port in Europe. The national airline, Royal Dutch Airlines (KLM), flies to more than a hundred cities throughout the world and is the world's oldest operating airline.

People. The Dutch are a Germanic people of three different basic strains: Friesian, Saxon and Frankish in the south. In the 8th and 9th centuries the Franks conquered and absorbed an indigenous people, the Batavians, whose physical characteristics of fair hair, blue eyes and sturdy frames agreed with the generally held idea of "the Dutch".

The country is one of the most densely populated in the world. About a third of its population is concentrated in the low-lying western region known as *Randstad Holland*, which has many of the largest and most historic towns, including Amsterdam, The Hague and Rotterdam. The Dutch language is West-Germanic in origin. Dialects of it are spoken in the north-western province of Friesland (Friesian) and in northern (Flemish) Belgium, and it is the basis of the South African language Afrikaans.

Education in The Netherlands is free and compulsory between the ages of 6 and 15. There are six universities, including those of Leiden (1575), Groningen (1614) and Utrecht (1636), and five "high schools" that award degrees.

The two largest religious communities are the Roman Catholic Church (40 per cent) and Dutch Reformed Church (23 per cent). There are also other Protestant Churches, an Old Catholic Church and Jewish communities. The royal family belongs to the Dutch Reformed Church. Traditionally the north of the country is Protestant and the south Catholic, but this distinction is fading as is the mutual mistrust between Protestant and Catholic.

Public life in The Netherlands is conducted in an orderly but informal way: the only lavish state occasion of the year is the opening of parliament. Although the country is a monarchy, there is little court ceremonial and the sovereign, Queen Juliana, is held in deep affection and respect.

Dutch meals tend to be substantial, but it is customary to have only one hot meal a day – usually in the early evening. Popular dishes include *erwtensoep* (pea soup with smoked sausages) and *uitsmijter*, a snack made from fried eggs on bread with slices of ham or beef. Indonesian dishes are common in restaurants, introduced by immigrants from the former Dutch East Indies (Indonesia). Sport has an

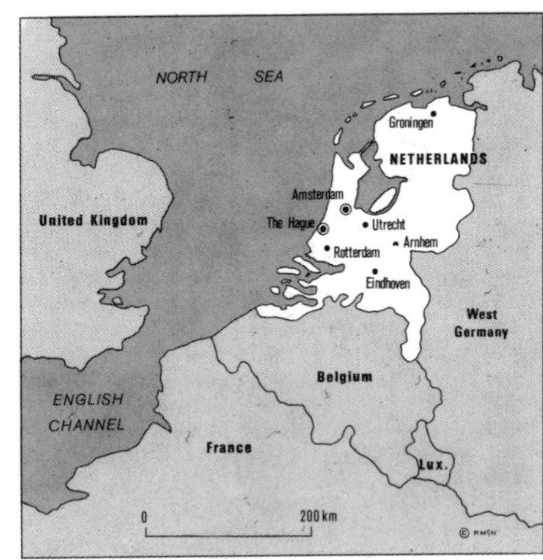

important place in Dutch life, the chief spectator sports being soccer and bicycle racing. Ice-skating has long been popular and, if the canals freeze in winter, school children may be given a holiday to go skating. In particularly severe winters, thousands of skaters take part in the 200km (124-mile) *Elfstedentocht* (Eleven-Towns Race) in Friesland.

The Dutch have a long and rich artistic and scholastic tradition, but little of its literature has been translated. The greatest figure in Dutch scholarship was probably Desiderius Erasmus – the apostle of common sense. The philosopher Baruch Spinoza was also a Dutchman. But it is through painting that the culture of The Netherlands is known to the world. Famous Dutch artists include not only Rembrandt but also Pieter de Hooch, Jan Vermeer, Frans Hals, Jacob van Ruisdael and, in more recent times, Vincent van Gogh.

Government. The country is a constitutional and

The brilliant bulb fields of the Netherlands in spring are a celebrated sight for tourists and make an important contribution to Dutch exports.

In The Netherlands some children learn to skate at the same time as they learn to walk, and in a hard winter almost everybody takes to the canals.

Nouméa is the capital and industrial centre of New Caledonia which, despite its tiny population, is the world's third largest producer of nickel.

hereditary monarchy. Female heirs may succeed to the throne if there are no male heirs. Legislative power is vested in the sovereign and the *Staten-Generaal* (States-General), which consists of two houses: the *Eerste Kamer* (First Chamber) of 75 members elected by the Provincial States – the councils of the 11 provinces; and the *Tweede Kamer* (Second Chamber) of 150 members elected directly for four-year terms by all citizens (with a few exceptions) more than 18 years of age. Bills may be proposed by the sovereign and the Tweede Kamer. The Eerste Kamer may approve or reject them but may not amend them.

History. In the 15th century the county of Holland and the dukedom of Brabant, together with Zeeland and Gelderland, became the property of the dukes of Burgundy and, through them, of the Hapsburgs. Charles of Burgundy (1500-58) became King of Spain and, as Charles V, also became Holy Roman Emperor. In 1555 he gave the Low Countries (now The Netherlands and Belgium) to his son Philip, who ruled Spain as Philip II. The attempts of Philip II to regulate the affairs of the Low Countries as an appendage of Spain and to introduce the Inquisition to stamp out the growing tide of Calvinism met with fierce opposition, which developed into open revolt in 1562.

In the northern provinces, the champion of Netherlandish independence was William I "the Silent", Prince of Orange. In 1576, in the Pacification of Ghent, he persuaded Catholics and Protestants to unite to expel the Spaniards. Spanish armies put down the revolt in the southern (Catholic) provinces, but the seven northern Protestant provinces, behind their river barriers, were able to hold out. In 1579 they formed the Union of Utrecht and two years later declared their independence, with William as *stadhouder* (governor). William was assassinated in 1584 and was succeeded by his son Maurice of Nassau. In 1609 the Spanish gave grudging recognition to the aspirations of The Netherlands, but fighting was resumed 12 years later and Dutch independence was not finally assured until the Peace of Westphalia (1648) ending the Thirty Years War. The whole period of conflict from 1568 to 1648 is known as the Eighty Years War.

During the early 17th century, despite their involvement in war, the Dutch were developing a great commercial empire. In 1602 the Dutch East India Company was formed to trade in Asia, and Dutch settlements were established in Ceylon (Sri Lanka), the Spice Islands (the Moluccas) and Malaya. Later, Dutch colonists began to settle in southern Africa. Amsterdam and other Dutch ports replaced Antwerp in importance, and Dutch ships carried much of the world's trade. This growing strength led to wars with England and with France, but the wars proved inconclusive. Within The Netherlands, the hereditary office of stadhouder was for a time abolished (1667) by a faction led by Jan de Witt, but after de Witt's murder the stadhouder William III (later also King of England) was restored to his position. On his death, the office was again abolished, to be re-established in 1747.

The power of The Netherlands declined throughout the 18th century. Revolutionary armies from France invaded the country in 1794-95 and set up the Batavian Republic. Napoleon created a new Kingdom of The Netherlands (1806) for his brother, Joseph, but later incorporated The Netherlands into France. Dutch independence was restored by the Congress of Vienna (1815), and The Netherlands, Belgium and Luxembourg were united under one king, called William I. But in 1830 the Belgians broke away, and in 1890 Luxembourg also became a separate state when a ten-year-old princess, Wilhelmina, succeeded to the Dutch throne; her mother, Queen Emma, acted as regent.

During World War I, The Netherlands remained neutral, but in World War II it was invaded by Germany and had to surrender within days. Queen Wilhelmina fled to London and a Dutch government-in-exile was set up there. When the Japanese entered the war they seized The Netherlands East Indies, the main Dutch possession in Asia.

After the war the Dutch had the task of rebuilding their country. New industry was established, and The Netherlands became one of the most prosperous lands in Western Europe. In 1949 the country was forced to relinquish control of the Netherlands East Indies, which became the independent republic of Indonesia. In 1976 and again in 1977 South Moluccan terrorists operating within The Netherlands held people hostage to publicize their claim for independence from Indonesia. Map 18.

Netherlands – profile

Official name Kingdom of The Netherlands
Area 41,160sq km (15,892sq miles)
Population (1976 est.) 13,796,000
Density 344 per sq km (890 per sq mile)
Chief cities Amsterdam (capital) 1,036,000; Rotterdam, 686,600; The Hague (seat of government) 550,600
Government Constitutional hereditary monarchy
Religions Roman Catholic (40%), Protestant (30%)
Language Dutch
Monetary unit Gulden
Gross national product (1974) £28,239,300,000
Per capita income £2,047
Agriculture Dairy products (particularly cheese), cereals, potatoes, sugar-beet, fruit, vegetables, flowers
Industries Steel, motor vehicles, aircraft, electrical and electronic goods, building materials, ships, chemicals, textiles, printing, processed foods
Minerals Petroleum, natural gas, salt
Trading partners Other members of the EEC, USA

Netherlands Antilles is an autonomous group of five main islands (and part of a sixth) in the West Indies in the Caribbean Sea. The capital is Willemstad on the main island, Curaçao; others include Aruba, Bonaire, Saba, Saint Eustatius and the southern half of St Maarten. The official language is Dutch and most of the people are Roman Catholics. Curaçao's economy depends mainly on the refining of petroleum from nearby Venezuela; other income derives from tourism and the export of salt and phosphates. The islands were visited in the 1490s by Christopher Columbus, Alonso de Ojeda and Amerigo Vespucci. Curaçao was settled by the Spanish in 1527 and the Antilles captured by the Dutch in 1634. Severe race riots occurred there in 1969. Area: 993sq km (383sq miles). Pop. (1975 est.) 242,000. Map 74.

Nevada. *See* UNITED STATES.

New Britain. *See* PAPUA NEW GUINEA.

New Brunswick. *See* CANADA.

New Caledonia (Nouvelle Calédonie) is a French overseas territory in the south-western Pacific Ocean, about 1,210km (750 miles) east of Australia. The territory takes its name from the main island, location of the capital, Nouméa. Other islands include the Belep, Chesterfield, Huon and Loyalty groups and the Isle of Pines and Walpole Island. The chief agricultural products of the islands are copra, coffee and cotton; ores of nickel, iron, manganese, cobalt and chromium are mined – some of the richest deposits in the world [*see* PE p.*133*]. New Caledonia was discovered in 1774 by Capt. James Cook, and a French Roman Catholic mission was established there in 1843. It was used as a penal colony from 1864-94 and became a French overseas territory in 1946. Area: 18,342sq km (7,082sq miles). Pop. (1975 est.) 136,000. Map 62.

Newfoundland. *See* CANADA.

New Guinea is a large island 2,415km (1,500 miles) long in the south-western Pacific Ocean, about 150km (95 miles) north of Australia. The western part of the island, Irian Barat (West Irian), is a province of Indonesia and the eastern half is occupied by the nation of Papua New Guinea. It has a tropical climate, high mountain ranges and mangrove swamps on the coasts. It was probably discovered in 1511 by the Portuguese navigator Antonio d'Abreu and later visited by various European explorers. The western part was annexed by the Dutch in 1828 (and called Dutch New Guinea). In 1884 the Germans took possession of the north-east (German

Coconut palms, source of copra, are also used as building materials in the islands of the New Hebrides, which are administered by Britain and France.

Wellington, named after the Iron Duke in 1840, has grown over the last century into the communications centre and chief port of New Zealand.

The North Island mountains, less sheer than those to the south, are all volcanic in origin and all but one are still classified as active.

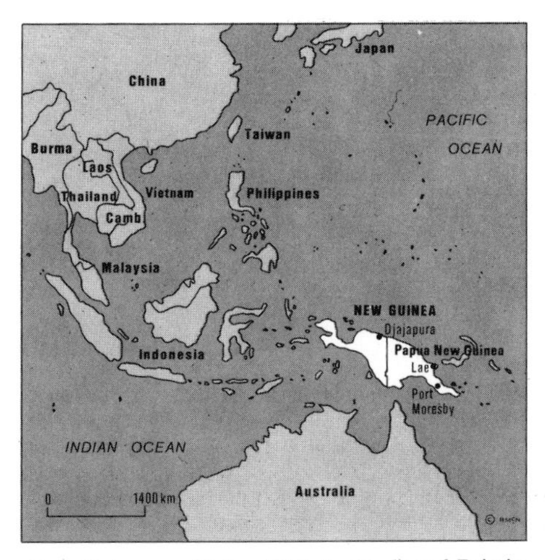

New Guinea, or Kaiser-Wilhelmsland) and Britain occupied the south-east (British New Guinea). For the subsequent history of the island *see* INDONESIA; PAPUA NEW GUINEA. Area: 885,780sq km (342,000sq miles). Pop. (1973 est.) 3,600,000.
Map 60.

New Hampshire. *See* UNITED STATES.

New Hebrides (Nouvelles Hébrides) is a group of coral and volcanic islands in the south-western Pacific Ocean, about 2,300km (1,430 miles) east of Australia, governed jointly by Britain and France. It is made up of about 80 islands that form a chain extending for 725km (451 miles). The chief islands are Efate (location of the capital, Vila), Espiritu Santo, Malekula, Malo, Pentecost and Tanna. Industries include the production of copra, fishing, farming and mining. The New Hebrides were discovered in 1606 by the Portuguese navigator Pedro Fernandez de Queirós and explored by Louis de Bougainville in 1768 and Capt. James Cook in 1774. British and French settled there in the early 1800s, and for many years the local people were taken as slaves to work on sugar-cane plantations in Queensland and Fiji. Joint Anglo-French naval control was established in 1887, being replaced by a condominium giving joint administrative control in 1906. Area: 14,760sq km (5,699sq miles). Pop. (1975 est.) 97,000.
Map 62.

New Ireland. *See* PAPUA NEW GUINEA.

New Jersey. *See* UNITED STATES.

New Mexico. *See* UNITED STATES.

New South Wales. *See* AUSTRALIA.

New York. *See* UNITED STATES.

New Zealand is an independent island nation in the south-western Pacific Ocean, 1,930km (1,200 miles) south-east of Australia. In spite of its distance from its major customers, it has developed a prosperous economy based on the ability to sell meat, wool, and dairy products on the international market. Its racially harmonious society was formed in the mid-19th century from a mixture of British settlers and indigenous Maoris. Today New Zealanders tend to be egalitarian, resourceful and conservative, more competitive in sport than in business. The country offers good opportunities for outdoor recreation, and the people enjoy excellent housing, education, and health and welfare services. The capital is Wellington, on North Island.

Land and climate. New Zealand consists of two rugged main islands – North Island and South Island (separated by the Cook Strait) – and Stewart Island to the south, as well several smaller islands [*see* PE p.57]. Few inland areas are more than 130km (80 miles) from the sea. A land of contrasts, it is geologically young and forms part of the circum-Pacific volcanic rim. Earth-tremors are frequent, although there have been fewer than 20 destructive earthquakes since European settlement.

On the North Island a narrow northerly peninsula with dairy farms and orchards extends southwards to Auckland, the largest city, straddling the isthmus between the Pacific Ocean and the Tasman Sea. South of this is fertile sheep and dairying land in the basin of the Waikato, at 435km (270 miles) the longest of more than 70 major rivers. It flows seawards from Lake Taupo (606sq km; 234sq miles) in the volcanic plateau at the centre of the North Island. The thermal area near Lake Taupo is a tourist centre. To the west lie the dairying plains of Taranaki. On the east coast the long sweep of the Bay of Plenty, another dairying centre, extends to the hilly grazing land of the East Cape and the sheep country inland of Poverty Bay and Hawke Bay. The Hawke's Bay area also has fertile lowlands in which farmers grow fruit and vegetables. South of the Manawatu River, mountain ranges run to Wellington, sited on a deep, enclosed harbour.

The slightly larger but less populated South Island is dominated by the Southern Alps, with permanent icefields rising to the country's highest peak, Mt Cook (3,764m; 12,349ft above sea-level). The Alps feed a series of deep lakes and wide glaciers and plunge steeply to the narrow, densely forested West Coast [*see* PE p.116]; on the east they descend into the broad Canterbury Plain, a major area for growing cereal crops and raising lambs. Beyond the rolling hill country of Otago the rich grasslands of Southland extend southwards to Invercargill. In the south-west the indented coast has spectacular sounds and peaks, and the Marlborough Sounds at the north of the island are formed from drowned valleys.

New Zealand's climate is moist and equable with warm summers and mild winters. Temperatures, often moderated by westerly sea breezes, average about 23°C (73°F) in January (summer) and 8°C (46°F) in July (winter), with only light frosts in the north. Although in the South Island winters are colder, snowfalls on low country are rare. Average rainfall throughout the country is 635 to 1,525mm (25–60in) per year, although the mountains cause wide variations, especially between the west coast of the South Island (annual average in the Southern Alps is 7,600mm; 300in) and the drier east at Otago (average 330mm; 13in). In most districts there are 100 to 150 days of rain each year and about 2,000 hours of sunshine.

Physical resources. Apart from its ample supplies of water and its large areas of grassland and forest, New Zealand has comparatively few natural resources. There is some coal and iron ore, natural gas has been discovered, and limestone is quarried. One of the world's largest man-made forests, 400,000 hectares (nearly a million acres) of radiata pines at Kaingaroa, is the centre of an expanding wood pulp and paper industry in the centre of the North Island.

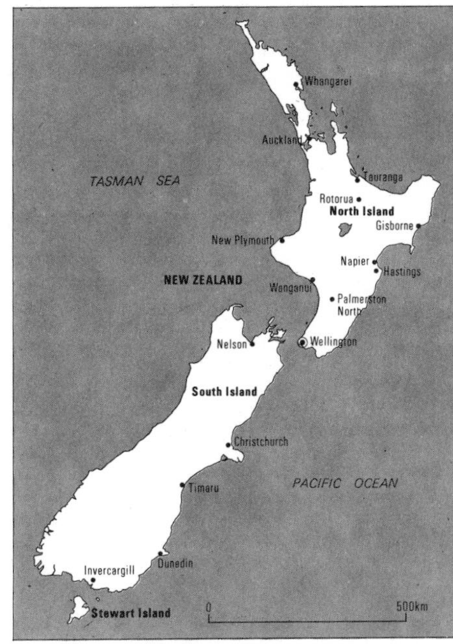

Hydroelectricity is the major source of energy. There are nine power stations on the Waikato River, but the country's greatest output comes from the southern lakes and rivers, particularly the Clutha and Waitaki rivers. From Benmore on the Waitaki power is taken to the North Island by a 500-kilovolt submarine cable across Cook Strait. The storage capacity of lakes Manapouri and Te Anau provides electric power to process Australian bauxite into aluminium. At Wairakei, in the thermal

Mixed evergreen forest is New Zealand's indigenous vegetation, but settlement of the country has left it with only a few areas of dense bushland.

Hot springs, geysers and boiling mudpools surround Rotorua, a town situated in the heart of the thermal belt running through the North Island.

The physicist Ernest Rutherford, who was born in New Zealand, made a major contribution to science with his discovery in 1911 of the atomic nucleus.

area around Taupo, a geothermal power station (second to be established in the world) produces 220,000 kilowatts [*see also* MM p.73].

The chief minerals produced in New Zealand are coal (2.5 million tonnes per year), limestone and building materials. Since the opening of the Kapuni field in Taranaki in 1970, natural gas has become an important energy source and the nearby offshore Maui field, developed in the late 1970s, is one of the world's largest. Deposits of ironsand on the western coast of the North Island contain an estimated 500 million tonnes of ore and are mined for export and for steelmaking [*see* PE p.134].

A small fishing industry operates in the shallow waters over the continental shelf around New Zealand, landing catches of snapper and other species, mainly for the local market. Rock lobsters provide the main export earnings and there are large catches of mussels, scallops, oysters and other shellfish. Big-game fishing grounds off the east of the North Island are being commercially explored for tuna.

Agriculture. Farming employs only 12 per cent of the workforce yet its products are the mainstay of the economy, earning 80 per cent of income from exports. Pastures covering a third of the total land area support 60 million sheep, 6 million beef cattle and 3 million dairy cattle (farm animals outnumber people by more than 20 to 1.) Grasslands are enriched by applications of super-phosphate fertilizer, much of it sprayed from aircraft [*see* PE p.*163, 171*]. Scientific research aimed at increasing yields, plus the advantage of being able to farm throughout the year, has given New Zealand farmers the highest productivity per man in the world.

The dairy industry is organized on co-operative lines, with centralized marketing and guaranteed prices designed to even out seasonal market fluctuations. An average farm has 100–120 cows (using only contract seasonal labour) on a 40–60 hectare property. In the mid-1970s New Zealand's annual production was 240,000 tonnes of butter, 100,000 tonnes of cheese and 200,000 tonnes of milk powder and casein (used for making plastics).

Sheep farming produces income from wool (320,000 tonnes per year), lamb and mutton. Millions of fat lambs are killed during November and December for the British market and total meat production (including beef) is 1 million tonnes a year. An average farm in hilly country has 12 hectares (30 acres) for every sheep, whereas lowland farms (fat lambs) have 25 sheep to every hectare (10 per acre). *See also* PE p.231.

Wheat, barley, maize and sugar-beet are grown, chiefly in the South Island. Fruit production is important and half the crop of apples and pears is exported. Nelson, Hawke's Bay, Central Otago and Gisborne are major orchard areas, as is Northland with its distinctive semi-tropical fruits such as kiwi-fruit and tamarillos.

Industry. Although it has a small population, New Zealand has a surprisingly wide range of industry. Historically, industry has been concerned mainly with producing basic consumer goods for the domestic market and with processing farm products for export. There was a rapid expansion of manufacturing under import controls during World War II, and again in the 1960s, but today the tendency is towards the exploitation of local resources and skills – yarn, carpets, steel, wood pulp and paper, chemicals and agricultural machinery. Tourism is another growing industry.

Most enterprises are on a small scale, and yet manufacturing industry employs a quarter of the workforce. It is strongly concentrated in Auckland (which produces a third of the total output), the Hutt Valley-Wellington area and Christchurch, followed by Hamilton, Dunedin and Rotorua. Food processing makes the chief contribution (a third of output), followed by textiles and clothing, forest products, machinery and transport equipment, and the fast-developing metals industry, with a steel mill at Glenbrook (near Auckland) and an aluminium smelter at Bluff.

Trade. New Zealand's heavy dependence on trade gives it a vested interest in removing international barriers against agricultural products. Until 1954 all its meat and dairy produce went to Britain under bulk purchase agreements, encouraging a complacency that was upset in the early 1960s when Britain began moving towards memebrship of the European Economic Community (EEC), with its protectionist agricultural policies. As intensive efforts were made to develop alternative markets, especially in Japan, the percentage of exports sent to Britain fell from 51 per cent in 1960 to 27 per cent in 1973. But Britain still buys much of New Zealand's butter, cheese and lamb and transitional arrangements have been made with the EEC for continuing this trade. Meat, wool and dairy products make up 70 per cent of all exports, and there is an expanding trade in forest products.

Britain is also the major source of imports, although Australia is gaining in importance, followed by the United States and Japan. Machinery, transport equipment, and other manufactured goods account for two-thirds of imports. New Zealand and Australia are linked in a limited free trade area to encourage rationalization of industrial development and component manufacture.

Transport and communications. New Zealand's narrow, rugged islands have presented formidable transport problems. But the country has an extensive road system (95,000km; 59,000 miles) with some first-class highways financed by petrol tax, all of which goes to the National Roads Board. The state-owned railway system, which crosses Cook Strait by ferry, is less extensive and has to compete with the internal airways network operated by the National Airways Corporation, supplemented by minor private firms. Air New Zealand operates services throughout south-eastern Asia and across the Pacific Ocean to the United States. Five other international airlines also use the main airports at Auckland, Christchurch and Wellington. There are good harbours at Auckland, Wellington, Lyttleton, Dunedin, other general ports at Bluff, Napier and New Plymouth, and specialist ports at Tauranga (for timber) and Whangarei (which serves the country's only oil refinery). Container services to Britain and the United States started in 1972 and an Australian-New Zealand firm uses small container ships and roll-on roll-off freighters to carry goods across the Tasman Sea. The transport industry employs 9 per cent of New Zealand's workforce.

Apart from a few private radio stations, broadcasting is controlled by a single corporation with some government participation. The New Zealand Broadcasting Corporation (NZBC) operates Radio New Zealand and two television services. The television channels carry some advertising and nearly 90 per cent of homes have television sets.

People. The population of just more than 3 million is overwhelmingly of British stock; Maoris make up about eight per cent. Other significant groups shown in the 1971 census were Pacific Islanders (45,000; mostly in Auckland), Dutch (26,000) and Chinese (13,000), with fewer numbers of Indians and Yugoslavs. The youthful Maori population, now estimated at 250,000 people, has recovered rapidly from a low figure of 40,000 in 1900. Early in the 20th century there was a revival of Maori spirit, led by such men as Apirana Ngata and Peter Buck. Today intermarriage with Europeans is more common than marriage to another Maori. As a result, an eventual homogenous society is envisaged, furthering a multi-racial policy that is practised as well as preached (although there is some racial friction in Auckland).

The main waves of British immigration were in the 1850s and 1870s, although British people continued to emigrate each year, sometimes with part of the fare paid by the New Zealand government. Recently entry has been restricted to those with useful skills and there is relatively less freedom of entry for Pacific Islanders. The emphasis in immigration policy on ease of assimilation has prevented a more varied racial mixture. Population growth since the 1950s has fluctuated between one and two per cent annually, but economic setbacks produced a record net outflow of 13,839 people in 1976–77.

Before World War II many people who wished to get to the top of their profession left New Zealand to complete their education, gain experience or work abroad, often in Britain. Such New Zealanders who achieved international acclaim include the physicist Ernest Rutherford (1871–1937) and the surgeon Archibald McIndoe. With increasing population and more opportunities at home, this tendency is becoming less marked.

Cultural life and recreation. Events since World War II have loosened New Zealand's strong cultural ties to Britain and led to more interest in what is distinctive at home because of the country's isolated position. Thus, although English is the only language widely spoken, there have been recent attempts to expand the teaching of Maori and to re-

Most Maori artistic traditions, being oral, were submerged in colonial culture; but the craft of decorative design is still widely practised.

The Southern Alps are fold mountains which stretch for 480km (300 miles). They contain Mt Cook – at 3,764m (12,349ft) New Zealand's highest mountain.

A surfer comes ashore at Hawke Bay on the east coast of New Zealand. Surfing, one of the most popular sports, is more often seen on the west coast.

vive Maori culture, with its traditions of carving and oral poetry. American influences have become more marked, especially in Auckland, overlaying a culture that has until now been deeply imitative of British customs, institutions and attitudes.

Education is free and compulsory between the ages of 6 and 15, and there is generous financial aid for most students who attend one of the country's six universities or two agricultural colleges. The standard of state education is high, but there is also a tradition of private schooling, particularly among Roman Catholics, who form the third largest religious group after Anglicans and Presbyterians.

Many of New Zealand's best writers, artists, musicians, actors and dancers were once drawn irresistibly to Britain, many never to return. Although this drain of talent continues in some fields it has been stemmed by a strengthening sense of national identity, more generous government assistance through the Arts Council, the establishment of a professional orchestra (the NZBC Symphony Orchestra) and theatre, and a growing appreciation of the works of local writers, artists and craftsmen, especially potters. The writers best known outside New Zealand are four women: Katherine Mansfield (1888–1923), Ngaio Marsh (1899-), Sylvia Ashton-Warner (1918-) and Janet Frame (1924-). A new wave of novelists has recently emerged, but New Zealand's self-image has been formed mainly by its short-story writers, notably Frank Sargeson (1903-) and by the spare, direct and witty verse of its poets. Among the most influential have been R.A.K. Mason (1905-), Charles Brasch (1909-76), Alan Curnow (1911-), Denis Glover (1912-) and more recently James K. Baxter (1926-72) and C.K. Stead (1932-).

Living close to the sea and mountains, New Zealanders have outstanding facilities for sailing, surfing, swimming, golfing, fishing, skiing and mountain climbing (Sir Edmund Hillary, conqueror of Everest, is a national hero). Team sports are dominated in winter by the national passion, rugby union. Always formidable, the New Zealand national side, the All Blacks, won 15 test matches against all-comers in the 1960s. Recent players who have won fame in all the rugby-playing countries include Don Clarke (1935-), Ian Kirkpatrick (1946-), Brian Lochore (1941-), Colin Meads (1936-) and Wilson Whineray (1935-). Soccer is increasing in popularity.

There is a wide variety of summer sports, including cricket and golf. One of the best-known cricketers was Martin Donnelly (1917-), who played for the national side and English county clubs, as well as once playing rugby for England. In golf, Bob Charles (1932-) was the first left-hander to win the Open championship (1963). New Zealand has been strikingly successful in middle-distance running. Its Olympic gold medallists include Jack Lovelock (1936), Murray Halberg (1960), Peter Snell (1960; 1964) and John Walker (1976). New Zealand is also among the world's strongest rowing nations. Denis Hulme (1936-) won the 1967 world motor racing championship and Ronnie Moore (1933-), Barry Briggs (1935-) and Ivan Mauger (1939-) long dominated world speedway championships. New Zealand's stud farms produce fine racehorses; a state-run off-course betting system is the main outlet for gambling.

Constitution, politics and law. New Zealand is a one-chamber parliamentary democracy within the Commonwealth. The nominal head of state is the British monarch, represented by a governor-general, but full sovereignty is vested in the House of Representatives, whose 87 members are elected every three years by voters aged 20 or over. Maoris may stand for any electorate, and four seats are reserved exclusively for them. The majority party forms the executive Cabinet, headed by a prime minister. The unwritten constitution – like the legal system with its Appeals Court, Supreme Court and lower magistrates' courts – relies largely on English precedent. Local government is vigorously decentralized, with county councils, borough councils, town boards and a wide range of other elected bodies. Auckland, however, has a regional authority to deal with services for its metropolitan area.

Since 1935 New Zealand has been governed by either the National Party or the Labour Party. Industrial workers tend to vote Labour and the larger business and farming groups tend to favour the National Party. Political differences are not strongly marked, however; both parties advocate a mixed economy with emphasis on social security and state aid for industry and agriculture.

History. New Zealand lay undiscovered until the 9th or 10th century AD, when Polynesian mariners arrived in open canoes, probably carried by wind and current from the Society Islands more than 2,400km (1,490 miles) to the north-east. Maori legend tells of a purposeful migration of a whole fleet of canoes in the 14th century. The archaeological record shows that soon afterwards there was a transition from a simple society which hunted the moa – a large flightless bird, now extinct [*see* PE p.*144*] – to a tribal society skilled in the arts of carving, weaving, building and agriculture and with a poetic and heroic mythology.

It was this society that rebuffed the first approach of European civilization in 1642 when Abel Tasman, of the Dutch East India Company, sighted the South Island. Tasman abandoned an attempt to land after Maoris killed four of his men. He retreated after charting the west coast and giving the name Nieuw Zeeland to what he thought was the rim of a southern continent.

Capt. James Cook, the next European to arrive, circumnavigated the country in 1769, charted it accurately and established relations with a Maori race he noted to be intelligent, warlike and hospitable. After the founding of a British colony in Australia in 1788, New Zealand became a landing place for sealers, whalers, traders and eventually missionaries and settlers. The Maoris, then numbering perhaps 250,000, found Europeans interesting and profitable. One Maori chief, Hongi Hika, visited England in 1820. He returned with muskets and quickly upset the delicate balance of power among feuding Maori chiefs; ruthless tribal warfare swept the country, adding to the ravages made by European diseases.

By 1839, alarmed by the decline of the Maoris and pressured by both settlers and missionaries, Britain decided on annexation. Capt. William Hobson persuaded many North Island chiefs to sign the Treaty of Waitangi (1840), under the terms of which sovereignty passed to the British Crown in return for protection and guaranteed Maori land rights. The moral guardianship implied in the treaty, a product of the Evangelical movement in England, was to make the colonization of New Zealand relatively enlightened.

The prime mover in British migration was Edward Gibbon Wakefield, an adventurer and visionary, who aimed to transplant to New Zealand a "vertical slice of decent British society". Wakefield settlements were established at Wellington, Nelson, Wanganui, New Plymouth, Canterbury (under the auspices of the Church of England) and Otago (by the Free Church of Scotland). In the almost empty South Island pastoralism soon flourished, and settlement was encouraged by the discovery of gold in Otago (1861) and the West Coast. But in the north, Maori unrest, brought under control by the firm governorship of George Grey (1845-53), flared up again in 1860. In Taranaki, the right of a paramount chief to veto sales of tribal land was ignored, promoting a revolt that was not crushed until 1865. The last guerrilla leader, Te Kooti, retreated in the early 1870s into the "King Country", an enclave named after the Maori king elected in 1857 in the Waikato. Elsewhere, private land sales that ignored Maori tribal rights soon put the remaining good land under European ownership.

The Constitution Act of 1852 led to a large measure of self-government, at first carried out by six provincial councils. By 1876 full control had passed to the central legislature in Wellington, which had replaced Auckland as the capital in 1865. Prime Minister Julius Vogel, who abolished the provincial councils, launched an expansionary programme with borrowed money, extending services and encouraging immigration (the population rose to half a million by 1880). An ensuing slump lasting until the 1890s was overcome by better land settlement and aided by the development of refrigerated ships which enabled New Zealand to send dairy products and meat to Britain, in addition to wool.

An adventurous Liberal administration under the leadership of Richard Seddon made New Zealand a pioneer of social legislation, including votes for women (1893), industrial conciliation and arbitration (1894) and old-age pensions (1898). Although it became an independent dominion in 1907, New Zealand expressed its fierce loyalty to Britain in World War I, sending nearly 10 per cent of its entire

Whakatane, a seaside resort and sawmill town has, like almost every town in New Zealand, its McKenzies shop, the largest chain store in the country.

A Victorian bandstand, typical of the New Zealand colonial style, stands in the centre of Blenheim, which grew rapidly after the discovery of gold in 1864.

Waikato River, the longest river in New Zealand, feeds seven artificial lakes which are the main source of the country's hydroelectric power.

population to fight in Gallipoli and France and losing 17,000 men. The war awakened a sense of national identity, as did the trusteeship in 1920 of Western Samoa (which New Zealand eventually guided to independence in 1962).

Land speculation under the "farmers" Reform government was followed by a postwar depression. Despite government action, including the establishment of a Reserve Bank and farm credits by the minister responsible for unemployment, Gordon Coates, the slump deepened with the onset of worldwide depression. Cuts in wages and public spending led to serious unemployment, strikes and riots. In 1935 the country overcame its fear of socialism (represented by the Labour Party) and elected the first Labour government under Michael Savage, a popular and moderate leader whose dynamic cabinet included Walter Nash (finance) and Peter Fraser (education and health).

During the next 14 years, under Savage (who died in 1940) and Fraser, Labour created a comprehensive welfare state, with guaranteed farm prices, guaranteed minimum wages, state-rental houses, national health services and eventually food subsidies and family and special benefits. The government was helped by an economic recovery that began in 1935 and continued during World War II, when food from New Zealand contributed greatly to Britain's survival. Wartime economic stabilization also encouraged local manufacturing.

Impatient with restrictive controls, a majority of electors turned in 1949 to the National Party under Sidney Holland, which strengthened its position in 1951 by breaking the power of militants in the waterfront (dockers') union. The National Party, adopting Labour's commitment to full employment and most of its welfare policies, has been predominant ever since. Labour's two subsequent terms of office, in 1957-60 and 1972-75, were both clouded by economic recession, and the death in office of Norman Kirk in 1974 removed the only powerful Labour politician to emerge in the postwar era. After the long, benign leadership of Keith Holyoake, prime minister from 1960-72, the National Party soon found an aggressive successor in Robert Muldoon, prime minister from 1975.

In foreign affairs, the postwar period has been marked by a steadily lessening reliance on Britain in defence and trade. New Zealand recognized the United States as its protector in 1941 when Britain's defeat at Singapore left the southern Pacific open to Japanese invasion. (The government decided not to recall the New Zealand Division from Europe where it fought in Greece, North Africa and Italy.) New Zealand joined Australia and the United States in a mutual defence alliance, ANZUS, in 1951 and became a member of the South-East Asia Treaty Organization (SEATO) in 1954, honouring these commitments by sending a small force to South Vietnam in 1965. New Zealanders earlier fought in Korea in 1950 and in Malaya from 1957 to 1960. Defence co-operation with Malaysia and Singapore continues through the Five Power Defence Arrangements of 1971, in which Australia and Britain also participate.

New Zealand has played a significant role among small nations in the United Nations, in the Colombo Plan, and in other aid schemes in south-eastern Asia and the Pacific area. It administers NIUE, the Tokelau islands and a large area of territory in Antarctica, the Ross Dependency, site of American scientific bases which are supplied from Christchurch. New Zealand also guides the COOK ISLANDS, a Pacific Ocean territory that became self-governing in 1965.

Trade with the United States, Australia, Japan and the Pacific area has increased ever more since Britain joined the EEC. The need to diversify exports is an over-riding economic concern. New Zealand has had to recognize that it cannot depend on a few types of food products to finance imports of raw materials for protected local industries in a welfare state. Falls in overseas prices quickly produce balance of payment deficits, import controls and business recessions. New Zealand accordingly does not have the expansionary confidence of its neighbour, Australia. Map 62.

Prime Ministers of New Zealand

Henry Sewell (1856)
William Fox (1856; 1861–62; 1869–72; 1873)
Edward Stafford (1856–61; 1865–69; 1872)
Alfred Domett (1862–63)
Frederick Whitaker (1863–64; 1882–83)
Frederick Weld (1864–65)
George Waterhouse (1872–73)
Julius Vogel (1873–75; 1876)
Daniel Pollen (1875–76)
Harry Atkinson (1876–77; 1883–84; 1884; 1887–91)
George Grey (1877–79)
John Hall (1879–82)
Robert Stout (1884; 1884–87)
John Ballance (1891–93)
Richard Seddon (1893–1906)
William Hall-Jones (1906)
Joseph Ward (1906–12; 1928–30)
Thomas Mackenzie (1912)
William Massey (1912–25)
Francis Bell (1925)
Gordon Coates (1925–28)
George Forbes (1930–35)
Michael Savage (1935–40)
Peter Fraser (1940–49)
Sidney Holland (1949–57)
Keith Holyoake (1957; 1960–72)
Walter Nash (1957–60)
John Marshall (1972)
Norman Kirk (1972–74)
Wallace Rowling (1974–75)
Robert Muldoon (1975–)

New Zealand – profile

Official name New Zealand
Area 268,676sq km (103,736sq miles)
Population (1977 est.) 3,130,000
Density 12 per sq km (30 per sq mile)
Chief cities Wellington (capital) (1976 est.) 139,300; Christchurch, 171,800; Auckland 152,600; the largest metropolitan area is Auckland (744,000)
Government Constitutional monarchy with a one-chamber parliamentary government
Religions Anglican, Presbyterian, Roman Catholic
Language English
Monetary unit New Zealand dollar
Gross national product (1974) NZ$9,296,700,000 (£5,299,100,000)
Per capita income NZ$2,970 (£1,693)
Agriculture Meat, wool, dairy products, fruit, wheat, vegetable, wine
Industries Food processing, textiles, forest products, machinery and transport equipment, metal products, electrical goods
Minerals Coal, iron ore, natural gas, limestone
Trading partners Britain, USA, Japan Australia

Nicaragua is the largest republic in Central America, between Honduras (to the north) and Costa Rica. It has shorelines on the Caribbean Sea to the east and the Pacific Ocean to the west. The capital is Managua. It has poor communications and a sparse population, most of whom work in agriculture producing a variety of crops such as cotton, coffee, maize, rice, sugar cane and tobacco. Mestizos (people of mixed European and American-Indian descent) make up most of the population, which is mainly Spanish-speaking and Roman Catholic. The land has four main regions: a large triangular area of folded faulted structures in the north-centre; a low coastal plain, called Miskito (or Mosquito), in the east; a lowland area extending from the Gulf of Fonseca south-eastwards to the Costa Rican border, including the large lakes Nicaragua and Managua; and a narrow highland region on the east coast. About half the land is forested, yielding fibres, resins and various kinds of timber. Other industries include sugar refining, textiles and the production of cigars, soap, cement and leather goods.

In 1502 Christopher Columbus "discovered" the coast of Nicaragua and found it inhabited by Indian tribes called Miskitos. Fernándes de Córdoba founded the cities of Granada and León in 1524. In the mid-17th century the Mosquito coast in Nicaragua and Honduras was counted a British dependency, but in 1786 Britain acknowledged Span-

Traditional motifs in a modern working lend a distinctive appearance to the Niger National Museum at Niamey; ornamental gardens surround it.

The Benue River, largest tributary of the River Niger, and the main carrier of goods in Niger, overflows its banks annually in the rainy season.

Ibadan, Nigeria's second-largest city, has little industry and few high-rise buildings, yet at its centre has a population density of 1,550 per sq km.

ish claims to the Caribbean coast and in 1821 Nicaragua proclaimed independence. It became part of the Mexican Empire and, for 13 years, a member of the Central American Federation before achieving total independence in 1838. Managua was made the official capital in 1855 to settle rival claims by Granada and León for that status. The American William Walker made himself president in 1855 but was deposed two years later. Thirty years of uneventful conservative rule ended in 1894 with the liberal presidency of José Zelaya, which lasted until 1909. A civil war broke out in 1912 and for 21 years US troops occupied the country. Anastasio Somoza García became president in 1937 and ruled (except for one interlude from 1947 to 1950) until he was assassinated in 1956. His son Anastasio Somoza Debayle was elected president in 1967, but he resigned in 1972, and was then re-elected in 1974. Area: approx. 130,000sq km (50,190sq miles). Pop. (1975 est.) 2,155,000. Map 74.

Niger, official name Republic of Niger, is the largest nation in Western Africa. It is an arid, landlocked and poor country. In 1974 the average annual income was only £51. The capital is Niamey.

Land and climate. Most of the land consists of flat plateaus and plains. The chief highland areas are the Aïr massif, which reaches 1,900m (6,234ft) above sea-level in the central north, and the high plateau in the north-east. The north is mainly hot desert and the centre rough pastureland; agriculture is confined to the far south, where the rainfall averages 560mm (22in) per year. But the Aïr massif is cooler and the rainfall, which exceeds 250mm (10in) per year, supports some pasture. The only permanent river is the Niger, in the south-west.

Economy. Livestock-rearing (cattle, goats and sheep) is an important occupation and live animals, hides and skins form about 18 per cent of all exports. But prolonged droughts in the late 1960s and early 1970s seriously depleted the animal population. Groundnuts account for 56 per cent of the exports and the chief food crop, grown in the south, is millet. Some uranium and cassiterite (tin ore) are mined, although manufacturing industry is extremely limited. Niger's remoteness and poor communications are obstacles to development.

People. The Berber Tuaregs are a pastoral group in the north. But most people live in the south, the largest groups being the Hausa, the Djerma-Songhai and the Fulani (or Peul). Islam is the religion of 85 per cent of the people, 14.5 per cent practise ethnic religions and the remainder are Christians. The literacy rate is 11 per cent.

History and government. The region was known to the Ancient Egyptians and there were several city states there in the 12th to 15th centuries. In the early 16th century it became part of the Songhai Empire of Gao. France occupied the area between 1897 and 1900 and ruled it as part of French West Africa until 1922, when it was made a separate French colony. In 1960 Niger became an independent republic and President Hamani Diori ruled with a cabinet and a National Assembly with 60 elected members. But a military group overthrew Diori in 1974 and Lt.-Col. Seyni Kountché became president. Kountché suspended the constitution, abolished the National Assembly, banned political parties and ruled Niger with a Supreme Military Council. He formed a new government with a majority of civilian members in 1976. Later in the year loyal troops foiled an attempted military coup, whose leaders were arrested and executed. Map 32.

Niger – profile

Official name Republic of Niger
Area 1,267,000sq km (489,189sq miles)
Population (1975 est.) 4,600,000
Chief city Niamey (capital) (1975) 130,299
Government Supreme Military Council
Religions Islam, ethnic, Christianity
Language French (official)
Monetary unit CFA franc
Agriculture Millet, groundnuts, cotton, rice
Chief mineral Uranium

Nigeria, official name Federal Republic of Nigeria, has more people than any other African country, although it is only fourteenth largest in area. It has tremendous potential for development. The discovery of petroleum in 1958 was particularly significant and, by the mid-1970s, Nigeria had become Africa's leading petroleum producer. Much of the revenue from petroleum is being invested in ambitious development projects. Lagos, in the south-west, is the present capital, but a new capital territory is being established near Abuja in central Nigeria.

Land and climate. A Y-shaped depression, occupied by the Niger and Benue rivers, divides Nigeria into three main regions. Northern Nigeria contains high plains and plateaus, flanked by the Sokoto basin in the north-west and the Chad basin in the north-east. South-eastern Nigeria includes the Cross River plains and plateaus that rise to 2,042m (6,700ft) above sea-level in the mountains bordering Cameroon. Most of south-western Nigeria is an upland between 300 and 600m (984-1,969ft) above sea-level. The vast, swampy Niger delta dominates the central part of the coast. Astride the delta the coast is mostly lined by sand spits, which enclose lagoons.

The climate is hot and humid, especially in the south where temperatures average 27°C (80°F) all the year round. In the north, however, temperatures sometimes drop as low as 10°C (50°F) in the coldest month and are more than 38°C (100°F) in the warmest month. The rainfall averages 1,000mm (39in) per year in the south-west and 2,500mm (98in) in the south-east, decreasing inland. The north has between 250 and 1,000mm (10-39in) per year. Mangrove swamps border much of the coast. Rain forests in the interior merge into woodland savanna in central Nigeria. Much of the north is savanna, but the Chad basin is semidesert.

Economy. Petroleum has recently dominated the economy, accounting for more than 80 per cent of all exports. Tin is also exported and some coal and iron ore are mined. But the traditional commercial products – cocoa, groundnuts, palm kernels, rubber and timber – remain important. About three-quarters of the people still depend on farming and Nigeria is almost self-sufficient in food. Manufacturing is developing quickly, especially at Enugu, Ibadan, Lagos, Port Harcourt and Sapele.

People. Nigeria has about 250 ethnic and language groups, the largest being the Hausa and Fulani in the north and centre, the Yoruba in the south-west and the Ibo in the south-east. Islam is

Slum clearance, to alleviate over-crowding and poor sanitation, produced the redevelopment of housing in central Lagos by the 1970s.

Inside the 15th-century walls of the city of Kano, in Nigeria, the inhabitants continue the life-style handed down to them by their ancestors.

The imposing Parliament House and grounds at Stormont in Belfast remain as a memorial of the time when Northern Ireland was self-governing.

the religion of about 44 per cent of the people, Christianity is practised by 22 per cent and ethnic religions by the remainder. The literacy rate is about 40 per cent.

The exact population figure for Nigeria is a matter of dispute. The 1973 census indicated 79,700,000 people, but allegations were made that the returns from some states were inflated for political reasons. The government therefore decided to set aside the 1973 census until another could be held. The United Nations estimate of Nigeria's 1975 population was 62,925,000.

Government. Nigeria became independent in 1960 as a federation of three states, North, West and East. Then in 1963 Nigeria became a federal republic and a fourth region, the Mid-West, was created from part of the West. Ethnic and cultural differences and rivalries between the regions, including fears of Northern domination, led to political instability. In 1966 two coups occurred and the

country came under the rule of a Supreme Military Council (SMC). The SMC consists of the heads of the armed forces and police and the military governors of the states. The chairman of the SMC is head of state, and the presidency and all political activity is banned. The only civilian representation is on a Federal Executive Council to which some matters are referred by the SMC. In 1976 the country was further reorganized into 19 states and the SMC announced plans to restore civilian rule by 1979.

History. Nigeria was a collection of individual states when Europeans – Portuguese, Dutch and British – first arrived in the late 15th and early 16th centuries. There were a series of civil wars, often connected with religion (Islam), in the early 19th century. The British annexed Lagos in 1861 and used it as an anti-slavery base. By 1885 Britain controlled most of the coast and between 1888 and 1897 its influence was extended inland. In 1900 Britain proclaimed its rule over Southern Nigeria and, between 1901 and 1903, it conquered the north. In 1914 the north and south, formerly administered as two protectorates, were united as the Colony and Protectorate of Nigeria.

In 1960 Nigeria became independent; in 1966 a military regime under Gen. Yakubu Gowon was established. In 1967 Gowon reorganized Nigeria into 12 states, although the people of the former Eastern Region attempted to secede by establishing their own nation, Biafra. The bitter civil war which followed ended in 1970 when the secessionists surrendered and their leader, Col. Ojukwu, went into exile. After the war the government sought to restore unity and the nation rapidly increased its prosperity through petroleum production. Then in 1975 another coup brought Gen. Murtala Mohammed to power, but he was assassinated in a military coup in 1976. He was replaced by Lt.-Gen. Olusegun Obasanjo. Map 32.

Nigeria – profile

Official name Federal Republic of Nigeria
Area 923,768sq km (356,667sq miles)
Population (1975 est.) 62,925,000
Density 68 per sq km (176 per sq mile)
Chief cities Lagos (capital) (1975 est.) 1,060,850; Ibadan 847,000; Ogbomosho 432,000; Kano 399,000
Government Military council
Religions Islam, ethnic, Christianity
Languages English (official), Hausa, Ibo, Yoruba
Monetary unit Naira
Gross national product (1974) £7,662,400,000
Per capita income £122
Agriculture Cassava, cocoa, cotton, groundnuts, maize, millet, palm kernels, rice, rubber, sorghum, timber, tobacco
Industries Brewing, cement, processed food, forest products, petroleum products, rubber, textiles, fishing
Minerals Petroleum and natural gas, coal, iron
Trading partners USA, Britain, Netherlands, France, West Germany

Niue (also known as Savage Island) is a coral island in the south-central Pacific Ocean, 2,150km (1,335 miles) north-east of New Zealand, to which it belongs. Copra and bananas are exported from the port and chief town, Alofi. Niue was discovered in 1774 by Capt. James Cook who named it Savage Island because of the hostility of the Polynesian inhabitants. It became a British protectorate in 1900 and was annexed to New Zealand a year later. Area: 260sq km (100sq miles). Pop. (1975 est.) 4,000. Map 62.

Norfolk is an island in the south-western Pacific Ocean, about 1,450km (900 miles) east of Australia, to which it belongs. Its large pine forests and warm climate attract many tourists. Farmers raise livestock and grow beans for export. Norfolk was discovered in 1774 by Capt. James Cook and made a British penal colony from 1788 to 1855. In 1844 it was annexed to Tasmania, in 1896 made a dependency of New South Wales, and in 1913 became a territory of Australia. Area: 34sq km (13sq miles). Pop. (1975 est.) 2,000. Map 62.

North Carolina. *See* UNITED STATES.

North Dakota. *See* UNITED STATES.

Northern Ireland, an integral part of the United Kingdom, was established in 1920 when the rest of Ireland broke away from the Union (*see* IRELAND, REPUBLIC OF). It consists of six counties (Antrim, Armagh, Down, Fermanagh, Londonderry and Tyrone) and two county boroughs (Belfast and Londonderry) of the province of Ulster, in the north-east of Ireland. The capital is Belfast. Northern Ireland has long had a troubled existence, and since 1968 has been disrupted by riots and terrorism encouraged by extremists intent on polarizing the differences between the Protestant majority and the large Roman Catholic minority. British troops have had the task of trying to keep the peace in an atmosphere of mutual hatred and mistrust.

Land and climate. Northern Ireland is a land of rolling, fertile plains surrounded by low mountains, mainly near the coast. The highest peak is Slieve Donard, which stands 852m (2,796ft) above sea-level in the Mourne Mountains in County Down in the south-east. The Sperrin Mountains rise to 683m (2,240ft) in the north-west, and along the north-east coast lie the Mountains of Antrim, a low basalt plateau rising to more than 550m (1,800ft). A feature of the northern coast of Antrim is the Giant's Causeway. This remarkable formation of some 40,000 polygonal blocks of basalt, some as much as 6m (20ft) high, stretches along the coast for nearly 10km (6 miles).

In the centre of Northern Ireland is Lough Neagh, the largest lake in the British Isles, with an area of 396sq km (153sq miles). Another large lake, Lough Erne, stretches across the centre of Fermanagh in the south-west. There are several large bays cutting into the coast, and these provide excellent harbours for cities such as Belfast and Londonderry. The climate of Northern Ireland is mild, averaging 15°C (59°F) in summer and 4.5°C (40°F) in winter. Rainfall varies from about 750mm (30in) on the lowlands to 1,500mm (60in) on some of the higher ground.

Economy. Northern Ireland is mainly agricultural, and more than 80 per cent of the land is suitable for farming. About two-thirds of this is devoted to mixed farming, and is owned mainly by small family concerns. The chief crops are potatoes and barley.

Some of the largest ocean-going vessels in the world have been built in the shipbuilding yards of the Harland and Wolff Company in Belfast.

Bangor, in Northern Ireland, has many attractions for tourists as a seaside resort, particularly during the annual yachting regatta.

Londonderry, on the River Foyle, is the second city of Northern Ireland and (then called Derry) originally grew up around an abbey founded by St Columba.

There are more than 1½ million cattle, nearly 1 million sheep, more than ½ million pigs and 12 million chickens.

Industry continues to increase in importance, and about 40 per cent of the working population is employed in manufacturing or construction. Engineering and allied industries are important, including shipbuilding and aircraft manufacture. The textile industry has diversified from the traditional linen into man-made fibres, and now contributes a third of Britain's output of synthetic fibres. Other important industries include clothing, footwear, food, drink and tobacco. Most of Northern Ireland's overseas trade is either with the rest of Britain or routed through Britain – some 84 per cent of exports and 75 per cent of imports. About 12 per cent of trade (import and export) is with the Republic of Ireland. Belfast is the chief port.

People. About two-thirds of the people of Northern Ireland are descended from English and Scottish Protestants, and most of the rest are of Irish Catholic descent. In the 1971 census, 53 per cent professed Protestant affiliations (26% Presbyterian, 22% Church of Ireland and 5% Methodist) and 31 per cent were Roman Catholics. Many Protestants belong to the Orange Order and are called Orangemen. This organization dates back to 1795 and relates to the defeat in 1690 by the Protestant William of Orange of the Roman Catholic ex-king James II in the struggle for Ireland. Orangemen celebrate William's victory in the Battle of the Boyne on Orange Day, 12 July.

Nearly all the people speak English, the official language, and a few also speak Gaelic. Education is administered by the Department of Education and five local boards. The Department is also responsible for community relations. Education is free and compulsory between the ages of 5 and 15. About a third of the children of secondary-school age attend grammar schools, the rest go to secondary (intermediate) schools. There are two universities, the Queen's University of Belfast (about 5,500 students) and the New University of Ulster at Coleraine (1,600 students).

Soccer is Northern Ireland's most popular sport, and this country of about 1½ million people has produced some of the world's finest footballers, such as George Best, Danny Blanchflower and Peter Doherty. Combined international sides with the Republic of Ireland are fielded in rugby union and some other sports.

Government. Northern Ireland is a part of the United Kingdom, with 12 representatives in the House of Commons at Westminster. It was once almost completely autonomous, but the separate parliament (Stormont, with a 52-member House of Commons and a 26-member Senate) and executive government established for Northern Ireland in 1921 was suspended in 1973. Direct rule (from Westminster) has been in force since 1972. Responsibility for the government rests with the secretary of state for Northern Ireland. The chief political parties are the United Ulster Unionist Coalition (Protestant), the Social Democratic Labour Party (Catholic), the Alliance Party and the New Unionist Party of Northern Ireland (Protestant). Local government in Northern Ireland is carried out by a system of 26 district and borough councils.

History. The political problems of Northern Ireland date from extensive settlement of Scottish and English Protestants in the reign of James I (1603–25). Ulster was the last pocket of resistance against the "planting" of Protestants in Ireland, and when the large landowners fled in 1607 – the "flight of the earls" – their vast estates were seized by the Crown and "planted". Catholic tenants retaliated with the "massacres of 1641", but Cromwell brutally crushed later rebellions and renewed planting.

The restoration of Charles II brought a relaxation of religious persecution in Ireland, although the Catholics still had few rights. They were gradually squeezed out of the Irish parliament, and two years after the Battle of the Boyne were completely ousted (1692). A series of laws deprived them of rights of property ownership, the franchise, education and office. Various measures of reform passed in the 19th century – such as admission to political office (1829) and disestablishment of the Irish Church (1869) – failed to arrest the demand from Catholics for the repeal of the Union. Home Rule was finally granted, over the protests of the Ulster Protestants, in 1914; but its implementation was delayed by the outbreak of World War I.

The Government of Ireland Act was passed in 1920 dividing Ireland into two separate countries, each with a certain amount of autonomy. This was rejected by the south, which eventually became the Irish Free State (1921), but was accepted by Ulster Protestants, and the province of Northern Ireland came into being. Relations between north and south improved, but many Roman Catholics in Northern Ireland opposed partition. A series of weak and sometimes blatantly anti-Catholic governments in Northern Ireland made little effort to win them over, and made matters worse by introducing electoral reforms that tended to operate against the Catholic minority: proportional representation was abolished in 1929, increasing the standing of the Unionists; the local government franchise favoured the Protestant majority; and the fixing of wards, or "gerrymandering", also had the effect of further separating the two groups.

Such a situation was ripe for exploitation by the IRA (Irish Republican Army), a now outlawed organization dedicated to the end of partition; they made frequent terrorist attacks in Northern Ireland from 1956 to 1962. The police (the Royal Ulster Constabulary) were armed and an all-Protestant volunteer force (the Ulster Special Constabulary) formed. Capt. Terence O'Neill, who became prime minister in 1963, attempted reforms, both electoral and economic, and tried to improve community relations. But before he resigned in 1969, frustrated as much by the reactionary elements in his own Unionist Party as by Catholic impatience, Ulster had plunged into an even deeper crisis.

The Northern Ireland Civil Rights Association (formed 1967) organized a series of protest marches in 1968. It resisted attempts to restrict their routes, resulting in police violence which in turn led to the formation at Queen's University of another protest movement, the People's Democracy. This escalation of events led to further riots, with extremists from both sides (such as Bernadette Devlin of the People's Democracy and the Protestant leader the Rev. Ian Paisley) advocating militancy.

Riots occurred in several towns in 1969, culminating in fierce fighting in Londonderry and Belfast in August. As a result, British troops were sent to Ulster, and the failure of the IRA's attempts to protect Catholic interests led to the formation of the Provisional IRA, a violent breakaway group. The army was at first welcomed by Catholics, but use of CS gas in the 1970 riots alienated them, and support grew for the Provisionals. Between 1969 and 1976 some 1,600 people were killed as a result of rioting and terrorism, including the spread of IRA bombing to England.

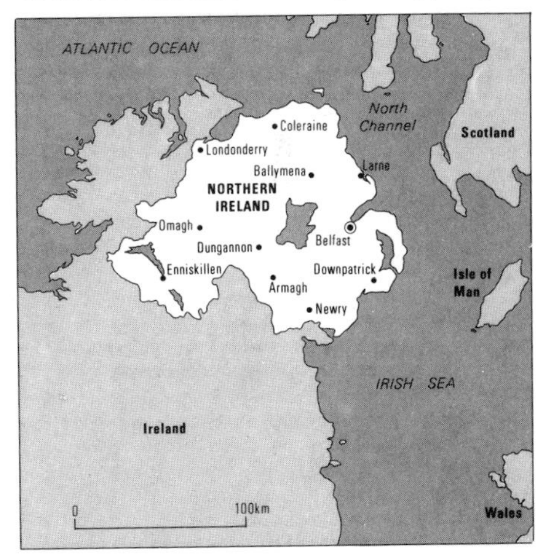

An intensification of the bloodshed in 1972 forced Britain to introduce direct rule and dissolve Stormont – which, despite Prime Minister Brian Faulkner's proposed reforms, had become powerless to reduce the violence. Protestant reaction was immediately provoked and ex-minister William Craig launched the Vanguard Movement, another extremist organization. The Ulster Defence Association (UDA), a Protestant para-military organization formed earlier in the year, began to prepare to do battle. In 1973 the British government established a 78-member Assembly (elected by proportional representation) and an Executive. Elections produced a Unionist coalition majority in the Assembly and Faulkner headed the power-sharing

Carrickfergus, whose castle was built in 1178, was formerly the assize town of Co. Antrim, Northern Ireland, until superseded by Belfast.

The ice-free port of Narvik, in northern Norway, is used for the shipping of iron ore from Sweden, whose harbours are frozen in the winter season.

Most Norwegian farms are small and farmers eke out a living from mixed farming by exploiting the timber resources which are usually part of their land.

Prime Ministers of Northern Ireland

Lord Craigavon (Sir James Craig)	1921–40
John Andrews	1940–43
Lord Brookeborough (Sir Basil Brooke)	1943–63
Capt. Terence O'Neill	1963–69
Maj. James Chichester-Clark	1969–71
Brian Faulkner	1971–72

Executive. He resigned, however, in 1974 when Protestant extremists organized a general strike in protest against power-sharing and the newly created Council of Ireland (a Dublin-Belfast axis). As a result, direct rule was resumed.

Further attempts to introduce power-sharing by means of a Constitutional Convention in 1975 and 1976 were unsuccessful. In July 1976 the newly appointed British ambassador to Dublin, Christopher Ewart-Biggs, was assassinated. In August two Roman Catholic women founded the Women's Peace Movement and held massive peace rallies. This and the failure in May 1977 of Protestant extremists to organize another general strike (this time in an effort to force the government to take stronger measures against the IRA) gave hope that the Northern Ireland controversy could eventually be settled by peaceful means. Map 88.

Northern Ireland – profile

Official name Northern Ireland
Area 14,121sq km (5,452sq miles)
Population (1975 est.) 1,537,000
Density 109 per sq km (282 per sq mile)
Chief cities Belfast (capital) (1972) 362,400; Londonderry 51,617
Government Parliamentary (suspended 1973; under direct rule by the British parliament)
Religions Protestant (53%), Roman Catholic (31%)
Languages English, Gaelic
Monetary unit Pound sterling
Agriculture Cattle, pigs, sheep, poultry, potatoes, barley, vegetables
Industries Engineering, dairy products, wool and textiles, food products
Major trading partners Britain, Republic of Ireland

Northern Territory. *See* AUSTRALIA.

Northwest Territories. *See* CANADA.

Norway (Norge), official name Kingdom of Norway, is a European nation that occupies part of the Scandinavian peninsula. It has a long, indented coastline, and most of the people live on or near the coast. The country has few natural resources, yet its people enjoy one of the highest standards of living in the world. The capital is Oslo.

Norway has a long tradition as a seafaring nation, and many of the country's heroes have been associated with the sea or with exploration. The great Viking leader Leif Ericsson reached North America some 500 years before Columbus "discovered" it; Fridtjof Nansen explored the North Polar Basin in the 1890s and later became a statesman, oceanographer and winner of the 1922 Nobel Peace Prize; Roald Amundsen became the first man to reach the South Pole (1911); and the anthropologist Thor Heyerdahl made remarkable Pacific expeditions in his balsa-wood raft *Kon-Tiki* in the late 1940s.

Land and climate. Most of Norway is a mountainous plateau, the only lowland areas being around Trondheim in the centre and Oslo in the south-east. The plateau consists mostly of bare rock, made smooth by the action of ancient glaciers. A striking feature of Norway is its jagged coastline, deeply indented by hundreds of fiords [*see* PE p.117], some of which penetrate farther inland. Norway also includes some 150,000 islands, most of which are small and uninhabited. The chief inhabited islands are Vesterålen, Lofoten Islands and Svalbard and Jan Mayen. The land rises steeply from the coast, and between the plateaus and the mountain ranges are deep valleys and swift-flowing rivers, and there are many mountain lakes. Norway's highest mountain is Galdhöpiggen, in the south, 2,468m (8,097ft) above sea-level. More than a third of the country lies north of the Arctic Circle, yet winters are unusually mild and the coast ice-free because of the warming effect of the Gulf Stream in the west. Summers are short and mild. Winds bring considerable rainfall in the west, and in winter there is snow cover for at least three months. Norway has more daylight in summer than any other inhabited region of the world; in the north the people have the light of the mid-night sun for 24 hours a day from mid-May to the end of July.

Economy. The geography and climate of Norway are not well suited to farming, and less than 3 per cent of the land is cultivated. Nor is the country rich in natural resources – except fast-flowing water, which has been utilized since 1900 to provide abundant hydroelectric power, and timber (about three-quarters of the country is covered in forests). For these reasons Norwegians have turned to the sea for a living. Norway's merchant fleet, with 2,800 ships (nearly 27 million gross tonnes), is the fourth largest in the world; it has been the major factor in closing the country's trade gap. Norwegian fishermen take about 5 per cent of the world's total catch (cod, mackerel, haddock, herring and capelin), and fish processing is an important industry. There have been rapid attempts to develop the reserves of petroleum and natural gas discovered beneath the North Sea in the late 1960s. Ships, fish and crude oil are all major exports.

Manufacturing and engineering are important industries; chief products include aluminium [*see* MM p.34], iron and steel, machinery, and paper and other timber products. Norway negotiated free-trade agreements with the EEC in 1973. Its chief customers are Britain (24%), Sweden (16%) and West Germany (10%); its chief suppliers are Sweden (19%), West Germany (16%) and Britain (10%).

People. The people are chiefly of Nordic origin, closely related to the Danes and Icelanders. About 20,000 Lapps live in the far north, where there are also people of Finnish ancestry. The established religion is the Evangelical Lutheran Church, to which 96 per cent of the people belong. Two forms of Norwegian are recognized equally as official languages. About 80 per cent of schoolchildren learn the older form, Bokmål (also called Riksmål) as their chief language, whereas the other 20 per cent learn the new form, Landsmål (or Nynorsk). The long-standing dispute about language will probably be settled in favour of a modified form of Bokmål (book language). Education is free and compulsory between the ages of 7 and 16, and in the early 1970s nearly half of all 18-year-olds received full-time schooling. All schools for the 17-19 age group are being made comprehensive (begun in 1976). Norway has four universities; the largest, Oslo (founded in 1811), has about 20,000 students.

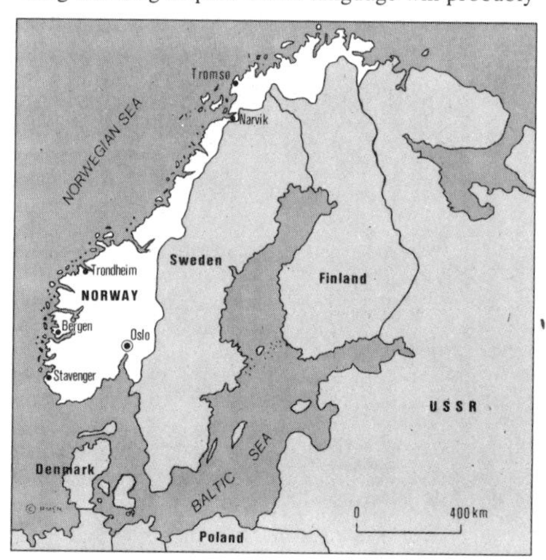

Norwegians enjoy a particularly high standard of living, with modern homes, state pension and health schemes, and many cultural and sporting facilities. Norway has contributed to most fields of culture and the arts. In literature the playwright Henrik Ibsen (1828–1906) was a great innovator and is often referred to as the "father of modern drama", and other writers include the Nobel prize-winners Bjørnstjerne Bjørnson (1832–1910), Knut Hamsun (1859–1952) and Sigrid Undset (1882–1949). In music the composer Edvard Grieg

Alesund, like Bergen, was destroyed by fire in 1904 and has been rebuilt entirely in stone. Today it is one of the largest fishing ports in Norway.

Bergen, for centuries one of Norway's important ports, was four times destroyed by fire before 1855, since when wooden buildings have been banned in the city.

Oman, like many of its neighbouring Arab nations, derives almost its entire wealth from petroleum; increasingly, more is being refined locally.

(1843–1907) was influenced by Norwegian folk-music; he wrote the incidental music to Ibsen's *Peer Gynt*. Painter Edvard Munch (1863–1944) was one of the forerunners of Expressionist art, and sculptor Gustav Vigeland (1869–1943) laid out Frogner Park in Oslo and exhibited many of his works there.

Most Norwegians are performers at sport rather than spectators. Skiing is the national sport – particularly ski-jumping, at which they excel. Many enthusiasts ski all the year round, and skiing is also an essential means of transport in parts of the country. Many people also enjoy walking and climbing in the mountains, skating, swimming, boating and fishing. Sport is predominantly amateur, but one famous Norwegian who turned professional was Olympic and world figure-skating champion Sonja Henie, who achieved film stardom in Hollywood.

Government. Norway is a constitutional and hereditary monarchy (with direct male-line ascent). Legislative power is vested in parliament (the *Storting*), which has 155 members elected to four-year terms by popular vote using proportional representation. The members choose a quarter of their number to form the *Lagting* (upper house), the rest forming the *Odelsting* (lower house). The king has nominal executive power, exercised through the Council of State (cabinet), headed by the prime minister. The king may veto a bill only twice. The country is divided into 19 counties *(fylker)* for local administration.

History. Norway's great seafaring tradition began with the Vikings, who sailed the seas in their long-ships and terrorized people on the coasts of western Europe from the late AD 700s. Norwegian Vikings colonized Iceland in the late 9th century. Western Norway was first united as a kingdom in 872 by Harold Fairhair. But when the male line of the monarchy ended in 1319, Norway began nearly 600 years of domination by its neighbours Sweden and Denmark.

United first with Sweden, Norway was soon devastated by the plague (1349–50), losing two-thirds of its population. Denmark joined the union in 1397, and when Sweden seceded in 1523 Denmark remained the dominant partner until 1814. It then had to hand Norway over to Sweden as part of the Peace of Kiel at the end of the Napoleonic Wars, but Norway refused to recognize the treaty, adopted a constitution of its own, and chose a Danish prince as king. Sweden refused to grant Norway independence, and used force to secure the union.

It was not until 1905 that the Norwegians finally persuaded Sweden to agree to their independence and a Danish Prince Charles, (Carl), became Haakon VII of Norway. The country had already started to utilize its hydroelectric potential, and the economy began to expand rapidly. – Norway's Industrial Revolution took place between 1905 and 1914. The country remained neutral during World War I, but its merchant fleet carried cargo for the Allies and suffered considerable losses. In World War II Norway was quickly occupied by Germany (in 1940), and King Haakon set up a government-in-exile in London. Despite the efforts of the Nazi Norwegian leader Vidkun Quisling to establish close collaboration with the Germans, the Norwegians continued courageously to defy the Nazis. At the end of the war the retreating Germans virtually destroyed the northern counties of Finnmark and Troms, an area the size of Austria.

Norway made a good economic recovery after the war, thanks to American aid and the determination of its people. Norwegian statesman Trygve Lie became the first secretary-general of the United Nations (1946), and Norway was a founder member of NATO (1949) and EFTA (1960). In the 1965 elections the Labour Party, dominant for 30 years, was defeated by a non-Socialist coalition, so beginning a period of minority governments. In a referendum in 1972 the Norwegian electorate rejected the proposed entry to the EEC. The development of Norway's North Sea oilfields in the 1970s promised to boost the economy. Map 16.

Norway – profile

Official name Kingdom of Norway
Area 323,886sq km (125,053sq miles)
Population (1976 est.) 4,027,000
Density 12 per sq km (32 per sq mile)
Chief cities Oslo (capital), 465,300; Bergen, 214,000; Trondheim, 134,000
Government Head of state, King Olav V (succeeded 1957)
Religion Lutheran
Language Norwegian
Monetary unit Norwegian krone (plural kroner)
Gross national product (1974) £9,064,300,000
Per capita income £2,251
Agriculture Forestry, fishing, livestock, cereals
Industries Petroleum, iron and steel, aluminium processing, timber products, dairy products, shipping, chemicals
Trading partners Sweden, Britain, West Germany, USA, Denmark

Nova Scotia. *See* CANADA.

Nyasaland. *See* MALAWI.

Ohio. *See* UNITED STATES.

Oklahoma. *See* UNITED STATES.

Oman ('Umān), until 1970 Muscat and Oman and official name Sultanate of Oman, is an independent Arab nation on the south-eastern Arabian peninsula bordering on the Arabian Sea and the Gulf of Oman. The Hajar Mountains, highest point Jebel Sham (3,018m; 9,900ft), run parallel to the gulf, and the interior of the country is a gently sloping plain. The capital is Muscat. Oil is the major export; other important industries are fishing and agriculture, with sugar cane, dates, olives and cereals as the main crops. Most of the people are Arabs; there are minority groups of Baluchis, Persians and Indians.

The Portuguese settled the port of Muscat in 1508 and held the region until it was taken by Turkey in 1659. In 1741 Oman was re-taken by Ahmad Ibn Said of Yemen, who founded the present royal line. The country was one of the most powerful of the Arabian states in the early 19th century, but in 1856 its control over Zanzibar and most of the coast of Iran and Baluchistan was lost and Oman became politically and economically dependent on Britain. British influence was reaffirmed by treaties in 1939 and 1951; a rebellion against the sultanate (1954–57) was suppressed with aid from Britain. The end of formal British influence came in 1965

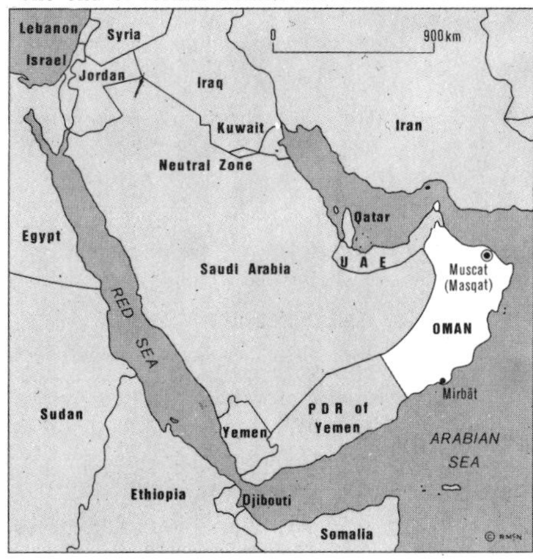

following demands by the United Nations. In 1970 Qabus bin Said deposed his father, Sultan Said bin Timur, and claimed by the end of 1975 to have quelled rebel activity. In 1971 Oman joined the Arab League and the United Nations. Area: 212,457sq km (82,030sq miles). Pop. (1975 est.) 766,000. Map 38.

Ontario. *See* CANADA.

Orange Free State. *See* SOUTH AFRICA.

Oregon. *See* UNITED STATES.

Pacific Islands Trust Territory includes about 2,000 islands and islets in a large area of the east-central Pacific Ocean. The major island groups are the Caroline, Marianas and Marshall islands. The territory is held by the United States under a trusteeship of the United Nations, and administered from

Islamabad is a new city constructed since 1960 to replace Karachi as the capital of Pakistan; the sun beats down on the National Assembly Building.

Pakistan's extensive building programme creates a great demand for ceramics (bricks and tiles) and, as shown here, quantities of cement.

The palace and mausoleum of the emperor Jahangir at Lahore are among many historic monuments in the city that contains Pakistan's oldest university.

Saipan in the Marianas. Copra, sugar cane, coffee and citrus fruits are produced by the Micronesian inhabitants. There are also deposits of phosphates and manganese ores. Once belonging to Germany, the islands were taken by Japan in 1914 and formally mandated to it by the League of Nations in 1922. The United States was given the trusteeship in 1947 after having occupied the territory during World War II. Area: 1,860sq km (718sq miles). Pop. (1975 est.) 120,000. Map 62.

Pakistan, official name the Islamic Republic of Pakistan, is a country of Asia in the northwestern corner of the Indian subcontinent. Until 1947 it was part of British-ruled India; in that year it was established as a separate homeland for India's Muslims, who feared Hindu dominance when India became independent. Pakistan is a poor country with few natural resources, and its economy has suffered from political upheavals and from the loss in 1971 of its eastern province – now the independent state of BANGLADESH – which had some of the most profitable industry. The name *Pakistan* means "land of the pure"; the capital is Islamabad.

Land and climate. Pakistan's north-eastern frontier is in Jammu and Kashmir, a territory whose possession it disputes with India. It has four provinces, each of them a land with a long and individual tradition: Punjab, North-West Frontier, Sind and Baluchistan. Much of the country is wild and inhospitable, although there is a great variety of terrain – mountains, plains, swamps and deserts. The north-west is overshadowed by the immense peaks of the Himalaya, Hindu Kush and Karakoram ranges. At the foot of the Hindu Kush, on the border with Afghanistan, is the stony Khyber Pass, the route followed by many invaders of India. The great River Indus, with its many large tributaries, flows south-westwards through the country, watering the fertile plains of the Punjab ("Five Rivers"); at its mouth it forms an alluvial lowland. East of the river is the barren and sandy Thar Desert. Pakistan's climate is monsoonal, but ranges from cool and wet in some mountainous regions to extremely hot and dry in the deserts.

Economy. Pakistan's natural resources include large deposits of natural gas and small amounts of other minerals. Among them are coal, iron ore, chromite, gypsum, limestone, sulphur and antimony. Some petroleum is produced.

About two-thirds of the people live by farming. Although the fertility of a great deal of land has been improved by skilful irrigation and drainage [*see* MM p. *197*], agricultural output is generally low because of the shortage of fertilizers and the use of old and inefficient farming techniques. Pakistan is generally self-sufficient in food. Its chief food crops are wheat and rice, but maize, sugar cane, millet and barley are also grown and parts of Baluchistan have long been known for their dates and other fruits. Cotton is a major crop, both for domestic needs and for export, and some tobacco is grown. The sea, lakes and fish farms provide a variety of fish for a large home market and for export.

The chief manufacturing industry is the making of cotton textiles. Food processing and leather working employ many people, and cement, paper, metal goods and fertilizers and other chemicals are produced. Cottage industries include the making of carpets, cloth and pottery. The railways are well developed, and road connections are also good between the major cities. But in mountainous and desert regions communications are difficult. Pakistan International Airlines (PIA) operates on domestic routes and to cities in many other countries.

People. The population is mixed in ancestry, being descended from the various invaders who, during many centuries, have made their way through the mountains into the subcontinent; among them have been Aryans, Greeks, Arabs, Persians, Afghans and Mongols. Today there are several distinct ethnic groups, however, the largest being the Punjabis, who make up more than 60 per cent of the population. Others include the Sindhis (13 per cent), and the Pathans, Baluchis and Kashmiris. The population is unevenly distributed, most people living in the Indus lowlands; the heaviest concentration is around Karachi in the south. Some of the desert and highland regions have only a scattering of people.

The official language of the country is Urdu, but English is recognised as a common language for business and government, and many people speak such other languages as Punjabi, Sindhi and Pushtu. Primary education is free (the period of schooling has, for the time being, been fixed at five years) and there are seven universities. About 16 per cent of the population is literate. Almost all the people (98 per cent) are Muslims, and Islam has a central place in the state. But the rights of religious minorities are protected, and there are small numbers of Hindus, Buddhists, Christians and Parsees.

Pakistan's largest city is the seaport of Karachi. Other important and historic cities include Lahore, Rawalpindi, Multan and Quetta. The new and well-planned capital city of Islamabad, near Rawalpindi, has as yet a population of less than 250,000.

Government. Under the Constitution of 1973, the country has a federal parliamentary system, with an elected National Assembly and a Senate. The head of state is the president, and the chief executive is the prime minister.

History. The political and cultural history of Pakistan until 1947 is part of the history of INDIA. The new country came into being on 14 August 1947 with Mohammed Ali Jinnah – the president of the Muslim League who had fought so hard for Pakistan's creation – as governor-general, and Liaquat Ali Khan as prime minister. The country then consisted of two parts, West Pakistan and East Pakistan (formed from East Bengal and the Sylhet District of Assam), separated by 1,450km (900 miles) of Indian territory. It remained a member of the Commonwealth of Nations with the British sovereign as head of state. The transfer of population (Muslims to Pakistan and Hindus to India) which followed the partition of the old India resulted in communal fighting in which thousands of people lost their lives. Later, Pakistan was involved in bitter fighting with India over the status of Kashmir – a dispute that is still unresolved.

In 1956 the country became a republic; under its new constitution the president was required to be a Muslim. The first president was Gen. Iskander Mirza who, two years later, annulled the constitution and dismissed the government. Within a few

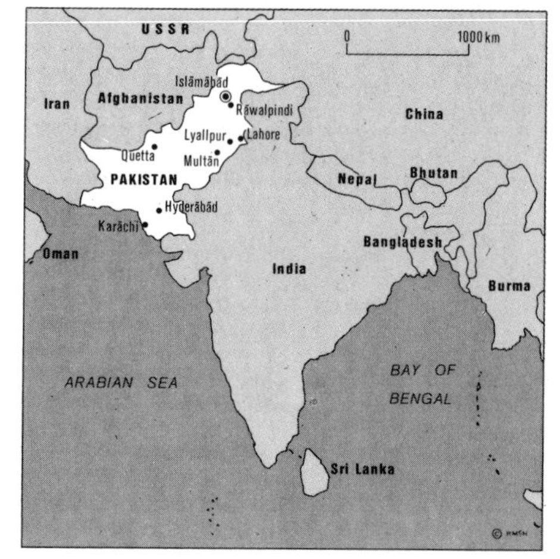

Pakistan – profile

Official name Islamic Republic of Pakistan
Area 803,943sq km (310,402sq miles)
Population (1975 est.) 70,260,000
Density 87 per sq km (227 per sq mile)
Chief cities Islamabad (capital) 250,000; Karachi, 3,442,000; Lahore, 1,985,000
Government Republic; federal parliamentary system
Religion Islam
Languages Urdu, English
Monetary unit Pakistan rupee
Gross national product (1974) £3,747,900,000
Per capita income £53
Agriculture Wheat, rice, cotton, maize, sugar cane, dates
Industries Textiles, fishing, processed foods, building materials, chemicals, carpets
Minerals Natural gas, petroleum, coal, iron, gypsum
Trading partners Britain, Japan, USA, West Germany

The capital of Panama, Panama City, still has on its fringes people who eke out a living selling local produce, such as bananas and fish.

Swampland still covers a large part of Panama, despite schemes of land reclamation, and only seven per cent of the land area is suitable for farming.

Asunción, the capital of Paraguay, is the country's main industrial centre and port; its buildings have a mixture of modern and Spanish-style architecture.

weeks, Mirza was replaced as president by the commander-in-chief of the army, Gen. Mohammed Ayub Khan, who governed by decree. In 1969, after a prolonged outbreak of riots and strikes, Ayub Khan handed over to Gen. Yahya Khan, who proclaimed martial law and put down disorder.

General elections were held at the end of 1970, the intention being that a new constitution should afterwards be agreed. But the successful parties in East and West Pakistan could not reach agreement: the Awami League demanded autonomy for the East, and a civil war gradually built up between West Pakistan forces and East Pakistan guerrillas. The Indian Army intervened, successfully, on the side of the guerrillas, and in March 1971 East Pakistan became the independent country of Bangladesh. President Zulfikar Ali Bhutto, who had succeeded Yahya Khan, announced in 1972 that Pakistan had left the Commonwealth in protest against Britain's recognition of Bangladesh.

In 1974 Pakistan accepted the independence of Bangladesh, and President Bhutto succeeded in restoring good relations with India. Widespread civil disorder followed a general election in 1977 in which Bhutto was seemingly confirmed in office; his opponents accused him of having falsified the electorial returns and in July he was overthrown by a military coup. Map 40.

Palestine. *See* ISRAEL.

Panama (Panamá), official name Republic of Panama, is an independent nation in Central America on the narrowest part of the isthmus joining North and South America. The capital is Panama City. Panama is governed by a military junta and its economy is almost entirely dependent on the CANAL ZONE, an area astride the Panama Canal administered by the United States.

Land and economy. There is a marked contrast between the Caribbean and Pacific sides of the isthmus. The Caribbean coastal strip, running east-west, produces most of the country's food, and the plains on the Pacific coast have fertile valleys in which farmers grow cereal crops (principally rice), bananas and raise livestock. Veraguas, a region of dense rain forests, produces coffee. The Caribbean plains west of the Canal Zone produce cocoa and rubber, and the lowlands in the east yield bananas. Two mountain systems, Sierra de Chiriquí and Cordillera de Veraguas, include volcanic peaks; the highest is Chiriquí, 3,475m (11,400ft). In 1968 a United Nations survey team discovered deposits of copper ore in the provinces of Colón and Darien.

People. Most Panamanians are Spanish-speaking Roman Catholics descended from Spanish colonists, immigrant West Indians and local Indians. Many work in agriculture, but the country produces only 40 per cent of its food requirements. Industries include the manufacture of cigarettes, clothing and food processing. Political parties were abolished in 1969 and the government rules by decree.

History. All Spain's major explorers had contact with Panama: Rodrigo de Bastidas explored it in 1501, Christopher Columbus sighted it in the same year, Vasco Balboa landed there on his way to the discovery of the Pacific Ocean, and Francisco Pizarro went there on his way to Peru in 1531. Panama remained under Spanish rule until 1821, when it joined the Confederation of Greater Colombia (its southern neighbour). It achieved independent status in 1903, principally as a result of its strategic position – made even more important after the building of the Panama Canal (1904-14). The United States (and its armed forces) played a major role in the development of the country (*see also* MM pp. 196–7). From the 1940s to the 1960s, there was a stormy period of rapid changes of government, rioting, assassinations and political intrigue. A military coup overthrew the presidency of Arnulfo Arias in 1968, political parties were abolished a year later, and a new constitution enacted in 1972.

In 1974 Panama and the United States agreed to discuss plans aimed at giving Panama more revenue and eventual jurisdiction over the Panama Canal. A new constitution was planned for 1978. Area: 75,651sq km (29,209sq miles). Pop. (1976 est.) 1,719,000. Map 74.

Panama Canal Zone. *See* CANAL ZONE.

Papua New Guinea is an independent nation of the western Pacific Ocean, consisting of the eastern half of NEW GUINEA, and the neighbouring islands of the Bismarck Archipelago (including New Britain, New Ireland and the Admiralty group) and Bougainville. The capital is Port Moresby. Most of the terrain is mountainous and there are many volcanoes, particularly in the Bismarck Archipelago. The only large lowland areas are to the north and south of the Central Highlands in New Guinea. Situated just south of the Equator, the island has a hot, wet climate throughout the year, giving rise to the natural vegetation of rain forest. Some of the world's most primitive people, with cultures resembling those of the Stone Age, live in Papua New Guinea and because of the difficult terrain isolated tribes are still being discovered. Farming is basically subsistence agriculture, with coffee, cocoa and copra as the main cash crops. The forests are now being exploited for hardwoods, and a large deposit of copper ore has been discovered on Bougainville.

New Guinea was charted by Portuguese and Spanish explorers, but little was known about the island until German traders began operating there in the 1870s. Germany claimed the north-eastern region (Kaiser-Wilhelmsland) in 1884, Australia occupied it during World War I, and the whole eastern part of the island became an Australian mandate in 1920. In 1973 Papua New Guinea became self-governing. In 1975 it became a fully

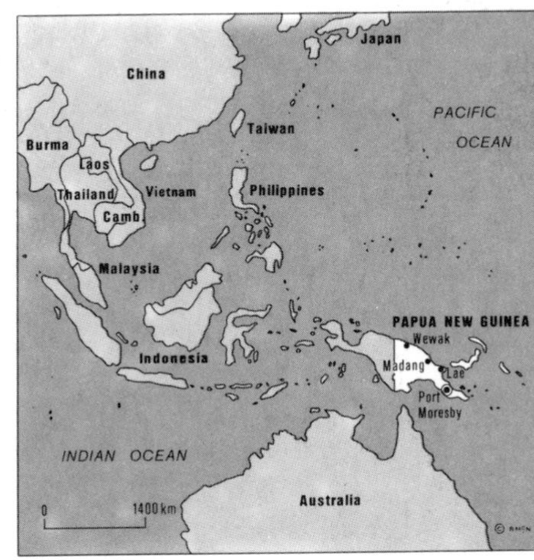

independent state within the Commonwealth. Government is through a House of Assembly which is elected by every citizen who is more than 18 years old. Area: 475,369sq km (183,540sq miles). Pop. (1975 est.) 2,756,000. Map 60.

Paraguay, official name Republic of Paraguay, is an independent South American nation. It is a land-locked country whose capital, Asunción, was founded in 1537. Nearly all Paraguayans are of mixed Spanish and American-Indian (Guaraní) descent, living in a society based on an agricultural economy. Spanish is the official language, although 90 per cent of the population speaks Guaraní. Most of the people are Roman Catholics, and the literacy rate is about 30 per cent.

Land and economy. Paraguay is surrounded by Argentina, Brazil and Bolivia and its agricultural products reach the Atlantic Ocean by means of a

The city of Arequipa, Peru, has many fine examples of Spanish colonial style architecture, most of which had to be restored after a severe earthquake in 1868.

Machu Picchu, the "lost city" of the Incas, was discovered in 1911 80km from Cuzco. It is the best surviving evidence of the ancient culture of Peru.

Fishing for anchoveta is a particularly important industry in Peru, and at one time catches of this fish were the largest of any species caught in the world.

river system flowing through Argentina. The Paraguay River, 2,550km (1,585 miles) long, divides the country into two regions: a temperate eastern zone of rolling hills, forest and grasslands, and a western area of scrub forest, uncertain rivers and open plains with little rainfall (the Chaco). There are few minerals; 90 per cent of exports are meat, timber, cotton, coffee and tobacco. Cattle-raising is the chief form of agriculture.

History. The Spaniard Alejo Garća was probably the first European to reach Paraguay, in about 1525. Reports of his expedition into the Inca Empire encouraged the explorations of Sebastian Cabot between 1526 and 1529. Other explorers followed the courses of Paraguay's rivers in an attempt to find a route across the continent. Paraguay became part of the Spanish province of Río de la Plata in the 1550s but was split from Argentina in 1617. Jesuit missionaries established agricultural colonies and tried to Christianize the Indians. In 1721 José Antequera Castro led a revolt, proclaimed independence from Spain and ruled the country for ten years. The Jesuits were expelled in 1767 but nine years later Paraguay was again incorporated into the viceroyalty of Río de la Plata. A bloodless revolution in 1811 finally resulted in lasting independence.

There followed a series of dictatorships until the War of the Triple Alliance against Argentina, Brazil and Uruguay (1865–70), in which more than half the population of Paraguay was killed. Short-lived dictatorships filled the political vacuum that ensued until the Chaco War of 1932–35, when Paraguay fought Bolivia over a border dispute. Paraguay won but at a great cost to its economy and the country again had a series of short-term governments. In 1954 Gen. Alfredo Stroessner led a successful military coup and after 1958 was re-elected for four consecutive five-year terms.

Maps 76, 78.

Paraguay – profile

Official name Republic of Paraguay
Area 406,752sq km (157,047sq miles)
Population (1976 est.) 2,742,000
Density 6.7 per sq km (17 per sq mile)
Chief cities Asunción (capital) (1975 est.) 574,000; Villarrica, 38,050; Encarnación 240,200
Government Republic; head of state President Alfredo Stroessner
Religion Roman Catholic
Languages Spanish (official), Guaraní (national)
Monetary unit Guaraní
Gross national product (1974) £512,800,000
Per capita income £188
Agriculture Maize, wheat, beans, groundnuts, tobacco, citrus fruits, beef cattle, timber
Industries Meat products, leather, timber products, tannin extract, vegetable oils
Minerals Metal ores
Trading partners Argentina, Brazil, USA, West Germany, Netherlands, Britain

Pennsylvania. *See* UNITED STATES.

Peru (Perú), official name Republic of Peru, is the third-largest nation of South America and home of the ancient Inca Empire. The country is ruled from Lima, the capital, by a military junta and has an economy based mainly on the exploitation of its rich deposits of minerals.

Land and economy. Peru shares boundaries with five other South American countries – Chile, Bolivia, Brazil, Colombia and Ecuador – and has a coast 2,250km (1,400 miles) long on the Pacific Ocean. Its immense mountain system, the Cordillera de los Andes, divides the country into three geographical zones. The western, coastal region (costa) is a dry strip of desert that produces 50 per cent of the country's income and where 40 per cent of the people live. The central region is a dry and cold mountain chain that hinders transport and communications. It contains rich mineral deposits and has 50 per cent of the country's population. Mt Huascarán, 6,678m (22,205ft), the highest point in Peru, is in the north of this region, and the southern end extends as far as Lake Titicaca, at 3,810m (12,500ft) above sea-level the highest body of navigable water in the world [*see* PE p.65]. The third division (montaña) is in the east. It is a hot, moist region of tropical forests and largely unexplored jungles; prospectors search for oil in the montaña.

Since the time of the Incas, Peru has been known for its abundant minerals – ores of lead, copper, zinc, iron, silver, cadmium, tin and gold as well as coal, marble, limestone and now oil. Half the population is employed in agriculture, and cotton, rice, coffee and sugar are exported. Fishing – particularly of anchoveta – is important [*see* PE p.241].

People. Descendants of the Incas make up 40 per cent of Peru's population; 11 per cent are Caucasian and 43 per cent are mestizo (of mixed European and American-Indian descent). Roman Catholicism is the state religion, practised by 90 per cent of the people. Spanish is the official language but many Indians speak Quechua and Aymara. Literacy is estimated at 60 per cent. The forest Indians are among the most primitive people in the world today.

History. When the Spanish landed in 1531, Peru was the centre of an advanced Inca civilization. Francisco Pizarro, in search of Inca treasure, conquered the country after 1532; his numerically inferior troops had the advantage of horses and firearms, neither of which had been seen in the land before. By 1542 Peru had become a major source of Spanish wealth and power. An attempted revolt by the Indians in the 1780s was crushed by the colonial power. In 1820–24 José de San Martín and Simón Bolívar led a successful War of Independence and declared the country free from Spain. Peru formed a confederation with Bolivia in 1836, but Chile regarded the union as a threat and attacked the confederation, which was disbanded after the Battle of Yungay in 1839. Civil unrest followed until the presidencies of Gen. Ramón Castilla (1844-50 and 1855-62). A republic was proclaimed in 1860 but Spain continued to try to reconquer the country until defeated in 1866; formal recognition of Peru's independence came in 1879. Chile again declared war on Peru in that same year, but this time Chile won and in 1883 took some of the southern provinces, although formal ownership was not settled until 1929.

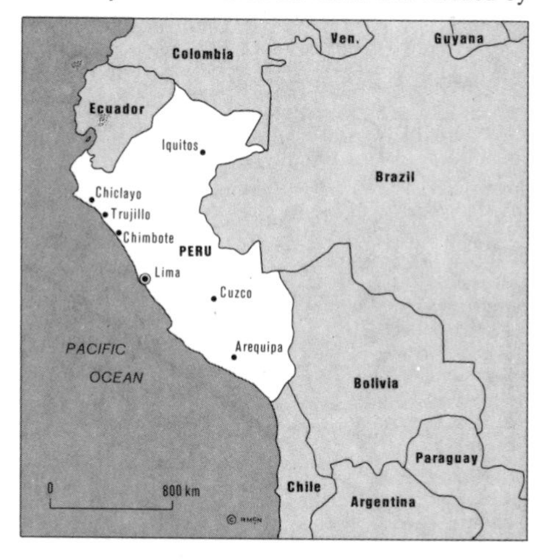

The first half of the 20th century was a time of political rivalry and insurrection. The military seized power in 1963 and held elections, but in 1968

Quechua Indians of Peru who have been fortunate enough to retain their land practise subsistence agriculture and exchange crops at local markets.

The urbanization of the Philippines and the growth of a modern transport network has made travel by traditional dug-out boats more a pleasure than a necessity.

The rebuilding of Warsaw after its destruction in World War II has been accomplished with careful regard to restore its ancient appearance.

the new government was overthrown by a military junta, which dissolved parliament and suspended the constitution. It undertook a programme of agricultural reform, nationalization of essential industry, and reorganization of education. The military government had to take firm action in 1976 against a wave of strikes (particularly in the anchoveta industry). Map 76.

Peru – profile

Official name Republic of Peru
Area 1,285,216sq km (496,222sq miles)
Population (1975 est.) 16,090,000
Density 12.5 per sq km (32 per sq mile)
Chief cities Lima (capital) (1972) 2,833,609; Callao, 335,400; Arequipa, 194,700
Government Military junta led by Francisco Bermúdez
Religion Roman Catholic
Language Spanish, Quechua (both official)
Monetary unit Sol
Gross national product (1974) £4,559,800,000
Per capita income £283
Agriculture Beef cattle, cotton, sugar cane, rice, coffee, sheep, fish, maize, tobacco
Industries Fish meal, processed foods, textiles, chemicals, metal products, cars
Minerals Copper, iron ore, lead, petroleum
Major trading partner USA

Philippines (Filipinas), official name Republic of the Philippines, is an island nation of south-eastern Asia in the south-western Pacific Ocean between the South China Sea and the Philippine Sea. It consists of more than 7,000 islands and islets; the new capital, Quezon City, is near the old capital of Manila on the largest island, Luzon. The official language is English but most of the people – predominantly a Roman Catholic farming community – speak Tagalog, also known as Filipino (Pilipino).

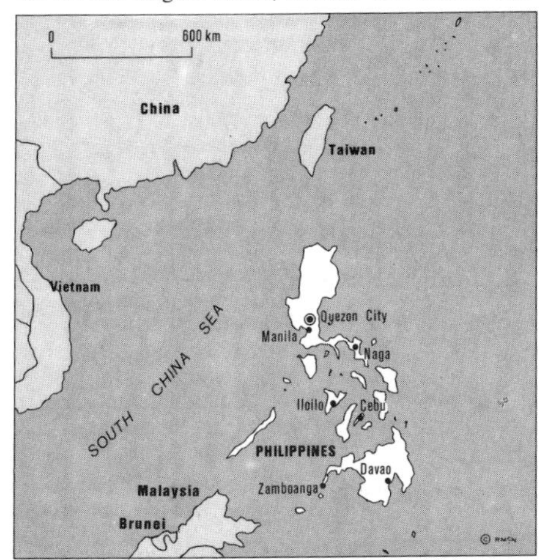

Land and economy. The two largest islands of the Philippines, Luzon in the north and Mindanao in the south, are separated by a number of smaller islands known as the Visayan group. Most of the islands are mountainous (some are volcanic), with the densest population concentrated in mountain plains. The climate is tropical and about 40 per cent of the land area is covered with forest; timber is the principal export. The mountains are rich in mineral deposits, many still undeveloped. The country's economy is based on agriculture, with rice as the staple food; the Philippines is the world's largest producer and principal exporter of coconuts and coconut products. The government has encouraged the development of manufacturing industry.

People and government. Most Filipinos are descended from people who travelled to the islands from south-eastern Asia or Indonesia. The short, black-skinned Aetas (Balugas) are the only remaining aboriginals. The republican government consists of an executive branch with a president, elected for a four-year term; a legislative branch made up of a House of Representatives of not more than 120 members elected every four years; and a Senate of 24 members elected every six years. There is also a judicial branch headed by a Supreme Court.

History. Muslims arrived in the late 15th century, but the first mass contact with the outside world was in the 1500s with the arrival of the Spanish. Ferdinand Magellan discovered the islands in 1521 and claimed them for Spain. The first permanent settlement was established on Cebu in 1565, and Manila was settled in 1571. Spanish soldiers methodically conquered the major islands, and the friars who accompanied them systematically converted the people to Roman Catholicism. Spain did little but exploit the natural resources until the 1800s when, plagued by European troubles, their grip was relinquished. The opening of the Suez Canal in 1869 improved access to European markets and boosted the economy. In 1899, after the United States had defeated Spain at the Battle of Manila Bay, the Philippines were ceded to the United States. Although promised independence, the Philippines were not established as a republic until 1946 because of delays caused by the American government and the islands' involvement in World War II. In 1972 President Ferdinand Marcos declared martial law and under his leadership the Philippines have loosened their ties with the United States. Map 50.

Philippines – profile

Official name Republic of the Philippines
Area 300,000sq km (115,830sq miles)
Population (1976 est.) 43,751,000
Density 146 per sq km (378 per sq mile)
Chief cities Quezon City (capital) (1973) 896,200; Manila (former capital), 1,435,500; Davao, 463,700
Government Head of state, president Ferdinand Marcos
Religion Christianity (87% Roman Catholic)
Languages Pilipino, English (official)
Monetary unit Peso
Gross national product (1974) £5,568,300,000
Per capita income £127
Industries Petroleum products, tobacco, plywood, veneers, paper
Agriculture Rice, maize, coconuts, sugar cane, timber, sweet potatoes
Minerals Silver, gold, copper, zinc
Trading partners (major) Japan, USA

Pitcairn is one of a group of four volcanic islands in the southern Pacific Ocean half-way between New Zealand and Panama. It is a British possession, administered by New Zealand. Its residents are descendants of ten mutineers (led by Fletcher Christian) from the British ship HMS *Bounty* who together with six men and 12 women from Tahiti landed on Pitcairn in 1790. (The remains of their ship were found off the south-eastern end of the island in 1957.) In 1831 the island became overpopulated and the inhabitants were removed to Tahiti; they returned in 1832. Britain took formal possession of Pitcairn in 1838. The island again became overpopulated in 1856 and the entire colony was transferred to NORFOLK Island, but within two years some families had returned. Today the islanders, who all live in Adamstown, grow and export fruit. Public revenue comes from the worldwide sale of postage stamps. The other, uninhabited, islands in the group are Ducie, Henderson and Oeno. Area: 6.5sq km (2.5sq miles). Pop. (1976) 67.

Poland (Polska), official name Polish People Republic, is a country of central Europe. It belongs to the so-called "Eastern bloc" of Communist countries but has retained some of the fierce individuality that has marked Polish life through the ages. Most of its farms, for example, are still privately owned, and the Roman Catholic Church (to which the majority of Poles belong) is still one of the dominant forces in the country. Throughout history Poland has been one of Europe's most troubled lands; according to one old joke, an essential part of every European peace treaty was a clause altering Poland's frontiers. In the opening stages of World War II in 1939, Poland was partitioned between Germany and the USSR. During the war the country was devastated and more than one-fifth of its population perished; at the war's end some 180,000sq km (69,500sq miles) of Polish territory were incorporated into the victorious USSR, Poland's ally, and Poland was given in return about 104,000sq km (40,150sq miles) of the territory of defeated Germany. The capital is Warsaw.

Poland

Steel production is one of Poland's major heavy industries; it uses locally mined iron ore and coal, and its products are used by Polish manufacturing industry.

The Leon Wyczokowski Museum on the canal at Bydgoszcz in Poland was built in the 17th century as a granary. The canal was constructed in the 18th century.

Wawel Cathedral and Castle, at Kraków, was built in the early 11th century, rebuilt in the 14th, and finally given a Gothic renovation in the 18th.

Land and climate. The greater part of Poland lies in the Central European Plain: the country's name comes from that of a Slav tribe called the *Polians* or "plain dwellers". Its coastline on the Baltic Sea has fine, sandy beaches backed by shallow lagoons. The northern part of the plain, particularly the area between the German border and the River Vistula (Wisla), has thousands of small lakes; to the east of the Vistula there are some vast and desolate regions of marshland. The southern part of the plain consists of rolling downland, interspersed with low rocky hills and marshy valleys. Farther south the plain rises slowly to the Sudete Mountains in the south-west, the Tatra in the south and the Beskids in the south-east. The Beskids are part of the Carpathian Range. Four-fifths of the country lies in the basin of the River Vistula, which is navigable for most of its course. Poland's western frontier with Germany follows the line of the Oder (Odra) and Neisse (Nysa Luzicka) rivers; this frontier, decided by the Allies at the Potsdam Conference in 1945, has been a subject of much controversy.

Poland's climate is generally equable. It tends to be warm and wet in summer and mild in winter, with average annual temperatures of 6°C (43°F) in the north and 8°C (46°F) in the south-west. Rainfall is 800 to 1,200mm (31–47in) in the mountains and averages 450mm (18in) in the lowlands.

Economy. The chief mineral resouce is coal, and Poland is one of the largest producers in Europe. The main coalfields are in Upper Silesia near the border with Czechoslovakia. Much of the coal is of high quality, and lignite (brown coal) is also mined. There are large reserves of copper, lead and zinc and smaller deposits of iron ore, sulphur and oil.

Agricultural output is high for central Europe. Almost 80 per cent of agricultural land is privately owned, in strong contrast to Poland's Communist neighbours, where only a small percentage of land is in private hands. The government aims at eventual collectivization of all land and encourages the formation of new collective farms by offering economic inducements. But so far the only regions that have been extensively socialized are the former German lands in East Prussia, Pomerania and Silesia where the estates of the Junkers were broken up at the end of World War II. The average size of a private farm is about 4ha (10 acres). The chief crops are potatoes and rye, of which Poland produces more than any other country except the USSR. In lower Silesia, however, where the soil is better than on the vast plains of the north and centre, wheat and sugar-beet are important. Other major crops are oats, barley, vegetables, fruits and flax. Dairy farming and pig-breeding are common in most parts of the country.

Nearly all manufacturing industries are state owned, and the country has recovered its prewar industrial strength. Steel production is important, and several kinds of engineering industries are associated with it. In recent years the chemical industry has greatly expanded; its products include fertilizers, petrochemicals and plastics. Other chief industries include the manufacture of textiles, building materials and electrical goods. Gdańsk and Szczecin have shipyards.

Although there is an extensive road network, many roads are poorly surfaced. Most heavy freight travels by rail, and many major rail routes have been electrified. The rivers Vistula and Oder are navigable, but water transport is of only minor importance. The Polish airline LOT operates on domestic and international routes.

People. The Poles are Slavs but also have the blood of Germanic and other peoples who have settled in Poland over the centuries. The country has a few small ethnic minorities, chiefly Ukrainians and Belorussians. The German population of the lands east of the Oder-Neisse line was expelled at the end of the war. The Polish language belongs to the West Slavonic group; it uses a Latin script. The country is traditionally Roman Catholic and today, despite the Communist regime's hostility to religion, four out of five Poles are Catholics. There are twice as many churches as before the war and there is a Roman Catholic university. The Polish Orthodox Church has about half a million members, and there are smaller numbers of Uniates and Lutherans. Education in Poland is free and compulsory between the ages of 7 and 15. There are ten universities, including those of Warsaw, Kraków, Wroclaw and Lódź.

The greatest density of population is in the south particularly in Upper Silesia. The growth of industry has attracted increasing numbers of people into the cities, although efforts are made to limit city expansion. At the end of the war in 1945 many cities were in ruins; 85 per cent of Warsaw had been destroyed, 70 per cent of Wroclaw and more than half of Poznań, Gdańsk and Szczecin. The smaller towns were also in ruins. Wherever possible town centres have been painstakingly rebuilt to look as they were before the war.

The development of the arts is, as in most Communist countries, a matter of government policy and has an earnest political and social purpose. The rigorous constraints that follow are to some extent balanced by the fact that state funds are made available to artistic organizations. Music and the theatre are probably the main beneficiaries. Among Polish composers of the past, the best-known is Frédéric Chopin (1810-49); another Polish composer of international repute is Stanislaw Moniuszko (1819-72). Ignace Paderewski (1866-1941) was not only one of the world's greatest concert pianists but was also Poland's prime minister and, later, its president-in-exile. In literature, Poland's best-known figure is the poet Adam Mickiewicz (1798-1855). A number of Polish novelists have won Nobel prizes for literature.

The most popular sport in Poland is soccer, and some clubs have international reputations. The next most common spectator sports are bicycle racing and motor racing.

Government. The Constitution is modelled on the 1936 Constitution of the USSR. The legislative body is the *Sejm,* whose members are elected for four-year terms. The Sejm appoints a council of State (whose chairman acts as head of state) and a Council of Ministers. The largest political party is the Polish United Workers' Party; others are the United Peasants' Party and the Democratic Party. All belong to the Communist-controlled National Unity Front. Effective political power is in the hands of the Politburo of the Central Committee of the Polish United Workers' Party.

History. In the 11th century a Polish principality created by the Polians was made into a kingdom and Boleslaw the Brave became king (1025). In the 13th century the Teutonic Knights seized the Polish province of Pomorze (Pomerania), and were not ejected until after their defeat at Grünwald (Tannenberg) in 1410. In 1683 one of the most triumphant events in Polish history occurred when King

John III – John Sobieski – led the Polish and imperial armies to victory over the Turks at Vienna.

The first of many partitions was forced on Poland in 1772 when Russia, Austria and Prussia each seized pieces of Polish territory, and in 1793 Russia and Prussia added further areas to their conquests. Despite the efforts of the patriot Thaddeus Kościuszko to rally his countrymen, a third partition by Russia, Austria and Prussia in 1795 erased all that was left of Poland. Napoleon reconstituted a Polish state as the Grand Duchy of Warsaw; and the Congress of Vienna in 1815 created a nominally independent Poland in personal union with the Tsar. Polish insurrections against Russian rule in 1830 and 1863 resulted in rigorous Russianization of the Polish way of life.

At the end of World War I (1918) Poland declared itself a republic, with Gen. Josef Pilsudski as head of state. Pilsudski resigned in 1922, but in 1926

The International Trade Fair at Poznań has been held since 1921. The city is one of Poland's leading academic, artistic and scientific centres.

Portuguese agriculture has long been in decline, partly from the prevalence of small farms, partly from the failure to mechanize production.

Leiria, Portugal, has a castle begun by Alfonso I in 1135; it changed hands many times in the wars with the Moors and is today a major tourist attraction.

returned after a coup and established an authoritarian regime that was continued after his death in 1935 by Marshal Edward Rydz-Śmigly. In 1939 Adolf Hitler demanded the return to Germany of the Port of Danzig (Gdańsk), the end of the "Polish Corridor" that gave Poland access to the Baltic but cut off East Prussia from the rest of Germany. Despite defensive guarantees to Poland by Britain and France, Germany invaded Poland on 1 September 1939 – the act that precipitated World War II. Poland was divided by Germany and the USSR (then an ally of Germany). During the war six million Poles were murdered, including three million Jews.

In 1944, as the Soviet army drove the Germans back through Poland, a Communist regime was established in the country. Much bitterness had been caused by the Soviet refusal to help the Polish uprising in Warsaw (February 1943) because it was under non-Communist leadership. The Polish People's Republic, with a new constitution, was established in 1952. The Communist government suppressed all opposition, but there was some lessening of the rigours of the regime after anti-Russian riots broke out in several cities. From 1956 to 1970 Wladyslaw Gomulka was the first secretary of the Communist Party. He resigned in 1970 when there were riots in Gdańsk and elsewhere over proposed increases in food prices. He was succeeded by Edward Gierek. In 1976 there were again riots and strikes about proposed increases in the price of food, and the government was forced to reconsider. Map 18.

Poland – profile

Official name Polish People's Republic
Area 312,676sq km (120,724sq miles)
Population (1976 est.) 34,481,000
Density 110 per sq km (286 per sq mile)
Chief cities Warsaw (capital) (1975 est.) 1,436,000; Łódź, 787,000; Kraków, 668,300
Government Communist republic; power is effectively in the hands of the Central Committee of the United Workers' Party
Religion Roman Catholicism
Language Polish
Monetary unit Zloty
Gross national product (1974) £35,230,800,000
Per capita income £1,022
Agriculture Potatoes, rye, wheat, sugar-beet, oats, barley, vegetables, fruits, flax
Industries Iron and steel, engineering products, fertilizers, petrochemicals, plastics, textiles, electrical goods, building materials
Minerals Coal, lignite, lead, zinc, copper, iron, sulphur, petroleum
Trading partners USSR, East Germany, West Germany, Czechoslovakia, Britain

Portugal, official name Portuguese Republic, is a small country in south-western Europe which shares the Iberian Peninsula with Spain. Today Portugal is one of Europe's poorest countries, but once it was the centre of a great colonial empire that included Brazil. In the 20th century Portugal underwent many changes: it ceased to be a kingdom, spent many years under a near-dictatorship, relinquished its overseas possessions and, in the 1970s, experienced a military coup that led to the foundation of a Socialist state. Portugal is popular with holiday-makers from other countries, largely because of its sunny Atlantic beaches. The capital is Lisbon.

Land and climate. Unlike Spain, Portugal has a coastline only on the Atlantic Ocean. The seaboard is low-lying, and so are inland areas in the south of the country. The north is generally high and has mountain ranges interspersed with lofty broken plateaus. This part of Portugal is really the western edge of the great central plateau of Spain; it is wild and wooded, with some spectacular scenery. The country has several sizeable rivers; the four largest – the Minho, Douro, Tagus (Tejo) and Guadiana – rise in Spain. The climate is predominantly Mediterranean in type, although winters can be severe in the mountains.

Economy. Portugal's economy is still predominantly agricultural. Most farmers live on small peasant holdings, although in recent years large commercial farms have been established in Alentejo and elsewhere, most of which produce cereals. The main crops are wheat, maize, rye, oats and barley. Other important crops are olives and rice. Some of the river valleys in highland areas are terraced with vineyards, the most celebrated of which are those of the Douro valley, which grow grapes for port; port (a fortified wine) is named after the seaport of Oporto (Pôrto, in Portuguese), from which it is shipped. In the south of the country citrus fruits, figs and almonds are grown. Portuguese fishermen catch sardines, cod and tuna; sardines are the most important catch and are a major export.

There are some mineral resources of importance, including ores of copper and tungsten (wolframite); iron ore, coal, tin and kaolin are also extracted. The forests are a major natural resource. Apart from timber and wood pulp they provide resins, turpentine and cork, of which Portugal is the world's largest producer.

Manufacturing industry is concerned chiefly with food processing, but there are growing textile, chemical and engineering industries, including shipbuilding and motor vehicle assembly. Road and rail communications are generally good, particularly between the larger cities in the northern part of the country. In country areas mules and horses are still used for transport.

People. The people of Portugal are descended from Iberian tribes who, over the centuries, intermarried with many invaders, including Phoenicians, Celts, Greeks, Romans, Goths and Moors. The Portuguese language belongs to the Romance group – those derived principally from Latin; although similar to Spanish, it has a quite distinct sound. About 70 per cent of the population is literate. Education, run by the state, is free and compulsory between the ages of 7 and 12. There are three universities; Coimbra University (founded in Lisbon in 1290) is one of the oldest in Europe. Most of the people are Roman Catholics.

The greatest density of population is in the north-west, where several of the larger cities are situated. The country's biggest city is Lisbon, built on terraced hills overlooking the estuary of the River Tagus; it is also a major port. Relics of the Moorish presence can be seen in many towns, particularly in the Algarve. The northern regions have many fortified villages – once a typical feature of the Portuguese countryside.

In the arts, Portugal is chiefly distinguished in literature. The most famous figure in its literary history is the poet Luís Vaz de Camões (1524–80),

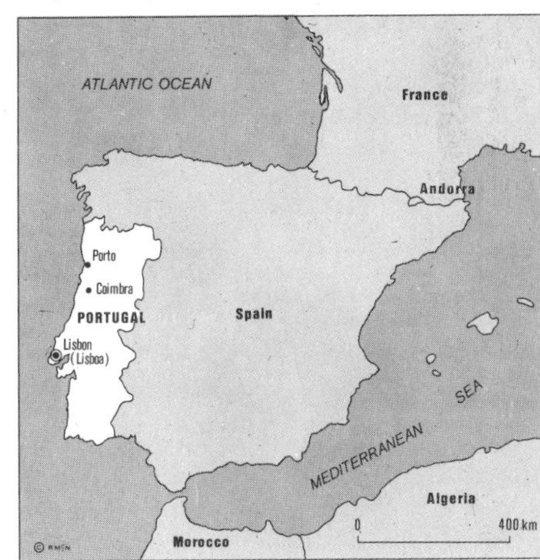

whose greatest work *Os Lusíadas* (The Lusiads) is the national epic. It tells the story of Vasco da Gama's voyage to India.

Bullfighting is popular in Portugal, as it is in Spain, although the bull is not killed. Soccer is the principal sport, with millions of followers.

Government. Portugal has had a confused political history since the Armed Forces Movement seized power in 1974. The country is basically a democracy, with an elected legislative assembly, a president and a military Revolutionary Council which has the function of advising the president and guaranteeing the country's democratic institutions. Constitutionally, Portugal is described as a unitary state that aims at establishing a Socialist society. Its territory is mainland Portugal and the two autonomous (island) regions of Madeira and the Azores.

History. In the 1st century AD central Portugal was included in the Roman province of Lusitania.

A grand 18th-century staircase takes thousands of pilgrims each Whitsun up to the Church of Bom Jesus dem Monte at Braga in Portugal.

Evora, an agricultural town in south-central Portugal, is a classic example of the Iberian domestic style – white stucco walls and red tiled roofs.

Qatar, like many of its neighbouring nations in the Persian Gulf, derives most of its income from oil and, through its nationalized refineries, petrochemicals.

Portugal – profile

Official name Portuguese Republic
Area (including Madeira and the Azores) 91,640sq km (35,382sq miles)
Population (1975 est.) 8,762,000
Density 96 per sq km (248 per sq mile)
Chief cities Lisbon (capital) 782,266; Oporto, 310,437
Government Democracy, with some degree of military control; President Gen. António dos Santos Ramalho Eanes
Religion Roman Catholic
Language Portuguese
Monetary unit Escudo
Gross national product (1974) £5,953,000,000
Per capita income £679
Agriculture Cereals, olives, citrus fruits, figs, almonds
Industries Wine, processed foods, textiles, fishing, chemicals, engineering goods
Minerals Copper, tungsten, coal, tin, kaolin
Trading partners EEC, Spain, USA, Angola

The Visigoths overran most of the Iberian Peninsula in the 5th century, and in the 8th century the Moorish conquest began. During the Christian reconquest, the Castilian king (in about 1094) gave Coimbra to one of his supporters, Henry of Burgundy. Henry was later called count of Portucalense or Portugal, a title derived from *Portus Cale*, the Roman name for Oporto. Henry's son, Alfonso Henriques, made himself King of Portugal and established his capital at Lisbon.

In the 15th century Portuguese explorers were among the most adventurous and successful in the great Age of Discovery. One of the king's sons, Prince Henry "the Navigator", devoted his life to the study of geography and the encouragement of exploration. In 1486 Bartholomew Diaz sailed round the Cape of Good Hope, and in 1498 Vasco da Gama reached India. Another Portuguese, Pedro Álvares Cabral, landed in Brazil (1500) and during an attempted journey to India made landfalls at Madagascar and Mozambique. Portugal gradually established an overseas empire.

In 1580 the ruling dynasty, the House of Aviz, died out and King Philip II of Spain seized the Portuguese throne. But the country regained its independence in 1640, and the Duke of Braganza became king as John IV. In the 18th century much was done to modernize the country by the autocratic Marquês de Pombal (1699–1782), the "power behind the throne". During the Napoleonic Wars, the French invaded Portugal (1807) and King John VI fled to Brazil; he returned in 1821. In the following year the Brazilians, dissatisfied with what they regarded as oppressive government, declared their independence of Portugal.

There was also a rising tide of dissatisfaction within Portugal, culminating in the assassination of King Carlos and the crown prince in Lisbon in 1908. Two years later a republic was declared and the new king fled. The republic was plagued by dissension and violence and a gradually worsening economic position. When Gen. António Carmona took over the government following a coup in 1926, he appointed António de Oliveira Salazar to manage the country's finances. Salazar became premier in 1932 and established a rigorously authoritarian *Estado Novo* (New State). He held power for 36 years. His successor, Marcello Caetano, was overthrown by a revolution staged by the Armed Forces Movement in 1974, and Gen. António de Spínola became president; he resigned after a few months. In two years, Portugal had six successive governments, and for a time it seemed that a Communist regime would be established. But in elections in 1976 the Communist candidates were resoundingly defeated. A major concern of the post-1974 governments was the dismantling of Portugal's remaining colonial empire, including MOZAMBIQUE and ANGOLA. Map 22.

Portuguese Guinea. *See* GUINEA BISSAU.

Portuguese Timor. *See* INDONESIA.

Portuguese West Africa. *See* ANGOLA.

Prince Edward Island. *See* CANADA.

Puerto Rico, formerly Porto Rico and official name Commonwealth of Puerto Rico, is a self-governing island in union with the United States, located in the West Indies about 1,600km (1,000 miles) southeast of Florida. Its territory includes the off-shore islands of Culebra, Mona and Vieques. It is ruled from San Juan, the capital, by an elected governor. The people, mostly Spanish-speaking Roman Catholics, have United States citizenship; they are descended from Spanish settlers and the original American-Indian inhabitants. The island is overpopulated, there is much unemployment, and many Puerto Ricans have emigrated to the United States. Sugar and its products, formerly the mainstay of the economy, have become less significant in recent years as tax incentives have attracted more industry to the island. Tobacco, coffee, pineapples and maize are also grown and dairy farming is important. Industries include cement making, canning, manufacturing and, increasingly, tourism. One of the world's largest radio telescopes is located in Puerto Rico [*see* SU p.*172*].

Puerto Rico was visited by Christopher Columbus in 1493 and Spanish colonization began 17 years later. It was ceded to the United States after the Spanish-American War in 1898. It was made a United States territory in 1917 and the present commonwealth proclaimed in 1952. It uses American currency (US dollar). Area: 8,870sq km (3,425sq miles). Pop. (1975 est.) 3,087,000. Map 74.

United States
0 700km
ATLANTIC OCEAN
Bahamas
Cuba
Dominican Republic
Cayman Islands
San Juan
Jamaica
Haiti
PUERTO RICO
CARIBBEAN SEA
Nicaragua
Colombia

Qatar, official name State of Qatar, is a small independent Arab nation on a peninsula that extends northwards into the Persian Gulf from the Arabian mainland. The capital is Doha (Ad-Dawhah). The economy is dominated by oil and natural gas, which have been in commercial production since 1949. Qatar has a hot climate; there is little rainfall and only sparse vegetation. Most of the people are Wahabi Muslims who live in villages along the coast; about half are indigenous Qataris and many of the remainder are Palestinian refugees. Fishing is the main activity of those who do not work in the oil industry. A maritime treaty of 1868 allowed Britain to gain predominance in the area and a further agreement in 1916 gave Britain control over Qatar's defence and foreign affairs. Qatar became an independent state in 1971 and joined the Arab League

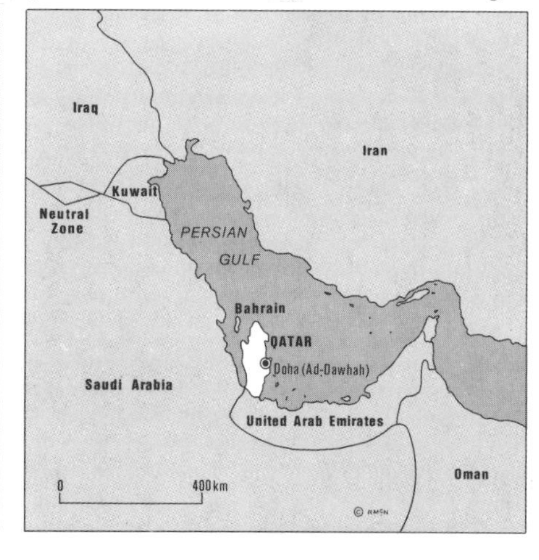

Constanta is the principal seaport of Romania and its broad sandy beach at Mamaia makes it the most popular of the Black Sea resorts.

Satalui open air museum in Bucharest, erected in 1936, re-creates for visitors, with authentic houses and artefacts, a typical Romanian early village.

A statue of Stephen the Great in Iaşi honours the 15th-century creator of the unified Moldavian state, which makes up the eastern half of present-day Romania.

and the United Nations. By 1975, under Sheikh Khalifah ibn Hamad al-Thani, the government had nationalized all the major independent oil companies. Area: 11,000sq km (4,247sq miles). Pop. (1975 est.) 180,000. Map 38.

Quebec. *See* CANADA.

Queensland. *See* AUSTRALIA.

Republic of Ireland. *See* IRELAND.

Reunion (Réunion) is an overseas départment of France comprising one of the Mascarene Islands in the Indian Ocean about 700km (435 miles) east of Madagascar. The capital is St Denis. There is an active volcano in the centre of the island and most of the people, descended from French settlers and their Asian slaves, live in the coastal lowlands and work on sugar-cane plantations. Other agricultural products are maize, vanilla and tobacco. Reunion was uninhabited when discovered by the Portuguese in 1513; it was claimed by France and used as a penal colony from about 1640. The island was British territory temporarily from 1810 to 1814. It changed status from a colony to an overseas département in 1947. Area: 2,510sq km (969sq miles). Pop. (1975 est.) 501,000. Map 34.

Rhode Island. *See* UNITED STATES.

Rhodesia, an African nation that is officially a self-governing British colony. In November 1965 the Rhodesian government, led by Prime Minister Ian Smith, made a unilateral declaration of independence (UDI), an action that the British government stated was illegal. Increasingly in recent years the African independence movement within Rhodesia – and other nations in the world – have referred to the country by the African name Zimbabwe. For a detailed account of the nation, *see* ZIMBABWE.

Rhodesia, Northern. *See* ZAMBIA.

Romania, official name Socialist Republic of Romania, is a Communist country in south-eastern Europe. It was once a Roman province called *Dacia,* and its name means (the land) "of the Romans". Until World War II Romania was a kingdom; at the end of the war the USSR aided the Communist Party in taking over the government. But since then Romania has remained neutral in the dispute between the two large Communist countries, the USSR and China, and has also acted as a bridge between the Communist nations of eastern and central Europe and the Western countries. The country's name is sometimes spelled *Rumania* or *Roumania.* The capital is Bucharest (Bucureşti).

Land and economy. The centre of Romania is occupied by the great mountain ranges of the Carpathians. They form an arc enclosing the plateau basin of Transylvania. The Eastern Carpathians extend south-eastwards about 400km (250 miles) from the Ukrainian border until they turn sharply west as the Southern Carpathians, sometimes known as the Transylvanian Alps. These, in turn, give way to the Western Carpathians, a range 320km (200 miles) long. The mountains are flanked by broad plains: the Moldavian plain in the east, the Wallachian plain in the south and the edge of the Tisza River basin in the west. The Carpathians are heavily forested and are cut by many valleys and ravines. The River Danube (Romanian, *Dunărea*) and its tributaries drain the greater part of the country; the Danube forms most of the country's southern boundary with Bulgaria. The climate is generally warm in summer, except in the mountains, but most winters are severe. The average annual temperature ranges from 7°C (45°F) in the north to 11°C (52°F) in the south. Rainfall in the mountains may be as much as 1,400mm (55in) per year, although some places are considerably drier and the national average is about half that amount.

Economy. Romania is rich in mineral resources, particularly oil and natural gas. The most important oilfields are in the south and east, and the highest output of natural gas is from places on the Transylvanian plateau. Other valuable minerals include salt, iron ore and coal, and Romania's forests are also a source of natural wealth. Most of the cultivatable land has been made into collective farms, and agricultural output is high. In the lowlands the principal crops are cereals – wheat, maize and barley. Sugar-beet, potatoes, sunflower seeds and hemp are other important crops on the plains. In the foothills farmers grow fruit, including grapes for wine.

Romania's manufacturing industries have expanded greatly since World War II. Industries include steel-making and the manufacture of machines (including tractors and other vehicles), chemicals, electrical equipment and textiles. Road and rail networks are extensive, but many roads are poorly surfaced. The state airline Transporturi Aeriene Române (TAROM) operates domestic and international routes.

People. The Romanians consider themselves a basically Latin people in contrast to their many Slav neighbours. Their language, Romanian, belongs to the Romance group (although with many Slavic and Turkish words) but they have the blood of many invaders and other settlers in addition to that of the Dacians of the Roman Empire. The country has two large minorities: about 1,600,000 Hungarians, who live mainly in Transylvania, and about 400,000 Germans. Education is free and compulsory between the ages of 6 and 16. There are six universities, of which the oldest is the University of Iaşi (1860). The largest religious body is the Romanian Orthodox Church, and there are minorities of Roman Catholics, Calvinists (chiefly among the Hungarians) and Lutherans (chiefly among the Germans). Religious practices are officially discouraged. More than 40 per cent of the population lives in the cities and towns, of which Bucharest is the only one to have a population of more than a million. Bucharest, once the capital of Wallachia, is the country's chief cultural and industrial centre. In modern times the most famous Romanian, apart from political figures, was probably the musician George Enesco (1881-1955), a composer and one of the world's great violinists. Romanian athletes have achieved considerable international success.

Government. The supreme organ of state, the Grand National Assembly, is elected for a term of five years; it meets for brief sessions twice a year. Between sessions its powers are exercised by the State Council, headed by the president. The chairman of the Council of Ministers is the head of government. Power is effectively in the hands of the Central Committee (and the Permanent Bureau) of the Romanian Communist Party.

History. In the Middle Ages the princes of Moldavia and Wallachia were independent rulers, although by the 16th century their autonomy was maintained only at the price of paying tribute to the Ottoman sultan. In the early 18th century the princes allied themselves with the tsar of Russia; as a result the sultan appointed *hospodars* (governors) in their place. Most of the hospodars were Phanariots a class of rich Byzantine Greeks who served the sultan as diplomats and viceroys. Their rule was oppressive. In 1859 Alexander John Cuza was elected prince of an autonomous Moldavian and Wallachian union, from then on know as *Romania.* Cuza was forced to abdicate in 1866 and was succeeded by a German prince, Carol I. In 1881 Romania was declared a kingdom.

During World War I Romania, under King Ferdinand I, remained neutral at first but entered the war on the Allied side in 1916. When Ferdinand died in 1927 he was succeeded by his five-year-old grandson, Michael I. But in 1930 Michael's father,

Romania has a great mixture of cultures; this traditional open-air grate comes from the Black Sea region of Dobruja, named after the 14th century Prince Dobrotich.

Tall rugged mountains and steep-sided valleys are typical of the terrain of Rwanda, the most densely populated country in Africa and one of its poorest.

Tristan da Cunha, in the British colony of Saint Helena, has a population of only about 300 people who had to be evacuated when the island's volcano erupted in 1961.

who had previously renounced his right of succession, returned to Romania and was proclaimed king as Carol II. He established an authoritarian regime but did not succeed in curbing the Fascist Iron Guard; he alienated many people who disapproved of his relationship with Elena (Magda) Lupescu.

At the start of World War II in 1939 Romania was forced to give up territory to the USSR, Bulgaria and Hungary. Carol abdicated in 1940 and went into exile; in 1947 in Brazil he married Magda Lupescu, who became Princess Elena. Carol was succeeded by his son, who became king as Michael I. The government leader, Gen. Ion Antonescu, established a dictatorship and Romania threw in its lot with the Axis powers. Michael succeeded in overthrowing Antonescu in 1944, however, and Romania declared war on Germany. In 1947 he was forced by the Communist-led and Soviet-supported coalition government formed in 1945 to abdicate, and Romania was declared a People's Republic.

In the postwar years there was at first close Soviet control of economic developments but during the 1950s and 1960s the country often found itself at odds with its Communist neighbours. In 1977 an earthquake killed 1,500 people in Bucharest and left 35,000 families homeless. Map 26.

Romania – profile

Official name Socialist Republic of Romania
Area 237,499sq km (91,698sq miles)
Population (1975 est.) 21,245,000
Density 89 per sq km (232 per sq mile)
Chief cities Bucharest (capital) (1975 est.) 1,707,000; Cluj, 203,000
Government Communist republic; head of state President Nicolae Ceauşescu
Religions Romanian Orthodox, Roman Catholic, Calvinist; religion is officially discouraged
Language Romanian
Monetary unit Leu
Gross national product (1974) £18,020,000
Per capita income £848
Agriculture Cereals, sugar-beet, potatoes, sunflower seeds, hemp, grapes (for wine) and other fruit
Industries Iron and steel, engineering products, chemicals, electrical equipment, textiles
Minerals Petroleum, salt, iron, coal, natural gas
Trading partners USSR, West Germany, East Germany, Britain, Czechoslovakia

Ruanda-Urundi. *See* BURUNDI; RWANDA.

Russia (full name, Russian Soviet Federated Socialist Republic) is the largest and most important constituent member of the Union of Soviet Socialist Republics (USSR). But, like the term *Soviet Union,* the name is often used for the whole of the USSR, and before 1917 the Russian Empire was also generally known simply as *Russia.* For an account of the history of the empire and a description of the modern republic, *see* UNION OF SOVIET SOCIALIST REPUBLICS. Map 30.

Russian Soviet Federated Socialist Republic. *See* UNION OF SOVIET SOCIALIST REPUBLICS.

Rwanda, official name Republic of Rwanda, is a small, landlocked but densely populated nation in east-central Africa. It was formerly part of the Belgian mandate of Ruanda-Urundi. In 1962 this territory was divided into Rwanda and BURUNDI. Rwanda is one of the world's poorest nations, its people having an average annual income of only £34 (in 1974). The capital is Kigali.

Land and climate. Eastern Rwanda consists of plateaus between 1,520 and 1,830m (4,987-6,004ft) above sea-level. To the west rugged mountains border the western arm of the African Rift Valley, which encloses Lake Kivu. The eastern plateaus have an average annual temperature of 20°C (68°F) and rainfall of 1,140mm (45in) per year. Moist woodlands once covered the plateaus, but most of them have been cleared. The mountains are cooler, with an average annual temperature of 17°C (63°F) and more than 1,520mm (60in) of rain per year. The Rift Valley floor has an average annual temperature of 23°C (73°F) and 760mm (30in) of rain per year.

Economy. About 95 per cent of the people are farmers, mostly at subsistence level. Food crops include bananas, beans, maize, peas, sorghum and sweet potatoes. The chief commercial crop and export is coffee. Cotton, pyrethrum and tea are also important. Some cassiterite (tin ore) and wolframite (tungsten ore) are mined, but manufacturing is on only a small scale.

People. About 90 per cent of the people are Bantu-speaking Hutu, most of whom are crop farmers. The Tutsi (9 per cent of the population) are tall Nilotic pastoralists, who dominated the Hutu and the small Twa (pygmy) minority for about 400 years untill their rule was ended in 1959. About half of the people are Christians; the remainder follow ethnic religions.

History and government. Formerly a Tutsi state and then a German colony (part of German East Africa), Ruanda-Urundi was governed by Belgium from 1916 to 1962. The Belgians maintained Tutsi feudal rule but in 1959 the Hutu rebelled. The *mwami* (king) and many Tutsi fled. In 1962 Rwanda became an independent republic, although communal tension and killings of Tutsi continued. A military coup in 1973 installed Maj.-Gen. Juvénal Habyalimana as president, heading a military government. For a few months in 1976 President Idi Amin of Uganda, because of a dispute with Kenya, refused to allow lorries carrying petrol and other fuels to cross Uganda to reach Rwanda, and the country's industry came to a halt. Map 34.

Rwanda – profile

Official name Republic of Rwanda
Area 26,338sq km (10,169sq miles)
Population (1976 est.) 4,321,000
Chief city Kigali (capital), 60,000
Religions Christianity, ethnic
Languages French, Kinyarwanda (both official)
Monetary unit Rwanda franc

Sabah. *See* MALAYSIA.

Saint Helena, together with the dependencies of Ascension and Tristan da Cunha, is a British island colony in the southern Atlantic Ocean. The capital is Jamestown on Saint Helena, which lies 1,930km (1,200 miles) west of Africa. The islanders produce vegetables such as sweet potatoes. Saint Helena is a mountainous volcanic island and was uninhabited when it was discovered in 1502 by the Portuguese explorer João de Nova Castella. It was claimed by the Dutch in 1633 and annexed by the British East India Company in 1659; it became a Crown colony in 1834. Napoleon I was exiled to Saint Helena in 1815 (and died there six years later) and the island was used as a prison for Boers from South Africa from 1900 to 1902.

Ascension, about 1,125km (700 miles) north-west of Saint Helena, was also discovered by João de Nova Castella (on Ascension Day, 1501). It has been a British possession since 1815 and was made a dependency (of Saint Helena) in 1922. Most of the small population work at the international cable ter-

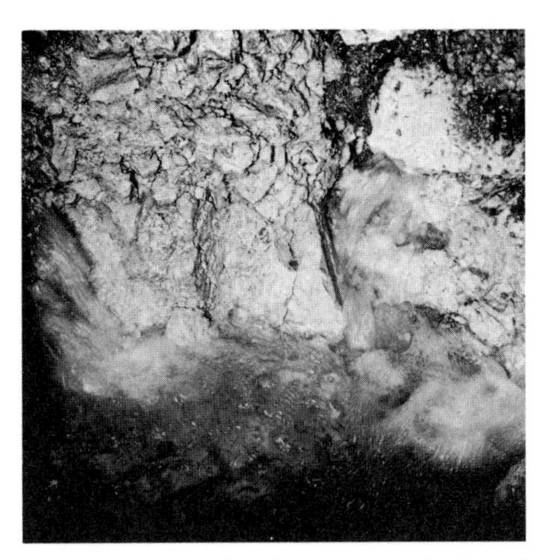

Salt extraction is one of the few industries on the island of Saint Kitts (Saint Christopher) in the West Indian state of Saint Kitts-Nevis.

San Marino, among independent nations second only to Naru in size, is located on the slopes of the Apennines close to the Adriatic Sea coast of Italy.

The world's largest exporter of crude oil, Saudi Arabia has along its Red Sea coast oil refineries that produce fuels, lubricants and petrochemicals.

minal at Georgetown or at the nearby American satellite tracking station.

Tristan da Cunha is the largest of four volcanic islands about 2,575km (1,600 miles) south-west of Saint Helena. The other islands, all uninhabited, are Gough, Inaccessible and Nightingale. The islands were discovered by the Portuguese admiral Tristan da Cunha in 1506 and annexed to Britain in 1816. They joined the Saint Helena dependency in 1938. In 1961 the long-dormant volcano on Tristan da Cunha erupted and the entire population was evacuated to England; almost all the people returned two years later. Total area: 412sq km (159sq miles). Total pop. (1975 est.) 6,438.

Saint Kitts-Nevis is a self-governing state of the Leeward Islands in the West Indies. it includes the islands of Saint Christopher (generally called Saint Kitts), Nevis and Sombrero. Basseterre, the capital, is on Saint Kitts and the chief town on Nevis is Charlestown. The islands are volcanic and mountainous, and their fine climate and scenery attract many tourists. Other income derives from the export of sugar, cotton, salt and coconuts.

Saint Kitts and Nevis were discovered in 1493 by Christopher Columbus. Saint Kitts was occupied by the British in 1623 and by French settlers in 1624. The British first went to Nevis in 1628. Anglo-French disputes over possession were settled in Britain's favour in 1783 by the terms of the Treaty of Paris. The islands were part of the colony of the Leewards from 1871 to 1956 and became members of the west Indies Federation from 1958 until its dissolution in 1962. They joined with ANGUILLA to form the self-governing state of Saint Kitts-Nevis-Anguilla in 1967, but Anguilla left the union after a few months. Area: 311sq km (120sq miles). Pop. (1976 est.) 48,000. Map 74.

Saint Lucia is a volcanic island of the Windward group in the West Indies, between the Atlantic Ocean and the Caribbean Sea. The capital is Castries. Bananas are grown for export in the fertile volcanic soil, coconuts yield copra and trees on the mountain slopes provide a variety of hardwoods. Saint Lucia was inhabited by Carib Indians when Christopher Columbus visited it in 1502. In the 1600s first British and then French settlers arrived and both nations contested ownership until 1803, when British possession was established; it was ceded to Britain in 1815. The people still speak a French patois (dialect). It joined the West Indies Federation in 1958 as part of the British colony of the Windward Islands until the colony was dissolved a year later. Saint Lucia achieved self-government in 1967 when it became one of the six Associated States of the West Indies. Area: 616sq km (238sq miles). Pop. (1975 est.) 112,000. Map 74.

Saint Pierre and Miquelon is a self-governing French territory consisting of nine small islands in the Gulf of St Lawrence, 16km (10 miles) south-west of Newfoundland. The largest island is Miquelon but the capital, St Pierre, is on the eastern coast of the island of the same name. Rocky, barren and often foggy, the islands are nevertheless a desirable possession because of their nearness to the Atlantic fishing grounds on the Grand Banks. The territory was claimed for France by Jacques Cartier in 1535 and, held by the British several times, was made a French possession in 1815. It was granted self-government in 1935. Area: 241sq km (93sq miles). Pop. (1975 est.) 5,000. Map 64.

Saint Vincent is a self-governing island state of the Windward group in the West Indies. It is made up of the island of Saint Vincent (location of the capital, Kingstown) and the smaller Grenadine Islands farther south. The highest point on Saint Vincent is the volcano Soufrière (1,234m; 4,048ft), whose eruption in 1902 destroyed much of the island. The chief source of income is tourism and the export of arrowroot [*see* PE p.*187*], bananas, copra and cotton. The islands were probably discovered by Christopher Columbus in 1498 but were not settled by Europeans until the British arrived in 1762. Apart from brief French occupation from 1779 to 1783, Saint Vincent remained in British hands, became part of the Barbados and Windward Islands colony in 1883, and remained with the Windward Islands when they received separate colonial status in 1885. From 1958 to 1962 it was a member of the West Indies Federation and it achieved self-government in 1969. Area: 389sq km (150sq miles). Pop. (1975 est.) 100,000. Map 74.

Sakhalin. *See* UNION OF SOVIET SOCIALIST REPUBLICS.

San Marino, one of the world's smallest independent republics, lies in north-eastern Italy, 18km (11 miles) south-west of Rimini. The capital is San Marino. It claims to be the oldest state in Europe. Most of its income derives from tourism, the sale of postage stamps, and the export of wine and wool. Nearly all the people are Italian-speaking Roman Catholics, most of whom work in agriculture. The country is named after its traditional founder, a Christian refugee called Marino who established a community there in the 4th century AD. San Marino is ruled by a ten-member Council of State and two regents, elected by the larger Grand Council. Women were not given the vote until 1960 and only since 1973 have they been able to hold public office. Area: 62sq km (24sq miles). Pop. (1976 est.) 20,000. Map 24.

São Tomé and Príncipe is an independent island republic in the Gulf of Guinea, on the equator off the west coast of Africa. The capital is São Tomé, located on the island of the same name which lies about 240km (150 miles) north-west of Gabon; the other major island, Príncipe, is about 145km (90 miles) north-east of São Tomé. Most of the people, who speak Portuguese, are descended from Portuguese settlers and African labourers who worked on the islands' plantations growing cocoa, coffee, bananas and coconuts – which are still the major products and whose export provide the republic's only income. The islands were uninhabited when discovered by the Portuguese navigators Pedro Escobar and João de Santarem in 1471. They were officially claimed by Portugal in 1522, although they were held for 100 years up to 1740 by the Dutch. São Tomé and Príncipe became independent in 1975 and established trade agreements with China and other Communist countries. Area: 964sq km (372sq miles). Pop. (1975 est.) 80,000. Map 34.

Sarawak. *See* MALAYSIA.

Sardinia. *See* ITALY.

Sark. *See* CHANNEL ISLANDS.

Saskatchewan. *See* CANADA.

Saudi Arabia, official name Kingdom of Saudi Arabia, is an Arab nation that occupies much of the Arabian Peninsula in south-western Asia. Until petroleum was discovered there in the 1930s it was an impoverished desert country, populated by nomadic tribesmen and known mainly as the birthplace of Islam. The exploitation of petroleum has vastly increased the country's wealth. Today Saudi Arabia is the world's third largest producer of crude oil (after the USSR and United States), but it is the leading exporter and has nearly one-fifth of the known reserves. The capital is Riyadh.

By the mid-1970s Saudi Arabia was receiving annually from petroleum sales more than twice the amount of revenue it needed to pay for its imports. Much of the country's wealth is being used to develop and diversify the economy and improve services, such as communications and education. Saudi Arabia has also invested money in developed countries, such as Italy and Japan, and has provided assistance to developing nations, especially in the Arab world.

Land. The Arabian Peninsula is an ancient land mass, highest in the west and sloping downwards to the east. There are four main regions. The narrow Tihama, or western coastal plain, borders on the Red Sea. Its greatest width is about 60km (37 miles) inland from Jidda.

The western highlands are composed of Precambrian crystalline rocks. In the north, the highlands are called *Hejaz* (meaning "barrier"). The highlands are less pronounced in the west-centre, but rise again to the south in the Asir highlands. These contain Jebel Abha, the country's highest peak, 3,133m (10,279ft) above sea-level.

East of the mountains lies the vast interior of Saudi Arabia, a region of monotonous rocky plateaus and great sandy deserts. The Nafud desert in the north covers 67,000sq km (25,900sq miles).

The Eastern Province of Saudi Arabia has the country's largest concentration of oases, fed by artesian wells and used to grow cereals and fruit.

The Haram, or Great Mosque, at Mecca is the goal of thousands of pilgrims every year, although Mecca was a religious centre even before Islamic times.

The twin towers of New College, the theological school in the University of Edinburgh, add the splendour of Victorian Gothic to the city's skyline.

In the low east of the country the Precambrian rocks are covered by more recent ones. The Al Hasa plain, on the Persian Gulf, contains Jurassic and Cretaceous strata which yield petroleum.

Climate. Saudi Arabia lacks surface water, although the many wadis are briefly filled after sudden storms. Average annual rainfall ranges from 370mm (15in) in the Asir highlands to only 75mm (3in) at Riyadh. But parts of the interior have practically no rain at all and the only settlements are at oases. Average temperatures in summer reach more than 40°C (104°F), although the nights can be cold, with frosts in highland areas. Winter temperatures average 23°C (73°F) on the Red Sea coast and 14°C (57°F) at Riyadh.

Economy. Petroleum dominates the economy, providing more than 90 per cent of the nation's revenue. The main oilfields are near the Persian Gulf coast and the chief producer is the Arabian American Oil Company (Aramco), in which the government has a 60 per cent share. But few Saudis work in the automated petroleum industry and 75 per cent of the people still live in rural areas where they depend on farming or nomadic pastoralism. Less than one per cent of Saudi Arabia is cultivated and the country imports most of its food. Farm products include butter, cereals, dates and hides. The country has about three million sheep and 300,000 cattle. The desperate need for water has led to the drilling of new wells, the building of reservoirs for irrigation, and the desalination of sea water. Manufacturing is increasing, especially petrochemical industries which are based on local natural gas and petroleum supplies.

People. Most of the people are Arabs, 90 per cent of whom are Sunni Muslims, although some Shiite Muslims live in the east and some Negroes in the west. Saudi Arabia's need for skilled labour has been partly met by substantial foreign minorities, including Arabs from other countries, Indians, Iranians, Pakistanis and Americans. The literacy rate has increased quickly in recent years and is now estimated at 30 per cent. Educational facilities, including opportunities for women (who have long had inferior status) are expanding.

Government. The king of Saudi Arabia, Khaled, is also prime minister. He heads the Council of Ministers which governs the country; the Council's chief decisions are issued as royal decrees. Elections occur only at municipal level and appointed officials, many of whom belong to the royal family, largely control local government. There are no political parties and Communists are forbidden entry to the country. Saudi Arabia is a member of the Arab League and has acted as negotiator in disputes within the Arab world.

History. The Prophet Mohammed was born in Mecca in about AD 570 and Saudi Arabia became the heartland of Islam. Internal rivalries later led to the breaking up of the Muslim Empire. The Ottoman Turks took Hejaz and other areas in the early 1500s and remained in the region until 1916, although the Muslim reformist Wahhabi group controlled much of the interior from the late 1700s and early 1800s.

A Wahhabi leader, Ibn Saud, gradual took over the whole region between 1902 and 1925, and in 1932 Saudi Arabia was established as a monarchy with the king as the effective absolute ruler. After World War II Saudi Arabia made gradually progress as revenue from petroleum increased. Ibn Saud died in 1953 and was succeeded by his son King Saud, one of whose acts was the formal abolition of slavery in 1962. Saud was deposed in 1964 and Prince Faisal became king until he was assassinated in 1975. He was succeeded by King Khaled. Map 38.

Saudi Arabia – profile

Official name Kingdom of Saudi Arabia
Area 2,149,690sq km (829,995sq miles)
Population (1975 est.) 8,966,000
Density 4 per sq km (11 per sq mile)
Chief cities Riyadh (capital) (1974) 660,800; Jidda, 300,000; Mecca, 250,000
Government Monarchy; head of state King Khaled
Religion Islam (official)
Language Arabic (official)
Monetary unit Riyal
Gross national product (1974) £7,132,500,000
Per capita income £796
Agriculture Cereals, dates, hides, vegetables
Industries Petroleum, cement, fertilizers, iron and steel, petrochemicals
Trading partners Japan, Italy, Netherlands, France, Britain, USA

Scotland, a country within the United Kingdom of Great Britain and Northern Ireland makes up just more than a third of the total area of Britain but has less than a tenth of its population. The capital is Edinburgh. Most Scots live in the Lowlands, a small area in central Scotland that has nearly all the industry and the largest cities. Few people live in the Highlands and islands, and the government is trying to develop these regions. But it is from the Highlands that Scotland gets much of its proud tradition – the clans, Highland dancing and colourful tartans. The Highlands also have some of the finest scenery in the British Isles, with rugged mountains, deep glens and many lochs (lakes), among which are Loch Lomond, and Loch Ness.

At one time Anglo-Scottish relations were uneasy and there is a long history of border skirmishes. Today many Scots desire a greater measure of independence, and the proposal for a separate Scottish Assembly with governmental control devolved from Westminster came under renewed discussion in the mid-1970s. After World War I Scotland became one of the most economically depressed areas of the United Kingdom with poverty and unemployment much higher than the national average. But the discovery of oil and natural gas off the Scottish coast offered possibilities of a change for the better in the country's fortunes.

Land and climate. Scotland can be considered as three broad regions: the Southern Uplands, the Central Lowlands and the Highlands. The Southern Uplands are mainly rolling moorlands. They include the Cheviot Hills which form most of the border with England and rise in places to more than 800m (2,625ft) above sea-level.

The Central Lowlands are dissected by the valleys of the rivers Tay, Forth and Clyde; they contain Scotland's finest agricultural land and its largest cities. The Tay, Scotland's longest river (190km; 118 miles), is famous for its salmon. The Clyde provides access for shipping to Glasgow and is the most important river commercially.

The Highlands are divided into two ranges, the Grampians and the North-West Highlands, by Glen More (the Great Glen). The highest point in the British Isles, Ben Nevis, rises to 1,343m (4,406ft) in the central Grampians.

Scotland's coastline is indented by several wide estuaries, called firths. Another feature of the country is its lochs, which are either lakes or coastal fiords. Glen More is a large cleft with a chain of long lakes (including Loch Ness) running across the country from Loch Linnhe to the Moray Firth, linked by the Caledonian Canal.

Hundreds of islands lie off the Scottish coast. The Hebrides, a group of about 500 (100 inhabited) off the west coast, include Harris and Lewis, the Uists, Mull and Skye. The Orkneys (90 islands, about 30 inhabited) and Shetlands (100 islands, 18 inhabited) lie in the Atlantic Ocean north-east of Scotland.

For such a northerly country, Scotland's climate is relatively mild. The west coast is warmed by the

Fishing has long been a key industry in Scotland. Here a fishing boat passes through the northern end of the Caledonian Canal after having left Loch Ness.

Bagpipes, whose origin is Mesopotamian and which were popular in ancient Greece and Rome, have in the recent past been primarily associated with Scotland.

Modern Glasgow is a creation of the Industrial Revolution, particularly in shipbuilding; half of Scotland's people live within 35km of the city.

Gulf Stream, but westerly winds bring more rain to that side of the country. Summers are cool and winters cold, especially inland. Annual rainfall varies from 635 to 1,525mm (25-60in) along the east coast to 2,540mm (100in) in the western mountains.

Economy. The economy of Scotland is linked to that of the United Kingdom as a whole. Once a major area of industrial growth, the country has suffered since World War I from the decline of its heavy industries – shipbuilding, engineering, steel-making and mining. Scotland has a higher rate of unemployment than the rest of Britain, resulting in a steady emigration of people looking for jobs elsewhere. To counter this trend, the government has attracted industries to Scotland, particularly light manufacturing such as electronics. The development of North Sea oilfields in the 1970s stimulated supply and support industries in Scotland, particularly in the Shetlands and Orkneys. Traditional industries still flourish. Paisley is famous for its cotton thread, the Shetlands for woollens, and Lewis and Harris for tweeds.

Scotland has few remaining natural resources, apart from some coal in the Central Lowlands, although its Highland rivers produce much of Britain's hydroelectric power. About 60 per cent of Scotland is rough grazing land and another 20 per cent is arable or permanent pasture. The chief agricultural activity is livestock raising. The mountain areas are used for grazing sheep, and Scottish farmers have developed many breeds of cattle, such as Aberdeen-Angus and Ayrshire [*see* PE pp.226, *231*, *235*]. The chief crops include oats and potatoes. Fishing is important, with Aberdeen as the chief fishing port [*see* PE p.*243*]. Scotland's chief export is Scotch whisky – most of which goes to the United States [*see* PE p.206].

People. About three-quarters of the people live in the Central Lowlands, most of them in the large cities. English is the official language, spoken (as in England) with a wide variety of accents. Gaelic is also spoken in some areas, particularly in the Hebrides; in 1971 there were 88,000 Gaelic speakers. The Church of Scotland is Presbyterian, and about 25 per cent of the people belong to it. About 15 per cent of the people are Roman Catholics.

Education has developed separately from that of England, but has recently undergone a similar reorganization with the introduction of comprehensive schools. Out-of-school activities have been widely developed, and many schoolchildren participate in such pursuits as mountaineering and sailing. There are eight universities; the oldest, St Andrews, was founded in 1411, and the largest are Glasgow and Edinburgh, each with more than 10,000 students.

The Scots are famous for their ceremony and tradition, much of which derives from the clans, groups of families with common ancestors and the same name. The most distinctive feature of the clans is their dress, which includes a plaid kilt (each clan has its own tartan pattern), sporran and other accoutrements. Other symbols of Scotland include the bagpipes and the national dish, haggis, made from oatmeal and the innards of a sheep or a calf mixed with onions and boiled in a bag made from a sheep's stomach.

Soccer is the favourite sport, especially in Glasgow where the two leading teams, Celtic and Rangers, are supported with a fanatical rivalry. Celtic were the first British side to win the European Cup. Scotland's international soccer team is also well supported and plays its home games at Hampden Park, Europe's largest stadium. Golf was first played in Scotland in about the 12th century, and famous courses include St Andrews, home of the Royal and Ancient Golf Club (which established the laws of the game). Curling also originated in Scotland and other winter sports, particularly skiing, have a growing following. Rugby, sailing and canoeing are also popular, and Scotland is well known for its salmon and trout streams, grouse moors and deer forests. There are numerous Highland gatherings in summer at such places as Braemar and Galashiels, with Highland dancing, pipe bands and special athletic events such as tossing the caber.

Culture. The golden age of Scottish culture began in the 18th century. Several writers emerged, including the national poet Robert Burns (1756-96), who wrote in the Scottish vernacular; romantic novelist Sir Walter Scott (1771-1832), the first great master of the historical novel; diarist and biographer James Boswell (1740-95); moral philosopher Adam Smith (1723-90), whose *Wealth of Nations* is still regarded as a masterpiece of classical economics; and philosopher David Hume (1711-76), who has had a profound influence on the development of modern thought.

Much of this cultural activity was centred on the city of Edinburgh. The New Town, begun in the 1760s, is an outstanding example of town planning, with its broad streets and elegant squares. Two great Scottish architects and designers were also flourishing at this time. Robert Adam (1728-92) originated the "Adam style", aided by his brother James (1730-94). They revolutionized English design not only in architecture but also in furniture, carpets and other aspects of interior decoration. Later outstanding Scottish writers include essayist and historian Thomas Carlyle (1795-1881), novelist Robert Louis Stevenson (1850-94) and playwright Sir James Barrie (1860-1937).

A 20th-century revival of Scottish literature, written mainly in the Scottish poetic dialect of Lallans, was led by poet Hugh McDiarmid (1892-). The Edinburgh Festival, launched in 1946 and held annually in August and September, has become a major international festival of music and drama.

Over the years, Scotsmen have also made outstanding contributions to the world of science. They include the mathematician John Napier (1550-1617), who invented logarithms; the anatomist and surgeon John Hunter (1728-93); James Watt (1736-1819), who developed the steam engine; bridge-builder Thomas Telford (1757-1834); physicist Lord Kelvin (William Thomson) (1824-1907); and inventors Alexander Graham Bell (1847-1922), John Logie Baird (1888-1946) and Sir Robert Watson-Watt (1892-1973). The last three made a major contribution to communications, inventing the telephone, mechanical television and radar. Other famous Scottish scientists were James Clerk Maxwell (1831-79), who developed the mathematical theory of radio, and Sir Alexander Fleming (1881-1955), who discovered penicillin.

Government. As part of the United Kingdom, Scotland sends 71 members of parliament to Westminster, and one member of the cabinet (the secretary of state for Scotland) is responsible for Scottish affairs and heads the Scottish Office, in London. Various departments of the Scottish Office, chiefly in Edinburgh, are responsible for day-to-day administration. A proposal for the establishment of

a separate Scottish Assembly of 142 members with powers of government devolved from Westminster was announced in November 1975, and has been a major topic of parliamentary discussion. The Scottish National Party, which enjoyed a revival in the 1960s, has been instrumental in persuading other political parties to take a keener interest in Scottish

The decline of the crofting industry in north-east Scotland has left the countryside dotted with abandoned farms which have fallen into disrepair.

Crayfish pots on the Isle of Skye are a sign of the increasingly large part which shellfish and crustaceans are playing in the Scottish fishing industry.

The railway bridge over the Firth of Forth, designed by Benjamin Baker, was completed in 1890. It was the world's first cantilever bridge.

affairs. Scottish local government was reorganized in 1975, creating nine regions on the mainland and three island authorities. The regions are further divided into a total of 53 districts.

Judiciary. The legal system of Scotland, unlike that of England and Wales, is based on civil law, and the organization and working of its courts differ considerably from those of England. The supreme criminal court is the High Court of Justiciary. Cases are tried by one judge (or more) and a jury of 15 people. Sheriff courts, presided over by a sheriff-principal or a sheriff (in criminal cases with a jury of 15, in civil of seven), has limited jurisdiction. District courts, with jurisdiction over minor offences, are presided over by lay magistrates known as justices. Scottish courts permit three verdicts for criminal offences: guilty, not guilty and not proven.

History. Settlers from the mainland of Europe began to arrive in Scotland 6,000 years ago. About 2,000 years ago the Celts, who had arrived in about 200 BC, were the dominant people; the area was divided among a number of tribes which were frequently at war with each other. From *c.* AD 84, when the Roman general Agricola won the battle of Mons Graupius, the Romans occupied parts of Scotland but were never able to subdue the Picts, (as they called them) – even after building Hadrian's wall (AD 127) right across the northern part of England – and finally abandoned the attempt at conquest in 211. Missionaries began to spread Christianity in the area from the late 4th century.

By the 7th century there were four kingdoms: the Picts, in the north; the Scots, a Celtic people who began crossing from Ireland in the 4th century and settled in Argyll and the western islands (the kingdom of Dalriada); the Britons, a Celtic (Welsh) people who settled in the south-west (Strathclyde); and the Angles, a Germanic people who migrated from England and colonized the Scottish Lowlands in the late 6th century. The four kingdoms engaged each other in many bitter struggles.

In the late 8th century Vikings began to raid the coast establishing settlements and gaining control of the main islands. This influx eventually had the effect of unifying Scotland, because the weakened Picts accepted Kenneth MacAlpin, (King of the Scots) as their ruler in 844. He created the kingdom of Alba (most of Scotland north of the Forth and Clyde). The southern kingdoms were added in the 11th century, and King Duncan became king of all Scotland in 1034.

Margaret, the wife of Malcolm III, was an English princess who had a great influence on the Scots in the late 11th century, introducing Church reforms and English customs and language. Their son David I (reigned 1124-53) introduced Norman nobles to Scotland, and with them the feudal system. David founded several burghs (towns), including Aberdeen and Edinburgh, as well as many monasteries. The reign of Alexander III (1249-86) was a peaceful and prosperous time when agriculture and trade flourished, roads were built, a border between Scotland and England was established and the Norsemen were expelled from the Hebrides (they still remained in the Orkneys and Shetlands).

After Alexander's reign the struggles with England that had disturbed the 150 years before it erupted again, and were to continue intermittently for another 300 years. Edward I of England intervened in a dispute for the Scottish throne and in 1292 recognized the weak John Balliol as king, but under his overlordship. When Balliol revolted and made a treaty with France (the start of an alliance that lasted 250 years), Edward invaded Scotland, defeated the Scots at Dunbar (1296), seized the Stone of Scone (the sacred coronation stone of the Scots) and declared himself King of Scotland.

Most Scots resented English rule, and William Wallace became the first popular Scottish hero by leading an army that crushed the English at Stirling Bridge (1297). He expelled the English from the country and ruled as "guardian". But Edward I returned from France, defeated Wallace at Falkirk (1298), re-occupied the country, and eventually executed Wallace (1305). The next great Scottish hero was Robert Bruce who, after many initial setbacks, led a brilliant guerrilla campaign against the English and decisively defeated them in 1314 at the Battle of Bannockburn. The Scots declared their independence in 1320, and Edward III finally recognized Bruce as King of Scotland in 1328.

Scottish independence did not run smoothly, however. A whole succession of kings came to the throne as minors and died young. The period from the succession of Robert II (the first of the Stuarts, then spelled Stewart) in 1371 to the abdication of Mary, Queen of Scots in 1567 was marked by wars with England, internal intrigue and rebellions by nobles. Nevertheless, Scotland made important advances during this period – in trade, education, government and culture. The Stuarts maintained close ties with the French, and to help them James IV went to war with England and was defeated and killed in the Battle of Flodden Field (1513).

When James V died in 1542 his new-born daughter Mary became queen. When the Scots refused an offer to marry her to Edward, son of Henry VIII of England, the English began ravaging southern Scotland. Mary, then aged five, was sent to France for safety, where she married the heir to the French throne in 1558 (he became King Francis II in 1559). Meanwhile the Protestant Reformation (led by John Knox) was taking hold in Scotland; when Mary returned in 1561 – a widow and a devout Catholic – it was as queen of a Protestant country. In 1567 she was forced to abdicate in favour of her infant son James VI. She was later imprisoned in England by Elizabeth I for 19 years before she was eventually executed in 1587.

On the death of Elizabeth in 1603 James, who was a descendant of Margaret Tudor (wife of James IV) and who had kept on friendly terms with his cousin Elizabeth, succeeded to the English throne as James I. He ruled the countries as separate kingdoms. They were eventually united in 1707 by the Action of Union.

The history of Scotland since 1603 is merged with that of the United Kingdom as a whole (*see* UNITED KINGDOM). But for the first 150 years or so Scotland's own story continued as a violent one, with bitter quarrels concerning the Church, the defiance of Cromwell's Commonwealth, the Jacobite Rebellion of 1715, and the defeat of "Bonnie Prince Charlie" (Charles Edward Stuart, the Young Pretender), at Culloden in 1746.

It was the Culloden defeat that marked the beginning of the end for the Highland clan system. The Highlanders were fearsome warriors, a powerful force when united against a common enemy. The government introduced measures to destroy the pattern of Highland life. The Highlands did not enjoy the same development of commerce and industry as the Lowlands, and many people left the area; crofters were dispossessed by landlords who wanted the land for raising sheep (the Highland Clearances of about 1780-1860). From about 1840 the population of the Highlands began to decline, and it has not recovered even today.

The Highlands and Islands Development Board was set up in 1965 to improve economic and social conditions. In 1975 a Scottish Development Agency was established to further the development of the

Rulers of Scotland

Scotland became united in 1034

Malcolm II	1005–34
Duncan I	1034–40
Macbeth (usurper)	1040–57
Lulach	1057–58
Malcolm III (Canmore)	1058–93
Donald Bane	1093–94
Duncan II	1094
Donald Bane (restored)	1094–97
Edgar	1097–1107
Alexander I	1107–24
David I	1124–53
Malcolm IV	1153–65
William I (the Lion)	1165–1214
Alexander II	1214–49
Alexander III	1249–86
Margaret, Maid of Norway	1286–90
Interregnum	1290–92
John Balliol	1292–96
Interregnum	1296–1306
Robert I (the Bruce)	1306–29
David II	1329–71
House of Stewart (Stuart)	
Robert II	1371–90
Robert III	1390–1406
James I	1406–37
James II	1437–60
James III	1460–88
James IV	1488–1513
James V	1513–42
Mary, Queen of Scots	1542–67
James VI	1567–1625

For subsequent rulers, *see* UNITED KINGDOM (*Rulers*).

Dundee, known as the "Scottish Geneva" because of its prominent part in the Scottish Reformation, is today a major jute-processing centre.

Dakar is Senegal's commercial and administrative centre; the city is also renowned for its scholarly institutes and international conferences.

In the Freetown market-place the Creole descendants of freed slaves who originally inhabited Sierra Leone are now outnumbered by Temne and Mende people.

economy and improve the environment, and the government also published proposals for the establishment of a Scottish Assembly with certain powers devolved from Westminster. Map 10.

Regions of Scotland

Region	**Area sq km**	**[sq miles]**	**Population (1975 est.)**
Borders	4,671	[1,803]	99,400
Central	2,631	[1,016]	269,300
Dumfries & Galloway	6,371	[2,460]	143,700
Fife	1,305	[504]	336,300
Grampian	8,705	[3,361]	448,800
Highland	25,130	[9,703]	182,000
Lothian	1,756	[678]	754,000
Strathclyde	13,727	[5,300]	2,504,900
Tayside	7,665	[2,959]	402,000
Island Authority			
Orkney	881	[340]	17,700
Shetland	1,427	[551]	18,500
Western Isles	2,901	[1,120]	29,600

Scotland – profile

Official name Kingdom of Scotland
Area 78,764sq km (30,411sq miles)
Population (1975 est.) 5,206,200
Density 67 per sq km (175 per sq mile)
Chief cities Edinburgh (capital) 475,042; Glasgow, 905,032; Aberdeen, 212,237
For additional information, *see* UNITED KINGDOM (profile).

Senegal (Sénégal), official name Republic of Senegal, is the most westerly nation in Africa. GAMBIA is a separate independent country that occupies an enclave in south-western Senegal. Much of the culture and organization of Senegal reflects its 300-year association with France. The capital, Dakar, stands on the Cape Verde peninsula.

Land and climate. Most of the country is flat and low-lying, although plateaus reach about 400m (1,312ft) above sea-level in the south-east. The coast is warm, although not as hot as might be expected from its latitude, and the annual rainfall averages between about 580mm (23in) in the north and 1,600mm (63in) in the south. The interior is hot and arid. The chief rivers are the Casamance, Gambia, Saloum and Senegal.

Economy. About 75 per cent of the people are farmers and the average annual income in 1974 was £141. Groundnuts and groundnut products account for about 80 per cent of exports. Fishing is also important and phosphates are mined and exported. The country had an early lead in industrial development and a good deal of light industry has been established. Dakar, formerly capital of French West Africa (which included the colony of Senegal), is the major industrial centre.

People and government. Most of the people are Negroid, the largest group being the Wolof; about 90 per cent are Muslims. The literacy rate is about 5 per cent. Senegal is governed by a president and a government council, headed by a prime minister appointed by the president. The National Assembly has 80 members, drawn from three political parties.

History. French influence began in the 1600s, although the country did not become a French colony until 1855. An attempt at federation with French Sudan (Mali) in 1959-60 failed, as did attempts to incorporate Gambia. Senegal became an independent republic in 1960, with Léopold Sédar Senghor, a noted poet and thinker, as president. Map 32.

Senegal – profile

Official name Republic of Senegal
Area 197,161sq km (76,124sq miles)
Population (1976 est.) 5,085,400
Chief city Dakar (capital), 799,000
Religiona Islam, Christianity, ethnic
Languages English (official), Krio
Monetary unit Leone

Seychelles is an independent island republic in the Indian Ocean about 970km (600 miles) north of Madagascar. It consists of 85 small islands of which the largest is Mahé, location of the capital, Victoria. The chief products of the Seychelles are copra, cinnamon, tea and fish. Tourism is an increasing source of income – the scenery and unique wildlife attract many visitors, particularly from South Africa. English and French are the official languages, although most of the people speak a French patois (dialect). The islands were uninhabited when visited in 1502 by Vasco da Gama (they were probably discovered by the Arabs earlier). They were claimed by the French in 1756 and colonized 12 years later by plantation owners and their African and Indian slaves. In 1814 they were ceded to Britain by the terms of the Treaty of Paris and in 1903 they were made a Crown colony. Seychelles gained self-government in 1975 and in 1976 was granted independence within the Commonwealth. Area: 269sq km (105sq miles). Pop. (1976 est.) 60,000. Map 34.

Sicily. *See* ITALY.

Sierra Leone, official name Republic of Sierra Leone, is a country on the coast of western Africa that was founded by Britain in 1787 for rescued slaves. It became a republic in 1971.

Land and climate. Most of the coastal region is swampy, although a lofty volcanic peninsula overlooks a fine harbour on which the capital (Freetown) stands. Inland a broad plain gradually rises to the interior uplands, which reach 1,948m (6,390ft) above sea-level. The climate is tropical and Freetown has an average annual rainfall of

Sierra Leone – profile

Official name Republic of Sierra Leone
Area 71,740sq km (27,699sq miles)
Population (1975 est.) 2,729,000
Chief city Freetown (capital) (1974) 274,000
Religions Islam, Christianity, ethnic
Languages English (official), Krio
Monetary unit Leone

3,360mm (132in). The interior is hot and wet, with most rain falling in summer.

Economy. Most people are subsistence farmers and the average annual income in 1974 was only £81. Coffee, palm kernels and cocoa are major cash crops, but minerals are more important economically. Diamonds, iron ore and rutile account for more than 75 per cent of all exports.

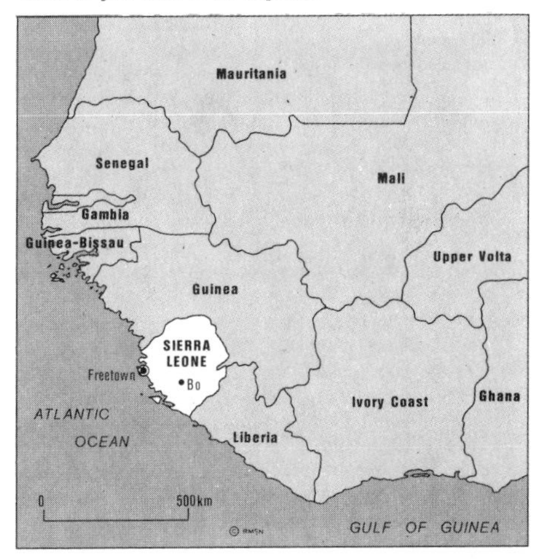

People. Sierra Leone has 18 main ethnic groups, the largest being the Mende and Temne. Most speak Krio (a form of English) and English is the official language. An important minority is the 42,000-strong Creole group, whose members are descended from former slaves. The chief religions are Islam (33 per cent), Christianity (5 per cent) and ethnic religions (62 per cent).

History and government. The Portuguese were the first Europeans to make contact with the region, in the 1460s. In 1787 the British founded a settlement at Freetown for ex-slaves and an Act of Parliament of 1807 made the peninsula a colony. In 1896 the interior was proclaimed a British protectorate, and the region became the headquarters from which British rule was extended to other parts of western Africa. Sierra Leone became independent in 1961. A military group ruled in 1967–68, but civilian government was restored. Map 32.

Once almost entirely rain forest, Singapore is now one of the world's most densely populated countries and an important centre of commerce.

The desert which covers most of Somalia supports little vegetation other than grass and shrubs, which have to provide feed for the country's abundant livestock.

Cape Town, legislative capital of South Africa, is expanding constantly; many of the shops and offices in the harbour area are on reclaimed land.

Sikkim. *See* INDIA.

Singapore, official name Republic of Singapore, is an island nation in south-eastern Asia at the southern end of the Malay Peninsula, between the Indian Ocean and the South China Sea. It consists of the island of Singapore and 60 smaller nearby islands. Its parliamentary form of government is headed by a president (Benjamin Sheares) and prime minister. The land was once covered by a tropical rain forest, but now more than 60 per cent of it has been cleared to accommodate the rapidly urbanizing economy. Less than 25 per cent of the land is used for agriculture; vegetables, tobacco, fruits, rubber and coconuts are the chief products.

Singapore is the largest importer in south-eastern Asia, one of the world's greatest commercial centres and has one of its busiest harbours. Principal industries include shipping, shipbuilding, tourism, food processing and steel products. More than 75 per cent of its two million inhabitants live in and around the capital city of Singapore. Of the total population, 76 per cent is Chinese; Malay, Chinese, Tamil and English are all spoken. The island has one of the highest standards of living in Asia, a high literacy rate, and excellent health facilities. It is the location of the University of Singapore (created in 1963 from the former University of Malaya, founded in 1949) and Nanyang University (1959).

Singapore was ceded in 1819 to the British East India Company by the Sultan of Johore through the efforts of T. Stamford Raffles, who founded the city of Singapore in the same year. It became a British possession in 1824 and grew, with the influx of Chinese and Malay merchants, into a major exporter of rubber and tin. It was one of the Straits Settlements from 1826 to 1946. After their abolition, Singapore joined Christmas Island and Cocos-Keeling islands as a British Crown colony; it became a self-governing state in 1959. In 1963 it merged with Malaya, Sarawak and Sabah to form the Federation of MALAYSIA but owing to internal and racial strife it agreed to separate in 1965 and became an independent republic. Area: 580sq km (224sq miles). Pop. (1976 est.) 2,278,000. Map 52.

Society Islands. *See* FRENCH POLYNESIA; TAHITI.

Solomon Islands. *See* BRITISH SOLOMON ISLANDS.

Somalia, official name Somali Democratic Republic, is a poor arid country on the "horn" of Africa on the east of the continent. The capital, Mogadisho, has a population of 285,000.

Land and climate. The narrow northern coast is backed by highlands. In the south, the broad coastal plain is crossed by the Juba and Shebeli rivers; inland are plateaus. The north is hot and dry, with less than 250mm (10in) of rain per year. The rainfall increases towards the south and Mogadisho has an average of 400mm (16in) of rain per year.

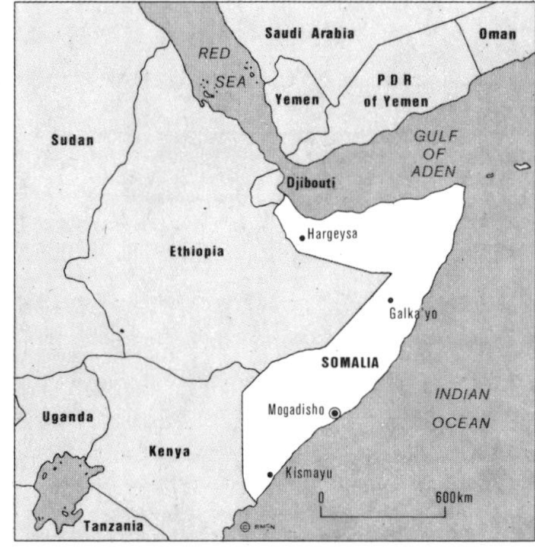

Economy. The average annual income in 1973 was only £33, and 80 per cent of the people are nomadic pastoralists. Animals and animal products account for 66 per cent of all exports. Crop farming is confined to the south, especially around the rivers; bananas are a major crop. Mining is unimportant, and there are a few processing industries. A severe drought in 1974 led to a famine in 1975 and the need for massive foreign aid.

People. Nearly all the people are Somalis, who are Sunni Muslims; the official language is Somali, a Cushitic tongue. Unlike many African nations, Somalia has been untroubled by ethnic divisions. But Somalis also live in DJIBOUTI, ETHIOPIA and KENYA, and the Somali desire for unification has led to disputes between Somalia and its neighbours.

History and government. Somalia began as a series of Arab trading stations, established along the coast from the 7th century. Northern Somalia became a British protectorate in 1887 and Italy took the south in 1905. In 1960 the two parts became independent and united. The army took control in 1969 and ruled the country through a Supreme Military Council. Somalia became a one-party state in 1976 and it is now ruled by the Political Bureau of the Somali Socialist Revolutionary Party, headed by the president, Maj. Gen. Mohammad Siyad Barrah. Area: 637,657sq km (246,199sq miles). Pop. (1976 est.) 3,258,000. Map 38.

South Africa (Suid-Afrika), official name Republic of South Africa, is a nation that occupies the southern end of the African continent. Its coastline faces the Atlantic Ocean in the west and the Indian Ocean in the east. To the north, South Africa is bordered by NAMIBIA (South West Africa), Botswana and Zimbabwe (Rhodesia). The area immediately round Walvis Bay is an enclave of South Africa within Namibia. To the north-east are Mozambique and Swaziland. South Africa also surrounds the small kingdom of LESOTHO.

The nation of South Africa was created in 1910 when two British territories, Cape Colony and Natal, were united with the two Afrikaner states Orange Free State and Transvaal. These territories became provinces within the Union of South Africa, which retained close ties with Britain – the British monarch being the head of state. Then in 1961 South Africa became a republic and left the British Commonwealth. South Africa has two capitals: Cape Town, in the south-west, is the seat of the legislature; Pretoria, in Transvaal, is the seat of the government.

Within this complex, multiracial society people of European descent (called Whites) constitute only 17.5 per cent of the population; they have control of the government and the country's economy. Since 1948 South Africa has pursued a rigorous segregationist policy towards the non-Whites – the Black Africans, Coloureds (the official name for people of mixed racial origin) and Asians. This policy, which is known as "separate development" or *apartheid,* has provoked criticism from many parts of the world. Yet despite international hostility, South Africa has made tremendous progress towards its goal of economic self-sufficiency. It is the richest nation in Africa in mineral resources (apart from petroleum) and is by far the continent's most industrialized country.

Land. Physically, South Africa consists of two main regions: the interior plateau and the marginal areas. The interior plateau is saucer-shaped. It is the southern extension of the African continental shield and includes the Orange and Limpopo drainage basins. The Orange River system rises in the highest part of the plateau – the Drakensberg in the east – and it drains westwards into the Atlantic Ocean. The Limpopo system drains the north-eastern part of South Africa, flowing eastwards

Towering over Cape Town is Table Mountain, whose flat summit is often clouded by a dense white mist known locally as the "tablecloth".

South Africa has a large iron and steel industry which uses haematite ore mined in Thabazimbi, Transvaal; here an electric arc furnace is being opened.

Mine wastes at Johannesburg show the devastation which mining for gold and diamonds has wreaked on parts of the South African countryside.

from the plateau into the Indian Ocean. Parts of the plateau rim, or Great Escarpment, have local names: from west to east the Great Escarpment includes the Roggeveldberge, the Nuweveldberge, the Sneeuberge, the Stormberge and the lofty Drakensberg, which reaches 3,482m (11,424ft) above sea-level in Lesotho. *(Berge* is the Afrikaans for *mountains.)*

The marginal region is between 60 and 240km (37-149 miles) wide. It includes the coastlands as well as the uplands and plateaus that rise to the foot of the Great Escarpment. The coastal plain is broadest in Zululand, near Mozambique and Swaziland, where it reaches about 64km (40 miles) in width. Nearly everywhere else it is narrow. Inland of the coastal plain the land rises in steps. In the south, the Langeberge and the Outeniekwaberge separate the coastal plain from the interior plateau. This plateau, the Little Karoo, extends inland to the Groot Swartberge, a range of mountains that separates it from a higher plateau (the Great Karoo), which extends to the foot of the Great Escarpment. In the south-west the Cape Ranges are fold mountains – the only true fold mountains in the continent, apart from the Atlas range, far away in north-western Africa.

Geologically, South Africa is underlaid by ancient Precambrian rocks, which are rich in mineral resources. They include, for example, the gold reefs of the Witwatersrand system. But later sedimentary rocks cover two-thirds of the surface. The Karoo system, consisting of Carboniferous to Jurassic rocks, contains Africa's largest coal deposits.

Climate. Most of South Africa lies south of the Tropic of Capricorn and the climate is temperate, although there are many regional contrasts; for instance, the east coast is subtropical because it is warmed by the southwards-flowing Mozambique current. The west coast, on the other hand, is chilled by the cold northwards-flowing Benguela current. As a result, Durban in the east has an average annual temperature of 21°C (70°F) whereas Port Nolloth on the west coast has an average annual temperature of only 14°C (57°F).

The height of the land also affects the climate. Cape Town (at sea-level) has an average annual temperature range of 13-22°C (55-72°F), whereas Johannesburg, which is 1,753m (5,751ft) above sea-level, has an average annual temperature range of 10-21°C (50-70°F). Yet Johannesburg is more than seven degrees of latitude north of Cape Town and would therefore be much warmer than the Cape were it not for its altitude. Much of the southern Transvaal experiences frosts, which may occur on more than 100 nights in every year.

About half of South Africa is arid or semi-arid and only 10 per cent of the country has more than 760mm (30in) of rain per year. In the interior the wettest areas are around the eastern and south-eastern plateau rim. The rainfall decreases westwards and the west coast is desert. Most of South Africa has summer rain and dry winters. But the Cape region in the south-west has a typical Mediterranean-type climate, with winter rain and summer drought.

Major regions. The interior plateau contains four main climatic regions. The Bushveld in north-western Transvaal is dry savanna country. It is lower and distinctly warmer than the High Veld in southern Transvaal and eastern Orange Free State, which is a grassland region with between 500 and 600mm (20-24in) of rain per year. The wettest area is the cool Lesotho highland zone, which contains large areas of mountain grassland. West of Bloemfontein and Mafeking, however, the western plateau becomes increasingly dry and is mostly desert or semi-desert scrub.

The marginal zone has seven main climatic regions. The eastern coast is warm and moist and supports subtropical vegetation and some forests. The eastern uplands, rising to the Drakensberg, contain grassland and forest. West of Port Elizabeth, the south coast has rain all the year round; it has large forests. Inland, however, the Karoo region is dry grassland. The south-western Cape has a Mediterranean climate and maquis-type vegetation, called *fynbos.* Behind the Cape is the dry south-west, a rain-shadow region which merges into the Karoo. The west coast is desert, a southward extension of the Namib Desert.

Economy. South Africa has passed through three stages in economic development. Before 1870 agriculture was the mainstay of the economy. But the discovery of diamonds in 1867 and gold in 1886 led to a rapid growth in mining and, by the early 1900s, it far exceeded agriculture in value. After World War I, however, manufacturing steadily increased in importance. By 1973 manufacturing contributed 23 per cent of the gross domestic product, mining and quarrying 13 per cent, and forestry, hunting and fishing only 8 per cent.

In the 1960s and 1970s South Africa's economy expanded rapidly by between 6 and 7 per cent per year. In the mid-1970s inflation, the world economic recession and the high cost of petroleum caused a slowing of economic growth and it was predicted that the growth rate in the late 1970s would be about 3 per cent per year. Even so, South Africa's economy remains basically sound.

Manufacturing. South Africa is the most industrialized nation in Africa. The chief manufacturing areas are in the southern Transvaal, especially the Witwatersrand (commonly called the Rand) from Randfontein through Johannesburg to Springs, and in the chief ports – Cape Town, Durban and Port Elizabeth. The oldest industries are food processing and canning. South Africa also has a large iron and steel industry; engineering, metal working, vehicle assembly, farm equipment, textiles, chemicals and diamond cutting are all important.

Mining. The most valuable mineral is gold, which was valued in 1974 at R2,403 million (£1,512 million). The industry employed about 404,000 workers, 91 per cent of whom were non-whites, including many immigrant workers. In 1974 South Africa produced 76 per cent of the new gold in the non-Communist world. Most of it, mined mainly in Transvaal and Orange Free State, is sold to the United States. Uranium and thorium are valuable by-products of the gold-mining industry.

South Africa also leads the world in the production of gem diamonds; the industry is centred on Kimberley [*see* PE p.*98*]. Coal is mined around the Witwatersrand and in northern Natal. Much of it is used to generate electricity (because South Africa lacks hydroelectricity). South Africa contains about 97 per cent of Africa's known coal reserves and also possesses abundant reserves of many other minerals, including asbestos, chromite, copper, iron ore, manganese, platinum, tin and zinc.

Agriculture, forestry and fishing. Arable land covers only about 5 per cent of South Africa, but grazing land of varying quality makes up another 80 per cent. Sheep (30 million) and goats (6 million) are raised on the drier pastures, such as those in the Karoo. Wool from Merino sheep is an especially important product. There are more than 10 million cattle; dairy farming is practised around the industrial zones, although most of the animals are reared for beef. Hides, skins and mohair are other major products derived from livestock.

The chief African food crop is maize. Most African farmers live at subsistence level and farming standards in the Bantustans (African homelands) are low. The methods of white farmers are, however, scientific and highly productive. Their crops vary from region to region. The Natal coast produces such subtropical crops as bananas, mangoes and sugar cane. Citrus fruits, cotton and tobacco thrive on irrigated farms in the Bushveld region of northern Transvaal, whereas the High Veld is a maize-growing and cattle-rearing area. The south-western Cape has a flourishing wine industry – a

Like every industrialized nation, South Africa has a well developed chemical industry, the most important product of which (shown here) is sulphuric acid.

Bantu people walk in the typical landscape of grassy savanna and steep-walled valleys which riverine erosion has formed in the Transkei.

The Nguni people of the Transkei and Cape Province make colourful tapestries depicting their traditional beadwork, weapons and tribal ritual.

characteristic of regions with a Mediterannean-type climate – and wheat is also important.

Forests occupy only about 1 per cent of South Africa's land area but the timber output satisfies 90 per cent of the country's needs. South Africa is also a major fishing nation and about 90 per cent of the catch is exported. The chief fish are anchovy, maasbanker, mackerel and pilchard, with the main fishing ports at Cape Town, East London, Mossel Bay and Port Elizabeth. Whaling is carried out from Durban and from the Donkergat station.

Tourism. South Africa has a warm climate, fine scenery and superb nature reserves and national parks, especially the Kruger National Park which covers more than 19,000sq km (7,336sq miles) of the north-eastern Transvaal. Such features attract tourists, and in 1973 about 609,000 people visited South Africa, making tourism an important factor in the economy.

Trade. South Africa is a major trading nation. Its trade figures usually include those of Namibia (South West Africa), together with those of Botswana, Lesotho and Swaziland – three nations which are linked with South Africa in a customs union. Excluding gold, the chief export of this group of nations is manufactured goods, which accounted for 28 per cent of all exports in 1974. Other important exports were food and livestock (25%), inedible raw materials (18%), diamonds (6%) and machinery and transport equipment (5%).

Gold makes by far the largest contribution to trade. In 1974 gold earned two-and-a-half times as much revenue as did manufactured goods. Imports in 1974 included machinery and transport equipment (44%), manufactured goods (22%) and chemicals (12%). South Africa's chief trading partners, not including gold sales, are Britain, West Germany and the United States.

Transport and communications. South Africa has progressed greatly since the early days when the ox-wagon was the chief form of transport. The discovery of minerals led to the building of an extensive railway network, which today totals about 35,800km (22,250 miles); the country also has a good network of roads. The main ports are Durban, Cape Town, Port Elizabeth and East London. South African Airways runs regular internal and international services.

About 500 newspapers, periodicals and journals are published in South Africa. The South African Broadcasting Corporation provides programmes in all the main local languages and television was introduced in 1976. In 1975 there were more than 1.9 million telephones.

People. South Africa is a multiracial society, with four main groups of people. According to the 1970 census Europeans (Whites) made up 17.5 per cent of the population, the largest concentration of white people on the African continent. Black Africans, also called Bantu after the languages they speak, constituted 70.2 per cent of the population; Coloureds formed 9.4 per cent; and Asians made up 2.9 per cent. These percentages are steadily changing because the rates of population growth among non-Whites is greater than that among Whites. In 1976 a South African report stated that the Black African population was increasing by an average of 2.72 per cent each year, the Coloureds by 2.69 per cent, and the Asians by 2.56 per cent. The annual rate of increase among whites was only 2.04 per cent.

The Whites. The descendants of Europeans in South Africa are divided into two main groups: the Afrikaners (who speak Afrikaans, a language derived principally from Dutch) and people who speak English. Afrikaans and English are both official languages. In 1975 an estimated 55 per cent of Whites spoke Afrikaans as their first language and 38 per cent spoke English, although bilingualism is becoming more common. South Africa also has sizeable minorities of people of German, Greek, Italian and Portuguese origin.

Most of the Afrikaners are concentrated in the Transvaal and Orange Free State, whereas the English-speaking Whites live mostly in Cape Province and Natal. Another difference between the communities is that, whereas 30 per cent of Afrikaners live in rural areas, most of the English-speaking Whites live in urban areas. Education for Whites is compulsory.

The Blacks. Nearly all South African Blacks speak Bantu languages, although there are still a few Khoisan (Bushmen and Hottentots). The chief Bantu-speaking groups are the Zulu (4.02 million in 1970), the Xhosa (3.93 million), the Tswana (1.72 million), the Sepedi, or Northern Sotho (1.6 million) and the Seshoeshoe, or Southern Sotho (1.42 million) [*see* MS p.*255*]. The literacy rate among Black Africans is estimated to be nearly 60 per cent. About 43 per cent live in Bantusans, where they have their own government and institutions. The remainder live and work in European-designated areas, most of them as unskilled labourers. A substantial number of migrant workers from the neighbouring countries of Botswana, Lesotho, Malawi, Mozambique and Swaziland also work in South Africa.

The Coloureds and Asians. The Coloureds, 87 per cent of whom live in Cape Province, are descended from Hottentots and other peoples who entered the Cape in the early days of South African history. The Cape Malays are a group who are descended from Muslims introduced from Asia by the Dutch East India Company. Most Coloureds work as labourers; their literacy rate is 75 per cent.

The Asians, 83 per cent of whom live in Natal, are descendants of Asian workers taken to Natal between 1860 and 1890. Today many work as factory hands, farmers and traders; their literacy rate is about 85 per cent.

Religion. Most whites are Protestants. Their chief church is the Dutch Reformed Church (Nederduits Gereformeerde Kerk), which includes among its members about 40 per cent of the white population (mostly Afrikaners). The other main denominations are Anglicans (11%), Methodists (9%) and Roman Catholics (8%); there is also a small Jewish community.

About 64 per cent of non-Whites are Christians, and about 25 per cent of Black Africans practise ethnic religions. Most of the Asians are Hindus or Muslims.

Cultural life and leisure. South Africa's culture is extremely varied. Black Africans have a rich tradition of music, dancing and oral literature, including prose and poetry, and some write in English.

Afrikaans literature includes much poetry, and writers such as Jan Celliers and Eugene Marais have reflected the national feelings of Afrikaners. Some English-language writers, such as Alan Paton (author of *Cry the Beloved Country*) and Laurens van der Post have become world famous.

South Africa's fine climate results in a strong emphasis on outdoor sports and nearly every game is played. The country's teams have achieved international fame in cricket and rugby union (in which both teams are known as Springboks), with such world-class players as Edie Barlow, Graeme Pollock, Mike Proctor and Barry Richards (cricket) and Jan Ellis, John Gainsford, Benjamin Osler, Erik du Preez and Davie de Villiers (rugby). Champion South African golfers include Bobby Locke and Gary Player. In the early 1970s the country's racial policy led to its sportsmen being excluded from many international events. Several South African cricketers went to Britain to play in English county sides.

Government. South Africa became a republic in 1961 following the results of a referendum held in 1960 among white voters. The head of state is the president, who is elected for a seven-year term by an electoral college formed from members of the Senate and the House of Assembly. Legislative power resides in the president, the Senate (which has 54 members, including four from Namibia) and the House of Assembly (which has 171 members, with six from Namibia). All members of the legislature are Whites, elected by White voters aged 18 years or more. The prime minister, appointed by the president, is leader of the majority party. The Coloureds are represented by a Coloured Persons Representative Council, which has 40 elected and 40 nominated members. It advises the government on affairs concerning the Coloured community. Since 1974 the advisory South African Indian Council has had 15 nominated and 15 elected members.

History. In 1488 the Portuguese navigator Bartholomew Diaz became the first European to reach the Cape of Good Hope. He was succeeded by Vasco da Gama, who rounded the Cape in 1497 and opened up a new trade route to Asia. The first European settlement was not established until 1652, when a Dutch surgeon, Jan van Riebeeck, founded a depot on the site of Cape Town to provide supplies for Dutch ships on their way to and from Asia.

Most of the early settlers were Dutch, although there were also Germans, Frenchmen and others.

Durban is the largest city and major port of Natal; this Tudor style building contrasts with the many modern office blocks of the commercial district.

Car assembly plants employ many thousands of people in Cape Province, especially in Port Elizabeth and Uitenhage; this is Chrysler's car works.

Pretoria is primarily an administrative centre although the city has major industries; iron and steel are produced and there is an oil refinery.

The Europeans made contact with the local Hottentots, who rapidly declined in numbers, partly as a result of succumbing to white man's diseases, partly through armed conflict and partly through intermarriage. Workers from Asia and other parts of Africa were also introduced for farms around the Cape or in the interior.

Pastoral farmers gradually pushed farther into the interior. By 1770 the Little Karoo had been settled and Europeans had spread to the foot of the Great Escarpment. To the north, the settlers met Kalahari Bushmen, but to the east they faced a much greater threat – the Bantu-speaking peoples. These had been gradually migrating southwards from their original home (in what is now Cameroon) for more than 2,000 years, displacing the indigenous peoples as they went. The first clash between Europeans and the Bantu-speaking people occurred in 1779-81. Fighting continued in the 1800s, especially with the Zulus, the most powerful African nation. The British finally defeated the Zulus in 1879.

The Europeans in the Cape, who numbered about 15,000 at the end of the 1700s, had evolved a distinctive way of life. They had their own language (Afrikaans), which was also called Cape Dutch. The people were called Boers (farmers) or Afrikaners (Africans).

In 1795 the British took the Cape and held it until 1803, when they returned it to the Dutch. But the British re-occupied the Cape three years later and in 1814 the region was formally ceded to Britain as a colony. The arrival of British settlers, administrators and missionaries was much resented by the Afrikaners, who feared total British domination.

In the Great Trek (1835-36) some of the Boers, who became known as *Voortrekkers* (advance pioneers), first left the Cape and began to move into the interior. In the years that followed, they were succeeded by hundreds of others who moved inland to establish the Orange Free State and the Transvaal, far beyond the extent of British influence. The Transvaal, then the South African Republic, was recognized by Britain in 1852 and the Orange Free State in 1854. But Britain occupied Natal and, finally, made it a separate colony in 1856. In developing Natal, the British introduced Asian labourers to work on the plantations.

Rivalry between the British and the Boers continued – especially after Britain claimed that Kimberley, site of a major diamond find, was a part of Cape Colony and not of the Orange Free State. In 1877 Britain took the South African Republic and this act led to the first Anglo-Boer War (or War of Freedom) of 1880-81. After the war the South African Republic regained its independence. But tension continued, particularly after gold was discovered on the Witwatersrand in 1886. Thousands of people, many of them British, flocked to the area. The Boers feared British domination and so refused to give the so-called *Uitlanders* (foreigners) any political rights. An attempted British invasion of the South African Republic in support of the Uitlanders failed in 1895. It was led by Leander Starr Jameson, a friend of the British politician, colonialist and businessman Cecil Rhodes.

The second Anglo-Boer War (also called the South African War) finally broke out in 1899. The Boers surrendered in 1902, but Britain restored self-government in the Orange Free State and the Transvaal five years later. In 1910 the four parts of South Africa were united to form the Union of South Africa.

In World Wars I and II South Africa fought on the side of the Allies and, in 1920, the League of Nations mandated South Africa to rule South West Africa (Namibia) – formerly a German territory. The inter-war years saw a great expansion of the economy, especially in manufacturing. In 1948 the predominantly Afrikaner National Party won a general election and, under the prime minister Daniel François Malan, it embarked on a programme to implement apartheid. This policy was continued by successive administrations, particularly that of Hendrik Verwoerd. Legislation was enacted to set up homelands for Black Africans, but the rights of Black Africans in White areas were continually eroded. This policy aroused much international opposition. South Africa became a republic in 1961 and withdrew from the Commonwealth.

In 1966 Balthazar Johannes Vorster became prime minister. He continued South Africa's policies of apartheid, but also sought détente in southern Africa and, to this end, began discussions with those leaders of Black African nations who were prepared to talk. In the early 1970s, Vorster met Black African leaders from neighbouring states to try to solve the problems facing ZIMBABWE.

Provinces and Bantustans. South Africa has four provinces: Cape Province, Natal, the Orange Free State and the Transvaal. Each has an elected Provincial Assembly, but any ordinances must be approved by the state president before they become law. Within the provinces certain areas, called Bantustans or Homelands, have been designated for occupation by the Bantu-speaking peoples. One of these, Transkei, was granted independence in 1976.

Cape Province, official name the Province of the Cape of Good Hope, was formerly known as Cape Colony. Its capital is Cape Town and it is the largest of South Africa's provinces, covering 59 per cent of the country. Excluding Bantu-speaking people in the Bantustans, the make-up of the population is Whites (26%), Black Africans (32.1%), Coloureds (41.4%) and Asians (0.5%).

Cape Province produces nearly all of South Africa's wine, together with a wide range of farm products. The chief industrial centres are Cape Town, East London and Port Elizabeth. The colony was first established by the Dutch in 1652, but it was formally ceded to Britain in 1814 and still retains many British traditions. Parliamentary government was granted in 1850, and in 1910 the colony became a province of the Union. Area: 721,224sq km (278,465sq miles). Pop. (1970) 6,731,820.

Natal, official name the Province of Natal, is South Africa's smallest province. Its capital is Pietermaritzburg. Excluding the Bantu-speaking people in its Bantustans, the population consists of Whites (20.7%), Black Africans (52.2%), Asians (24%) and Coloured (3.1%).

The major crop of farms along the warm coast is sugar cane; inland, livestock and cereals are important. Coal is the chief mineral and Durban is the leading industrial centre. Natal was named by the Portuguese, who first sighted it on Christmas Day 1497. It became a British colony in 1843 but was annexed to Cape Colony in 1844. In 1856 it again became a separate colony. Responsible government was granted in 1893 and in 1910 it became a province of the Union of South Africa. Area: 86,967sq km (33,578sq miles). Pop. (1970) 4,236,770.

Orange Free State (Oranje-Vrystaat), official name the Province of the Orange Free State, is a landlocked province with its capital at Bloemfontein. Only a small part has been allocated for Bantustans. The population is composed of Whites (17.9%), Black Africans (79.9%) and Coloureds (2.2%); the 1970 census recorded only five Asians in the province.

Eastern Orange Free State is part of the High Veld, where maize is grown and livestock reared; the west is arid. The province's wealth lies in its minerals, especially gold. Orange Free State was settled by Europeans in the 1810s and its population rapidly increased in the 1830s. In 1848 it was proclaimed a British possession, but became independent in 1854. In 1900, during the South African War, it was annexed by Britain and named the Orange River Colony. As Orange Free State it became a province of the Union of South Africa in 1910. Area: 129,153sq km (49,866sq miles). Pop. (1970) 1,716,350.

Transvaal, official name the Province of the Transvaal, was formerly called the South African Republic. It is South Africa's second largest province and has more people than the others. The capital is Pretoria, although the largest city is Johannesburg. Excluding Bantu-speaking people in the Bantustans, its population consists of Whites (29.6%), Black Africans (66.8%), Coloureds (2.4%) and Asians (1.2%).

Livestock is important and maize grows on the High Veld. In the lower, northern Bushveld citrus fruits and tropical crops are cultivated. In the Witwatersrand, Transvaal contains one of the world's great mining complexes. The southern Transvaal also produces about 75 per cent of South Africa's manufactured goods. The region was colonized by the Boers in the 1830s and Britain recognized its independence in 1852. A year later it took the name South African Republic, but Britain annexed the area in 1877, an action which led to a Boer uprising in 1880. In 1881 Britain granted the area internal self-government, but continuing rivalries led to the South African War (1899-1902). After the war, Transvaal became a British colony but achieved in-

The discovery at Kimberley in 1870 that the "blue ground" of volcanic pipes was diamond-bearing opened up the prospect of large-scale diamond mining.

The urban development of the eastern coast of Cape Province has made the region the most densely populated part of South Africa.

Alhambra, a group of buildings overlooking Granada, was built in the 13th and 14th centuries; it is the finest example of Moorish architecture in Spain.

South Africa – profile

Official name Republic of South Africa
Area (including Transkei) 1,221,042sq km (471,444sq miles)
Population (1975 est.) 25,471,000
Density 21 per sq km (54 per sq mile)
Chief cities Pretoria (administrative capital) (1975 est.) 614,400; Cape Town (legislative capital), 818,100; Johannesburg, 1,498,700; Durban, 721,265
Government Republic; head of state President Nicolaas Diederichs
Religions Christianity (whites and non-whites); ethnic (non-whites); Hinduism, Islam (Asians)
Languages Afrikaans, English (both official)
Monetary unit Rand
Gross national product (1974) R19,824,165,000 (£12,482,900,000)
Per capita income R778 (£490)
Agriculture Barley, cattle, cotton, fruits, groundnuts, kaffir-corn, maize, oats, rye, sheep, sorghum, sugar cane, tobacco, wheat, wine, wool
Industries Car assembly, electric motors, fertilizers, food and beverages, tobacco, furniture, machinery, paper, plastics, steel, textiles, wine
Minerals Antimony, asbestos, chromite, coal, copper, diamonds, gold, iron, lead, manganese, platinum, silver, tin, uranium, vanadium, vermiculite, zinc
Trading partners Britain, USA, West Germany, Japan, Italy, France

ternal self-government again in 1907. In 1910 it became a province of the Union of South Africa. Area: 286,065sq km (110,450sq miles). Pop. (1970) 8,717,530.

Bantustans. The policy for setting up Bantustans as homelands for South Africa's Bantu-speaking peoples was put forward in 1951. According to the government, the aim of the apartheid policy ("separate development") was to enable Black Africans to maintain and develop their own traditions and institutions in their own areas. In 1977 there were ten Bantustans: Basotho-Qwaqwa (for the Seshoeshoe group); Bophuthatswana (Tswana), scheduled for independence in December, 1977; Ciskei (Xhosa); Gazankulu (Shangaan); Kwazulu (Zulu); Lebowa (Sepedi); South Ndebele (Ndebele); Amanswazi (Swazi); Transkei (Xhosa), which became independent in October 1976; and Vhavenda (Venda). Together they occupy a combined area of 157,391sq km (60,769sq miles) – that is, less than 13 per cent of the country's area. The South African government has announced its aim of making all the Bantustans independent by 1979.

Critics of the Bantustans claim that they are not viable units. First, they are fragmented; second, they depend on assistance from the South African government because their essentially subsistence farming economies make them very poor; and third, critics argue that less than half of the Bantu-speaking people actually live in the homelands. The remainder work in white areas and many of them, who by law are considered as citizens of Bantustans, have never even visited them.

Transkei, capital Umtata, was the first Bantustan to become independent, amid considerable controversy. Most of its people are Xhosa, and Transkei is the least fragmented of the Bantustans, being divided into three parts with a total area of 39,008sq km (15,061sq miles). According to the 1970 census the population was 1,733,931, but another 1,323,442 Transkeians were recorded as living in white areas. On independence, these people became citizens of Transkei and lost their South African citizenship.

Transkei's parliament has 150 members – 75 elected and 75 nominated; the prime minister is Paramount Chief Kaiser Mantanzima. The United Nations has declared, however, that it will not accept Transkei as a truly independent nation, and the only country to recognize it as such has been South Africa. Map 36.

South Australia. *See* AUSTRALIA.

South Carolina. *See* UNITED STATES.

South Dakota. *See* UNITED STATES.

Southern Yemen. *See* YEMEN, PEOPLE'S REPUBLIC OF.

South West Africa. *See* NAMIBIA.

Soviet Union. *See* UNION OF SOVIET SOCIALIST REPUBLICS.

Spain (España), official name the Spanish State, is a kingdom occupying four-fifths of the Iberian Peninsula in south-western Europe. It is a land of many peoples, from the fiercely independent Basques in the mountainous north to the Andalusians of the southern coast. Once a thriving province of the Roman Empire, Spain became the greatest power in Europe and had its own, short-lived, empire in Latin America and Africa. It spread its language, culture and religion to hundreds of millions of people. A violent civil war in the late 1930s led to a Fascist dictatorship under Gen. Francísco Franco for nearly 40 years. As it became industrialized in the 1960s, Spain gradually returned to prosperity, although there are still great contrasts between wealth and poverty. Franco made concessions to democracy a few years before his death in 1975, and free elections were held again in 1977. The capital city is Madrid.

Land and climate. Three-quarters of Spain is a broad, barren plateau – the Meseta – sloping to the south and east and broken up by a series of high rocky hills and river valleys from east to west. The height of the plateau averages more than 600m (2,000ft) above sea-level; it is largely bounded by mountain ranges. The Andalusian Mountains in the south are the highest, rising to Mulhacén (3,478m; 11,411ft) in the Sierra Nevada. The Pyrenees are the highest of the northern ranges, reaching 3,404m (11,169ft) in the Pico de Aneto and forming a formidable barrier between Spain and France. A coastal plain on the east extends southwards from Barcelona along the whole east coast, broken only by a spur of the Sierra Nevada near Cape Nao. The indented north-west provides several fine harbours. Most of Spain's large rivers, such as the Tagus and the Guadalquivir, rise in the Meseta and flow into the Atlantic Ocean. An exception is the Ebro, which rises in the Cantabrian Mountains in the

north and empties into the Mediterranean Sea.

Spain has a generally dry climate, with hot summers and cold winters. Temperature extremes are most marked on the Meseta, where the summer sun makes it hot and dusty, with July temperatures averaging 22°C (72°F). In January the temperatures drop to an average 5°C (41°F), with little rain and much wind. The average annual rainfall on the plateau is only 450mm (18in). Northern Spain is wetter – 1,500mm (60in) – with cooler summers and milder winters. The southern and eastern coasts also have milder winters; July temperatures average 23°C (73°F) and rainfall is rarely more than 350mm (14in) a year.

Economy. Spain is predominantly an agricultural country, although the proportion of workers employed on the land fell from 41 per cent to 25 per cent between 1960 and the mid-1970s. About two-fifths of the land is under cultivation, the remainder being

Spain is the home of fine sherry and the third largest wine producer in Europe; here workers pick the grape harvest at Logrono, which is famous for its Rioja wine.

Bilbao, in the industrial north-west of Spain, is the leading town of the Basque population, which in the 1970s mounted a campaign for independence.

A rich store of buildings of the 17th and 18th centuries and the summer resorts of the adjacent sierras have made Cordoba a major tourist town.

too arid or too mountainous. About three-quarters of the arable land is pasture, and livestock includes more than 15 million sheep, 4 million cattle and nearly 9 million pigs. The chief cereal crops are wheat, barley and maize, and potatoes, sugar-beet, tomatoes and onions are the main vegetables. Spain is famous for its fruit, especially oranges. Olives and olive oil are also important products. Vines are cultivated on about 8 per cent of the arable land, and Spain is one of the world's leading wine-producing nations [*see* PE pp.200, 202]; it is especially famous for sherry. Cork is a leading forest product, and the most important fish are sardines, tuna and cod.

Spain is rich in minerals, especially coal, lignite, iron ore and potash, and is the world's leading producer of mercury. Other important minerals are the ores of zinc, tin and lead, and there are deposits of copper, uranium, silver, phosphates and sulphur.

Industry's share in Spain's gross national product has risen dramatically since the early 1960s, and 40 per cent of the country's workforce is now employed in manufacturing industries. Chief products include machinery, cotton and woollen goods, shoes, paper, motor cars, ships (fourth in world production), cement, steel and pig iron. Spain's long-standing trade deficit is largely offset by invisible earnings, especially from tourism.

Manufactured goods account for about a quarter of Spain's exports, and fruit and vegetables for about an eighth. Other important exports include machinery, chemicals, ships, cars and shoes. EEC countries take nearly half of Spain's exports (France 14 per cent, West Germany 11 per cent) and the United States takes about 10 per cent. Crude petroleum accounts for over a fifth of Spain's imports, and about half of this is supplied by Saudi Arabia. The United States is Spain's leading supplier (16 per cent), and EEC countries account for more than 35 per cent (West Germany 10 per cent).

Transport has long been a problem in Spain, with its rugged mountains and shallow rivers. Since the 1960s there has been a marked improvement, however, due both to the demands of the tourist industry and to Spain's own motor-car boom (6½ million cars in the mid-1970s), and rail, road and air travel are now well developed.

People. Most Spaniards are Roman Catholics, but the country has a number of languages and customs that vary sharply from region to region. Castilian is the language spoken by three-quarters of the people. In the north the Basques, who cling to their own traditions and rarely marry outside their group, speak their own language, which is not related to Spanish. Catalans in the north-east and Galicians in the north-west also have their own dialects. All of these regional tongues, together with Valencian (in the east), were officially recognized for legal purposes in 1976.

Many of Spain's regional minorities, particularly the Basques and Catalans, resent centralized government and want a much greater measure of autonomy. Spain is divided into 12 educational districts, each with a university. Education is free and compulsory between the ages of 6 and 14. The illiteracy rate is low – about 8 per cent. There are 20 universities, with half the students in Madrid and Barcelona.

Spain's traditional spectator sport is bullfighting and all the ceremony and ritual that goes with it. The Basques are famous for pelota, one of the fastest games in the world. The national sport is soccer, and its club sides rank with the world's best (Real Madrid won the European Cup five times between 1956 and 1960). Cycling also has an enthusiastic following, and tennis, basketball, skiing and athletics are becoming increasingly popular. Spanish cuisine is largely regional. Veal is the preferred meat, and there is a great variety of fish and seafood, especially in Basque country. Valencia is famous for *paella*, a rice dish, and Galicia for its stews. Wine is taken with most meals.

Government. The monarchy was restored in 1975 after the death of Gen. Franco, and in 1976 electoral reforms approved by a referendum provided for a new, two-chamber parliament *(Cortes)*, consisting of a 350-member Congress of Deputies and a 207-member Senate, each elected for four-year terms. Deputies are elected by universal suffrage (proportional representation) and senators on a regional basis. The king is head of state; he appoints the prime minister and has the right to call a referendum on important national issues regardless of parliamentary decisions. The abolition in 1976 of the National Movement (Spain's only political party under Franco) saw the emergence of several political parties (the Communist Party was legalized in 1977). Spain is divided into 50 provinces (including the Balearics and Canaries), and there are two small African possessions (Ceuta and Melilla). There are more than 8,000 municipalities.

Spain's armed forces consist of more than 300,000 men (220,000 army, 45,000 navy, 35,000 air force), with a paramilitary Guardia Civil of 65,000. Military service (from 16 to 24 months) is compulsory. Spain has a defence agreement with the United States, which uses air and naval bases in return for military and financial aid.

History. The expressive art of Stone Age men who lived on the Iberian Peninsula about 15,000 years ago is preserved in the cave paintings of Altamira, in north-western Spain. The Phoenicians began to establish colonies in Spain about 1100 BC, and later the peninsula was invaded by Greeks and Carthaginians. The latter were forced out in 202 BC by the Romans, who established political unity, built a vast network of roads and spread their language (Latin), from which Spanish developed. During their rule, Christianity was introduced and spread throughout the country.

In the 5th century Spain was overrun by Germanic tribes, first the Vandals and then the Visigoths, who were in turn conquered by the Moors (early AD 700s). Spain became a Muslim country except for some Christian kingdoms in the north. The Moors greatly influenced Spanish life and culture, building cities and magnificent mosques (as at Córdoba) and introducing an efficient system of irrigation. The Christian reconquest took some 800 years. The power of the Moors began to break up in the 11th century with the fragmentation into petty kingdoms, and by 1276 only the southern state of Granada remained in Moorish hands.

The various Christian kingdoms were united in the 15th century, the union of the two most powerful coming about in 1469 when Ferdinand of Aragón married Isabella of Castile. Their determination to make Spain a Roman Catholic country led in 1478 to the start of the infamous Inquisition and the later expulsion of the Jews and Muslims. The year 1492 was a significant one, for the Spanish finally drove the Moors out of Granada and Christopher Columbus obtained the finance for his voyage that led to the discovery of the New World. The Spaniards soon established a vast empire, which included most of South America, large parts of North America, the Philippines and parts of Africa.

Then in 1588 the Spanish Armada was defeated by the English and, already weakened by dynastic and religious wars, Spanish power began to decline. Driven out of Portugal in 1640, they lost further European territory in the War of the Spanish Succession in 1714, including Gibraltar to Britain. They prospered for a while, but joined France in the Napoleonic Wars, had their sea power destroyed by the British fleet at Trafalgar (1805), and then enlisted British help to drive the French out of Spain (1813). In the early 19th century their American colonies won their independence, and they lost the rest of their New World empire in the Spanish-American War of 1898.

Spain – profile

Official name The Spanish State
Area 504,750sq km (194,884sq miles)
Population (1976 est.) 35,972,000
Density 71 per sq km (185 per sq mile)
Chief cities Madrid (capital) (1976 est.) 3,751,000; Barcelona, 1,745,000; Valencia, 654,000
Government Head of state, King Juan Carlos I (acceded 1975)
Religion Roman Catholicism
Language Spanish
Monetary unit Peseta
Gross national product (1974) £29,337,600,000
Per capita income £816
Agriculture Wheat, barley, maize, livestock vegetables, fruit
Industries Textiles, ship-building, mining, iron and steel, fishing, tourism
Trading partners USA, West Germany, France, Britain and other EEC countries

Traditional handicrafts in Spain have been given special government protection to prevent their being smothered by modern mass production.

Anuradhapura in northern Sri Lanka has several impressive Buddhist monuments, among them this domed stupa for housing religious relics.

Classical dance in Sri Lanka plays a prominent role on ceremonial and festive occasions. These elaborately costumed men perform to the rhythm of a barrel drum.

Despite revolutions and disorder at home, Spain's economy flourished, and the country remained neutral during World War I. After the war Catalonia increased its efforts to attain more self-government, and a strong anarchist movement developed in Barcelona. In 1931 the people voted for republican candidates, King Alfonso XIII left Spain, and a republic was declared. In 1936 an army revolt precipitated the Spanish Civil War. It raged for 2½ years, with the USSR and the International Brigade (organized by the world Communist movement) supporting the Loyalists (Republicans) and Nazi Germany and Fascist Italy the Rebels (Nationalists), led by Gen. Francísco Franco. Spain became a battleground for the two extreme ideologies, Communism and Fascism. The war ended in April 1939 with victory for Franco, who set up a harsh authoritarian state, with the Fascist Falange (later the National Movement) the only legal party.

Spain remained neutral in World War II, and Franco continued to rule largely with the support of the army, but with increasing opposition from discontented workers, students and Basque Nationalists. In 1969 Franco designated Don Juan Carlos (grandson of Alfonso XIII) to succeed him as chief of state, and in 1973 he relinquished his post as prime minister to Admiral Carrero Blanco, who was assassinated by Basque Nationalists later that year. Following Franco's death in 1975 Juan Carlos was proclaimed king. In 1976 he replaced premier Carlos Arias Navarro with Adolfo Suárez González, and charged the new cabinet to prepare for elections, which were held in June 1977. Map 22.

Spanish Guinea. *See* EQUATORIAL GUINEA.

Spanish Sahara. *See* WESTERN SAHARA.

Sri Lanka, official name Republic of Sri Lanka and former name Ceylon, is an independent island state in the Indian Ocean, 32km (20 miles) off the south-eastern coast of India. The people are of Indian origin. The capital is Colombo.

Land and economy. The island of Sri Lanka, a continuation of the Indian continental shelf, is separated from the mainland by Palk Strait. A mountainous massif dominates the south-centre of the island, trailing off to coastal plains to the west, north and east. The climate is tropical, and 70 per cent of the island is covered by forest and natural grassland. The annual rainfall varies with location between 1,016 and 5,080mm (40–200in). The economy is primarily agricultural and relies heavily on the export of tea, rubber and coconuts. High-quality graphite is the principal mineral export. Industry, once of minor importance, is growing significantly under governmental direction and the use of foreign aid. The population has more than doubled since 1950, causing serious problems of food supply.

People. Most of the people are of Indian descent – Sinhalese and Tamil. The Sinhalese are Buddhists and make up more than 9 million of the population; Tamils, Hindu in belief, account for nearly 2¼ million. There are two groups of Tamils, Ceylonese and Indian (imported as labourers by the British in the late 1800s). Christians and Muslims constitute two minority groups. Schooling is available to all children, and most of the people are literate.

Sri Lanka – profile

Official name Republic of Sri Lanka
Area 65,610sq km (25,332sq miles)
Population (1975 est.) 13,986,000
Density 213 per sq km (552 per sq mile)
Chief cities Colombo (capital) (1974 est.) 592,000; Jaffna, 101,000; Kandy, 78,000; Galle, 73,000
Government Republic, head of state President William Gopallawa
Religions Buddhism, Hinduism
Languages Sinhalese (official), English
Monetary unit Sri Lanka rupee
Gross national product (1974) £764,900,000
Per capita income £55
Agriculture Rice, tea, coconuts, rubber
Industries Rice milling, cement, pharmaceuticals, petroleum products
Major minerals Graphite, salt
Trading partners Britain, China, USA, Australia

Government. Sri Lanka is a member of the Commonwealth. The 1972 constitution established a National State Assembly of 157 members elected every five years. The president serves as head of state and the prime minister heads the cabinet. Local government is invested in provinces and districts. The judicial system consists of a Supreme Court and many lesser courts. There are four major political parties, frequently necessitating coalition governments.

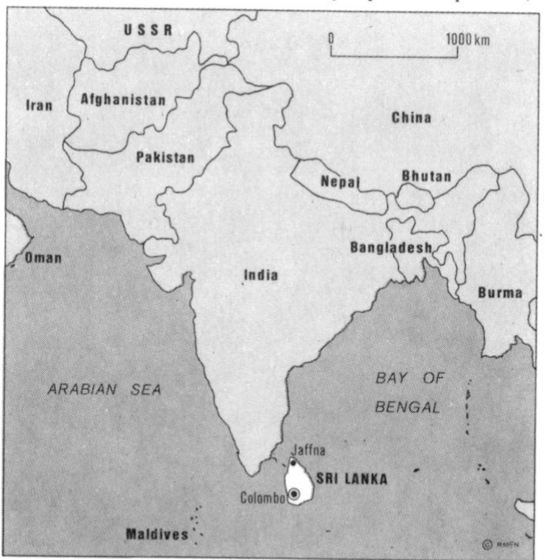

History. Sri Lanka has had a recorded history for more than 2,000 years. The island was settled in the 5th century BC by the Sinhalese from India and controlled by their Buddhist kings until the arrival of the Portuguese in 1505. By 1619 the Portuguese governed the entire island, only to be driven out by the Dutch (with the help of Sinhalese kings) in 1648. It became a British colony in 1798 and Britain continued its dominance until a series of disturbances (beginning in 1915) led to the granting of independence in 1948. The major figure in the early years of independence was S.W.R. Bandaranaike, who was assassinated in 1959. His widow, Sirimavo Bandaranaike, governed from 1959 to 1965 and became prime minister again in 1970. Communal rioting in 1976 caused a state of emergency. Map 40.

Sudan, official name Democratic Republic of the Sudan, is Africa's largest country. But much of the land is arid and thinly populated, and most of the people are poor. Khartoum is the capital city.

Land and climate. Most of Sudan is an undulating plateau. It includes much of the Upper Nile basin, and the chief highlands are around the country's borders. Sudan is a hot country and in the central region temperatures of 38°C (100°F) occur throughout the year. The northern third is desert, with less than 100mm (4in) of rain per year. Central Sudan contains some densely populated areas along the banks of the River Nile and around oases. In the south there are large areas of *sudd* – land flooded by the White Nile and covered by floating plants [*see* NW p.*232*]. In the far south, the highlands bordering Uganda have 1,520mm (60in) of rain per year.

Economy. Most people are subsistence farmers or nomadic pastoralists and their average annual income is about £50. Cotton and cotton products make up 65 per cent of all exports; groundnuts, gum arabic and sesame are other leading products. Min-

Nubia, an ancient region of north-eastern Africa, extends from the First Cataract of the River Nile to near Khartoum, the capital of the Sudan.

A Swazi woman carries water in a pot on her head in the age-old manner. Much of the lower parts of the country are dry and provide only poor crop yields.

Stockholm is said to be one of the world's finest cities architecturally, with wide streets, parks, well-planned houses and modern shopping precincts.

ing is unimportant and manufacturing is on only a small scale. About 80 per cent of Sudan's trade passes through Port Sudan on the Red Sea.

People. The people in the north and centre are a mixture of Arab, Hamitic and Negroid people with a Muslim culture. They differ greatly from the Negroid southern peoples, some of whom are Christians although most follow ethnic religions [*see* MS p.*193*]. These differences led to a civil war from 1964 to 1972. The literacy rate is between 10 and 15 per cent.

History and government. In about 2000 BC Egyptians colonized northern Sudan, an area then called Nubia. By the 8th century AD the land was made up of two Christian states, which over the following centuries became converted to Islam. From 1899 Sudan was ruled jointly by Britain and Egypt as a condominium. Sudan became independent in 1956 as a constitutional republic; then in 1958 a military group seized control. Another period of civilian rule began in 1964 but in 1969 Maj.-Gen. Gaafar Nimeiry gained power after a military coup. In 1972 the ruling ten-man Revolutionary Council ended the civil war. It granted the southern provinces a measure of self-government, with a People's Regional Assembly. An attempted coup, backed by Libya, failed in July 1976 and led to a strengthening of the army and a defence pact with Egypt. Map 32.

Sudan – profile

Official name Democratic Republic of Sudan
Area 2,505,813sq km (967,494sq miles)
Population (1975 est.) 17,757,000
Chief cities Khartoum (capital) (1972 est.) 300,000; Omdurman, 258,530
Government Military Revolutionary Council, led by Maj.-Gen. Gaafar Nimeiry
Religions Islam, ethnic, Christianity
Language Arabic (official)
Monetary unit Sudanese pound

Sumatra. *See* INDONESIA.

Surinam (Suriname), formerly known as Dutch Guiana or Netherlands Guiana, is an independent nation on the Atlantic coast of north-eastern South America between French Guiana (to the east) and Guyana. Paramaribo is the capital city. The president is head of state, although the country is run by a premier through a 39-member legislative council. Surinam consists of three major regions: the Guinea Highlands Plateau, a flat coastal plain, and an inland forest area that covers 80 per cent of the country. Fast-flowing rivers provide hydroelectric power. The main agricultural products are rice, bananas, sugar cane, groundnuts, coffee, coconuts, timber and citrus fruits. Bauxite is the chief export and mainstay of the economy. Other industries include food processing and timber products. Most of the people are descended from Creoles, Indonesians, Indians or other Asians; they have complete religious freedom, with denominations of Hindus, Roman Catholics, Muslims, Protestants and Confucians. The Guiana coast was visited in 1499 by the Spanish explorer Alfonso de Ojeda and the Dutch founded the first colony there in 1616. Britain gained the GUYANA region in 1815 by the terms of the Congress of Vienna; the Dutch retained control of Surinam. The country was awarded internal autonomy in 1954 and in 1975 gained full independence from The Netherlands. Many Dutch and Hindu Surinamese have left the country, and the new government has tried to halt this emigration. Also there is a continuing drift of population from country districts to the cities, leaving insufficient people to work the land. Area: 163,265sq km (63,037sq miles). Pop. (1975 est.) 350,000. Map 76.

Swaziland, official name Kingdom of Swaziland, is a small landlocked nation in southern Africa enclosed by Mozambique and South Africa. A mountainous country, it is often called "the Switzerland of Africa". The capital is Mbabane.

Land and climate. Swaziland is made up of four regions running roughly north-south. The western highveldt lies between 910 and 1,830m (2,985–6,005ft) above sea-level, and to the east are the middleveldt and the lowveldt, which rise between 150 and 300m (490–985ft). The Lubombo plateau overlooks the lowveldt. Climate varies with altitude. The highveldt has a temperate climate, with between 1,140 and 1,900mm (45–75in) of rain per year. Temperatures increase and rainfall decreases to the east – the lowveldt is almost tropical and dry, with between 500 and 760mm (20–30in) of rain per year.

Economy. In the 1960s and 1970s Swaziland's

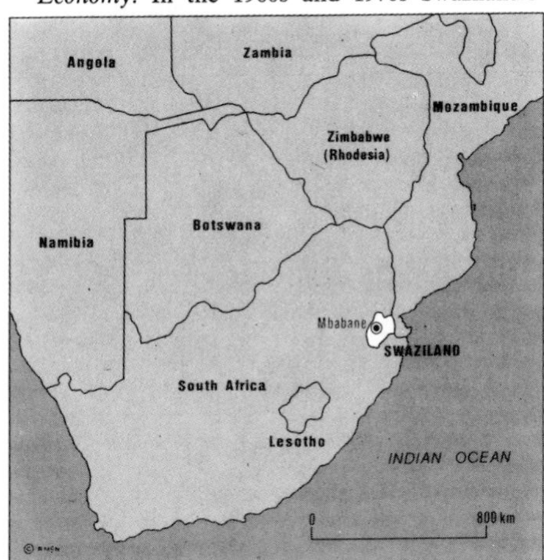

economy expanded rapidly, reducing its dependence on South Africa. About 8,000 Swazis still work in South Africa, but the money they earn is now much less important to the economy. The average annual income in 1973 was £135, although much wealth is in the hands of the small European minority, who own about 44 per cent of the land. About three people out of every four depend on agriculture for a living. Sugar and timber are leading products, and cattle and sheep are also reared. Minerals include iron ore, asbestos and coal. But the chief iron mine, near Mbabane, will probably be exhausted by 1980.

People. About 90 per cent of the people are Swazis. Minorities include Shangaans, Tongas and Zulus and there are some white people (mostly South Africans) and people of mixed origin. More than 60 per cent of the people are Christians, and most of the remainder follow ethnic religions. The literacy rate is 25 per cent.

History and government. Swazis migrated to their present region in the early 1800s to escape attacks by Zulus. Swaziland was made a British protectorate after the South Africa War of 1899–1902 and became an independent monarchy in 1968. King Sobhuza II was the first head of state and executive power rested in the Cabinet and the House of Assembly. But political unrest led the king to repeal the constitution in 1973. He banned all political groups and ruled by proclamation. Relations with South Africa became strained and Swaziland's exports passed through Mozambique. Map 36.

Swaziland – profile

Official name Kingdom of Swaziland
Area 17,366sq km (6,705sq miles)
Population (1976 est.) 482,000
Chief cities Mbabane (capital), 21,000; Manzini, 6,081
Religions Christianity, ethnic
Languages English, siSwati (both official)
Monetary unit Lilangeni (plural Emalangeni)

Sweden (Sverige), a constitutional monarchy occupying the eastern part of the Scandinavian peninsula, is a land of rivers, lakes and forests. The fast-flowing rivers provide abundant hydroelectric power, there are about 96,000 lakes, and forests cover more than half the land. From being a poor, agrarian country at the end of the 19th century, Sweden has developed into the most prosperous country in Europe – highly industrialized and with one of the best state welfare systems in the world. The country has low unemployment, negligible illiteracy, and the world's highest longevity rate and lowest infant mortality rate. The Swedes have maintained neutrality since 1814 and have been prominent in the United Nations, providing the second

More than half of Sweden's land area is forested; the country's softwoods are made into wood pulp for manufacturing paper – particularly newsprint.

The bride and groom pose for photographs after a traditional-style wedding at Seglora, not far from Gothenburg, Sweden.

Malmö is the third largest city of Sweden and stands on the bank of the Öresund opposite Copenhagen, the capital of Denmark.

secretary-general, Dag Hammarskjöld (1953–61). The capital is Stockholm.

Land and climate. Sweden is a long, narrow country, the northern part of which lies within the Arctic Circle. The country may be divided broadly into four main regions. Norrland, which occupies more than half the area in the north, is drained by many swift rivers flowing south-eastwards to the Gulf of Bothnia. In their upper courses they widen into long lakes. Most of Norrland is covered with great forests of pine and spruce. The Kjölen Mountains along the border with Norway include Sweden's highest peak, Kěbnekaise (2,123m; 6,965ft), and there are hundreds of small glaciers on their higher slopes [*see* PE p.116].

Svealand forms the central lowlands and also contains numerous lakes; the largest, Vänern, covers 5,545sq km (2,141sq miles). The lakes are linked by rivers and canals to form a complex system of waterways. In the south, Götaland consists of two regions: a low plateau (centred on Småland) covered with forests, lakes and rivers, and the fertile lowland of Scania (Skåne).

Apart from Scania, which has sandy beaches, the coasts of Sweden are mostly rocky and fringed with groups of small islands. There are also two large islands in the Baltic, Gotland and Öland, both flat, unlike the mainland.

Sweden's climate varies widely from north to south. The average February temperature in Kiruna, in Lapland, is -12°C (10°F), compared with -3°C (27°F) in Stockholm. The difference in summer temperatures is less marked, averaging about 13°C (55°F) in the north and 17°C (63°F) in the south. The annual precipitation (rain and snow) varies between 400 and 650mm (16–26in), being heaviest in the uplands.

Economy. Sweden is rich in mineral resources, especially high-grade iron ore, and in hydroelectric potential and timber. As a result, with a population of only 8 million, it has become a major industrial power. About 29 per cent of the country's 4 million workforce is engaged in manufacturing and mining industries, compared with only 7 per cent in agriculture, forestry and fishing. Sweden produces about 5 per cent of the world's iron ore, which is the basis of its domestic heavy industry even though more than 90 per cent of it is exported. The major deposits are located north of the Arctic Circle. Other important minerals include ores of copper, lead and zinc.

Sweden's chief products include machinery, road vehicles, ships (third in world production), and electronic and telecommunications equipment. The country also has an international reputation for the quality of its furniture, porcelain and glass. The chief agricultural products derive from the forests – timber, wood for fuel, pitch and raw materials for the paper and rayon industries. Only 7 per cent of the land is arable, the chief crops being barley, wheat, oats, potatoes and sugar-beet. Less than 2 per cent of the land is meadow or pasture. There are about 2 million cattle and 2½ million pigs, and dairy products account for about 30 per cent of farming output.

Sweden's chief exports include machinery, paper, motor vehicles [*see* MM p.*95*], iron and steel, wood pulp, timber, arms and ships. Its leading customers are Norway (11%), Britain (11%) and West Germany (10%). Crude oil and petroleum products account for some 16 per cent of Sweden's imports. Other imports include machinery, chemicals and transport equipment, and their chief suppliers are West Germany (19%), Britain (11%) and the other Scandinavian countries.

Sweden has an excellent railway system of more than 12,000km (7,450 miles) of track, largely state-owned, of which more than a half is electrified. Swedish State Railways also operate a ferry service to Denmark and West Germany. Sweden has a merchant fleet of some 7 million gross tonnes, and its chief ports are Gothenburg (Göteborg) and Stockholm. Lapland ore is shipped from Luleå on the Gulf of Bothnia in the summer, but in winter it is carried across the mountains to the ice-free port of Narvik, on Norway's Atlantic coast. On land, almost as much freight is carried by road as by rail.

People. There are few minority groups in Sweden. In the north there are about 10,000 Lapps and the country has 250,000 foreign workers, nearly half of whom are from Finland. Most of the people speak Swedish, the official language, although the Lapps and immigrant Finnish workers speak their own languages. Most of the people live in the southern lowlands, three-quarters of them in towns and cities. Almost all belong to the Evangelical Lutheran Church, the established state religion.

Sweden's schools were reorganized as comprehensives in the 1960s. Many children under the age of seven attend private kindergartens. Attendance at *grundskolan*, between the ages of 7 and 16, is free and compulsory. It is divided into three schools: lower, middle and senior. English, Sweden's second language, is compulsory in middle school and 90 per cent of pupils elect to learn it also in upper school. After comprehensive school, children may enter integrated upper secondary schools, with a wide choice of academic and vocational courses. There are six state universities, including Uppsala (founded in 1477) with 15,000 students and Stockholm with more than 20,000.

Sweden has an advanced system of social security and health schemes, with pensions and benefits. The people enjoy a high standard of living. They eat various kinds of fish and sausages, and much frozen food; coffee is a favourite beverage. They are famous for their smorgasbord, an elaborate and elegantly displayed cold table, sometimes with hundreds of different dishes.

The Swedes are an outdoor people, and hunting and fishing are popular recreations. They also enjoy cross-country skiing, ice hockey and athletics. Every March about 5,000 skiers set off on the 90km (56 mile) Vasa Race, which commemorates the 16th-century Swedish hero Gustavus Vasa. Gymnastics, part of the school curriculum, is also widely practised by adults. Soccer is mainly an amateur sport (played in summer) although the Swedes have a good international record – they reached the final of the 1958 World Cup.

Government. Sweden is a representative and parliamentary democracy, with a hereditary king as head of state but with only formal powers. Executive power is vested in a cabinet, headed by the prime minister, and responsible to the 349-member, one-house parliament (the *Riksdag*). The members are elected for three-year terms, 310 directly from the 28 constituencies and the remaining seats being distributed proportionately. The chief political party, the Social Democratic Party, was in power almost without interruption from 1932 to 1976, when it was ousted by a non-socialist coalition. For local administration, Sweden is divided into 24 counties consisting of 278 municipalities. There is a small regular army, with a permanent force at the disposal of the United Nations. National service is between 7½ and 15 months long, and there is a total reserve strength of 750,000 men.

History. People first began to settle in the southern tip of Sweden about 8,000 years ago, moving farther north as the climate improved. About 2,000 years ago the people were trading with the Romans, and in about AD 100 the Roman historian Tacitus wrote of the "Suiones", or Svear, from whom the country got its name. After a long period of conflict the Svear and another tribe, the Gotar (who had settled in the south), united in the 6th century.

In the 13th and 14th centuries there were constant struggles between the rulers and the nobles. The country was further weakened by a trading alliance between its merchants and the German Hanseatic League. An effort to attain strong rule led to the Union of Kalmar (1397), which united Sweden with Denmark and Norway under Queen Margaret

Stockholm is built on several islands and peninsulas connected by bridges; besides being a flourishing port, it is sometimes called the "Venice of the North".

A major international winter sports centre, Davos in Switzerland has one of the world's finest ski runs: it is also a fashionable health resort.

Lucerne is one of Switzerland's most popular tourist resorts with facilities for sailing, horse racing, show jumping and various winter sports.

of Denmark. It was an uneasy alliance, however, and the Swedes finally succeeded in breaking away in 1523, led by Gustavus Vasa (who defeated the Danes). As Gustavus I, he laid the foundations of modern Sweden, setting up a strong army, centralizing power, and encouraging industry, trade, and the spread of Lutheranism.

For the next 200 years Sweden fought a series of wars with Denmark, Poland and Russia to gain control of the Baltic. Under their great military leader King Gustavus II Adolphus (reigned 1611–32), they won much territory as he led Sweden and the Protestant cause to sterling victories in the Thirty Years War. Sweden became a powerful and respected nation, but their empire came to an abrupt end when Charles XII (1697–1718) (who had won many victories) invaded Russia – the Swedish army was annihilated by Peter the Great at Poltava (1709).

Sweden became involved in the Napoleonic Wars against France in the early 19th century. They lost Finland (which had long been a province of Sweden) to Russia in 1809, and were seriously weakened as a result. Parliament then elected Jean Baptiste Bernadotte, one of Napoleon's generals, as heir apparent to the childless Charles XIII. Before he became king in 1818 as Charles XIV he fought the last war in Sweden's history, to ensure union with Norway in 1814. (Norway eventually broke away peaceably in 1905.) Bernadotte, the ancestor of the present-day royal family, was the originator of Swedish neutrality.

The 19th century was a period of industrial growth, but agriculture could not keep pace with the rapidly increasing population. The resulting poverty and hardship led to a wave of emigration, especially to the United States (about half a million people left between 1865 and 1885). After World War I, socialist leaders Hjalmar Branting and Per Albin Hansson abandoned doctrinaire Marxism, opting for social democracy by reform rather than revolution. The Social Democrats championed the welfare state, and guided the country through the years of the Depression to a position of unprecedented prosperity.

With the long-serving Tage Erlander as prime minister (1946–69), they consolidated this position after World War II (in which Sweden remained neutral). Sweden's international reputation for stability, however, has not been won without hard work and sacrifice. The high cost of social welfare (about a third of the national budget) is met by taxes and compulsory contributions, and the 1970s saw unofficial strikes and a new militance against high taxation and a steeply rising cost of living. There is also increasing concern that Sweden's collective society has become too bureaucratic, a trend that the conservative coalition that came to power in 1976 promised to reverse. Map 16.

Sweden – profile

Official name Kingdom of Sweden
Area 449,964sq km (173,731sq miles)
Population (1976 est.) 8,226,000
Density 18 per sq km (47 per sq mile)
Chief cities Stockholm (capital), 665,200; Gothenburg, 444,650; Malmö, 243,600
Government Head of state, King Carl XVI Gustaf (succeeded 1973)
Religion Evangelical Lutheran
Languages Swedish, English, Finnish
Monetary unit Krona (plural kronor)
Gross national product (1974) £23,440,200,000
Per capita income £2,850
Agriculture Forestry, cereals, potatoes, sugar-beet
Industries Mining, iron and steel, machinery, wood pulp, paper, electronic equipment, fishing
Trading partners West Germany, Britain, Denmark, Norway, Finland, USA

Switzerland, a small, landlocked mountainous republic in southern Europe, has three official names; *Schweiz* (German), *Suisse* (French) and *Svizzera* (Italian). It is a confederation of 25 states called *cantons* and *demi-cantons*. The country is known for its alpine scenery, its clocks and watches and its importance as a centre of finance. It has a tradition of neutrality in war: the Geneva Conventions, which provide for the humane treatment of wounded and prisoners in wartime, are named after one of Switzerland's chief cities; and the Red Cross, which was founded in Switzerland, uses as its emblem the Swiss flag with the colours reversed. The country is sometimes called the *Helvetic Confederation* after Switzerland's original inhabitants, the Helvetii. The capital is Bern.

Land and climate. Switzerland is the most mountainous country in Europe: three-quarters of it consists of mountain ranges. The Jura Mountains, rising along the north-western border with France, are separated by the plateau of the Mittelland from snow-capped ranges of the Alps that occupy the whole southern part of the country [*see* PE pp.*44,45*]. Alpine peaks in Switzerland include the Jungfrau (4,158m; 13,642ft), the Eiger (3,975m; 13,040ft) and, on the Swiss-Italian border, the Matterhorn (Mte Cervino) (4,478m; 14,691ft) [*see* PE p.*117*]. Two great rivers rise in the Swiss Alps: the Rhine, which originates in two headstreams flowing into Lake Constance, and the Rhône, which has its source in a glacier in the south-west. Some of Switzerland's most beautiful scenery is found around its lakes, the largest of which are Constance (shared with West Germany and Austria, where it is called Boden See), Zürich, Lucerne (Vierwaldstätter See), Neuchâtel and Geneva (shared with France, where it is called Lac Léman). The climate varies: on south-facing slopes it is generally milder than elsewhere. The path of the föhn, a warm south-westerly wind, includes the Mittelland in the spring.

Economy. There are few natural resources except for the forests, some salt deposits and the swift-flowing rivers that can be utilized for the generation of hydroelectricity. Switzerland has one of the highest living standards in Europe, and its prosperity is based mainly on the skill and industry of its people. The chief form of agriculture is dairy farming; much of the cultivatable land is too high or too difficult to reach for growing crops. In the Alps cattle and sheep spend winter in the sheltered valleys, and in spring and summer are driven to pastures above the tree line. Swiss dairy products include Gruyère and Emmental cheeses [*see* PE p.226–229]. Milk is also used in the production of chocolate, for which Switzerland is well known. Fruit-growing is important in lakeside areas and in the valleys of the foothills, where there are vineyards and orchards of apples, pears, cherries and plums.

Swiss industry concentrates to a large extent on

small, high-precision products, although there is an important steel industry that uses imported iron ore. Associated with it is the manufacture of industrial machinery, rolling-stock, turbines and other heavy equipment. Among the major precision industries are the manufacture of electrical machinery, scientific instruments and optical goods. The pharmaceutical industry has an international reputation. The making of clocks and watches – for which Switzerland is internationally famous – ranks only third or fourth in economic importance within the country. It is an assembly industry, with more than a thousand small factories each making a particular component. Tourism is a major source of income, and so are banking and insurance; Switzerland is one of the world's chief banking centres.

The country's roads and railways are among the best in Europe. The Alps are crossed by road and rail passes and tunnels, including the St Bernard,

Simplon Pass connects Switzerland with the Lake District of Northern Italy; the road through the pass was built by Napoleon I in the early 1860s.

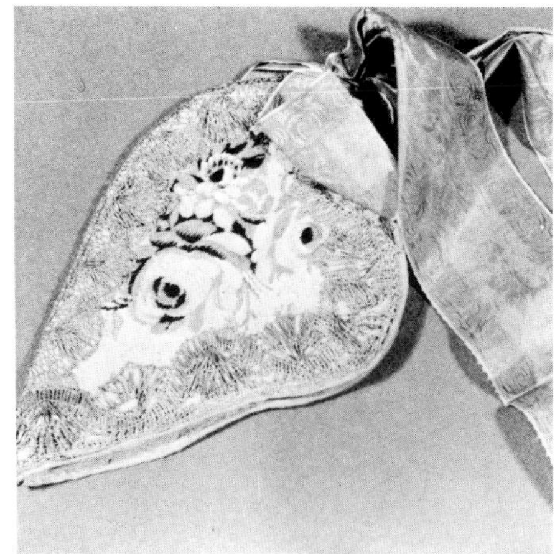

Appenzell, a sparsely populated region of Switzerland, retains many of its ancient customs and is famous as an embroidery centre.

Among the attractions in Lausanne, Switzerland, is the famous Gothic Cathedral of Notre Dame; there are several notable museums and a university too.

Simplon [*see* MM p. 188] and St Gotthard. Swissair, the national airline, flies on internal and inter national routes.

People. The Swiss are sometimes said to be the most European of people because their country is like a miniature Europe: it has three official languages – German, French and Italian – and a fourth, Romansch, has the status of a "national" language. The most widely used is German; it is the first language of more than two-thirds of the people and of 16 of the 22 cantons. Switzerland's particular form of German is called *Schwyzerdütsch.* It has many variations, such as *Baslerdütsch* (spoken in the canton of Basel) and *Barndütsch* (spoken in Bern). One person in five speaks French, and about one in eight Italian. Romansch is spoken by about 50,000 people in the cantons of Ticino and Graubünden, and has several dialects of which the best-known is Ladin.

Switzerland has the highest proportion of foreign residents of any country in Europe – about one-sixth of the total population. Education is free and compulsory at primary level, but its form varies from canton to canton. The oldest of the seven universities is the University of Basel (1460). There is complete religious freedom (except that the Society of Jesus – the Jesuit Order – is banned), and no one has to pay taxes that help to maintain a creed to which he does not belong. The population is almost equally divided between Roman Catholics (49 per cent) and Protestants (48 per cent). Central Switzerland, except for Bern, is mostly Catholic; the rest, is mainly Protestant.

Government. Legislative and governmental powers are shared between the cantons and the federal authority. The federal parliament consists of two houses: the *Ständerat* (Council of States) of 44 members, two for each canton; and the *Nationalrat* (National Council) of 200 deputies directly elected for four-year terms. Parliament elects the federal government, the *Bundesrat* (Federal Council), which consists of seven members from seven cantons, elected for four years. The Bundesrat elects a president to hold office for one calendar year; while in office he is the President of the Confederation.

History. The Helvetii, a Celtic people, were conquered by Julius Caesar in about 58 BC. In the 5th century AD the territory was invaded by the Burgundii and the Alemanni; the River Sarine (Saane), the boundary between the territories occupied by these invaders, still forms a rough dividing line between French and German Switzerland. Later the region became part of the Holy Roman Empire.

On 1 August 1291 the men of Schwyz, Uri and Nidwalden formed an "Everlasting League" to prevent Hapsburg encroachments on their liberty; this event is taken as the foundation of the Swiss Confederation, and 1 August is the country's national day. According to legend one of the heroes of the alliance was William Tell. During the next centuries the Swiss had many times to resort to arms to maintain their independence from the Austrians. They defeated Austrian armies at Morgarten in 1315, at Sempach in 1386 and at Näfels in 1388. The original cantons were joined by Luzern, Zürich, Glarus, Zug and Bern in the 14th century, by Fribourg and Solothurn in 1481, by Schaffhausen and Basel in 1501 and by Appenzell in 1513. Switzerland became a considerable military power and made conquests of its own, but was decisively defeated by the French at Marignano in 1515. The Swiss policy of neutrality dates from this defeat.

The Reformation split the cantons into Catholic and Protestant camps. The Protestants were defeated in battle, and only a vague and precarious sense of unity survived. In the 1540s John Calvin established his austere theocracy in Geneva. Switzerland's independence of the Holy Roman Empire was formally recognized by the Peace of Westphalia in 1648.

In 1798, during the Napoleonic Wars, French armies invaded the country. Napoleon united the cantons into a single state, the Helvetic Republic, which was later reorganized as a federation. After Napoleon's defeat the old union of cantons was restored and the Congress of Vienna (1815) guaranteed Switzerland's neutrality. In 1847, however, a civil war broke out between the cantons that wished for a more formal confederation and those that were opposed to change. The confederates won, and the Swiss Confederation was formed in 1848.

In the 20th century, Switzerland has avoided Europe's worst troubles: it remained neutral in both World Wars. In 1920 Geneva was made the headquarters of the League of Nations. After World War II Switzerland decided – in order not to prejudice its position of neutrality – not to join in the United Nations, although it has become a member of some of the specialized agencies. Many other international organizations have their headquarters in Switzerland – for example, the International Labour Organization, the World Health Organization and the scientific organization CERN [*see* SU p.95]. In 1959 the canton of Vaud made Swiss history: it granted women voting rights. In 1971 all women got the right to vote. Map 20.

Switzerland – profile

Official name Switzerland, or the Swiss Confederation
Area 41,288sq km (15,941sq miles)
Population (1976 est.) 6,333,000
Density 157 per sq km (397 per sq mile)
Chief cities Bern (capital) 162,405; Zürich, 422,640; Basel, 212,857; Geneva, 173,618
Government Federal democracy
Religions Roman Catholic (49%); Protestant (48%)
Languages German, French, Italian, Romansch
Monetary unit Swiss franc
Gross national product (1974) £18,423,100,000
Per capita income £2,909
Agriculture Dairy products, including cheese; fruit, including grapes (for wine)
Industries Electrical machinery, scientific instruments, optical equipment, watches, steel, rolling-stock, turbines, confectionery
Trading partners EEC, Austria, Sweden

Syria (As-Sūrīyah), official name Syrian Arab Republic, is an independent nation in the Middle East. It is dominated politically by the avowedly socialist Ba'ath administration, and its economy has suffered because of preoccupation with the continual Arab-Israeli conflicts. The capital is Damascus.

Land and economy. Syria is situated at the eastern end of the Mediterranean Sea, with its dominant geographical features parallel to the coast. The Anti-Lebanon and Alawite mountains run along the coast from Israel to Turkey, the valley of the River Euphrates crosses the country from north to south-east, the Jebel al-Druze mountains (which include the Golan Heights) rise in the south and the south-east is a desert plateau. The climate is mainly dry and years of drought, lack of foreign investment, and heavy military expenditure have been drawbacks to the Syrian economy, which is primarily dependent on agriculture and stock-raising. There is sufficient arable land for its people, and about 65 per cent of the population is dependent on the soil. Cotton is the major export, with cereal crops second. The most developed industry is textiles. Petroleum reserves are being exploited.

People. Most of the people are Arabs, but there are also minorities of Kurds, living in the north along the Turkish border, and Armenians, most of whom live in towns and cities. Probably the only

The 12th century Byzantine Citadel in Aleppo, Syria, occupies an imposing sight on the city's skyline; its present form dates from the 17th century.

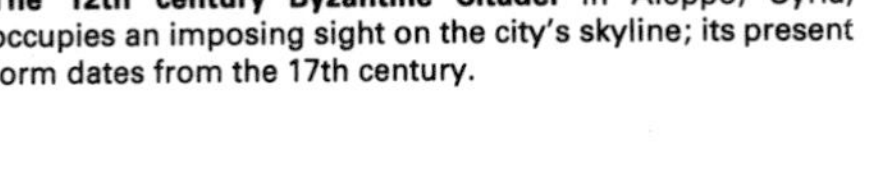

The old quarter of Damascus, the capital of Syria, contrasts with the modern part of the city, which is the administrative and communications centre of the country.

The twin peaks of Mount Kilimanjaro in Tanzania are permanently covered in snow and ice; coffee is grown on the lower slopes of the mountain.

indigenous people are the Alawis, a Muslim sect living in the province of Latakia.

Government. The 1973 constitution provides for a 186-member People's Council, with most power in the hands of the president.

History. Located where three continents merge Syria has held a strategic position since about 2500 BC. Dominated by a series of rulers, it fell into Muslim hands in the AD 630s, had its cities sacked by the Mongols in 1401, and was under Turkish rule for 400 years after 1516. It was a French League of Nations mandate after World War I and declared itself a republic in 1941. Full independence came in 1944. For a short time (1958–61), Syria joined Egypt in the United Arab Republic. In 1963 the socialist Ba'ath Party seized power, becoming the only legal party. In the 1967 Arab-Israeli War, Syria lost the strategic Golan Heights, overlooking Israel. In 1975–76 Syria's involvement in the war in Lebanon alienated most other Arab nations. Map 38.

Syria – profile

Official name Syrian Arab Republic
Area 185,123sq km (71,476sq miles)
Population (1976 est.) 7,585,000
Density 41 per sq km (106 per sq mile)
Chief cities Damascus (capital) (1975 est.) 1,049,500; Aleppo, 639,361
Government Socialist one-party system; head of state Gen. Hafez al-Assad
Religions Islam, Christianity
Language Arabic (official)
Monetary unit Syrian pound
Gross national product (1974) £1,487,200,000
Per capita income £196
Agriculture Cotton, barley, wheat, fruits, vegetables, sugar-beet, sheep
Industries Textiles, flour milling, oil refining, cement, tobacco products, glassware, brassware, soap
Mineral (major) Petroleum
Trading partners USSR, China, Lebanon, France, Italy, West Germany, Britain

Tadzhikistan. *See* UNION OF SOVIET SOCIALIST REPUBLICS.

Taiwan, official name Republic of China and also known as Nationalist China, is an independent island nation on the Tropic of Cancer 145km (90 miles) off the south-eastern coast of mainland CHINA. It consists of the island of Taiwan (formerly Formosa) and the Pescadores, Quemoy and Matsu islands. The capital is T'aipei, on Taiwan island. The nation was established, with support from the United States, by Gen. Chiang Kai-shek's Kuomintang government after Communists gained control of the mainland in 1949.

Taiwan's climate is semi-tropical and it lies in a typhoon and earthquake belt. A range of mountains runs the length of the island. In the east the coastal plains are narrow; in the west, they broaden out. Chief crops are rice, tea, sugar, sweet potatoes and bananas. Mineral resources include coal, oil and natural gas. Light manufacturing industry has replaced agriculture as the dominant factor in the economy, with the emphasis on textiles, clothing and electrical goods. The Taiwanese are descendants of aboriginal Philippine tribes or early immigrants from the mainland, or some of the two million people who have travelled there from mainland China since 1949. The principal religion is Buddhism-Taoism; the literacy rate 84 per cent.

The history of Taiwan is one of a series of migrations, beginning with people from the mainland in the 7th century. It was known to the Portuguese in the 1590s, occupied by the Dutch (1624–62) and then seized, after 1683, by several warring Chinese dynasties. After the Sino-Japanese War Taiwan was awarded to Japan in 1895. It was returned to China after World War II and Chiang Kai-shek withdrew there in 1949. In the 1950s Quemoy and Matsu were shelled by Communist forces, whose threatened invasion was discouraged by a treaty between Taiwan and the United States. After the death of Chiang Kai-shek in 1975 power passed to his son Chiang Ching-kuo. Area: 35,962sq km (13,885sq miles). Pop. (1976 est.) 16,324,000. Map 48.

Tanganyika. *See* TANZANIA.

Tanzania, official name United Republic of Tanzania, is an eastern African country consisting of the former territories of Tanganyika and Zanzibar. It is a poor nation and more than 90 per cent of the people live in rural areas. The only large city is Dar-es-Salaam, which is also the capital, chief port and major industrial centre. By 1985 the government plans to make Dodoma the capital.

Land and climate. Inland from the narrow coastal plain lies a series of plateaus separated by highlands and the African Rift Valley. The most spectacular highland zone is in the north where Mt Kilimanjaro, Africa's highest peak, towers 5,895m (19,340ft) above sea-level [*see* MS p.*216*]. The eastern arm of the Rift Valley extends north-south, separating the Southern Highlands, the south-eastern plateau and the Masai steppes from the high interior plateau; the western arm encloses Lake Tanganyika. In most places the climate is tropical, but the hot, humid coastlands contrast with the cooler, drier high plateaus. Only a quarter of Tanzania has a rainfall of more than 760mm (30in) per year.

Economy. Most people are subsistence farmers, and beans, maize and millet are leading food crops. In the 1970s the government grouped people on the mainland in *ujamaa* villages (*ujamaa* is Swahili for *familyhood,* and the villages are organized along communal lines). The inhabitants enjoy facilities that they lacked in scattered bush settlements.

The main commercial crops are coffee, cotton and sisal which, with cloves from Zanzibar, together account for half of all exports. Cashew nuts, pyrethrum, tea and tobacco are also important. The chief mineral, diamonds, accounted for four per cent of exports in 1974. Manufacturing is based mainly on Dar-es-Salaam. Tanzania's national parks are attracting more and more tourists.

People. Tanzania has about 125 ethnic and language groups, nearly all of whom speak Bantu languages (which include Swahili, the lingua franca). Islam, Christianity and ethnic religions are all practised. The literacy rate is about 30 per cent.

Government. Tanzania is a one-party state. After union with Zanzibar in 1964, the two areas retained their own political institutions but in 1977 they merged into the *Chama cha Mapindozi* (Revolutionary Party). The National Assembly contains members from the mainland and ZANZIBAR, but Zanzibar retains its own legislature. A new constitution, introduced by President Nyerere, made the Revolutionary Party supreme.

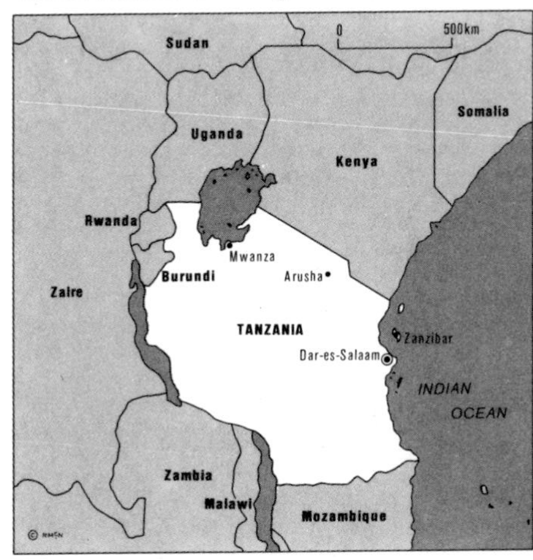

History. The remains of some of man's earliest ancestors, dating from about 1.75 million years ago, have been found in Olduvai Gorge in Tanzania [*see* MS pp.22–24]. Coastal trading posts were established in the early centuries AD. The first European to visit the region was the Portuguese navigator Vasco da Gama, who landed on the coast in 1498. German influence on the mainland began in the 1880s and by 1907 the Germans had subjugated the interior, calling the territory German East Africa. Zanzibar came under British protection. In 1920 the League of Nations mandated Britain to rule also the mainland region, Tanganyika.

Tanganyika became independent in 1961 and a republic in 1962. In 1963 Zanzibar also became independent and in 1964 it united with Tanganyika (which adopted the name Tanzania). The United

Ngorongoro National Park is situated in the crater of an extinct volcano; it is Tanzania's major conservation area and a popular tourist attraction.

Making kites in Thailand is an activity engaged in by all ages. Kites originated in Asia and have been a favourite pastime in the region for many centuries.

Traditional Thai dancers await their cue to start a performance. Thai dance is complex and is based on symbolic postures and characterizations centuries old.

Republic of Tanzania has pursued socialist policies, and differences with Kenya and Uganda have led to the gradual breakdown of the East African Community, by means of which the countries shared common services. Instead, Tanzania has been co-operating increasingly with its southern neighbours Mozambique and Zambia. Map 34.

Tanzania – profile

Official name United Republic of Tanzania
Area 945,087sq km (364,898sq miles)
Population (1975 est.) 15,312,000
Density 16 per sq km (42 per sq mile)
Chief city Dar-es-Salaam (capital) (1975 est.) 517,000
Government One-party republic; head of state President Julius Nyerere
Religions Islam, Christianity, ethnic
Languages Swahili, English (both official)
Monetary unit Tanzanian shilling
Gross national product (1974) £880,300,000
Per capita income £57
Agriculture Cashew nuts, cloves, coconuts, coffee, hides, meat, maize, sisal, sugar, tea, tobacco
Industries Cement, food processing, petroleum products, sugar refining, textiles, tanning
Minerals Diamonds, gold, salt, tin
Trading partners Britain and other members of the EEC, China

Tasmania. *See* AUSTRALIA.

Tennessee. *See* UNITED STATES.

Texas. *See* UNITED STATES.

Thailand (Prathet Thai), formerly known as Siam and official name Kingdom of Thailand, is an independent nation of south-eastern Asia. The capital is Bangkok. Thailand has a fast-developing economy but is still dependent on agriculture, with rice as the chief crop.

Land and economy. Thailand is surrounded on the west, north and east by three other countries: Burma, Laos and Cambodia. To the south is a coastline on the Gulf of Siam, with a narrow neck of land down to the southern border with Malaysia. There are four major land divisions: a central fertile region watered by the Chao Phraya River and irrigation canals; a large plateau in the north-east, which has poor soil and suffers from frequent droughts and floods; an area of forested mountains and fertile valleys in the north; and the rain-forested isthmus on the Malayan peninsula. The whole country has a tropical monsoon climate. Exports of rice account for 20 per cent of foreign earnings, followed in importance by rubber, maize and tin (of which Thailand is the world's third-largest producer). Tourism is gaining in importance.

People. Most of the population is descended from Thai stock, and includes about three million Chinese (mainly in the towns), about a million Malay-speaking Muslims and minority groups of hill tribes and Vietnamese. The rural population is concentrated in the fertile valleys. Thai is the official language, and many people speak English. There is compulsory education and the literacy rate is about 70 per cent. Most of the people are Buddhists.

History. The Thais originally ruled a kingdom in what is now Yunnan, China, and migrated to Thailand about a thousand years ago, encouraged by the Mongol invasion of southern China. Contact with the West began with visits by the Portuguese in the 16th century. Burmese conquerors in the 18th century were driven out by Rama I, founder of the present Thai ruling family. As the European colonizing powers grew stronger, successive rulers modernized Thailand in an attempt to allow it to survive as a nation. The country was occupied by the Japanese from 1941 until the end of World War II. The victories of Communist forces elsewhere in south-eastern Asia forced Thailand to modify its pro-Western policies in the mid-1970s. There have been several changes of government, and student riots in October 1976 were followed by much bloodshed, martial law and a change of constitution. Map 50.

Thailand – profile

Offical name Kingdom of Thailand
Area 514,000sq km (198,455sq miles)
Population (1976 est.) 43,569,000
Density 85 per sq km (220 per sq mile)
Chief city Bangkok (capital) (1976 est.) 4,349,500
Government Constitutional monarchy, under martial law
Religion Buddhism
Languages Thai (official), English
Monetary unit Baht
Agriculture Cassava, rice, rubber, maize, coconuts, tobacco, pepper, groundnuts, beans, cotton, jute
Industries Forestry, fishing, tapioca, car assembly, pharmaceuticals, textiles, electrical goods
Minerals Tin, iron, manganese, tungsten, antimony
Trading partners Japan, USA, Malaysia, Singapore, Hong Kong, West Germany, Britain

Timor. *See* INDONESIA.

Togo, official name Republic of Togo, is a small country in western Africa. The capital, Lomé (pop. 148,443), is situated on the coast. The country extends about 550km (342 miles) north from the Gulf of Guinea to its border with Upper Volta. The east-west distance varies between 145km (90 miles) in the centre to only 64km (40miles) in the south. The Togo-Atacora mountains cross central Togo, low plateaus cover the north, and tablelands and fertile plains are in the south. The average annual rainfall on the coast is 740mm (29in), with 1,780mm (70in) on the mountains.

Most Togolese are farmers and the average income in 1974 was £107. The chief cash crops are cocoa and coffee, but phosphates are the most valuable export. Manufacturing is on only a small scale.

Togo has about 30 ethnic groups and French is the official language. Christianity is practised by 24 per cent of the people and Islam by 7 per cent; the remainder follow ethnic religions. Togoland was a German colony from 1884 until World War I. After the war, it was mandated to Britain and France. British Togo, in the west, was incorporated into the Gold Coast and became independent as part of GHANA in 1957; French Togo became the independent republic in 1960, having voted in 1956 to remain autonomous within the French Union. In 1967 a military group led by Gen. Gnassingbe Eyadema seized power. Area: 56,000sq km (21,622sq miles). Pop. (1975 est.) 2,222,000. Map 32.

Tonga, formerly called the Friendly Islands and official name Kingdom of Tonga, is an independent island nation in the south-western Pacific Ocean about 2,200km (1,370 miles) north-east of New Zealand. It consists of about 150 islands and islets in three groups: Tongatapu (to the south), Vava'u (north) and Ha'apai (centre). The capital is Nuku'alofa on Tongatapu island. Native Tongans are Polynesians, most of whom work in fishing or on farms growing coconuts or bananas. There is compulsory education and a high literacy rate. The islands were charted between 1616 and 1643 by the Dutch and they were visited in the 1770s by Capt. James Cook (who named them the Friendly Islands). English missions were established in 1797 and British power gradually increased until the islands became a self-governing British protectorate in 1900 under King George Tupou II. Tonga became completely independent in 1970 under King Taufa'ahau Tupou IV, who succeeded to the throne in 1965 on the death of his mother, Queen Salote Tupou III. Area: 699sq km (270sq miles). Pop (1975 est.) 102,000. Map 62.

Transkei. *See* SOUTH AFRICA.

Transvaal. *See* SOUTH AFRICA.

Trinidad and Tobago is an independent nation consisting of the two southernmost islands of the West Indies, separated from Venezuela's north-eastern coast by the Gulf of Paria and the Serpent's Mouth channel. The islands of Trinidad and Tobago are separated by the Dragon's Mouth channel. The cap-

Shanty towns on the outskirts of Port of Spain, Trinidad, reflect the resourcefulness of the poor in the harsh economic climate that exists in many West Indian islands.

The city of Tunis, which has been the capital of the region since medieval times, dates back to before the rise of Carthage, whose ruins are near Tunis.

The Saharan desert, which covers much of the southern half of Tunisia, is made passable by a number of oases, including this one at Gabes, on the coast.

ital, Port of Spain, is on Trinidad. Major industries include sugar production, petroleum and tourism. Almost half the population is of African descent, with East Indians making up a third; the remainder are European, Middle Eastern or Chinese. The official language is English, although many people speak a French patois (dialect). Trinidad was visited in 1498 by Christopher Columbus and, after temporary Dutch and French occupation, taken by the British in 1797 and ceded to the Crown in 1802. Tobago was settled in 1632 by the English, who were driven away by local Carib Indians. Again the Dutch and French occupied the island before it was retaken by Britain in 1803. The two islands formed a union in 1888 and were members of the West Indies Federation from 1958 to 1962, when they became an independent member of the Commonwealth with a parliamentary form of government. Area: 5,128sq km (1,980sq miles). Pop. (1974 est.) 1,070,000. Map 74.

Tristan da Cunha. *See* SAINT HELENA.

Trucial States. *See* UNITED ARAB EMIRATES.

Tunisia, official name Republic of Tunisia, is a small nation in northern Africa. The country's sunny climate, fine beaches and historic remains – such as the ruins of Carthage near Tunis (the capital) – attract many visitors. In 1974 some 716,000 people visited Tunisia and the tourist industry provides about 20 per cent of the country's foreign earnings.

Land and climate. Physically, northern Tunisia is a complex region of mountain ranges, representing the eastward limit of the folded Atlas mountains. These extend from Morocco, through Algeria and into Tunisia where the two main chains are separated by the fertile valley of the River Mejerda. The highest peak in Tunisia is Jebel Chambi, 1,554m (5,220ft) above sea-level in the High Tell. The Chott Djerid depression occupies central Tunisia, and Saharan plateaus cover the south.

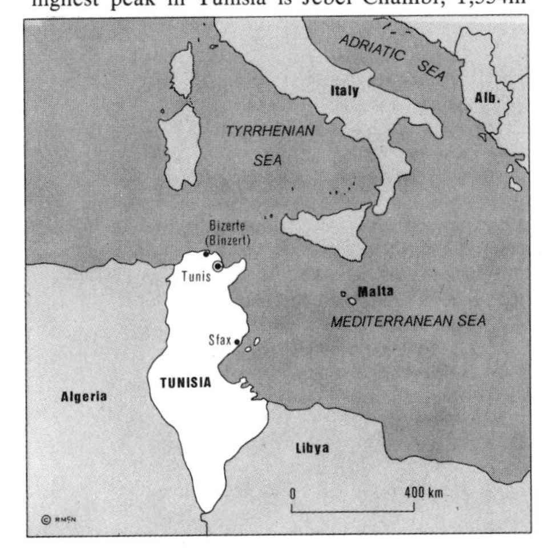

Tunisia – profile

Official name Republic of Tunisia
Area 164,150sq km (63,378sq miles)
Population (1975) 5,588,200
Density 34 per sq km (88 per sq mile)
Chief cities Tunis (capital), 642,000; Sfax, 250,000
Government Republic; head of state President Habib Bourguiba
Religion Islam
Languages Arabic (official), French
Monetary unit Tunisian dinar
Gross national product (1974) £1,316,200,000
Per capita income £236
Agriculture Almonds, cereals, citrus and other fruits, livestock
Industries Construction materials, leather goods, olive oil, petroleum products, processed foods, textiles, wine
Minerals Iron ore, lead, petroleum, phosphates
Trading partners France, Italy, USA, West Germany

Northern Tunisia has a Mediterannean climate with cool winters (when most of the rain falls) and hot, dry summers. The average annual temperature range is between 10°C (50°F) in January and 27°C (81°F) in July. Rainfall averages between 510 and 1,020mm (20–40in) per year. To the south summer temperatures increase, and the rainfall steadily decreases. The Chott Djerid has an average of only 150mm (6in) of rain per year; the Saharan plateaus in the far south are even drier.

Economy. Farming is the most important sector of the economy, providing employment for about 60 per cent of the people. The five main agricultural regions are the fertile northern valleys and plains; the north-eastern Cap Bon peninsula, the leading area for citrus fruits; the eastern coastal plains, called the Sahel, which are known for their olive groves; the central uplands, which provide good pasture; and the oases in the south. The main food crops are barley and what. The chief farm exports are olive oil, wine, fruits and vegetables, which together account for 26 per cent of all exports (Tunisia is the world's largest producer of olive oil). The livestock-rearing industry in the central uplands has been modernized; the country has about 3,200,000 sheep, 680,000 cattle and 460,000 goats.

The most valuable exports, however, are minerals, especially petroleum. Although Tunisia accounts for only 0.3 per cent of the petroleum produced by the members of the Organization of Petroleum Exporting Countries (OPEC), petroleum makes up about 30 per cent of Tunisia's exports and there is a refinery at Bizerte. Phosphates account for 14 per cent of the exports and ores of iron and lead for another 4½ per cent. Manufacturing is steadily increasing and traditonal craft industries still flourish.

People. Most of the people are Muslims who speak Arabic, but Berber is spoken in parts of the south. Some Europeans, mostly Frenchmen and Italians, live in Tunisia and French is a second language in the cities. With a high birth-rate (a net population increase of 2.4 per cent per year) and a youthful population (64 per cent of Tunisians are under 25), Tunisia faces considerable difficulties in providing sufficient educational facilities and jobs. Unemployment is the chief source of discontent in this generally politically stable country.

Government. A former French protectorate, Tunisia became independent in 1956 as a monarchy, the head of state being the bey. But the bey was deposed in 1957 and Tunisia became a republic; the president is also head of the government. The National Assembly has 90 members and the only party is the Destourian Socialist Party.

History. Carthage, the leading Phoenician trading city, was founded in Tunisia in the 9th century BC. It became a major Mediterranean power and pitted its strength against Rome in the Punic Wars between 264 and 146 BC. After Carthage was destroyed, the area was ruled successively by Romans, Vandals and Byzantines. The Arabs invaded Tunisia in AD 647–669 and converted it to Islam. The Turks took Tunisia in the mid-16th century and made it part of the Ottoman Empire. But in 1881 France occupied the country and made it a protectorate. Nationalists opposed French rule and their leader from 1934 was Habib Bourguiba. Guerrilla warfare in the early 1950s led France to withdraw, and Tunisia became independent in 1956. Bourguiba was elected president in 1957 and re-elected in 1959, 1964 and 1969. In 1974 he was elected president for life. Map 32.

Turkey (Türkiye), official name Republic of Turkey, is an independent nation located between Asia and Europe. It became a republic in 1923, and is governed from Ankara, the capital.

Land and economy. The strategically important Turkish straits (the Bosporus, Sea of Marmara, and the Dardanelles) connect the Black Sea and the Mediterranean Sea. The mild, narrow coastal plain supports a variety of crops, including tea and cotton. Wheat is grown on the western part of the central Anatolian plateau; the eastern part is mountainous and has severe winters. The south-west is treeless, with some mountains 3,050m (10,000ft) above sea-level and little population. The rivers Tigris and Euphrates rise in the east and flow southwards to the Persian Gulf. In 1973 a suspension bridge was completed linking Europe and Asia across the Bosporus [*see* MM p.*193*].

Turkey has an agricultural economy, the main

A masterpiece of Ottoman architecture is the Suleiman mosque, built between 1550 and 1557 by Sinan, one of the great Islamic builders.

Kars is on a plateau 1,750m (5,740ft) above sea-level in Turkey near the border of the USSR; it is an important centre for livestock.

People and pipes disappear into the side of the Ruwenzori Mountains in the quest for copper ore at a mine in Kilembe, Uganda; the country is a major copper producer.

crops being cotton, tobacco and cereals. About 65 per cent of the working population is employed in farming and allied occupations. About half of the economy depends on state-owned or state-controlled enterprises. Opium production was halted in 1971 and resumed for the pharmaceutical industry in 1974.

People. Most Turks (99 per cent Sunni Muslim) live on the Anatolian peninsula. At one time many people lived in small villages, but villagers have moved to the towns and most cities now have shanty-town communities surrounding them. There is no officially recognized religion and no legal discrimination against the minority groups of Greeks and Armenians. The largest ethnic minority, the Kurds, live in primitive conditions in the remote areas of the east and south-east. Primary education is free and compulsory; the literacy rate is 65 per cent in towns but less in the country districts.

Government. The 1961 constitution provides for a

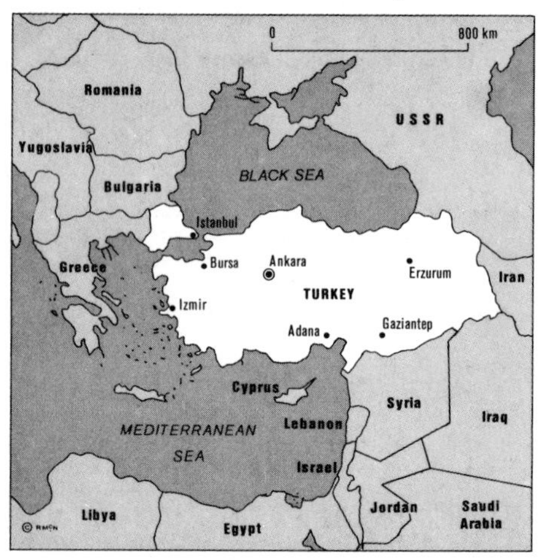

president and a two-chamber legislature. The prime minister is chosen from the majority party.

History. In classical times a centre of Greek civilization, the region now occupied by Turkey was subsequently under the Roman, Byzantine and Ottoman empires. When the 600-year-old rule of the Ottoman Empire collapsed after fighting as one of Germany's allies in World War I, Nationalism grew and the trappings of the old empire were abolished; Turkey's history dates from this time. Under the leadership of Kemal Atatürk, Turkey became a republic in 1923, with Atatürk its first president. It turned away from imperial traditions and became Westernized, with social and economic reforms that were the basis of modern Turkey.

Turkey joined the Allies near the end of World War II and under the Truman Doctrine received military and economic aid from the United States. In 1950 Atatürk's party was defeated and the Democratic Party gained power until 1960 when they were overthrown by a military coup. A return to civil government came in 1961. Tension with neighbouring Greece has been a constant factor in foreign affairs. In 1974 this led to a Turkish invasion of CYPRUS, and the partitioning of the Turkish and Greek communities on the island. Another continuing dispute with Greece concerns rights to any minerals (particularly petroleum) that might be discovered in the bed of the Aegean Sea. There was rioting in Turkey between left-wing and right-wing factions throughout 1976. Map 6.

Turkey – profile

Official name Republic of Turkey
Area 780,574sq km (301,380sq miles)
Population (1975) 40,198,000
Density 52 per sq km (133 per sq mile)
Chief cities Ankara (capital) (1974 est.) 1,522,350; Istanbul, 2,487,100; Izmir, 619,150
Government Republic; head of state, President Fahri Korutürk
Religion Islam
Languages Turkish, Kurdish, Arabic
Monetary unit Turkish lire
Gross national product (1974) £11,453,000,000
Per capita income £285
Agriculture Tobacco, cereals, cotton, olives, livestock, fruits, sugar-beet, opium, forestry
Industries Olive oil, fibres, opium, iron and steel, leather goods, furniture, cement, paper, glass
Minerals Antimony, borate, copper, chronium, manganese, lead, zinc, coal, iron ore, petroleum
Trading partners EEC, USA

Turkmenistan. *See* UNION OF SOVIET SOCIALIST REPUBLICS.

Turks and Caicos Islands is a British dependency in the West Indies. It includes more than 30 islands and islets, six inhabited, that form a south-eastern continuation of the Bahamas. The capital is on Grand Turk island. The chief exports are salt, sponges and shell fish; most of the people are of African descent. The islands were visited in 1512 by the Spanish explorer Juan Ponce de León, settled by British Loyalists from the United States in the 1780s and annexed to Jamaica in 1874. Since 1962 the dependency has been governed by an administrator appointed by Britain. Area: 430sq km (166sq miles). Pop. (1975 est.) 6,000. Map 74.

Uganda, official name Republic of Uganda, is a landlocked nation in east-central Africa. The capital and largest city is Kampala. The country was a British protectorate between 1893 and 1962, and since January 1971 it has been ruled as a military dictatorship by Gen. Idi Amin. Pursuing a policy of Africanization, in 1972 he expelled Ugandan Asians who held British passports. He has used the army ruthlessly to suppress any attempts at opposition to his regime.

Land and climate. Most of Uganda lies on a plateau between 1,070 and 1,370m (3,510–4,495ft) above sea-level. Water covers about 15 per cent of the country. Lake Victoria, Africa's largest lake and source of the White Nile, is shared with neighbouring Kenya and Tanzania, and Lake Victoria occupies a shallow depression in the plateau. To the west, Uganda is bordered by the western arm of the deep African Rift Valley, which contains lakes Idi Amin Dada (formerly Lake Edward) and Mobutu Sese Seko (formerly Lake Albert). The eastern rim of the Rift Valley has been uplifted to form the Ruwenzori, a block mountain range which reaches 5,109m (16,762ft) above sea-level. Mountains also border Uganda to the north and east. Mt Elgon, a volcanic massif, rises to 4,321m (14,176ft) on the border with Kenya. The three main land regions are the thickly populated Lake Victoria lowlands (including the Lake Kyoga region to the north), the plains of the northern savanna plateau and the Rift Valley zone.

The climate is mostly hot and humid all the year round. As a result, the country never attracted many permanent European settlers, who preferred the dry heat of the nearby Kenyan highlands. The average annual rainfall varies between 760 and 2,000mm (30–79in). It is greatest on the northern shores of Lake Victoria and on the mountains; the northern plateau is the driest region.

Economy. More than 90 per cent of the people live in rural areas and most are subsistence farmers. Important food crops include cooking-bananas (plantains), cassava, maize, millet and sweet potatoes. Fish is a useful source of protein for people who live near the lakes. The chief cash crops are coffee, cotton and tea. In 1973 coffee accounted for 59 per cent of all exports, cotton for 16 per cent and tea for 5 per cent. Copper, the chief mineral, made up a further 5 per cent of the exports. Other important products include animal foodstuffs and hides and skins. Manufacturing has developed at Jinja, Kampala and Tororo. Most factories use electricity from the Owen Falls hydroelectricity station at Jinja [*see* MM p.199]. But manufacturing and mining contribute only about 10 per cent of the gross domestic product, as opposed to agriculture, which contributes nearly 50 per cent. Uganda has three superb national parks – Kabalega Falls (formerly Murchison Falls), Kidepo and Ruwenzori. But tourism has been slow to develop because of the political situation.

People. Uganda has about 40 ethnic and language groups. About two-thirds of the people are Bantu-speaking Negroids, the largest group being the

The Owen Falls Dam and hydroelectric project on the Victoria Nile, completed in the 1950s, turned Jinja, Uganda, into a major industrial centre.

Until 1972 many Ugandan shops were owned and run by Asians, after whose expulsion the shops were re-opened under African management.

The Great Palace of the Moscow Kremlin, a complex of buildings occupying a large site in the city centre, overlooks the Moscow River.

Baganda (16 per cent of the total population), the Banyankore, the Basoga and the Bakiga. One-sixth are Nilotic, including the Lango and Acholi, and another sixth are Nilo-Hamitic, mainly Iteso. There are about 10,000 Europeans and refugees from Rwanda, Sudan and Zaire. Christianity is practised by nearly 50 per cent of the people and Islam by 6 per cent; the remainder follow ethnic religions. The literacy rate is estimated at 25 per cent.

Government. Uganda was a constitutional republic until 1971 when the army commander Gen. Idi Amin seized power. He dissolved parliament, suspended parts of the constitution and centralized government through a Defence Council under his chairmanship. Amin's authority is virtually absolute and in 1976 he was made president for life.

History. Before the arrival of Europeans, the Bantu-speaking peoples were organized in small kingdoms. The most powerful was Buganda – the land of the Baganda. This kingdom was visited by

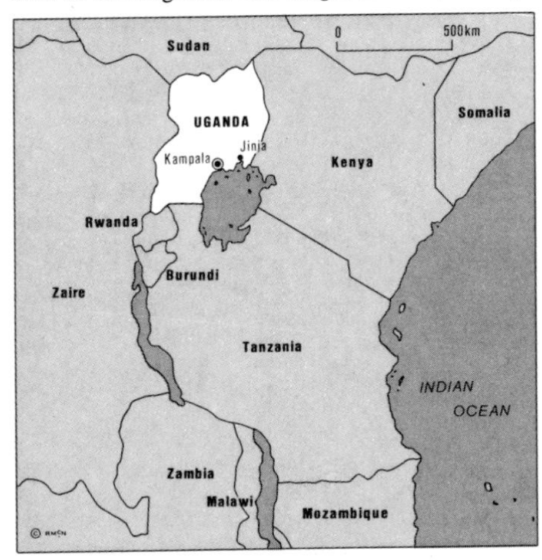

the British explorer John Hanning Speke in 1862. In 1894 Buganda became a British protectorate, but the traditional monarchy was retained. By 1914 all of Uganda was under British rule and the railway from Mombasa in Kenya to Uganda made economic development possible. The labourers on the railway were Asians, and many of them settled in Uganda after the railway was completed. They then played an important part in developing trade.

As Uganda approached independence, difficulties arose concerning the status of Buganda, the most developed part of the country. Following independence in 1962 it was granted a special federal relationship with the rest of the country. In 1963, in an attempt to integrate Buganda more closely into the rest of the country, the *kabaka* (king) of Buganda, Mutesa II, was made president. However, the prime minister, Apollo Milton Obote, deposed the kabaka in 1966 and, under a new constitution of 1967, became president. But in 1971 Obote was overthrown by Gen. Idi Amin. In 1972 President Amin expelled most of the country's Asians. Strained relations with neighbouring Kenya came to a head in 1976 when Kenya insisted that goods passing through on route to Uganda be paid for in Kenyan currency.

Maps 32, 34.

Uganda – profile

Official name Republic of Uganda
Area 236,036sq km (91,133sq miles)
Population (1975 est.) 11,549,000
Density 49 per sq km (127 per sq mile)
Chief city Kampala (capital) (1975 est.) 542,000
Government Military dictatorship
Religions Christianity, ethnic, Islam
Language English (official)
Monetary unit Ugandan shilling
Gross national product (1974) £760,700,000
Per capita income £66
Agriculture Bananas, cassava, coffee, cotton, hides and skins, maize, millet, oil seeds, sisal, sugar, sweet potatoes, tobacco
Industries Cement, chemicals, cigarettes, copper smelting, food processing, textiles
Minerals Copper, tin
Trading partners Britain, USA, West Germany, Japan

Ukraine. *See* UNION OF SOVIET SOCIALIST REPUBLICS.

Union of Soviet Socialist Republics (USSR) is the largest country in the world. It occupies half of Europe and one-third of Asia, and stretches for more than 8,000km (5,000 miles) from its western borders in Europe to its eastern coastline on the Pacific Ocean. At its most north-easterly point, on the Bering Strait, it is only 90km (56 miles) from Alaska in the United States.

The USSR is a federation of 15 republics, inhabited by people of many different cultures. Three-quarters of its territory is in Asia, but most of its population and important cities – including the capital, Moscow – are in Europe.

The name of the USSR in Russian is *Soyuz Sovyetskikh Sotsialisticheskikh Respublik* – SSSR. In their Cyrillic alphabet SSSR is written CCCP. The country is often called the *Soviet Union* because its Constitution provides for a system of government by *soviets* (councils). Many people also call the country *Russia*, and more than a half of its people are Russians, from the republic of Russia (Russian Soviet Federalist Republic; RSFSR).

Land and climate. The general landscape of the USSR is that of an enormous amphitheatre in which high mountains along the eastern and southern borders slope down to vast plains or steppes similar to the great prairies of North America. The East European Plain extends from the Polish border in the west to the Ural Mountains, a range that runs north-south and marks the boundary between Europe and Asia. The Urals are densely forested except in the north.

Two other great lowland areas lie east of the Urals: the West Siberian Lowland in the north and the Turanian Plain in the south. The first of these is the Siberia of folklore, poetry and the novel; it skirts the Arctic Ocean and extends as far east as the Yenisey River. The largest continuous lowland in the world, it has an area of about 2,500,000sq km (970,000sq miles). Its northern zone is tundra, with stunted vegetation and ground that is never completely free from frost. Farther south is the taiga, a great belt of dense forest. The Turanian Plain, in Soviet Central Asia, lies south of the Aral Sea. Much of it consists of stark deserts: the *Kara Kum* (Dark Sands) and the *Kyzyl Kum* (Red Sands).

At the eastern end of the Siberian plain are the Central Siberian Uplands which rise to about 900m (2,950ft) above sea-level. Still farther east are the wild and lofty East Siberian Highlands, which extend along the Kamchatka Peninsula. Their highest peaks rise to about 4,500m (14,800ft); there are about 30 active volcanoes in the peninsula.

The highest mountains are along the southern borders of the USSR. In the south-west the Caucasus range has peaks of up to 5,500m (18,000ft). In the Pamir Knot, near the border with Afghanistan, Communism Peak rises to 7,495m (24,590ft), the highest point in the country. The high southern ranges extend eastwards with the Tien Shan, Altai, Sayan and Yablonovyy mountains.

Three of the longest rivers are in Siberia and flow northwards into the Arctic Ocean. They are the Ob (3,410km; 2,120 miles long), the Yenisey (3,200km; 1,990 miles) and the Lena (4,800km; 2,980 miles). The Amur (4,500km; 2,800 miles) flows along the border with China and empties into the Gulf of Sakhalin. The European part of the USSR is also watered by great rivers. The Volga (3,750km; 2,330 miles), Europe's longest river, flows into the Caspian Sea. The broad, leisurely Don (2,000km; 1,240 miles) rises near Moscow and winds southwards to the Black Sea. The Dnieper, or Dnepr (2,200km; 1,370 miles), the chief river of the Ukraine, also empties into the Black Sea.

The Caspian Sea, which lies to the east of the Caucasus, covers about 427,000sq km (164,870sq miles) and is the largest body of inland water in the world. On its north-eastern shore is the Karagiye Depression, 130m (430ft) below sea-level and the lowest point in the USSR. The Caspian and the nearby Aral Sea – also landlocked – have salt water. Europe's largest freshwater lake, Lake Ladoga (near Leningrad), covers an area of 4,400sq km (1,700sq miles). Lake Baykal, in the Central Siberian Uplands, is the deepest freshwater lake in the

Union of Soviet Socialist Republics

world; its greatest depth is about 1,750m (5,740ft).

The Soviet climate is as varied as the terrain. In the central areas of the European zone, January temperatures are between −10°C (14°F) and −16°C 3°F), but in the Lena River basin in Siberia winter temperatures average −40°C (−40°F), falling to −72°C (−98°F) in Yakutia. Summer temperatures in the European zone average about 20°C (68°F) and from 4°C (40°F) to 10°C (50°F) in the far north. In central Asian regions the climate is torrid, with mean temperatures rising as high as 50°C (122°F).

Precipitation is lowest in the Turanian Plain and in the tundra regions, which have less than 250mm (10in) of rain or snow a year. The heaviest precipitation is on the eastern shores of the Black Sea with 2,440mm (96in) per year. The East European Plain and the Pacific coastal lands have an average of between 510 and 1,020mm (20–40in) per year.

Physical resources. The USSR probably has the world's greatest store of mineral wealth. Much of it, however, is difficult to extract because it is located in remote areas in the far north and east of the country. Full exploitation of recently discovered mineral deposits in Siberia would require considerable – and not always available – resources of equipment, money and manpower.

The USSR is the world's largest producer of petroleum, with an annual output of nearly 500 million tonnes. The most important oilfields are in the Caucasus, the Volga–Ural region, Central Asia and the far eastern regions. The country's natural gas deposits are also the largest in the world. They are estimated at 19 billion cu m (671 billion cu ft).

Soviet coal deposits account for an estimated 68 per cent of the world's total reserve; using modern technology, about one-third of the coal is recoverable. The chief coalfields are in the basins of the Donets and the Pechora in the European part of the country and in Kuznetsk and Kazakhstan in Asia. Annual production is about 700 million tonnes. Some 40 per cent of the world's known resources of iron ore are in the USSR. Most of the mining areas are in the European part around Krivoi Rog and Kursk. There are also important deposits in the Urals, Kazakhstan and Siberia.

Kazakhstan also has copper, lead, zinc, nickel and titanium. Aluminium is extracted from bauxite mined in the Urals. The USSR is a major producer of gold from deposits in eastern Siberia, as well as of precious and semi-precious stones such as diamonds, emeralds, rubies, jasper and malachite. The forests are also a source of wealth, although vast areas of forest in Asia are at present under-exploited because of their inaccessibility. The most prosperous forest industries are located west of the Urals.

Agriculture. Soviet agriculture is based on either the collective farm (*Kolkhoz*) or state farm (*sovkhoz*). Collective farms operate as autonomous units, apart from obligatory deliveries of certain amounts of produce to the government; employees are paid on a profit-sharing basis. On state farms employees receive wages, and management is directly responsible to the Ministry of Agriculture. Each employee on either type of farm is allowed a plot of land of up to a hectare (2.5 acres) for growing fruit and vegetables for his own use or for sale, and for keeping livestock. There are about 29,000 collective farms and 18,000 state farms. The larger farms occupy about 6,000 hectares (14,820 acres) and support about 400 families.

Cereals – particularly wheat – are the chief crops of Soviet agriculture; wheat has an average annual yield of 90 million tonnes. Smaller quantities of oats, barley, rye and maize are produced. Other food crops of major importance are potatoes, fruit, tea and sugar-beet. Soviet production of sugar-beet is the largest in the world, and the USSR is also the largest grower of flax. Cotton is the chief crop in watered areas of Soviet Central Asia and in parts of Transcaucasia; in some other areas of the south tobacco is grown. Cattle (beef and dairy), sheep and pigs are raised in most farming regions; in some dry places and in mountainous areas livestock raising is the chief source of income.

Although the USSR devotes more land to agriculture than does any other country in the world, Soviet farming is not as efficient as farming in the United States, Canada or even some small countries such as Britain. Often the climate is the reason for low yields; periodic droughts cause disastrously bad harvests, and grain has then to be imported from North America. The major grain-producing areas – Ukraine, the Volga region and Kazakhstan – are the most liable to drought, and poor harvests occur about once every four years. The government has embarked on large-scale irrigation projects in these areas. By contrast the Baltic region and the heart of the RSFSR have too much water, and massive drainage schemes are being brought into operation.

Fishing provides employment for many thousands of people ; the annual catch is surpassed only by those of Peru and Japan. Soviet deep-sea trawlers operate in many of the world's fishing grounds. Smaller vessels fish in the Black Sea, the Caspian Sea and the Baltic Sea [see PE p.244].

Industry. There is no private industry. All enterprises are either state-owned or operate on a collective or co-operative basis. Industrial planning is centralized by the various ministries under five-year plans. In the past, strong emphasis was laid on the development of heavy industry. But since the 1960s there has been a marked increase in the production of consumer goods.

The USSR produces about one-fifth of the world's steel and in engineering manufactures it ranks second after the United States. All types of metal-cutting machine tools are made, including ultra-precision lathes and automated production lines. The output of motor cars rose from 524,000 in 1960 to 1,964,000 in 1976, and about a million motor cycles and scooters are produced annually.

The importance given to railway transport in the USSR is shown in the number of locomotives manufactured, about 1,800 a year – more than any other country. Similarly the strenuous efforts being made to improve agricultural efficiency are mirrored in the output figures for farm machinery: about half a million tractors a year and 85,000 combine harvesters.

The Soviet chemical industry is second only to that of the United States. Its major products include fertilizers and synthetic fibres and resins. The development of the consumer goods industry is highlighted by the large increase in the production of household equipment. In 1960, for example, 529,000 refrigerators were made. By 1975 this figure had increased to 5,506,000, while the number of washing machines manufactured in the same period grew from 895,000 to 3,284,000.

The output of electricity in 1976 was almost five times that of 1960. Coal, once the main source of power, is progressively being replaced by oil and gas, which together now account for 60 per cent of energy used, compared with 20 per cent in 1950.

Trade and economics. The USSR trades with more than 100 countries, and ranks ninth among the countries of the world in trade turnover. Its chief exports are metals and ores, which comprise more than 30 per cent of the total. Next comes machin-

The Russian steppes, with almost half of the world's area of fertile black earth, are increasingly being turned into arable farming lands.

Novaya Zemla, an archipelago of mountainous islands which are a continuation of the Urals, has been used by the USSR for testing nuclear weapons.

Issyk-Kul, a resort area in the Ala-Tau range of central USSR, is the world's second-largest mountain lake and reaches a depth of 702m (2,303ft).

ery, followed by fertilizers and other chemicals.

Most of the foreign trade, some 56 per cent, is with the Communist countries of Europe. Another 31 per cent is with the EEC countries and North America: West Germany, Japan, the United States, Finland, Italy and Britain are the main non-Communist trading partners. The USSR's imports consist chiefly of industrial equipment, machinery and manufactured goods.

As in other sectors of the economy, trade is conducted by the state. There is, however, a limited but important form of private trade in towns where collective farmers sell produce from their family plots.

Transport and communications. Railways transport about 60 per cent of the country's freight, and also carry more than 300 million passengers a year. The railway system extends over 140,000km (87,000 miles) of track, one-tenth of the world's total. Water transport is also important, but many of the large and otherwise navigable rivers are iced up for part of the year. There are about 19,000km (11,800 miles) of canals and 180 million passengers travel on them each year. Buses and other motor vehicles carry about 40 per cent of passenger traffic (compared with 50 per cent on the railways); they are used chiefly within cities and for short-distance journeys. The Soviet merchant fleet ranks sixth in the world. It has about 1,600 vessels, 80 per cent of which are less than 20 years old.

Because of the vastness of the country, the ruggedness of its terrain and the severity of the climate in many places, aircraft play an important role in internal travel, particularly to the more remote regions. Aeroflot, the government-owned airline, operates about 580,000km (360,500 miles) of routes inside the country; it also flies to many cities in other countries. The USSR was the first to build a supersonic transport plane, the Tu-144, although the Anglo-French *Concorde* was the first to enter regular international service.

There are three national radio channels, broadcast from Moscow, and 130 television stations (some with more than one channel). Individual republics broadcast up to three local radio programmes. Throughout the country there are nearly 60 million radio receivers and more than 50 million television sets. All of the country's publishing houses are licensed by the State Committee for the Press. Publications include more than 4,000 periodicals and 6,000 newspapers, of which the most influential are *Pravda* and *Izvestia*.

Constitution and government. The USSR is a federation of 15 republics (see table). Within the union there are 20 so-called autonomous republics, 16 of which are in the RSFSR.

The federal legislature is the Supreme Soviet, a two-chamber parliament consisting of an elected Soviet of the Union and an elected Soviet of the Nationalities. It normally meets only twice a year for about a week at a time; a Presidium carries out legislative functions between meetings of the Supreme Soviet. The chairman of the Presidium acts as head of state. There is also an executive consisting of a Council of Ministers (formally called the People's Commissars), headed by a chairman who ranks as premier.

Every citizen aged 18 years or more is entitled to vote in elections. The Communist Party, which has about 15 million members, is the only political group allowed. It plays a crucial role in government because, in practice, it is the sole policy-maker and the Supreme Soviet has only to give its formal consent to the measures proposed. The highest organ of the Communist Party is its Central Committee in plenary session. In day-to-day affairs, however, effective power lies with an inner committee, the Political Bureau (Politburo), in which the first secretary is a key figure with a political authority exceeding that of the premier. In 1977 there were 15 members of the Political Bureau.

Each of the republics has a Constitution similar to that of the federation: each has its own Supreme Soviet and party organization. A republic has the right to secede from the federation, but the close integration of executive functions — particularly at party level — makes secession unlikely, if not almost impossible.

The highest court is the Supreme Court of the USSR; each republic also has a Supreme Court. Judges are elected for terms of five years. There is no trial by jury. Cases are heard by a judge and two elected officials called assessors. Minor cases, comprising some 90 per cent of judicial business, are heard by People's Courts elected by local voters. The maximum prison sentence is 15 years. Capital punishment can be imposed for murder and for what the Constitution describes as "encroachments upon the foundations of the socialist system".

Armed services and police. The USSR maintains a large standing defence force with several branches: the Strategic Rocket Force, Air Defence, and the army, air force and navy. All Soviet male citizens fit for military service undergo training for a period of from two to five years. No official statistics are issued regarding the strength of the armed services, but it has been estimated that the total Soviet military manpower is about 3,500,000. Of these about 1,800,000 are in the army, which has more than 100 mechanized divisions, 50 tank divisions and seven airborne divisions. The navy has about 475,000 men and its equipment includes hundreds of submarines, many of them nuclear-powered, as well as powerful surface vessels. The air force has more than 5,000 combat aircraft. The armoury of the Strategic Rocket Force includes intercontinental ballistic missiles, self-launching missiles and long-range bombers. The USSR is a founder-member of the Warsaw Pact military alliance, along with Poland, Bulgaria, Czechoslovakia, Romania and East Germany.

Day-to-day policing of the USSR is carried out by the militia, assisted by volunteer public order squads whose members spend a few hours of their free time each month in patrol duties. The secret police force specializes in state security and operates in the field of espionage and military intelligence. It reports to a committee in the Ministry of Internal Affairs called the *Komityet Gosudarstvenny Bezopasnosti* (KGB; Committee for Public Safety).

People. The largest ethnic group of people in the USSR are the Russians, who make up about 55 per cent of the population. They are Slavs, as are the Ukrainians (18 per cent) and the Belorussians or "White Russians" (3.5 per cent). There are smaller numbers of other European peoples, including Latvians, Estonians, Lithuanians, Poles and Germans. Rather less than a quarter of the population is Asian. The Uzbeks (3.5 per cent) are the largest of the Asian groups; others include Tatars, Kazakhs, Turkmenians, Yakuts and Chukchi. South of the Caucasus Mountains the Georgians, Armenians and Azerbaijanians each have their own republic. The Jewish population numbers 2,200,000 — about one per cent of the total. Altogether there are some 100 ethnic groups, ranging down to the Aleuts who number only about 400 people.

The population is unevenly distributed; two-thirds of the Soviet people live in the East European Plain, and one-fifth of the entire population lives within 500km (310 miles) of Moscow. Siberia and the eastern regions, which together make up more than half the country, have only one-tenth of the people, and most of those are in the south.

Most city dwellers live in flats. Rents are the lowest in the developed world, but by Western standards the flats are small and sparsely furnished. Many apartment houses are built by local authorities. Some factories have blocks of flats for their workers, and there are some private apartments, usually constructed by co-operative housing associations. Where accommodation is scarce in cities, local authority flats are allocated according to a system of priorities: factors taken into account include health, overcrowding and disability.

In smaller towns and in the countryside, family houses are common; again, they are usually small. Some of the richer people – particularly city dwellers – have private *dachas* (summer cottages).

The social security system provides retirement pensions for men at age 60 and for women at age 55. There is no unemployment benefit because, in the official view, nobody is without work. The disabled are given special help; blind and deaf people receive pensions and are exempt from income tax.

Cultural life and leisure. The official language of the federation is Russian, which is the native tongue of about 60 per cent of the population. But more than 60 other languages are spoken in various parts of the country. They include Ukrainian, Belorussian, Lithuanian, Latvian, Estonian, Moldavian, Yiddish, Georgian, Armenian, Azerbaijani and Kazakh. The state encourages the use of these languages and the preservation of the cultures they represent. But where Russian is not the language in everyday use, it is taught in schools.

According to the Constitution "freedom of religious worship and anti-religious propaganda is per-

The Cathedral of St Sophia, built within the Kremlin at Novgorod in the 11th century, is one of the best examples of early Russian architecture.

Mikhail Lomonosov, whose bust stands in a Leningrad square, was as scientist, grammarian and poet – a commanding figure in the Russian Enlightenment.

The Caucasus, whose mountains have formed an historic barrier between Europe and Asia, came under Russian sway from the 16th to the 19th centuries.

mitted to all citizens". In practice, however, membership of a religious body or any kind of religious observance attracts official disapproval. The Russian, Georgian and Armenian Orthodox Churches are the largest Christian denominations. The Roman Catholic Church is strongest in the western Ukraine, the Baltic republics and parts of Belorussia. The largest Protestant groups are the Evangelical Christian Baptists and the Lutherans. There are Muslim communities in the Asian republics, as well as some Buddhists. Every large city where there is a Jewish community has a synagogue.

The educational system is geared to the political and social objectives of the state, and Marxist-Leninist teachings are an essential part of all curricula. Education is free and compulsory between the ages of 7 and 15 or 16; primary and secondary schools have a total of about 50 million pupils. Provision is also made for secondary schooling for workers. There are thousands of technical schools and other institutions offering specialized training, and there are 58 universities, of which the oldest is Moscow University (founded in 1755).

The people of the USSR are proud of their long musical, literary and dramatic traditions. Russian composers of the past include Alexander Borodin, Modest Mussorgsky, Peter Ilyich Tchaikovsky and Nikolai Rimsky-Korsakov. Since the Revolution, other famous names have been added, including Sergei Prokofiev, Aram Khachaturyan and Dmitri Shostakovich.

Russian literature is equally rich in names: Alexandr Pushkin, Nikolai Gogol, Mikhail Lermontov, Ivan Turgenev, Fyodor Dostoevsky, Leo Tolstoy, Anton Chekhov and Maxim Gorky. Writers of the Soviet era include Mikhail Sholokhov and Ilya Ehrenburg. Contemporary writers are expected to confine themselves to themes that stress the collective approach to problems encountered on the road to achieving Communism and Socialism, and those who fail to do so incur official disfavour. But such writers (known outside the USSR as dissidents) have grown in number since the death of Stalin and have presented a different vision of Soviet life. Two of them, Boris Pasternak and Aleksandr Solzhenitsyn, have been awarded Nobel prizes for literature.

In the performing arts, the USSR has produced many internationally celebrated figures. They include the ballet dancers Galina Ulanova, Maya Plisetskaya and Rudolf Nureyev, the violinist David Oistrakh, the cellist Mstislav Rostropovich and the pianists Vladimir Ashkenazy, Sviataslav Richter and Emil Gil. There are more than 550 professional theatres throughout the country, of which about 40 are devoted to opera and ballet. The Bolshoi in Moscow has a permanent staff – apart from the performers – of more than a thousand, the largest of any theatre in the world. The Soviet film industry has some 40 studios, producing more than 200 full-length feature films a year. Such great directors as Sergei Eisenstein and Mark Donskoi have contributed greatly to the art of cinema.

Sport plays an important part in Soviet life, and receives considerable encouragement from the state. This help is rewarded by the success of Soviet athletes in international events, particularly in the Olympic Games, where they do well at gymnastics, athletics, basketball, volleyball, fencing, handball, canoeing, wrestling and weightlifting. At the Winter Olympics, too, Soviet athletes are often successful in cross-country skiing, biathlon, figure skating, speed skating and ice hockey. Chess is also widely played, and millions of people enthusiastically follow tournaments.

History. The Slav peoples originated in the region round the River Elbe, spread eastwards and southwards and by AD 800 had founded the trading city of Novgorod. They were a peaceful people and invited the Varangians, a Viking tribe, to govern and defend their city and its surrounding region, which was known as the land of the Rus.

The region prospered and became known as Russia. Mongol and Tatar hordes occupied it in the 13th century, to be displaced finally by Ivan IV, "the Terrible", in the late 1400s; Ivan became "Tsar of all the Russias". Western influence began to be felt during the reign of Peter I, "the Great" (1689–1725), who also added Estonia and Latvia to the tsardom and built a new capital at St Petersburg (now Leningrad). Russia expanded even more under Catherine II (1762–96), also known as "the Great". But the ordinary people still lived as serfs in conditions not far removed from slavery. Catherine's son, Paul I, was murdered in a palace revolution. His successor, Alexander I, promised reforms but they were not carried out. During his reign (1801–25) Russia defeated Napoleon's Grand Army and became a major world power.

Alexander II (reigned 1855–81) introduced the first social reforms of any consequence, but they were too little and too late. His emancipation of the serfs in 1861 did not satisfy the nihilists and anarchists and he was killed by a terrorist bomb in St Petersburg. The assassination resulted in another period of repression that lasted until 1917, when the tsarist system collapsed.

In 1905 Russia, which had extended its empire far into Asia during the 1800s, was humiliatingly defeated in a war with Japan. The tsar, Nicholas II (reigned 1894–1917), was a feeble ruler and had been led into the war by his advisers in the belief that Japan could not win. The defeat caused waves of revolutionary unrest in Moscow and St Petersburg. Tsarist troops opened fire on a peaceful demonstration of workers in St Petersburg, and more than 100 men, women and children were killed in what came to be known as "Bloody Sunday". A series of strikes then swept the country. Nicholas II was forced to promise to grant a constitution with an elected parliament, and between 1906 and 1914 four assemblies ("Dumas") were elected.

In 1914 Russia entered World War I as an ally of Britain and France against Austria-Hungary and Germany. After early successes, the Russian army – the great "Steam Roller", as the world thought of it – was slowly but inexorably pushed back. Sickness, hunger, poor equipment and shortage of ammunition combined to reduce it to an ineffective rabble. At the same time the people of St Petersburg rioted because of bread shortages. In March 1917, in the face of distrust on all sides, Nicholas was forced to abdicate and hand over to a provisional government, at first under Prince Lvov and then under Aleksandr Kerensky.

Throughout all this, the revolutionary Vladimir Lenin had been marshalling his Bolshevik Party and waiting for the right time to strike. Lenin had returned from Switzerland to Russia with German help. He insisted that all power should be given to the soviets, the workers' councils that had been set up during the revolutionary ferment. Lenin also promised to take Russia out of the war, and this proved to have an irresistible appeal to the masses of the people [*see* MS p.*279*].

In October 1917, as the country still floundered on with the war, Bolshevik troops brought down the provisional government and Lenin came to power. Civil war broke out but, despite military intervention from France, Japan, Britain and the United States on the side of the "White Russian" moderates and monarchists, the Red Army was victorious.

The civil war left the new state in economic chaos. The situation was made worse when droughts caused bad harvests, and about 7 million people starved to death. Once more there were signs of revolt and discontent among the people. Lenin modified his economic plans and introduced the New Economic Policy. This allowed a considerable restoration of free enterprise in commerce, but left heavy industry firmly in the hands of the state. This policy was maintained until 1928.

The USSR was officially proclaimed in 1922. Lenin died two years later and was succeeded by a triumvirate consisting of Joseph Stalin, Lev Kamenev and Grigori Zinoviev. Stalin was determined to rule by himself. With great skill he packed the party with his supporters and had his two rivals removed from office. They were eventually executed during the purges of the 1930s in which millions of Stalin's opponents, real or imagined, perished. Lev Trotsky, the activist who had organized the Red Army, was banished and eventually assassinated in 1940 while living in exile in Mexico City.

Stalin carried through the enforced collectivization of agriculture during the 1920s and 1930s. In 1928 he introduced the first five-year plan and swept away private enterprise, such as the *kulaks,* the relatively prosperous peasants who refused to join the collectives.

In foreign policy Stalin always made the security of the Soviet state his major priority, rather than thinking first of furthering the cause of world revolution as Trotsky had advocated. When Japan invaded China in 1937, Stalin urged the Chinese Communists to support their erstwhile foe, the Nationalist leader Chiang Kai-shek. This policy

The Dnieper River, the chief river of the Ukraine, has been since the construction of the Dnieproges Dam in 1932 navigable for almost its full course.

Extensive irrigation of the lowlands in the Tashkent mountains has allowed the development of rice, cotton and some wheat farming in the area.

The Georgian city of Tbilisi, taken by the Russians in 1801 has become a manufacturing centre for machine tools and electric locomotives.

contributed in part to the mutual suspicion that 20 years later was to divide the two great Communist nations, the USSR and China.

Stalin also tried to develop a system of collective European security against the growing power and ambition of Hitler's Germany. He failed, and in 1939 he signed a non-aggression treaty with Germany, taking a part of Poland when Germany invaded that country and precipitated World War II.

Despite the treaty, Germany attacked the USSR in 1941 and advanced rapidly to the gates of Moscow. But as had happened to Napoleon's army, the Russian winter and the determination of the people vanquished the invaders. Leningrad withstood a long and terrible siege, and the city of Stalingrad was defended street by street and house by house. The battle of Stalingrad in 1942 was the turning point and by 1943 the German army was in full retreat.

After the war, Eastern Europe came under the influence and domination of the USSR. People's democracies were set up in several countries, and their economic, political and military destinies were tied to those of their powerful neighbour. Soviet domination of Eastern Europe led to the so-called Cold War with the West. There were many flash-points, such as the Soviet blockade of Berlin and the Korean War.

Stalin died in 1953. Lavrenti Beria, chief of the secret police, tried unsuccessfully to seize power. He was arrested, tried in secret and executed. A new collective leadership was set up with Georgi Malenkov as premier and Nikita Khrushchev as First Secretary of the Communist Party.

In 1955 Malenkov was replaced by Nikolai Bulganin, but Khrushchev began to emerge as the dominant force. Under his leadership, signs of greater tolerance began to appear; to some extent these were perhaps forced by simmering discontent among the subject peoples, which led to risings in Poland and Hungary in 1956. In 1957 Krushchev denounced the tyranny of Stalin, and promised the Soviet people less austerity in their lives.

Khrushchev's downfall in 1964 followed a confrontation two years earlier with the United States over the setting up of Soviet missile sites in Cuba. At the last moment Khrushchev relented and agreed to remove the rockets. His party colleagues forced him out of office because of what they considered his recklessness in foreign affairs. He was also blamed for Soviet agriculture's poor performance and for the widening rift with China.

Khrushchev was succeeded as First Secretary by Leonid Brezhnev, with Alexei Kosygin as premier. Under Brezhnev the USSR sought to develop closer ties with the West. Like Khrushchev before him, Brezhnev visited the United States and had talks with President Nixon. The world's two most powerful nations also tried to agree on reducing their deadly stockpiles of strategic nuclear weapons.

From the late 1950s the USSR pioneered space travel – in rivalry with the United States – and registered some notable achievements [*see* MM pp. 156–158]. In 1957 it launched *Sputnik*, the first artificial Earth satellite, and two years later it landed a space probe on the Moon. It also succeeded in sending a spacecraft to circumnavigate the Moon and obtained the first pictures of its hidden side.

An even more notable achievement occurred in 1961, when the Soviet cosmonaut Yuri Gagarin [*see* MM p.158] became the first man to orbit the Earth. This success was paralleled in 1963 when Valentina Tereshkova became the first woman to perform the same feat. In 1965 a Soviet rocket landed on the planet Venus. [*See also* SU p.268.]

The 1960s and 1970s saw a great improvement in the standard of living of the Soviet people. It was also a time in which dissent began to be voiced. For a time greater freedom of expression and criticism was allowed to writers. Aleksandr Solzhenitsyn published *A Day in the Life of Ivan Denisovich,* a novel that dealt with the rigours of forced labour camps in the Stalin era. Such books were permitted as part of the "destalinization" process introduced by Khrushchev. This tolerance was, however, only temporary. Solzhenitsyn's later works were proscribed, as were those of other writers who refused to accept the offical line for authorship. Solzhenitsyn was banished from the country in 1974.

In spite of repression the "dissidents" maintained their pressure on the Soviet authorities and actively enlisted the support of international opinion. The USSR also came under pressure from abroad for its alleged discrimination against Soviet Jews wishing to emigrate to Israel. International criticism led the Soviet authorities sometimes to yield, in order not to prejudice good relations with the West. Within the USSR there was an obviously swelling tide of liberal opinion. Map 30.

Republics of the USSR

Armenia Capital Yerevan; area 29,800sq km (11,505sq miles). Pop. (1976) 2,842,000
Azerbaijan Capital Baku; area 86,600sq km (33,435sq miles). Pop. (1976) 5,700,000
Belorussia Capital Minsk; area 207,600sq km (80,155sq miles). Pop. (1976) 9,384,000
Estonia Capital Tallin; area 45,100sq km (17,415sq miles). Pop. (1976) 1,438,000
Georgia Capital Tbilisi; area 69,700sq km (26,910sq miles). Pop. (1976) 4,965,000
Kazakhstan Capital Alma-Ata; area 2,717,300sq km (1,049,150sq miles). Pop. (1976) 14,406,000
Kirgizia Capital Frunze; area 198,500sq km (76,640sq miles). Pop. (1976) 3,372,000
Latvia Capital Riga; area 63,700sq km (24,595sq miles). Pop. (1976) 2,499,000
Lithuania Capital Vilnius; area 65,200sq km (25,175sq miles). Pop. (1976) 3,317,000
Moldavia Capital Kishinev; area 33,700sq km (13,010sq miles). Pop. (1976) 3,858,000
Russian Soviet Federal Socialist Republic (RSFSR; Russia) Capital Moscow; area 17,361,400sq km (6,703,235sq miles). Pop. (1976) 134,485,000
Tadzhikistan Capital Dushanbe; area 143,100sq km (55,250sq miles). Pop. (1976) 3,400,000
Turkmenistan Capital Ashkhabad; area 488,100sq km (188,455sq miles). Pop. (1976) 2,582,000
Ukraine Capital Kiev; area 445,000sq km (171,815sq miles). Pop. (1976) 49,075,000
Uzbekistan Capital Tashkent; area 447,400sq km (172,740sq miles). Pop. (1976) 14,090,000

USSR – profile

Official name Union of Soviet Socialist Republics
Area 22,402,200sq km (8,649,489sq miles)
Population (1976 est.) 255,413,000
Density 11 per sq km (30 per sq mile)
Chief cities Moscow (capital) 7,734,000; Leningrad, 4,372,000; Kiev, 2,013,000; Tashkent, 1,643,000; Baku, 1,406,000
Government Communist federation. Effectively, political power is exercised by the Central Committee of the Communist Party
Religions Christianity (Russian, Georgian, Armenian Orthodox, Roman Catholic, Baptist); Islam; Judaism
Languages Russian (official); ethnic minorities have their own languages
Monetary unit Rouble
Gross national product (1974) £248,183,800,000
Per capita income £972
Agriculture Cereals, potatoes, fruit, tea, sugar-beet, flax, cotton, tobacco
Industries Iron and steel, machine tools, motor cars, aircraft, rolling stock, farm machinery, chemicals, foodstuffs, household equipment, optical equipment, watches, carpets
Minerals Petroleum, coal, ores of iron, copper, lead, zinc, nickel, titanium and aluminium, natural gas, gold, precious stones
Trading partners Communist countries of Eastern Europe, EEC, Japan, USA, Finland

United Arab Emirates (Ittihād al-Imārāt al-'Arabīyah), formerly the Trucial States, is a federation of seven emirates (Dubai, Ajman, Abū Dhabi,

Many English monarchs have been buried in London's Westminister Abbey near the shrine of Edward the Confessor or in Henry VII's Chapel.

Snowdon, the highest mountain in England and Wales (1,085m; 3,560ft), forms part of the Snowdonia National Park, a popular tourist centre.

Loch Ness is a lake in the Great Glen, Scotland; it is best known for the "Loch Ness Monster" which supposedly rises above the surface of the lake.

Ras al-Khaimah, Fujairah, Sharjah, and Umm al-Qaiwain) on the Persian Gulf. The temporary capital is Abū Dhabi [*see* MS p. *317*]. The land is flat, consisting mainly of sand and salt-flat desert; only in the east does the land become hilly. Fewer people follow the traditional nomad existence as wealth from oil production attracts people to the towns of the Gulf. Agriculture is limited to the hilly region, oases and areas where irrigation is provided; the main crops are dates and vegetables. Oil production is the economic mainstay of the Union; first produced in 1962 from Abū Dhabi, it now comes also from Dubai and off-shore sites. Government is in the hands of the Supreme Council, which is made up of the royal families of the emirates; they elect the president (Sheikh Zaid ibn Sultan am-Nahayan in 1976). From the 19th century to 1971 Britain was responsible for the area's defence and for the foreign relations of the Trucial States (so called because they each had truces with Britain); the UAE was formed when Britain withdrew from the region. Area: 83,600sq km (32,278sq miles). Pop. (1975 est.) 656,000. Map 38.

United Kingdom, official name United Kingdom of Great Britain and Northern Ireland, is the island nation of north-western Europe that consists of ENGLAND, WALES, SCOTLAND, NORTHERN IRELAND (Ulster), the CHANNEL ISLANDS and the ISLE OF MAN. Often it is called Great Britain or, as in this book, merely Britain. **For a description of the land, economy and people of the constituent countries, see their separate articles.**

The United Kingdom is a constitutional monarchy. Parliament, the supreme legislative power, has two houses at Westminster, in the capital city, London: the House of Lords, with about 1,000 members drawn from the hereditary peerage, life peers and the episcopate; and the House of Commons, with 635 members elected for a maximum of five years. Wales was united with England in 1536 and Scotland joined the Union in 1707; both send representatives to Parliament, but plans for devolution could in time result in a greater degree of autonomy for both countries. Northern Ireland, represented in the House of Commons by 12 members, also had its own Parliament at Stormont (Belfast) from 1921 to 1972, when all powers were assumed by the Parliament in Westminster because of sectarian violence in Ulster. The Channel Islands and the Isle of Man are Crown dependencies with their own legislatures. The Channel Islands, as part of the old Duchy of Normandy, came under the control of the English Crown in 1066. The Isle of Man became subject to England in 1346.

History. The history of the countries of the United Kingdom before 1603 is described in the articles on ENGLAND, SCOTLAND and WALES. The kingdoms of Great Britain (England and Wales) and Scotland were under one rule from 1603, when James VI of Scotland ascended to the English throne as James I. The realms were not formally united until 1707, but the history of Great Britain begins with James' accession.

The relations between Parliament and James I – and his son Charles I – were stormy, because Parliament's control of taxation thwarted many of the Crown's plans. The struggle led to the first Civil War (1642-46) and then the second Civil War in 1648, which was followed by the execution of Charles I and the proclamation of the Commonwealth in 1649, headed by Oliver Cromwell. The Restoration in 1660 brought Charles II to the throne. His brother James II acceded in 1685 and was opposed for his Roman Catholicism, particularly after the birth of a male heir. William of Orange, husband of James' Protestant daughter Mary, arrived in 1688 in the Glorious Revolution, and as William III and Mary ascended to the throne in 1689. After William's death his sister-in-law Anne took the crown which, by the Act of Settlement, passed to the house of Hanover in 1714.

The early years of the Hanoverians were devoted to consolidation at home, while Britain was becoming a great colonial and maritime power abroad. The loss of the American colonies in the Revolutionary War was balanced by gains in India. Demands for political reform were stimulated by the American and French revolutions but the governing class resisted attempts at reform. Rebellion in Ireland in 1798 was followed in 1800 by the union of Ireland and Britain. Political changes received new impetus with the extension of the franchise in the Reform Act of 1832.

United Kingdom – monarchs

House of Stuart		
James I (VI of Scotland)		1603–25
Charles I		1625–49
Council of State		1649–53
Oliver Cromwell		1653–58
Richard Cromwell		1658–59
Charles II		1660–85
James II		1685–89
Mary II	ruled jointly	1689–94
William III	ruled jointly	1689–1702
Anne		1702–14
House of Hanover		
George I		1714–27
George II		1727–60
George III		1760–1820
George IV		1820–30
William IV		1830–37
Victoria		1837–1901
House of Saxe-Coburg		
Edward VII		1901–10
House of Windsor		
George V		1910–36
Edward VIII		Jan.-Dec. 1936
George VI		1936–52
Elizabeth II		1952-

The 19th century was a time of development for Britain as the nation continued its rapid industrialization and became the world's foremost power, with a huge overseas empire. Politically, the long reign of Queen Victoria was marked by the emergence of national political parties. The Liberal Party was headed by William Gladstone, and the Conservatives were led by Benjamin Disraeli and Lord Salisbury. Further extensions of the franchise, the growth of trade unions, and reforms in Ireland were major domestic issues. In the early 20th century, the Liberals' influence declined and the new Labour Party increased its power in Parliament.

The great events of the 20th century all had major effects on Britain. The nation was slow to recover from the cost of fighting in World War I, which had asked a great effort of the population. At the end of 1918 there were 8 million men and 1 million women serving in the armed forces or working in munitions factories. Women came out the war strengthened in their resolve to win the right to vote, which they gained in 1918. In 1922 southern Ireland at last won its fight for independence and became the Irish Free State (*see* IRELAND, REPUBLIC OF).

Severe unemployment after 1919 helped to bring the Labour Party its first government, a short-lived minority one, in 1924. Two years later the General Strike, begun by a miners' dispute, ended after nine

Giant's Causeway is the name given by Irish folk tradition to the basalt columns of irregular hexagons which lie off Antrim in Northern Ireland.

Palm trees dominate the landscape at Miami in Florida, the southernmost American state and the only one with a subtropical vegetation.

Brooklyn Bridge, completed in 1883, is one of several suspension bridges in New York. It spans 486m and joins Brooklyn to Manhattan Island.

days in defeat for the strikers. Labour was again in office in 1929, but the dislocation of jobs and industry brought by the Great Depression led to the formation of a National government (an all-party coalition) which stayed in office from 1931 to 1945. In 1936 the monarchy was shaken by Edward VIII's abdication; confidence in the Crown was restored by George VI, who reigned from 1936 to 1952.

After World War II, most of Britain's colonies achieved independence. Britain's influence in the world steadily declined. At home the years 1945-50 saw the establishment of the welfare state. Coal, steel and the railways were nationalized and comprehensive free medical treatment organized in the National Health Service. Since the early 1960s both Labour and Conservative governments have struggled to overcome the pressure of inflation and the relative stagnation of British industry. In 1973 Britain took a major step away from her traditional constitution by joining the European Economic Community. The decision was supported by the population in a referendum held in 1975. Map 8.

United Kingdom – profile

Official name United Kingdom of Great Britain and Northern Ireland
Area 244,046sq km (94,226sq miles)
Population (1975 est.) 56,147,000
Density 230 per sq km (596 per sq mile)
Chief cities London (capital) (1976) 7,028,200; Birmingham, 1,013,400; Glasgow, 905,032
Government Monarchy and parliament
Religions Anglican (England), Presbyterian (Scotland), Roman Catholic
Languages English, Welsh, Gaelic (Scottish)
Monetary unit Pound sterling
Gross national product (1974 est.) £59,568,700,000
Per capita income £1,061
Agriculture Cereals, potatoes, sheep, cattle
Industries Coal, iron and steel, textiles, industrial chemicals, oil refining, cement, paper, motor vehicles, petrochemicals, fishing
Minerals Coal, iron ore, petroleum
Trading partners USA, West Germany, Netherlands, France

United States of America, the richest country in the world, is the acknowledged leader of the so-called Western bloc of nations, as distinct from the countries of the Communist bloc. This mighty nation – often called simply "America" – came into being only 200 years ago in 1776, when it consisted of 13 states with a total population of about 2½ million. Today there are 50 states and a federal district (District of Columbia) and the population is more than 200 million. Dependencies of the United States include the CANAL ZONE, GUAM, PUERTO RICO and the VIRGIN ISLANDS OF THE US. Few people trace their ancestry back to the original inhabitants of the mainland, the North American Indians. Most are descendants of European immigrants or settlers, especially from Britain, although a large minority are Negroes descended from slaves taken from Africa between the 17th and 19th centuries.

In the 20th century the United States became an affluent and powerful nation. Isolationism was strong, but overcome as the country was eventually drawn into World Wars I and II. The United States then became involved in further wars in an attempt to prevent the spread of Communist regimes. At the same time it entered an arms race with the USSR, and each country has accumulated sufficient nuclear weaponry to obliterate the other (and the rest of the world) several times over.

The American people have revised their attitudes in the last two decades – in the light of the Civil Rights movement, the consequences of their involvement in Vietnam and Cambodia, the Watergate scandal, and such problems as pollution, the increase in crime and the build-up of nuclear weaponry. Nevertheless the nation is justifiably proud of its many remarkable achievements, accomplished in such a brief space of time.

Land and climate. The United States mainland (excluding Alaska) may be divided into several major regions, each with its own characteristics, but more broadly into three: the western mountain area, the eastern highlands, and the vast plains in between. The western mountain area covers more than a third of the country. It is dominated by two major mountain ranges: the Pacific Mountain System in the west, separated from the ocean by only a narrow coastal plain, and the Rocky Mountains, which in places extend almost 1,600km (1,000 miles) inland.

The Rockies in the United States have several peaks exceeding 4,270m (14,000ft) in height. The Pacific mountains consist of the Cascade Range in the north and the Sierra Nevada in the south, each with peaks also exceeding 4,270m. Mt Whitney, at 4,418m (14,495ft) the loftiest mountain in the United States (excluding Alaska), lies in California, in the southern part of the Sierra Nevada.

Between the Rockies and the Cascade-Sierra Nevada systems lie the Intermontane Plateaus and basins. The northern part, the Columbian Plateaus, is a volcanically formed area dissected by deep river canyons. The southern part, the Colorado Plateaus, is partly volcanic in origin and partly the result of block faulting. It also is dissected by deep canyons, and includes the Grand Canyon of the Colorado River, one of the natural wonders of the world; at its deepest it is more than 1,900m (6,230ft) below the main surface of the plateau [*see also* PE pp. *124,125*]. Between the two plateau areas lies the Great Basin, a region of high-lying desert and mountains, into which the waters of the surrounding ranges drain [*see* PE p.*114*]. Its rivers run fast in wet weather, but dry out when it is hot. Most of the lakes in this region, such as the great Salt Lake, are shallow and extremely saline.

East of the Rocky Mountains lie the Interior Plains, stretching from the Great Lakes to the border with Mexico and covering half the country. This heartland of the United States contains much fertile land. The region is watered by the Mississippi River and its major tributaries, the Missouri, Arkansas and Ohio. The Missouri, North America's longest river, flows 4,380km (2,722 miles) from its source in Montana before it joins the Mississippi. The Mississippi-Missouri is the third-longest river system in the world, flowing for 6,050km (3,760 miles) into the Gulf of Mexico [*see* PE p.100].

To the north of the Interior Plains lies a small region around Lake Superior, the Laurentian Uplands, which has little agricultural value but is rich in mineral deposits. The Great Lakes are important North American waterways; only Lake Michigan lies wholly within the United States. Lake Superior is the second largest body of fresh water in the world (after the Caspian Sea), covering an area of 82,413sq km (31,820sq miles). To the south of the Plains region is the Ozark Plateau, an area of low mountains unsuitable for much farming but containing mineral deposits.

The eastern margin of the Plains is marked by the Appalachian Highlands, a strip of mountain and plateau lands running south-westwards from the Canadian border in the north-east to within 600km (nearly 400 miles) of the Gulf of Mexico. The eastern part of the Appalachian Highlands is formed by the Blue Ridge Mountains, which are continued farther north by the Green Mountains and the White Mountains. The western part is bounded by the Appalachian Plateau and stretching down the centre is the Great Valley, a series of valleys running from New York to Tennessee. The Coastal Lowlands include an eastern coastal region and most of the southern states. The eastern part of this region, extending from New England in the north to Florida in the south, is the Atlantic Coastal Plain; it is narrow in the north (where the Appalachians approach the coast) but becomes wider in the south. Between the Blue Ridge Mountains and the coastal plain is the Piedmont, a low plateau sloping down from the foothills of the mountains to the plain. In the south is the wide, low-lying region of the Gulf Coastal Plain, much of which is formed by the floodplain and delta of the Mississippi River.

The western half of the United States has an average rainfall of less than 500mm (20in) a year, although in the far north-west the Olympic Mountains have the highest rainfall (3,550mm; 140in) in the country. Nevada, which is mostly desert, has less than 190mm (7.5in) a year. The driest place is Death Valley, California, with an annual average of less than 50mm (2in) of rain. The natural vegetation throughout most of the west is grassland, suitable for grazing, or desert scrub. The eastern half of the country has a much higher rainfall, most of it bet-

The Petrified Forest National Park, Arizona, has the world's largest display of jasper and agate "stone trees", which date from the Triassic period.

ween 400 and 1,200mm (16-48in) with the south having up to 1,600mm (64in). The southern part of the country generally has mild or warm winters and hot summers; the rest of the country has cold or extremely cold winters and warm to hot summers.

Alaska lies much farther north and west than the rest of the country and is extremely mountainous. To the south of the Arctic coastal plain is a treeless region in the tundra belt with permafrost extending 300m (985ft) below the surface; it includes the Brooks Range. Part of the Rocky Mountain System, it rises to 2,745m (9,000ft) in the east and has steep peaks cut by numerous glaciers [*see* PE p.116]. A large central tract of Alaska is drained by the Yukon River and its tributaries, and has low, rolling hills and broad, swampy valleys; about two-thirds of it is forested. To the south lies the great Alaska Range, containing the highest mountain in North America, Mt McKinley (6,194m; 20,320ft). This region of rugged mountains, deep valleys and fiords extends to the Pacific; it has cool summers and comparatively mild winters, with heavy annual precipitation. Extending south-west from the Alaska Range is the Aleutian Range, which forms the backbone of the long Alaska Peninsula and the Aleutian Islands, stretching a total of some 2,500km (1,550 miles) into the Pacific. It contains several active volcanoes [*see* PE p.*28*].

Hawaii, 3,860km (2,400 miles) west of the mainland, consists of a chain of 122 islands extending over 2,600km (1,610 miles) of the Pacific Ocean. Seven of the eight main islands are inhabited, and four-fifths of the people live on the third-largest island, Oahu. The islands are volcanic in origin and there are still active volcanoes on the largest, Hawaii, which gives its name to the state [*see* PE pp.30, *31, 57*]. Temperatures are warm in summer and winter, and rainfall varies from 250 to 12,700mm (10-500in).

Physical resources. The natural resources of the United States have resulted in its being both the world's greatest industrial nation and one of the world's major food producers. Away from the mountains and the deserts, most of the land is suitable for some form of agriculture. Vast herds of cattle and sheep are raised on ranches in the grasslands of the west, and the rich soils farther east support many different kinds of crops. The forests contain a wealth of timber, both softwoods and hardwoods, and although much of the forest resources has been squandered, careful conservation and reafforestation is helping to maintain this valuable asset.

Mineral resources in the United States provide most of the metals and other minerals required for modern industrial production, although the country's consumption is so large that even some of these requirements have also to be imported. The country is particularly rich in coal, iron ore, natural gas and crude oil. Petroleum comes principally from Texas, Louisiana and California. Coal is fairly widely distributed, with the chief fields lying in the east-central part of the country, especially West Virginia [*see* PE p. 148], Pennsylvania and Ohio. The main deposits of iron ore are close to lakes Superior and Michigan and in Alabama, Pennsylvania and New York. Other metallic minerals produced in quantity include copper, gold, lead, magnesium, molybdenum, silver, tungsten, uranium, vanadium and zinc. The United States leads the world in the production of several minerals, including coal and natural gas, although it has been overtaken by the USSR in the production of crude oil.

Economy. The economic strength of the United States is due partly to its large volume of domestic consumption — about ten times the volume of trade with other countries. This enables exporting industries to operate more economically, and also cushions all industries against fluctuations in international commerce. The value of American exports in the mid-1970s was about four times as great as in the mid-1960s. Some of this increase was due to worldwide inflation, but at least part was due to a steady increase in trade.

Despite a recession in the early 1920s, the balance of trade throughout the 20th century has generally been in favour of the United States. The pattern of commodities has varied over the years; in the 1970s farm produce became the main export by value, followed by transport equipment (including motor cars). Ten years earlier, manufactured goods were the main American exports. The chief imports are crude oil and other fuels, machinery and cars. Canada is the chief trading partner, taking about 20 per cent of America's exports and supplying nearly 23 per cent of its imports. Japan is next (9% exports and 12% imports), followed by West Germany (both 5%).

Agriculture. Only about 4 per cent of the working population is engaged in agriculture, but with modern farming methods and machinery this comparatively small force cultivates nearly half the total land area [*see* PE p.*157*]. The leading crop is maize (corn), cultivated mainly in the north-central part of the country, the so-called Corn Belt; in the southern states cotton is the main crop. The United States leads the world in the production of maize. Of the other cereal crops, wheat is the most important and the United States ranks second in world output, as it does for oats and tobacco. Other major crops include groundnuts, millet, sorghum, potatoes, rye, soya beans [*see* PE p.184], and barley. The country is a leading producer of oranges, lemons, grapefruit and limes, and leads the world in tomato production.

More than half the agricultural land is given over to livestock. Beef cattle are raised in the mid-west and on the great ranges of the west, whereas dairy cattle are common in the north-east. The country is a leading producer of cheese (ranking first in the world), milk (second) and butter. The total number of cattle on American farms and ranches in the mid-1970s was about 130 million. There were also about 50 million pigs, mainly in the Corn Belt. Sheep are raised principally in Texas and the western states, and the number has been declining steadily from 30 million in the 1950s to fewer than half that number in the mid-1970s. Most American sheep are raised for meat rather than wool. The United States is the world's largest producer of meat. [*See also* PE p.226–233.]

Industry. About a quarter of American workers are employed in manufacturing industry, which contributes about a quarter of the gross national product. The main industrial region lies in the north-east of the country, close to the major sources of coal and iron ore. A secondary belt of industrial development has been established in the south, running through Alabama, Georgia, the Carolinas, Tennessee and Virginia, and a third area is developing along the Pacific coast. Modern industrial development tends to be in regions where land and labour are most readily available.

The United States ranks first in the world in several manufacturing industries, including cars, steel, aluminium [*see* MM p.34] and synthetic rubber. The manufacture of arms is a massive industry, and in addition to its own requirements the United States supplies military equipment to most of the countries of the free world. Food processing is widely distributed.

Transport and communications. The great waterways such as the Mississippi River (which helped to open up the United States in pioneer days) still play an important part in the transport network. The St Lawrence Seaway [*see* MM pp.195–197] which the United States shares with Canada, allows ocean-going shipping to reach Chicago and other Great Lakes ports. The Chicago River is linked by canal to the Mississippi, and traffic can sail from the Great Lakes to the Gulf of Mexico [*see* MM p.*196*]. In 1975 the United States had a merchant fleet of 612 cargo ships and 279 tankers. The busiest ports are New York and New Orleans. Because of the great size of the country, air transport also plays a vital role, with more than 200 million passengers carried annually on internal flights. American airliners fly more passenger-kilometres than any other country's, and Chicago's O'Hare airport is the world's busiest [*see* MM p.187].

Rail, road and water transport are the chief carriers of freight. The United States has more than 6,100,000km (3,600,000 miles) or roads, including more than 56,000km (35,000 miles) of motorways, with about 7,600km (4,725 miles) of toll roads. There are about 130 million motor vehicles of all kinds (more than three-quarters of which are cars), an average of more than one vehicle for every two people. Despite the financial difficulties that they share with railways throughout the world, America's railways are the main carriers of freight. There is a dense network of lines in the east, and four major routes link the east and west coasts. With a total of more than 320,000km (199,000 miles), the United States has nearly 30 per cent of the world's track.

Government. The Constitution of the United States

The Hale Observatory at Mt Palomar, California, was developed by the astronomer George Hale in 1928. It has a 200-inch reflecting telescope.

The Adobe architecture of the Pueblo Indians, whose New Mexican culture dates back 700 years, represents the highest Indian civilization north of Mexico.

San Francisco's cable cars, originally the only solution to the problem of providing public transport in such a hilly city, are now kept running mainly as tourist attractions.

was ratified in 1788, and has since been amended several times; a 27th amendment, giving equal rights to women, was still awaiting ratification by a majority of states in the late 1970s. The Constitution provides for a federal form of government. Each of the 50 states has its own constitution, legislature, governor and judicial system, which have considerable powers within the states. The Constitution defines the powers that the federal government can exercise, but the state governments have authority to do anything not specifically reserved for the federal government or otherwise prohibited in the Constitution. The American political system works on the two-party principle; there are only two major political parties, although there are also several minor parties.

Defence and police. The president is the commander-in-chief of the armed forces, which are organized by the Department of Defense (one of the 11 departments of the executive). Military service is voluntary. The military strength of the United States numbers more than 2 million personnel, including an estimated (late 1970s) average strength for the army of 780,000, navy 530,000, marines 196,000 and air force 570,000. Defence is organized on a global basis, and the United States is a member of NATO, SEATO and ANZUS.

The forces are equipped with an extensive array of nuclear weapons, including more than a thousand inter-continental ballistic missiles in underground silos [*see* MM pp. *169, 180*] and 656 submarine-launched ballistic missiles. The Strategic Air Command has nearly 400 long-range bombers. The United States also has a number of orbiting spacecraft carrying out reconnaissance and a battery of early-warning radar systems.

Education. Elementary and secondary education is in general the responsibility of the separate states. Free education is available for 12 years, including kindergarten. There are three basic plans for schooling in the public (free) schools after kindergarten: eight elementary grades (years) plus four high school; six elementary and six high; or six elementary plus three junior high and three senior high school grades. All lead to graduation from high school (usually at age 17 or 18). Schooling is compulsory in all states except Mississippi, generally from the ages of 6 or 7 to 16, although in some states the lower limit is 8 and in several the upper limit is 17 or 18.

In the mid-1970s there were more than 50 million children at school (including kindergarten), representing more than a quarter of the population, with nearly 9 million enrolled in higher education. There are about 2,750 colleges and universities in the United States, ranging in size from fewer than 100 students to the 225,000 of the City University of New York. The oldest is Harvard University (or College), founded in 1636.

People and culture. Nearly 88 per cent of United States citizens are white and 11 per cent are black. There are also about 800,000 American Indians (less than 0.5 per cent of the population). Nearly half the Indians live in urban areas, and 28 per cent reside on the 115 major reservations. The largest tribe is the Navajo (Navaho), nearly 100,000 strong. Only about 5 per cent of American citizens are of foreign birth but, according to a sample survey held in 1972, about half claim specific origins by country, including British and Irish (22½%), German (12½%), Spanish (mostly Mexican and Puerto Rican, 4½%), Italian (4%), Polish (2½%) and Russian (1%).

The ethnic diversity of the country is due chiefly to the large-scale immigration that occurred before about 1920. Today the law restricts immigration to 120,000 people a year from the Western Hemisphere and 20,000 per country (maximum 170,000) from the Eastern Hemisphere (including Europe). Spouses, children and parents of US citizens are exempt from these regulations. Between 380,000 and 400,000 immigrants are admitted each year, including 60,000-70,000 from Mexico and 20,000-30,000 each from Cuba, Korea and the Philippines.

A population census is carried out every ten years. In 1790 the United States had nearly 4 million inhabitants (5 per cent urban), in 1880 more than 50 million (28 per cent urban), and in 1970, 203 million (73.5 per cent urban). There are 35 metropolitan areas with a population of 1 million or more, including six cities. About 17 per cent of the population live in a belt lying between Boston and Washington, DC, a region sometimes referred to as the "megalopolis", and altogether more than half the people of the United States live within 80km (50 miles) of the coast or the Great Lakes.

A strong unifying factor in American life is the use of English as the official language. Although many immigrants continue to speak their mother tongue in private, they are encouraged to learn English, and their children are taught in that language. Spanish is, however, the preferred language of sizeable minorities in New York City (Puerto Ricans), Florida (Cuban refugees) and along the border with Mexico. Religious freedom was one force that induced people to emigrate to the United States, and today 62 per cent of the people have some religious affiliation, including 34 per cent Protestants, 23 per cent Roman Catholics and 3 per cent Jews.

Americans in general enjoy a high standard of living, and are proud of their ultra-modern homes — containing many labour-saving devices. America is renowned for the variety and size of its hamburgers and hot-dogs; the country was also one of the first to make extensive use of canned, frozen and concentrated foods. But Americans are becoming increasingly conscious of the nutritional value of what they eat and of the dangers of the so-called "junk foods" their children were traditionally fed.

The national sports are American football and baseball, and Americans play most other sports. Tenpin bowling and pool are played throughout the country, and horse racing and motor racing are popular spectator sports. Hunting and fishing are popular leisure pursuits.

The arts in the United States have their origins in Europe, the ancestral home of most of its people, but the country has also been responsible for some major contributions of its own. Art forms virtually created in the United States include the film, the stage musical, and jazz, which is largely the contribution of the Negroes. The home of American theatre is Broadway, in New York City, although the enormous cost of staging a new show is limiting the kind of production that can be mounted with any anticipation of a profit. Similar financial problems plague the "big four" opera companies — the Metropolitan and the City in New York and the Chicago Lyric and San Francisco company — as well as smaller companies. Orchestras also require heavy subsidies, mostly from private funds. Despite this handicap, the United States possesses many of the world's finest orchestras, and its composers, such as Charles Ives (1874–1954), Walter Piston (1894–), Virgil Thomson (1896–), George Gershwin (1898–1937), Aaron Copland (1900–), Samuel Barber (1910–) and Leonard Bernstein (1918–), are world famous.

In its comparatively short history the United States has produced many great writers, who have made a prolific and fresh contribution to English literature. Among early distinguished pioneers of literature in America were Washington Irving (1783–1859) and James Fennimore Cooper (1789–1851), the transcendentalists Ralph Waldo Emerson (1803–82) and Henry Thoreau (1817–62), and the so-called "Boston Brahmins", who included the poet Henry Wadsworth Longfellow (1807–82) and poet and satirist Oliver Wendell Holmes (1809–94).

Other leading 19th-century writers include the short-story writers Edgar Allan Poe (1809–49) and Nathaniel Hawthorne (1804–64); novelists Herman Melville (1819–91), Mark Twain (1835–1910) and Henry James (1843–1916); and poets Walt Whitman (1819–92) and Emily Dickinson (1830–86). America's first major dramatist was Eugene O'Neill (1888–1953), and others have included Thornton Wilder (1897–1975), Tennessee Williams (1914–) and Arthur Miller (1915–).

History. The history of the United States dates from 4 July 1776, when 13 British colonies signed the Declaration of Independence. The original inhabitants of the land were the American Indians (or Amerindians), who are related to the Mongoloid peoples of central Asia. Their ancestors reached the continent from Asia across the Bering Strait, which was bridged either by land or by ice.

The true colonial period began in 1608, when a group of about 100 men from England founded a settlement at Jamestown, Virginia. They were followed in 1620 by a group of Puritans, known ever since as the Pilgrim Fathers.

In the 1760s relations between Britain and its American colonies began to turn sour. Trouble was provoked largely by the British government's attempts to tax the colonies. The colonists resisted

Salt Lake City, Utah, an urban oasis in the mid-West plains, is the centre of world Mormonism and home of the massive Mormon Tabernacle.

The domestic manner of Frank Lloyd Wright – peaked roof and broken surface area – has become a standard American architectural style in the 20th century.

The rolling and wooded countryside of Vermont, in New England, makes it a favourite state for American holiday-makers, both in winter and in summer.

taxation vehemently, particularly in 1773 when the government refused to repeal the tax on tea in order to give the ailing British East India Company a trading advantage in that commodity. One night a group of colonists dressed as Indians boarded East India Company ships in Boston harbour and threw the cargo of tea overboard – the historic "Boston Tea Party". Severe British legislation in reprisal – known as the Intolerable Acts – led 12 of the colonies to hold a Continental Congress in 1774, protesting against the Acts.

In April 1775 British troops tried to seize a colonial arms store at Concord, Massachusetts. They were met by armed resistance at nearby Lexington, shots were fired, and the American War of Independence (or American Revolution) had begun. After a year of desultory fighting a Second Continental Congress realized that no settlement was possible, and issued the Declaration of Independence. The war dragged on until 1783, when Britain finally recognized the new country.

The first government of the new United States proved to have inadequate powers to deal with the problems facing the country. A new conference was called in 1787 and its delegates agreed on a different form of government, the powers of which were set forth in the United States Constitution. The Constitution came into force in 1789 and in the following year the first president, George Washington (commander-in-chief in the War of Independence), took office. All the original colonies lay east of the Appalachian Mountains, but the terms of the 1783 agreement with Britain gave the United States land westwards to the Mississippi River. Spain nominally controlled lands farther west and south, but in 1800 it ceded the region between the Mississippi and the Rocky Mountains – the Louisiana Territory – to France. Three years later the French ruler Napoleon Bonaparte, unable to defend the territory, sold it to the United States for $15 million.

A new clash with Britain occurred in 1812, caused by the continuing British war with France. British attempts to stop American ships from carrying supplies to France provoked an American declaration of war; fighting lasted for 2½ years. American objections to European interference were stated clearly a few years later in the Monroe Doctrine (1823), in which President James Monroe declared that the United States would oppose any meddling by European nations in the affairs of any independent country of the Western Hemisphere.

The first half of the 19th century emerged as a period of rapid expansion for the United States. In 1836 American settlers in Texas revolted against Mexican rule and proclaimed a republic. Texas was admitted as a state of the Union in 1845. In 1846, the Americans and British reached agreement over the Oregon Territory: the boundary line was drawn along the 49th parallel of latitude, with the exception of the southern part of Vancouver Island. Tension with Mexico grew after the annexation of Texas, and the United States offered to buy all the territory north of the Rio Grande. Mexico refused, and the Americans declared war and seized the disputed lands. They later paid for them, and the "Gadsden Purchase" of 1853 finally established the present frontier with Mexico.

During this period, industrialization proceeded apace. The American inventions of the cotton gin, the reaper and the electric telegraph were paralleled by the development of European inventions such as the railway and the steamboat. The great political

States of the United States

State	Area sq km	[sq miles]	Population (1975 est)	No. Reps. in House	Capital	Joined Union
Alabama	133,667	[51,609]	3,614,000	7	Montgomery	1819
Alaska	1,518,769	[586,397]	352,000	1	Juneau	1959
Arizona	295,023	[113,908]	2,224,000	4	Phoenix	1912
Arkansas	137,539	[53,104]	2,116,000	4	Little Rock	1836
California	411,013	[158,692]	21,185,000	43	Sacramento	1850
Colorado	270,000	[104,247]	2,534,000	5	Denver	1876
Connecticut	12,973	[5,009]	3,095,000	6	Hartford	1788
Delaware	5,328	[2,057]	579,000	1	Dover	1787
Florida	151,670	[58,560]	8,357,000	15	Tallahassee	1845
Georgia	152,488	[58,876]	4,926,000	10	Atlanta	1788
Hawaii	16,705	[6,450]	865,000	2	Honolulu	1959
Idaho	216,412	[83,557]	820,000	2	Boise	1890
Illinois	146,075	[56,400]	11,145,000	24	Springfield	1818
Indiana	93,993	[36,291]	5,311,000	11	Indianapolis	1816
Iowa	145,790	[56,290]	2,870,000	6	Des Moines	1846
Kansas	213,094	[82,276]	2,267,000	5	Topeka	1861
Kentucky	104,623	[40,395]	3,396,000	7	Frankfort	1792
Louisiana	125,674	[48,523]	3,791,000	8	Baton Rouge	1812
Maine	86,026	[33,215]	1,059,000	2	Augusta	1820
Maryland	27,394	[10,577]	4,098,000	8	Annapolis	1788
Massachusetts	21,386	[8,257]	5,828,000	12	Boston	1788
Michigan	150,779	[58,216]	9,157,000	19	Lansing	1837
Minnesota	217,735	[84,067]	3,926,000	8	St Paul	1858
Mississippi	123,584	[47,716]	2,346,000	5	Jackson	1817
Missouri	180,455	[69,674]	4,763,000	10	Jefferson City	1821
Montana	381,086	[147,137]	748,000	2	Helena	1889
Nebraska	200,017	[77,227]	1,546,000	3	Lincoln	1867
Nevada	286,297	[110,539]	592,000	1	Carson City	1864
New Hampshire	24,097	[9,304]	818,000	2	Concord	1788
New Jersey	20,295	[7,836]	7,316,000	15	Trenton	1787
New Mexico	315,113	[121,665]	1,147,000	2	Santa Fe	1912
New York	128,401	[49,576]	18,120,000	39	Albany	1788
North Carolina	136,523	[52,712]	5,451,000	11	Raleigh	1789
North Dakota	183,022	[70,665]	635,000	1	Bismarck	1889
Ohio	106,764	[41,222]	10,759,000	23	Columbus	1803
Oklahoma	181,089	[69,918]	2,712,000	6	Oklahoma City	1907
Oregon	251,180	[96,981]	2,288,000	4	Salem	1859
Pennsylvania	117,412	[45,333]	11,827,000	25	Harrisburg	1787
Rhode Island	3,144	[1,214]	927,000	2	Providence	1790
South Carolina	80,432	[31,055]	2,818,000	6	Columbia	1788
South Dakota	199,551	[77,047]	683,000	2	Pierre	1889
Tennessee	109,411	[42,244]	4,188,000	8	Nashville	1796
Texas	692,405	[267,338]	12,237,000	24	Austin	1845
Utah	219,931	[84,915]	1,206,000	2	Salt Lake City	1896
Vermont	24,887	[9,609]	471,000	1	Montpelier	1791
Virginia	105,710	[40,815]	4,967,000	10	Richmond	1788
Washington	176,616	[68,191]	3,544,000	7	Olympia	1889
West Virginia	62,629	[24,181]	1,803,000	4	Charleston	1863
Wisconsin	145,438	[56,154]	4,607,000	9	Madison	1848
Wyoming	253,596	[97,913]	374,000	1	Cheyenne	1890

Las Vegas, the largest city in Nevada, derives most of its revenue from a large number of nightclubs and gambling casinos, which have made it world-famous.

New Mexico Indians enjoy the intertribal festival which is held each August at Gallup, the centre of a region inhabited by Navajo, Zuni and Hopi Indians.

Glacier National Park, on the Montana border with Canada, is a Rocky Mountain wilderness which bears many traces of its ice-age formation.

issue was slavery. Although the importation of slaves was prohibited from 1807, there were about 4 million in the 15 southern "slave states" by 1860, nearly a third of their population. The northern states had banned slavery, but the southern states, which needed cheap labour, clung to it.

The dispute between north and south came to a head after the acquisition of the Mexican territory: the two sides could not agree whether the new territories should have slaves. The final split came when Abraham Lincoln, an opponent of slavery, was elected president in 1860. Seven southern states – Alabama, Florida, Georgia, Louisiana, Mississippi, South Carolina and Texas – seceded from the Union and formed the Confederate States of America early in 1861. Four more states – Arkansas, North Carolina, Tennessee and Virginia – joined the Confederacy. The terrible civil war that followed lasted until 1865; it ended in victory for the north, and the restoration of the Union. About 750,000 soldiers died in the war and the southern states were devastated. Slavery was finally abolished.

The second half of the 19th century was a period of reconstruction after the war, and saw great industrial growth. The development of industry led to the emergence of big business as a power in the land. The United States also bought more territory, this time Alaska, purchased from Russia for about 2 cents an acre. About 4½ million people had migrated to the country between 1800 and 1860. The civil war halted the flow for a time, but in the 1880s a new wave began, and in the period 1870-1916 more than 25 million migrants arrived (nearly all from Europe). This great influx of new blood, which helped to swell the country's population from about 40 million to 100 million, enabled the United States to exploit and develop its natural resources quickly. And by 1890 the Western Frontier stood on the Pacific coast.

Important technical developments during this period of American history include the invention of the telephone, motor-car, electric light and phonograph, while new industrial processes and methods helped the nation's factories to produce an ever-increasing flow of goods. Such rapid development brought its own problems: squalid working conditions, exploitation, and corruption in local and national government. But such abuses led to a wave of reform, to improve conditions for the poor and for workers generally.

In 1898 came another war, when the United States intervened to help the Cubans against an oppressive Spanish administration. The Spanish-American War not only brought the United States more territory – Guam, the Philippines, and Puerto Rico – but also made them a world power. The same year they annexed Hawaii, which was already controlled by American businessmen.

The outbreak of World War I in Europe in 1914 found the United States neutral, and determined to remain so. But Germany's unrestricted submarine warfare (designed to cut off supplies to its enemies) involved American shipping, and in 1917 the United States declared war and sent troops to help in the struggle and the final defeat of Germany. America's president, Woodrow Wilson, played a major role in the peace talks that followed the war, and proposed the formation of an international body – the League of Nations – to try to prevent future wars. But Congress sabotaged his efforts and also the League, by barring American participation, thus leaving the League with insufficient power and the United States again pursuing a policy of isolation.

The 1920s were at first a boom period – the time of the cinema, cars, gramophones, the Charleston, "bright young things" and, above all, Prohibition. This ban on drinking alcohol came into effect in 1920, but far from curbing drunkenness and crime it had the reverse effect. Ordinary, law-abiding citizens broke the Prohibition laws to drink, and gangsters provided the liquor for them. This period of euphoria was soon doused. The boom atmosphere had led people to invest unwisely, and in 1929 a fall in share prices began which quickly escalated into a panic; millions of dollars were written off shares, investors were ruined, and in the next three years the Great Depression set in. It led to the unemployment of 12 million people, the failure of more than 5,000 banks and 30,000 businesses, and widespread misery and hardship. In 1933 Franklin D. Roosevelt, a Democrat, took office as president and pledged a "New Deal" to end the Depression. The New Deal included funds for public works and help for farmers and manufacturers, and it led to a restoration of confidence.

The outbreak of World War II in Europe in 1939 had the effect of helping the United States to recover from the slump. War orders produced a huge demand for American products, so that when the United States was itself precipitated into the war in December 1941 (by a Japanese attack on Pearl Harbor naval base in Hawaii), the country was already geared to war production. A long and costly struggle ended in victory for the Allies in the summer of 1945. Final victory over Japan was secured by the use of the atomic bomb, secretly made in the United States. This great advance in technology was the start of a scientific revolution that has been going on in the United States ever since, culminating in the "space race" with the USSR in the 1960s, and the landing of the first man – American Neil Armstrong – on the Moon in 1969 [*see* SU p.268].

The country was determined not to make the same mistake in international affairs after the second world conflict as it had after the first. By means of the Marshall Plan (named after US Secretary of State George C. Marshall) the United States financed the postwar rebuilding of Europe; American occupation of Japan enabled that country's economy to recover; and the United States has led the way in providing economic aid and technical advice to developing countries. It has also played a major role in the work of the United Nations – which succeeded the League of Nations – and in 1950 responded to a UN call by helping South Korea to combat aggression by Communist-dominated North Korea. The Korean War lasted until 1953. A similar intervention in Vietnam (1965-73) was unsuccessful, and American troops were eventually pulled out with the war going against them.

At home the Civil Rights movement, demanding true equality for the country's Negro population, was the dominant issue throughout the 1960s and 1970s. Legislation passed by Congress aimed to provide this equality, and to ensure integration of black and white children in schools. There were two major events involving the presidency. In November 1963 President John Kennedy was assassinated, and in 1974 Richard Nixon became the first president to resign (over the Watergate scandal). Nixon's successor, Gerald Ford, helped to restore confidence in the office of president, and in 1977 President James Carter began new diplomatic moves designed to ease world tensions. Map 66.

United States of America – profile

Official name United States of America
Area 9,363,123sq km (3,614,343sq miles)
Population (1976 est.) 215,135,000
Density 23 per sq km (60 per sq mile)
Chief cities (1975 est.) Washington, DC (capital), 716,000; New York, 7,567,000; Chicago, 3,173,000; Los Angeles, 2,747,000; Philadelphia, 1,862,000; Detroit, 1,387,000; Houston, 1,320,000
Government Head of state, James E. Carter, president (elected 1976)
Religions Protestant (34%), Roman Catholic (23%), Jewish (3%)
Monetary unit US dollar
Gross national product (1974) $1,406,610,000,000 (£601,115,400,000)
Per capita income $6,540 (£2,794)
Agriculture Maize, wheat, sorghum, oats, potatoes, soya beans, groundnuts, cotton, tobacco, fruits, livestock, forestry
Industries Manufacturing, mining, iron and steel, meat and dairy products, metal processing, cars and lorries, cement, paper, rubber and plastics, chemicals, fishing, tourism
Exports (major, excluding arms) Machinery, grain, food products, motor vehicles, chemicals, electrical and electronic equipment
Trading partners (major) Canada, Japan, West Germany, Britain, Mexico

Upper Volta (Haute-Volta), official name Republic of Upper Volta, is a landlocked nation in western Africa. It is one of the world's poorest countries — the average annual income in 1974 was only £38. In

Montevideo is a major fishing station for the South Atlantic fleets; almost all Uruguay's imports and exports pass through the city.

Almost half the population of Uruguay lives in the capital, Montevideo, despite the country's agricultural and pastoral based economy.

St Peter's Square is a large open space facing the church in the Vatican City; a red granite obelisk stands in the centre of the piazza.

the 1960s the average rate of population growth — 2.1 per cent per year — exceeded the annual economic growth rate and incomes decreased. The capital is Ouagadougou (pop. 110,000).

Land and climate. Most of Upper Volta is a flat plateau, about 305m (1,000ft) above sea-level. It is crossed by several rivers: the Black Volta, Red Volta, White Volta and tributaries of the Niger. The average rainfall in this hot, mostly infertile, country varies between 1,170mm (46in) in the south-west to 500mm (20in) in the north; it is unreliable everywhere.

Economy and people. Money sent home by migrant workers in the Ivory Coast and Ghana and earnings from transit trade are the chief sources of income. Subsistence farming occupies 90 per cent of the people. Animals and animal products account for nearly half of the exports, followed by cotton, groundnuts and sesame seeds. Mining and manufacturing are relatively unimportant. The largest group among the Black population is the Mossi; French is the official language. Ethnic religions are practised by 75 per cent of the people, Islam by 20 per cent and Christianity by 5 per cent.

History and government. For several centuries Upper Volta existed as a collection of powerful states. The French ruled the country from the 1890s, although the Mossi peoples continued to resist foreign domination until 1902. In 1960 it became an independent republic; a military group took over in 1966. Civilian rule was restored in 1971 and then another military regime took control in 1974. The president, Gen. Sangoulé Lamizana, dissolved the Assembly and ruled with a Government of National Renewal. Area: 274,200sq km (105,869sq miles). Pop. (1975) 6,144,000. Map 32.

Uruguay, official name Eastern Republic of Uruguay, is an independent nation on the eastern coast of South America between Brazil (to the north) and Argentina (west and south). It is a small country with state ownership of major utilities and of some industry; it has been fighting inflation to maintain its high standard of living and social welfare programmes. The capital is Montevideo.

Uruguay – profile

Official name Eastern Republic of Uruguay
Area 177,508sq km (68,536sq miles)
Population (1976 est.) 3,101,000
Density 17 per sq km (45 per sq mile)
Chief city Montevideo (capital) 1,230,000
Government Socialist republic (with some military control)
Religion Roman Catholic
Language Spanish
Monetary unit Peso
Agriculture Wheat, corn, rice, cattle, sheep, forestry, citrus fruits, oats
Industries Meat products, wool, hides, construction materials, chemicals, wine
Trading partners Western European countries, Argentina, Brazil, USA

Land and economy. Uruguay has large areas of grasslands with ample rainfall and a temperate climate, making livestock raising — particularly cattle and sheep — the mainstay of the economy. Wheat, rice and flax are the chief crops grown in the northern agricultural areas. Wool, which with meat makes up 35-40 per cent of exports, has declined in importance in recent years with the drop in world market prices. During the same period guerrilla activity by leftist Tupamaros has discouraged foreign investment in the country.

People. More than a third of Uruguay's population lives in and around Montevideo. Spanish, both in language and culture, predominates, although 25 per cent of the population is of Italian origin. Most people are Roman Catholics. Primary education is compulsory, higher education is free, and the literacy rate is 95 per cent.

History. The Spanish were the first Europeans to settle in Uruguay, more than 100 years after the initial exploration in the region of the Río de la Plata (River Plate). Struggles with Spain and Portugal, and then with Brazil and Argentina, marked its history until independence was achieved in 1828. Even so, civil wars and foreign intervention continued to plague the country until the end of the 19th century. Since then Uruguay has been known for its stability as a democracy whose pattern of political and social reform was begun by President José Batlle y Ordóñez in 1903. Except for a coup in 1933 and a military council formed in 1973 to fight the Tupamaros, its government has remained democratic. In 1976 rule passed to a military-civilian council, headed by President Aparicio Méndez. Map 78.

USA. *See* UNITED STATES.

USSR. *See* UNION OF SOVIET SOCIALIST REPUBLICS.

Utah. *See* UNITED STATES.

Uzbekistan. *See* UNION OF SOVIET SOCIALIST REPUBLICS.

Vatican City (Città del Vaticano), official name State of the Vatican City, is an independent sovereign state – the smallest in the world – existing as an enclave within Rome; it is also known as the Holy See. The Vatican is the official home of the pope and the centre of the Roman Catholic Church, with its own passports, currency and postage stamps. Its government is based on canon law, apostolic constitutions and papal laws.

People. Most of the population of the Vatican City are Italian or Swiss born, and citizenship is granted only to people who hold office or are employed within the Vatican, such as apostolic delegates and the pope's spiritual staff. Seventy countries have diplomatic representatives in the Holy See. Italian is the chief language; official acts are written in Latin.

History. Once a boggy swamp and a charioteers' burial ground, the Vatican was made a garden area by Nero in AD 59. Popes held sovereignty over mid-Italy (the Papal States) until 1861, when conquests caused much of the papal dominion to be moved to the Kingdom of Sardinia; the pope's sovereignty was confined to Rome. By the terms of a 1929 treaty, the Holy See and the Italian government agreed to full independence for the Vatican, granted special status to the Church, and provided compensation for lands taken. Area: 44 hectares (109 acres). Pop. (1974 est.) 1,000. Map 24.

Venezuela, official name Republic of Venezuela, is an independent nation of northern South America.

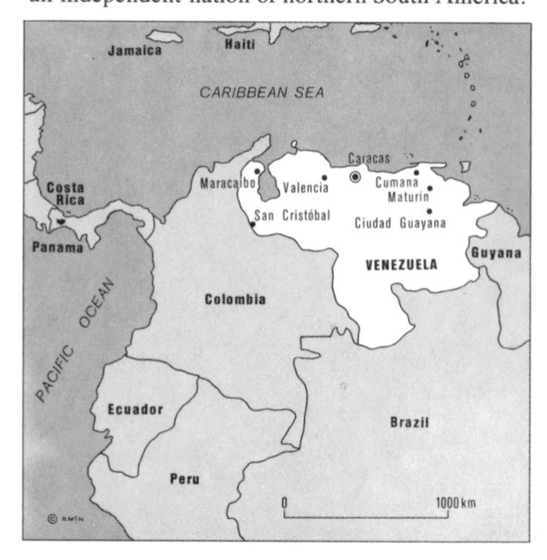

The River Orinoco, a main transport route, forms a boundary between Venezuela and Colombia and flows for 2,150 km (1,336 miles) before entering the Atlantic Ocean.

Hanoi, after the peace agreement in 1973, still for a time displayed the scars of nearly two decades of war; fighting thereafter was farther south.

European architecture may still be seen in Saigon (briefly called Ho Chi Minh City), although few Europeans stayed after the Vietnam War.

It includes 72 islands, the largest of which is the dependency of Margarita Island. Its petroleum deposits have made it the world's fifth largest producer of oil and one of the wealthiest countries in Latin America. The capital is Caracas.

Land and economy. Venezuela can be divided into four geographical regions: the Orinoco basin, the mountains at the northern extent of the Andes, the Guiana Highlands and the coastal lowlands around Lake Maracaibo [*see* PE p.*64*]. Angel Falls, the highest waterfall in the world (980m; 3,215ft), is in the Guiana Highlands. A quarter of the gross national product and 80 per cent of the country's income come from petroleum, although foreign investment in this resource has been discouraged by the government's plan to take over control of all petroleum assets after 1983. Next to oil the chief exports are iron ore, coffee, cocoa, rice, cotton, steel products, sugar, fish and fruit. Agricultural reforms have increased the area of cultivated land and resettled 100,000 families on their own farms. Venezuela's rivers have made it the fourth largest producer of hydroelectricity in Latin America.

People. Venezuela is a country of contrasting peoples. The population is drifting from a rural to an urban society; it is based on descendants of South American Indians and Spanish colonials, and yet has had a large influx of post-World War II immigrants from Italy, Portugal and Spain. Most of the people are Spanish-speaking and Roman Catholic. Education is free and the literacy rate is estimated as 80 per cent. Voting is compulsory for all citizens more than 18 years old. The 1961 constitution guarantees religious freedom and a strong central government elected by universal suffrage.

Venezuela – profile

Official name Republic of Venezuela
Area 912,050 sq km (352,143sq miles)
Population (1976 est.) 12,361,000
Density 14 per sq km (35 per sq mile)
Chief cities Caracas (capital) (1976 est.) 2,576,000; Maracaibo, 690,400
Government Democracy, head of state President Carlos Pérez
Religion Roman Catholic
Language Spanish (official)
Monetary unit Bolívar
Gross national product (1974) £8,474,400,000
Per capita income £686
Agriculture Coffee, cocoa, citrus fruits, sugar cane, rice, tobacco, bananas, cotton, maize, cattle
Minerals Petroleum, gold, copper, coal, salt, nickel, manganese, asbestos
Industries Petrochemicals, iron, paper products, canned fish, steel, textiles, tyres, shoes, dairy products
Trading partners USA, West Germany, Japan

History. Venezuela was sighted in 1498 by Christopher Columbus and the coastline explored in the following year by Alonso de Ojeda and Amerigo Vespucci. It was under Spanish domination until 1821 when Simón Bolívar, Venezuela's national hero, finally won independence from Spain. Venezuela became part of Greater Colombia, but broke away in 1830 under José Páez. There followed a long period of civil wars and unstable dictatorships and only after World War II did an elected president serve a full term (Rómulo Betancourt, 1959–64). Even then there were several uprisings by left- and right-wing groups, guerrilla activity and disputes with Colombia and Guyana. Map 76.

Vermont. *See* UNITED STATES.

Victoria. *See* AUSTRALIA.

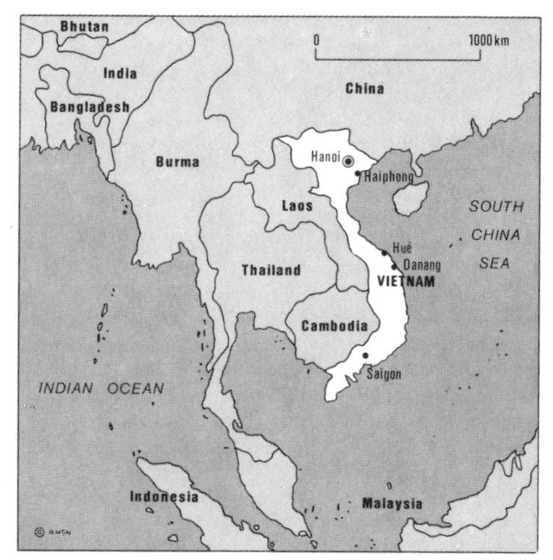

Vietnam, official name the Socialist Republic of Vietnam, is an independent nation of south-eastern Asia. Under Chinese and then French influence for centuries, it was in 1954 divided into two countries (North and South Vietnam) and reunited in 1975 after the Vietnam War, which resulted in a victory of the Communist forces in the south (assisted by the north). The capital is Hanoi.

Land and economy. Vietnam is located in the Indochina peninsula. Most of the northern part is covered by thick, mountainous jungle. Rice is the main crop in the heavily populated and cultivated Red River delta. The climate is monsoonal, causing frequent flooding. The southern part of Vietnam is dominated by a flat, marshy, coast – the Mekong River system. A tropical climate and rich soil result in abundant rice harvests. Both industry and agriculture came to a standstill during the war years.

People. For more than a thousand years the Vietnamese were subservient to the might of China. This dependence is still reflected in language and art, and in the importance of the family, knowledge and maturity – the Confucian ethic. Scientific Socialism is now the official creed, although Buddhism is still tolerated, especially among the older generation. About a tenth of the people are Roman Catholics. Science and vocational training are emphasized in the government-controlled school system. Chinese and Russian languages are taught and literacy is estimated at 95 per cent in the north, 65 per cent in the south. Vietnamese is the official language. The government is Communist, with a Provisional Revolutionary Government in the south.

Vietnam – profile

Official name Socialist Republic of Vietnam
Area 338,392sq km (130,653sq miles)
Population (1975 est.) 45,211,000
Density 134 per sq km (346 per sq mile)
Chief cities Hanoi (capital) (1976 est.) 1,443,500; Saigon (Ho Chi Minh City), 3,500,000; Danang, 437,700
Government Communist republic
Religions Taoism, Buddhism, Roman Catholicism
Languages Vietnamese, French, English
Monetary unit Dong (north), new piastre (south)
Gross national product (combined 1974 figures of former North and South Vietnam) £2,735,100,000
Per capita income £60
Agriculture Rice, rubber, forestry, livestock, cereals, tea, coffee, sweet potatoes, tobacco, sugar cane
Industries Shellac, processed foods, textiles, fishing, rubber products
Minerals Coal, zinc, tin
Trading partners USSR, China, Japan

History. Originally the Vietnamese people lived in the region of China's Yellow River valley, but they were driven south to the Red River delta. They were under Chinese rule from the 2nd century BC until they revolted in 939 and founded their own empire. They remained independent until the mid-19th century, when the French took control in Indochina. Inspired by the success of the Chinese freedom drive under Sun Yat-sen, a nationalist movement staged an uprising in 1930 against the French. In the same year Ho Chi Minh organized the Indochinese Communist Party.

Japan occupied Vietnam during World War II, and in 1945, with the war over, a Communist-led revolt in Hanoi proclaimed the Democratic Republic of Vietnam. Ho Chi Minh led the Communists in an eight-year guerrilla war against the French, who were finally defeated in 1954. Vietnam was divided at the 17th parallel of latitude into two countries – the Communist north and the Nationalist south.

The cliff railway runs down Constitution Hill into Aberystwyth, a large resort in Cardigan Bay and seat of the University of Wales.

The mining villages in the valleys of south Wales are known for their terraced houses, their strong community spirit and their male-voice choirs.

Llandaff Cathedral in south Wales, during restoration after World War II, was given a great arch surmounted by Jacob Epstein's statue of Christ.

Ngo Dinh Diem, prime minister in the south, faced a ruined economy, refugee problems and conflicting religious and political factions. The Communists established agricultural reforms, rebuilt industry and embarked on a campaign to overthrow the southern regime. In 1961 the United States supplied its first military advisers, and its involvement gradually increased. The South Vietnamese government (who were backed by the United States) was unable to defeat the insurgent guerrillas (backed by North Vietnam). American air strikes against North Vietnam began in 1965 and eventually the United States committed more than half a million troops to the Vietnam War [*see* MM p.*181*]. Strong opposition to the war within the United States influenced the withdrawal of troops after 1969, when peace talks began. A cease-fire agreement was signed in Paris in 1973. Pressure against South Vietnam continued, however, and in 1975 the southern (Saigon) regime collapsed, and the country fell to the Communists.

The new unified Vietnam became the most powerful military nation in south-eastern Asia, but the economy lay in ruins (especially in the south) and the government began moving unemployed people from city areas to work on the land. Map 52.

Virginia. *See* UNITED STATES.

Virgin Islands. *See* BRITISH VIRGIN ISLANDS; VIRGIN ISLANDS OF THE UNITED STATES.

Virgin Islands of the United States is the official name for the group of 68 islands of the Lesser Antilles which are administered by the American Department of the Interior. Another 36 neighbouring islands to the north-east constitute the British colony called the BRITISH VIRGIN ISLANDS. The chief islands of the United States group are St Croix and St Thomas (location of the capital, Charlotte Amalie), which are used for raising livestock and growing sugar cane, and St John, most of which is given over to the Virgin Islands National Park. The Danish West Indies Company began to colonize St Thomas in 1672. Denmark claimed St John in 1683 and bought St Croix from France in 1733; the group became a Danish Royal colony in 1754. The United States purchased their islands from Denmark in 1917, because of their strategic position near the Panama Canal. Their residents were granted United States citizenship in 1927. Area: 344sq km (133sq miles). Pop. (1975 est.) 92,000. Map 74.

Wales (Cymru), a principality within the United Kingdom of Great Britain and Northern Ireland, occupies a broad peninsula on the west coast of Great Britain. It is more closely linked with ENGLAND than are NORTHERN IRELAND and SCOTLAND, particularly in relation to local government, education, banking and the justiciary. For these reasons, England and Wales are often regarded as a single entity. The national identity and culture of the people of Wales, however, are strong, and about a fifth of them speak Welsh, which has equality with English as an official language. Most of the people live in the south, the main industrial area.

Land and climate. Two-thirds of Wales is covered by the Cambrian Mountains. About a quarter of the land is more than 300m (985ft) above sea-level. The only lowlands are the north and south coastal plains, and the river valleys. Snowdon, in the north-west, is the highest peak in England and Wales, rising to 1,085m (3,560ft) in Snowdonia. There are many large, grassy plateaus in the centre, cut by deep valleys and gorges, with lakes and waterfalls. The chief mountains in the south are the Brecon Beacons, rising to 886m (2,907ft). The Severn, Britain's longest river, rises in the central mountains.

The only large island is Anglesey, separated from the north-west coast by the narrow Menai Strait. It has an area of 715sq km (276sq miles) and, unlike the mainland, is generally low-lying. Much of the coast of Wales is lined with cliffs, and there are numerous natural bays and harbours.

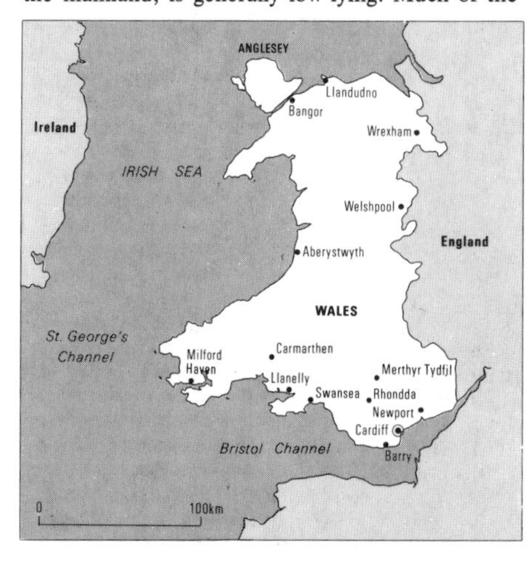

Wales has a maritime climate with mild, wet winters and cool, moist summers. The country is often covered with a layer of cloud. The average annual rainfall is about 1,270mm (50in) and is heaviest in the mountain areas – as much as 5,000mm (200in) on Snowdon.

Economy. Wales's chief contributions to the economy of the United Kingdom come from coal mining and iron and steel manufacture. The main industrial area is in the south, which produces a quarter of Britain's output of crude steel; there are steel mills at Ebbw Vale and in the Port Talbot area. More than a quarter of Britain's aluminium is made in South Wales; tin-plate manufacture is also important, and other metals processed include zinc, copper and nickel.

The South Wales coalfield, once the world's chief coal-exporting region, is still a major producer and provides nearly all of Britain's limited supplies of anthracite. Large dock areas at Cardiff, Swansea and Newport grew up in the late 19th century for the export of coal, and expanded for the import of iron ore when the Welsh supply ran out; a large new harbour was opened at Port Talbot in 1970. The fine natural harbour at Milford Haven, developed as an oil terminal, handles nearly a third of Britain's petroleum imports, much of which is refined nearby. There is a second industrial area in North Wales, centred on Wrexham and based on the smaller North Wales coalfield. Spinning for textiles is another leading industry, especially using man-made fibres such as rayon and nylon. There are also many small potteries and wool-mills. Limestone and slate are quarried in the mountains.

Although more than 80 per cent of the land is used for farming, less than 15 per cent is arable – the rest is rough grazing land or permanent pasture. Sheep are reared on the hills [*see* PE p.231], cattle on the better pastures. The main crops are fodder crops.

One of Wales's chief natural resources is the water from its many rivers. Reservoirs have been created by damming and flooding the deep river valleys, and these produce hydroelectric power as well as providing water for Wales and English cities such as Birmingham and Liverpool.

People and culture. About two-thirds of the people live in the industrial south-east, in the valleys or on the coast. At the time of the 1971 census, 21 per cent of the population spoke Welsh (a Celtic language), compared with 25 per cent in 1961 and 29 per cent in 1951 – a decline of about 4 per cent in ten years. In 1971, 1 per cent of the population spoke only Welsh, compared with 4 per cent in 1931. The main Welsh-speaking areas are in the west of the country, where three-quarters of the people speak it.

In the predominantly Welsh-speaking areas, Welsh is the main language of instruction in primary schools, and there are bilingual schools in many other places. The University of Wales, with colleges in Aberystwyth, Bangor, Cardiff, Lampeter and Swansea, has more than 17,000 students. The

Welsh counties

	Area sq km	Area sq miles	Population (1975 est.)
Clwyd	2,426	[937]	374,800
Dyfed	5,765	[2,226]	321,700
Gwent	1,376	[531]	440,100
Gwynedd	3,866	[1,493]	224,200
Mid-Glamorgan	1,019	[393]	540,100
Powys	5,077	[1,960]	100,800
South Glamorgan	416	[161]	391,600
West Glamorgan	816	[315]	371,700

The Norman gatehouse on the bridge over the Monnow provides evidence of the antiquity of Monmouth, south Wales, where Henry V was born.

Yachting and bathing from the sandy beaches are the main preoccupations of holiday-makers who go to Abersoch, near the tip of the Lleyn Peninsula.

Tobago is one of the southernmost islands of the West Indies; like most of the other islands it derives much of its income from the sale of coconuts and copra.

Church in Wales was disestablished from the Church in England in 1914 and Wales formed into a separate province. Most of the people are Protestants, mainly Methodists.

The Welsh are proud of their cultural traditions, in particular of their poetry, music and literature. These traditions are seen at their best at the many *eisteddfodau*, the national bardic festivals held every year. They include competitions for poetry, writing and singing, mainly in Welsh. At the National Eisteddfod, lasting a week and held alternately in North and South Wales, the most important competitions are for the Chair (for a poem in strict Welsh metre) and the Crown (for less formal verse). Another famous eisteddfod (now international) is held at Llangollen, where the idea was first revived in 1858. In music, the male-voice choirs of the valley towns are justly famous.

Welsh sport is dominated by rugby union, which is played and followed with enthusiastic fervour. Soccer is also popular (Welsh clubs play in the English leagues), as are boxing and outdoor activities such as hunting and climbing.

Government. As part of the United Kingdom, Wales sends 36 members of parliament to Westminster and a member of the cabinet (the secretary of state for Wales) is responsible for Welsh affairs. The Welsh Office, centred in Cardiff, is responsible for most day-to-day administration. The Welsh national party, Plaid Cymru, seeks ultimate independence for Wales. Plans for the establishment of a separate Welsh Assembly came under serious discussion in the mid-1970s. Welsh local government was reorganized in 1974; there are eight counties made up of a total of 37 districts.

History. The Roman invasion of Britain had little effect upon the Britons in Wales, which the Romans conquered in AD 78. Christianity was introduced during the Roman occupation. After the Romans left, Anglo-Saxons conquered much of Britain and drove the Britons into Wales, Cornwall and Strathclyde and gave them the name *Waelisc* (Welsh), meaning "foreign". By the early 600s they had isolated Wales from Cornwall and Strathclyde. For the first time the inhabitants of the Welsh peninsula began to call themselves *Cymry* (fellow-countrymen) and their land *Cymru*. In the 8th century Offa, King of Mercia, made inroads into Welsh territory; to prevent counter-attacks he constructed a boundary known as Offa's Dyke from the River Dee to the River Wye.

In the late 12th century a powerful Welsh prince, Llewelyn ap Iorwerth, freed most of Wales. His grandson Llewelyn ap Gruffydd also won control over much of the country in the 13th century, and in 1267 was recognized by Henry III of England as Prince of Wales. In return Llewelyn recognized Henry as his overlord, but later refused the same recognition to Henry's son Edward I and was killed in battle with English troops in 1282. Edward annexed Wales in 1284, and made his son Edward II Prince of Wales in 1301. He divided Llewelyn's territory into counties under English barons, who built strong castles to defend their lands.

The imposition of a "foreign" system of laws led to much local ill-feeling and there were minor rebellions, but the 14th century was a period of relative peace in Wales, during which poetry rose to great heights with Dafydd ap Gwilym. There was no serious resistance to English rule until 1400, when Owen Glendower (Owain Glyndŵr) led a successful revolt and drove out the English. But he gradually lost the land he had gained, and after about ten years much of Wales lay devastated and exhausted by the struggle. Eventually a Welsh family, the Tudors, succeeded to the English throne (Henry VII became king in 1485), and the Welsh gradually began to accept the idea of union with England. By the Acts of Union of 1536 and 1543 Henry VIII joined the two countries under the same system of government. Wales was divided into shires and given representation in the English parliament. English became the official language, and Welshmen enjoyed equal rights with Englishmen.

From this time on, the history of Wales is entwined with that of England. The Industrial Revolution affected Wales in much the same way as in England, although its consequences were even more dramatic because they were focused on a relatively small area in South Wales. A cultural revolution also took place, as proliferation of the Welsh language in books, schools and institutions awakened an awareness of Welsh heritage, stimulated by the revival in the 19th century of the eisteddfod.

Wales – particularly South Wales – suffered badly in the years after World War I. By 1932 unemployment affected nearly 250,000 people out of a population of 2½ million, and nearly 260,000 workers had migrated to England. After World War II Welsh nationalism underwent a revival and the government extended more authority to Wales. A secretary of state was appointed for Wales in 1964.

In 1975 a Welsh Development Agency was set up to expand the economy and improve the environment, and proposals were made for the establishment of a Welsh Assembly with certain powers devolved from Westminster. Map 8.

Wales – profile

Official name Principality of Wales
Area 20,761sq km (8,016sq miles)
Population (1975 est.) 2,765,000
Density 133 per sq km (344 per sq mile)
Chief cities Cardiff (capital), 287,000; Swansea, 188,350; Newport, 133,500
For further information, *see* UNITED KINGDOM (*profile*).

Washington. *See* UNITED STATES.

Western Australia. *See* AUSTRALIA.

Western Sahara, formerly Spanish Sahara, is a desert territory in north-western Africa. In early 1976 Spain withdrew its troops and the territory was partitioned; the northern two-thirds was taken by MOROCCO and the rest by MAURITANIA. But a nationalist group, the Popular Front for the Liberation of Saharan Territories (Polisario), proclaimed the territory independent as the Sahrawi Arab Democratic Republic. This declaration was supported by neighbouring Algeria, which aided Polisario in launching a guerrilla campaign against the Moroccans and Mauritanians. The capital is El Aaiún (pop. 24,500).

Nearly all of this hot territory has an average annual rainfall of less than 50mm (2in). The people are Muslims of Arab and Berber origin, who speak Arabic or Spanish; most work as pastoral farmers. The chief resource is the huge phosphate deposit at Bu Craa in the north. Phosphates, first exported in 1972, soon dominated the economy.

Spain ruled the territory between 1884 and 1976. The agreement to partition the territory was made by Spain, Morocco and Mauritania, in consultation with Saharan chiefs. Area: 266,000sq km (102,703sq miles). Pop. (1976 est.) 151,275. Map 32.

Western Samoa, official name The Independent State of Western Samoa, is an island nation in the southern Pacific Ocean made up of the western part of Samoa and nine other major islands, including Savai'i, Manono and Upolu, location of the capital, Apia. Many of the islands are the peaks of underwater volcanic mountains. They have a wet, tropical climate – ideal for growing yams, taro, bananas, breadfruit, cacao, papayas and coconuts. Farmers also raise pigs and poultry. Industries include food processing, furniture-making and, increasingly, tourism.

Most Western Samoans are Polynesians, living in a society based on the family. Some are Christians; others still follow traditional ethnic religions. Western influence, much of it through New Zealand, has brought about cultural and social changes. After 1899 the islands belonged to Germany, but were occupied by New Zealand in 1914. In 1921 a League of Nations mandate assigned them to New Zealand; the United Nations was awarded trusteeship in 1946. Western Samoa proclaimed independence in 1962. Area: 2,841sq km (1,097sq miles). Pop. (1976 est.) 151,275. Map 62.

West Indies is a group of islands that lie in an area extending from Florida to Venezuela, encircling the Caribbean Sea and separating it from the Atlantic Ocean. Geographically they form three major groups: the Bahamas, the Greater Antilles and the Lesser Antilles (which include the Leeward Islands and Windward Islands). Most are now independent, but they can also be grouped politically into former (or present) American, British, Dutch and Spanish possessions. Each of the nations and major islands has a separate article in this book.

Much of the eastern part of Yemen is taken up by the Rub al-Khali, an arid desert region with little vegetation and incapable of supporting agriculture.

Sarajevo, founded by the Turks, is now the capital of Bosnia-Hercegovina in Yugoslavia. Archduke Ferdinand's assassination here in 1914 precipitated WWI.

Skopje, the capital of Macedonia in Yugoslavia, lies on the Vardar River. In July 1963 an earthquake destroyed most of the city and killed more than a thousand people.

The first European contact with many of the islands was made in the 1490s by Christopher Columbus, and the first settlement (on Hispaniola) was Spanish. English, Dutch and French settlement followed and gave rise to many conflicts between the various colonial nations. The same countries imported thousands of slaves to work on plantations, most of which were given over to growing sugar cane. The short-lived West Indies Federation was formed in 1958 by ten former British possessions, chief of which were Barbados, Jamaica, and Trinidad and Tobago. In 1961 the latter two members left the federation, which was dissolved the following year. Map 74.

West Irian. *See* INDONESIA.

WestVirginia. *See* UNITED STATES.

Windward Islands are a group of islands in the WEST INDIES, the chief of which are DOMINICA, GRENADA, MARTINIQUE, SAINT LUCIA and SAINT VINCENT. Map 74.

Wisconsin. *See* UNITED STATES.

Wyoming. *See* UNITED STATES.

Yemen, official name Yemen Arab Republic, is an independent nation at the southern end of the Red Sea, on the Arabian Peninsula. It is sometimes known as North Yemen, to distinguish it from the neighbouring People's Democratic Republic of Yemen, or Southern Yemen. The capital is San'a. An absolute monarchy until 1962, it is now governed by an Army Council, which seized power in 1974. It is one of the world's poorest countries.

The land consists of interior highlands and a narrow coastal area called the *Tihamah.* Most of the people work in agriculture or in industries processing agricultural products. The chief crops are coffee (exports of which have declined in recent years) and *qat* (kat), a shrub whose leaves contain a narcotic and are used to make a type of tea. Local craft goods are exported. The population is predominantly Arab, with some people of mixed Arab and African descent in the Tihamah.

The region now occupied by the Yemen was the cradle of three major early civilizations: the Minaeans, the Sabaeans and the Himyarites. It was invaded by the Romans in the 1st century BC and, after subsequent Ethiopian conquests and the rise of Christianity and Judaism, Islam arrived in the 7th century. Following the break-up of Muslim rule, the Rassite dynasty gained power and evolved a political structure which survived until 1962.

From 1958 to 1961 Yemen was joined with Egypt and Syria in the nominal alliance called the United Arab States. A republican movement within the Yemen was supported by Egypt, while Saudi Arabia and Jordan sided with the royalists. A military junta proclaimed a republic and there were continuing clashes between royalists and opposing republican factions. By 1970 external support had ceased and the royalist cause had been defeated. In 1974 a military coup led by Col. Ibrahim al-Hamdi established a new regime, which suspended the constitution. Area: 195,000sq km (75,290sq miles). Pop. (1976 est.) 6,668,000. Map 38.

Yemen, People's Democratic Republic of (Al-Yamin ash-Sha'bīyah), formerly Southern Yemen, is an Arab nation at the southern end of the Arabian Peninsula, south and east of the Yemen Arab Republic. It includes the islands of Kamaran, Kuria Muria, Perim and Socotra; the capital is Aden. It was under various rulers, culminating in British occupation for more than 100 years until 1967, when it became independent.

Most of the land is hot and dry, rising from the southern coastal plain to mountains and highland plateaus averaging 1,980m (6,500ft). The economy is based on agriculture and relies heavily on foreign aid. Crops include cotton, tobacco, coffee, cereals and dates. The major industry, accounting for 75 per cent of exports, is petroleum processing at the refinery at Little Aden. Other industries include fishing, textiles, handicrafts, shipbuilding and furniture manufacture. Most of the people are Arabs, and there are some African and European influences. Some people in the north still pursue a nomadic way of life, although most of the people, in the south, live in towns and villages.

The area flourished under Minaean, Sabaean and Himyarite rule as part of the larger region called Al-Yaman. Muslim influence was established in the 7th century and it became part of the territory of the imams of Yemen in the Ottoman Empire by the 1500s. British occupation (beginning with Aden) dated from 1839; purchases of land and treaties with local rulers resulted in a British protectorate by 1914. In 1959 the various British interests were combined as the Federation of the Emirates of the South (renamed the Federation of South Arabia in 1963). The people of Aden opposed the union and violent campaigns against British control began in the 1960s. The National Liberation Front (NLF) forced the federation's collapse and in 1967 the country declared independence under the name Southern Yemen. The present name was adopted in 1970. Area: 287,683sq km (111,046sq miles). Pop. (1975 est.) 1,690,000. Map 38.

Yugoslavia, official name Socialist Federal Republic of Yugoslavia, is a mountainous country on the eastern shore of the Adriatic Sea. It has a Communist government, but has refused to accept the dominance of the USSR (unlike some Communist countries in central and eastern Europe). Part of its territory extends into the Balkan Peninsula, which was long considered to be the most politically explosive part of Europe. Yugoslavia itself – created at the end of World War I – is a land of deep, but suppressed, political and social tensions, chiefly those between the two largest of its constituent national groups, the Serbs and the Croats. Its name means "the country of the Southern Slavs". The capital is Belgrade (Beograd).

Land and climate. Yugoslavia's long coastline on the Adriatic Sea is fringed with small islands, many of them now popular as holiday resorts. Inland from the narrow coastal plain rises the Karst, a much-dissected barren limestone plateau which includes a number of mountain ranges; the most important are the Velebit Mountains and the Dinaric Alps. The the northern part of the Dinarics is heavily forested. In the north-eastern part of the country is the fertile Pannonian Lowland, drained by the River Danube (Serbo-Croat, Dunav) and its tributaries, the chief of which in this region are the Sava, Drava, Tisza and Morava. The mountainous regions of Yugoslavia suffer frequent earth tremors; an earthquake in 1963 caused much destruction in the city of Skopje. The mountain valleys contain lakes, the largest of which are Scutari and Ohrid on the Albanian border and Prespa on the border with Albania and Greece. The climate in coastal regions is mild and warm. In the Pannonian Lowland summers are often hot and humid and winters cold. In mountainous areas the climate also tends to extremes and there are heavy falls of rain or snow. The Dalmatian coastlands suffer from a cold wind called the *Bora.*

Economy. The country is comparatively rich in minerals, the chief sources being in the centre and south-east. The largest deposits are of low-quality coal and lignite (brown coal). Yugoslavia is among Europe's principal producers of ores of aluminium, antimony, lead and copper. Petroleum and natural gas have been of increasing importance in recent years. Mining of iron ore is the basis of much of the country's industry, and there are substantial deposits of mercury, manganese and chromium; gold and silver are also found.

The Drina River basin in central Yugoslavia is an area of agricultural wealth; at several points the river is dammed to provide hydroelectric power.

A rural atmosphere prevails in this residential area on the outskirts of Priština, Yugoslavia; the city, a manufacturing centre, produces jewellery and textiles.

Kinshasa, capital of Zaire, is a centre of trade, industry and communications despite being at the very edge of the country on the River Congo.

Most Yugoslavian farms are in the hands of peasant proprietors; such private holdings are limited to a maximum size of 10 hectares (25 acres), and the average farm is about 4 hectares (10 acres). Output on family holdings improved greatly after World War II through mechanization – mainly by means of co-operative ventures – and the increased use of fertilizers. The most productive land is in the Danubian basin. Farmers there grow maize, wheat, barley, rye, potatoes, sugar-beet and hemp. On the slopes to the south-west of the plain, grapes (for wine) and other fruits, including plums, are grown. The output of vegetables has been increased, and tobacco is an important crop. In mountainous areas livestock raising is the chief form of agriculture; dairy farming and horticulture are important in the coastal areas.

The main industrial region is the north-west, but government planners have sought to extend industry to other regions, too. Long-established industries include engineering and shipbuilding; the manufacture of chemicals, steel, textiles and paper are also important. Road and rail communications are well developed in the north and along the Adriatic coast; major highways include those linking Belgrade with Zagreb and Rijeka with Dubrovnik and Skopje. The Danube is an important waterway for passenger boats and barges, and there are many other navigable rivers and canals. The state airline, Jugoslovenski Aero Transport, operates domestic and international routes.

People. The overwhelming majority of the people of Yugoslavia are Slavs, although each of the main nationalities has its own distinctive culture. The Serbs are the most numerous; their chief city is Belgrade. They belong to the Serbian Orthodox Church and write their language, Serbo-Croatian, using the Cyrillic alphabet. The Croats, the next largest group, resent what they regard as Serbian dominance in the country's affairs. They are mainly Roman Catholic and also use the Serbo-Croatian language but write it with the Latin alphabet. They were for long associated with the Austrians and Hungarians and tend to be more Western in outlook than the Serbs, who were under Turkish domination until near the end of the 19th century. The Slovenes, who live in the most industrialized part of Yugoslavia, are also Roman Catholic and their language, Slovene, is written in Latin script. The Macedonians – who are numerous in Bulgaria as well as in Yugoslavia – have their own language, which uses Cyrillic characters. The Montenegrins formerly had an independent principality (after 1910, a kingdom) and once had a reputation for aggressiveness and lawlessness. Montenegro is named after the "Black Mountains", Mt Lovcen, in the south-west of the republic. Yugoslavia also has several small ethnic minorities, including Albanians, Magyars, Bulgarians and Turks.

Government. The Constitution describes the country as a federation composed of the socialist republics of Bosnia and Hercegovina, Crna Gora (Montenegro), Croatia, Macedonia, Serbia (and the autonomous provinces of Vojvodina and Kosovo) and Slovenia. The supreme organ of government is the Federal Assembly, consisting of the Federal Chamber and the Chamber of Republics and Provinces. Elections are based on universal suffrage for everyone more than 18 years old. Executive power is exercised by the Presidency (a body composed of representatives of the constituent republics) and the Federal Executive Council. The Constitution provides that each member of the Presidency in turn becomes president of Yugoslavia for a year. But Tito, first president of the Federal Republic, was made president for life.

History. Serbia became a kingdom in 1882 after gaining its independence from the Ottoman Empire. A quarrel between Serbia and Austria resulted in the murder in Sarajevo of the Austrian archduke Franz Ferdinand in 1914 – the act that precipitated World War I. After Austria-Hungary's defeat in the war, the Austrian or Hungarian territories of Croatia, Slovenia, Bosnia and Hercegovina were united with Serbia to form the Kingdom of the Serbs, Croats and Slovenes. Montenegro also joined, after deposing its king.

The Croats soon became dissatisfied with their position in the new state and demanded autonomy. In 1928 their leader, Stefan Radić, was shot and fatally wounded in Parliament. In the following year the king abolished the constitution and established a dictatorship to preserve the unity of the country, whose name was changed to Yugoslavia.

King Alexander was assassinated in Marseilles in 1934 by Croat nationalists. His 11-year-old son, Peter II, became king, but real power was in the hands of Alexander's brother, the regent Prince Paul. In March 1941, Prince Paul declared Yugoslavia's support for the Axis powers, but within two days he was ousted with King Peter II and a new government was formed. The Germans and Italians quickly occupied Yugoslavia, and Peter set up a government-in-exile in London. The Germans sponsored a puppet state in Croatia and Serbia headed by Ante Pavelić, the leader of the terrorist organization (the Ustachi) that had killed King Alexander; but many Croats fought with the anti-German resistance. Two groups of Yugoslav resistance fighters evolved: the royalist *Chetniks* led by Gen. Draza Mihailović and the Communist-dominated *Partisans* led by Josip Broz, called *Tito.*

By the end of the war the Partisans controlled most of Yugoslavia, and in an election in 1945 Tito gained an overwhelming victory. The country was declared a republic, and a Communist state was set up. The Yugoslavs soon ceased to follow the Soviet line, however; a decentralized governmental system was created and Tito became openly critical of many of the actions of the USSR. In the 1950s the two countries agreed on amicable acceptance of their differing viewpoints, although even then Yugoslavia gaoled some pro-Soviet "dissidents" in 1976. Map 24.

Yugoslavia – profile

Official name Socialist Federal Republic of Yugoslavia
Area 255,802sq km (98,756sq miles)
Population (1976 est.) 21,617,000
Density 85 per sq km (219 per sq mile)
Chief cities Belgrade (capital) (1975 est.) 870,000; Zagreb, 566,000; Skopje, 313,000
Government Federation of socialist (Communist) republics
Religions Orthodox (Pravoslav, Serbian, Macedonian); Roman Catholic
Languages Serbo-Croatian, Slovene, Macedonian
Monetary unit Dinar
Gross national product (1974) £10,867,500,000
Per capita income £503
Agriculture Cereals, potatoes, sugar-beet, hemp, grapes (for wine), plums, vegetables, tobacco, dairy products
Industries Steel, engineering products, ships, textiles, paper
Minerals Coal, lignite, aluminium, antimony, lead, copper, mercury, manganese, chrome, gold, silver, petroleum, natural gas
Trading partners USSR, EEC, Czechoslovakia

Zaire (Zaïre), official name Republic of Zaire, in west-central Africa, is Africa's second largest country. Before independence in 1960 Zaire was called the Belgian Congo, and between 1960 and 1971 it was generally known as Congo (Kinshasa) to distinguish it from its smaller neighbour, also called CONGO. Kinshasa is Zaire's capital. About 200 languages and dialects are spoken in Zaire and national unity has been difficult to achieve.

Land and climate. Most of Zaire occupies a depression in the African plateau, which forms the drainage basin of the River Zaire (better known as Congo) and its tributaries. Highlands and plateaus rise in the south and east. In the west, the River Zaire cuts through the rim of the depression to the Atlantic Ocean. Zaire's eastern border runs through the African Rift Valley, where lakes Idi Amin Dada (formerly Edward), Mobutu Sese Seko (formerly Albert) and Tanganyika are located. Central Zaire has an equatorial climate. Annual temperatures average between 25 and 27°C (77-81°F); the uplands are cooler. Rainfall varies between 1,250 and 2,030mm (50-80in) per year.

Economy. About 70 per cent of the people are subsistence farmers. Cattle are reared in the uplands, and the chief cash crops are coffee, cotton, palm products and rubber. The leading export is copper from Shaba (formerly Katanga) province; it

The Post Office at Lubumbashi in Zaire displays evidence of the time when the country was the Belgian Congo and the city was Elisabethville.

Much of the power used in the mining of copper, Zaire's most valuable export, is hydroelectricity produced by the numerous rivers of the country.

The railway line between Dar-es-Salaam in Tanzania and Lusaka in Zambia was finally completed at New Kapiri Mposhi, Zambia.

accounts for nearly 70 per cent of all exports. Industrial diamonds, cobalt, zinc, cassiterite (tin ore) and gold together account for another 16 per cent. Zaire has enormous hydroelectric potential and the Inga scheme, north of Kinshasa, will be the world's largest hydroelectric project when it is completed. Industries are developing, especially in Kinshasa and Lubumbashi.

People. There are about 200 ethnic and language groups. About two-thirds, including the Bakongo, Baluba, Balunda and Bamongo, speak Bantu languages. About 100,000 pygmies live in the forests and Hamites, Nilotes and Sudanese Negroes live in the north and north-east. About half of the people are Christians, mostly Roman Catholics, whereas the others practise ethnic religions. The literacy rate is about 40 per cent.

Government. Zaire has a presidential government. The supreme body, the Political Bureau of the only party, the Mouvement Populaire de la Révolution, is headed by the president. It has supremacy over the elected National Assembly and the government.

History. Henry Morton Stanley explored the River Zaire in 1874-77, and in 1884 the Conference of Berlin granted the territory to King Léopold II of Belgium as his personal property. But the Belgian government assumed responsibility in 1908. After independence in 1960 army mutinies, communal warfare and the attempted secession of the mineral-rich province of Shaba (then Katanga) created chaos. The United Nations helped to restore order in 1960-64. In 1965 the army took power and Gen Joseph-Désiré Mobutu (who later Africanized his name to Mobutu Sese Seko) became president, ruling with a civilian government. He put down revolts in 1966, 1967 and 1977. Map 34.

Zaire – profile

Official name Republic of Zaire
Area 2,345,409sq km (905,562sq miles)
Population (1975 est.) 24,902,000
Density 11 per sq km (28 per sq mile)
Chief cities Kinshasa (capital), 1,633,760; Katanga, 428,960; Lubumbashi, 357,369
Government Republic
Religions Christianity, ethnic
Language French (official)
Monetary unit Zaire
Gross national product (1974) £1,559,800,000
Per capita income £63
Agriculture Bananas, cassava, cacao, coffee, cotton, maize, mangoes, millet, palm products, rice, rubber, spices, sugar cane, tea
Industries Brewing, cement, food processing, mineral refining, palm products, soap, textiles
Minerals Cassiterite, cobalt, copper, gold, iron, manganese, silver, tungsten, uranium, zinc
Trading partners Belgium and Luxembourg, Italy, West Germany, USA

Zambia, official name Republic of Zambia, is a landlocked nation in south-central Africa. It was known as Northern Rhodesia until it became independent in 1964. Its economy depends on copper mining and a fall in world copper prices in the mid-1970s caused serious economic problems. The capital is Lusaka.

Land and climate. Most of Zambia is a tableland, between 1,070 and 1,525m (3,510-5,003ft) above sea-level. In the south-west a somewhat lower region is largely covered by Kalahari sands. In the south and east is the Zambezi-Luangwa trench, a depression associated with the African Rift Valley, which borders Zambia in the north. The climate varies with altitude. The seasons are dry and cool from May to August, hot and dry from September to November and hot and wet from December to April, when temperatures may reach as high as 38°C (100°F). Most rain falls between December and April. It exceeds 1,270mm (50in) per year in the north, but the south has between 510 and 760mm (20-30in) of rain per year. Most of the country is savanna and there is swamp vegetation around Lake Bengweulu, the largest lake entirely within Zambia.

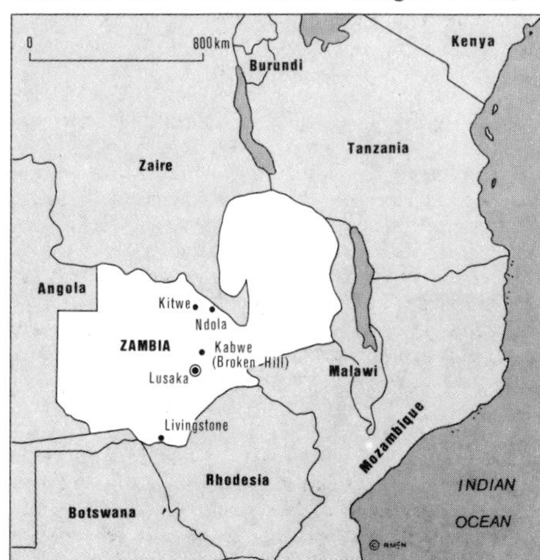

Economy. In 1973 copper accounted for 94 per cent of Zambia's exports, with zinc, lead and cobalt accounting for another 3½ per cent. Copper is mined in the Copperbelt, near the border with ZAIRE, and the government holds a 51 per cent interest in the industry. Other minerals come from Kabwe (Broken Hill). Coal deposits are being exploited. Both areas are served by the railway that runs from Maramba (Livingstone) through Lusaka to Kitwe, alongside which most economic activity takes place. This line is now linked to Dar-es-Salaam in Tanzania by the Uhuru (Freedom) Railway which was completed in 1975. About 60 per cent of Zambians are subsistence farmers. The chief cash crop, tobacco, accounts for 0.6 per cent of the exports. Tourism is developing rapidly and manufacturing is increasing in the towns.

People. Most people speak Bantu languages, of which there are six main ones and about 65 dialects. The largest group, the Tonga, lives in the south. The second largest group, the Bemba, lives in the north-east and many work on the Copperbelt. Most people follow ethnic religions, although about 500,000 are Christians. The literacy rate is between 15 and 20 per cent.

Zambia – profile

Official name Republic of Zambia
Area 752,614sq km (290,584sq miles)
Population (1975 est.) 4,896,000
Density 6.5 per sq km (17 per sq mile)
Chief cities Lusaka (capital) (1972) 448,000; Kitwe, 331,000
Government Republic
Religions Ethnic, Christianity
Language English (official)
Monetary unit Kwacha
Gross national product (1974) £987,200,000
Per capita income £202
Agriculture Cotton, dairy products, hides and skins, maize, meat, sugar cane, timber, tobacco
Industries Food processing, textiles, tobacco
Minerals Cobalt, copper, lead, zinc
Trading partners Britain and other members of the EEC, Japan

The first white man to see the Victoria Falls, now part of the Zambia-Zimbabwe border, was David Livingstone who named them after Queen Victoria.

Kariba Dam, completed in 1958, spans the mighty River Zambezi as it flows through the Kariba Gorge that separates Zimbabwe from Zambia.

Salisbury was named after the then British prime minister; it is now the capital of Rhodesia (Zimbabwe) and centre of communications and trade.

Government. The 1972 constitution made Zambia a one-party republic. The president heads the government and the vice-president leads the government in the elected National Assembly.

History. Early men probably lived in the region now occupied by Zambia more than a million years ago. By the 13th century some Bantu-speaking peoples had arrived and other ethnic groups entered the region over the following centuries. Cecil Rhodes' British South Africa Company entered the area in 1889. British protection was formalized in 1891, and in 1911 the territory became the British protectorate of Northern Rhodesia. From the 1920s the European settlers in Southern Rhodesia (Zimbabwe) pressed Britain to amalgamate the two Rhodesias. In 1953 the Federation of Rhodesia and Nyasaland (now MALAWI) was formed. But Black Africans opposed the Federation, fearing domination by the substantial European population of Southern Rhodesia. The Federation was dissolved in 1963 and Northern Rhodesia became the independent Republic of Zambia in 1964. But Zambia's economy remained closely tied to that of Southern Rhodesia. When Southern Rhodesia, then called Rhodesia, declared itself unilaterally independent in 1965, Zambia opposed this step and faced many problems. In the 1970s, Zambia's president Kenneth Kaunda was active with other African leaders in seeking a solution to the problem of Rhodesia (ZIMBABWE). Map 34.

Zanzibar, now part of Tanzania, was a British protectorate off the east coast of Africa. The capital is the city of Zanzibar and it includes two main islands: Zanzibar, which covers 1,658sq km (640sq miles), and Pemba, which covers 984sq km (380sq miles). The protectorate became independent in 1963 and in 1964 joined with Tanganyika to form the United Republic of TANZANIA, although Zanzibar retained its own government and legislature. Zanzibar is the world's leading producer of cloves. Pop. (1967) 354,360. Map 34.

Zimbabwe, in south-central Africa, remains legally the British self-governing colony of Rhodesia. But in Africa and, increasingly, outside Africa it is now becoming known as Zimbabwe – after the impressive stone ruins in the country, which some scholars believe are evidence of a major ancient African civilization. Since 1965 efforts have been made by Britain, South Africa, the United States and Black African nations in southern Africa to restore a legal regime to lead the country to independence under a constitution based on majority rule. The capital is Salisbury.

Land and climate. Central Zimbabwe consists of a plateau called the High Veld, most of which is between 1,220 and 1,525m (4,003-5,003ft) above sea-level. In the east, along the border with Mozambique, heights exceeding 2,440m (8,005ft) are reached. A deep trench in the north is occupied by the River Zambezi. The river has been dammed at Kariba and its waters have accumulated in Lake Kariba, which is about 280km (174 miles) long. Zimbabwe's border with Zambia runs along the Zambezi and through Lake Kariba. The Low Veld, in southern Zimbabwe, is mostly below 915m (3,002ft) above sea-level; it forms part of the Limpopo and Save river basins.

Altitude greatly affects the climate. The average annual temperature is 20°C (68°F) on the High Veld and 24°C (75°F) in the deep Zambezi trench. Rainfall averages 1,520mm (60in) per year on the eastern uplands and between 710 and 890mm (28-35in) per year on the central High Veld. The Low Veld has less than 400mm (16in) of rain per year.

Economy. The land is divided between African reserved areas (45.6 per cent), European reserved areas (45.6 per cent) and National Areas, for the preservation of wildlife (8.8 per cent). Most of the Africans are subsistence farmers, whose main food crop is maize. African cattle are mostly of poor quality. White farmers use scientific methods and get high yields. The chief cash crop is tobacco, and sugar cane, tea and fruits are also important. More than 2½ million cattle are kept on European farms. They are reared for beef, hides, and dairy products.

Mining and manufacturing are about twice as valuable as agriculture. Asbestos, chrome, coal and gold are mined. Manufacturing is important in Bulawayo, Gwelo, Galooma, Que Que, Salisbury and Umtali. Many industries are now powered by electricity from the Kariba hydroelectric power station. Zimbabwe is a landlocked country and relies on its neighbours to handle its trade. Since Mozambique closed its frontiers with Zimbabwe in 1976, most goods have passed through South Africa, mainly along the direct rail route through Beitbridge or the indirect rail route through Botswana. Trade figures have not been available since the United Nations imposed economic sanctions in 1965.

People. About 95 per cent of the people are Bantu-speaking Black Africans, many of whom belong to one of two major groups, the Shona in the north and the Ndebele (Matabele) in the south. There are about 275,000 white people (4.5 per cent of the population) and there are other minorities of Asians and people of mixed origin. Although exact figures are not available, many Black Africans have been converted to Christianity, whereas others still follow ethnic religions. The literacy rate among Black Africans is less than 30 per cent.

Government. The European-dominated government declared the country a republic in 1970. The cabinet, headed by prime minister Ian Smith, has included some African chiefs since 1976. About 94 per cent of the electorate are whites. The Senate contains ten Europeans, ten African chiefs and three others appointed by the president. The House of Assembly has 50 members elected by Europeans and 16 Black African members – eight elected by African voters and eight elected by African chiefs.

History. Bantu-speaking peoples began to arrive in the region in the 5th century AD, and the Zimbabwe ruins date from about this time. From 1898 Zimbabwe was officially ruled by a British high commissioner, based in South Africa. But the British South Africa Company and local European settlers effectively controlled internal affairs. In 1923 the territory became a British self-governing colony, called Southern Rhodesia to distinguish it from neighbouring Northern Rhodesia (now ZAMBIA). The government introduced discriminatory measures, the most significant of which was the Land Apportionment Act of 1930, which reserved much of the best land for European occupation only.

In 1953 Southern Rhodesia was federated with Northern Rhodesia and Nyasaland (now MALAWI). But Black Africans claimed that it extended white domination north of the Zambezi, and in 1963 the federation was dissolved. Southern Rhodesian Europeans wanted independence from Britain but no agreement was reached, largely because the Europeans were not prepared to lose their supremacy. In 1965 the government made a unilateral declaration of independence (UDI), an act declared illegal by Britain and the United Nations. But in spite of economic sanctions, mounting outside pressure for a government based on majority rule, and guerrilla warfare in the 1970s, Prime Minister Ian Smith maintained his regime. The gaining of independence by MOZAMBIQUE in 1975 and the closure of the Mozambique-Zimbabwe frontier in 1976 weekened the position of the minority regime. In 1976 Ian Smith accepted proposals made by the United States for independence under majority rule, but a subsequent conference failed to reach an agreement on how this was to be achieved. Map 34.

Zimbabwe – profile

Official name Rhodesia
Area 390,580sq km (150,080sq miles)
Population (1977 est.) 6,100,000
Density 16 per sq km (40 per sq mile)
Chief cities Salisbury (capital), 435,000; Bulawayo, 281,000
Government Internationally, the republican constitution and the government are considered illegal
Religions Christianity, ethnic
Monetary unit Rhodesian dollar
Gross national product (1974) £1,089,700,000
Per capita income £179
Agriculture Cassava, cotton, dairy products, maize, meat, millet, sorghum, sugar cane, tea, tobacco
Industries Chemical products, food processing, iron and steel, metal products, textiles
Minerals Asbestos, coal, chrome, copper, iron
Trading partner South Africa

Almanac

More and more in the modern world – in travel, science, agriculture and engineering – time is a key factor. Today a supersonic jet airliner can fly from London to Washington in about four hours. Yet on their arrival passengers see the hands on a clock in Washington recording a time 50 minutes earlier than their London departure time. This section of *The Modern World* explains the various methods of defining and measuring time, discusses time zones, and gives information about holidays and festivals around the world. The final page of the Almanac gives some world vital statistics and the composition of various international organizations.

Time measurement

Time is measured in terms of the rotation of the Earth about its axis (rotational time) or in terms of periodic phenomena occurring within atoms (atomic time). The various forms of rotational time and the relationship between them and atomic time are explained below.

Solar time The time taken for the Earth to complete one rotation on its axis, relative to the Sun, is a *solar day*. Because the Earth moves round the Sun in an ellipse rather than in a circle the *apparent solar day* is not constant throughout the year. The *mean solar day* averages out the differences over a period of one year. Mean solar time observed on the meridian of the telescope at the Royal Observatory at Greenwich is called *Greenwich Mean Time* (GMT). GMT is also known as *Universal time.*

Sidereal time The time taken for the Earth to complete one rotation on its axis, with reference to a point in the heavens called the Fixed Point of Aries, is a *sidereal day*. Because of the precession of the Earth's axis and various oscillations (mutation), astronomers use a *mean sidereal time* from which these fluctuations have been removed.

Ephemeris time A detailed study of the motions of the Sun, Moon and planets relative to the Earth has shown that there are irregularities in the Earth's rotation. A table giving the apparent daily positions of the Sun, planets, and Moon is called an ephemeris; the method of time measurement based on bringing the ephemeris into agreement with observed values is *ephemeris time*.

Atomic time Time can be measured to a high degree of accuracy using the interval between energy changes (quantum transitions) within an atom. Hydrogen, rubidium and caesium atoms have been used for this purpose. The SI unit of time is now the second as defined in terms of a specified transition within the caesium atom.

Relationship between rotational and atomic time Because of variations in the speed of rotation of the Earth, the *mean solar second* (1/86,400 of the mean solar day) is unsatisfactory for precision measurements. In 1955 the International Astronomical Union defined the second of ephemeris time as 1/31,566,925.9747 of the tropical year for 1 January 1900 at 12 hours ephemeris time. In 1956 this definition was accepted by the General Conference of Weights and Measures. In 1964 the Conference also defined an ephemeris second as 9,192,631,770 periods of radiation of a particular transition of the caesium-133 atom. The definition was adopted as the sole definition of the SI unit of time in 1967.

British Standard Time and British Summer Time (BST) During World War I an Act of Parliament decreed that during the summer months in Britain the legal time should be one hour in advance of GMT. During World War II the duration of summer time was extended, and in 1941-45 and again in 1947 double summer time (two hours in advance of GMT) was introduced. In 1968 the British Standard Time Act put the legal time one hour ahead of GMT for the whole year to bring Britain into line with Central European Time. As from 31 October 1971 Parliament restored the concept of summer time, to operate one hour in advance of GMT between 0200 hr GMT on the day following the third Saturday in March and 0200 hr GMT on the day following the fourth Saturday in October.

British Summer Time

	from		to
1977BST	from 20 March	to	23 October
1978	19 March		29 October
1979	18 March		28 October
1980	16 March		26 October
1981	22 March		25 October
1982	21 March		24 October
1983	20 March		23 October
1984	18 March		28 October
1985	17 March		27 October

Time Around the World In 1883 a system for dividing the world into separate longitudinal time zones was internationally agreed. These zones are based on Greenwich Mean Time, time in each zone usually differing from GMT by a whole number of hours. Each zone is approximately 15 degrees of latitude wide, but each country adopts a standard of time most convenient to it. Some large countries have several time zones; the United States, Canada, and the USSR are divided into zones approximately 7½ degrees of latitude wide on each side of central meridians. East of the Greenwich meridian the time is in advance of GMT, west of Greenwich the time is behind GMT. The date changes on the *International Date Line,* an imaginary line running along the 180th meridian (except that the line is adjusted locally so that it does not divide countries or groups of islands). West of the date line the date is one day ahead of that east of the line (*see map, below).*

The calendar

A calendar is a method of reckoning time for regulating religious, commercial and civil life, and for dating events. Calendars have been in use from earliest times – the Ancient Egyptians had a system based on movement of the star Sirius and on the seasons (determined by fluctuations of the River Nile). Like all calendars this was based on the astronomical motions of the Earth, Moon, and Sun; the lunar cycle produced the Moon's phases and the solar cycle the changing seasons. The units of time used in calendar reckoning are the day, the month and the year.

The day is the time taken for the Earth to make one complete turn on its axis. In modern calendars the day is reckoned from midnight to midnight and is divided into 24 hours. The hours of daylight vary throughout the year. They also vary with latitude; the farther north one gets the greater is the difference between daylight hours and hours of darkness.

The month is the time taken for the Moon to complete one full revolution around the Earth. There are different ways of defining this period; the one used in calendars is the *synodic month,* which is the time between successive new moons – a period of 29.53059 days. The month is used because it is a convenient number of days to count, and because the changes in the Moon's shape are easily recognized. The timing of many religious festivals, such as Easter, is based on the Moon. The *week* probably owes its origin to the four phases of the Moon.

Phases of the Moon The moon rises and sets as a result of the Earth's rotation and, in addition, it

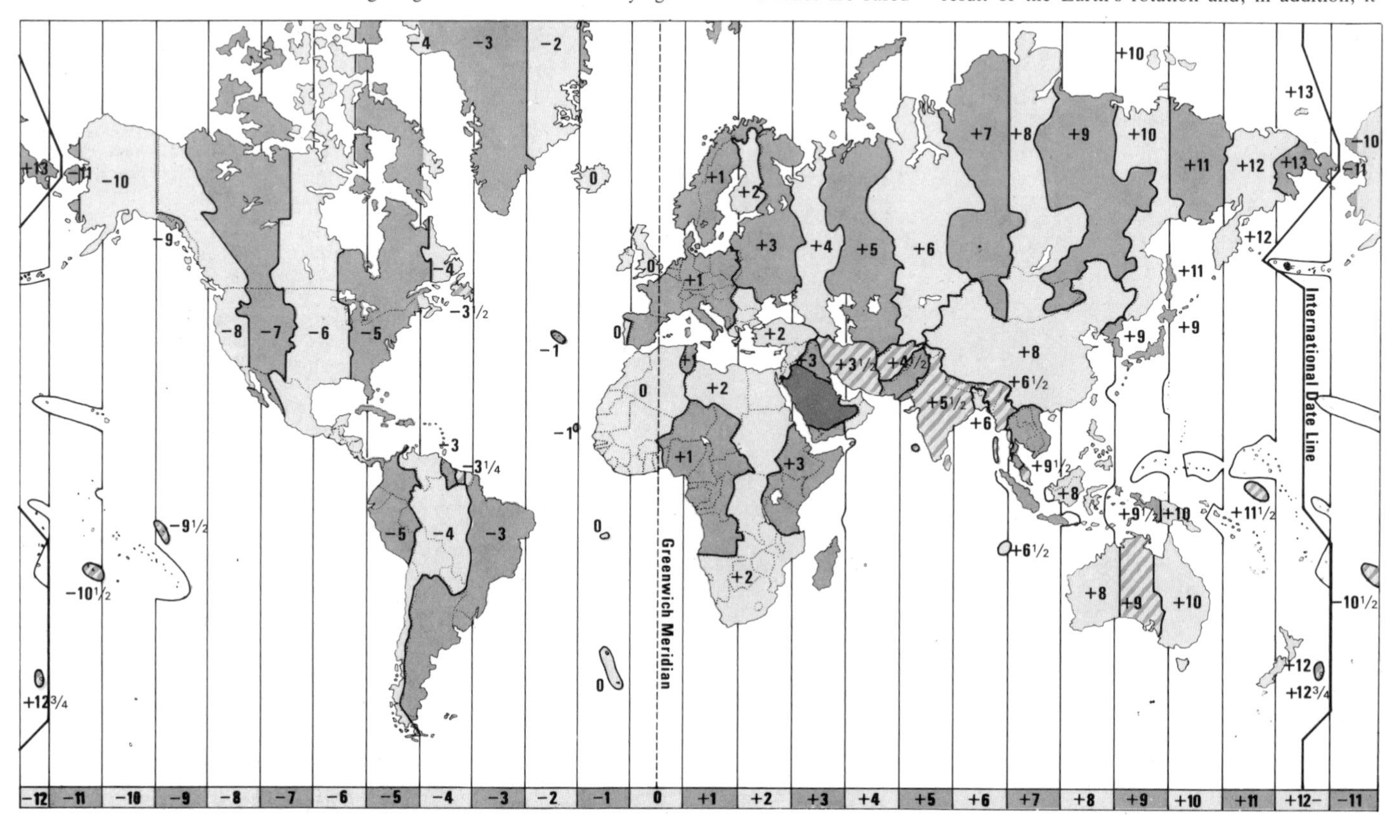

shows characteristic phases because of its movement around the Earth. At different points in its orbit different parts of the illuminated half are visible. At *new moon* the part illuminated by the Sun is hidden from the Earth and the Moon's disc is completely dark. In a lunar cycle the visible part grows (waxes) from a thin crescent, through a half moon (*first quarter*), to reach a *full moon* when the disc is completely illuminated. It then diminishes (wanes) through a second half moon (the *last quarter*) back to a new moon again. When the Moon is between a half moon and a full moon it is said to be *gibbous*. The full moon nearest to the autumnal equinox (22 September) is often called a *harvest moon;* the following full moon, in October, is a *hunter's moon.*

The tides The Moon and, to a lesser extent, the Sun cause tides by their gravitational pull on the water of the oceans. The effect of the Moon's attraction is to raise the water level at a point on the Earth closest to the Moon, thus distorting the Earth's water mass and producing a similar "bulge" on the opposite side of the Earth. These places have high tides; points on the "sides" of the Earth have low tides.

As the Earth rotates different places have high tides; most places have two tides every day separated by approximately 12 hours. The precise times of tides depend on the relative positions of the Moon and the Sun, and also on ocean currents. Each month there are two particularly high tides occurring at full and new moon, where the Sun and the Moon are acting in the same direction. These are known as *spring tides.* Midway between these two are the *neap tides*, with the lowest rise and fall, when the Sun and Moon act in opposite directions.

The year is the time taken for the Earth to complete one revolution around the Sun. Again there are different ways of defining it; the one used in calendars is the *tropical year*, which is the time taken for the Sun to complete a full cycle of its apparent motion north and south of the Equator. The tropical year is 365.242199 days.

The seasons occur because the Earth's axis is not at right-angles to a line between the Earth and the Sun, but is tilted at an angle. At one part of the orbit the Northern Hemisphere is tilted towards the Sun. At these times it is summer in the Northern Hemisphere, the hours of daylight exceed the hours of darkness, and the Sun is north of the Equator. As the Earth moves to the opposite side of the orbit, six months later, the Northern Hemisphere is tilted away from the Sun. It is then winter, the hours of darkness exceed the hours of daylight, and the Sun is south of the Equator.

The seasons are arbitrary divisions of the year into four periods. The generally accepted dates in Europe are:

Spring	21 March to 21 June
Summer	22 June to 22 September
Autumn	23 September to 22 December
Winter	23 December to 20 March

Equinoxes and solstices The equinoxes occur when the Sun crosses the Equator on its apparent journey north or south. At these times the Sun is directly overhead at the Equator and the hours of daylight equal the hours of darkness at all points on the Earth. The *vernal equinox* occurs when the Sun is moving northwards, about 21 March. The *autumnal equinox* occurs on about 22 September, when the Sun is moving southwards.

Midway between the equinoxes are the solstices, which occur when the Sun is at the extremes of its apparent motion north and south of the Equator. The *summer solstice* occurs on about 21 June, when the Sun is farthest north. In the Northern Hemisphere this is the longest day of the year. The *winter solstice*, on about 22 December, is the shortest day in the Northern Hemisphere and occurs when the Sun is farthest south.

Quarter days are four days marking quarters of the year; formerly they were days on which payments were made and accounts settled. In England, Wales and Northern Ireland the quarter days are:

Lady day	25 March
Midsummer day	24 June
Michaelmas	29 September
Christmas day	25 December

The days fall close to the equinoxes and solstices; note, however, that Midsummer day is not the longest day of the year.

In Scotland the quarter days are:

Candlemas	2 February
Whit Sunday	movable
Lammas	1 August
Martinmas	11 November

The modern calendar The main difficulty in compiling a calendar is that the day, the month and the year are not commensurate – the month is not an exact number of days and the year is not an exact number of months. For convenience, months and years are assigned a whole number of days, and extra days (called *intercalations*) are added at intervals to compensate. Thus the year, which is about 365¼ days, is taken to be 365 days. As time goes by the seasons get out of step with the dates – the solstices and equinoxes occur one day later every four years. To compensate, an extra day (29 February) is added every four years (leap years).

The calendar used at the present time is the Gregorian calendar, which is a modification of the earlier Julian calendar.

The Julian calendar was introduced by Julius Caesar in the first century BC and came from an earlier Roman system based on the Moon. The Julian calendar abandoned any attempt to keep the calendar months in step with the Moon's phases; as a result, the months became arbitrary divisions of time and full moon did not occur on the same day of the month every year. The calendar was based on a year of 365¼ days with an extra day every four years added to February. The calendar was modified by Augustus, who changed the lengths of the months to the number of days they now have.

The Gregorian calendar The value of 365.25 days of the Julian calendar is slightly longer than the true value and over the years the error mounted up. By 1582 there was a difference of 10 days between the dates of the solstices and their original dates. In this year a papal bull issued by Gregory XIII corrected the discrepancy by making 5 October into 15 October. The bull also modified the rule for leap years, introducing the rule that the first year of a century (1600, 1700, and so on) is a leap year only if it is divisible by 400 (thus 1600 was a leap year but 1700 was not).

The Gregorian system was not adopted immediately in all countries, in particular in Protestant countries. It was introduced in England only in 1752 by dating the day following 2 September 1752 as 14 September. In addition the start of the year was fixed at 1 January in 1752; from the 14th century the year had been reckoned from 25 March (the feast of the Annunciation).

The Gregorian calendar is often called the *New Style* calendar, to distinguish it from the Julian, *Old Style*, calendar. In converting dates between New and Old Styles it is necessary to subtract or add 11 days and to take into account the different starting dates for the year and the fact that 1700 was a leap year in the Julian calendar but not in the Gregorian calendar.

Leap years The modern rule for determining leap years is that leap years are divisible by four unless they are centennial years, in which case they are leap year if they are divisible by 400 except for the year 4000 and its multiples. These will not be leap years. Thus the year 2000 will be a leap year but 3000 and 4000 will not.

Table 1 Movable Christian Feasts

	1977	1978	1979	1980	1981	1982	1983	1984	1985
Ash Wednesday	23 Feb.	8 Feb.	28 Feb.	19 Feb.	4 Mar.	24 Feb.	16 Feb.	7 Mar.	20 Feb.
Easter Sunday	10 April	26 March	15 April	6 April	19 April	11 April	3 April	22 April	7 April
Ascension Day	19 May	4 May	24 May	15 May	28 May	20 May	12 May	31 May	16 May
Whit Sunday	29 May	14 May	3 June	25 May	7 June	30 May	22 May	10 June	26 May
Trinity Sunday	5 June	21 May	10 June	1 June	14 June	6 June	29 May	17 June	2 June
Advent Sunday	27 Nov.	3 Dec.	2 Dec.	30 Nov.	29 Nov.	28 Nov.	27 Nov.	25 Nov.	1 Dec.

Religious calendars

Religious calendars are cycles of festivals and seasons marking events or rituals of special significance. Often they are based on the Moon and do not follow the civil calendar. Nevertheless they have a wider importance – movable feasts such as Easter and Whitsuntide, in the Christian faith, are also national holidays.

The Christian calendar developed from the Jewish calendar, which was based on a lunar cycle. The festival of Easter developed from the Jewish Passover; Christmas was introduced later to be celebrated at the time of the winter solstice, which had long been marked by such pagan festivals as the Saturnalia.

The Christian year is set by the governing bodies of individual Churches and inevitably there are differences in practice – in particular, there are differences between the Eastern and Western Churches. The festivals dated here are those currently used in the Western Roman Catholic and Anglican Churches. There are two cycles of important dates: one of saints' days and other fixed dates, the other of movable feasts based on Easter. The fixed dates are given later in the table of Important Dates. Table 1 shows the main movable feasts.

Shrove Tuesday is the day before Ash Wednesday; Palm Sunday is the Sunday before Easter Sunday; and Maundy Thursday and Good Friday are the Thursday and Friday before Easter Sunday. Corpus Christi (RC) occurs on the Thursday after Trinity Sunday. Rogation Sunday is five weeks after Easter. The Festival of Christ the King (RC) is the Last Sunday after Pentecost. The Baptism of Christ is the first Sunday after Epiphany (6 Jan.).

The Jewish calendar is based on both lunar and solar cycles. Years are dated from the creation, supposed to be at the time of the autumnal equinox in 3760 BC, and designated AM (*anno mundi*). The Jewish year runs from September to September: 1 January 1977 AD occurred in 5737 AM.

Normal years are divided into 12 months which start at about the time of the new moon: these months have 30 and 29 days alternately, leading to a year of 354 days. In order to keep the months roughly in step with the seasons an extra month is included in certain years and changes are also made in the number of days in some months. Jewish fasts and festivals occur on fixed dates in the Jewish calendar, which of course means that they are movable in the civil calendar. Some important dates are given in Table 2.

The Muslim calendar is a purely lunar calendar consisting of 12 months of 30 or 29 days. An extra day is added to the last month at intervals to keep the months in step with the new moon, which occurs about the beginning of each month. The months do not keep in step with the seasons.

Table 2 Jewish Festivals

	1977	1978	1979	1980	1981	1982	1983	1984	1985
Jewish New Year	13 Sept.	3 Oct.	23 Sept.	12 Sept.	30 Sept.	19 Sept.	9 Sept.	28 Sept.	17 Sept.
Day of Atonement	22 Sept.	11 Oct.	1 Oct.	20 Sept.	8 Oct.	27 Sept.	17 Sept.	6 Oct.	25 Sept.
Feast of Tabernacles	27 Sept.	16 Oct.	6 Oct.	25 Sept.	13 Oct.	2 Oct.	22 Sept.	11 Oct.	30 Sept.
Purim	4 March	23 March	13 March	2 March	18 Feb.	9 March	27 Feb.	17 Feb.	7 March
Passover	3 April	22 April	12 April	1 May	19 April	8 April	29 March	17 March	6 April

Important Dates

The table below gives the dates of some of the main religious and secular anniversaries and festivals of the year. The saints' days marked "RC" are mainly celebrated in the Roman Catholic calendar; those marked "Ang" are in the Anglican *Book of Common Prayer*; unmarked saints are common to both. The saints' days are "fixed" – that is, they invariably fall on the designated date irrespective of the day of the week. In most countries, Independence Day (or National Day) is also fixed, although sometimes a government "moves" it to avoid its falling on a particular day (such as a Sunday). The dates in the table were those adopted in 1977.

Religious	Secular
January	
1 Circumcision; Solemnity of the BV Mary (RC)	New Year's Day
2	Scotland: New Year's Holiday
3 St Genevieve (RC)	
6 Twelfth Night	Iraq: Army Day
7 St John the Baptist (RC); Orthodox Christmas	Liberia: Pioneers Day
8 St Lucian (Ang)	USA: Battle of New Orleans
10 St Gregory (RC)	Bolivia: Oruro's Birthday
12	Tanzania: Revolution Day
13	Ghana: Redemption Day
17 St Anthony (RC)	
18 St Prisca (Ang)	
20 St Fabian (Ang)	Mali: Army Day
21 St Agnes	Dominican Republic: Altagrecía Day
22 St Vincent	Saint Vincent: Discovery Day
24 St Francis de Sales (RC)	
25 Conversion of St Paul	Uganda: Republic Day
26 SS Timothy and Titus (RC)	India: Republic Day
27 St Devoté (Monaco)	
28 St Thomas Aquinas (RC)	
30 St Basil (RC)	
31 St John Bosco (RC)	Australia: Australia Day
February	
2 Purification of the Virgin Mary	Scotland: Quarter Day
3 St Blasius (Ang)	Mozambique: Day of Martyrs
5 St Agatha	Tanzania: Afro-Shirazi Day
6	New Zealand: New Zealand Day
10 St Scholastica (RC)	
12	Burma: Union Day
14 St Valentine; SS Cyril and Methodius (RC)	Sri Lanka: Maha Sivarahri Day
18	Gambia: Independence Day
24 St Matthias	Ghana: National Liberation Day
29	Leap Year Day
March	
1 St David	Ethiopia: Battle of Adawa
2 St Chad (Ang)	Burma: Peasants Day
6	Ghana: Independence Day
7 St Perpetua (Ang)	Tasmania: Eight Hour Day
9	Australia: Canberra Day
12 St Gregory (Ang)	Mauritius: Independence Day
14	Australia: Labour Day
17 St Patrick (RC)	Northern Ireland: Bank Holiday
18 St Edward (Ang)	
19 St Joseph (RC)	
21 St Benedict (Ang)	Vernal Equinox
23	Pakistan: Pakistan Day
25 The Annunciation of the Virgin Mary	Lady Day (Quarter Day)
29	Malagasy: Memorial Day
April	
1	All Fools' Day; Cyprus: National Day
3 St Richard (Ang)	
4 St Ambrose (Ang)	Hungary: Liberation Day
5	Fiscal Year Ends
6	Hilary Law Sitting Ends
19 St Alphege (Ang)	Primrose Day; Venezuela: Independence Day
21 St Anselm (RC)	Israel: Independence Day
23 St George	
25 St Mark	Australia, New Zealand: Anzac Day
29 St Catherine (RC)	Japan: Emperor's Birthday
30	Netherlands: Queen Juliana's Anniversary
May	
1 SS Philip and James	Labour Day
3 St Athanasius (RC) Invention of the Cross	Japan: Constitution Day
6 Martyrdom of St John the Evangelist (Ang)	Denmark: Great Prayer Day
14 St Matthias (RC)	Malawi: Kamuzu Day
17	Norway: Independence Day
19 St Dunstan (Ang)	Botswana: President's Day
20	Cameroon: National Day
24	Commonwealth Day; Zambia: Africa Day
25 St Bede	Argentina, Jordan: Independence Day
26 St Philip Neri (RC); St Augustine (Ang)	Guyana: Independence Day
27 St Augustine (RC); St Bede (Ang)	Afghanistan: Independence Day
29	Britain: Oak Apple Day
31 The Visitation of our Blessed Lady (RC)	South Africa: Republic Day
June	
1 St Justin (RC); St Nicomede (Ang)	Brunei: Regiment Day
3 SS Chas. Lwanga and companions (RC)	Bahamas: Labour Day
5 St Boniface (Ang)	Denmark: Constitution Day
11 St Barnabas	
13 St Anthony (RC)	
14	Swaziland: Commonwealth Day
17 St Alban (Ang)	West Germany: National Memorial Day
19	Trinidad and Tobago: Labour Day
20 St Alban (RC)	
21 St Aloysius (RC)	Summer solstice: longest day
22 SS John Fisher and Thomas More (RC)	Congo: Army Day
24 St John the Baptist	Midsummer Day (Quarter Day); New Zealand: Labour Day
25	Mozambique: Independence Day
26	Malagasy: Independence Day
28 St Irenaeus (RC)	
29 SS Peter and Paul (RC); St Peter (Ang)	Seychelles: Independence Day
July	
1	Canada: Dominion Day; Ghana: Republic Day
2 Visitation of the Blessed Virgin Mary (Ang)	
3	Vietnam: Reunification Day
4	USA: Independence Day
10	Bahamas: Independence Day
11 St Benedict (RC)	Mongolia: National Day
12	Northern Ireland: Orangemen's Day
14	France: Bastille Day; Nicaragua: National Day
15 St Bonaventura (RC); St Swithin	
19	Burma: Martyrs Day
20 St Margaret (Ang)	Colombia: National Day
21	Belgium: National Day
22 St Mary Magdalen	Poland: National Day
24	Mozambique: National Day
25 St James	Puerto Rico: Commonwealth Day
26 SS Joachim and Anne (RC); St Anne (Ang)	Liberia: Independence Day
29 St Martha (RC)	
August	
1 St Alphonsus (RC); Lammas Day	Guyana: Commonwealth Day; Scotland: Quarter Day
4 St John Vianney (RC)	
6 Transfiguration of Jesus	Jamaica: Independence Day
7 Name of Jesus	
8 St Dominic (RC)	
10 St Lawrence	Ecuador: Independence Day
11 St Clare (RC)	Chad: Independence Day
14	Pakistan: Independence Day
15 The Assumption of Mary (RC)	India, South Korea: Independence Day
20 St Bernard	Hungary: Constitution Day
24 St Bartholomew	Liberia: Day of the Flag
27 St Monica (RC)	
28 St Augustin of Hippo (Ang)	
29 St John the Baptist	Hong Kong: Liberation Day
31	Malaysia: National Day; Trinidad and Tobago: Independence
September	
1 St Giles (Ang)	Libya: Revolution Day
3 St Gregory (RC)	Qatar: Independence Day
5	Canada, USA: Labour Day; South Africa: Family Day
6	Pakistan: Defence of Pakistan Day; Swaziland: Independence Day
7 St Evurtius (Ang)	Brazil: Independence Day
8 Nativity of the Virgin Mary	
13 St John Chrysostom (RC)	
14 Holy Cross Day	Bolivia: Cochabamba's Birthday
15 Our Lady of Sorrows (RC)	Costa Rica, El Salvador, Guatemala, Honduras: Independence Day
16 SS Cornelius and Cyrian (RC)	Mexico, Papua New Guinea: Independence Day
17 St Lambert (Ang)	Uganda: Remembrance Day
21 St Matthew	Malta: Independence Day
22	Autumnal Equinox
24 Feast of Our Lady of Mercy (Dominican Republic)	
26 St Cyprian (Ang)	Sri Lanka: Commemoration Day
27 St Vincent de Paul (RC)	
29 SS Michael, Gabriel and Raphael (RC); St Michael and All Angels (Ang)	Michaelmas (Quarter Day); Paraguay: Battle of Boqueron Day
30 St Jerome	Botswana: National Day
October	
1 St Teresa of the Child Jesus (RC); St Remigius (Ang)	China: National Liberation Day; Nigeria: Republic Day
4 St Francis (RC)	
6 St Faith (Ang)	Egypt: Army Day
7	East Germany: National Republic Day
9 St Denys (Ang)	Uganda: Independence Day
10	Canada: Thanksgiving Day; Fiji: Fiji Day; South Africa: Kruger's Day
12 Our Lady Fiesta Day (Spain)	Equatorial Guinea: Independence Day
13 St Edward the Confessor	
15 St Teresa of Avila (RC)	French Guiana: Cayenne Day
17 St Ignatius (RC); St Ethelreda (Ang)	Malawi: Mother's Day
18 St Luke	Jamaica: National Heroes' Day
20	Kenya: Kenyatta Day
24	United Nations Day; Zambia: Independence Day
25 St Crispin	
27	Zaire: Anniversary Day
28 SS Simon and Jude	
31 Hallowe'en	
November	
1 All Saints' Day	Algeria: Revolution Day; Antigua: State Day
2 All Souls' Day	
4 St Charles (RC)	Italy: Day of National Unity
5	Britain: Guy Fawkes' Day
7	Albania, USSR: Anniversary of the Revolution
10 St Leo (RC)	
11 St Martin	Scotland: Martinmas (Quarter Day)
13 St Britius (Ang)	
15 St Machutus (Ang)	Burma: National Day
16	West Germany: National Repentence Day
17 St Elizabeth of Hungary (RC)	Zaire: National Army Day
18 St Hugh (Ang)	Oman: National Day
20 St Edmund (Ang)	Mexico: Revolution Day
22 St Cecilia	
23 St Clement (Ang)	Japan: Labour Day
24	USA: Thanksgiving Day
25 St Catherine (Ang)	
30 St Andrew	Barbados, Yemen PDR: Independence Day
December	
3 St Francis Xavier (RC)	
6 St Nicholas (Netherlands)	Finland: Independence Day
7 St Ambrose (RC)	Ivory Coast: Independence Day
8 Immaculate Conception	
12	Angola: MPLA Day; Kenya: Independence Day
13 St Lucy (Lucia)	Malta: Republic Day
14 St John of the Cross (RC)	
16	Bahrain: National Day; South Africa: Day of the Covenant
21 St Thomas	
22	Winter solstice (shortest day)
25 Christmas Day	Quarter Day
26 St Stephen	Boxing Day
27 St John the Evangelist	
28 Holy Innocents' Day	
29 St Thomas Becket (RC)	Nepal: King's Birthday
31 St Silvester (Ang)	New Year's Eve; Scotland: Hogmanay

World facts and figures

CONTINENTS

Name	Area in sq km	[sq miles]
Asia	44,250,000	[17,084,900]
Africa	30,264,000	[11,684,900]
North America	24,398,000	[9,420,100]
South America	17,807,800	[6,875,600]
Antarctica	13,209,000	[5,100,000]
Europe	9,906,000	[3,824,700]
Australasia	8,842,400	[3,414,100]

ISLANDS

Name	Ocean	Area in sq km	[sq miles]
Greenland	Atlantic	2,175,600	[840,000]
New Guinea	Pacific	885,780	[342,000]
Borneo	Pacific	743,330	[287,000]
Madagascar	Indian	587,045	[226,658]
Baffin	Arctic	476,070	[183,810]
Sumatra	Indian	473,600	[182,860]
Honshu	Pacific	230,540	[89,010]
Great Britain	Atlantic	218,050	[84,190]
Ellesmere	Arctic	212,690	[82,120]
Victoria	Arctic	212,200	[81,930]
Sulawesi	Pacific	189,040	[72,990]
South Island [NZ]	Pacific	150,450	[58,090]
Java	Pacific	126,290	[48,760]
North Island [NZ]	Pacific	114,450	[44,190]

LAKES

Name	Country	Area in sq km	[sq miles]
Caspian	USSR/Iran	393,896	[152,083]
Superior	US/Canada	82,413	[31,820]
Victoria	Kenya/Uganda/ Tanzania	69,484	[26,828]
Aral	USSR	68,681	[26,518]
Huron	USA/Canada	59,596	[23,010]
Michigan	USA	58,015	[22,400]
Tanganyika	Zaire/Tanzania	32,893	[12,700]
Great Bear	Canada	31,792	[12,275]
Baykal	USSR	30,510	[11,780]
Nyasa	Malawi/Tanzania/ Mozambique	29,604	[11,430]

DESERTS

Name	Location	Area in sq km	[sq miles]
Sahara	North Africa	9 million	[3.5 million]
Gobi	Mongolia	1.3 million	[500,000]
Kalahari	Southern Africa	910,000	[351,000]
Rub al-Khali	Southern Arabia	650,000	[251,000]
Great Sandy	Australia	420,000	[162,000]
Taklamakan	Central Asia	325,000	[125,000]
Kara Kum	USSR	290,000	[112,000]
Thar	South Asia	260,000	[100,000]
Kyzyl Kum	USSR	230,000	[89,000]

VOLCANOES

Name	Location	Height in m	[ft]
Etna[6]	Sicily	3,340	[10,958]
Fuji[1]	Japan	3,778	[12,395]
Mauna Kea[1]	Hawaii	4,200	[13,779]
Mauna Loa[3]	Hawaii	4,160	[13,648]
Ngauruhoe[2]	New Zealand	2,290	[7,513]
Njamiagira[6]	Zaire	3,059	[10,036]
Nyiragongo[5]	Zaire	3,472	[11,391]
Pacaya[2]	Guatemala	2,546	[8,353]
Popocatepetl[6]	Mexico	5,456	[17,900]
Stromboli[6]	Italy	927	[3,041]
Tristan da Cunha[4]	Atlantic Ocean	2,062	[6,765]
Vesuvius[2]	Italy	1,278	[4,193]

1 Dormant
2 Steaming
3 Last erupted 1950
4 Last erupted 1961
5 Last erupted 1970
6 Last erupted 1971

WATERFALLS

Name	Location	Height in m	[ft]
Angel	Venezuela	980	[3,215]
Tugela	South Africa	949	[3,113]
Yosemite	USA	740	[2,428]
Cuquenán	Venezuela	610	[2,001]
Sutherland	New Zealand	581	[1,906]

MOUNTAINS

Name	Location	Height in m	[ft]
Everest	Nepal/Tibet	8,848	[29,029]
K2 [Godwin Austen]	Kashmir	8,616	[28,267]
Kanchenjunga	Nepal/Sikkim	8,591	[28,185]
Makalu	Nepal/Tibet	8,481	[27,824]
Dhaulagiri	Nepal	8,177	[26,827]
Nanga Parbat	Kashmir	8,131	[26,676]
Annapurna	Nepal	8,078	[26,502]
Gasherbrum	Kashmir	8,073	[26,486]
Gosainthan	Tibet	8,019	[26,309]
Nanda Devi	India	7,822	[25,662]
Rakaposhi	Kashmir	7,793	[25,567]
Kamet	India/Tibet	7,761	[25,462]
Namcha Barwa	Tibet	7,761	[25,462]

RIVERS

Name	Length in km	[miles]
Nile	6,669	[4,145]
Amazon	6,516	[4,050]
Mississippi-Missouri	6,050	[3,760]
Yangtze-Kiang	5,526	[3,434]
Ob-Irtysh	5,149	[3,200]
Amur	4,666	[2,900]
Zaire	4,373	[2,718]
Hwang Ho[Yellow River]	4,344	[2,700]
Lena	4,256	[2,645]
Mackenzie	4,240	[2,635]
Mekong	4,183	[2,600]
Niger	4,183	[2,600]
Yenisey	3,797	[2,360]

MEMBERSHIP OF INTERNATIONAL ORGANIZATIONS

Arab League Algeria, Bahrain, Egypt, Iraq, Jordan, Kuwait, Lebanon, Libya, Mauritania, Morocco, Oman, Palestine Liberation Organization, Qatar, Saudi Arabia, Somalia, Sudan, Syria, Tunisia, United Arab Emirates, Yemen, People's Democratic Republic of Yemen — founded 1945 to form and strengthen links between members

Benelux Economic Union Belgium, Luxembourg, Netherlands — formed 1958 to achieve economic union of its members

Colombo Plan Afghanistan, Australia, Bangladesh, Bhutan, Britain, Burma, Cambodia, Canada, Fiji, India, Indonesia, Iran, Japan, Laos, Malaysia, Maldives, Nepal, New Zealand, South Korea, Pakistan, Papua New Guinea, Philippines, Singapore, Sri Lanka, Thailand, United States — founded 1950 to aid developing countries of South and South-East Asia

Council for Mutual Economic Assistance [COMECON] Bulgaria, Cuba, Czechoslovakia, East Germany, Hungary, Mongolia, Poland, Romania, USSR — founded 1949 to co-ordinate and promote economic development of members

Commonwealth of Nations A free association of countries that were formerly ruled under the British empire and recognize the British monarch as head of the community — formed to further international co-operation and to strengthen links between members

European Economic Community [EEC] Belgium, Britain, Denmark, France, Ireland, Italy, Luxembourg, Netherlands, West Germany — formed 1957 to integrate members' economic development and eventually to achieve economic union

European Free Trade Association [EFTA] Austria, Finland, Iceland, Norway, Portugal, Sweden, Switzerland — formed 1960 to eliminate tariffs between members. Britain and Denmark left EFTA Dec. 1972 to join the EEC

Latin American Free Trade Association [LAFTA] Members are ten South American countries and Mexico — formed 1960 to work for a Latin American common market

Nordic Council Denmark, Finland, Iceland, Norway, Sweden — formed 1952 to foster legal, social and economic co-operation

North Atlantic Treaty Organization [NATO] Belgium, Britain, Canada, Denmark, France, West Germany, Greece, Iceland, Italy, Luxembourg, Netherlands, Norway, Portugal, Turkey, United States — formed 1949 as a major defensive military alliance

Organization of African Unity [OAU] Members are 42 independent African countries — formed 1963 to promote unity and co-operation in all spheres and to eliminate colonialism

Organization of American States [OAS] Argentina, Barbados, Bolivia, Brazil, Chile, Colombia, Costa Rica, Dominican Republic, Ecuador, El Salvador, Grenada, Guatemala, Haiti, Honduras, Jamaica, Mexico, Nicaragua, Panama, Paraguay, Peru, Trinidad and Tobago, United States, Uruguay, Venezuela — founded in 1948 to promote unity and understanding and to defend sovereignty of members

Organization of Petroleum-Exporting Countries [OPEC] Algeria, Ecuador, Gabon, Indonesia, Iran, Iraq, Kuwait, Libya, Nigeria, Qatar, Saudi Arabia, United Arab Emirates, Venezuela — formed in 1961 to administer a common policy for the sale of petroleum

South-East Asia Treaty Organization [SEATO] Australia, Britain, France, New Zealand, Philippines, Thailand, United States — founded in 1954 to oppose communist aggression, now also aims at economic and social co-operation

United Nations [UN] Organization of independent states formed in 1945 to promote peace, international co-operation and security. In 1977 the membership of the UN included the following nations: Afghanistan, Albania, Algeria, Angola, Argentina, Australia, Austria, Bahamas, Bahrain, Bangladesh, Barbados, Belgium, Benin, Bhutan, Bolivia, Botswana, Brazil, Bulgaria, Burma, Burundi, Byelorussia, Cambodia, Cameroon, Canada, Cape Verde Islands, Central African Empire, Chad, Chile, China, Colombia, Comoro Islands, Congo, Costa Rica, Cuba, Cyprus, Czechoslovakia, Denmark, Dominican Republic, Ecuador, Egypt, El Salvador, Equatorial Guinea, Ethiopia, Fiji, Finland, France, Gabon, Gambia, Germany, East, Germany, West, Ghana, Greece, Grenada, Guatemala, Guinea, Guinea-Bissau, Guyana, Haiti, Honduras, Hungary, Iceland, India, Indonesia, Iran, Iraq, Ireland, Republic of, Israel, Italy, Ivory Coast, Jamaica, Japan, Jordan, Kenya, Kuwait, Laos, Lebanon, Lesotho, Liberia, Libya, Luxembourg, Malagasy, Malawi, Malaysia, Maldives, Mali, Malta, Mauritania, Mauritius, Mexico, Mongolia, Morocco, Mozambique, Nepal, Netherlands, New Zealand, Nicaragua, Niger, Nigeria, Norway, Oman, Pakistan, Panama, Papua New Guinea, Paraguay, Peru, Philippines, Poland, Portugal, Qatar, Romania, Rwanda, Samoa, São Tomé and Príncipe, Saudi Arabia, Senegal, Seychelles, Sierra Leone, Singapore, Somalia, South Africa, Spain, Sri Lanka, Sudan, Surinam, Swaziland, Sweden, Syria, Tanzania, Thailand, Togo, Trinidad and Tobago, Tunisia, Turkey, Uganda, Ukraine, USSR, United Arab Emirates, United Kingdom, United States, Upper Volta, Uruguay, Venezuela, Yemen, Yemen, People's Democratic Republic of, Yugoslavia, Zaire, Zambia

Warsaw Pact [Eastern European Mutual Assistance Treaty] Bulgaria, Czechoslovakia, East Germany, Hungary, Poland, Romania, USSR — formed 1955 as a defensive military alliance [the Soviet bloc's equivalent to NATO]

Acknowledgements

The publishers wish to acknowledge the many contributions made to the text of *The Modern World*, particularly by the following: Norman Barrett BSc, Michael Darton, Keith Lye BA, FRGS, Daniel J. Sinclair MA (Edin.), Robert Stewart MA, DPhil, Christopher Tunney MA and, for authentication of pp.17-24, Dr Whitney Smith, Flag Research Center, Winchester, Mass., USA.

Illustrations are from the Fabbri Picture Library (Milan) and Mitchell Beazley Publishers Ltd, except for the following (pictures are numbered 1 to 3, left to right): Barnaby's Picture Library pp.180(1), 185(3), 186(1), 186(2); Colorpix 27(3), 122(1), 140(1), 156(3), 187(2), 187(3); J. Allen Cash 29(3), 30(3), 84(1), 84(2), 111(1); Keystone Press Agency 28(2), 30(1), 31(2), 43(2), 52(1), 52(3), 53(1), 54(3), 55(3), 56(3), 57(1), 57(2), 58(2), 59(1), 61(2), 62(1), 66(2), 66(3), 75(1), 75(2), 93(2), 93(3), 94(1), 101(3), 102(3), 103(1), 105(1), 105(2), 120(2), 121(3), 122(2), 126(2), 126(3), 127(3), 128(3), 129(3), 130(1), 130(2), 137(3), 138(1), 140(3), 150(2), 154(1), 154(3), 157(2), 157(3), 161(3), 162(1), 162(2), 162(3), 163(1), 165(2), 169(2), 181(2), 181(3), 182(1), 182(2), 182(3), 183(1), 186(3), 187(1); Spectrum Colour Library 29(2), 32(3), 58(3), 136(3), 137(1), 137(2), 168(3), 169(1).

Atlas of the world

Maps in the atlas are numbered from 2 to 80, and map references at the ends of articles in the main part of *The Modern World* refer to these numbers. For example, the reference to Map 18 at the end of the article on BELGIUM indicates that a political map of Belgium appears on page 18 of the atlas (Central Europe). Physical maps of the world's major oceans and land regions appear in *The Physical Earth,* one of the other volumes in *The Joy of Knowledge Library.*

Contents

Legend to maps in the atlas

Inhabited Localities

The symbol represents the number of inhabitants within the locality

At scales 1:4 000 000 to 1:9 000 000

- 0—10,000
- 10,000—25,000
- 25,000—100,000
- 100,000—250,000
- 250,000—1,000,000
- >1,000,000

At 1:16 000 000 scale

- 0—50,000
- 50,000—100,000
- 100,000—250,000
- 250,000—1,000,000
- >1,000,000

Urban Area (area of continuous industrial, commercial, and residential development)

The size of type indicates the relative economic and political importance of the locality

Écommoy — Lisieux — **Rouen**

Trouville — **Orléans** — **PARIS**

Jabrīn — Oasis

Capitals of Political Units

BUDAPEST — Independent Nation

Cayenne — Dependency (Colony, protectorate, etc.)

Lasa — State, Province, etc.

Alternate Names

MOSKVA
MOSCOW
Basel
Bâle

English or second official language names are shown in reduced size lettering

Volgograd
(Stalingrad)

Historical or other alternates in the local language are shown in parentheses

Political Boundaries

International (First-order political unit)

Demarcated and Undemarcated

Disputed de jure

Indefinite or Undefined

Demarcation Line (used in Korea)

Internal

State, Province, etc. (Second-order political unit)

MURCIA — Historical Region (No boundaries indicated)

PANTELLERIA (Italy) — Administering Country

Transport

Primary Road

Secondary Road

Minor Road, Trail

Railway

Canal du Midi — Navigable Canal

Bridge

Tunnel

Ferry

Miscellaneous Cultural Features

National Park or Monument

HOOVER DAM — Dam

Hydrographic Features

Shoreline

Undefined or Fluctuating Shoreline

Amur — River, Stream

Intermittent Stream

Rapids, Falls

Irrigation or Drainage Canal

Reef

The Everglades — Swamp

VATNAJÖKULL — Glacier

L. Victoria — Lake, Reservoir

Tuz Gölü — Salt Lake

Intermittent Lake, Reservoir

Dry Lake Bed

(395) — Lake Surface Elevation

Topographic Features

Mt. Kenya 5199 — Elevation Above Sea Level

76 — Elevation Below Sea Level

Mount Cook 3764 — Highest Elevation in Country

Khyber Pass 1067 — Mountain Pass

133 — Lowest Elevation in Country

Elevations are given in metres

The Highest and Lowest Elevation in a continent are underlined

Sand Area

Lava

Salt Flat

A N D E S / BODELE — Mountain Range, Plateau, Valley, etc.

KAMČATKA / CABO DE HORNOS — Peninsula, Cape, Point, etc.

BAFFIN ISLAND / ÎLE D'OUESSANT — Island

World

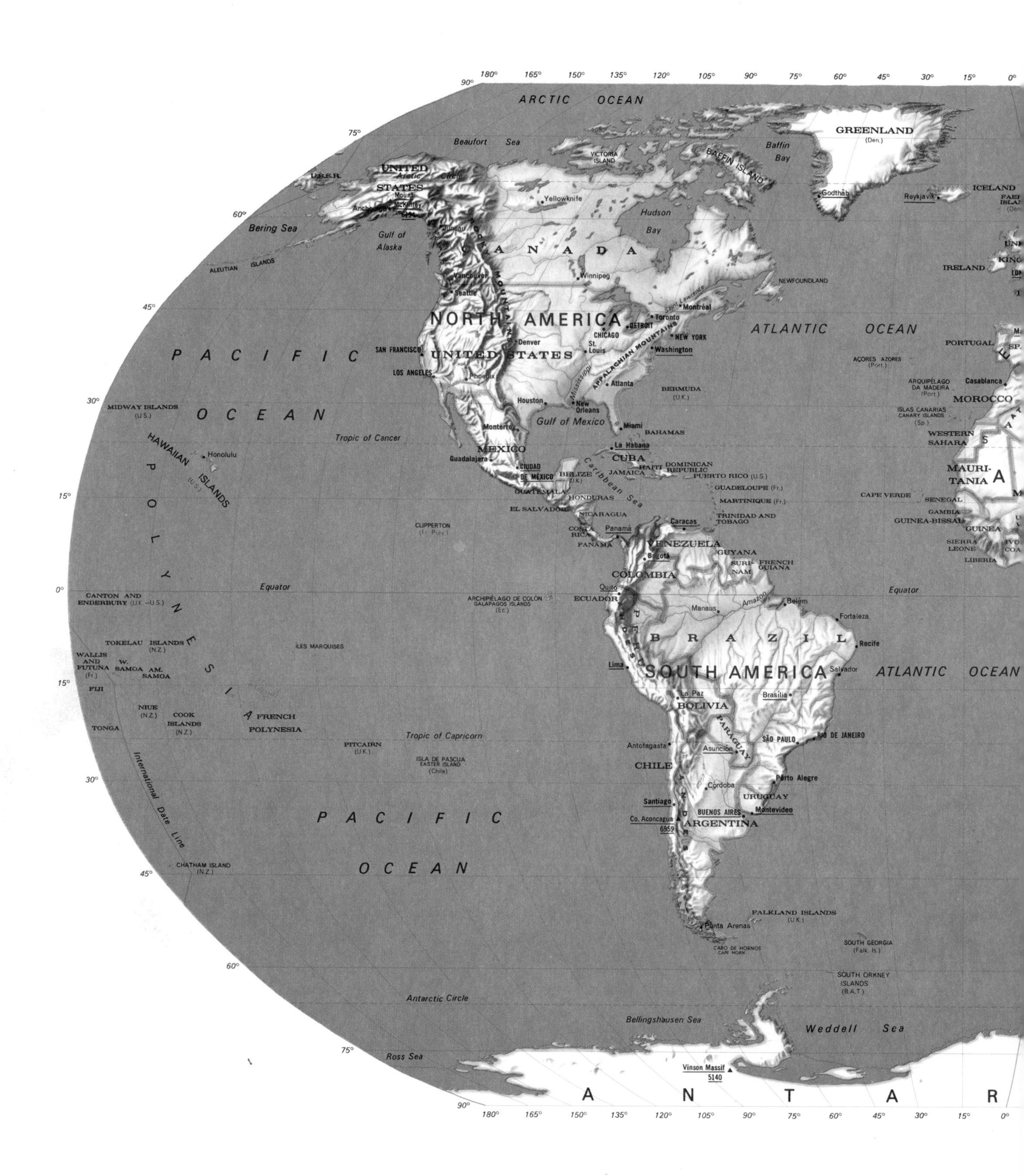

ARCTIC OCEAN
Beaufort Sea
VICTORIA ISLAND
BAFFIN ISLAND
Baffin Bay
GREENLAND
(Den.)
Godthåb
Reykjavík
ICELAND
U.S.S.R.
UNITED STATES
Arctic Circle
Mount McKinley
6194
Anchorage
Bering Sea
Gulf of Alaska
ALEUTIAN ISLANDS
Yellowknife
Hudson Bay
CANADA
Winnipeg
NEWFOUNDLAND
Vancouver
Seattle
ROCKY MOUNTAINS
Saint Lawrence
Montréal
Toronto
DETROIT
NORTH AMERICA
CHICAGO
NEW YORK
Washington
APPALACHIAN MOUNTAINS
SAN FRANCISCO
Denver
St. Louis
UNITED STATES
LOS ANGELES
Phoenix
Mississippi
Atlanta
BERMUDA
(U.K.)
ATLANTIC OCEAN
IRELAND
PORTUGAL
AÇORES AZORES
(Port.)
ARQUIPÉLAGO DA MADEIRA
(Port.)
Casablanca
MOROCCO
ISLAS CANARIAS CANARY ISLANDS
(Sp.)
WESTERN SAHARA
MAURITANIA
CAPE VERDE
SENEGAL
GAMBIA
GUINEA-BISSAU
GUINEA
SIERRA LEONE
LIBERIA
PACIFIC OCEAN
MIDWAY ISLANDS
(U.S.)
Houston
New Orleans
Gulf of Mexico
Miami
BAHAMAS
Tropic of Cancer
Monterrey
MEXICO
La Habana
CUBA
HAWAIIAN ISLANDS
(U.S.)
Honolulu
Guadalajara
CIUDAD DE MÉXICO
BELIZE
(U.K.)
Caribbean Sea
HAITI
DOMINICAN REPUBLIC
JAMAICA
PUERTO RICO (U.S.)
GUADELOUPE (Fr.)
MARTINIQUE (Fr.)
GUATEMALA
HONDURAS
EL SALVADOR
NICARAGUA
TRINIDAD AND TOBAGO
Caracas
CLIPPERTON
(Fr. Poly.)
COSTA RICA
Panamá
PANAMA
VENEZUELA
Bogotá
GUYANA
SURINAM
FRENCH GUIANA
COLOMBIA
POLYNESIA
Equator
Quito
ECUADOR
CANTON AND ENDERBURY (U.K.–U.S.)
ARCHIPIÉLAGO DE COLÓN
GALAPAGOS ISLANDS
(Ec.)
Manaus
Amazon
Belém
Fortaleza
Recife
PERU
BRAZIL
TOKELAU ISLANDS
(N.Z.)
ÎLES MARQUISES
WALLIS AND FUTUNA
(Fr.)
W. SAMOA
AM. SAMOA
FIJI
Lima
SOUTH AMERICA
Salvador
ATLANTIC OCEAN
La Paz
BOLIVIA
Brasília
NIUE
(N.Z.)
COOK ISLANDS
(N.Z.)
FRENCH POLYNESIA
TONGA
Tropic of Capricorn
PITCAIRN
(U.K.)
PARAGUAY
SÃO PAULO
RIO DE JANEIRO
Antofagasta
Asunción
ISLA DE PASCUA
EASTER ISLAND
(Chile)
International Date Line
CHILE
Córdoba
Porto Alegre
URUGUAY
Santiago
BUENOS AIRES
Montevideo
PACIFIC OCEAN
Co. Aconcagua
6959
ANDES
ARGENTINA
CHATHAM ISLAND
(N.Z.)
FALKLAND ISLANDS
(U.K.)
Punta Arenas
CABO DE HORNOS
CAPE HORN
SOUTH GEORGIA
(Falk. Is.)
SOUTH ORKNEY ISLANDS
(B.A.T.)
Antarctic Circle
Bellingshausen Sea
Weddell Sea
Ross Sea
Vinson Massif
5140
ANTAR
180° 165° 150° 135° 120° 105° 90° 75° 60° 45° 30° 15° 0°
90° 75° 60° 45° 30° 15° 0° 15° 30° 45° 60° 75° 90°

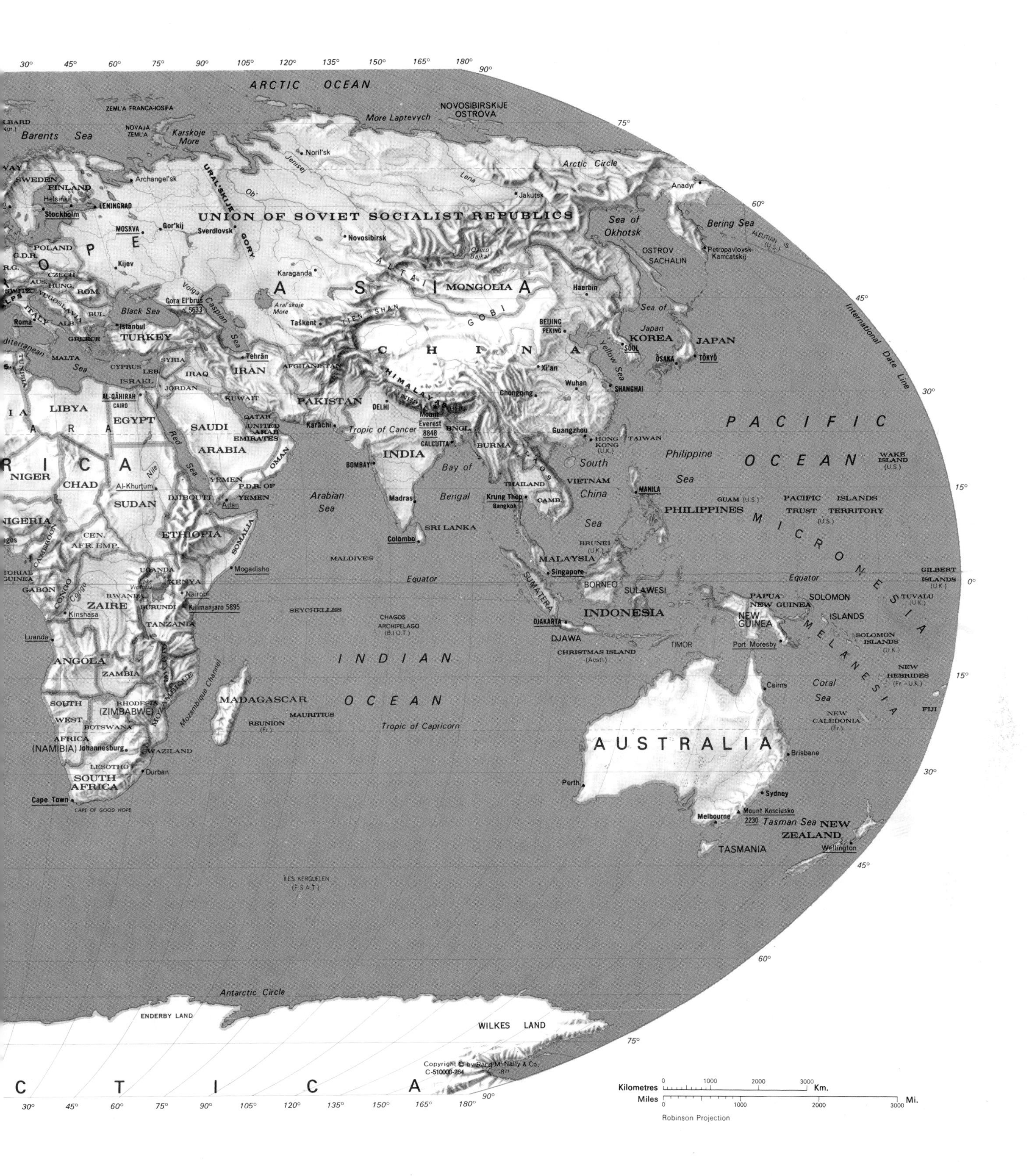

ARCTIC OCEAN
NOVOSIBIRSKIJE OSTROVA
More Laptevych
Barents Sea
NOVAJA ZEML'A
Karskoje More
Noril'sk
Jenisej
Lena
Arctic Circle
Anadyr'
Jakutsk
Archangel'sk
SWEDEN
FINLAND
Helsinki
LENINGRAD
Stockholm
UNION OF SOVIET SOCIALIST REPUBLICS
URAL'SKIJE GORY
Ob'
MOSKVA
Gor'kij
Sverdlovsk
Novosibirsk
Sea of Okhotsk
Bering Sea
OSTROV SACHALIN
Petropavlovsk-Kamčatskij
POLAND
Kijev
Karaganda
Ozero Bajkal
ALTAI
MONGOLIA
Haerbin
International Date Line
Volga
Gora El'brus 5633
Caspian Sea
Aral'skoje More
Black Sea
Istanbul
Taškent
TIEN SHAN
GOBI
BEIJING PEKING
Sea of Japan
KOREA
JAPAN
SOUL
ŌSAKA
TŌKYŌ
TURKEY
Roma
Tehrān
IRAN
AFGHANISTAN
CHINA
Xi'an
Yellow Sea
SHANGHAI
MALTA
CYPRUS
SYRIA
IRAQ
ISRAEL
JORDAN
KUWAIT
HIMALAYAS
Wuhan
Chongqing
LIBYA
AL-QĀHIRAH CAIRO
EGYPT
PAKISTAN
DELHI
Mount Everest 8848
SAUDI ARABIA
QATAR
UNITED ARAB EMIRATES
Karāchi
Tropic of Cancer
BNGL.
CALCUTTA
BURMA
Guangzhou
HONG KONG (U.K.)
TAIWAN
PACIFIC OCEAN
WAKE ISLAND (U.S.)
Philippine Sea
NIGER
CHAD
Nile
OMAN
INDIA
BOMBAY
Bay of Bengal
South China Sea
Red Sea
YEMEN
P.D.R. OF YEMEN
Aden
Al-Khurṭūm
SUDAN
DJIBOUTI
Arabian Sea
Madras
THAILAND
VIETNAM
MANILA
Krung Thep Bangkok
CAMB.
GUAM (U.S.)
PACIFIC ISLANDS TRUST TERRITORY (U.S.)
PHILIPPINES
MICRONESIA
NIGERIA
ETHIOPIA
SOMALIA
SRI LANKA
Colombo
CEN. AFR. EMP.
BRUNEI (U.K.)
MALAYSIA
MALDIVES
Mogadisho
Singapore
Equator
BORNEO
SUMATERA
SULAWESI
GABON
UGANDA
KENYA
Nairobi
ZAIRE
RWANDA
BURUNDI
Kilimanjaro 5895
SEYCHELLES
CHAGOS ARCHIPELAGO (B.I.O.T.)
DJAKARTA
INDONESIA
PAPUA NEW GUINEA
NEW GUINEA
SOLOMON ISLANDS
GILBERT ISLANDS (U.K.)
TUVALU (U.K.)
Kinshasa
TANZANIA
Luanda
DJAWA
TIMOR
Port Moresby
SOLOMON ISLANDS (U.K.)
MELANESIA
ANGOLA
ZAMBIA
INDIAN OCEAN
CHRISTMAS ISLAND (Austl.)
Cairns
Coral Sea
NEW HEBRIDES (Fr.-U.K.)
FIJI
Mozambique Channel
MADAGASCAR
RHODESIA (ZIMBABWE)
MOZAMBIQUE
SOUTH WEST AFRICA (NAMIBIA)
BOTSWANA
MAURITIUS
REUNION (Fr.)
Tropic of Capricorn
NEW CALEDONIA (Fr.)
Johannesburg
SWAZILAND
LESOTHO
Durban
SOUTH AFRICA
Cape Town
CAPE OF GOOD HOPE
AUSTRALIA
Brisbane
Perth
Sydney
Mount Kosciusko 2230
Melbourne
Tasman Sea
NEW ZEALAND
Wellington
TASMANIA
ÎLES KERGUELEN (F.S.A.T.)
Antarctic Circle
ENDERBY LAND
WILKES LAND
ANTARCTICA
Copyright © by Rand McNally & Co.
C-510000-264
Kilometres 0 1000 2000 3000 Km.
Miles 0 1000 2000 3000 Mi.
Robinson Projection

Arctic

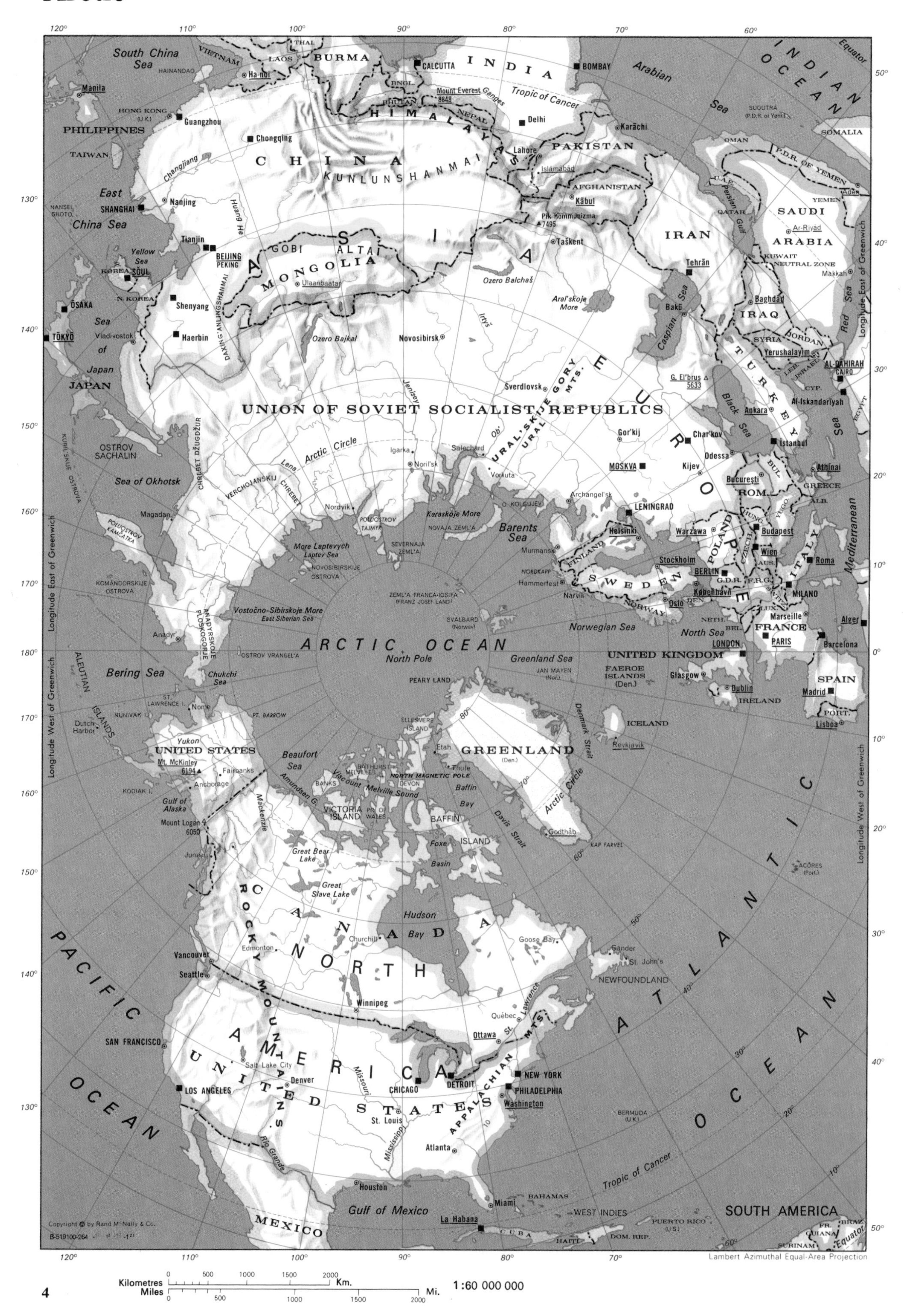

ARCTIC OCEAN
North Pole
ASIA
EUROPE
NORTH AMERICA
UNION OF SOVIET SOCIALIST REPUBLICS
CHINA
INDIA
MONGOLIA
JAPAN
PHILIPPINES
BURMA
PAKISTAN
AFGHANISTAN
IRAN
IRAQ
SAUDI ARABIA
TURKEY
SYRIA
JORDAN
ISRAEL
EGYPT
GREECE
SPAIN
FRANCE
UNITED KINGDOM
IRELAND
ICELAND
GREENLAND
CANADA
UNITED STATES
MEXICO
CUBA
SOUTH AMERICA
PACIFIC OCEAN
ATLANTIC OCEAN
INDIAN OCEAN
South China Sea
East China Sea
Yellow Sea
Sea of Japan
Sea of Okhotsk
Bering Sea
Chukchi Sea
Beaufort Sea
Arabian Sea
Caspian Sea
Black Sea
Red Sea
Mediterranean Sea
North Sea
Norwegian Sea
Greenland Sea
Barents Sea
Hudson Bay
Baffin Bay
Gulf of Mexico
Gulf of Alaska
HIMALAYA
KUNLUNSHANMAI
GOBI
ALTAI
ROCKY MOUNTAINS
APPALACHIAN MTS.
Arctic Circle
Tropic of Cancer
Equator
Mount Everest 8848
Mt. McKinley 6194
Mount Logan 6050
CALCUTTA
BOMBAY
Delhi
Karāchi
Lahore
Kābul
Tehrān
Baghdād
Ankara
İstanbul
MOSKVA
LENINGRAD
Kijev
Novosibirsk
Sverdlovsk
Taškent
BEIJING
SHANGHAI
Guangzhou
Chongqing
Tianjin
Shenyang
Haerbin
Ulaanbaatar
TŌKYŌ
ŌSAKA
SŎUL
Manila
Ha-noi
Stockholm
Helsinki
Oslo
København
BERLIN
Warszawa
Wien
Budapest
Roma
MILANO
PARIS
LONDON
Madrid
Lisboa
Barcelona
Dublin
Glasgow
Reykjavík
Godthåb
Anchorage
Fairbanks
Juneau
Vancouver
Seattle
SAN FRANCISCO
LOS ANGELES
Denver
Salt Lake City
Winnipeg
Edmonton
Churchill
CHICAGO
DETROIT
St. Louis
Atlanta
Houston
Miami
NEW YORK
PHILADELPHIA
Washington
Ottawa
Québec
Gander
St. John's
La Habana
Longitude East of Greenwich
Longitude West of Greenwich
Lambert Azimuthal Equal-Area Projection
Copyright © by Rand McNally & Co.
Kilometres 0 500 1000 1500 2000 Km.
Miles 0 500 1000 1500 2000 Mi.
1:60 000 000

Antarctic

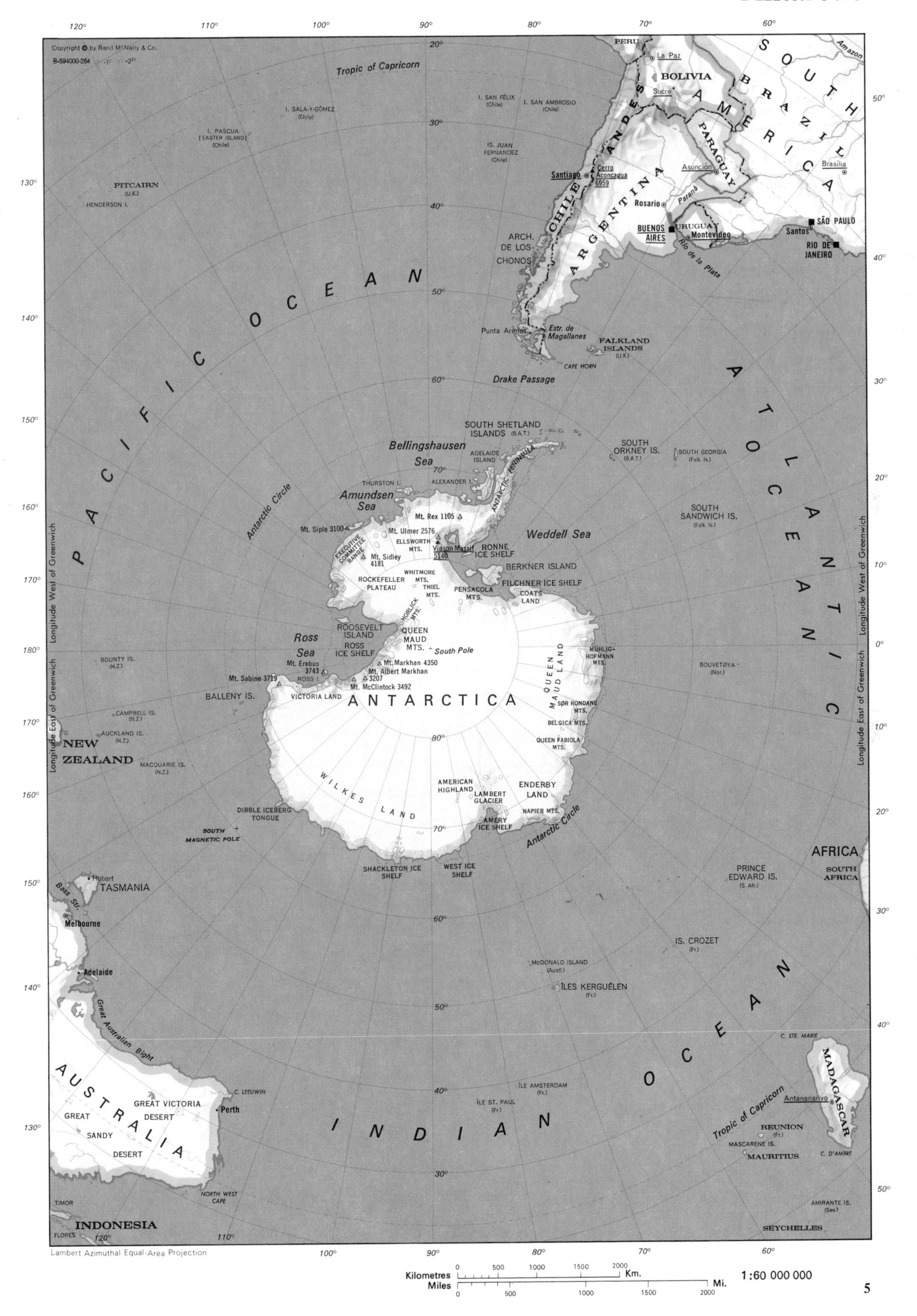

Copyright © by Rand McNally & Co.
Tropic of Capricorn
PERU
La Paz
BOLIVIA
Sucre
SOUTH AMERICA
BRAZIL
Amazon
ANDES
PARAGUAY
Asunción
Brasília
Santiago
Cerro Aconcagua 6959
CHILE
ARGENTINA
Rosario
Paraná
BUENOS AIRES
URUGUAY
Montevideo
Río de la Plata
SÃO PAULO
Santos
RIO DE JANEIRO
I. SAN FÉLIX (Chile)
I. SAN AMBROSIO (Chile)
I. SALA-Y-GÓMEZ (Chile)
I. PASCUA (EASTER ISLAND) (Chile)
IS. JUAN FERNANDEZ (Chile)
PITCAIRN (U.K.)
HENDERSON I.
ARCH. DE LOS CHONOS
PACIFIC OCEAN
Punta Arenas
Estr. de Magallanes
FALKLAND ISLANDS (U.K.)
CAPE HORN
Drake Passage
ATLANTIC OCEAN
SOUTH SHETLAND ISLANDS (B.A.T.)
Bellingshausen Sea
ADELAIDE ISLAND
ANTARCTIC PENINSULA
SOUTH ORKNEY IS. (B.A.T.)
SOUTH GEORGIA (Falk. Is.)
THURSTON I.
ALEXANDER I.
Amundsen Sea
Antarctic Circle
Mt. Rex 1105
Mt. Siple 3100
Mt. Ulmer 2576
ELLSWORTH MTS.
Vinson Massif 5140
RONNE ICE SHELF
Weddell Sea
SOUTH SANDWICH IS. (Falk. Is.)
EXECUTIVE COMMITTEE RANGE
Mt. Sidley 4181
BERKNER ISLAND
WHITMORE MTS.
ROCKEFELLER PLATEAU
THIEL MTS.
PENSACOLA MTS.
FILCHNER ICE SHELF
COATS LAND
Longitude West of Greenwich
HORLICK MTS.
ROOSEVELT ISLAND
Ross Sea
ROSS ICE SHELF
QUEEN MAUD MTS.
South Pole
QUEEN MAUD LAND
MÜHLIG-HOFMANN MTS.
BOUVETØYA (Nor.)
BOUNTY IS. (N.Z.)
Mt. Erebus 3743
ROSS I.
Mt. Markhan 4350
Mt. Albert Markhan 3207
Mt. McClintock 3492
Mt. Sabine 3719
BALLENY IS.
VICTORIA LAND
ANTARCTICA
SØR RONDANE MTS.
BELGICA MTS.
QUEEN FABIOLA MTS.
Longitude East of Greenwich
CAMPBELL IS. (N.Z.)
AUCKLAND IS. (N.Z.)
NEW ZEALAND
MACQUARIE IS. (N.Z.)
WILKES LAND
AMERICAN HIGHLAND
LAMBERT GLACIER
ENDERBY LAND
NAPIER MTS.
AMERY ICE SHELF
DIBBLE ICEBERG TONGUE
SOUTH MAGNETIC POLE
SHACKLETON ICE SHELF
WEST ICE SHELF
AFRICA
PRINCE EDWARD IS. (S. Afr.)
SOUTH AFRICA
Hobart
TASMANIA
Bass Str.
Melbourne
IS. CROZET (Fr.)
McDONALD ISLAND (Austl.)
ÎLES KERGUÉLEN (Fr.)
Adelaide
Great Australian Bight
C. STE. MARIE
MADAGASCAR
INDIAN OCEAN
C. LEEUWIN
ÎLE AMSTERDAM (Fr.)
ÎLE ST. PAUL (Fr.)
Antananarivo
AUSTRALIA
GREAT VICTORIA DESERT
Perth
GREAT SANDY DESERT
REUNION (Fr.)
MASCARENE IS.
MAURITIUS
C. D'AMBRE
NORTH WEST CAPE
TIMOR
AMIRANTE IS. (Sey.)
INDONESIA
FLORES
SEYCHELLES
Lambert Azimuthal Equal-Area Projection
Kilometres 0 500 1000 1500 2000 Km.
Miles 0 500 1000 1500 2000 Mi.
1:60 000 000

Europe

ICELAND
Reykjavík
Akureyri
Arctic Circle
NORWEGIAN SEA
NORWAY
Trondheim
Bergen
Oslo
Stavanger
SWEDEN
Göteborg
Stockholm
FAEROE ISLANDS (Den.)
Tórshavn
SHETLAND ISLANDS
Lerwick
ORKNEY ISLANDS
HEBRIDES
ATLANTIC OCEAN
NORTH SEA
DENMARK
København
Århus
Ålborg
Malmö
Baltic Sea
UNITED KINGDOM
SCOTLAND
ENGLAND
WALES
Aberdeen
Dundee
Glasgow
Edinburgh
Belfast
Newcastle
Teesside
Leeds
York
Kingston upon Hull
Manchester
Liverpool
Sheffield
Nottingham
Stoke
Birmingham
Norwich
Leicester
Coventry
Oxford
Swansea
Cardiff
Bristol
LONDON
Southampton
Portsmouth
Plymouth
IRELAND
Dublin
Cork
Limerick
Galway
Waterford
Londonderry
ISLE OF MAN
Irish Sea
English Channel
GUERNSEY
JERSEY
NETHERLANDS
Amsterdam
's-Gravenhage
The Hague
Rotterdam
Antwerpen
Bruxelles
BELGIUM
LUXEMBOURG
Lille
FEDERAL REPUBLIC OF GERMANY
GERMAN DEMOCRATIC REPUBLIC
Hamburg
Bremen
Hannover
BERLIN
Kiel
Lübeck
Rostock
Essen
Dortmund
Düsseldorf
Köln
Bonn
Frankfurt
Stuttgart
München
Nürnberg
Leipzig
Dresden
Halle
Magdeburg
Szczecin
Gdańsk
Gdynia
Bydgoszcz
Poznań
Wrocław
Katowice
Częstochowa
Praha
CZECHOSLOVAKIA
Brno
Ostrava
Bratislava
Wien
AUSTRIA
Linz
Graz
Innsbruck
Salzburg
Budapest
HUNGARY
Szeged
Pécs
FRANCE
PARIS
Le Havre
Rouen
Brest
Rennes
Le Mans
Nantes
Tours
Orléans
Reims
Metz
Nancy
Strasbourg
Mulhouse
Dijon
Lyon
Saint-Étienne
Grenoble
Limoges
Clermont-Ferrand
Bordeaux
Toulouse
Montpellier
Marseille
Nice
Toulon
Perpignan
La Rochelle
Bay of Biscay
MASSIF CENTRAL
SWITZERLAND
Genève
Lausanne
Bern
Basel
Zürich
ALPS
ITALY
Milano
Torino
Genova
La Spezia
Verona
Venezia
Padova
Trieste
Bologna
Firenze
Livorno
SAN MARINO
Ancona
Perugia
Roma
Pescara
Napoli
Salerno
Foggia
Bari
Taranto
Brindisi
MONACO
CORSE CORSICA (Fr.)
SARDEGNA SARDINIA (It.)
Cagliari
Tyrrhenian Sea
SICILIA SICILY
Palermo
Messina
Catania
Reggio di Calabria
Ionian Sea
Adriatic Sea
YUGOSLAVIA
Ljubljana
Zagreb
Rijeka
Split
Beograd
Sarajevo
Novi Sad
ALBANIA
MALTA
Valletta
SPAIN
Madrid
Barcelona
Valencia
Zaragoza
Bilbao
San Sebastián
Oviedo
Gijón
Vigo
La Coruña
Valladolid
Sevilla
Córdoba
Granada
Málaga
Murcia
Alicante
Cádiz
Gibraltar
ANDORRA
ISLAS BALEARES BALEARIC ISLANDS
Palma
MALLORCA
PORTUGAL
Lisboa
Porto
MEDITERRANEAN SEA
MOROCCO
Rabat
Casablanca
Tanger
Meknès
Fès
Oujda
ATLAS MOUNTAINS
ALGERIA
Alger
Oran
Constantine
Annaba
TUNISIA
Tunis
Sfax
Copyright © by Rand McNally & Co.
Kilometres
Miles
1:16 000 000

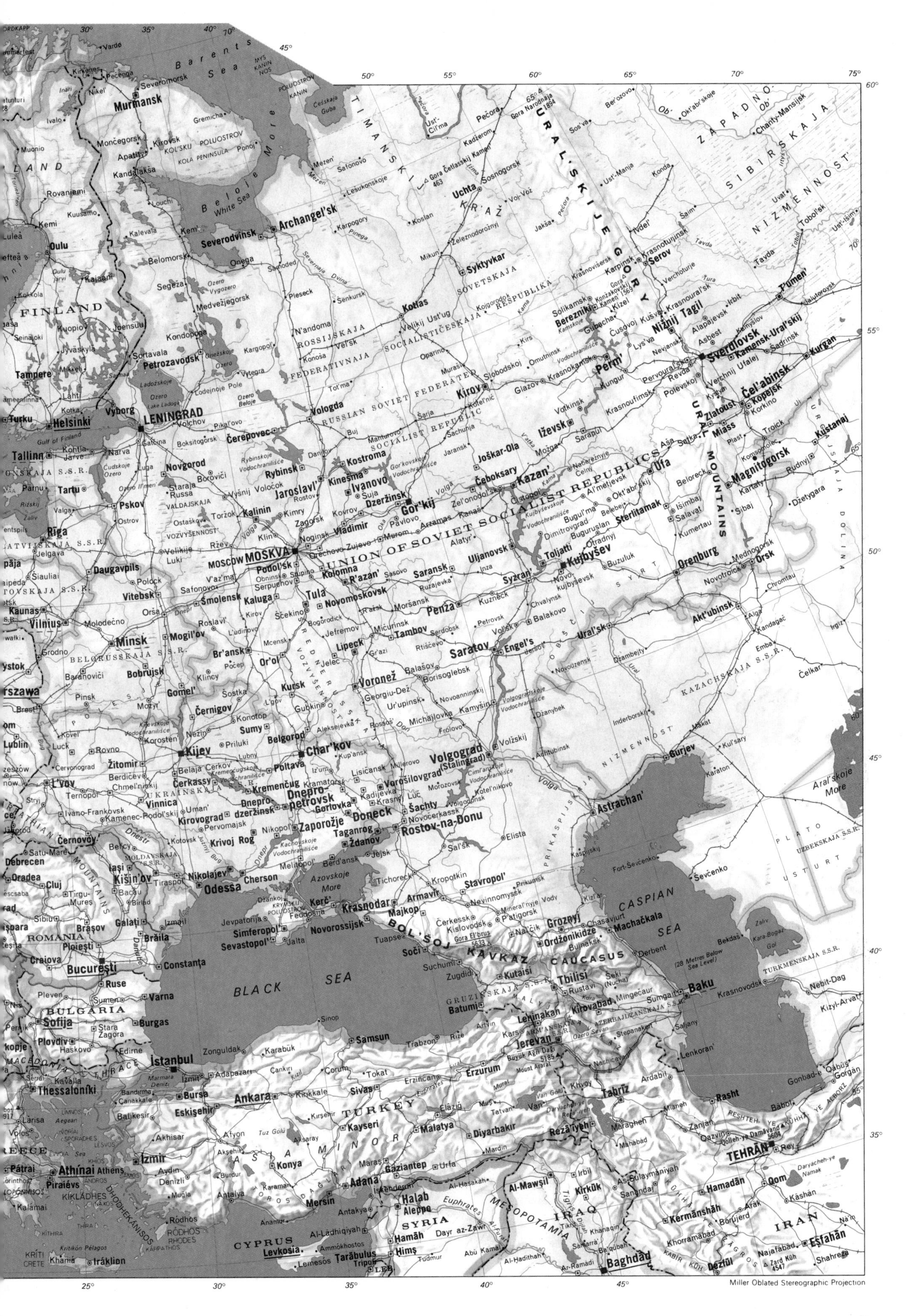

Barents Sea
Murmansk
Severomorsk
Kirovsk
KOL'SKIJ POLUOSTROV
KOLA PENINSULA
Beloje More
White Sea
Archangel'sk
Severodvinsk
Onega
Kem'
Belomorsk
Kandalakša
Mončegorsk
Apatity
Oulu
Kemi
FINLAND
Joensuu
Kuopio
Jyväskylä
Tampere
Helsinki
Turku
Vyborg
LENINGRAD
Petrozavodsk
Kondopoga
Medvežjegorsk
Segeža
Sortavala
Ladožskoje Ozero
Lake Ladoga
Onežskoje Ozero
Tallinn
Tartu
Pärnu
Riga
Narva
Novgorod
Pskov
Velikije Luki
Daugavpils
Vitebsk
Kaunas
Vilnius
Minsk
Grodno
Baranoviči
Bobrujsk
Pinsk
Brest
Gomel'
Mogil'ov
Smolensk
Orša
Moskva
Podol'sk
Kolomna
Tula
Kaluga
Br'ansk
Or'ol
Kursk
Voronež
Lipeck
Tambov
R'azan'
Novomoskovsk
Jaroslavl'
Rybinsk
Kostroma
Ivanovo
Vladimir
Gor'kij
Kalinin
Vologda
Čerepovec
Kotlas
Syktyvkar
Uchta
Pečora
Kirov
Perm'
Berezniki
Solikamsk
Iževsk
Kazan'
Čeboksary
Joškar-Ola
Uljanovsk
Toljatti
Kujbyšev
Syzran'
Penza
Saransk
Saratov
Engel's
Volgograd
(Stalingrad)
Ufa
Sterlitamak
Orenburg
Orsk
Magnitogorsk
Čel'abinsk
Sverdlovsk
Nižnij Tagil
Serov
Kurgan
Kustanaj
Tjumen'
Zlatoust
Miass
Ural'sk
Akt'ubinsk
Gur'jev
Astrachan'
Elista
Rostov-na-Donu
Taganrog
Ždanov
Doneck
Vorošilovgrad
Char'kov
Poltava
Kremenčug
Dnepropetrovsk
Zaporožje
Krivoj Rog
Kirovograd
Kijev
Černigov
Žitomir
L'vov
Vinnica
Černovcy
Kišin'ov
Odessa
Nikolajev
Cherson
Simferopol'
Sevastopol'
Kerč'
Krasnodar
Novorossijsk
Stavropol'
Groznyj
Machačkala
Ordžonikidze
Soči
Kutaisi
Tbilisi
Batumi
Jerevan
Kirovabad
Baku
Krasnovodsk
Azovskoje More
BLACK SEA
CASPIAN SEA
Aral'skoje More
BOL'ŠOJ KAVKAZ
CAUCASUS
Gora El'brus 5633
URAL'SKIJE GORY
URAL MOUNTAINS
TIMANSKIJ KR'AŽ
ZAPADNO-SIBIRSKAJA NIZMENNOST'
RUSSIAN SOVIET FEDERATED SOCIALIST REPUBLIC
ROSSIJSKAJA SOVETSKAJA FEDERATIVNAJA SOCIALISTIČESKAJA RESPUBLIKA
UNION OF SOVIET SOCIALIST REPUBLICS
UKRAINSKAJA S.S.R.
BELORUSSKAJA S.S.R.
KAZACHSKAJA S.S.R.
UZBEKSKAJA S.S.R.
TURKMENSKAJA S.S.R.
ROMANIA
Bucureşti
Constanţa
Braşov
Cluj
Galaţi
Brăila
Ploieşti
Craiova
BULGARIA
Sofija
Varna
Burgas
Ruse
Plovdiv
İstanbul
Edirne
Thessaloníki
Athínai Athens
Piraiévs
KRÍTI CRETE
Iráklion
TURKEY
ASIA MINOR
Ankara
Bursa
İzmir
Eskişehir
Konya
Kayseri
Sivas
Samsun
Trabzon
Erzurum
Malatya
Diyarbakir
Adana
Mersin
Gaziantep
Büyük Ağri Daği 5185
Mount Ararat
CYPRUS
Levkosia
SYRIA
Halab
Aleppo
Hamāh
Ḥimş
IRAQ
MESOPOTAMIA
Al-Mawşil
Kirkūk
Baghdād
IRAN
TEHRĀN
Tabrīz
Rasht
Qom
Hamadān
Eşfahān
Kermānshāh
Rezā'iyeh
Damāvand 5604
Euphrates
Tigris
Miller Oblated Stereographic Projection

British Isles

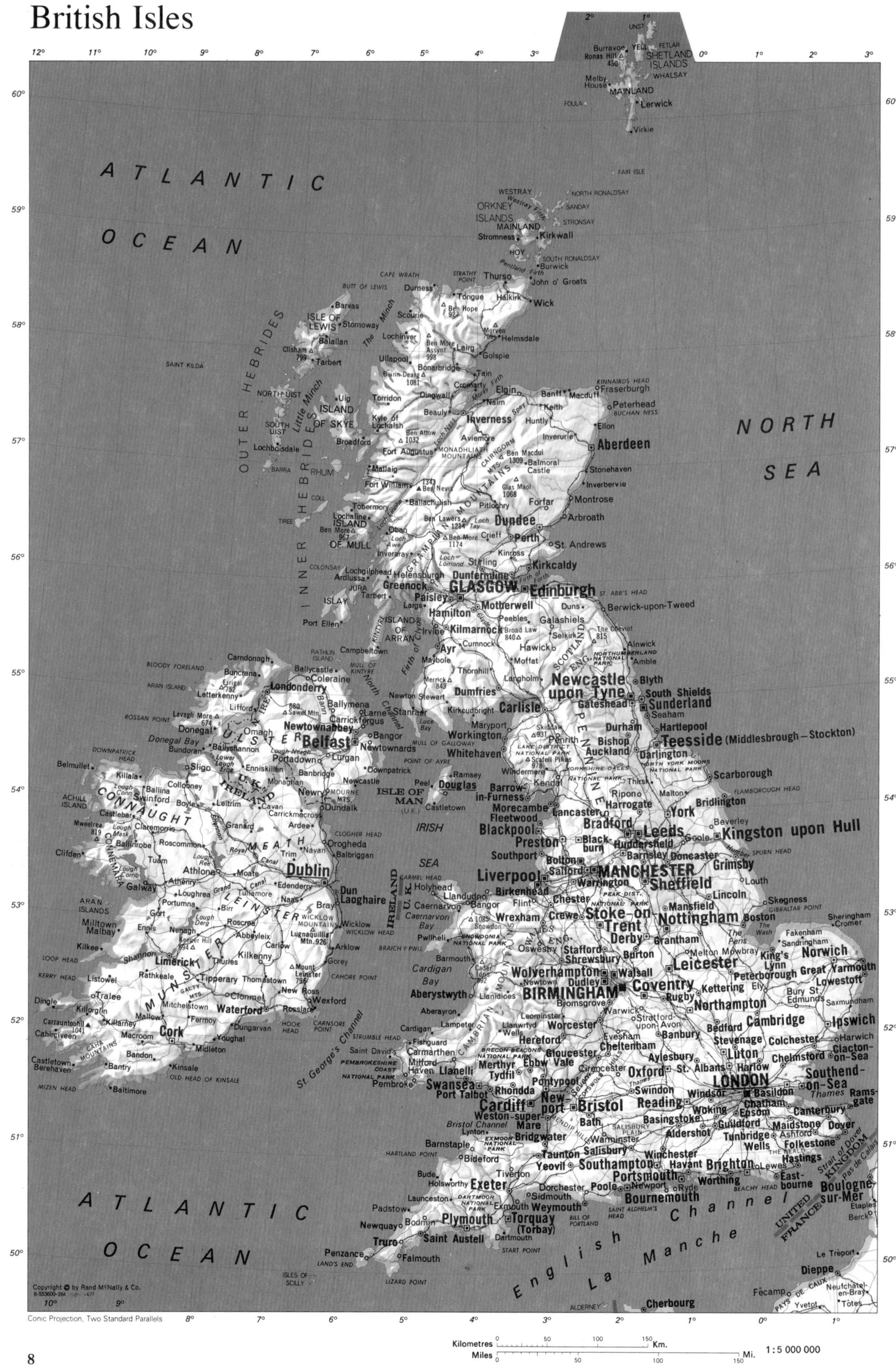

Ireland

ATLANTIC
OCEAN
North Channel
IRISH SEA
St. George's Channel
SCOTLAND
KINTYRE
Campbeltown
Southend
RATHLIN ISLAND
MULL OF KINTYRE
MALIN HEAD
Ballygorman
INISHTRAHULL
INISHOWEN
Culdaff
Carndonagh
Moville
Greencastle
Portrush
Bushmills
Ballycastle
Portstewart
Coleraine
Lough Foyle
FANAD HEAD
TORY ISLAND
HORN HEAD
Tory Sound
BLOODY FORELAND
Gortahork
Carrickart
Rathmullen
Ramelton
Gweedore
Errigal 752
Glendowan
Dungloe
ARAN ISLAND
GOLA
Rosses Bay
Gweebarra Bay
Loughros More Bay
Fintown
Maas
Glenties
Ardara
Glencolumbkille
Malin Beg
League 601
Killybegs
Dunkineely
Donegal
Donegal Bay
Ballintra
Ballyshannon
Bundoran
Belleek
Pettigo
Lower Lough Erne
Killeter
Drumquin
Castlederg
Newtown-stewart
Omagh
Stranorlar
Letterkenny
Raphoe
Lifford
Strabane
Plumbridge
Cloghan
Londonderry
Limavady
Dungiven
Sawel Mt. 680
Mullaghmore 554
Draperstown
Maghera
Magherafelt
Ballymoney
Finvoy
Dunloy
Ballymena
Cushendall
Trostan 551
GARRON POINT
Glenarm
Carncastle
Larne
ISLAND MAGEE
Whitehead
Agnews Hill 476
Antrim
Carrickfergus
Greenisland
Belfast Lough
North Down (Bangor)
Donaghadee
Newtownabbey
Belfast
Newtownards
Castlereagh
Lough Neagh
Cookstown
Stewartstown
Dungannon
Coalisland
Lisburn
Lurgan
Portadown
Hillsborough
Comber
Ballywalter
Strangford Lough
Portaferry
Strangford
BALLYQUINTIN POINT
Killyleagh
Crossgar
Downpatrick
Ballynahinch
Dromore
Banbridge
Gilford
Armagh
Moy
ULSTER
Sixmilecross
Dromore
Fintona
Trillick
Kesh
Augher
Fivemiletown
Slieve Beagh 383
Emyvale
Enniskillen
Lisnaskea
Monaghan
MONAGHAN
Clones
Castleblayney
Slieve Gullion 575
Newry
Warrenpoint
Kilkeel
MOURNE MTS.
Slieve Donard 850
Dundrum Bay
Newcastle
Castlewellan
Rathfriland
Poyntzpass
Middletown
Keady
Crossmaglen
Dundalk
Dundalk Bay
Castlebellingham
Dunleer
LOUTH
Drogheda
CLOGHER HEAD
Balbriggan
Ardee
Carrickmacross
Kingscourt
Shercock
Cootehill
Ballybay
Newbliss
Belturbet
Swanlinbar
Cavan
CAVAN
Ballieborough
Virginia
Ballyjamesduff
Butlers Bridge
Killashandra
Arvagh
Stradone
Lough Oughter
Ballyconnell
Lough Allen
Drumshanbo
Manorhamilton
Belcoo
Dromahair
Drumcliffe
Sligo
Strandhill
Sligo Bay
LEITRIM
DONEGAL
Lough Melvin
Lough Gill
Ballysadare
Collooney
Ballymote
SLIGO
Skreen
Easky
Dromore West
Inniscrone
Ballina
Killala
Killala Bay
Crossmolina
Lough Conn
Nephin 806
NEPHIN BEG RANGE
MAYO
BENWEE HEAD
DOWNPATRICK HEAD
Portacloy
Broad Haven
ERRIS HEAD
Glenamoy
Ballycastle
MULLET PENINSULA
Belmullet
Carrowmore Lake
Bangor Erris
Aghleam
Blacksod Bay
Gweesalia
Ballycroy
ACHILL HEAD
Dooagh
ACHILL ISLAND
Mallaranny
Newport
Castlebar
Clew Bay
CLARE ISLAND
Westport
Louisburgh
INISHTURK
Cregganbaun
INISHBOFIN
Mweelrea 819
Benwee 682
Leenaun
INISHSHARK
Letterfrack
Cleggan
The Twelve Pins 730
Clifden
Maam Cross
Ballyconneely
SLYNE HEAD
Roundstone
CONNEMARA
Oughterard
Kilkieran
Costelloe
Lettermullen
Kilkieran Bay
GORUMNA ISLAND
Galway Bay
ARAN ISLANDS
INISHMORE
INISHMAAN
INISHEER
BLACK HEAD
Moycullen
Galway
Oranmore
Athenry
Attymon
GALWAY
Lough Corrib
Headford
Shrule
Cong
Ballinrobe
Partree
Killavally
Claremorris
Balla
Kiltimagh
Swinford
Foxford
Charlestown
Tobercurry
Ballaghaderreen
Frenchpark
Boyle
Carrick-on-Shannon
Mohill
Dromod
Elphin
Strokestown
ROSCOMMON
Castlerea
Tulsk
Ballyhaunis
Ballindine
Ballymoe
Roscommon
CONNAUGHT
Kilmaine
Dunmore
Glenamaddy
Tuam
Ballygar
Caltra
Kiltoom
Athlone
Ballinasloe
Moate
Clara
Tullamore
Glasson
Lough Ree
Ballymahon
Ballinalack
Longford
LONGFORD
Drumlish
Granard
Castlepollard
Delvin
Mullingar
WESTMEATH
Kinnegad
Rochfort Bridge
Athboy
Trim
Navan
MEATH
Kells
Ceanannus Mór
Oldcastle
Ashbourne
Summerhill
Clonee
Swords
Lusk
Malahide
LAMBAY ISLAND
Baldoyle
Howth
Maynooth
DUBLIN
BAILE ÁTHA CLIATH
Dublin Bay
Clondalkin
Tallaght
Dun Laoghaire
DUBLIN
Bray
Greystones
Naas
Brittas
Droichead Nua
Kildare
Monasterevin
Rathangan
Robertstown
Edenderry
Daingean
Portarlington
Mountmellick
Portlaoighise
LAOIGHIS
Mountrath
Arderin 529
Stradbally
Athy
Ballitinglass
Dunlavin
KILDARE
LEINSTER
WICKLOW
Wicklow
WICKLOW HEAD
Lugnaquillia 926
Rathdrum
Aughrim
WICKLOW MOUNTAINS
MIZEN HEAD
Arklow
Tinahely
Shillelagh
KILMICHAEL POINT
Gorey
CARLOW
Carlow
Tullow
Muine Bheag
Bunclody
Mt. Leinster 796
Borris
Ferns
Ballycanew
CAHORE POINT
Enniscorthy
Blackwater
WEXFORD
Wexford
Wexford Harbour
New Ross
Taghmon
Rosslare
Rosslare Harbour
GREENORE POINT
CARNSORE POINT
SALTEE ISLANDS
HOOK HEAD
Bannow Bay
Waterford Harbour
Passage East
Waterford
Tramore
Kilmacthomas
WATERFORD
Dungarvan
Ringville
Cappoquin
Lismore
Tallow
Clashmore
Youghal
KNOCKMEALDOWN MOUNTAINS
Portlaw
Carrick-on-Suir
Clonmel
Newcastle
Clogheen
Slievenamon 722
Kilkenny
KILKENNY
Thomastown
Mullinavat
Callan
Fethard
Cashel
Tipperary
Golden
Dundrum
Thurles
Urlingford
Freshford
Durrow
Abbeyleix
Rathdowney
Templemore
Borrisoleigh
Roscrea
Borrisokane
Birr
Banagher
Ferbane
Cloghan
OFFALY
Portumna
Eyrecourt
Woodford
Lough Derg
Nenagh
Toomevara
Silvermines
Keeper Hill 695
Newport
Killaloe
Ballina
TIPPERARY
Cappamore
Limerick
SHANNON AIRPORT
Ballyneety
Croom
Rathkeale
LIMERICK
Newcastle West
Abbeyfeale
Knocklong
Kilmallock
Galtymore 920
GALTY MTS.
Kilfinane
Mitchelstown
Caher
New Inn
MUNSTER
Rath Luirc
Blackrock 519
Dromcolliher
Freemount
Buttevant
Kildorrery
Fermoy
Mallow
Rathcormack
Watergrasshill
Midleton
Cork
Cobh
Carrigaline
Cloyne
Passage West
Blarney
Coachford
Macroom
Ballyneety
CORK
Kinsale
Cork Harbour
OLD HEAD OF KINSALE
Courtmacsherry Bay
Clonakilty
Clonakilty Bay
Rosscarbery
GALLEY HEAD
Skibbereen
Baltimore
CLEAR ISLAND
CAPE CLEAR
FASTNET ROCK
Roaringwater Bay
MIZEN HEAD
Schull
Dunmanus Bay
Durrus
Bantry
Bantry Bay
Dunmanway
Ballineen
Drimoleague
Bandon
Crookstown
Ballingeary
Glengarriff
Adrigole
CAHA MOUNTAINS
Castletown Bere (Castletown Bearhaven)
BEAR ISLAND
DURSEY HEAD
Ardgroom
Kenmare River
Kenmare
Caherdaniel
Ballinskelligs Bay
Waterville
SKELLIG ROCKS
BOLUS HEAD
BRAY HEAD
VALENCIA ISLAND
Cahirciveen
Glenbeigh
Dingle Bay
SLEA HEAD
GREAT BLASKET ISLAND
Dingle
DINGLE PENINSULA
Brandon Mt. 953
BRANDON HEAD
Brandon Bay
Cloghane
Tralee Bay
Tralee
Castleisland
Castlemaine
Inch
Killorglin
Lough Leane
Killarney
Lakes of Killarney
Carrantuohill 1041
MACGILLYCUDDY'S REEKS
Mangerton Mt. 840
KERRY
The Paps 696
Rathmore
Millstreet
Ballymakeery (Ballyvourney)
Newmarket
Banteer
Abbeydorney
KERRY HEAD
Listowel
Ballybunion
Mouth of the Shannon
LOOP HEAD
Kilbaha
Kilrush
Kilkee
Tarbert
Foynes
R. Shannon
Kildysart
Newmarket-on-Fergus
Quilty
Miltown Malbay
Liscannor Bay
HAGS HEAD
Ennistymon
Corofin
Ennis
Tulla
Scarriff
CLARE
Crusheen
Gort
Lisdoonvarna
Ballyvaughan
Kinvara
Kilchreest
Loughrea
Craughwell
Derrybrien
Lambert Conformal Conic Projection
Copyright © by Rand McNally & Co.
B-551700-264
Kilometres
Miles
1:1 900 000

Scotland

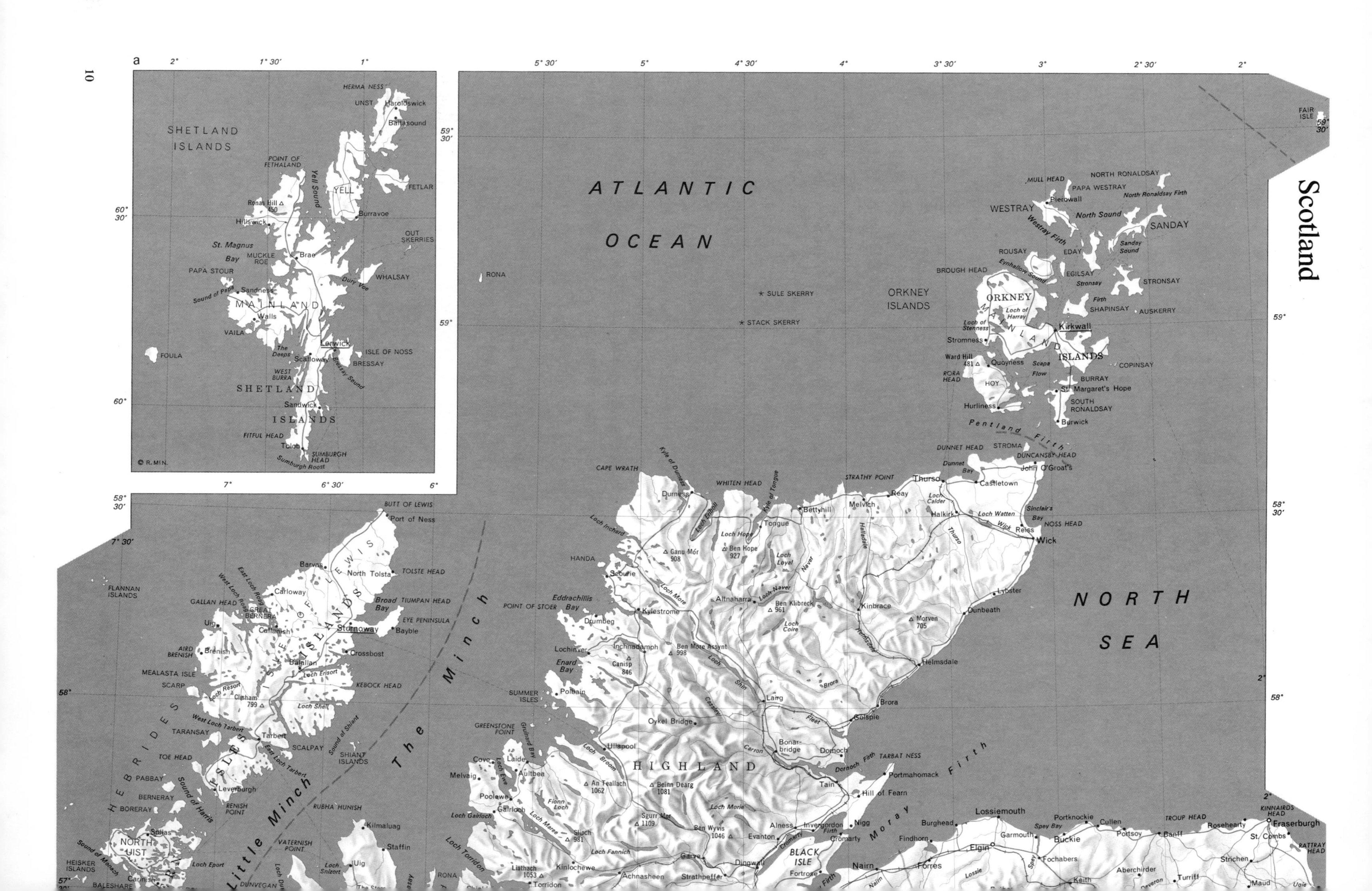

ATLANTIC
OCEAN
NORTH
SEA
ORKNEY ISLANDS
SHETLAND ISLANDS
HIGHLAND
The Minch
Little Minch
Pentland Firth
Moray Firth
WESTRAY
SANDAY
STRONSAY
ROUSAY
SHAPINSAY
HOY
Kirkwall
Stromness
Thurso
Wick
Dunbeath
Helmsdale
Brora
Golspie
Dornoch
Tain
Lairg
Durness
Ullapool
Kinlochewe
Gairloch
Stornoway
Tarbert
Leverburgh
Lerwick
Scalloway
Fraserburgh
Elgin
Forres
Nairn
Lossiemouth
Buckie
Banff
CAPE WRATH
DUNNET HEAD
SULE SKERRY
STACK SKERRY
RONA
FOULA
FAIR ISLE
UNST
YELL
FETLAR
WHALSAY
BRESSAY
SUMBURGH HEAD
BUTT OF LEWIS
FLANNAN ISLANDS
NORTH UIST
© R.MN.

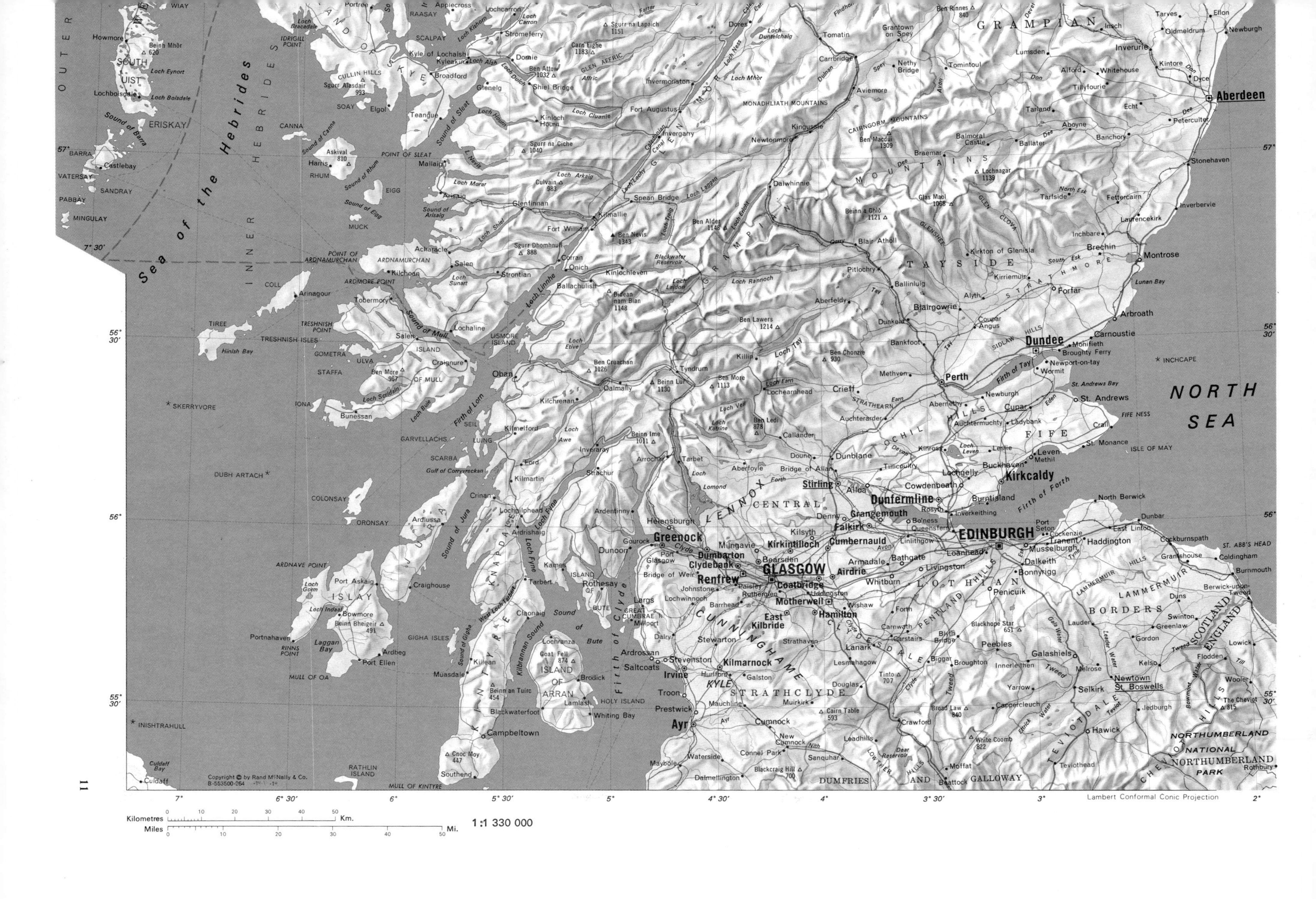

NORTH SEA
Sea of the Hebrides
OUTER HEBRIDES
INNER HEBRIDES
GRAMPIAN MOUNTAINS
MONADHLIATH MOUNTAINS
CAIRNGORM MOUNTAINS
TAYSIDE
CENTRAL
LOTHIAN
FIFE
BORDERS
STRATHCLYDE
DUMFRIES AND GALLOWAY
NORTHUMBERLAND
NORTHUMBERLAND NATIONAL PARK
SCOTLAND
ENGLAND
Aberdeen
Dundee
Perth
Stirling
EDINBURGH
GLASGOW
Kirkcaldy
Dunfermline
Greenock
Paisley
Ayr
Oban
Fort William
Montrose
Arbroath
St. Andrews
Firth of Forth
Firth of Tay
Firth of Clyde
Firth of Lorn
Sound of Mull
Sound of Jura
Loch Fyne
Loch Lomond
Loch Tay
Loch Ness
ISLAY
JURA
OF MULL
ISLAND OF ARRAN
KINTYRE
TIREE
COLL
SKYE
RHUM
EIGG
CANNA
SOUTH UIST
BARRA
Ben Nevis 1343
Ben Lawers 1214
Ben More 1113
Ben Macdui 1309
Copyright © by Rand McNally & Co.
Lambert Conformal Conic Projection
Kilometres
Km.
Miles
Mi.
1:1 330 000

Northern England and Wales

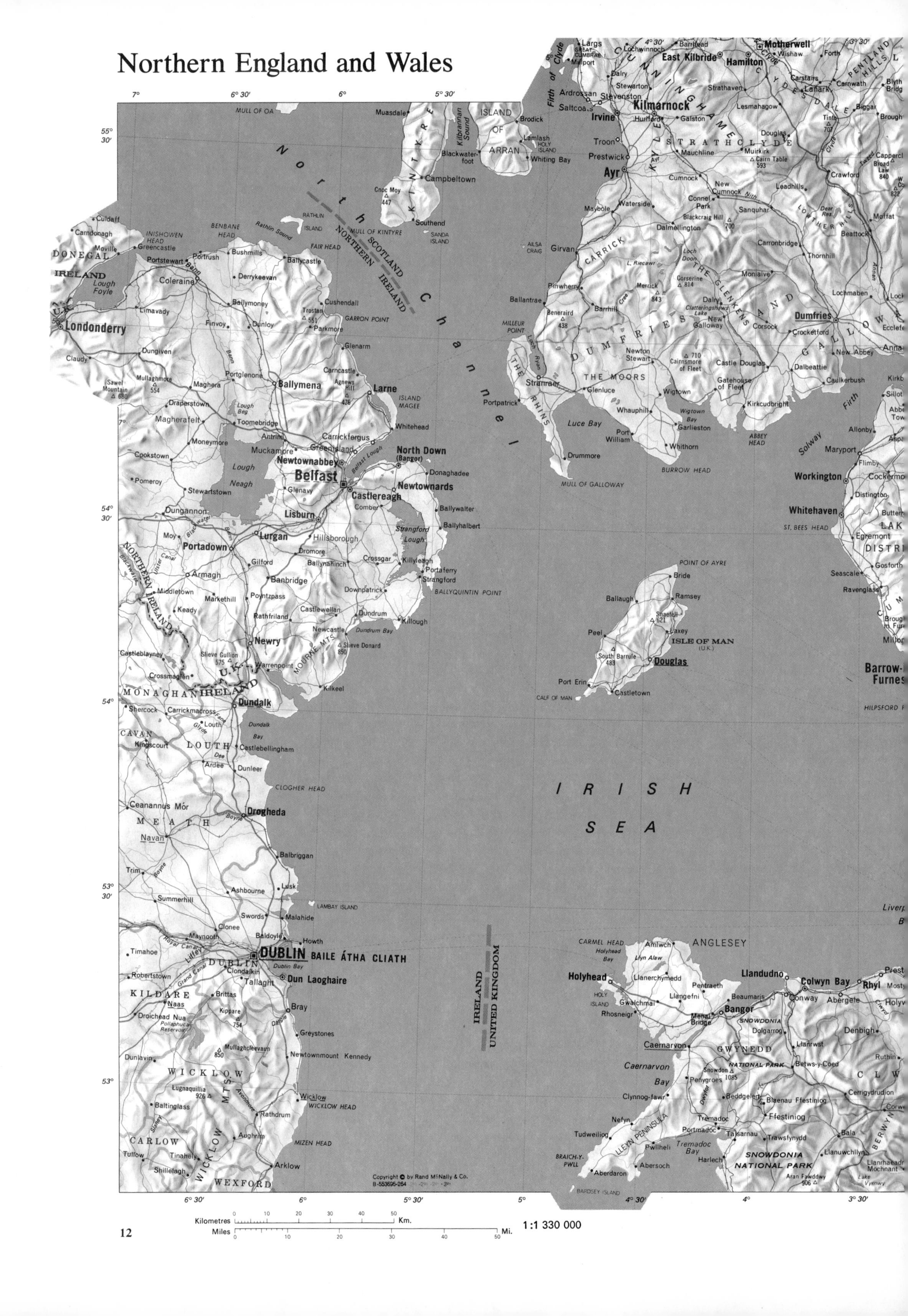

NORTH SEA
LAMMERMUIR
Duns
Berwick-upon-Tweed
Swinton
Lauder
Greenlaw
Gordon
Coldstream
HOLY ISLAND
FARNE ISLANDS
Galashiels
Melrose
Kelso
Flodden
Lowick
Belford
Seahouses
North Sunderland
Newtown St. Boswells
Selkirk
BORDERS
Wooler
Embleton
Jedburgh
The Cheviot
815
Longhoughton
Hawick
TEVIOTDALE
CHEVIOT HILLS
Whittingham
Alnwick
Alnmouth
Teviothead
Amble
Rothbury
Longframlington
Rochester
NORTHUMBERLAND NATIONAL PARK
NORTHUMBERLAND
Otterburn
Longhorsley
Ellington
Roan Fell
568
Newbiggin-by-the-Sea
SCOTLAND
ENGLAND
Newcastleton
Falstone
Cambo
Morpeth
Bellingham
Wansbeck
Blyth
Cramlington
Seaton Delaval
Roadhead
Colwell
Ponteland
Whitley Bay
Longtown
HADRIAN'S WALL
NEWCASTLE UPON TYNE
Gosforth
Tynemouth
South Shields
Gretna Green
Haltwhistle
Hexham
Haydon Bridge
Hebburn
Gateshead
TYNE AND WEAR
Brampton
Prudhoe
Sunderland
Harper Town
Allendale Town
Derwent Res.
Stanley
Ryhope
Carlisle
Cumwhinton
Consett
Chester-le-Street
Seaham
Thursby
Alston
Croglin
Hetton-le-Hole
Peterlee
Horden
High Hesket
Durham
Stanhope
Wearhead
DURHAM
Melmerby
Hartlepool
Cross Fell
893
Spennymoor
Ferryhill
Penrith
Bishop Auckland
Cow Green Reservoir
TEESDALE
Shildon
Redcar
Greystoke
Temple Sowerby
Mickle Fell
790
Middlesbrough
Newton Aycliffe
CLEVELAND
Loftus
Appleby
Barnard Castle
Stockton-on-Tees
Ormesby
Guisborough
Ullswater
Darlington
Whitby
Patterdale
Haweswater Res.
Brough
STAINMORE FOREST
Stokesley
Castleton
Egton
Robin Hood's Bay
LAKE DISTRICT NATIONAL PARK
Tebay
Kirkby Stephen
Richmond
Ambleside
High Seat
710
NORTH YORK MOORS NATIONAL PARK
NORTH YORK MOORS
Catterick
Windermere
Swale
Northallerton
Kendal
Sedbergh
Hawes
Leyburn
NORTH YORKSHIRE
Scalby
Garsdale Head
Aysgarth
Bedale
HAMBLETON HILLS
Scarborough
Newby Bridge
Whernside
736
YORKSHIRE DALES NATIONAL PARK
Helmsley
Pickering
Kirkby Lonsdale
Thirsk
Filey
Masham
Ampleforth
Grange-over-Sands
Kettlewell
Ripon
Thormanby
Malton
Norton
Foxholes
Burton Fleming
FLAMBOROUGH HEAD
Carnforth
Ingleton
Horton in Ribblesdale
Easingwold
Morecambe
Pateley Bridge
Boroughbridge
Bridlington
Bridlington Bay
Lancaster
Giggleswick
Grassington
Fridaythorpe
Driffield
Heysham
Knaresborough
Galgate
Wigglesworth
Hellifield
Nidd
Huntington
Stamford Bridge
Harrogate
Skipton
Middleton-on-the-Wolds
Hornsea
LANCASHIRE
York
Barnoldswick
Ilkley
Wharfe
Wetherby
HUMBERSIDE
Bowgreave
WEST YORKSHIRE
Tadcaster
Aldbrough
Clitheroe
Guiseley
Market Weighton
Beverley
Colne
Keighley
Riccall
Derwent
Cottingham (Haltemprice)
Blackpool
Longridge
Nelson
Pudsey
Selby
Kingston upon Hull (Hull)
Preston
Padiham
Burnley
Bradford
LEEDS
Ouse
Howden
Hedon
Withernsea
Accrington
Sowerby Bridge
Rothwell
Hessle
Lytham St Anne's
Blackburn
Todmorden
Halifax
Batley
Castleford
Goole
Brough
Patrington
Tarleton
Leyland
Rawtenstall
Dewsbury
Pontefract
New Holland
Barton-upon-Humber
Brighouse
Wakefield
Humber
Southport
Chorley
Ramsbottom
Rochdale
Huddersfield
Hemsworth
Thorne
Immingham Dock
Grimsby
SPURN HEAD
Ormskirk
Horwich
Bolton
Kirkburton
South Kirkby
Scunthorpe
Bury
Royton
Holmfirth
Adwick le Street
Cleethorpes
Wigan
Oldham
Barnsley
Brigg
Atherton
Thurnscoe
Doncaster
Caistor
Maghull
MANCHESTER
Ashton-under-Lyne
Penistone
Mexborough
Waddingham
North Somercotes
St. Helens
Salford
Stocksbridge
Swinton
Conisbrough
Bootle
Newton-le-Willows
Stretford
Glossop
Rotherham
Bawtry
Misterton
LINCOLNSHIRE WOLDS
Roby
Stockport
Ecclesfield
SOUTH YORKSHIRE
Gainsborough
Market Rasen
Louth
LIVERPOOL
Warrington
Hazel Grove
New Mills
Sheffield
Dinnington
Mablethorpe
MERSEYSIDE
Widnes
Wilmslow
Ladybower Res.
Worksop
LINCOLNSHIRE
Whaley Bridge
Dronfield
East Retford
Wragby
Mersey
Frodsham
Knutsford
Bollington
PEAK DISTRICT NATIONAL PARK
Alford
Helsby
Ellesmere Port
Northwich
Weaverham
Macclesfield
Buxton
Chesterfield
East Markham
Lincoln
NOTTINGHAMSHIRE
DERBYSHIRE
Bolsover
Horncastle
Skegness
CHESHIRE
Warsop
Sutton on Trent
Chester
Winsford
Congleton
Mansfield Woodhouse
Mansfield
Clay Cross
Matlock
Sutton-in-Ashfield
Witham
Wainfleet All Saints
GIBRALTAR POINT
Buckley
Tarporley
Sandbach
Biddulph
Coningsby
Crewe
Alsager
Kidsgrove
Leek
Newark-upon-Trent
Alfreton
Kirkby in Ashfield
Wirksworth
SHERWOOD FOREST
Southwell
Leadenham
STAFFORDSHIRE
Ripley
Hucknall
Arnold
Balderton
Holland Fen
Wrexham
Nantwich
Burslem
Belper
Butterwick
Newcastle-under-Lyme
Stoke-on-Trent
Ashbourne
Heanor
Eastwood
Carlton
Ancaster
Sleaford
Boston
Ruabon
Whitchurch
Upper Tean
Ilkeston
Nottingham
Bottesford
Grantham
The Wash
Derby
Beeston
West Bridgford
Donington
Market Drayton
Stone
Long Eaton
Folkingham
Holbeach Marsh
Chirk
Ellesmere
Uttoxeter
Chellaston
Waltham on the Wolds
Holbeach
Whittington
Wem
Hodnet
Eccleshall
Trent
Welland
Nene
Burton-upon-Trent
Melbourne
Kegworth
Colsterworth
Spalding
Sutton Bridge
King's Lynn
Oswestry
Ruyton-Eleven-Towns
Stafford
Swadlincote
Loughborough
Melton Mowbray
Bourne
Shawbury
Newport
Rugeley
Barton-under-Needwood
Shepshed
Stretton
Walpole Saint Peter
3°
2° 30′
2°
1° 30′
1°
0° 30′
0°
55° 30′
55°
54° 30′
54°
53° 30′
53°
Lambert Conformal Conic Projection

Southern England and Wales

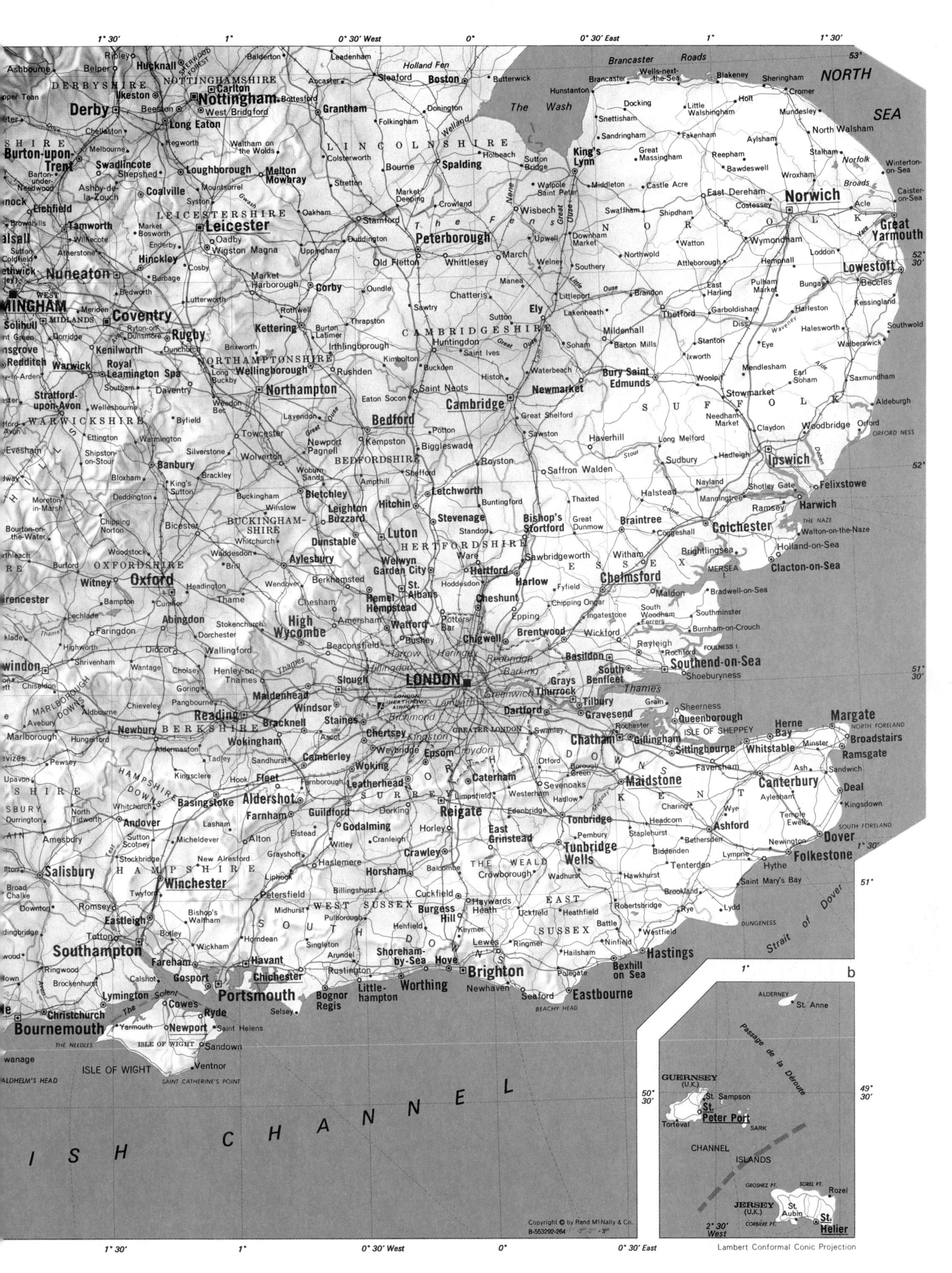
1° 30'
1°
0° 30' West
0°
0° 30' East
1°
1° 30'
NORTH SEA
The Wash
Brancaster Roads
DERBYSHIRE
NOTTINGHAMSHIRE
LINCOLNSHIRE
LEICESTERSHIRE
NORTHAMPTONSHIRE
CAMBRIDGESHIRE
NORFOLK
SUFFOLK
BEDFORDSHIRE
BUCKINGHAMSHIRE
HERTFORDSHIRE
OXFORDSHIRE
ESSEX
BERKSHIRE
GREATER LONDON
SURREY
KENT
HAMPSHIRE
WEST SUSSEX
EAST SUSSEX
WARWICKSHIRE
WEST MIDLANDS
The Fens
NORTH DOWNS
SOUTH DOWNS
THE WEALD
MARLBOROUGH DOWNS
HAMPSHIRE DOWNS
Derby
Nottingham
Carlton
West Bridgford
Ilkeston
Long Eaton
Grantham
Boston
Leicester
Loughborough
Melton Mowbray
Coalville
Hinckley
Nuneaton
Coventry
Rugby
Kettering
Corby
Wellingborough
Northampton
Peterborough
Spalding
King's Lynn
Wisbech
Ely
Cambridge
Newmarket
Bury Saint Edmunds
Thetford
Norwich
Great Yarmouth
Lowestoft
Ipswich
Felixstowe
Harwich
Colchester
Clacton-on-Sea
Chelmsford
Southend-on-Sea
Basildon
Bedford
Luton
Letchworth
Hitchin
Stevenage
Welwyn Garden City
Hertford
Harlow
St. Albans
Hemel Hempstead
Watford
Aylesbury
High Wycombe
Oxford
Abingdon
Banbury
Reading
Slough
Windsor
Maidenhead
LONDON
Bracknell
Newbury
Swindon
Salisbury
Winchester
Southampton
Portsmouth
Bournemouth
Basingstoke
Aldershot
Farnham
Guildford
Woking
Reigate
Crawley
Horsham
Worthing
Brighton
Eastbourne
Hastings
Tunbridge Wells
Tonbridge
Maidstone
Chatham
Gillingham
Dartford
Gravesend
Canterbury
Margate
Ramsgate
Broadstairs
Deal
Dover
Folkestone
Ashford
Chichester
Bognor Regis
Gosport
Fareham
Havant
Eastleigh
Newport
Ryde
Cowes
Sandown
Ventnor
ISLE OF WIGHT
ISLE OF SHEPPEY
Thames
ENGLISH CHANNEL
Strait of Dover
Copyright © by Rand McNally & Co.
Lambert Conformal Conic Projection
b
ALDERNEY
St. Anne
Passage de la Déroute
GUERNSEY (U.K.)
St. Sampson
St. Peter Port
SARK
CHANNEL ISLANDS
JERSEY (U.K.)
St. Aubin
St. Helier
Rozel
2° 30' West

Scandinavia

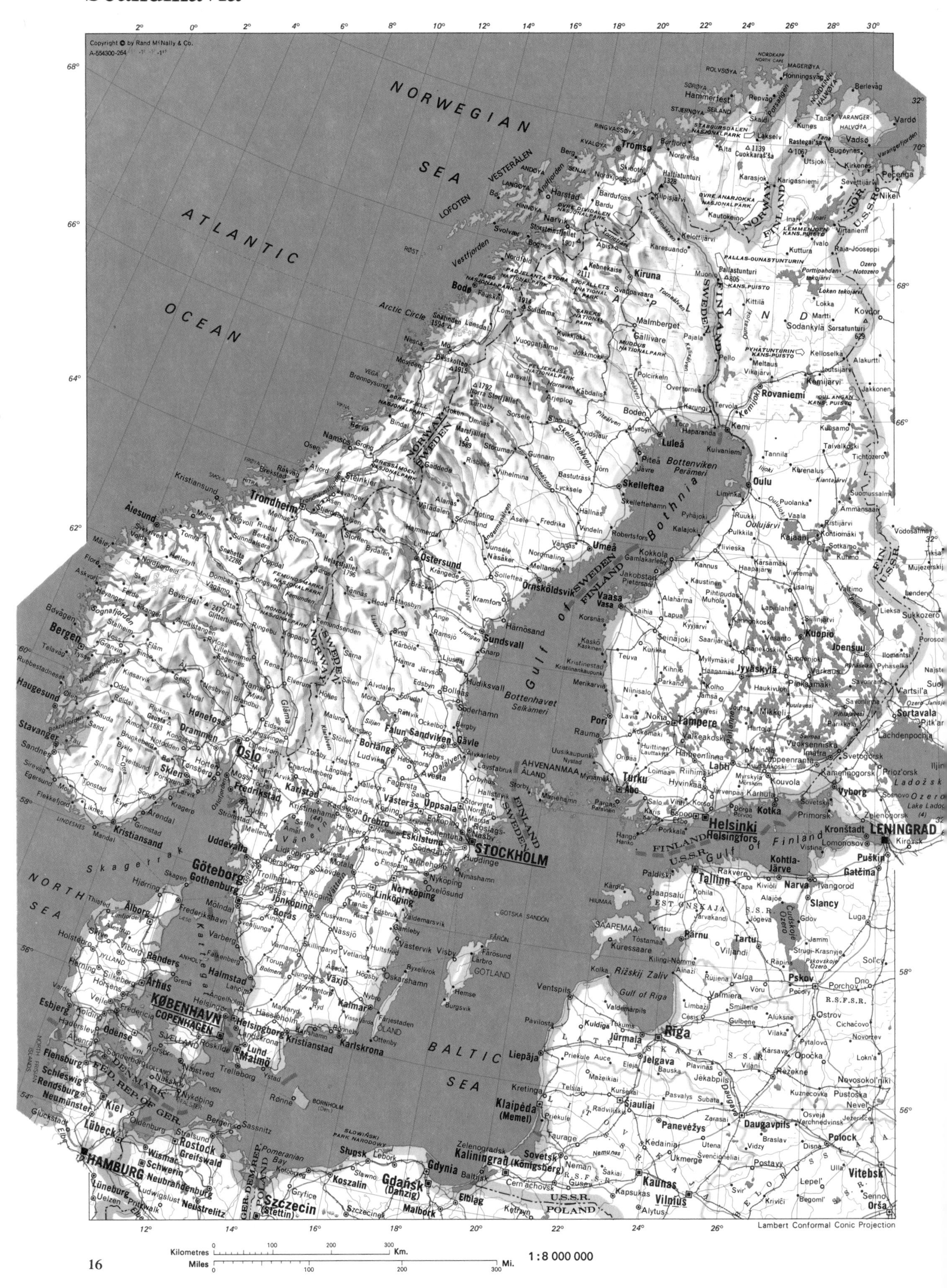

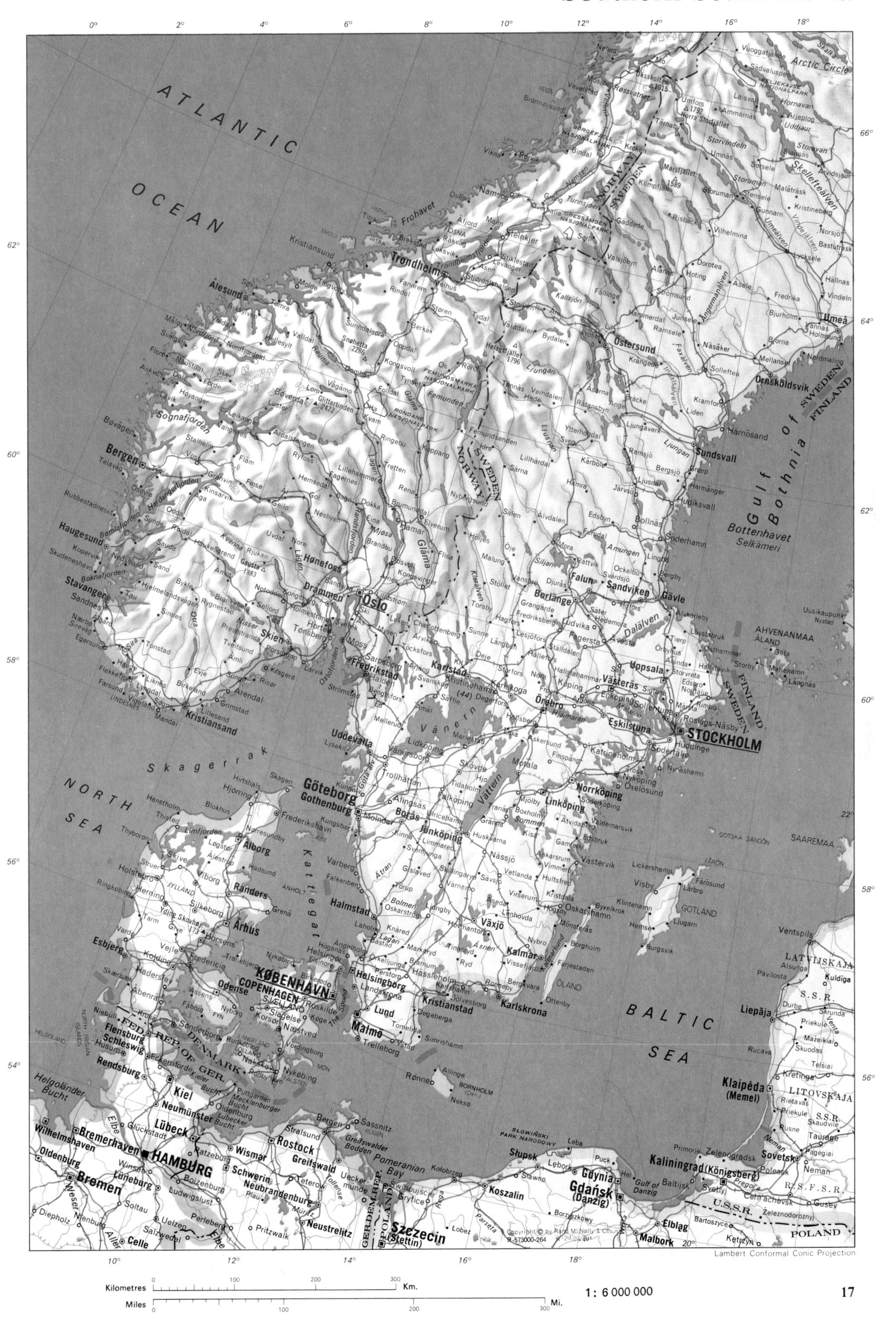

Kilometres 0 100 200 300 Km.

Miles 0 100 200 300 Mi.

1 : 6 000 000

Central Europe

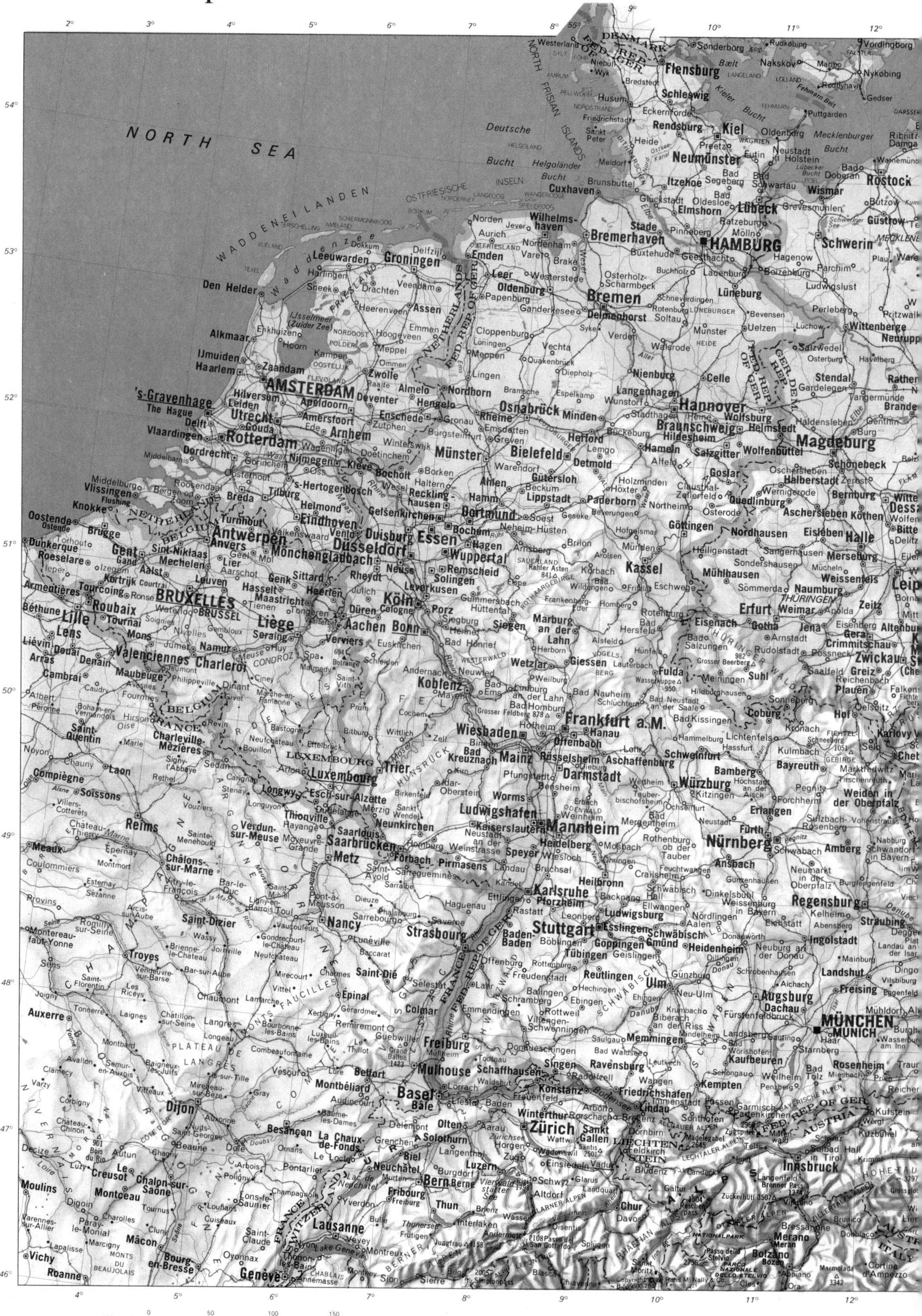

Kilometres 0 50 100 150 Km.
Miles 0 50 100 150 Mi.

1:4 000 000

onic Projection, Two Standard Parallels

France and the Alps

Kilometres 0 50 100 150 Km.
Miles 0 50 100 150 Mi.

1:4 000 000

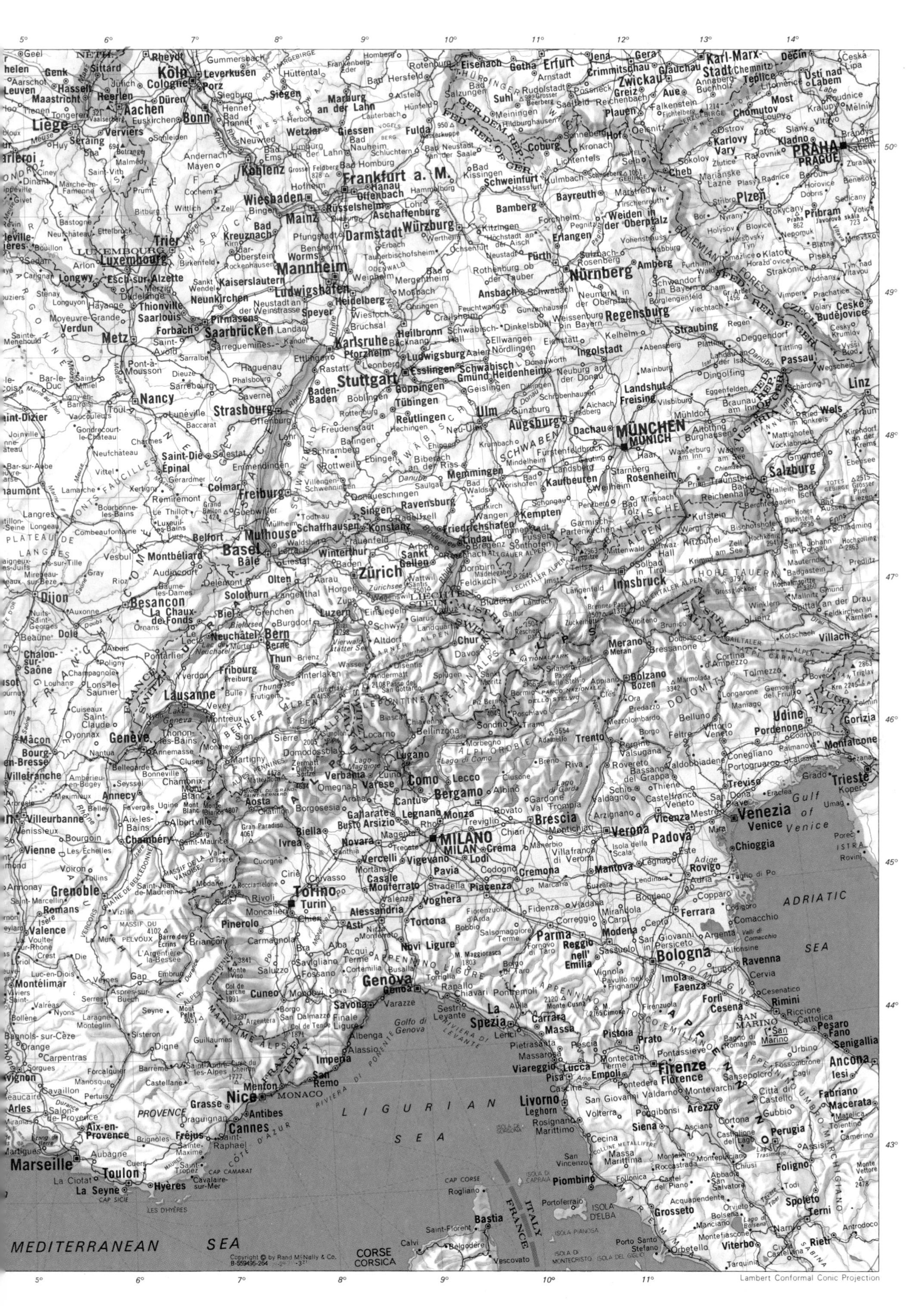

5° 6° 7° 8° 9° 10° 11° 12° 13° 14°
50° 49° 48° 47° 46° 45° 44° 43°
Köln
Cologne
Bonn
Aachen
Liège
Koblenz
Frankfurt a. M.
Wiesbaden
Mainz
Darmstadt
Würzburg
Mannheim
Heidelberg
Karlsruhe
Stuttgart
Nürnberg
Regensburg
Ingolstadt
Augsburg
Ulm
MÜNCHEN
MUNICH
Salzburg
Linz
Passau
Plzeň
PRAHA
PRAGUE
Karl-Marx-Stadt
Zwickau
Erfurt
Luxembourg
Metz
Nancy
Strasbourg
Saarbrücken
Freiburg
Mulhouse
Basel
Bâle
Zürich
Bern
Lausanne
Genève
Besançon
Dijon
Innsbruck
Bolzano
Bozen
Trento
Udine
Trieste
Venezia
Venice
Gulf of Venice
Padova
Verona
Brescia
Bergamo
Como
MILANO
MILAN
Torino
Turin
Genova
Grenoble
Lyon
Chambéry
Annecy
Bologna
Parma
Modena
Ferrara
Ravenna
Rimini
Firenze
Florence
Pisa
Livorno
Leghorn
Siena
Perugia
Ancona
Marseille
Toulon
Nice
Cannes
Monaco
PROVENCE
ADRIATIC SEA
LIGURIAN SEA
MEDITERRANEAN SEA
CORSE
CORSICA
Bastia
ISOLA D'ELBA
ALPS
APPENNINO
SCHWARZWALD
Bodensee
SAN MARINO
Copyright © by Rand McNally & Co.
Lambert Conformal Conic Projection

Spain and Portugal

GASCOGNE
ARMAGNAC
Toulouse
Montauban
Albi
Castres
Carcassonne
Narbonne
Béziers
Montpellier
Nîmes
Avignon
Arles
Aix-en-Provence
Marseille
Toulon
La Seyne
Hyères
Cannes
Nice
Antibes
MONACO
Grasse
Fréjus
Saint-Raphaël
PROVENCE
CAMARGUE
LANGUEDOC
Sète
Golfe du Lion
Perpignan
ROUSSILLON
Pau
Tarbes
Lourdes
PYRENEES
FRANCE
SPAIN
ANDORRA
Andorra
Pic d'Aneto 3404
Monte Perdido 3355
Huesca
Lérida
Gerona
Figueras
Vich
Manresa
Sabadell
Tarrasa
Mataró
Badalona
BARCELONA
Hospitalet
Villanueva y Geltrú
Reus
Tarragona
CATALUÑA
Zaragoza
Ebro
DELTA DEL EBRO
CABO DE TORTOSA
Tortosa
Amposta
San Carlos de la Rápita
Vinaroz
Benicarló
Castellón de la Plana
Teruel
Sagunto
Golfo de Valencia
Valencia
Alcira
Gandía
Denia
Alcoy
Elda
Alicante
Elche
Murcia
Cartagena
CABO DE LA NAO
CABO DE PALOS
ISLAS COLUMBRETES
MENORCA
MINORCA
Ciudadela
Mahón
CABO DE CABALLERÍA
MALLORCA
MAJORCA
Palma
Manacor
Inca
ISLA DRAGONERA
CABRERA
CABO DE SALINAS
ISLAS BALEARES
BALEARIC ISLANDS
IBIZA
Ibiza
San Antonio Abad
FORMENTERA
MEDITERRANEAN SEA
ALGER
ALGIERS
Blida
Médéa
Tizi-Ouzou
Bejaïa (Bougie)
Djidjelli
Sétif
El Eulma
El Asnam (Orléansville)
Mostaganem
Oran (Ouahran)
Arzew
Sig
Sidi bel Abbès
KABYLIE
ATLAS TELLIEN
MONTS DU HODNA
PLAINE DU HODNA
Bou Saâda
Batna
ALGERIA
Kilometres
Km.
Miles
Mi.
1:4 000 000

Italy

ADRIATIC SEA
LIGURIAN SEA
Gulf of Venice
BUDAPEST
Bratislava
WIEN VIENNA
MÜNCHEN MUNICH
Innsbruck
Salzburg
Linz
Graz
Zagreb
Ljubljana
Trieste
Rijeka
Pula
Zadar
Split
Sarajevo
Mostar
Dubrovnik
Venezia Venice
Padova
Verona
MILANO MILAN
Torino Turin
Genova Genoa
Bologna
Firenze Florence
Pisa
Livorno Leghorn
Ancona
Pescara
Perugia
Basel Bâle
Zürich
Bern Berne
Lausanne
Nice
CORSE CORSICA
Bastia

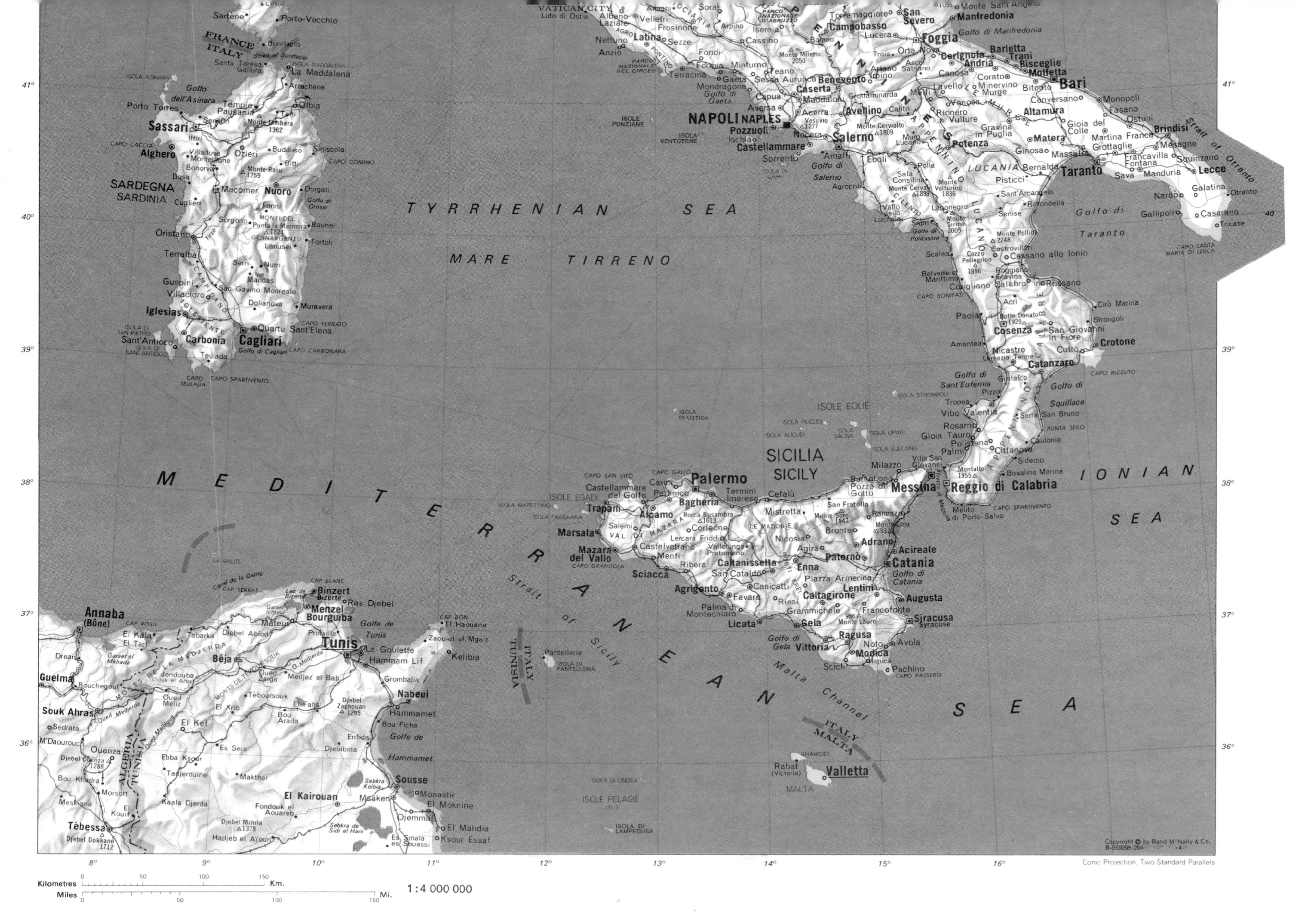

TYRRHENIAN SEA
MARE TIRRENO
MEDITERRANEAN SEA
IONIAN SEA
SICILIA
SICILY
SARDEGNA
SARDINIA
Strait of Sicily
Malta Channel
Strait of Otranto
Golfo di Taranto
FRANCE
ITALY
ITALY
TUNISIA
ITALY
MALTA
ALGERIA
TUNISIA
VATICAN CITY
NAPOLI NAPLES
Palermo
Catania
Messina
Reggio di Calabria
Bari
Taranto
Lecce
Brindisi
Foggia
Salerno
Potenza
Catanzaro
Cosenza
Crotone
Siracusa
Agrigento
Trapani
Marsala
Cagliari
Sassari
Nuoro
Oristano
Iglesias
Valletta
Tunis
Annaba (Bône)
Bizerte
Sousse
Nabeul
El Kairouan
Béja
Souk Ahras
Tébessa
Guelma
Monastir
El Mahdia
ISOLE EOLIE
ISOLE EGADI
ISOLE PELAGIE
ISOLE PONZIANE
Pantelleria
Copyright © by Rand McNally & Co.
Conic Projection, Two Standard Parallels
1:4 000 000
Kilometres
Miles
Km.
Mi.

Southeastern Europe

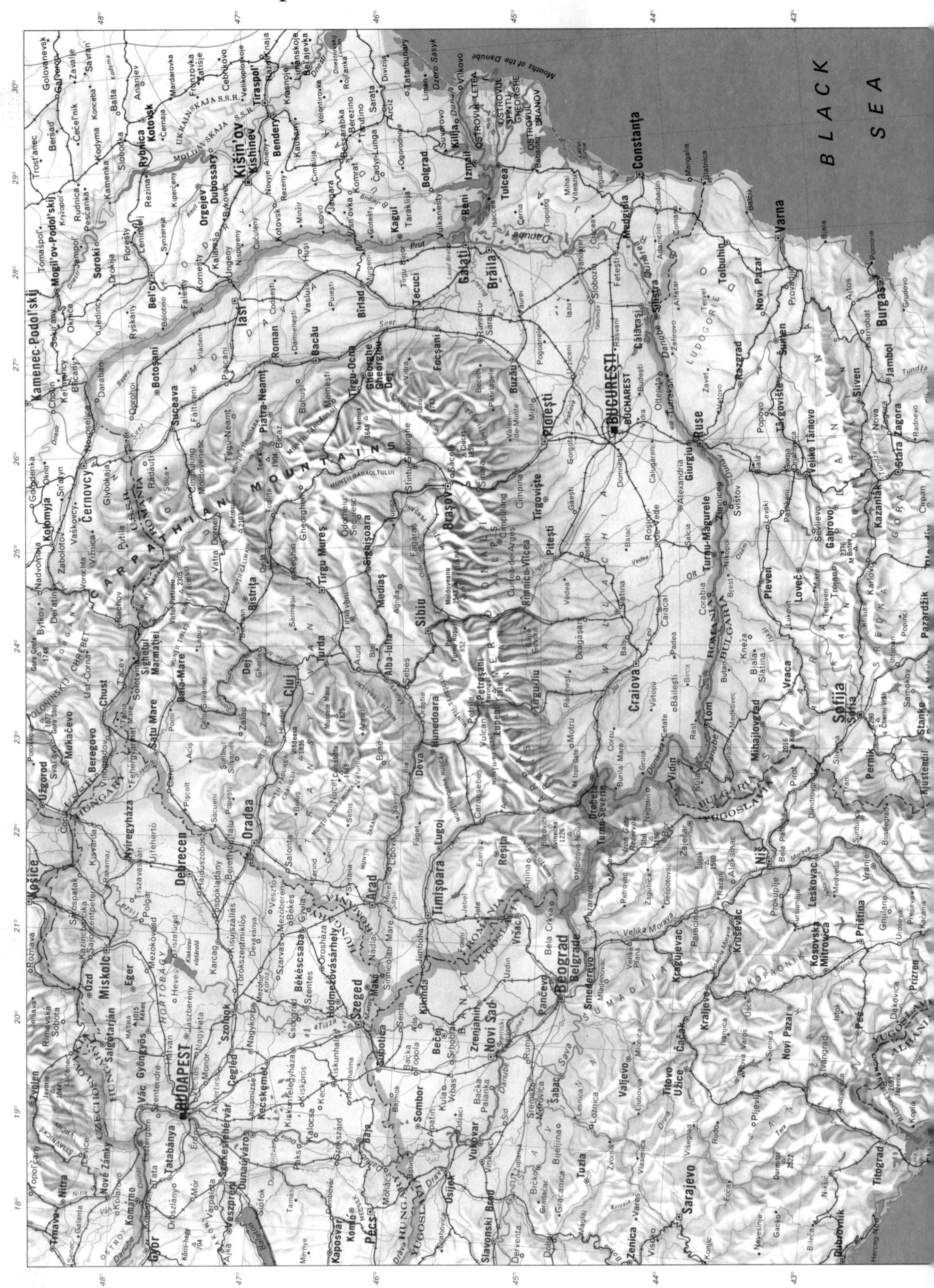

ADRIATIC SEA
Strait of Otranto
ITALY
Otranto
Tricase
IONIAN SEA
AEGEAN SEA
MEDITERRANEAN SEA
Marmara Denizi
Sea of Marmara
Thrakikón Pélagos
Kritikón Pélagos
Sea of Crete
Mirtóön Pélagos
Thermaïkós Kólpos
Saronikós Kólpos
Korinthiakós Kólpos
Gulf of Corinth
Kiparissiakós Kólpos
Lakonikós Kólpos
Argolikós Kólpos
Kerme Körfezi
RHODOPE MOUNTAINS
BULGARIA
GREECE
TURKEY
YUGOSLAVIA
ALBANIA
PELOPONNISOS
PELOPONNESUS
IÓNIOI NISOI
IONIAN ISLANDS
KIKLÁDHES
CYCLADES
DHODHEKÁNISOS
DODECANESE
KRÍTI
CRETE
RÓDHOS
RHODES
LÉSVOS
LESBOS
KHÍOS
CHIOS
SÁMOS
IKARÍA
NÁXOS
PÁROS
TÍNOS
ÁNDROS
MÍLOS
THÍRA
KÍTHIRA
KÁRPATHOS
KOS
ÉVVOIA
SKÍROS
VORÍAI SPORÁDHES
THÁSOS
SAMOTHRÁKI
SAMOTHRACE
GÖKÇEADA
KÉRKIRA
CORFU
LEVKÁS
KEFALLINÍA
ZÁKINTHOS
ÍSTANBUL
ATHÍNAI
ATHENS
Piraiévs
Thessaloníki
Salonika
Tiranë
Durrës
Elbasan
Berat
Vlorë
Korçë
Ohrid
Bitola
Prilep
Titov Veles
Skopje
Edirne
Kırklareli
Lüleburgaz
Tekirdağ
Çorlu
Çanakkale
Bandırma
Bursa
İnegöl
Mustafakemalpaşa
Balıkesir
Edremit
Bergama
Akhisar
Manisa
İzmir
Turgutlu
Salihli
Uşak
Ödemiş
Aydın
Nazilli
Denizli
Söke
Tire
Haskovo
Kârdžali
Smoljan
Komotiní
Xánthi
Kavála
Dráma
Sérrai
Alexandroúpolis
Kérkira
Ioánnina
Tríkala
Lárisa
Vólos
Lamía
Agrínion
Pátrai
Kalámai
Trípolis
Khalkís
Mitilíni
Khíos
Ródhos
Iráklion
Khaniá
Réthimnon
Kateríni
Véroia
Kozáni
Kastoría
Flórina
Préveza
Árta
Návpaktos
Mesolóngion
Kórinthos
Árgos
Návplion
Spárti
2917 Ólimbos
Mount Olympus
Conic Projection, Two Standard Parallels
Copyright © by Rand McNally & Co.
Kilometres
Km.
Miles
Mi.
1:4 000 000

Moscow, Leningrad and the Baltic

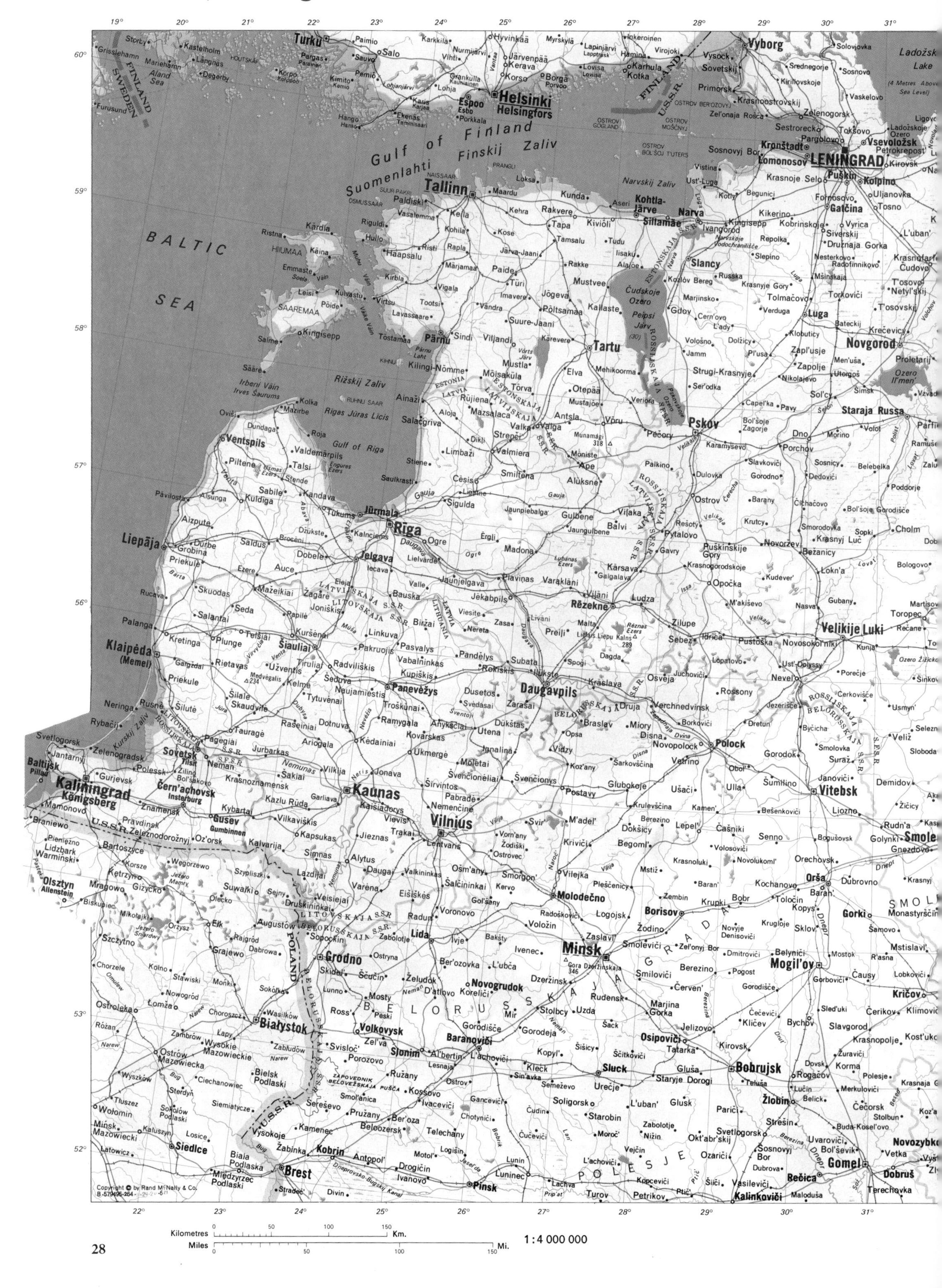

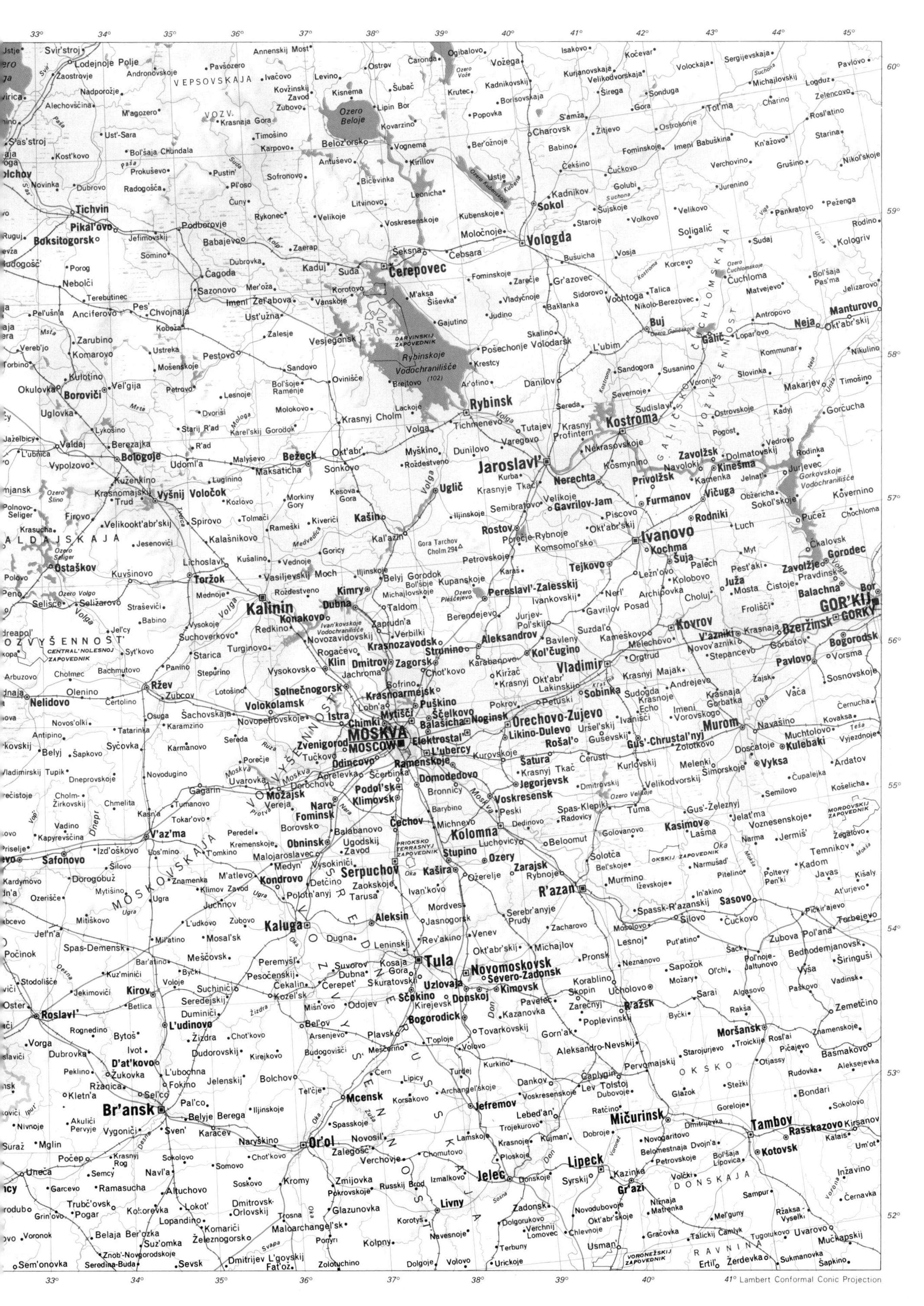
MOSKVA
MOSCOW
GOR'KIJ
GORKY
Jaroslavl'
Kalinin
Vologda
Čerepovec
Rybinsk
Kostroma
Ivanovo
Vladimir
Tula
Kaluga
R'azan'
Br'ansk
Or'ol
Lipeck
Tambov
Mičurinsk
Jelec
Tichvin
Sokol
Uglič
Rostov
Pereslavl'-Zalesskij
Kovrov
Dzeržinsk
Murom
Kolomna
Serpuchov
Novomoskovsk
Rybinskoje Vodochranilišče
Ozero Beloje
VEPSOVSKAJA
VALDAJSKAJA
MOSKOVSKAJA
SREDNERUSSKAJA
GALIČSKO-ČUCHLOMSKAJA VOZVYŠENNOST'
OKSKO DONSKAJA RAVNINA
Volga
Oka
Don
Lambert Conformal Conic Projection

Eastern Soviet Union

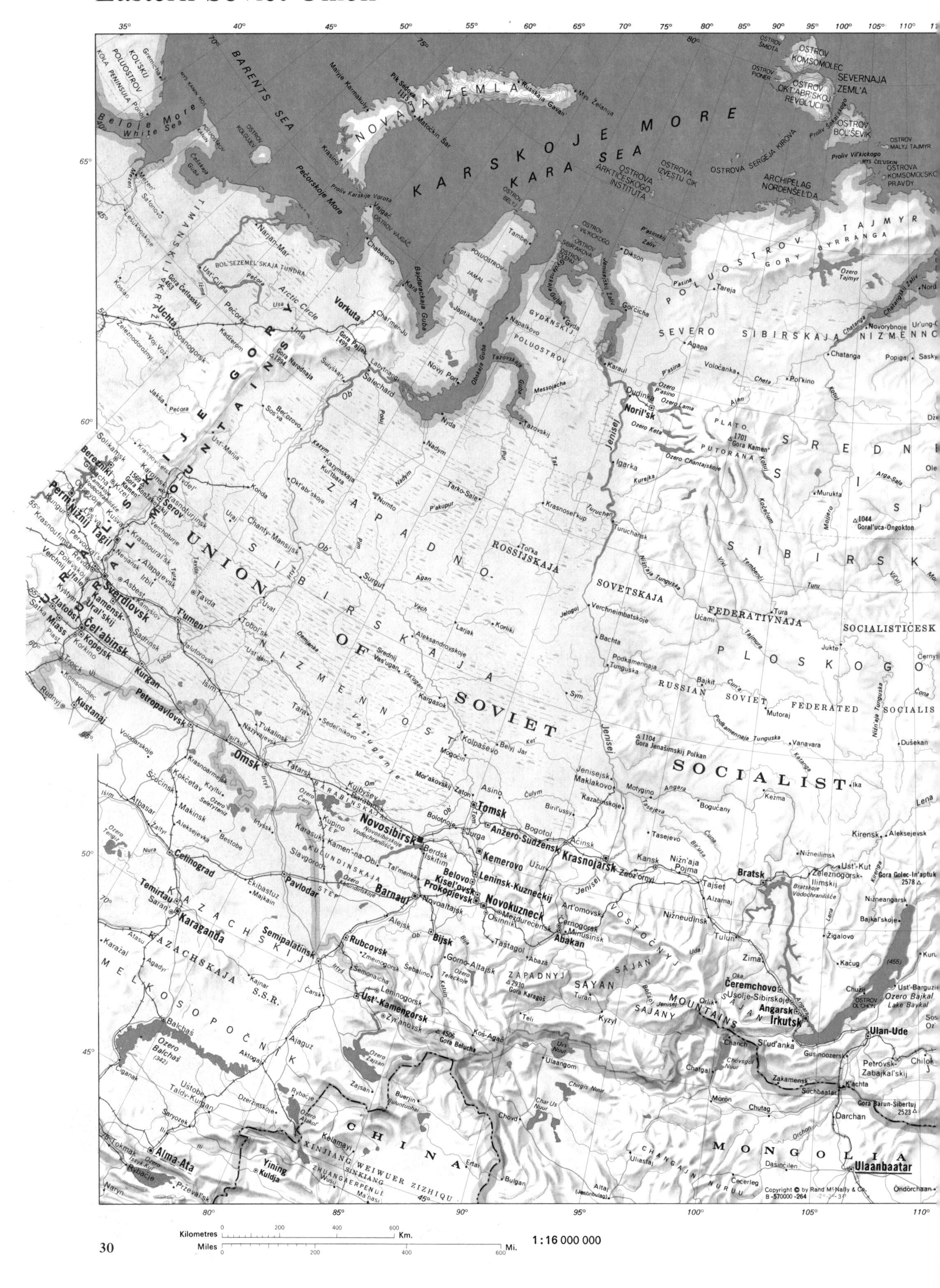

Kilometres 0 200 400 600 Km.
Miles 0 200 400 600 Mi.

1 : 16 000 000

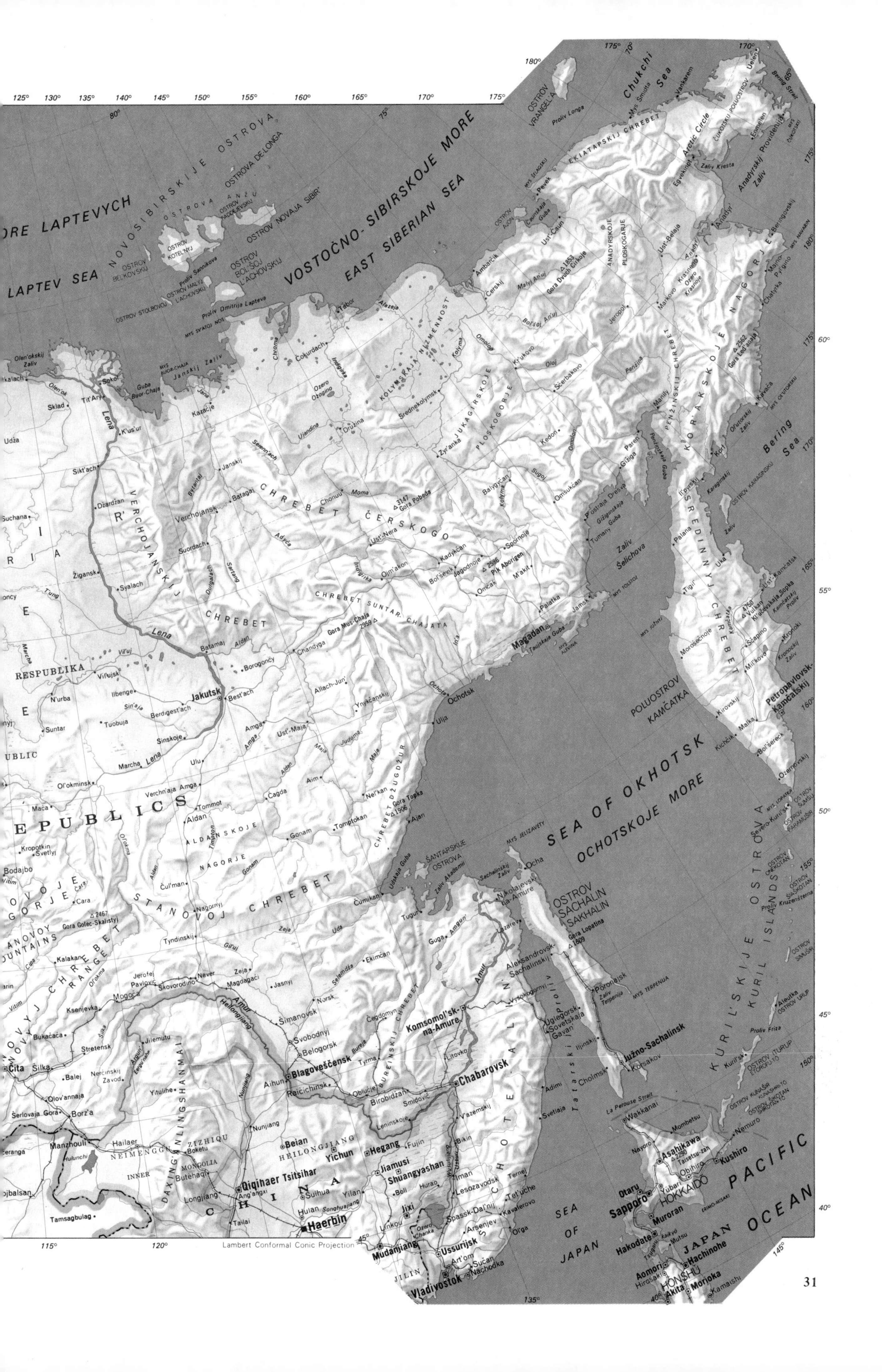

125° 130° 135° 140° 145° 150° 155° 160° 165° 170° 175°
180° 175° 170° 170°
80° 75° 70° 65°
ORE LAPTEVYCH
LAPTEV SEA
NOVOSIBIRSKIJE OSTROVA
OSTROVA DE-LONGA
OSTROVA ANŽU
OSTROV KOTEL'NYJ
OSTROV BEL'KOVSKIJ
OSTROV NOVAJA SIBIR'
OSTROV BOL'ŠOJ L'ACHOVSKIJ
OSTROV MALYJ L'ACHOVSKIJ
Proliv Sannikova
OSTROV STOLBOVOJ
Proliv Dmitrija Lapteva
MYS SV'ATOJ NOS
VOSTOČNO-SIBIRSKOJE MORE
EAST SIBERIAN SEA
OSTROV VRANGEL'A
Proliv Longa
Chukchi Sea
MYS Šmidta
Vankarem
Uelen
Bering Strait
ČUKOTSKIJ POLUOSTROV
Arctic Circle
Egvekinot
Zaliv Kresta
Enmelen
Providenija
Anadyrskij Zaliv
EKIATAPSKIJ CHREBET
MYS ŠELAGSKIJ
Pevek
OSTROV AJON
Čaunskaja Guba
Ust'-Čaun
ANADYRSKOJE PLOSKOGORJE
Ust'-Belaja
Anadyr'
Beringovskij
MYS NAVARIN
Markovo
Ozero Krasnoje
Majno-Pyl'gino
Chatyrka
Ambarčik
Čerskij
Malyj An'uj
Gora Dvuch Cirkov
Bol'šoj An'uj
Jeropol
Alazeja
Tabor
KOLYMSKAJA NIZMENNOST'
Kolyma
Omolon
Omolon
Čokurdach
Chroma
Indigirka
Ozero Ožogino
Srednekolymsk
Družina
Ujandina
JUKAGIRSKOJE PLOSKOGORJE
Zyr'anka
Kedon
Oloj
Penžina
Ščerbakovo
PENŽINSKIJ CHREBET
KORJAKSKOJE NAGORJE
Gora Ledanaja
Manily
Penžinskaja Guba
Kamenskoje
Korf
Ol'utorskij Zaliv
MYS OL'UTORSKIJ
Bering Sea
Oleņ'okskij Zaliv
Jana
Janskij Zaliv
MYS BUOR-CHAJA
Guba Buor-Chaja
Sokol
Ujandina
Tit'Ary
Lena
Sklad
Kazačje
Kus'ur
Udža
Sikt'ach
Sellach
Janskij
Verchojansk
Batagaj
Moma
Gora Pobeda
Chonuu
CHREBET ČERSKOGO
Ust'-Nera
Ojm'akon
Kadykčan
Bol'ševik
Jagodnoje
Pik Aborigen
Sporoje
Omčak
M'akit
Balygyčan
Sugoj
Omsukčan
Gižiga
Paren'
Gižiginskaja Guba
Zaliv Šelichova
Tumanyj
Palana
Ust'-Kamčatsk
SREDINNYJ CHREBET
Il'pyrskij
Karaginskij Zaliv
OSTROV KARAGINSKIJ
Zaliv
Tigil'
Kl'učevskaja Sopka
Kamčatskij Proliv
Ščapino
Kronockij Zaliv
Mil'kovo
Petropavlovsk-Kamčatskij
Kirovskij
Maiiko
Bol'šereck
Ozernovskij
POLUOSTROV KAMČATKA
Džardžan
VERCHOJANSKIJ CHREBET
Suordach
Žigansk
Syalach
Tung
Suchana
CHREBET SUNTAR-CHAJATA
Gora Mus-Chaja
Chandyga
Batamaj
Aldan
Borogončy
Vil'uj
Vil'ujsk
RESPUBLIKA
Jakutsk
Best'ach
Ilbenge
Sin'aja
Berdigest'ach
N'urba
Tuobuja
Suntar
Amga
Ust'-Maja
Sinskoje
UBLIC
Marcha
Lena
Ulu
Allach-Jun'
Ynykčanskij
Ochotsk
Ulja
Magadan
Taujskaja Guba
Palatka
Jamsk
MYS TOLSTOJ
MYS ALEVINA
Judoma
Maja
Ol'okminsk
Verchn'aja Amga
Čagda
Aim
Nel'kan
CHREBET DŽUGDŽUR
Gora Topko
Ajan
Tommot
Aldan
Tomptokan
Gonam
REPUBLICS
Maca
Kropotkin
Svetlyj
Bodajbo
Vitim
ALDANSKOJE NAGORJE
Timpton
Čul'man
Cara
Gora Golec-Skalistyj
STANOVOJ CHREBET
Nagornyj
Tyndinskij
Zeja
Uda
Čumikan
Tugur
Uda
ŠANTARSKIJE OSTROVA
Ul'banskij Zaliv Akademii
Sachalinskij Zaliv
Nikolajevsk-na-Amure
OSTROV SACHALIN SAKHALIN
Ocha
MYS JELIZAVETY
SEA OF OKHOTSK
OCHOTSKOJE MORE
Gora Lopatina
Guga
Amgun
Ekimčan
Selemdža
Kalakan
STANOVOY MOUNTAINS
STANOVOY RANGE
Jerofej Pavlovič
Skovorodino
Never
Zeja
Magdagači
Jasnyj
Norsk
Mogoča
Ksenjevka
Amur
Heilongjiang
Šimanovsk
Svobodnyj
Belogorsk
Čegdomyn
BUREINSKIJ CHREBET
Komsomol'sk-na-Amure
Lazarev
Aleksandrovsk-Sachalinskij
Poronajsk
Zaliv Terpenija
MYS TERPENIJA
Vysokogornyj
Ugleogorsk
Sovetskaja Gavan'
Južno-Sachalinsk
Il'jinskij
Korsakov
Cholmsk
Tatarskij Proliv
Bukačača
Stretensk
Šilka
Čita
Balej
Nerčinskij Zavod
Olov'annaja
Šerlovaja Gora
Borz'a
Jilemutu
Argun
Blagoveščensk
Raičichinsk
Aihun
Tyrma
Bureja
Litovko
Chabarovsk
Obluče
Birobidžan
Smidovič
Vjazemskij
SICHOTE-ALIN
Adimi
Svetlaja
La Perouse Strait
Wakkanai
Mombetsu
Nayoro
Asahikawa
Kushiro
Nemuro
Obihiro
HOKKAIDO
Otaru
Sapporo
Yubari
Muroran
ERIMO-MISAKI
Hakodate
Tsugaru Kaikyō
Aomori
Hachinohe
Hirosaki
Akita
Morioka
Kamaishi
HONSHU
JAPAN
PACIFIC OCEAN
SEA OF JAPAN
KURIL'SKIJE OSTROVA
KURIL ISLANDS
Severo-Kuril'sk
OSTROV PARAMUSIR
OSTROV ŠUMŠU
MYS LOPATKA
OSTROV ONEKOTAN
OSTROV ŠIAŠKOTAN
Proliv Kruzenšterna
OSTROV SIMUŠIR
Aleutka
OSTROV URUP
Proliv Friza
Kuril'sk
OSTROV ITURUP
OSTROV KUNAŠIR
OSTROV ŠIKOTAN
Yitulihe
Nunjiang
Nunjiang
Hailar
Manzhouli
Hulunchi
Boketu
NEIMENGGU ZIZHIQU
INNER MONGOLIA
DAXINGANLINGSHANMAI
Butehaqi
Qiqihaer Tsitsihar
Ang'angxi
Longjiang
Tailai
HEILONGJIANG
Beian
Yichun
Hegang
Fujin
Jiamusi
Shuangyashan
Bikin
Iman
Lesozavodsk
Terney
Tet'uche
Kavalerovo
Ol'ga
Suihua
Yilan
Boli
Hurao
Hulan
Songhuajiang
Jixi
Spassk-Dal'nij
Ozero Chanka
Arsenjev
Haerbin
Linkou
Mudanjiang
Ussurijsk
Art'om
Sučan
Nachodka
Vladivostok
JILIN
CHINA
Čoibalsan
Tamsagbulag
Erdeni
115° 120°
Lambert Conformal Conic Projection
45° 135° 145°
40° 45° 50° 55° 60°
150° 155° 160° 165° 170° 175° 180°

Western North Africa

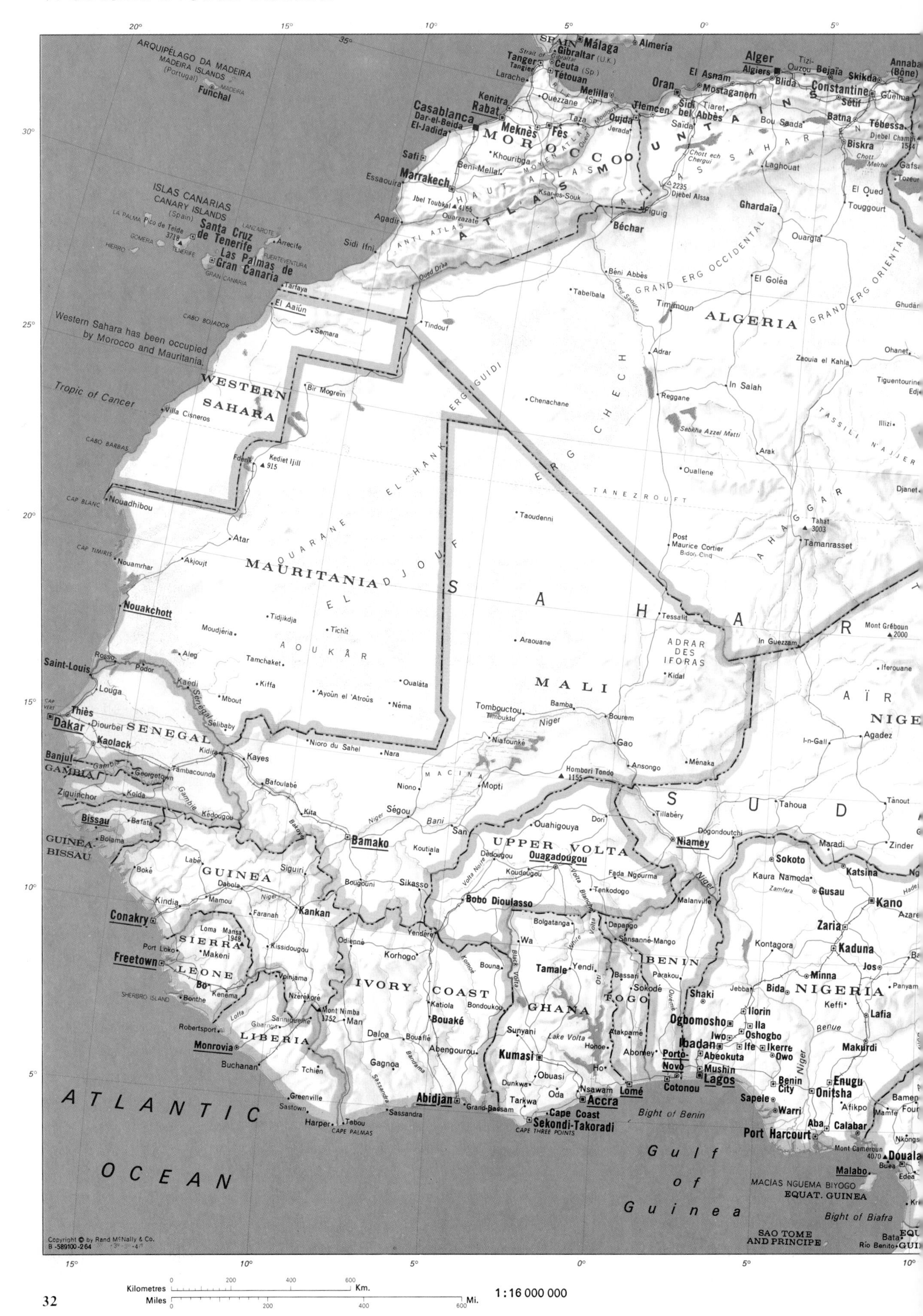

Kilometres 0 200 400 600 Km.

Miles 0 200 400 600 Mi.

1 : 16 000 000

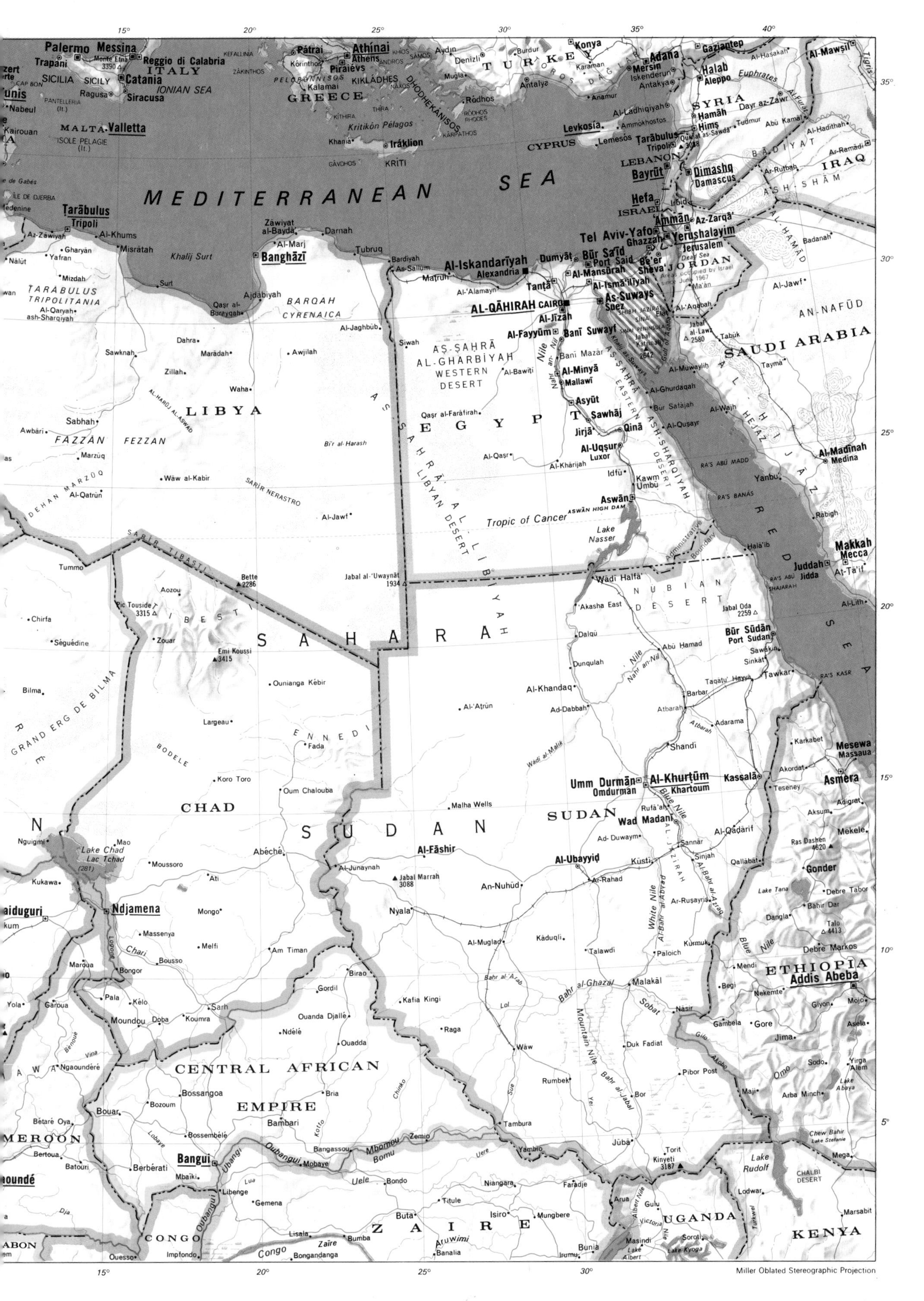

Miller Oblated Stereographic Projection

Southern Africa and Madagascar

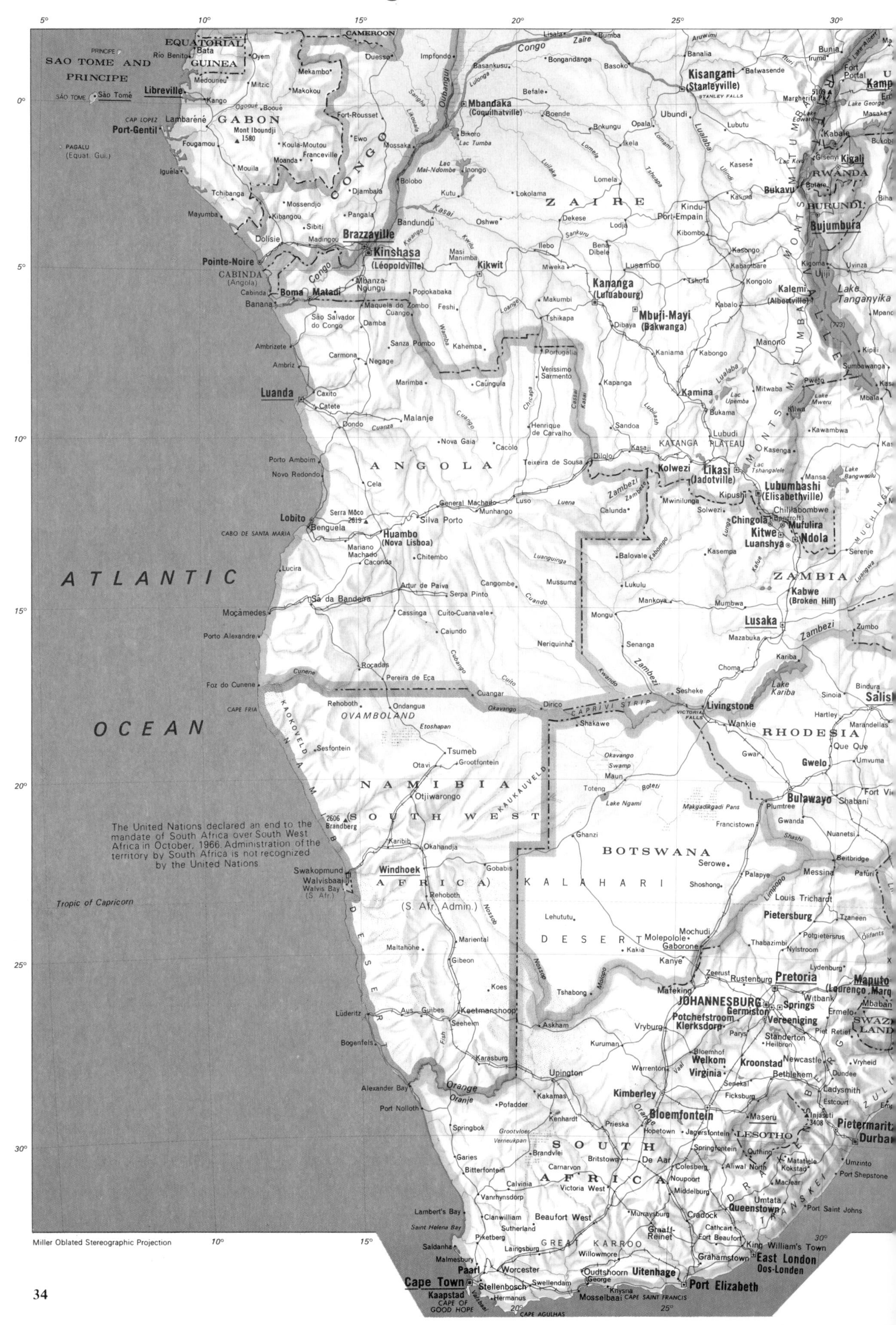

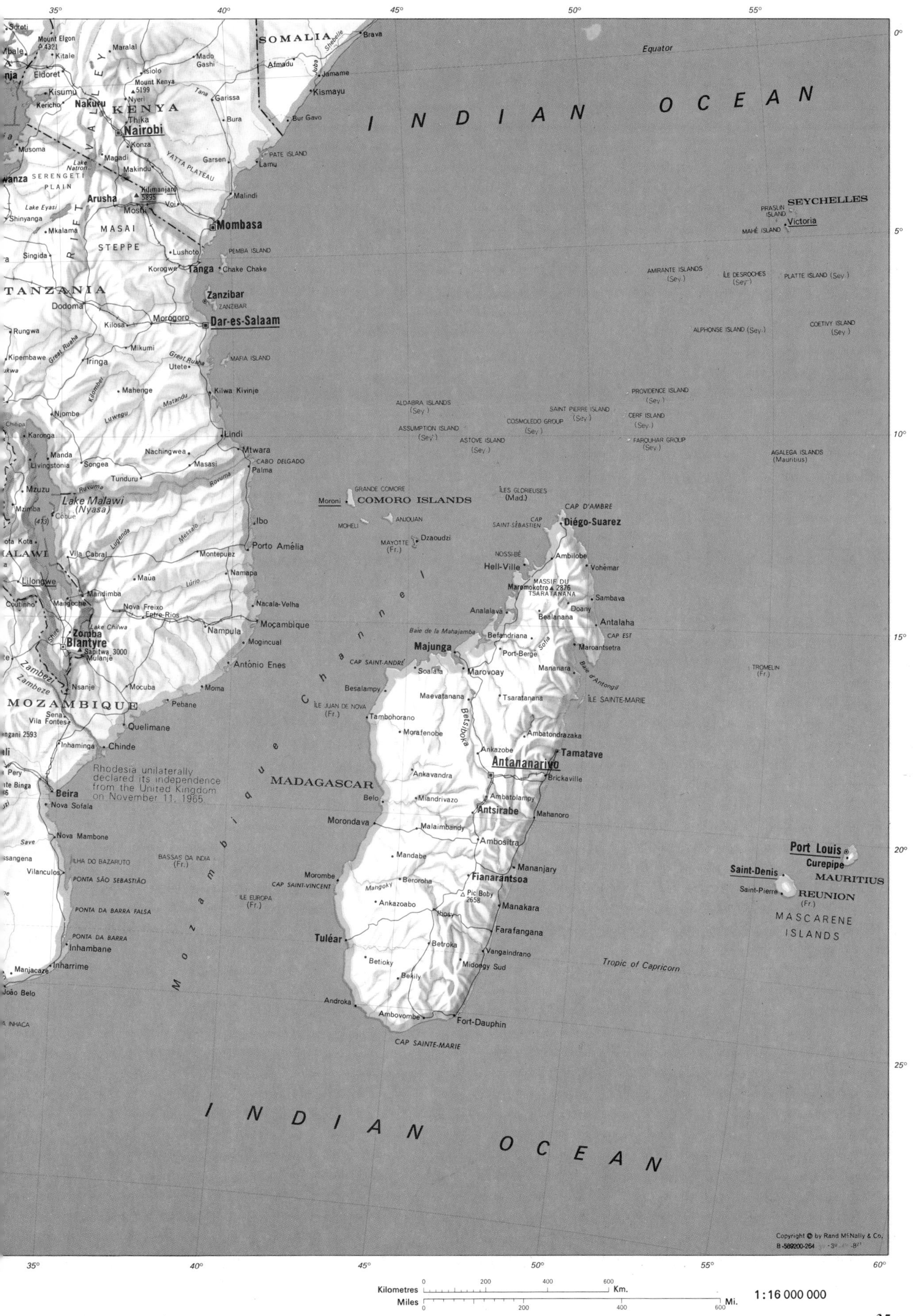

INDIAN OCEAN
Equator
SOMALIA
KENYA
Nairobi
Mombasa
TANZANIA
Dar-es-Salaam
Zanzibar
MOZAMBIQUE
MALAWI
Lake Malawi (Nyasa)
Lilongwe
Blantyre
Zomba
Beira
Moçambique
Quelimane
Mozambique Channel
COMORO ISLANDS
Moroni
MADAGASCAR
Antananarivo
Tamatave
Majunga
Diégo-Suarez
Fianarantsoa
Tuléar
Fort-Dauphin
Antsirabe
SEYCHELLES
Victoria
MAHÉ ISLAND
PRASLIN ISLAND
AMIRANTE ISLANDS (Sey.)
ALDABRA ISLANDS (Sey.)
PROVIDENCE ISLAND (Sey.)
AGALEGA ISLANDS (Mauritius)
TROMELIN (Fr.)
MAURITIUS
Port Louis
Curepipe
REUNION (Fr.)
Saint-Denis
Saint-Pierre
MASCARENE ISLANDS
Tropic of Capricorn
Rhodesia unilaterally declared its independence from the United Kingdom on November 11, 1965.
CAP SAINTE-MARIE
Kilometres
Miles
Km.
Mi.
1:16 000 000
Copyright © by Rand McNally & Co.

South Africa

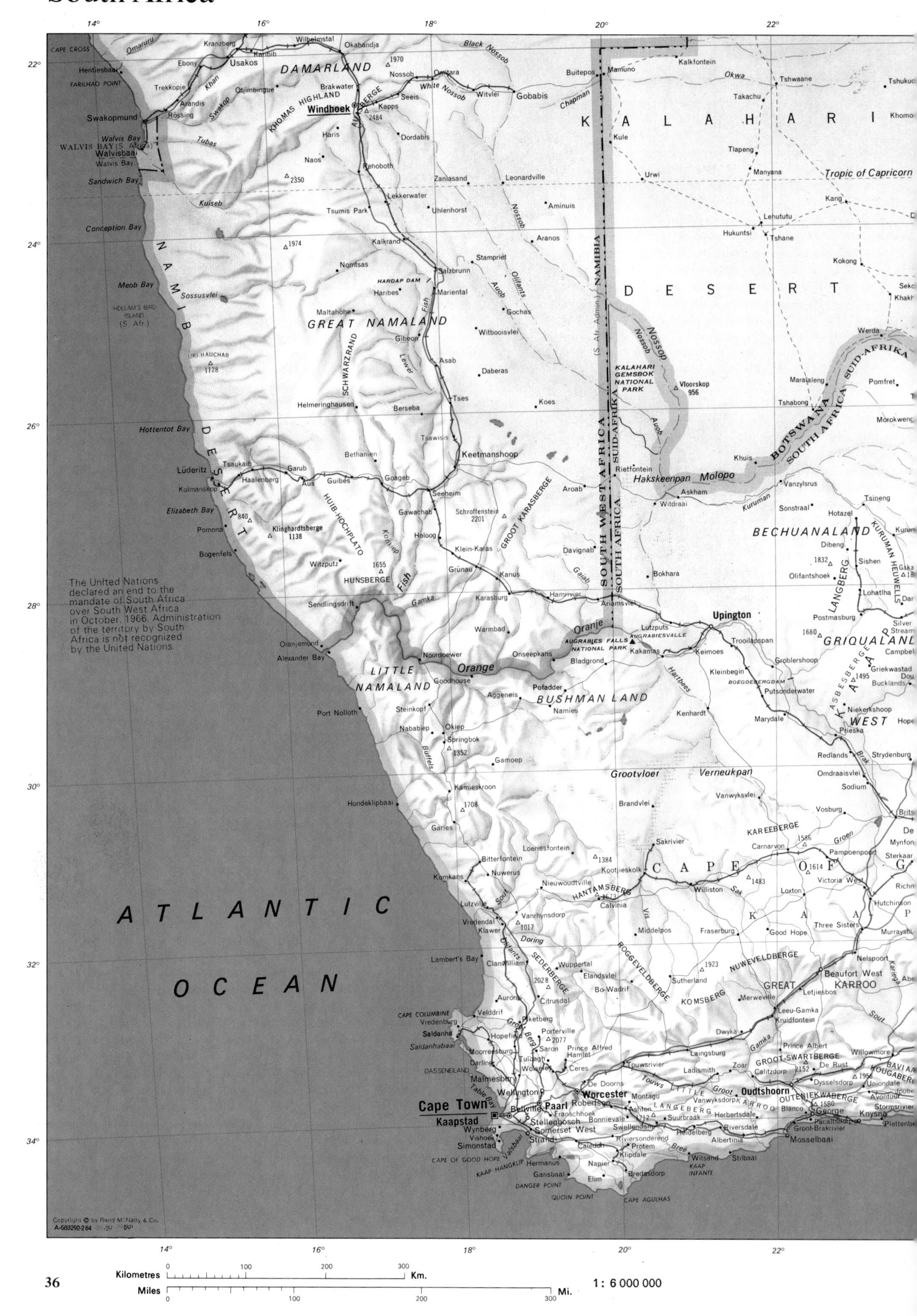

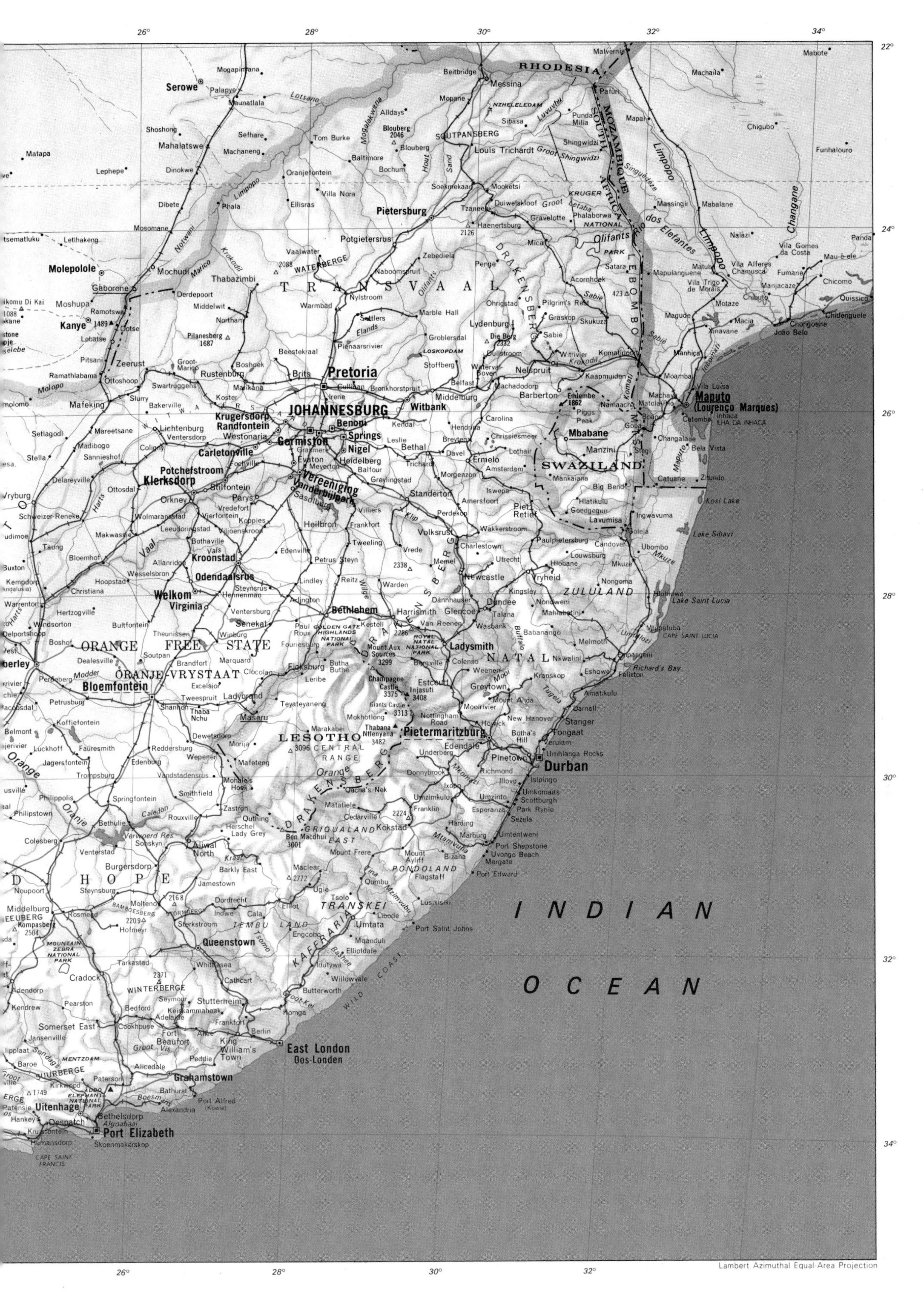

RHODESIA
MOZAMBIQUE
SOUTH AFRICA
TRANSVAAL
ORANGE FREE STATE
ORANJE-VRYSTAAT
LESOTHO
SWAZILAND
NATAL
ZULULAND
TRANSKEI
INDIAN
OCEAN
JOHANNESBURG
Pretoria
Maputo
(Lourenço Marques)
Mbabane
Maseru
Bloemfontein
Pietermaritzburg
Durban
East London
Oos-Londen
Port Elizabeth
Gaborone
Serowe
Molepolole
Kanye
Mafeking
Pietersburg
Witbank
Germiston
Benoni
Springs
Nigel
Krugersdorp
Randfontein
Westonaria
Carletonville
Potchefstroom
Klerksdorp
Vereeniging
Vanderbijlpark
Kroonstad
Odendaalsrus
Welkom
Virginia
Bethlehem
Ladysmith
Queenstown
Grahamstown
Uitenhage
Umtata
Messina
Louis Trichardt
Nelspruit
Lydenburg
Barberton
Newcastle
Vryheid
Richard's Bay
DRAKENSBERG
LEBOMBO
Limpopo
Olifants
Vaal
Orange
Tugela
Lambert Azimuthal Equal-Area Projection

Northeastern Africa and Arabia

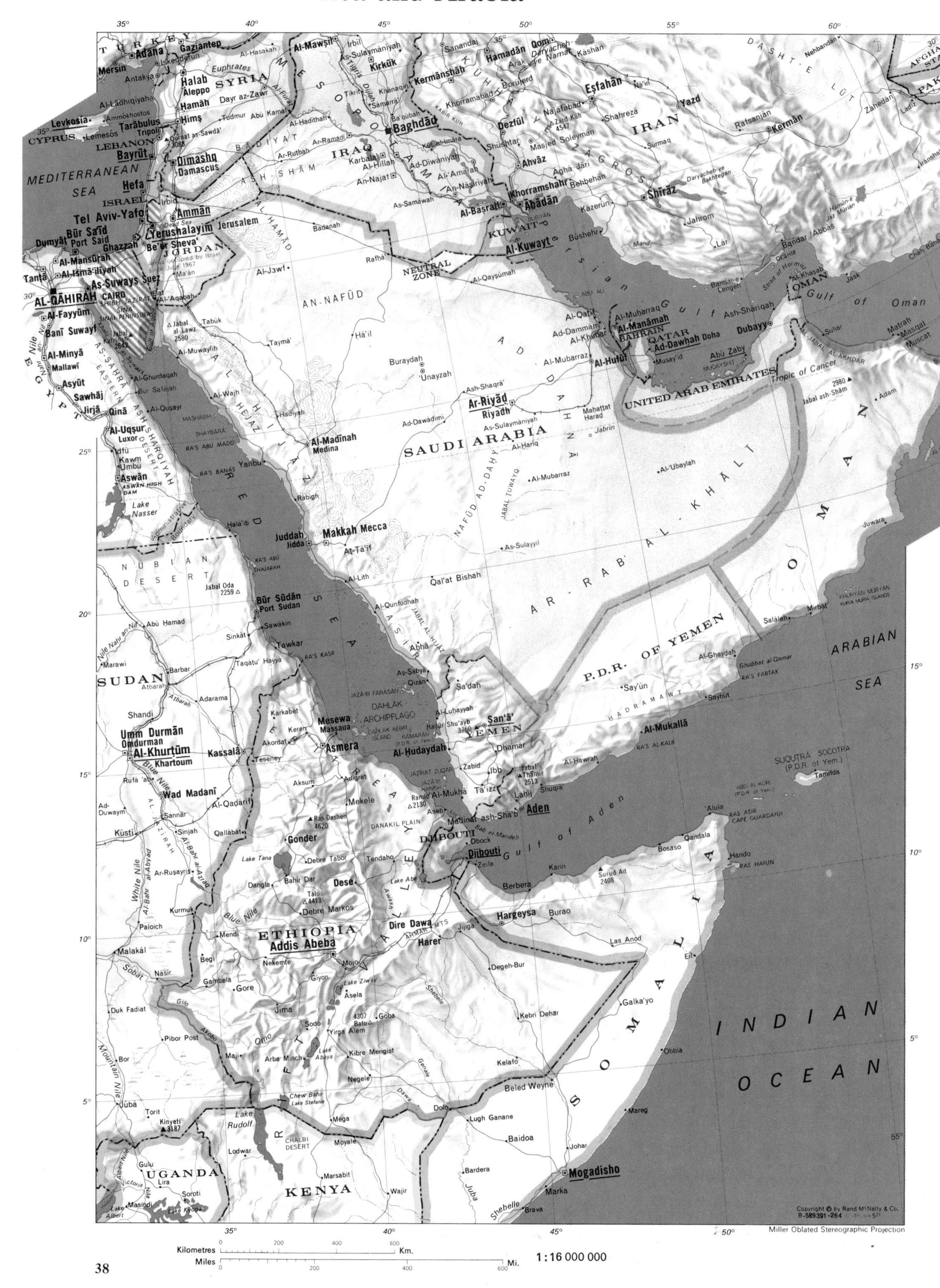

Miller Oblated Stereographic Projection

Kilometres 0 200 400 600 Km.
Miles 0 200 400 600 Mi.
1:16 000 000

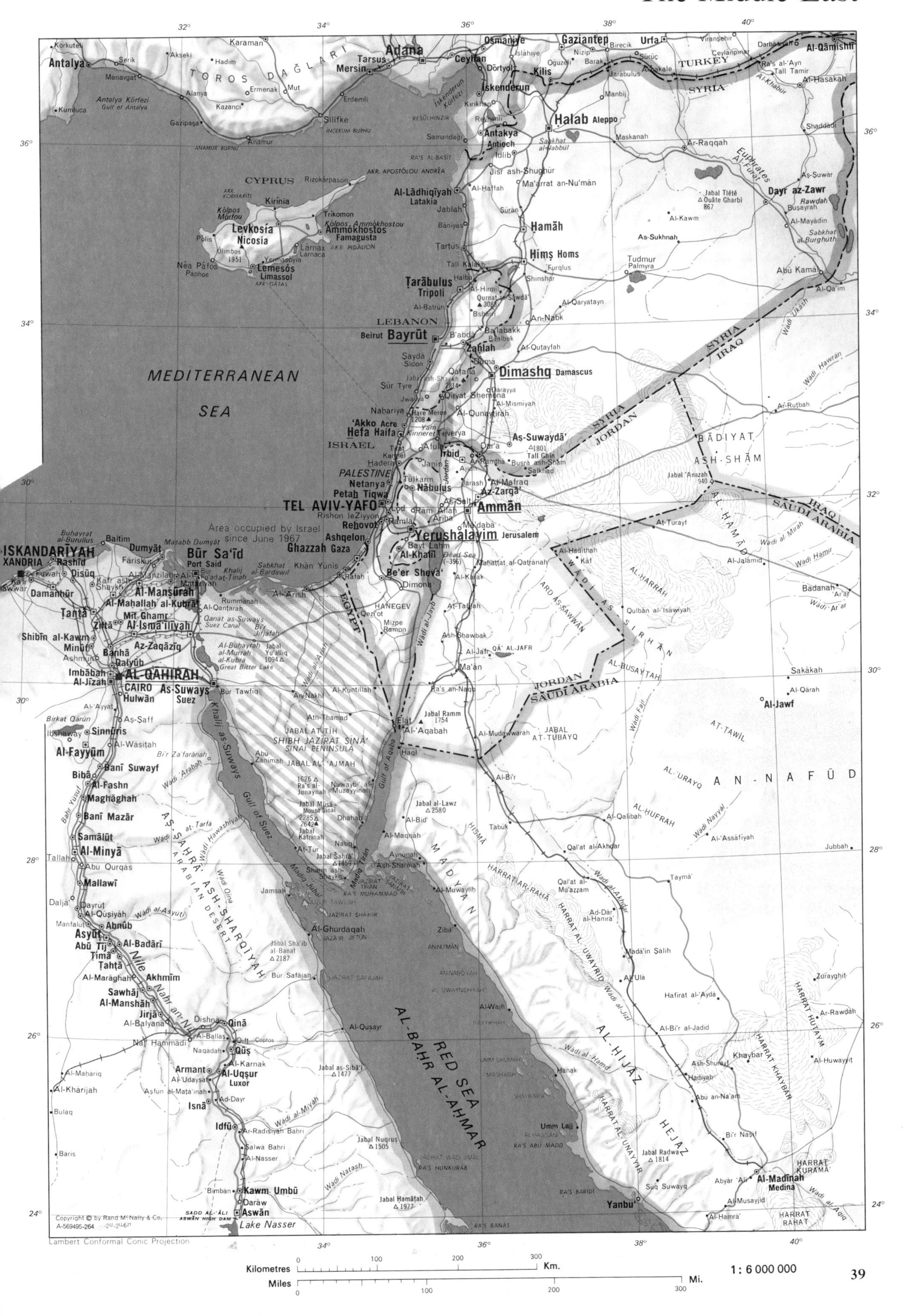

Kilometres 0 100 200 300 Km.

Miles 0 100 200 300 Mi.

1 : 6 000 000

India

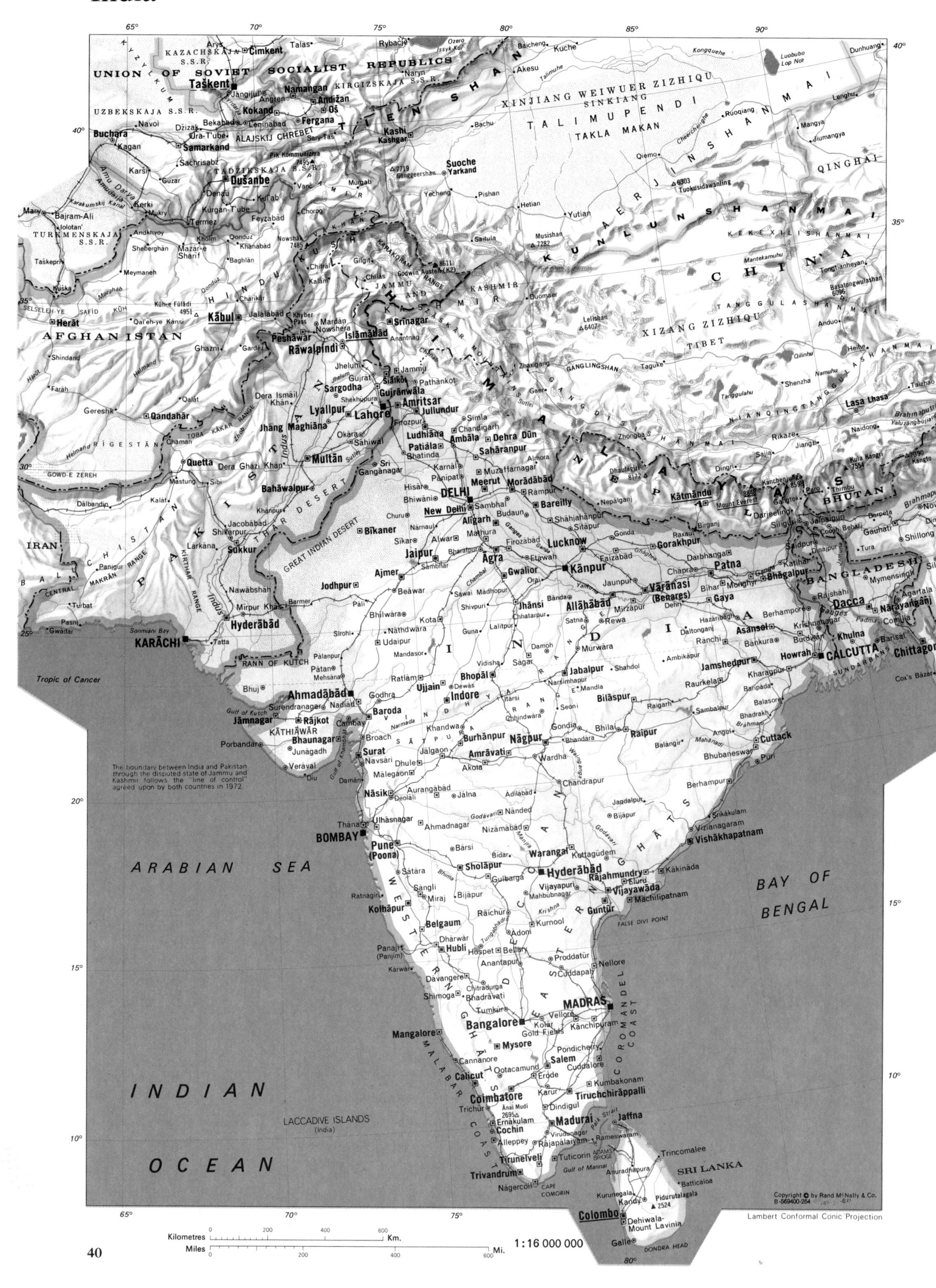

UNION OF SOVIET SOCIALIST REPUBLICS
KAZACHSKAJA S.S.R.
UZBEKSKAJA S.S.R.
KIRGIZSKAJA S.S.R.
TADZIKSKAJA S.S.R.
TURKMENSKAJA S.S.R.
TIEN SHAN
XINJIANG WEIWUER ZIZHIQU
SINKIANG
TALIMUPENDI
TAKLA MAKAN
KUNLUN SHAN
CHINA
XIZANG ZIZHIQU
TIBET
HINDU KUSH
KARAKORAM RANGE
JAMMU AND KASHMIR
HIMALAYAS
AFGHANISTAN
PAKISTAN
IRAN
NEPAL
BHUTAN
BANGLADESH
INDIA
SRI LANKA
THAR DESERT
GREAT INDIAN DESERT
RANN OF KUTCH
KATHIAWAR
VINDHYA RANGE
SATPURA RANGE
DECCAN
WESTERN GHATS
EASTERN GHATS
MALABAR COAST
COROMANDEL COAST
ARABIAN SEA
BAY OF BENGAL
INDIAN OCEAN
LACCADIVE ISLANDS (India)
Tropic of Cancer
Taškent
Samarkand
Dušanbe
Kabul
Herāt
Qandahār
Peshāwar
Rāwalpindi
Islāmābād
Srīnagar
Lahore
Amritsar
Multān
Quetta
KARĀCHI
Hyderābād
DELHI
New Delhi
Jaipur
Āgra
Lucknow
Kānpur
Vārānasi (Benares)
Patna
Kātmāndu
Dacca
CALCUTTA
Ahmadābād
Baroda
Surat
BOMBAY
Pune (Poona)
Nāgpur
Hyderābād
Vishākhapatnam
MADRAS
Bangalore
Mysore
Madurai
Colombo
Kandy
Jaffna
Trivandrum
Cochin
Calicut
Mangalore
Goa
Panaji (Panjim)
Mount Everest 8848
Godwin Austen (K2) 8611
CAPE COMORIN
The boundary between India and Pakistan through the disputed state of Jammu and Kashmir follows the "line of control" agreed upon by both countries in 1972.
Copyright © by Rand McNally & Co.
Lambert Conformal Conic Projection
Kilometres
Km.
Miles
Mi.
1:16 000 000

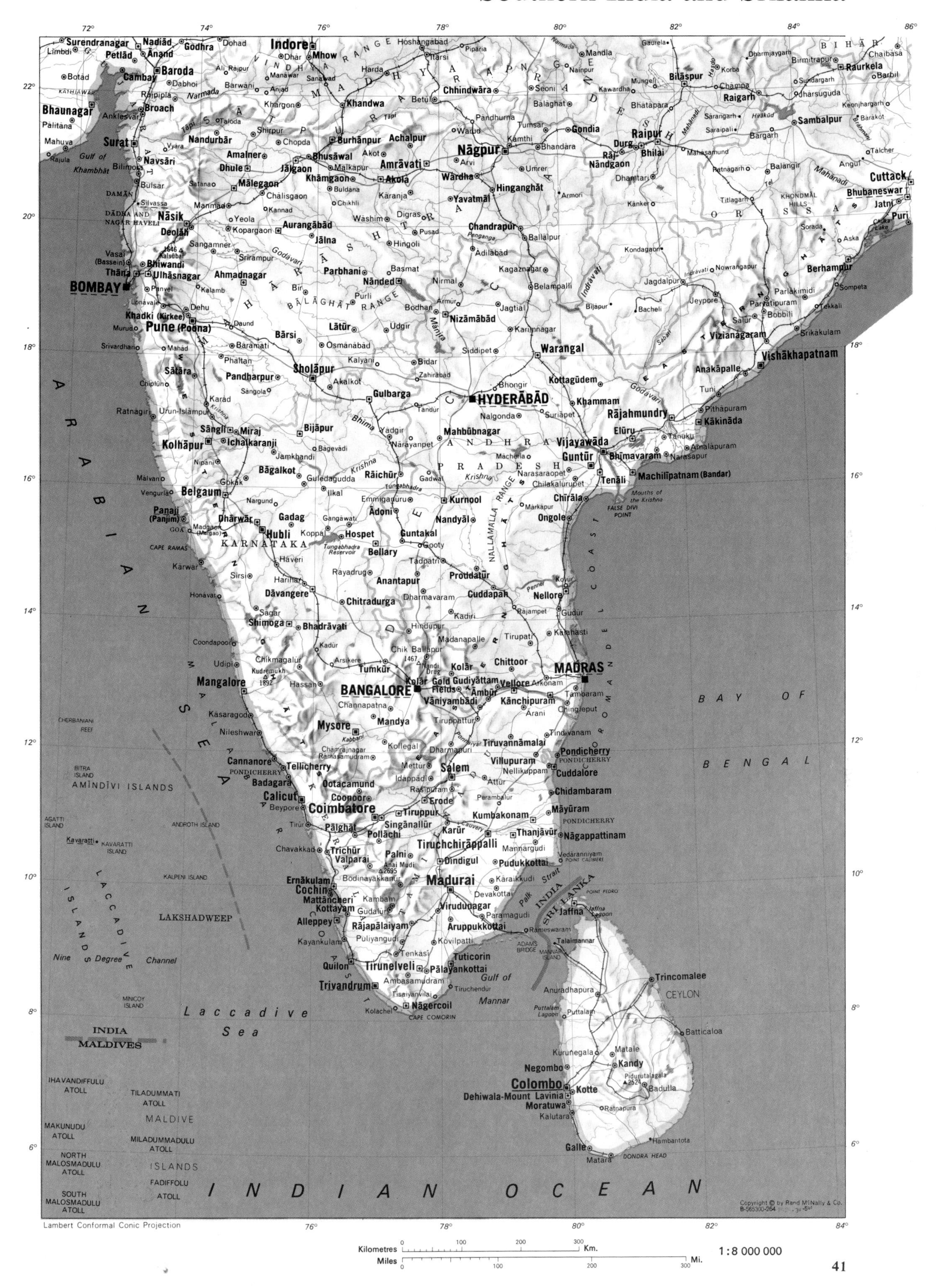

Kilometres 0 100 200 300 Km.

Miles 0 100 200 300 Mi.

1 : 8 000 000

Northern India and the Himalayas

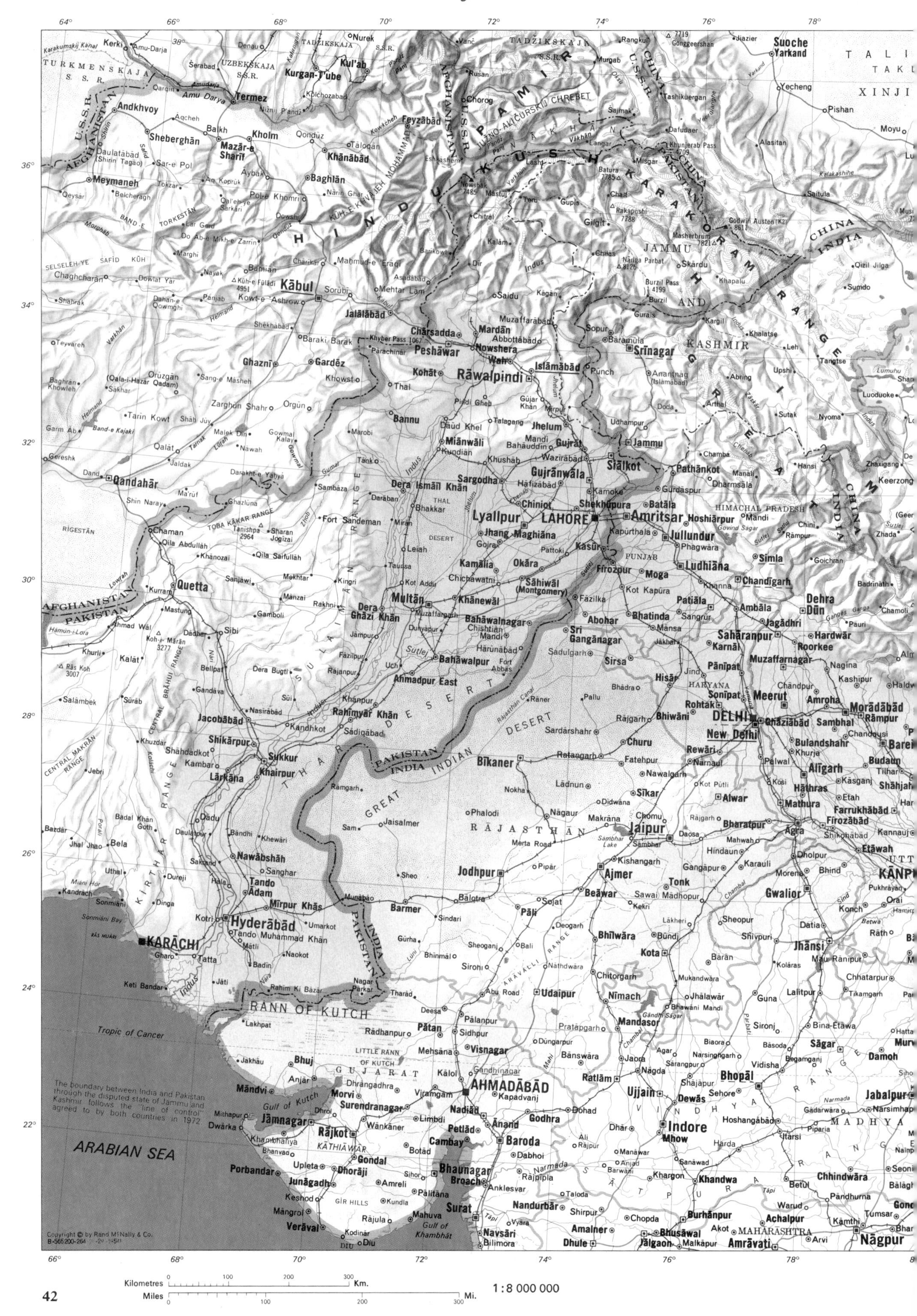

XIZANG ZIZHIQU
TIBET
WEIWUER ZIZHIQU
SINKIANG
QINGHAI
CHAIDAMU PENDI
KUNLUN SHAN
KEKEXILI SHAN
TANGGULA SHAN
NIANQINGTANGGULA SHAN
GANGDISI SHAN
HIMALAYA
Mount Everest
Zhumulangmafeng
Qiemo
Yutian
Andilangan
Dachaidan
Delingha
Golmud
Ge'ermu
Nangqian (Xiangda)
Lasa
Lhasa
Rikaze
Shigatse
Lazi
Jiangzi
Gyangze
Katmandu
Lalitpur
Bhaktapur
Pokhara
Darjeeling
Kalimpong
Siliguri
Jalpaiguri
Gangtok
SIKKIM
BHUTAN
Thimbu
Paro
Punakha
Tongsa Dzong
ARUNACHAL PRADESH
ASSAM
NAGALAND
MEGHALAYA
TRIPURA
MIZORAM
MANIPUR
BIHAR
ORISSA
WEST BENGAL
CHINA
NEPAL
INDIA
BANGLADESH
BURMA
Shillong
Gauhati
Imphal
Kohima
Agartala
Aijal
Silchar
Dibrugarh
Tinsukia
Digboi
Jorhat
Lucknow
Allahabad
Varanasi (Benares)
Gorakhpur
Patna
Gaya
Bhagalpur
Ranchi
Jamshedpur
Dhanbad
Asansol
Durgapur
Burdwan
Howrah
CALCUTTA
Kharagpur
Cuttack
Bhubaneswar
Puri
Raipur
Durg
Bhilai
Sambalpur
Dacca
Chittagong
Khulna
Mymensingh
Rajshahi
Comilla
Narayanganj
Sylhet
Cox's Bazar
Ganges
Brahmaputra
Padma
Jamuna
Hooghly
Mouths of the Ganges
BAY OF BENGAL
Tropic of Cancer
Lambert Conformal Conic Projection

China, Japan and Korea

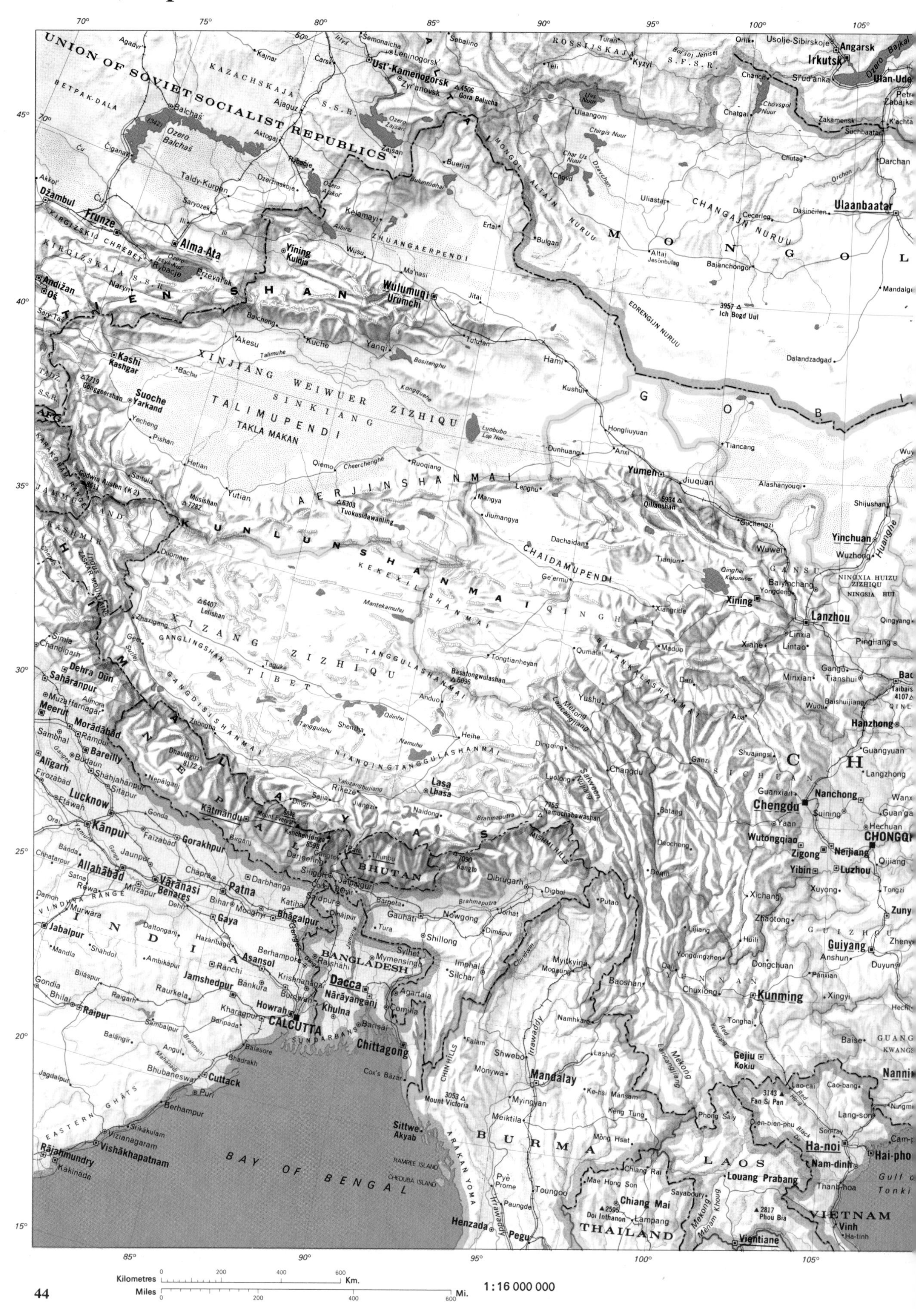

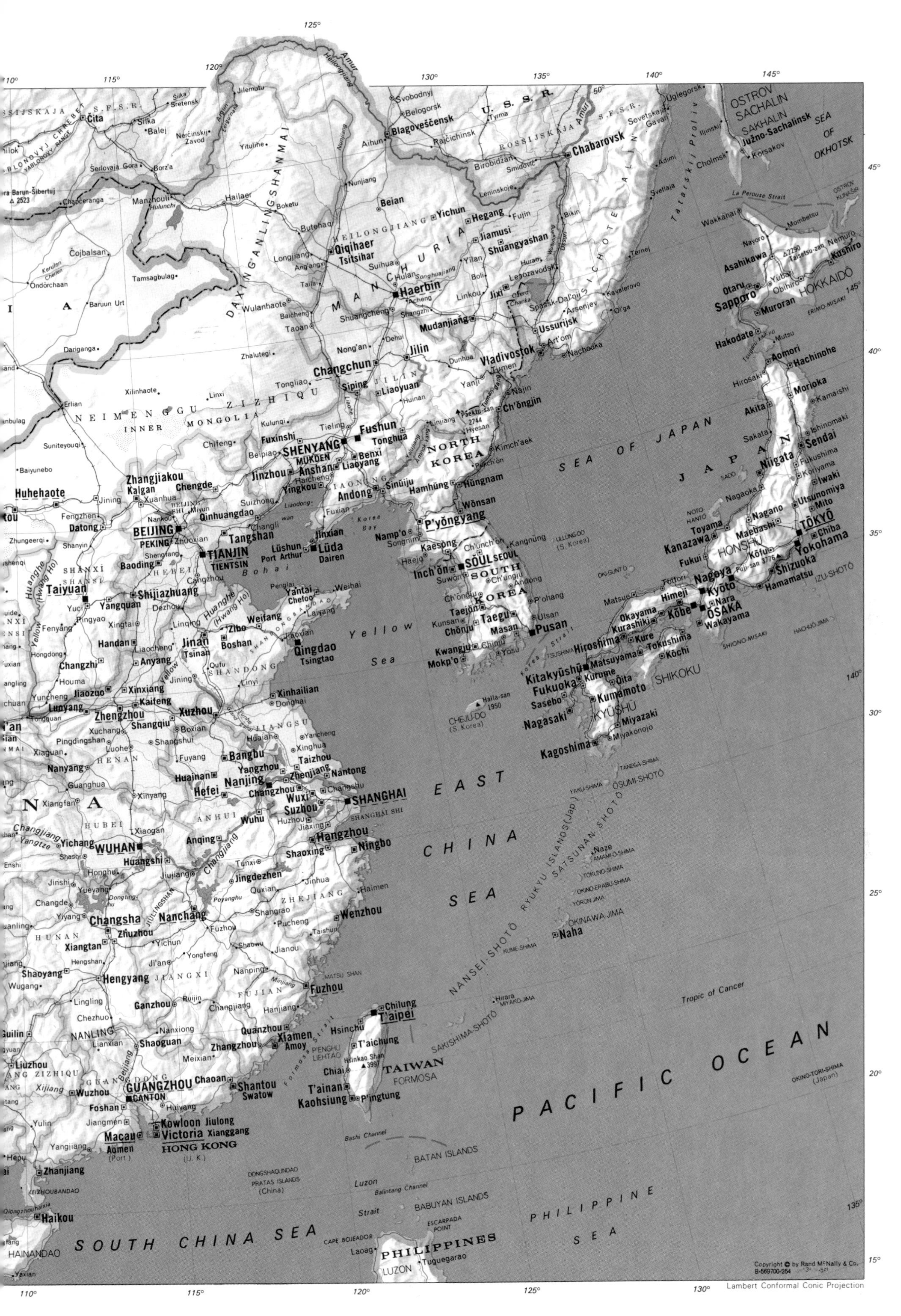

U. S. S. R.
MANCHURIA
HEILONGJIANG
DAXING'ANLINGSHANMAI
NEIMENGGU ZIZHIQU
INNER MONGOLIA
JILIN
LIAONING
HEBEI
SHANXI
SHANDONG
JIANGSU
HENAN
ANHUI
HUBEI
HUNAN
JIANGXI
ZHEJIANG
FUJIAN
GUANGDONG
NORTH KOREA
SOUTH KOREA
JAPAN
HOKKAIDŌ
HONSHU
SHIKOKU
KYŪSHŪ
TAIWAN
FORMOSA
PHILIPPINES
LUZON
HONG KONG
(U. K.)
SEA OF JAPAN
SEA OF OKHOTSK
OSTROV SACHALIN
SAKHALIN
Yellow Sea
Bohai
Korea Bay
EAST CHINA SEA
SOUTH CHINA SEA
PHILIPPINE SEA
PACIFIC OCEAN
Tatarskij Proliv
La Perouse Strait
Korea Strait
Formosa Strait
Bashi Channel
Balintang Channel
Luzon Strait
Tropic of Cancer
NANSEI-SHOTŌ
RYUKYU ISLANDS (Jap.)
SATSUNAN-SHOTŌ
ŌSUMI-SHOTŌ
SAKISHIMA-SHOTŌ
OKINAWA-JIMA
BATAN ISLANDS
BABUYAN ISLANDS
PRATAS ISLANDS
(China)
CHEJU-DO
(S. Korea)
Halla-san 1950
ULLŪNG-DO
(S. Korea)
OKINO-TORI-SHIMA
(Japan)
IZU-SHOTŌ
Amur
Heilongjiang
Huanghe
(Hwang Ho)
Changjiang
Yangtze
Čita
Blagoveščensk
Chabarovsk
Vladivostok
Južno-Sachalinsk
Harbin
Haerbin
Qiqihaer
Tsitsihar
Changchun
Jilin
Shenyang
Mukden
Fushun
Anshan
Benxi
Dandong
Lüda
Dairen
Lüshun
Port Arthur
BEIJING
PEKING
TIANJIN
TIENTSIN
Tangshan
Huhehaote
Datong
Taiyuan
Shijiazhuang
Baoding
Jinan
Tsinan
Qingdao
Tsingtao
Zibo
Zhengzhou
Kaifeng
Luoyang
Xuzhou
Nanjing
Hefei
Wuxi
Suzhou
SHANGHAI
Hangzhou
Ningbo
Wenzhou
WUHAN
Changsha
Nanchang
Fuzhou
Xiamen
Amoy
Shantou
Swatow
GUANGZHOU
CANTON
Kowloon
Victoria
Macau
Aomen
(Port.)
Zhanjiang
Haikou
HAINANDAO
T'aipei
Chilung
T'aichung
T'ainan
Kaohsiung
P'yŏngyang
Hŭngnam
Ch'ŏngjin
SŎUL
SEOUL
Inch'ŏn
Taejŏn
Taegu
Pusan
Kwangju
Mokp'o
Sapporo
Asahikawa
Hakodate
Aomori
Sendai
Niigata
TŌKYŌ
Yokohama
Nagoya
Kyōto
Ōsaka
Kōbe
Hiroshima
Kitakyūshū
Fukuoka
Nagasaki
Kagoshima
Naha
Fuji-san 3776
Copyright © by Rand McNally & Co.
B-569700-264
Lambert Conformal Conic Projection

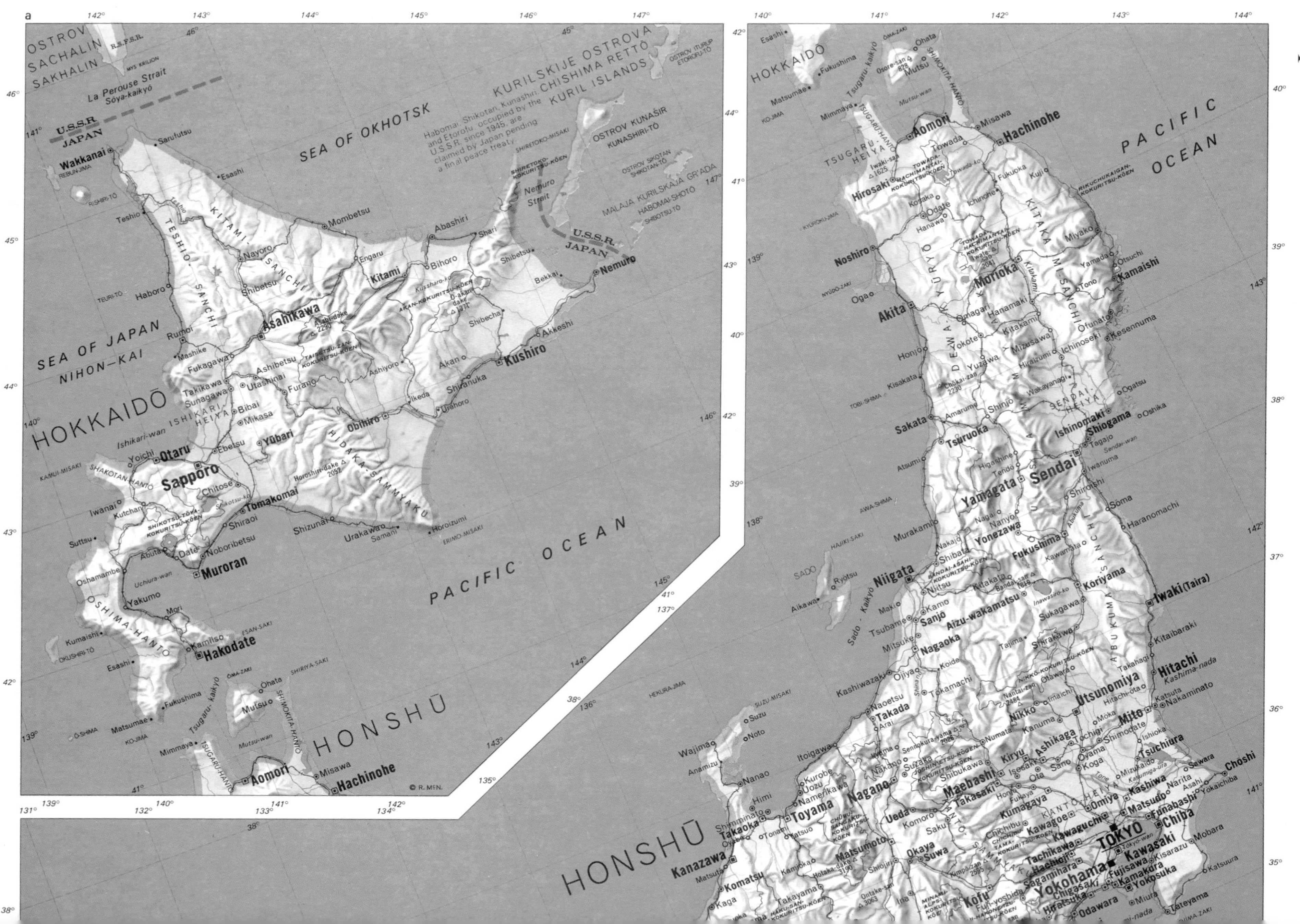
PACIFIC OCEAN
SEA OF OKHOTSK
SEA OF JAPAN
NIHON-KAI
HOKKAIDŌ
HONSHŪ
KURIL'SKIJE OSTROVA
CHISHIMA-RETTŌ
KURIL ISLANDS
OSTROV SACHALIN
SAKHALIN
La Perouse Strait
Sōya-kaikyō
U.S.S.R.
JAPAN
Wakkanai
Asahikawa
Sapporo
Otaru
Muroran
Hakodate
Kushiro
Obihiro
Kitami
Nemuro
Abashiri
Aomori
Hachinohe
Morioka
Akita
Sendai
Yamagata
Fukushima
Niigata
Toyama
Kanazawa
Nagano
Maebashi
Utsunomiya
Mito
Hitachi
Iwaki (Taira)
Kōriyama
TŌKYŌ
Yokohama
Kawasaki
Chiba
Chōshi
Kamaishi
Sakata
Tsuruoka
Noshiro
Hirosaki
Tsugaru-kaikyō
Sado-kaikyō
SADO
© R.MEN.

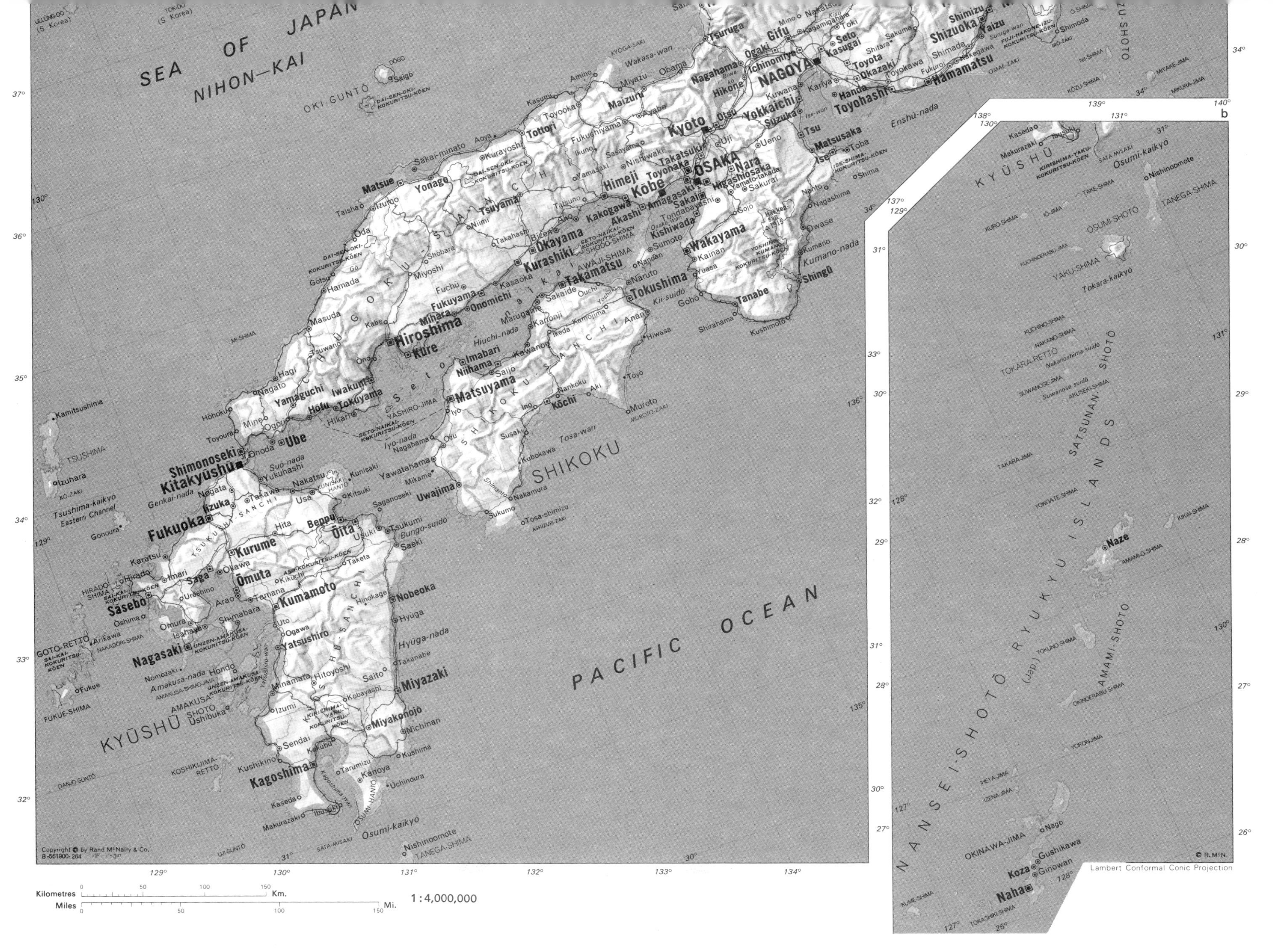
SEA OF JAPAN
NIHON-KAI
PACIFIC OCEAN
SHIKOKU
KYŪSHŪ
OKI-GUNTŌ
TSUSHIMA
NANSEI-SHOTŌ
RYUKYU ISLANDS (Jap.)
SATSUNAN-SHOTŌ
AMAMI-SHOTŌ
ŌSUMI-SHOTŌ
TOKARA-RETTŌ
OKINAWA-JIMA
Naha
Naze
NAGOYA
ŌSAKA
Kyōto
Kōbe
Hiroshima
Kitakyūshū
Fukuoka
Nagasaki
Kumamoto
Kagoshima
Miyazaki
Matsuyama
Kōchi
Tokushima
Takamatsu
Okayama
Wakayama
Hamamatsu
Shizuoka
Shimonoseki
Ube
Ōita
Beppu
Nobeoka
Sasebo
Kurume
Ōmuta
Himeji
Matsue
Tottori
Yonago
Kure
Fukuyama
Onomichi
Kurashiki
Tsu
Gifu
Toyohashi
Toyota
Okazaki
Nara
Yokkaichi
Kii-suidō
Tosa-wan
Enshū-nada
Hyūga-nada
Bungo-suidō
Ōsumi-kaikyō
Tsushima-kaikyō
Eastern Channel
1:4,000,000
Kilometres
Miles
Lambert Conformal Conic Projection
Copyright © by Rand McNally & Co.
B-561900-264

Eastern China, Hongkong and Taiwan

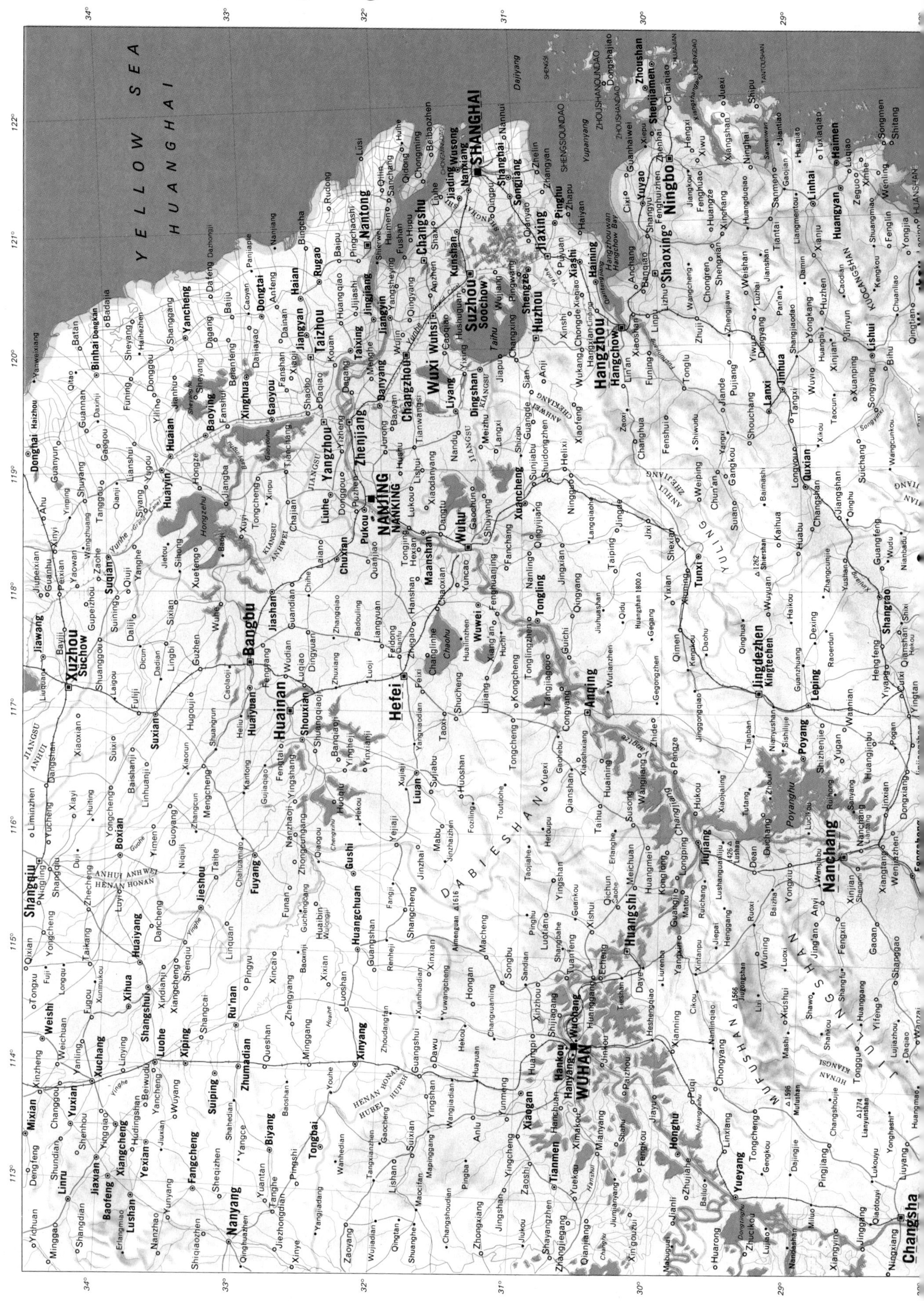

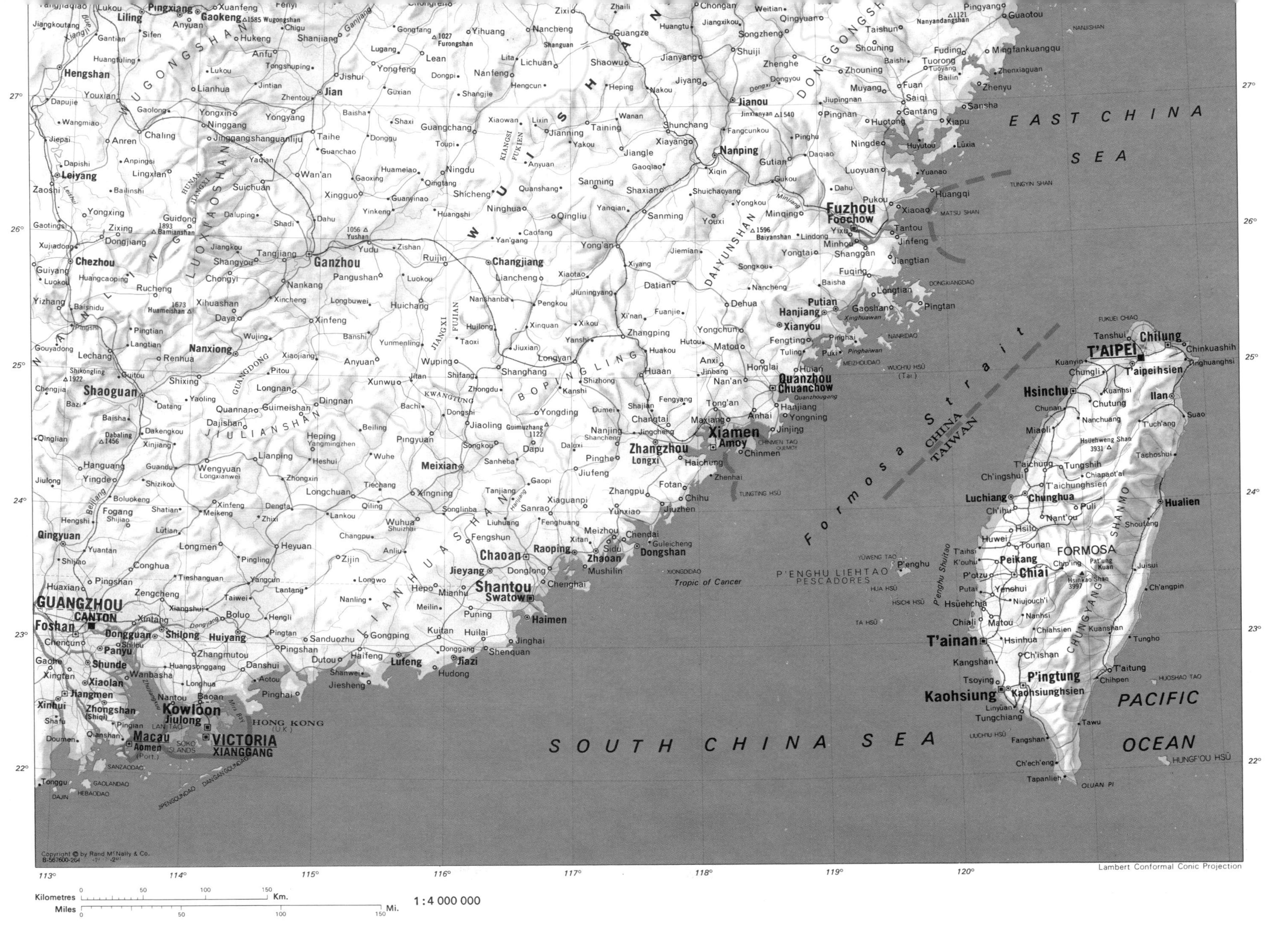
EAST CHINA SEA
SOUTH CHINA SEA
PACIFIC OCEAN
Formosa Strait
CHINA TAIWAN
FORMOSA
CHUNGYANG SHANMO
T'AIPEI
Chilung
Hsinchu
Hualien
Chunghua
Chiai
T'ainan
Kaohsiung
P'ingtung
P'ENGHU LIEHTAO
PESCADORES
Tropic of Cancer
Fuzhou
Foochow
Putian
Quanzhou
Chuanchow
Xiamen
Amoy
Zhangzhou
Shantou
Swatow
Chaoan
Meixian
DAIYUNSHAN
JIULIANSHAN
GUANGZHOU
CANTON
Foshan
Macau
Aomen
HONG KONG
(U.K.)
VICTORIA
XIANGGANG
Kowloon
Shaoguan
Ganzhou
Jian
Hengshan
Chezhou
Nanping
Jianou
Sanming
Changjiang
Lambert Conformal Conic Projection
Kilometres
Miles
1:4 000 000
Copyright © by Rand McNally & Co.

Southeastern Asia

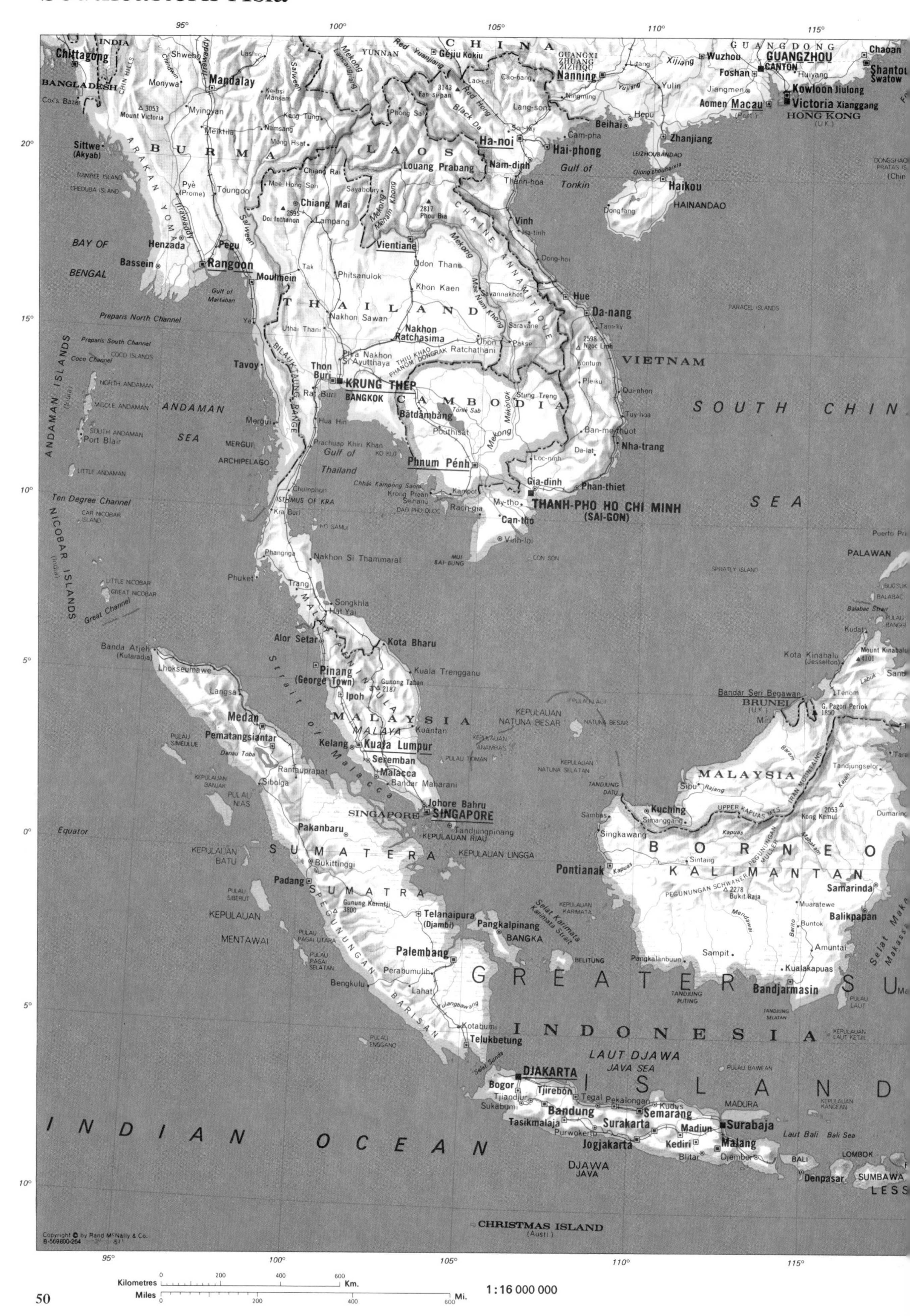

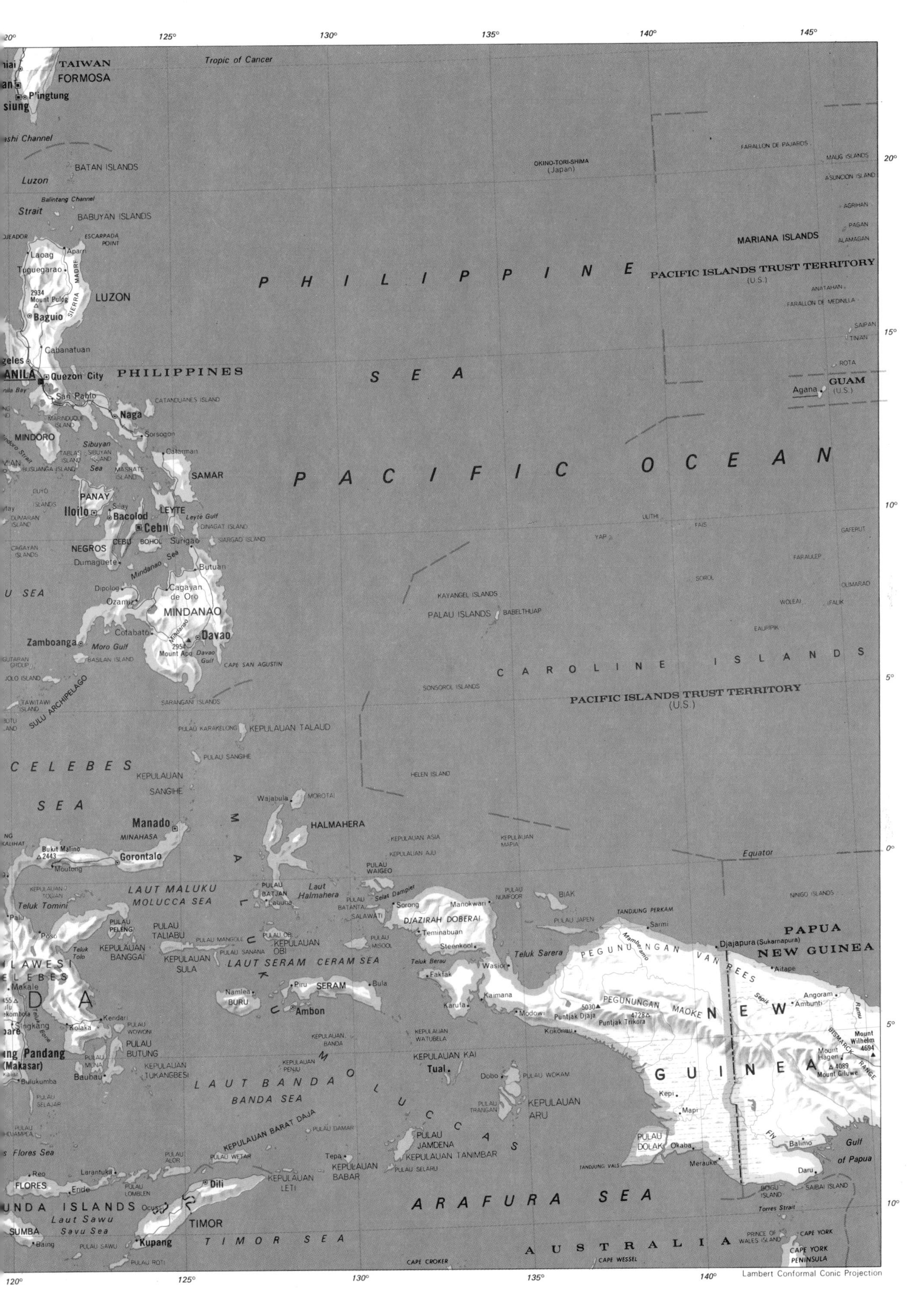
TAIWAN
FORMOSA
P'ingtung
Tropic of Cancer
BATAN ISLANDS
Luzon
Strait
BABUYAN ISLANDS
LUZON
Baguio
Mount Pulog
Cabanatuan
Quezon City
PHILIPPINES
San Pablo
Naga
MINDORO
Sibuyan Sea
SAMAR
PANAY
Iloilo
Bacolod
LEYTE
Leyte Gulf
Cebu
NEGROS
CEBU
BOHOL
Surigao
Dumaguete
Mindanao Sea
Butuan
Dipolog
Ozamiz
Cagayan de Oro
MINDANAO
Zamboanga
Moro Gulf
Cotabato
Davao
Mount Apo
Davao Gulf
CAPE SAN AGUSTIN
SULU ARCHIPELAGO
SARANGANI ISLANDS
KEPULAUAN TALAUD
CELEBES SEA
KEPULAUAN SANGIHE
Manado
MINAHASA
Gorontalo
Bukit Malino
MOROTAI
HALMAHERA
LAUT MALUKU
MOLUCCA SEA
Laut Halmahera
PULAU WAIGEO
Selat Dampier
Sorong
Manokwari
DJAZIRAH DOBERAI
Teminabuan
Steenkool
BIAK
PULAU JAPEN
TANDJUNG PERKAM
Sarmi
Djajapura (Sukarnapura)
PAPUA NEW GUINEA
Aitape
Teluk Sarera
PEGUNUNGAN VAN REES
PEGUNUNGAN MAOKE
Puntjak Djaja
Puntjak Trikora
Wasior
Fakfak
Kaimana
Kokonau
Mount Wilhelm
Mount Hagen
Mount Giluwe
BISMARCK RANGE
Angoram
Ambunti
Sepik
NEW GUINEA
Teluk Tomini
PULAU PELENG
PULAU TALIABU
KEPULAUAN BANGGAI
KEPULAUAN SULA
KEPULAUAN OBI
LAUT SERAM
CERAM SEA
SERAM
Bula
BURU
Namlea
Ambon
Kendari
Kolaka
PULAU BUTUNG
Baubau
KEPULAUAN TUKANGBESI
Ujung Pandang (Makasar)
Bulukumba
LAUT BANDA
BANDA SEA
MOLUCCAS
KEPULAUAN KAI
Tual
Dobo
KEPULAUAN ARU
Kepi
Mapi
PULAU DOLAK
Okaba
Merauke
Fly
Balimo
Gulf of Papua
Daru
Torres Strait
KEPULAUAN BARAT DAJA
PULAU JAMDENA
KEPULAUAN TANIMBAR
KEPULAUAN BABAR
KEPULAUAN LETI
Dili
TIMOR
Kupang
Flores Sea
FLORES
Ende
Larantuka
SUMBA
SUNDA ISLANDS
Laut Sawu
Savu Sea
TIMOR SEA
ARAFURA SEA
AUSTRALIA
CAPE YORK
CAPE YORK PENINSULA
CAPE WESSEL
CAPE CROKER
PHILIPPINE SEA
PACIFIC OCEAN
OKINO-TORI-SHIMA (Japan)
MARIANA ISLANDS
PACIFIC ISLANDS TRUST TERRITORY (U.S.)
GUAM (U.S.)
Agana
ROTA
TINIAN
SAIPAN
CAROLINE ISLANDS
PALAU ISLANDS
KAYANGEL ISLANDS
BABELTHUAP
YAP
ULITHI
SONSOROL ISLANDS
HELEN ISLAND
Equator
Lambert Conformal Conic Projection

Indochina

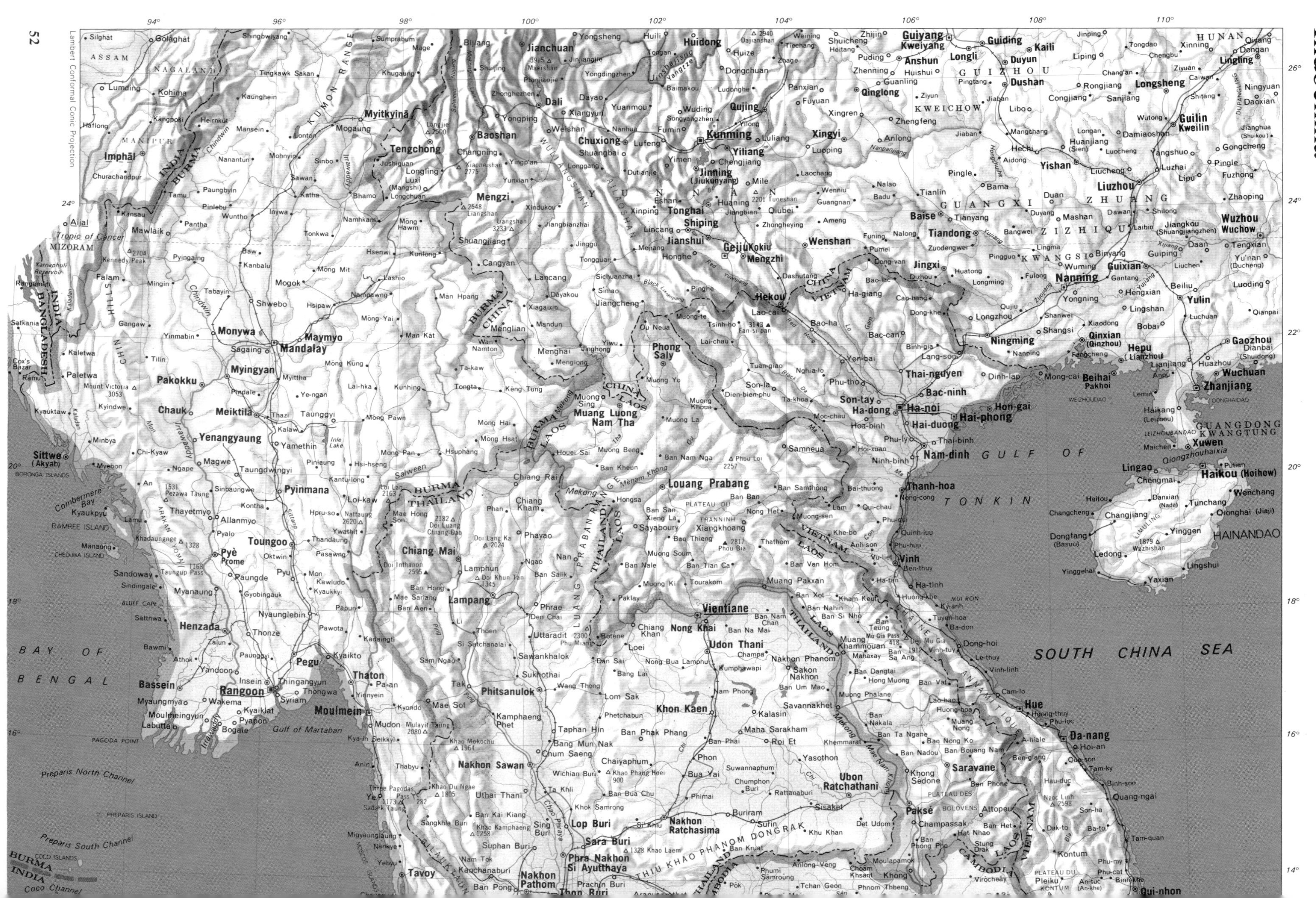

Lambert Conformal Conic Projection

Kilometres 0 100 200 300 Km.

Miles 0 100 200 300 Mi.

1:8 000 000

Australia

INDIAN OCEAN
INDONESIA
Timor Sea
Tasikmalaja
Jogjakarta
Surakarta
Madiun
Kediri
Blitar
Malang
Djember
DJAWA JAVA
BALI
Denpasar
LOMBOK
SUMBAWA
LESSER SUNDA ISLANDS
FLORES
SUMBA
Laut Sawu
Baing
PULAU ROTI
SAWU
Ocussi
TIMOR
Kupang
MELVILLE ISLAND
BATHURST ISLAND
Beagle Gulf
Van Diemen Gulf
Darwin
Rum Jungle
ARNHEM LAN
Katherine
CARTIER ISLAND (Austl.)
CAPE LONDONDERRY
Joseph Bonaparte Gulf
SCOTT REEF
BONAPARTE ARCHIPELAGO
Wyndham
Victoria
Birdum
LYNHER REEF
BUCCANEER ARCHIPELAGO
Collier Bay
KIMBERLEY PLATEAU
CAPE LEVEQUE
King Sound
KING LEOPOLD RANGES
936
Derby
Mount Ord
Wave Hill
ROWLEY SHOALS
Broome
Fitzroy
Fitzroy Crossing
Halls Creek
NORTHER
La Grange
EIGHTY MILE BEACH
TANAMI DESERT
TERRITOR
GREAT SANDY DESERT
DAMPIER ARCHIPELAGO
Port Hedland
Roebourne
Dampier
Marble Bar
Lake White (Dry)
Barrow
BARROW ISLAND
MUIRON ISLANDS
NORTH WEST CAPE
Onslow
Exmouth Gulf
Fortescue
Nullagine
THROSSELL RANGE
Lake Auld (Dry)
Lake Mackay (Dry Salt Lake)
HAMERSLEY RANGE
1227
Mount Bruce
Lake Disappointment (Dry Salt Lake)
Mount Leisler
1006
Mount 1510
MACDON
WESTERN
GIBSON DESERT
Lake Amadeus (Dry)
Tropic of Capricorn
1106 Mount Augustus
910 Mount Essendon
Mount Olga 1069
A U S T
Lake McLeod
Carnarvon
BERNIER ISLAND
Gascoyne
Lake Carnegie (Dry)
Mount Aloysius 987
MUSGRAVE RANGES
1440
Mount Wood
Shark Bay
Wooramel
DIRK HARTOG ISLAND
Murchison
Meekatharra
Wiluna
AUSTRALIA
GREAT VICTORIA DESERT
Lake Austin (Dry)
Sandstone
Mount Magnet
Mount Redcliffe 576
HOUTMAN ROCKS
Mullewa
Geraldton
Dongara
Three Springs
Leonora
Malcolm
Lake Carey (Dry Salt Lake)
Lake Barlee (Dry Salt Lake)
SOU
Ooldea
Forrest
Dalwallinu
Kalgoorlie
Coolgardie
Rawlinna
NULLARBOR PLAIN
Eucla
Moora
Bullfinch
Southern Cross
Lake Cowan (Dry Salt Lake)
CAPE ADIEU
Northam
DARLING RANGE
Perth
Beverley
Norseman
Newdegate
Wagin
Geographe Bay
Bunbury
Ravensthorpe
Esperance
CAPE ARID
Great Australian Bight
INVEST
CAPE NATURALISTE
Busselton
ARCHIPELAGO OF THE RECHERCHE
Augusta
Pemberton
CAPE LEEUWIN
Albany
CAPE VANCOUVER
POINT D'ENTRECASTEAUX
INDIAN OCEA
110°
115°
120°
125°
130°
105°
10°
15°
20°
25°
30°
35°
40°
Kilometres
0
200
400
600
Km.
Miles
0
200
400
600
Mi.
1:16 000 000

Sea
Gulf of Port Moresby
PAPUA NEW GUINEA
Papua
NEW GUINEA
Torres Strait
PRINCE OF WALES ISLAND
CAPE YORK
WESSEL ISLANDS
THE ENGLISH COMPANY'S ISLANDS
CAPE ARNHEM
Gulf of Carpentaria
GROOTE EYLANDT
Limmen Bight
SIR EDWARD PELLEW GROUP
MORNINGTON ISLAND
WELLESLEY ISLANDS
BARKLY TABLELAND
CAPE YORK PENINSULA
Weipa
Coen
Musgrave
Cooktown
Mareeba
Cairns
Ravenshoe
Burketown
Normanton
Croydon
Forsayth
Ingham
Townsville
Home Hill
Bowen
Charters Towers
Camooweal
Mount Isa
Cloncurry
Richmond
Hughenden
Duchess
Winton
Mackay
Blair Athol
QUEENSLAND
Longreach
Barcaldine
Emerald
Rockhampton
Blackall
Gladstone
GREAT ARTESIAN BASIN
SIMPSON DESERT
Yaraka
Windorah
Theodore
Bundaberg
Hervey Bay
FRASER ISLAND
Maryborough
Gayndah
Charleville
Quilpie
Mitchell
Roma
Gympie
Kingaroy
Mount Kiangarow
Nambour
Dalby
Toowoomba
Ipswich
Brisbane
NORTH STRADBROKE ISLAND
Oodnadatta
Lake Eyre (North) (Dry Salt Lake)
Coopers Creek
Innamincka
STURT DESERT
GREY RANGE
Thargomindah
Cunnamulla
Saint George
Goondiwindi
Warwick
Murwillumbah
Lake Eyre (South)
Marree
AUSTRALIA
Warrina
Tenterfield
Lismore
Milparinka
Moree
Walgett
Bourke
Inverell
Glen Innes
Grafton
The Round Mountain
Armidale
Coffs Harbour
Lake Torrens (Dry Salt Lake)
Lake Frome (Dry Salt Lake)
Woomera
Lake Gairdner (Dry Salt Lake)
Saint Mary Peak
FLINDERS RANGES
GAWLER RANGES
Port Augusta
Broken Hill
Wilcannia
Darling
Nyngan
Tamworth
Port Macquarie
Taree
Dubbo
NEW SOUTH WALES
Whyalla
Peterborough
Port Pirie
EYRE PENINSULA
Mount Hope
Spencer Gulf
Lincoln
Elizabeth
Adelaide
Gulf Saint Vincent
Investigator Strait
KANGAROO ISLAND
Encounter Bay
Cessnock
Maitland
Newcastle
Orange
Mildura
Hay
Griffith
Lachlan
SYDNEY
Wollongong
Wagga Wagga
Goulburn
Canberra
A.C.T.
Swan Hill
Murray
Albury
Bordertown
Horsham
Naracoorte
Shepparton
Wangaratta
Bendigo
VICTORIA
Cooma
Mount Kosciusko
Bombala
Ararat
Ballarat
GREAT DIVIDING RANGE
Mount Gambier
Geelong
MELBOURNE
Orbost
CAPE HOWE
Portland
Warrnambool
Morwell
Sale
NINETY MILE BEACH
CAPE OTWAY
SOUTH EAST POINT
WILSONS PROMONTORY
KING ISLAND
Bass Strait
FLINDERS ISLAND
FURNEAUX GROUP
HUNTER ISLAND
Smithton
Burnie
Banks Strait
Scottsdale
Devonport
Launceston
Mount Ossa
Zeehan
Saint Marys
Strahan
TASMANIA
New Norfolk
Hobart
SOUTH WEST CAPE
SOUTH BRUNY
SOUTH EAST CAPE
Popondetta
TROBRIAND ISLANDS
Losuia
D'ENTRECASTEAUX ISLANDS
Kulumadau
WOODLARK ISLAND
OWEN STANLEY RANGE
Esa-Ala
Samarai
LOUISIADE ARCHIPELAGO
MISIMA I.
TAGULA ISLAND
ROSSEL ISLAND
Solomon Sea
VELLA LAVELLA
CHOISEUL
SANTA ISABEL
Gizo
NEW GEORGIA
RENDOVA
VANGUNU
SOLOMON ISLANDS (U.K.)
GUADALCANAL
Honiara
Mt. Popomanasiu
RENNELL ISLAND
INDISPENSABLE REEFS
GREAT BARRIER REEF
OSPREY REEF
Coral Sea
BOUGAINVILLE REEF
CAPE GRAFTON
HOLMES REEFS
WILLIS ISLETS (Austl.)
CORINGA ISLETS (Austl.)
LIHOU REEFS
TREGOSSE ISLETS (Austl.)
MELLISH REEF
HINCHINBROOK ISLAND
Halifax Bay
ÎLES CHESTERFIELD (N. Cal.)
ÎLES DE SABLE (N. Cal.)
CUMBERLAND ISLANDS
CAPE PALMERSTON
SWAIN REEFS
KENN REEFS
SAUMAREZ REEF
CAYE DE L'OBSERVATOIRE (N. Cal.)
BELLONA REEFS
WRECK REEFS
CURTIS I.
CATO ISLAND
Tropic of Capricorn
PACIFIC OCEAN
MIDDLETON REEF
ELIZABETH REEF
LORD HOWE ISLAND (N.S.W.)
Tasman Sea
Lambert Conformal Conic Projection

Southeastern Australia

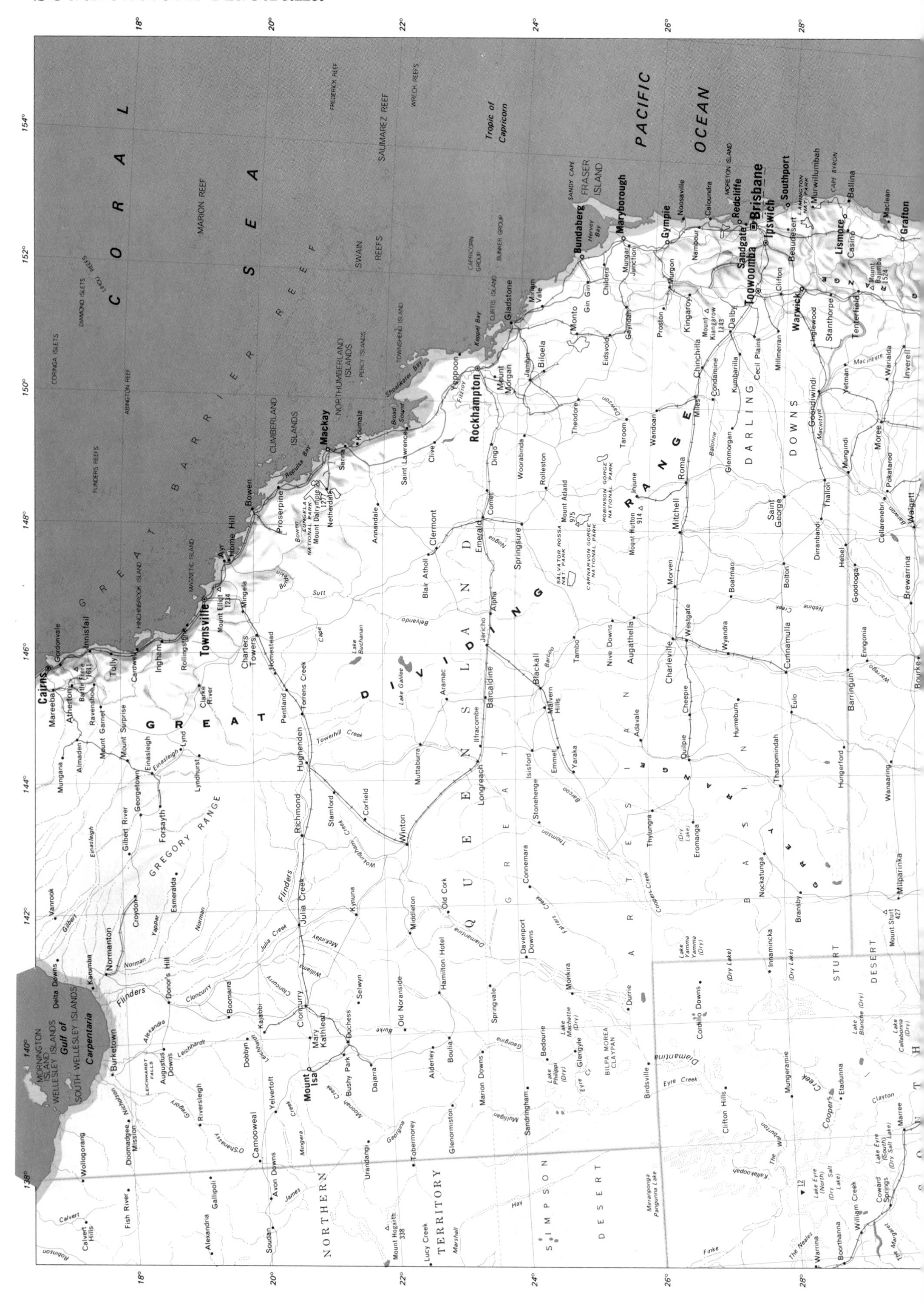

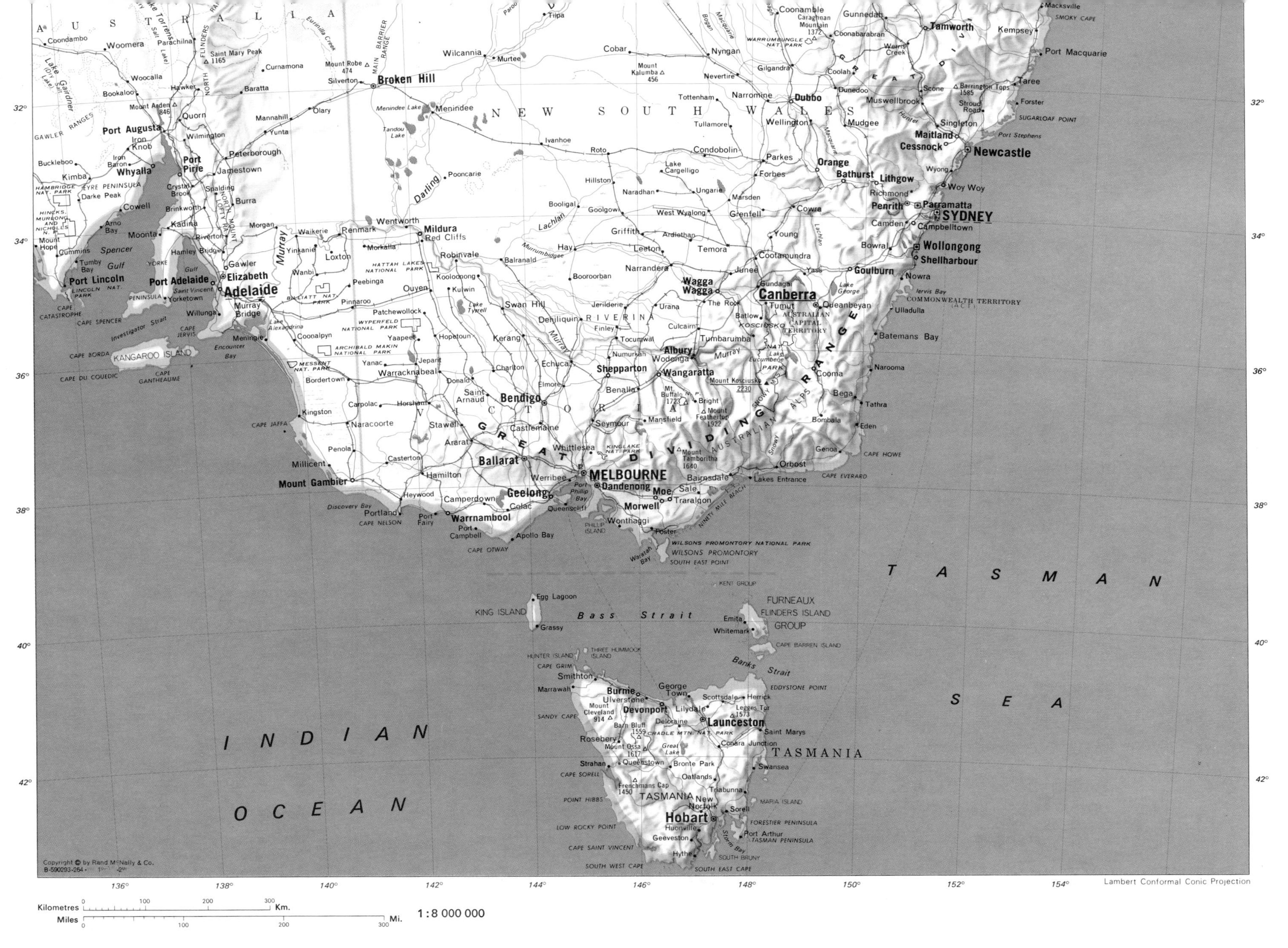

NEW SOUTH WALES
VICTORIA
TASMANIA
AUSTRALIAN CAPITAL TERRITORY
GREAT DIVIDING RANGE
INDIAN OCEAN
TASMAN SEA
Bass Strait
Banks Strait
KANGAROO ISLAND
KING ISLAND
FURNEAUX
FLINDERS ISLAND
GROUP
SYDNEY
MELBOURNE
Adelaide
Canberra
Hobart
Newcastle
Wollongong
Broken Hill
Port Augusta
Port Pirie
Whyalla
Port Lincoln
Port Adelaide
Elizabeth
Mildura
Bendigo
Ballarat
Geelong
Warrnambool
Mount Gambier
Shepparton
Wangaratta
Albury
Wagga Wagga
Dubbo
Orange
Bathurst
Lithgow
Goulburn
Tamworth
Launceston
Devonport
Burnie
Morwell
Moe
Dandenong
Mount Kosciusko
2230
RIVERINA
Spencer Gulf
Gulf Saint Vincent
Investigator Strait
Encounter Bay
Port Phillip Bay
WILSONS PROMONTORY
CAPE OTWAY
CAPE HOWE
Lambert Conformal Conic Projection
Kilometres
Km.
Miles
Mi.
1 : 8 000 000

Southwestern Australia

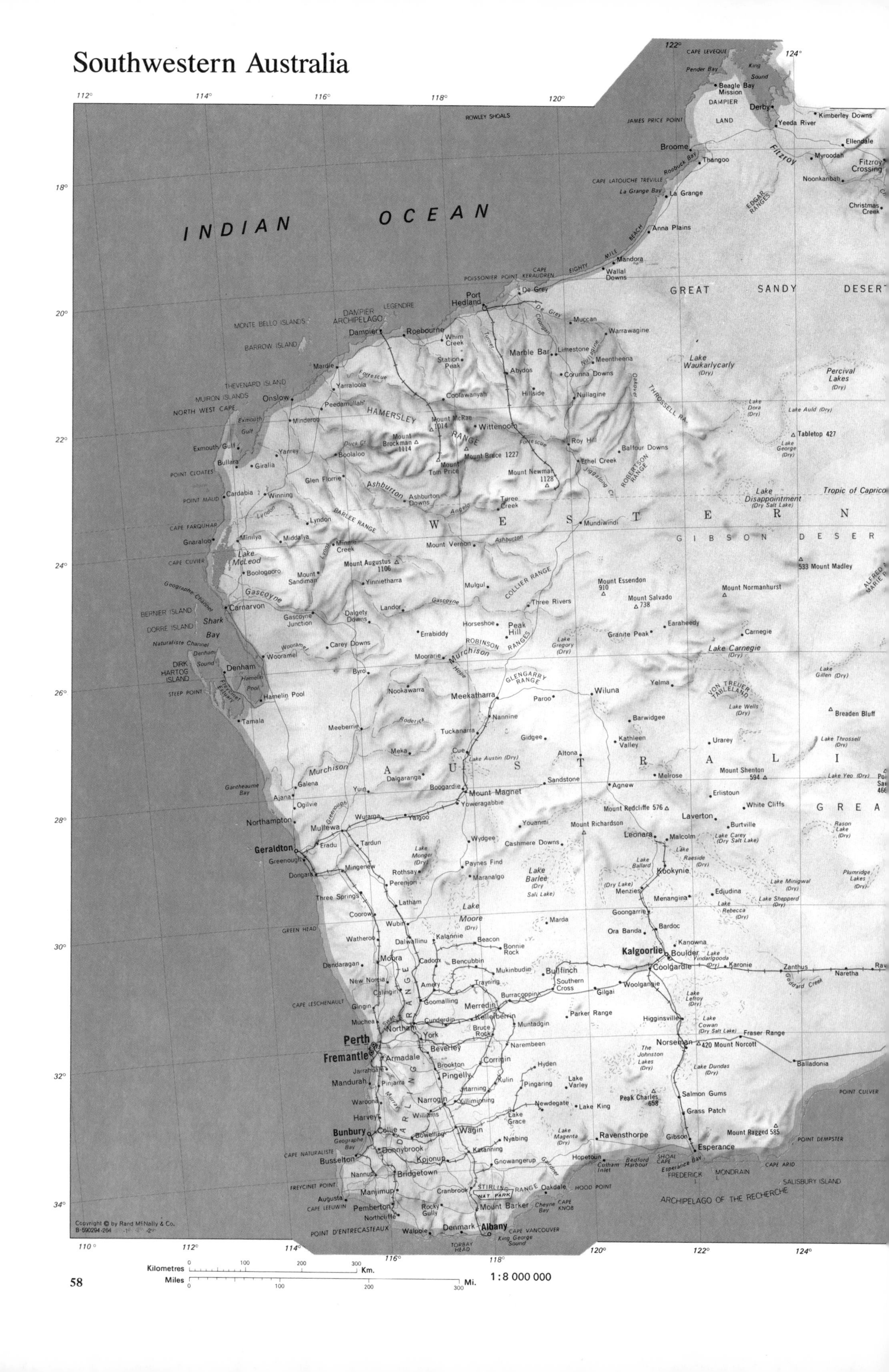

Kilometres 0 100 200 300 Km.

Miles 0 100 200 300 Mi.

1 : 8 000 000

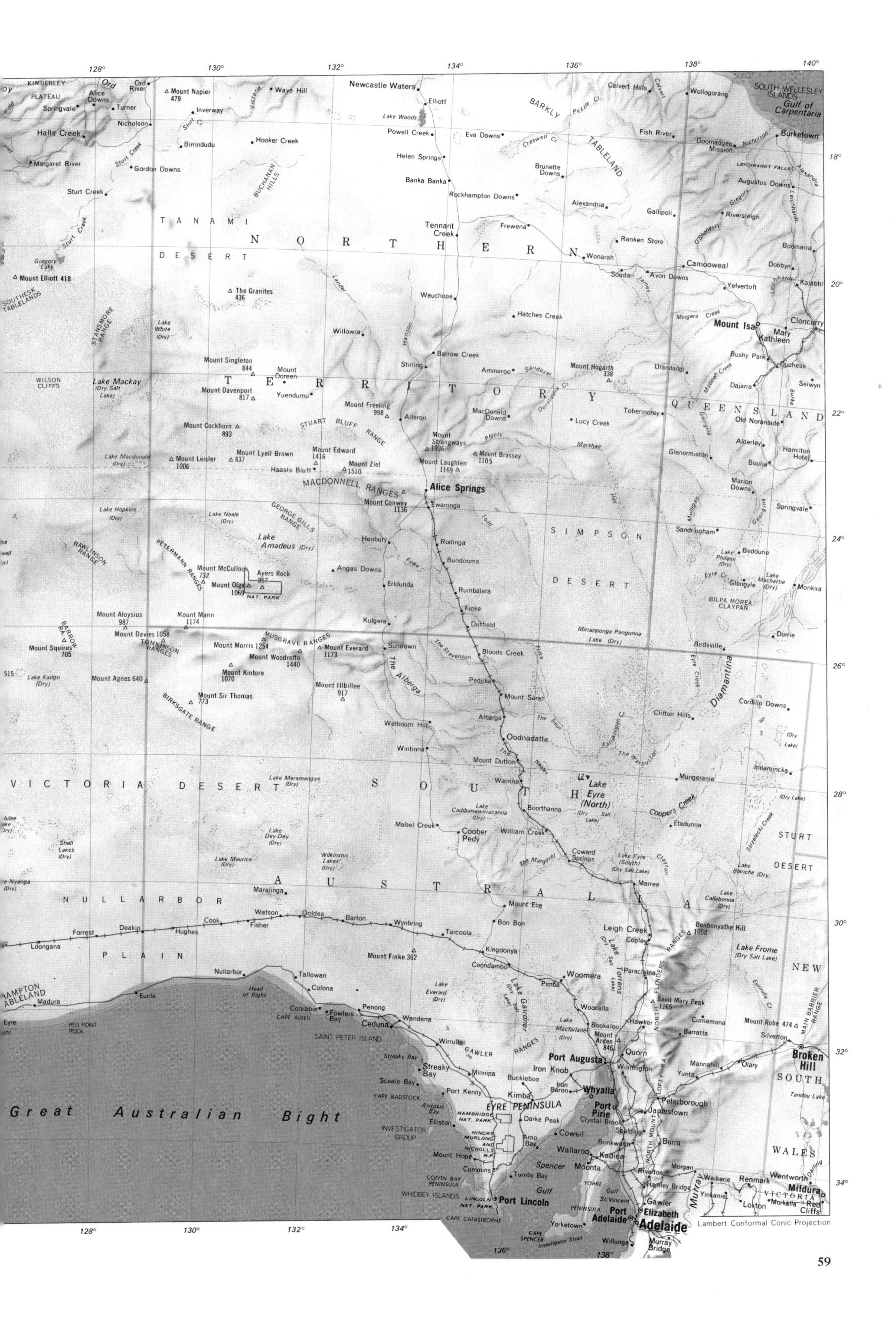
128°
130°
132°
134°
136°
138°
140°
18°
20°
22°
24°
26°
28°
30°
32°
34°
KIMBERLEY
PLATEAU
Ord
Ord River
Alice Downs
Springvale
Turner
Nicholson
Halls Creek
Mount Napier 479
Inverway
Victoria
Wave Hill
Sturt Cr.
Birrindudu
Hooker Creek
Sturt Creek
Margaret River
Gordon Downs
BUCHANAN HILLS
Sturt Creek
T A N A M I
D E S E R T
Gregory Lake
Mount Elliott 418
SOUTHESK TABLELANDS
The Granites 436
STANSMORE RANGE
Lake White (Dry)
WILSON CLIFFS
Lake Mackay (Dry Salt Lake)
Mount Singleton 844
Mount Doreen
Mount Davenport 817
Yuendumu
Mount Cockburn 893
Lake Macdonald (Dry)
Mount Leisler 1006
Mount Lyell Brown 837
Haasts Bluff
Mount Edward 1416
Mount Ziel 1510
Lake Hopkins (Dry)
Lake Neale (Dry)
GEORGE GILLS RANGE
Lake Amadeus (Dry)
RAWLINSON RANGE
PETERMANN RANGES
Mount McCulloch 732
Mount Olga 1069
Ayers Rock 867
NAT. PARK
Mount Aloysius 987
Mount Mann 1174
BARROW RA.
Mount Davies 1058
TOMKINSON RANGES
Mount Squires 705
Mount Morris 1254
MUSGRAVE RANGES
Mount Everard 1173
Mount Woodroffe 1440
Lake Kadgo (Dry)
Mount Agnes 640
Mount Kintore 1070
BIRKSGATE RANGE
Mount Sir Thomas 773
Mount Illbillee 917
Newcastle Waters
Elliott
Lake Woods
Powell Creek
Eva Downs
Helen Springs
Banka Banka
Rockhampton Downs
Tennant Creek
Frewena
N O R T H E R N
T E R R I T O R Y
Lander
Wauchope
Hatches Creek
Willowra
Hanson
Barrow Creek
Stirling
Ammaroo
Sandover
Mount Hogarth 338
Mount Freeling 998
Aileron
STUART BLUFF RANGE
MacDonald Downs
Lucy Creek
Mount Strangways 1036
Mount Laughlen 1169
Mount Brassey 1105
Plenty
MACDONNELL RANGES
Alice Springs
Mount Conway 1136
Ewaninga
Todd
Henbury
Rodinga
Finke
Bundooma
Angas Downs
Erldunda
Rumbalara
Finke
Kulgera
Duffield
Sundown
The Stevenson
Bloods Creek
The Alberga
Pedirka
Mount Sarah
Alberga
Welbourn Hill
Oodnadatta
Wintinna
Mount Dutton
Warrina
BARKLY TABLELAND
Calvert Hills
Calvert
Wollogorang
Puzzle Cr.
Creswell Cr.
Brunette Downs
Fish River
Alexandria
Gallipoli
Ranken Store
Wonarah
Soudan
James
Avon Downs
Camooweal
S I M P S O N
D E S E R T
Hay
Mirranponga Pangunna Lake (Dry)
SOUTH WELLESLEY ISLANDS
Gulf of Carpentaria
Burketown
Doomadgee Mission
Nicholson
LEICHHARDT FALLS
Alexandra
Augustus Downs
Leichhardt
Gregory
Riversleigh
O'Shanassy
Boomarra
Dobbyn
Yelvertoft
Kajabbi
Mingera Creek
Mount Isa
Cloncurry
Mary Kathleen
Bushy Park
Duchess
Urandangi
Moonah Creek
Selwyn
Dajarra
Burke
Tobermorey
Q U E E N S L A N D
Old Noranside
Georgina
Alderley
Glenormiston
Boulia
Hamilton Hotel
Marion Downs
Mulligan
Springvale
Georgina
Sandringham
Lake Philippi (Dry)
Bedourie
Eyre Cr.
Glengyle
Lake Machattie (Dry)
Monkira
BILPA MOREA CLAYPAN
Durrie
Birdsville
Diamantina
Eyre Creek
Cordillo Downs
Clifton Hills
Kallakoopah Cr.
The Warburton
Innamincka
Mungeranie
Lake Eyre (North) (Dry Salt Lake)
Coopers Creek
Etadunna
Strzelecki Creek
STURT
DESERT
Lake Blanche (Dry)
Lake Callabonna (Dry)
Clayton
Lake Eyre (South) (Dry Salt Lake)
Marree
V I C T O R I A
D E S E R T
Lake Meramangye (Dry)
S O U T H
A U S T R A L I A
Shell Lakes (Dry)
Lake Dey-Dey (Dry)
Lake Maurice (Dry)
Wilkinson Lakes (Dry)
Lake Cadibarrawirracanna (Dry)
Boorthanna
Mabel Creek
Coober Pedy
William Creek
The Margaret
Coward Springs
N U L L A R B O R
P L A I N
Maralinga
Mount Eba
Bon Bon
Watson
Ooldea
Barton
Cook
Fisher
Wynbring
Forrest
Deakin
Hughes
Loongana
Tarcoola
Mount Finke 362
Kingoonya
Coondambo
Leigh Creek
Copley
Benbonyathe Hill 1058
Lake Frome (Dry Salt Lake)
Lake Torrens (Dry Salt Lake)
FLINDERS RANGES
NORTH FLINDERS RANGES
Parachilna
Woomera
Pimba
Lake Gairdner (Dry Salt Lake)
Lake Everard (Dry)
Woocalla
Saint Mary Peak 1165
Eurinilla Cr.
Bookaloo
Hawker
Curnamona
Mount Robe 474
MAIN BARRIER RANGE
Lake Macfarlane (Dry)
Mount Arden 846
Barratta
Silverton
Quorn
Broken Hill
Port Augusta
Wilmington
Mannahill
Olary
NEW
SOUTH
WALES
Tandou Lake
Nullarbor
Tallowan
Colona
Head of Bight
HAMPTON TABLELAND
Eucla
Madura
RED POINT ROCK
Coorabie
Fowlers Bay
CAPE ADIEU
Penong
Ceduna
Wandana
SAINT PETER ISLAND
Wirrulla
GAWLER RANGES
Streaky Bay
Streaky Bay
Minnipa
Buckleboo
Iron Knob
Iron Baron
Whyalla
Sceale Bay
CAPE RADSTOCK
Port Kenny
Kimba
EYRE PENINSULA
Port Pirie
Crystal Brook
Peterborough
Jamestown
Yunta
LOFTY RA.
Anxious Bay
HAMBRIDGE NAT. PARK
Darke Peak
Elliston
INVESTIGATOR GROUP
HINCKS MURLONG AND NICHOLLS N.P.
Arno Bay
Cowell
Spalding
Brinkworth
Burra
NORTH MOUNT LOFTY RANGES
Mount Hope
Wallaroo
Kadina
Spencer Gulf
Cummins
Tumby Bay
Moonta
YORKE PENINSULA
Riverton
Morgan
Waikerie
Renmark
Wentworth
Darling
Mildura
COFFIN BAY PENINSULA
Hamley Bridge
Murray
Yinkanie
Loxton
VICTORIA
Morkalla
Red Cliffs
WHIDBEY ISLANDS
LINCOLN NAT. PARK
Port Lincoln
Gulf St. Vincent
Gawler
Elizabeth
Port Adelaide
Adelaide
CAPE CATASTROPHE
Yorketown
CAPE SPENCER
Investigator Strait
Willunga
Murray Bridge
Great Australian Bight
Lambert Conformal Conic Projection

Northern Australia

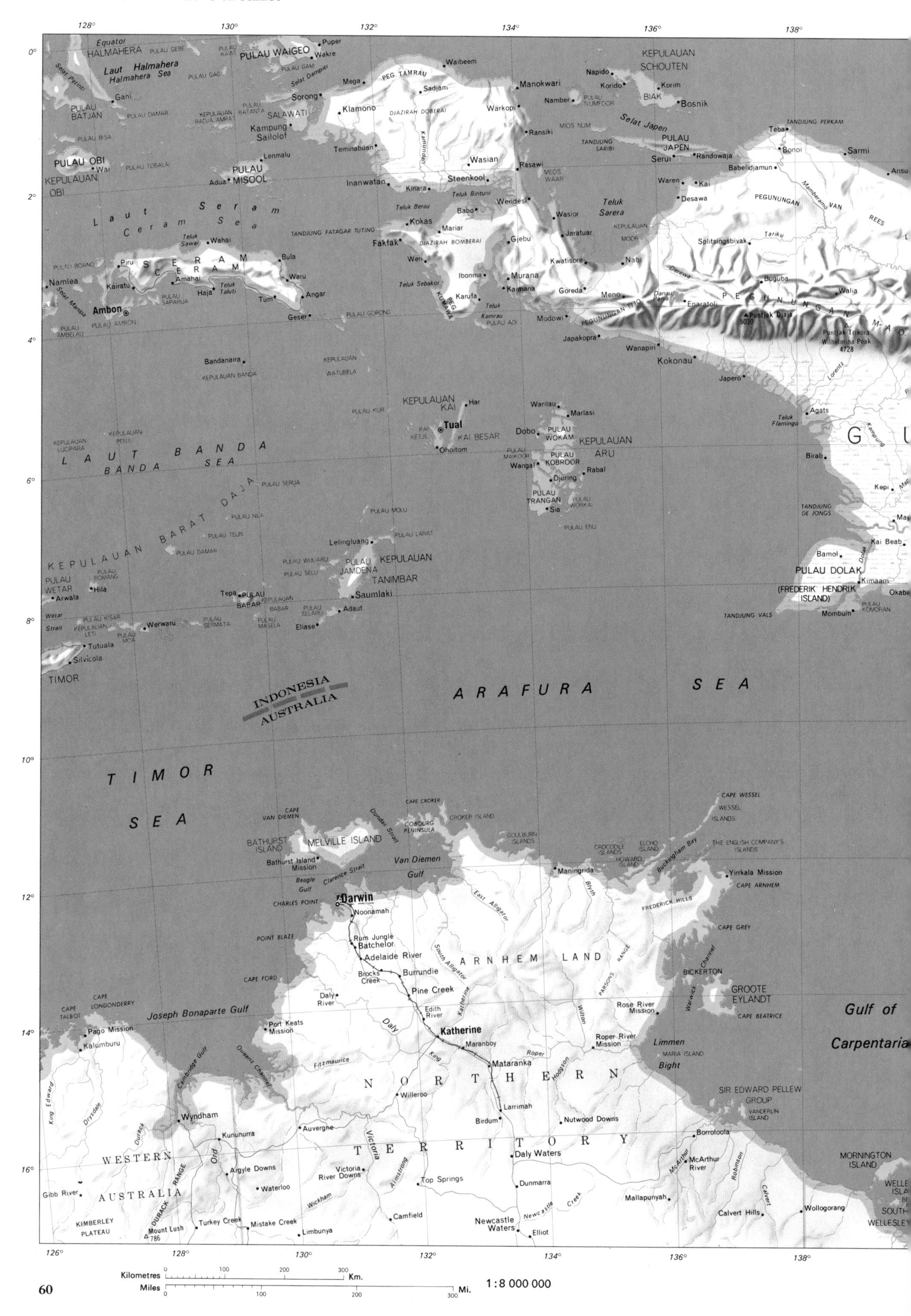

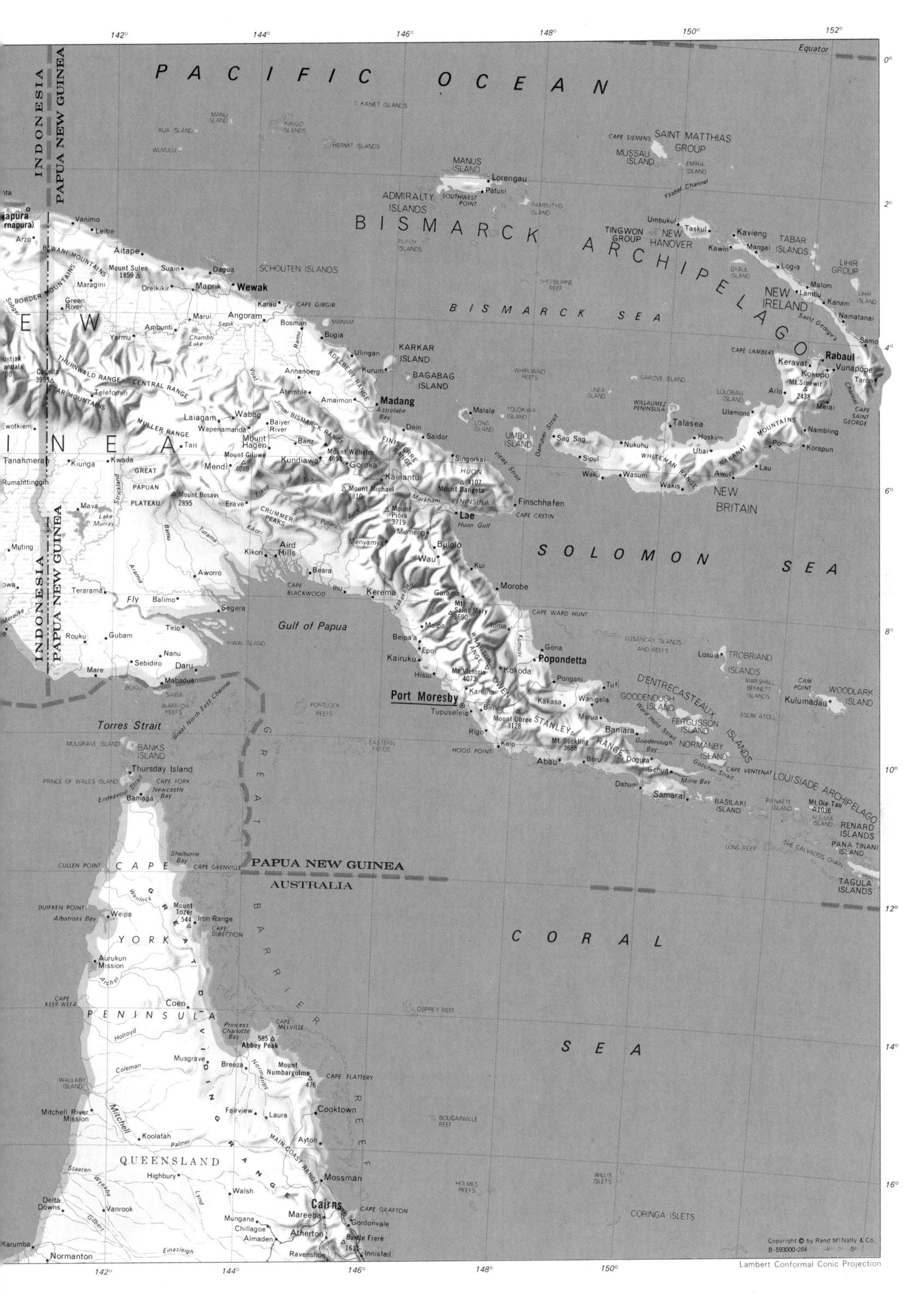

PACIFIC OCEAN
INDONESIA
PAPUA NEW GUINEA
BISMARCK ARCHIPELAGO
BISMARCK SEA
SOLOMON SEA
CORAL SEA
Gulf of Papua
Torres Strait
MANUS ISLAND
Lorengau
ADMIRALTY ISLANDS
SAINT MATTHIAS GROUP
MUSSAU ISLAND
NEW HANOVER
Kavieng
TABAR ISLANDS
LIHIR GROUP
NEW IRELAND
Rabaul
Kokopo
NEW BRITAIN
Talasea
Kimbe
KARKAR ISLAND
BAGABAG ISLAND
Madang
Wewak
SCHOUTEN ISLANDS
Aitape
Vanimo
Angoram
Mount Hagen
Goroka
Kainantu
Mount Wilhelm 4694
Lae
Huon Gulf
HUON PENINSULA
Finschhafen
Wau
Bulolo
Morobe
Kerema
Daru
Port Moresby
Kokoda
Popondetta
OWEN STANLEY RANGE
GOODENOUGH ISLAND
FERGUSSON ISLAND
NORMANBY ISLAND
D'ENTRECASTEAUX ISLANDS
TROBRIAND ISLANDS
WOODLARK ISLAND
Samarai
LOUISIADE ARCHIPELAGO
TAGULA ISLANDS
RENARD ISLANDS
GREAT PAPUAN PLATEAU
Fly
Aird Hills
PAPUA NEW GUINEA
AUSTRALIA
GREAT BARRIER REEF
CAPE YORK PENINSULA
GREAT DIVIDING RANGE
QUEENSLAND
Thursday Island
BANKS ISLAND
Weipa
Coen
Cooktown
Mossman
Cairns
Mareeba
Atherton
Innisfail
Normanton
Karumba
CORINGA ISLETS
Equator
Copyright © by Rand McNally & Co.
Lambert Conformal Conic Projection

New Zealand

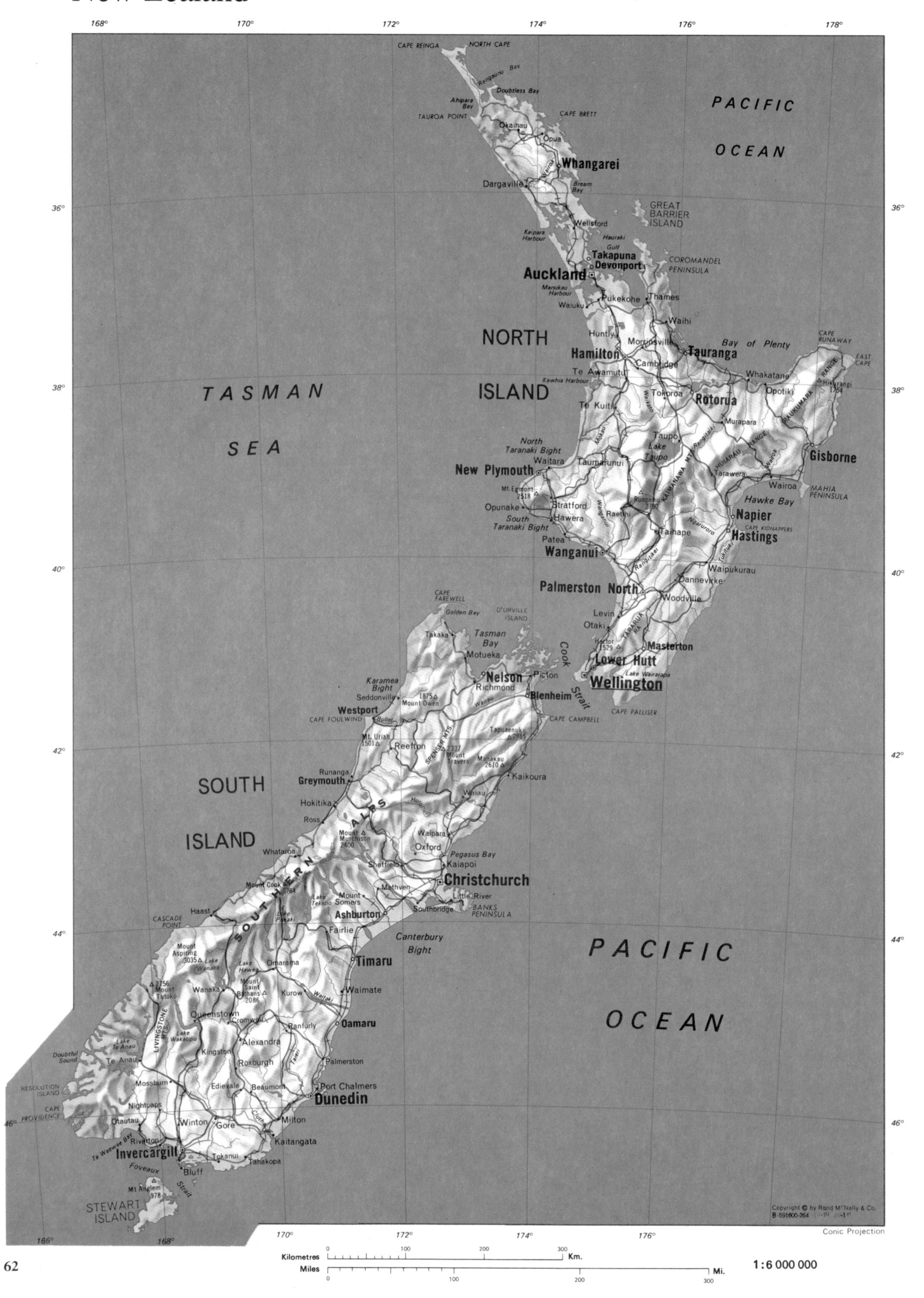

PACIFIC OCEAN
TASMAN SEA
NORTH ISLAND
SOUTH ISLAND
STEWART ISLAND
CAPE REINGA
NORTH CAPE
Whangarei
Dargaville
GREAT BARRIER ISLAND
Takapuna
Devonport
Auckland
Hamilton
Tauranga
Bay of Plenty
Rotorua
Gisborne
New Plymouth
Napier
Hastings
Hawke Bay
Wanganui
Palmerston North
Masterton
Lower Hutt
Wellington
Cook Strait
Nelson
Blenheim
Westport
Greymouth
Hokitika
Kaikoura
Christchurch
Pegasus Bay
Ashburton
Timaru
Canterbury Bight
Oamaru
Dunedin
Invercargill
Foveaux Strait
SOUTHERN ALPS
Conic Projection
Kilometres
Miles
1:6 000 000

Pacific Islands

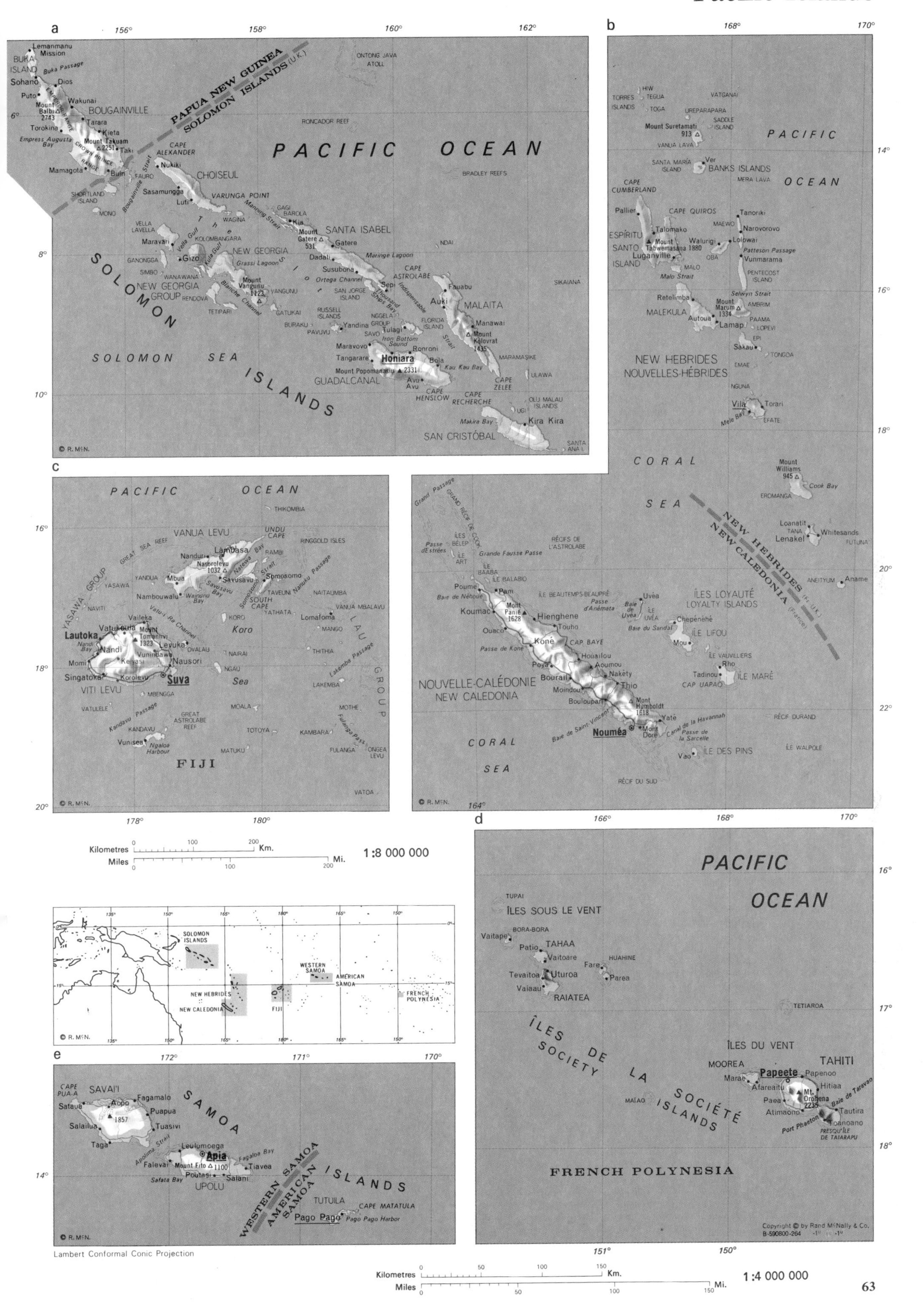

Canada

Beaufort Sea
PACIFIC OCEAN
Gulf of Alaska
UNITED STATES
CANADA
ALASKA RANGE
BROOKS RANGE
ROCKY MOUNTAINS
COAST MOUNTAINS
MACKENZIE MTS.
SELWYN MOUNTAINS
CASCADE RANGE
SIERRA NEVADA
NORTH WEST TERRITORIES
BRITISH COLUMBIA
ALBERTA
SASKATCHEWAN
MANITOBA
WASHINGTON
OREGON
IDAHO
MONTANA
WYOMING
NEVADA
UTAH
COLORADO
NORTH DAKOTA
SOUTH DAKOTA
NEBRASKA
MINNESOTA
IOWA
BANKS ISLAND
VICTORIA ISLAND
MELVILLE ISLAND
PRINCE OF WALES ISLAND
VANCOUVER ISLAND
QUEEN CHARLOTTE ISLANDS
Great Bear L.
Great Slave Lake
Lake Athabasca
Lake Winnipeg
Anchorage
Fairbanks
Whitehorse
Dawson
Yellowknife
Edmonton
Calgary
Saskatoon
Regina
Winnipeg
Vancouver
Victoria
Seattle
Spokane
Portland
Boise
Salt Lake City
Sacramento
SAN FRANCISCO
Minneapolis
Bismarck
Fargo
1:16 000 000
Kilometres 0 200 400 600 Km.
Miles 0 200 400 600 Mi.

GREENLAND
(Denmark)
Baffin
Bay
Davis
Strait
Labrador
Sea
ATLANTIC
OCEAN
ATLANTIC
OCEAN
Hudson
Bay
Foxe
Basin
PÉNINSULE
D'UNGAVA
Ungava
Bay
QUÉBEC
LABRADOR
NEWFOUNDLAND
ONTARIO
James
Bay
BAFFIN
ISLAND
CUMBERLAND
PENINSULA
SOUTHAMPTON
ISLAND
DEVON ISLAND
Lancaster Sound
CAPE CHIDLEY
RESOLUTION ISLAND
Frobisher Bay
HALL PENINSULA
MELVILLE PENINSULA
PRINCE CHARLES ISLAND
MANSEL ISLAND
COATS ISLAND
BELCHER ISLANDS
AKIMISKI ISLAND
CAPE HENRIETTA MARIA
All islands within Hudson Bay, James Bay, and Ungava Bay lie within Northwest Territories
Arctic Circle
KAP FARVEL
CAPE HARRISON
Goose Bay
Churchill
Battle Harbour
St. Anthony
Gulf of St. Lawrence
ÎLE D'ANTICOSTI
Sept-Îles
Chicoutimi
Quebec
Trois-Rivières
MONTREAL
Ottawa
Sherbrooke
NEW BRUNSWICK
Fredericton
Moncton
Saint John
Halifax
NOVA SCOTIA
Sydney
Glace Bay
CAPE BRETON ISLAND
PRINCE EDWARD ISLAND
Charlottetown
ST. PIERRE AND MIQUELON
(Fr.)
St. John's
CAPE RACE
Corner Brook
Gander
CAPE SABLE
SABLE ISLAND
APPALACHIAN MTS.
NOTRE DAME MTS.
Bangor
Portland
BOSTON
Providence
New Bedford
Hartford
New Haven
NEW YORK
PHILADELPHIA
Albany
Schenectady
Worcester
Manchester
Concord
Burlington
Montpelier
ADIRONDACK MOUNTAINS
Syracuse
Rochester
Buffalo
Utica
Binghamton
Scranton
Wilkes-Barre
Harrisburg
Trenton
Newark
Poughkeepsie
Elmira
Erie
Cleveland
Youngstown
Toledo
Windsor
DETROIT
Flint
Lansing
Saginaw
Bay City
Sarnia
London
Kitchener
Guelph
Hamilton
TORONTO
Oshawa
Peterborough
Kingston
Watertown
Brockville
Cornwall
Pembroke
North Bay
Sudbury
Sault Ste. Marie
Timmins
Kirkland Lake
Rouyn
Thunder Bay
Marquette
Escanaba
Green Bay
Appleton
Oshkosh
Milwaukee
Racine
Kenosha
Madison
La Crosse
Dubuque
Rockford
CHICAGO
South Bend
Kalamazoo
Grand Rapids
Muskegon
MICHIGAN
WISCONSIN
ILLINOIS
NEW YORK
PENNSYLVANIA
NEW JERSEY
OHIO
MAINE
Lake Superior
Lake Huron
Lake Michigan
Lake Erie
Lake Ontario
Moosonee
Fort Albany
Attawapiskat
Fort Severn
Winisk
Copyright © by Rand McNally & Co.
Lambert Conformal Conic Projection

United States

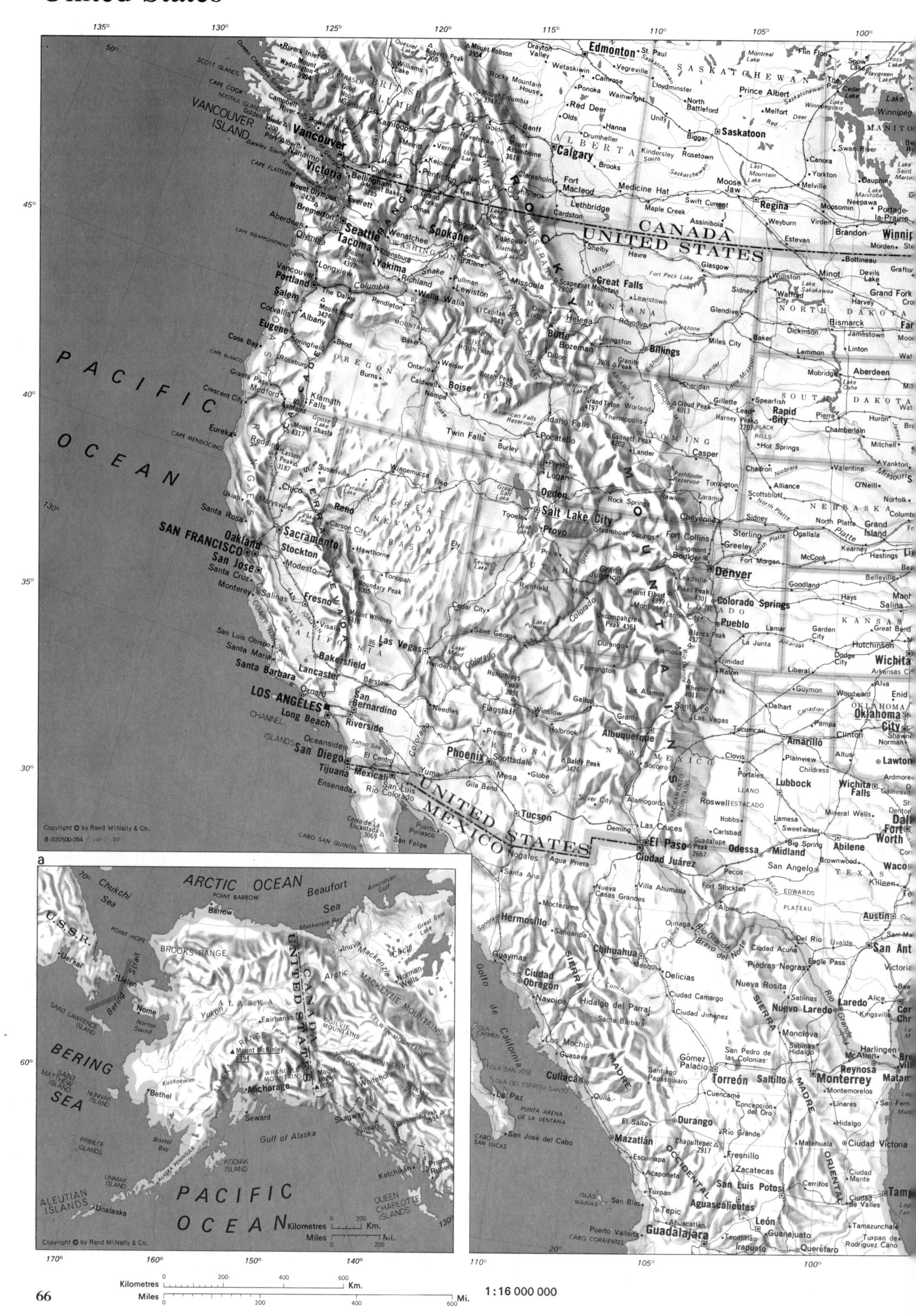

Kilometres 0 200 400 600 Km.
Miles 0 200 400 600 Mi.
1:16 000 000

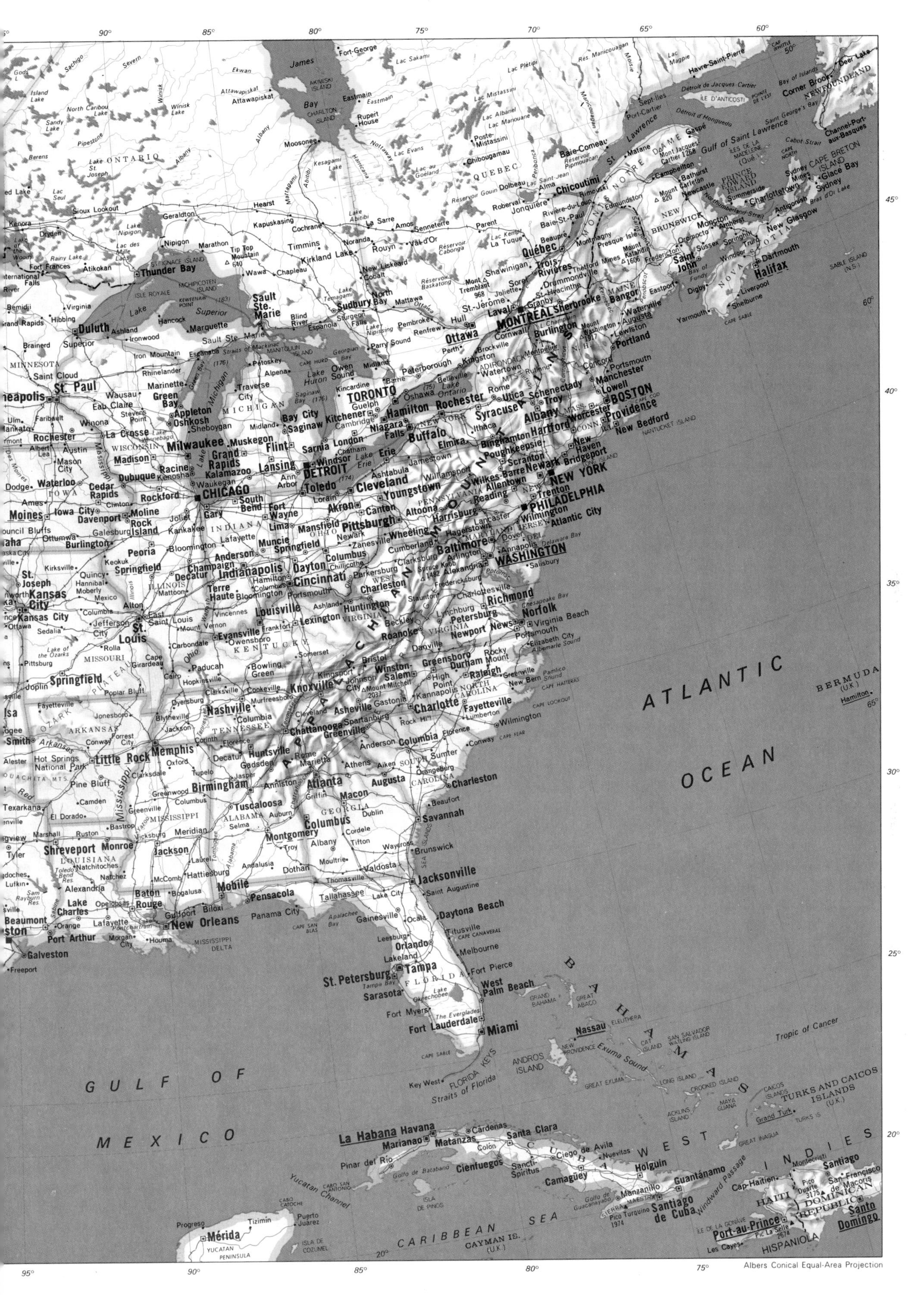
ATLANTIC
OCEAN
GULF OF
MEXICO
CARIBBEAN SEA
WEST INDIES
BAHAMAS
APPALACHIAN MOUNTAINS
NOTRE DAME MTS.
Gulf of Saint Lawrence
Lake Superior
Lake Michigan
Lake Huron
Lake Erie
Lake Ontario
James Bay
QUEBEC
ONTARIO
NEW BRUNSWICK
NOVA SCOTIA
PRINCE EDWARD ISLAND
NEWFOUNDLAND
MINNESOTA
WISCONSIN
MICHIGAN
IOWA
ILLINOIS
INDIANA
OHIO
KENTUCKY
TENNESSEE
MISSOURI
ARKANSAS
LOUISIANA
MISSISSIPPI
ALABAMA
GEORGIA
FLORIDA
NORTH CAROLINA
VIRGINIA
WEST VIRGINIA
PENNSYLVANIA
NEW YORK
MAINE
MONTREAL
Québec
Ottawa
TORONTO
DETROIT
CHICAGO
NEW YORK
PHILADELPHIA
WASHINGTON
BOSTON
Baltimore
Pittsburgh
Cleveland
Buffalo
Cincinnati
Indianapolis
Louisville
St. Louis
Milwaukee
Minneapolis
St. Paul
Duluth
Thunder Bay
Memphis
Nashville
Atlanta
Birmingham
Charlotte
Richmond
Norfolk
Charleston
Savannah
Jacksonville
Tampa
Orlando
Miami
New Orleans
Mobile
Houston
Galveston
Halifax
Saint John
Sydney
Nassau
La Habana Havana
Santiago de Cuba
Port-au-Prince
Santo Domingo
HAITI
DOMINICAN REPUBLIC
HISPANIOLA
TURKS AND CAICOS ISLANDS (U.K.)
BERMUDA (U.K.)
Hamilton
CAYMAN IS. (U.K.)
Yucatan Channel
Straits of Florida
Tropic of Cancer
Mérida
YUCATAN PENINSULA
Albers Conical Equal-Area Projection

Northeastern United States

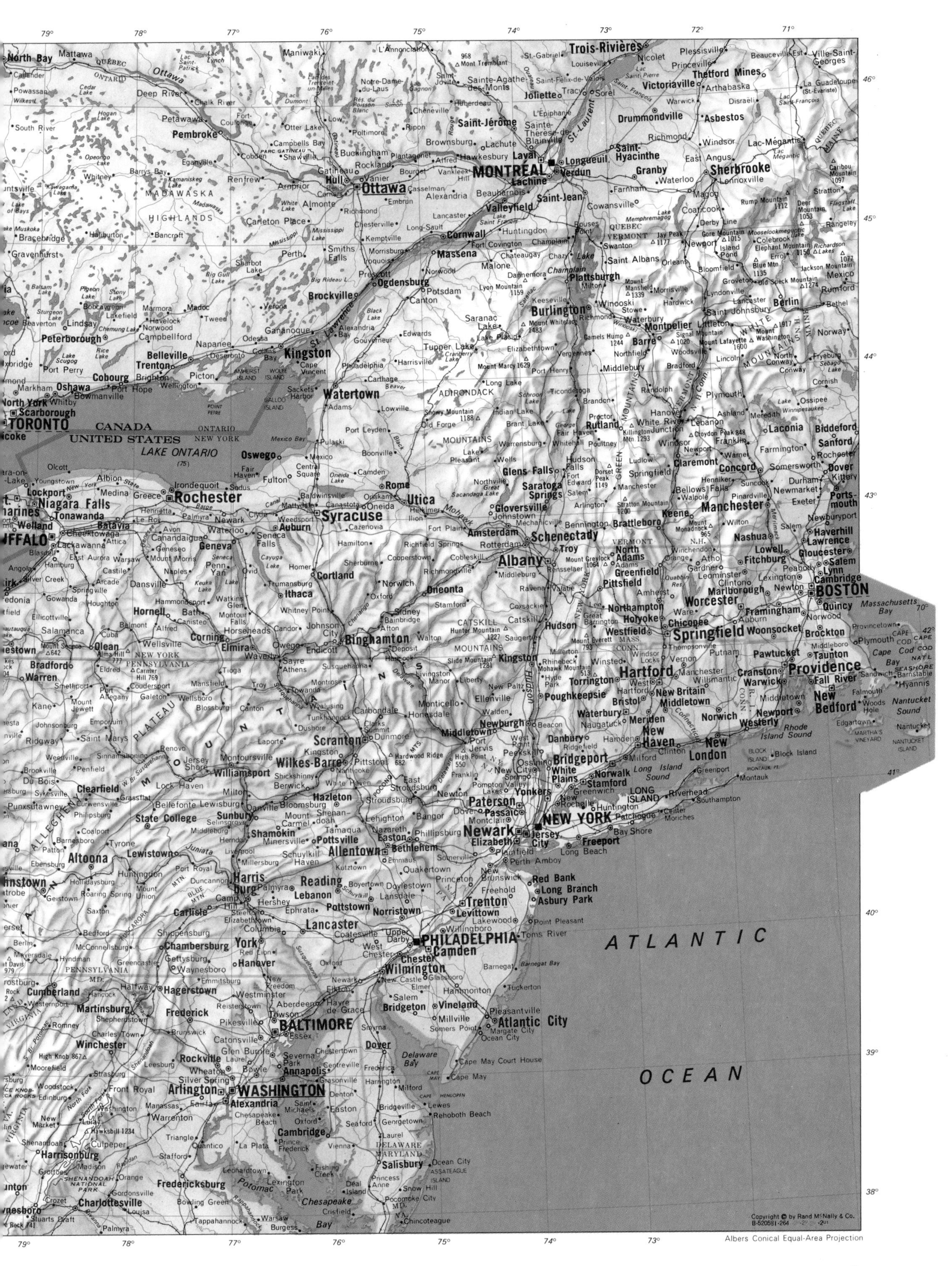

Albers Conical Equal-Area Projection

California, Nevada and Hawaii

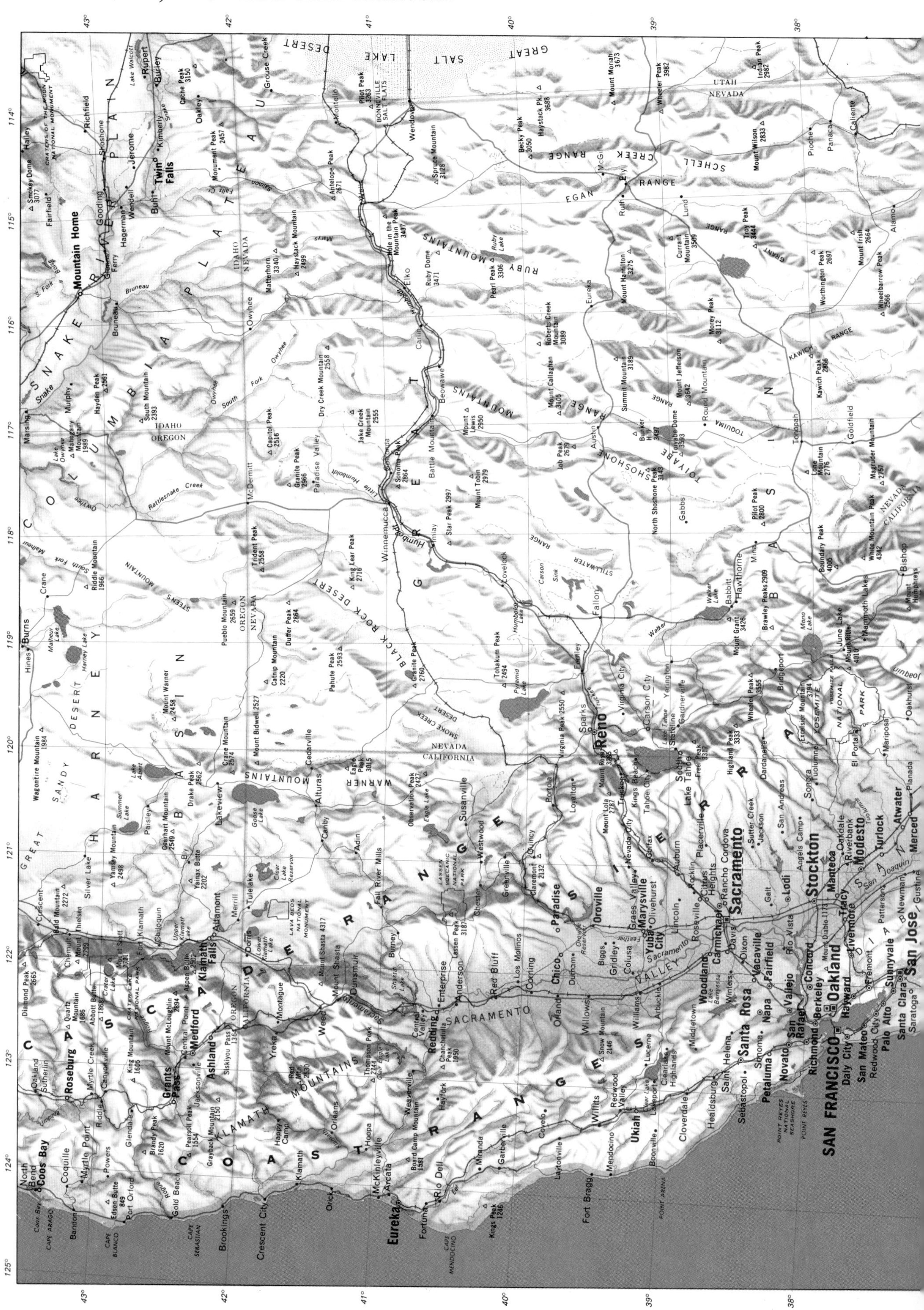

PACIFIC OCEAN
Los Angeles
San Diego
Las Vegas
Fresno
Bakersfield
Santa Barbara
San Bernardino
Riverside
Pasadena
Long Beach
Santa Ana
Tijuana
Mexicali
Ensenada
Yuma
Salinas
Monterey
San Luis Obispo
Lompoc
Ventura
Oxnard
Lancaster
Barstow
Victorville
Needles
Kingman
Palm Springs
Indio
El Centro
Calexico
Brawley
San Luis Rio Colorado
Visalia
Porterville
Tulare
Hanford
Delano
Santa Maria
Oceanside
Escondido
Chula Vista
Santa Monica
Torrance
Anaheim
MOJAVE DESERT
SIERRA NEVADA
COAST RANGES
SAN JOAQUIN VALLEY
CHANNEL ISLANDS
DESIERTO DE ALTAR
Golfo de California
Salton Sea
IMPERIAL VALLEY
ARIZONA
NEVADA
CALIF.
U.S.
MEXICO
Colorado
Lake Mead
Honolulu
OAHU
KAUAI
NIIHAU
MOLOKAI
LANAI
MAUI
KAHOOLAWE
HAWAII
Hilo
Kailua
Kaneohe
Wahiawa
Lihue
Kahului
Wailuku
Lahaina
Kaunakakai
PACIFIC OCEAN
Kaulakahi Channel
Kauai Channel
Kaiwi Channel
Alenuihaha Channel
1:4 000 000
Kilometres
Miles
Albers Conical Equal-Area Projection
Copyright © by Rand McNally & Co.

Mexico

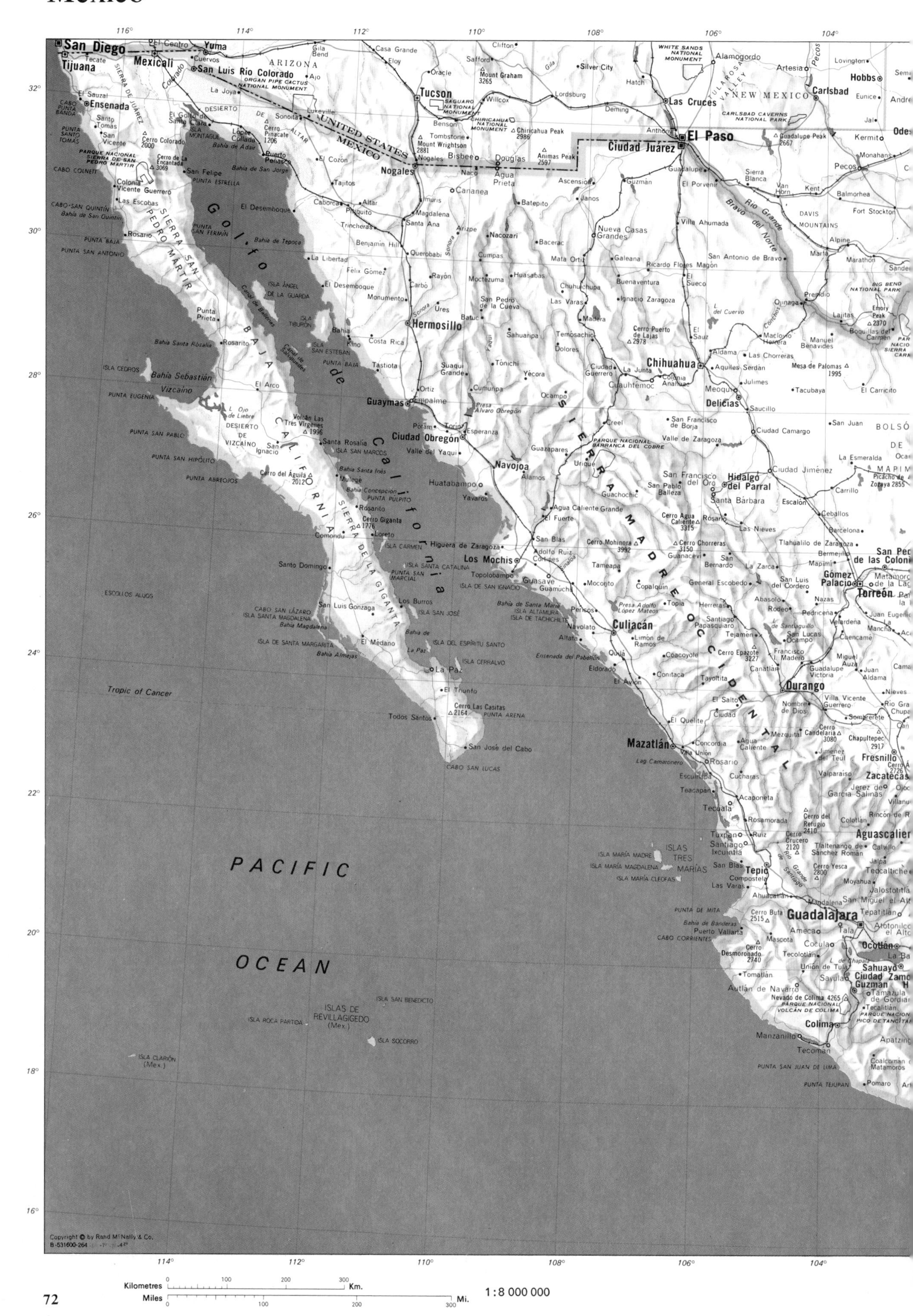

Kilometres 0 100 200 300 Km.
Miles 0 100 200 300 Mi.

1 : 8 000 000

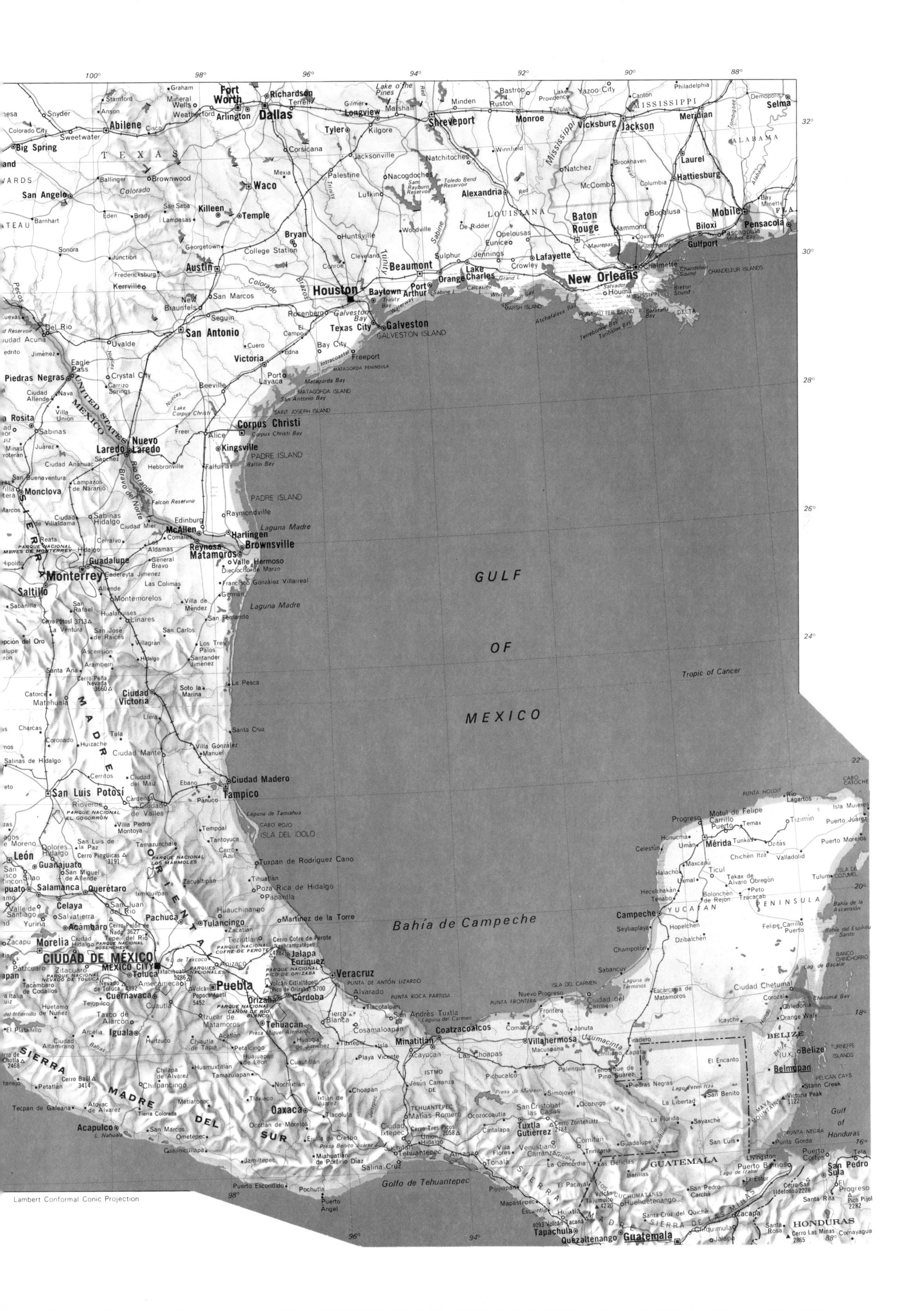

100°
98°
96°
94°
92°
90°
88°
32°
30°
28°
26°
24°
22°
20°
18°
16°
GULF
OF
MEXICO
Bahía de Campeche
Tropic of Cancer
Golfo de Tehuantepec
Gulf of Honduras
T E X A S
LOUISIANA
MISSISSIPPI
ALABAMA
FLA.
UNITED STATES
MEXICO
YUCATAN PENINSULA
BELIZE
GUATEMALA
HONDURAS
SIERRA MADRE ORIENTAL
SIERRA MADRE DEL SUR
Fort Worth
Dallas
Arlington
Richardson
Abilene
Big Spring
Snyder
Sweetwater
Waco
Killeen
Temple
Bryan
Austin
San Antonio
Houston
Baytown
Beaumont
Port Arthur
Orange
Galveston
Texas City
Victoria
Corpus Christi
Kingsville
Laredo
Nuevo Laredo
McAllen
Harlingen
Brownsville
Tyler
Longview
Kilgore
Corsicana
Palestine
Lufkin
Nacogdoches
Huntsville
San Angelo
Brownwood
Del Rio
Eagle Pass
Uvalde
Seguin
New Braunfels
San Marcos
College Station
Kerrville
Shreveport
Monroe
Ruston
Minden
Bastrop
Alexandria
Natchitoches
Lake Charles
Lafayette
Crowley
Jennings
Opelousas
Eunice
Baton Rouge
New Orleans
Houma
Chalmette
Vicksburg
Jackson
Natchez
McComb
Brookhaven
Hattiesburg
Laurel
Meridian
Columbia
Bogalusa
Hammond
Gulfport
Biloxi
Pascagoula
Mobile
Pensacola
Selma
Yazoo City
Canton
Philadelphia
Monterrey
Saltillo
Monclova
Piedras Negras
Reynosa
Matamoros
Guadalupe
Linares
Montemorelos
Ciudad Victoria
Ciudad Mante
Ciudad Madero
Tampico
San Luis Potosí
Matehuala
León
Guanajuato
Salamanca
Querétaro
Celaya
Irapuato
Morelia
Acámbaro
Pachuca
Tulancingo
CIUDAD DE MÉXICO
MEXICO CITY
Toluca
Cuernavaca
Puebla
Tlaxcala
Jalapa Enríquez
Veracruz
Córdoba
Orizaba
Tehuacán
Iguala
Chilpancingo
Acapulco
Oaxaca
Salina Cruz
Tehuantepec
Juchitán
Minatitlán
Coatzacoalcos
Villahermosa
Campeche
Ciudad del Carmen
Mérida
Progreso
Valladolid
Chetumal
Belize
Belmopan
Tuxtla Gutiérrez
San Cristóbal las Casas
Comitán
Tapachula
Guatemala
Quezaltenango
Huehuetenango
Puerto Barrios
San Pedro Sula
Puerto Cortés
Poza Rica de Hidalgo
Tuxpan de Rodríguez Cano
Papantla
Martínez de la Torre
Alvarado
San Andrés Tuxtla
Cosamaloapan
Tierra Blanca
PADRE ISLAND
MATAGORDA ISLAND
GALVESTON ISLAND
ISLA DEL CARMEN
Laguna Madre
Lambert Conformal Conic Projection

Central America and the Caribbean

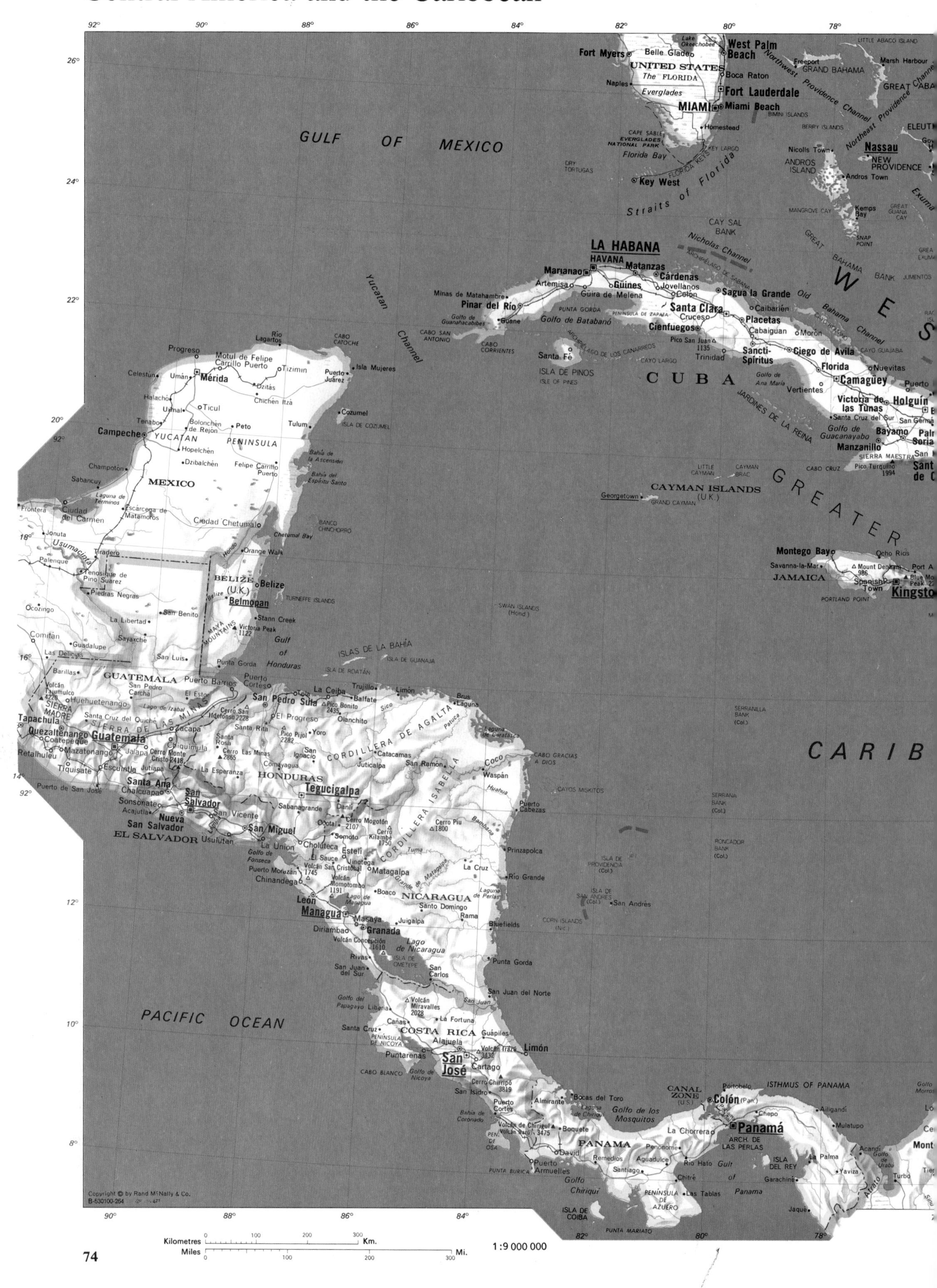

Kilometres 0 100 200 300 Km.
Miles 0 100 200 300 Mi.

1:9 000 000

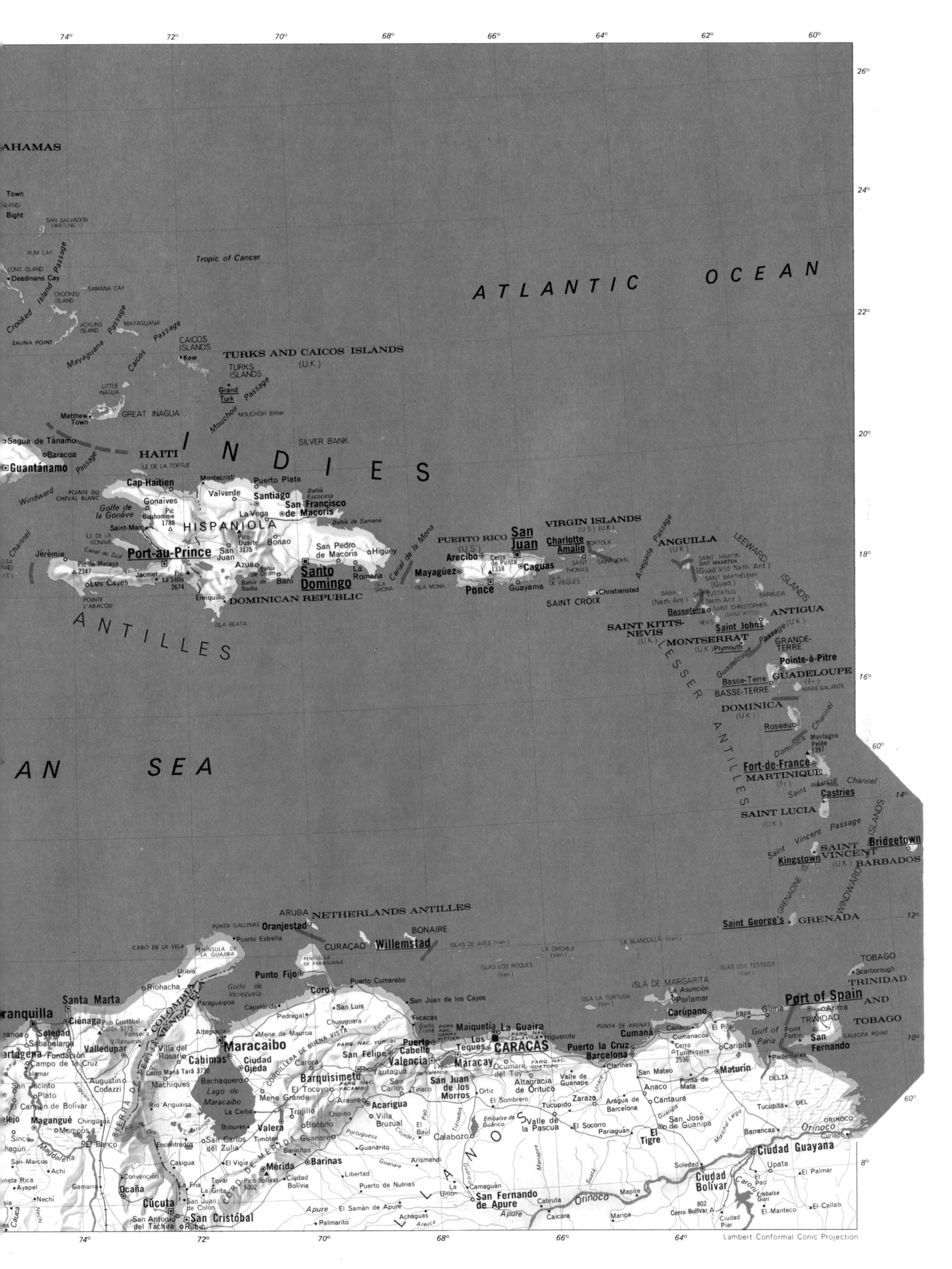

ATLANTIC OCEAN
INDIES
ANTILLES
AN SEA
BAHAMAS
Tropic of Cancer
Deadmans Cay
Crooked Island Passage
Mayaguana Passage
Caicos Passage
CAICOS ISLANDS
Kew
TURKS AND CAICOS ISLANDS
(U.K.)
TURKS ISLANDS
Grand Turk
Mouchoir Passage
GREAT INAGUA
Matthew Town
SILVER BANK
Sagua de Tánamo
Baracoa
Guantánamo
Windward Passage
HAITI
Cap-Haïtien
Gonaïves
Golfe de la Gonâve
Saint-Marc
HISPANIOLA
Jérémie
Port-au-Prince
Jacmel
Les Cayes
Montecristi
Puerto Plata
Valverde
Santiago
San Francisco de Macorís
La Vega
Bonao
San Juan
Azua
Baní
Santo Domingo
San Pedro de Macorís
La Romana
Higüey
Enriquillo
DOMINICAN REPUBLIC
Canal de la Mona
PUERTO RICO
(U.S.)
Arecibo
San Juan
Mayagüez
Caguas
Ponce
Guayama
VIRGIN ISLANDS
(U.S.) (U.K.)
Charlotte Amalie
Christiansted
SAINT CROIX
Anegada Passage
ANGUILLA
(U.K.)
LEEWARD ISLANDS
SAINT KITTS-NEVIS
(U.K.)
Basseterre
Saint Johns
ANTIGUA
MONTSERRAT
(U.K.) Plymouth
GRANDE-TERRE
Pointe-à-Pitre
GUADELOUPE
(Fr.)
Basse-Terre
BASSE-TERRE
Guadeloupe Passage
LESSER ANTILLES
DOMINICA
(U.K.)
Roseau
Dominica Channel
Fort-de-France
MARTINIQUE
(Fr.)
Saint Lucia Channel
Castries
SAINT LUCIA
(U.K.)
Saint Vincent Passage
SAINT VINCENT
(U.K.)
Kingstown
Bridgetown
BARBADOS
WINDWARD ISLANDS
Saint George's
GRENADA
ARUBA
NETHERLANDS ANTILLES
Oranjestad
BONAIRE
CURAÇAO
Willemstad
Puerto Estrella
Punto Fijo
Golfo de Venezuela
Puerto Cumarebo
Coro
San Juan de los Cayos
TOBAGO
Scarborough
TRINIDAD AND TOBAGO
Port of Spain
Arima
San Fernando
Gulf of Paria
ISLA DE MARGARITA
La Asunción
Porlamar
Carúpano
Cumaná
Barcelona
Puerto la Cruz
Güiria
Maturín
Santa Marta
Barranquilla
Ciénaga
Soledad
Cartagena
Valledupar
Riohacha
Uribia
COLOMBIA
VENEZUELA
Maracaibo
Cabimas
Ciudad Ojeda
Lago de Maracaibo
Machiques
Valera
Trujillo
Mérida
Barinas
Barquisimeto
Valencia
Maracay
Puerto Cabello
San Felipe
Maiquetía
La Guaira
Los Teques
CARACAS
Acarigua
Calabozo
Valle de la Pascua
El Tigre
Ciudad Guayana
Ciudad Bolívar
San Fernando de Apure
Cúcuta
San Cristóbal
Ocaña
Orinoco
Apure
LLANOS
Magangué
Mompós
El Banco
Lambert Conformal Conic Projection
74°
72°
70°
68°
66°
64°
62°
60°
26°
24°
22°
20°
18°
16°
14°
12°
10°
8°

Northern South America

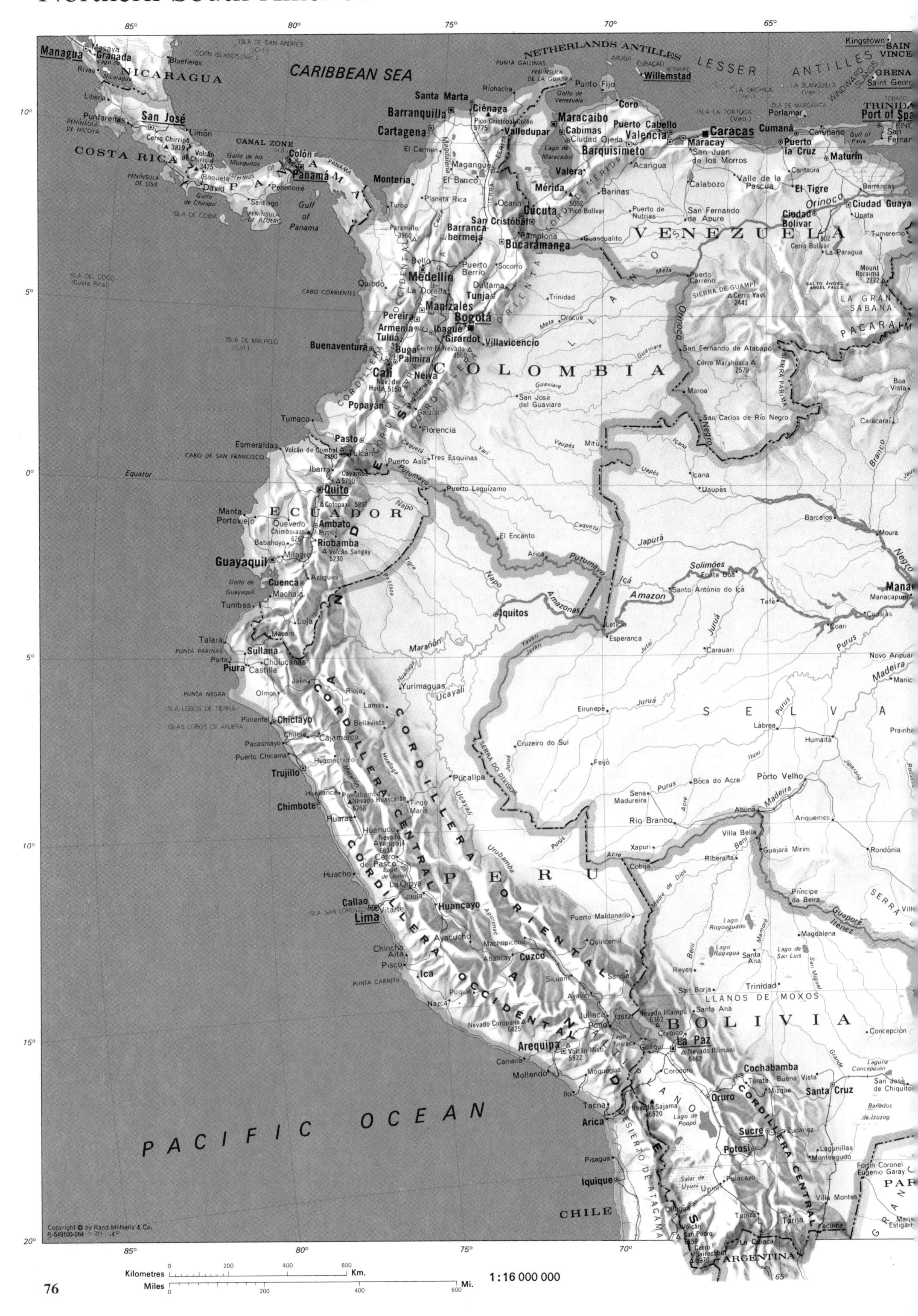

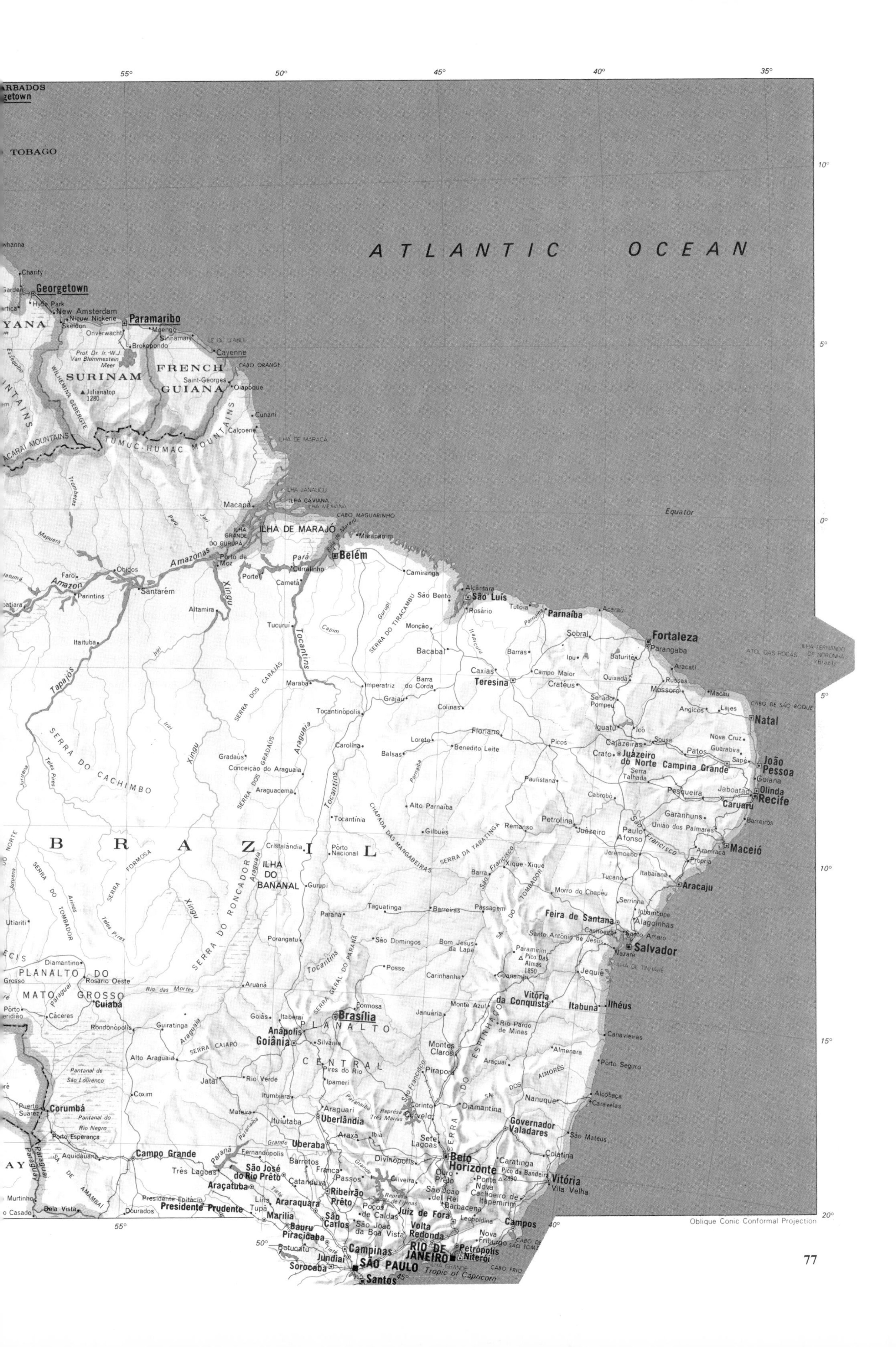

ATLANTIC OCEAN
Equator
Tropic of Capricorn
Oblique Conic Conformal Projection
55°
50°
45°
40°
35°
10°
5°
0°
15°
20°
BARBADOS
TOBAGO
Georgetown
New Amsterdam
Nieuw Nickerie
Paramaribo
SURINAM
FRENCH GUIANA
Cayenne
Saint-Georges
Oiapoque
ÎLE DU DIABLE
CABO ORANGE
Julianatop 1280
WILHELMINA GEBERGTE
TUMUC-HUMAC MOUNTAINS
ACARAI MOUNTAINS
Cunani
Calçoene
ILHA DE MARACÁ
ILHA JANAUCU
ILHA CAVIANA
ILHA MEXIANA
CABO MAGUARINHO
Macapá
ILHA DE MARAJÓ
ILHA GRANDE DO GURUPÁ
Amazonas
Amazon
Óbidos
Santarém
Parintins
Faro
Porto de Moz
Portel
Cametá
Curralinho
Pará
Belém
Marapanim
Camiranga
Alcântara
São Luís
São Bento
Rosário
Tutóia
Parnaíba
Acaraú
Fortaleza
Parangaba
Sobral
Ipu
Baturité
Aracati
Russas
Mossoró
Macau
ATOL DAS ROCAS
ILHA FERNANDO DE NORONHA (Brazil)
CABO DE SÃO ROQUE
Natal
Angicos
Lajes
João Pessoa
Campina Grande
Olinda
Recife
Caruaru
Jaboatão
Goiana
Maceió
Aracaju
Salvador
Feira de Santana
Alagoinhas
Ilhéus
Itabuna
Vitória da Conquista
Canavieiras
Pôrto Seguro
Alcobaça
Caravelas
São Mateus
Governador Valadares
Colatina
Vitória
Vila Velha
Teresina
Caxias
Campo Maior
Crateús
Barras
Bacabal
Monção
Imperatriz
Barra do Corda
Grajaú
Colinas
Floriano
Picos
Juàzeiro do Norte
Crato
Iguatu
Icó
Cajazeiras
Sousa
Patos
Quixadá
Senador Pompeu
Petrolina
Juàzeiro
Paulo Afonso
Garanhuns
Pesqueira
Barreiros
Propriá
Arapiraca
Tucano
Serrinha
Itabaiana
Jequié
Xique-Xique
Barra
Barreiras
Bom Jesus da Lapa
Tucuruí
Marabá
Tocantinópolis
Carolina
Balsas
Altamira
Itaituba
Tapajós
Xingu
Tocantins
Araguaia
São Francisco
Gurupi
SERRA DO CACHIMBO
SERRA DOS CARAJÁS
SERRA DOS GRADAÚS
SERRA DO RONCADOR
CHAPADA DAS MANGABEIRAS
SERRA DA TABATINGA
SERRA GERAL DO PARANÁ
SERRA DO ESPINHAÇO
SERRA DO TOMBADOR
B R A Z I L
ILHA DO BANANAL
PLANALTO DO MATO GROSSO
PLANALTO CENTRAL
Cuiabá
Cáceres
Rondonópolis
Corumbá
Campo Grande
Aquidauana
Três Lagoas
Dourados
Brasília
Goiânia
Anápolis
Goiás
Formosa
Montes Claros
Pirapora
Diamantina
Uberlândia
Uberaba
Araguari
Belo Horizonte
Juiz de Fora
Barbacena
Campos
Volta Redonda
Ribeirão Prêto
São José do Rio Prêto
Araçatuba
Presidente Prudente
Marília
Bauru
Araraquara
São Carlos
Piracicaba
Campinas
Jundiaí
Sorocaba
SÃO PAULO
Santos
RIO DE JANEIRO
Niterói
Petrópolis
Nova Friburgo
CABO FRIO
CABO DE SÃO TOMÉ

Southern South America

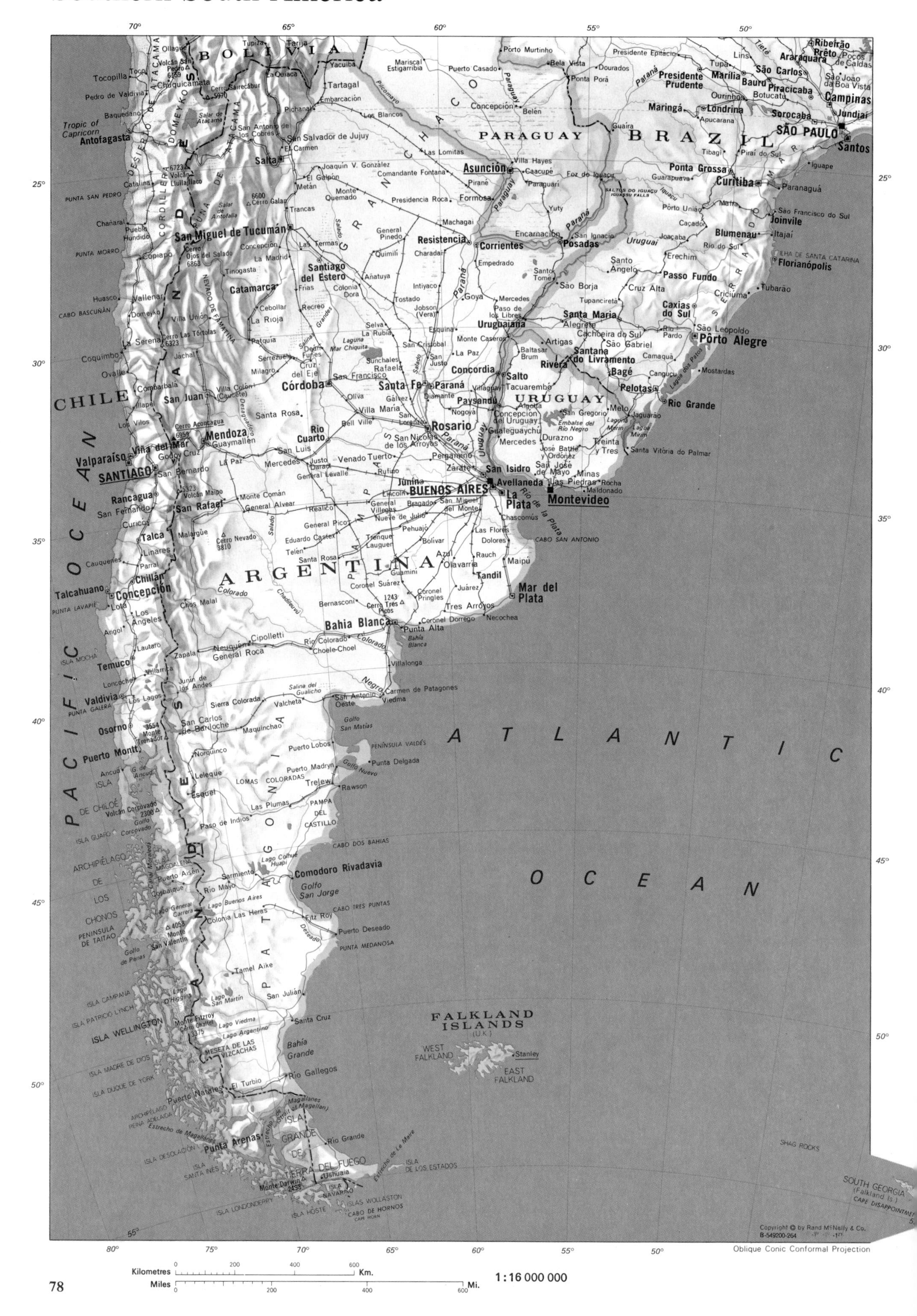

Central Argentina and Uruguay

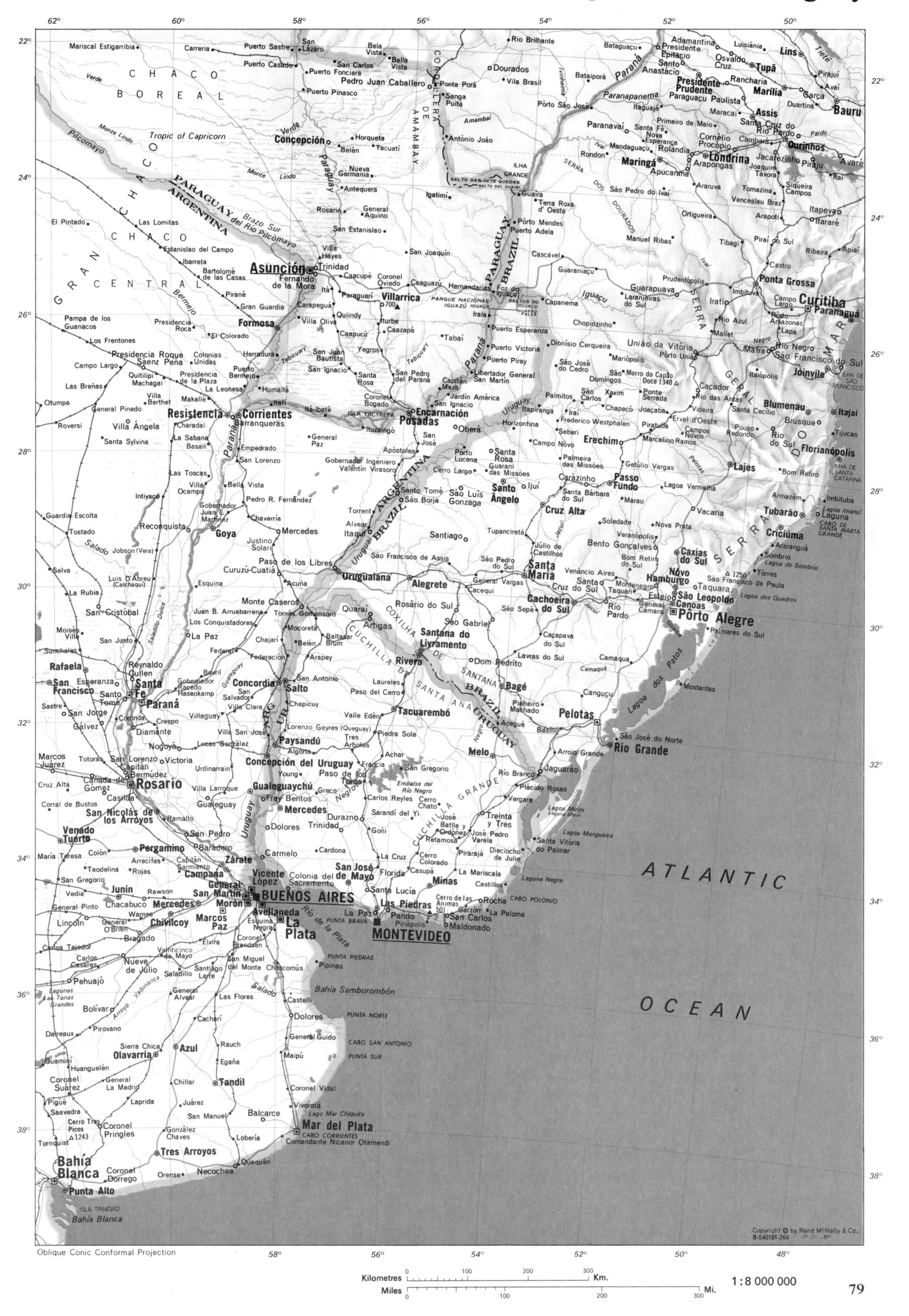

Southeastern Brazil

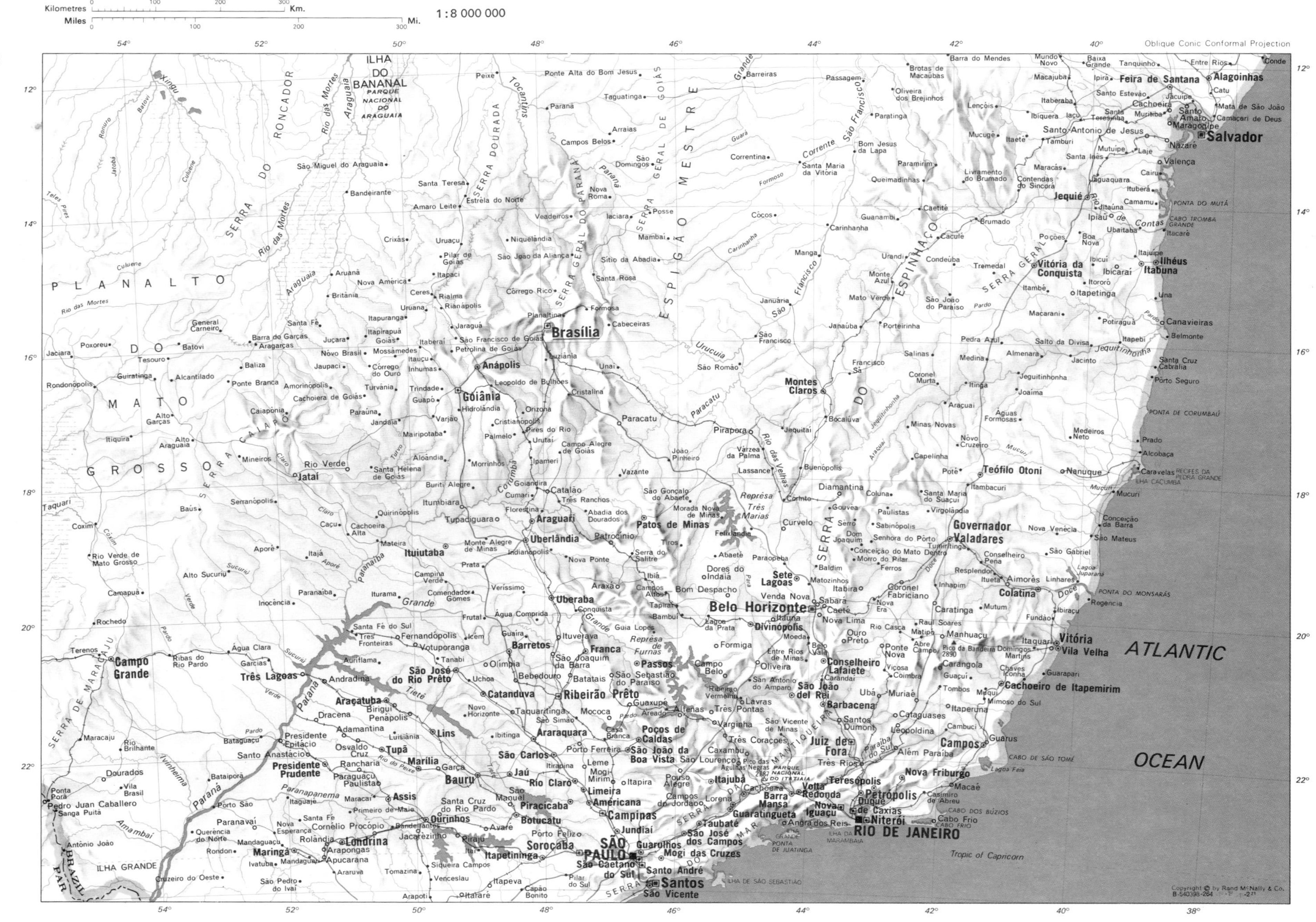

The Index includes in a single alphabetical list some 21,000 names appearing on the maps. Each name is followed by a page reference and the location of the feature on the map in coordinates of latitude and longitude. If a page contains several maps, a lowercase letter identifies the particular map. The page reference for two-page maps is always to the left-hand page.

Most map features are indexed to the largest-scale map on which they appear. Countries, mountain ranges, and other extensive features are generally indexed to the map that shows them in their entirety.

The features indexed are of three types: *point, areal,* and *linear*. For *point* features (for example, cities, mountain peaks, dams), latitude and longitude coordinates give the location of the point on the map. For *areal* features (countries, mountain ranges, etc.), the coordinates generally indicate the approximate center of the feature. For *linear* features (rivers, canals, aqueducts), the coordinates locate a terminating point—for example, the mouth of a river.

NAME FORMS Names in the Index, as on the maps, are generally in the local language and insofar as possible are spelled according to official practice. Most features that extend beyond the boundaries of one country have no single official name, and these are usually named in English. Many conventional English names are cross referenced to the primary map name. All cross references are indicated by the symbol →. A name that appears in a shortened version on the map due to space limitations is given in full in the Index, with the portion that is omitted on the map enclosed in brackets, for example, Acapulco [de Juárez].

TRANSLITERATION For names in languages not written in the Roman alphabet, the locally official transliteration system has been used where one exists. Thus, names in the Soviet Union and Bulgaria have been transliterated according to the systems adopted by the academies of science of these countries. Similarly, the transliteration for mainland Chinese names follows the Pinyin system, which has been officially adopted in mainland China. For languages with no one locally accepted transliteration system, notably Arabic, transliteration in general follows closely a system adopted by the United States Board on Geographic Names.

ALPHABETIZATION Names are alphabetized in the order of the letters of the English alphabet. Spanish *ll* and *ch*, for example, are not treated as distinct letters. Furthermore, diacritical marks are disregarded in alphabetization—German or Scandinavian *ä* or *ö* are treated as *a* or *o*.

The names of physical features may appear inverted, since they are always alphabetized under the proper, not the generic, part of the name, thus: "Gibraltar, Strait of ɰ." Otherwise every entry, whether consisting of one word or more, is alphabetized as a single continuous entity. "Lakeland," for example, appears after "La Crosse" and before "La Paz." Names beginning with articles (Le Havre, Den Helder, Al-Qāhirah, As-Suways) are not inverted. Names beginning "Mc" are alphabetized as though spelled "Mac," and names beginning "St." and "Sainte" as though spelled "Saint."

In the case of identical names, towns are listed first, then political divisions, then physical features. Entries that are completely identical (including symbols, discussed below) are distinguished by abbreviations of their official country names and are sequenced alphabetically by country name. The many duplicate names in Canada, the United Kingdom, and the United States are further distinguished by abbreviations of the names of their primary subdivisions. (See list of abbreviations below.)

ABBREVIATION AND CAPITALIZATION Abbreviation and styling have been standardized for all languages. A period is used after every abbreviation even when this may not be the local practice. The abbreviation "St." is used only for "Saint." "Sankt" and other forms of the term are spelled out.

All names are written with an initial capital letter except for a few Dutch names, such as 's-Gravenhage. Capitalization of noninitial words in a name generally follows local practice.

SYMBOL The symbols that appear in the Index graphically represent the broad categories of the features named, for example, ʌ for mountain (Everest, Mount ʌ). Superior numbers following some symbols in the Index indicate finer distinctions, for example, ʌ[1] for volcano (Fuji-san ʌ[1]). A complete list of the symbols and those with superior numbers is given on page 82.

LIST OF ABBREVIATIONS

	LOCAL NAME	ENGLISH
Afr.	—	Africa
Ala., U.S.	Alabama	Alabama
Alaska, U.S.	Alaska	Alaska
Alg.	Algérie	Algeria
Alta., Can.	Alberta	Alberta
Ang.	Angola	Angola
Antig.	Antigua	Antigua
Arg.	Argentina	Argentina
Ariz., U.S.	Arizona	Arizona
Ark., U.S.	Arkansas	Arkansas
Ar. Sa.	Al-'Arabīyah as-Sa'ūdīyah	Saudi Arabia
As.	—	Asia
Austl.	Australia	Australia
Ba.	Bahamas	Bahamas
B.C., Can.	British Columbia	British Columbia
Bel.	Belgique België	Belgium
Ber.	Bermuda	Bermuda
Bharat	Bhārat	India
Blg.	Bălgarija	Bulgaria
Bol.	Bolivia	Bolivia
Bra.	Brasil	Brazil
B.R.D.	Bundesrepublik Deutschland	Federal Republic of Germany
Calif., U.S.	California	California
Can.	Canada	Canada
Cay. Is.	Cayman Islands	Cayman Islands
Česko.	Československo	Czechoslovakia
Col.	Colombia	Colombia
Colo., U.S.	Colorado	Colorado
Comores	Comores	Comoro Islands
Conn., U.S.	Connecticut	Connecticut
C.R.	Costa Rica	Costa Rica
Dan.	Danmark	Denmark
D.C., U.S.	District of Columbia	District of Columbia
D.D.R.	Deutsche Demokratische Republik	German Democratic Republic
Del., U.S.	Delaware	Delaware
Dom.	Dominica	Dominica
Ec.	Ecuador	Ecuador
Eire	Eire	Ireland
Ellás	Ellás	Greece
El Sal.	El Salvador	El Salvador
Eng., U.K.	England	England
Esp.	España	Spain
Eur.	—	Europe
Falk. Is.	Falkland Islands	Falkland Islands (Islas Malvinas)
Fla., U.S.	Florida	Florida
Fr.	France	France
Ga., U.S.	Georgia	Georgia
Gam.	Gambia	Gambia
Gren.	Grenada	Grenada
Guat.	Guatemala	Guatemala
Guinée	Guinée	Guinea
Guy.	Guyana	Guyana
Guy. fr.	Guyane française	French Guiana
H.K.	Hong Kong	Hong Kong
Hond.	Honduras	Honduras
Idaho, U.S.	Idaho	Idaho
Ill., U.S.	Illinois	Illinois
Ind., U.S.	Indiana	Indiana
Indon.	Indonesia	Indonesia
I. of Man	Isle of Man	Isle of Man
Iowa, U.S.	Iowa	Iowa
Īrān	Īrān	Iran
'Irāq	Al-'Irāq	Iraq
It.	Italia	Italy
Jam.	Jamaica	Jamaica
Jugo.	Jugoslavija	Yugoslavia
Kans., U.S.	Kansas	Kansas
Kípros	Kípros Kıbrıs	Cyprus
Ky., U.S.	Kentucky	Kentucky
La., U.S.	Louisiana	Louisiana
Liber.	Liberia	Liberia
Lībiyā	Lībiyā	Libya
Lubnān	Al-Lubnān	Lebanon
Macau	Macau	Macau
Magreb	Al-Magreb	Morocco
Maine, U.S.	Maine	Maine
Mali	Mali	Mali
Malta	Malta	Malta
Man., Can.	Manitoba	Manitoba
Mass., U.S.	Massachusetts	Massachusetts
Md., U.S.	Maryland	Maryland
Méx.	México	Mexico
Mich., U.S.	Michigan	Michigan
Minn., U.S.	Minnesota	Minnesota
Miss., U.S.	Mississippi	Mississippi
Mo., U.S.	Missouri	Missouri
Mont., U.S.	Montana	Montana
Mya.	Myanma	Burma
N.A.	—	North America
N.B., Can.	New Brunswick	New Brunswick
N.C., U.S.	North Carolina	North Carolina
N. Dak., U.S.	North Dakota	North Dakota
Nebr., U.S.	Nebraska	Nebraska
Ned.	Nederland	Netherlands
Nepāl	Nepāl	Nepal
Nev., U.S.	Nevada	Nevada
Newf., Can.	Newfoundland	Newfoundland
N.H., U.S.	New Hampshire	New Hampshire
Nic.	Nicaragua	Nicaragua
Nig.	Nigeria	Nigeria
Nihon	Nihon	Japan
N. Ire., U.K.	Northern Ireland	Northern Ireland
N.J., U.S.	New Jersey	New Jersey
N. Mex., U.S.	New Mexico	New Mexico
Nor.	Norge	Norway
N.S., Can.	Nova Scotia	Nova Scotia
N.W. Ter., Can.	Northwest Territories	Northwest Territories
N.Y., U.S.	New York	New York
N.Z.	New Zealand	New Zealand
Ohio, U.S.	Ohio	Ohio
Okla., U.S.	Oklahoma	Oklahoma
Ont., Can.	Ontario	Ontario
Oreg., U.S.	Oregon	Oregon
Öst.	Österreich	Austria
Pa., U.S.	Pennsylvania	Pennsylvania
Pāk.	Pākistān	Pakistan
Pan.	Panamá	Panama
Para.	Paraguay	Paraguay
Perú	Perú	Peru
Pil.	Pilipinas	Philippines
Port.	Portugal	Portugal
P.R.	Puerto Rico	Puerto Rico
Que., Can.	Québec	Quebec
Rep. Dom.	República Dominicana	Dominican Republic
Réu.	Réunion	Reunion
Rh.	Rhodesia	Rhodesia
R.I., U.S.	Rhode Island	Rhode Island
Rom.	România	Romania
S.A.	—	South America
S. Afr.	South Africa Suid-Afrika	South Africa
Sask., Can.	Saskatchewan	Saskatchewan
S.C., U.S.	South Carolina	South Carolina
Schw.	Schweiz; Suisse; Svizzera	Switzerland
Scot., U.K.	Scotland	Scotland
S. Dak., U.S.	South Dakota	South Dakota
Sén.	Sénégal	Senegal
Sey.	Seychelles	Seychelles
S.L.	Sierra Leone	Sierra Leone
S.S.S.R.	Sojuz Sovetskich Socialističeskich Respublik	Union of Soviet Socialist Republics
Sūd.	As-Sūdān	Sudan
Sve.	Sverige	Sweden
S.W. Afr.	South West Africa	South West Africa
Taehan	Taehan-Min'guk	South Korea
T'aiwan	T'aiwan	Taiwan

Introduction to the Index

LIST OF ABBREVIATIONS CONT'D

	LOCAL NAME	ENGLISH
Tchad	Tchad	Chad
Tenn., U.S.	Tennessee	Tennessee
Tex., U.S.	Texas	Texas
Trin.	Trinidad and Tobago	Trinidad and Tobago
Tun.	Tunisie	Tunisia
Tür.	Türkiye	Turkey
U.K.	United Kingdom	United Kingdom
'Umān	'Umān	Oman
Ur.	Uruguay	Uruguay
U.S.	United States	United States
Utah, U.S.	Utah	Utah
Va., U.S.	Virginia	Virginia
Ven.	Venezuela	Venezuela
Vt., U.S.	Vermont	Vermont
Wales, U.K.	Wales	Wales
Wash., U.S.	Washington	Washington
Wis., U.S.	Wisconsin	Wisconsin
W. Va., U.S.	West Virginia	West Virginia
Wyo., U.S.	Wyoming	Wyoming
Yai.	Yaitopya	Ethiopia
Yukon, Can.	Yukon	Yukon
Zaïre	Zaïre	Zaire
Zam.	Zambia	Zambia
Zhg.	Zhongguo	China

KEY TO SYMBOLS

- ∧ **Mountain**
- ∧[1] Volcano
- ∧[2] Hill
- ∧∧ **Mountains**
- ∧∧[1] Plateau
- ∧∧[2] Hills
-)(**Pass**
- ∨ **Valley, Canyon**
- ≚ **Plain**
- ≚[1] Basin
- ≚[2] Delta
- ≻ **Cape**
- ≻[1] Peninsula
- ≻[2] Spit, Sand Bar
- I **Island**
- I[1] Atoll
- I[2] Rock
- II **Islands**
- II[1] Rocks
- ∸ **Other Topographic Features**
- ∸[1] Continent
- ∸[2] Coast, Beach
- ∸[3] Isthmus
- ∸[4] Cliff
- ∸[5] Cave, Caves
- ∸[6] Crater
- ∸[7] Depression
- ∸[8] Dunes
- ∸[9] Lava Flow
- ≏ **River**
- ≏[1] River Channel
- ≣ **Canal**
- ≣[1] Aqueduct
- ⌊ **Waterfall, Rapids**
- ʯ **Strait**
- C **Bay, Gulf**
- C[1] Estuary
- C[2] Fjord
- C[3] Bight
- ⊜ **Lake, Lakes**
- ⊜[1] Reservoir
- ☷ **Swamp**
- ᗭ **Ice Features, Glacier**
- ⍑ **Other Hydrographic Features**
- ⍑[1] Ocean
- ⍑[2] Sea
- ⍑[3] Anchorage
- ⍑[4] Oasis, Well, Spring
- ⩪ **Submarine Features**
- ⩪[1] Depression
- ⩪[2] Reef, Shoal
- ⩪[3] Mountain, Mountains
- ⩪[4] Slope, Shelf
- □ **Political Unit**
- □[1] Independent Nation
- □[2] Dependency
- □[3] State, Canton, Republic
- □[4] Province, Region, Oblast
- □[5] Department, District, Prefecture
- □[6] County
- □[7] City, Municipality
- □[8] Miscellaneous
- □[9] Historical
- ʊ **Cultural Institution**
- ʊ[1] Religious Institution
- ʊ[2] Educational Institution
- ʊ[3] Scientific, Industrial Facility
- ⊥ **Historical Site**
- ♠ **Recreational Site**
- ⊠ **Airport**
- ▪ **Military Installation**
- ⬩ **Miscellaneous**
- ⬩[1] Region
- ⬩[2] Desert
- ⬩[3] Forest, Moor
- ⬩[4] Reserve, Reservation
- ⬩[5] Transportation
- ⬩[6] Dam
- ⬩[7] Mine, Quarry
- ⬩[8] Neighborhood
- ⬩[9] Shopping Center

Index

Symbols against index entries represent categories identified in the key on page 82.

Symbols against index entries represent categories identified in the key on page 82.

15.18

Symbols against index entries represent categories identified in the key on page 82.

Symbols against index entries represent categories identified in the key on page 82.

33.19

Symbols against index entries represent categories identified in the key on page 82.

Symbols against index entries represent categories identified in the key on page 82.

51.42

Symbols against index entries represent categories identified in the key on page 82.

Symbols against index entries represent categories identified in the key on page 82.

70.3

Symbols against index entries represent categories identified in the key on page 82.

Symbols against index entries represent categories identified in the key on page 82.

88.15

Symbols against index entries represent categories identified in the key on page 82.

Symbols against index entries represent categories identified in the key on page 82.

Symbols against index entries represent categories identified in the key on page 82.

Symbols against index entries represent categories identified in the key on page 82.

Symbols against index entries represent categories identified in the key on page 82.

Symbols against index entries represent categories identified in the key on page 82.

142.36

Symbols against index entries represent categories identified in the key on page 82.

Symbols against index entries represent categories identified in the key on page 82.

160.48

Symbols against index entries represent categories identified in the key on page 82.

Symbols against index entries represent categories identified in the key on page 82.

Symbols against index entries represent categories identified in the key on page 82.

Symbols against index entries represent categories identified in the key on page 82.

Symbols against index entries represent categories identified in the key on page 82.

215.25

Symbols against index entries represent categories identified in the key on page 82.

Symbols against index entries represent categories identified in the key on page 82.

Symbols against index entries represent categories identified in the key on page 82.

Symbols against index entries represent categories identified in the key on page 82.

252.2

Symbols against index entries represent categories identified in the key on page 82.

Symbols against index entries represent categories identified in the key on page 82.

Symbols against index entries represent categories identified in the key on page 82.

Symbols against index entries represent categories identified in the key on page 82.

Symbols against index entries represent categories identified in the key on page 82.

Symbols against index entries represent categories identified in the key on page 82.

Symbols against index entries represent categories identified in the key on page 82.

Symbols against index entries represent categories identified in the key on page 82.

Symbols against index entries represent categories identified in the key on page 82.

Symbols against index entries represent categories identified in the key on page 82.

Symbols against index entries represent categories identified in the key on page 82.

Symbols against index entries represent categories identified in the key on page 82.

Symbols against index entries represent categories identified in the key on page 82.

Symbols against index entries represent categories identified in the key on page 82.

Symbols against index entries represent categories identified in the key on page 82.

Symbols against index entries represent categories identified in the key on page 82.

W

Symbols against index entries represent categories identified in the key on page 82.

Symbols against index entries represent categories identified in the key on page 82.

Symbols against index entries represent categories identified in the key on page 82.